Handbook of Experimental Pharmacology

Volume 139

Springer
Berlin
Heidelberg
New York
Barcelona
Hong Kong
London
Milan
Paris
Singapore
Tokyo

Retinoids

The Biochemical and Molecular Basis of Vitamin A and Retinoid Action

Contributors

A. Agadir, R.L. Beard, R.P. Bissonnette, W.S. Blaner, J. Buck, R. Cesario, R.A.S. Chandraratna, C. Chomienne, D. Crombie, L. Didierjean, E. Dmitrovsky, G. Eichele, M.M.A. Elmazar, M.V. Gamble, D.F.C. Gibson, F.E. Gomez, B. Harvat, C.E. Hayes, R.A. Heyman, K.A. Hoag, A.M. Jetten, R. Lotan, H.-C. Lu, X. Lu, C.-P. Lu, M. Maden, F.E. Nashold, K. Nason-Burchenal, H. Nau, N. Noy, A.I. Packer, M. Pfahl, R. Piantedosi, F.J. Piedrafita, R.K. Randolph, K. Roegner, J.C. Saari, J.H. Saurat, I.G. Schulman, G. Shao, B. Shroot, G. Siegenthaler, O. Sorg, A. Sykes, C. Thaller, E. Vakiani, S. Vogel, D.J. Wolgemuth, X-C. Xu, M.H. Zile

Editors

Heinz Nau and William S. Blaner

Springer

Professor Dr.Dr.h.c. HEINZ NAU
Department of Food Toxicology
Center of Food Science
School of Veterinary Medicine
Bischofholer Damm
D-30173 Hannover
GERMANY
e-mail: hnau@lebensmittel.tiho-hannover.de

WILLIAM S. BLANER, Ph.D.
College of Physicians and Surgeons
of Columbia University
Department of Medicine
Division of Preventive Medicine & Nutrition
630 West 168th Street
New York, NY 10032
USA
e-mail: wsb2@columbia.edu

With 81 Figures and 36 Tables

ISBN 3-540-65892-0 Springer-Verlag Berlin Heidelberg New York

Library of Congress Cataloging-in-Publication Data

Retinoids: the biochemical and molecular basis of Vitamin A and
retinoid action / editors, Heinz Nau and William S. Blaner;
contributors, A. Agadir . . . [et al.].
p. cm. — (Handbook of experimental pharmacology; v. 139)
Includes bibliographical references and index.
ISBN 3-540-65892-0 (alk. paper)
1. Retinoids—Physiological effect. 2. Retinoids—Therapeutic
use. I. Nau, Heinz, 1943– . II. Blaner, William S., 1950– .
III. Agadir, A. IV. Series.
QP905.H3 vol. 139
615′.1 s — dc21
[615′.328]

99-35527
CIP

Printed in Germany

Cover design: *design & production* GmbH, Heidelberg

Typesetting: Best-set Typesetter Ltd., Hong Kong

SPIN: 10575926 27/3020 – 5 4 3 2 1 0 – Printed on acid-free paper

Preface

In the future' the decade of the 1990s will likely be viewed as a Golden Age for retinoid research. There have been unprecedented research gains in the understanding of retinoid actions and physiology since the retinoid nuclear receptors were first identified and the importance of retinoic acid in developmental processes was first broadly recognized in the late 1980s. Between then and now, our knowledge of retinoid action has evolved from one of a near complete lack of understanding of how retinoids act within cells to one of sophisticated understanding of the molecular processes through which retinoids modulate transcription.

In this volume, we have tried to provide a comprehensive update of the present understanding of retinoid actions, with an emphasis on recent advances. The initial chapters of the volume, or Section A, focus on the physicochemical properties and metabolism of naturally occurring retinoids:

– NOY provides an uncommonly encountered view of retinoid effects from the perspective of the physiochemical properties of retinoids.
– VAKIANI and BUCK lend a perspective on the biological occurrence and actions of retro- and anhydro-retinoids.

Section B considers both the retinoid nuclear receptors and their mechanisms of action as well as synthetic retinoids that have been used experimentally to provide mechanistic insights into receptor actions and have potential therapeutic use for treating disease:

– PIEDRAFITA and PFAHL provide a comprehensive review of retinoid nuclear receptor biochemistry and molecular biology.
– Chapters senior-authored by CHANDRARATNA and HEYMAN, respectively, take pharmacologic perspectives in reviewing recent developments in the synthesis, characterization and use of both retinoic acid receptor- and retinoid X receptor-specific agonists and antagonists. These chapters point to, and provide a basis for, the future use of synthetic retinoids in clinical medicine.

Section C of this volume considers the actions of retinoids in modulating cellular differentiation and proliferation and in preventing and treating cancer:

– Chapters by HARVAT and JETTEN and DMITROVSKI and NASON-BURCHENAL, respectively, consider the mechanisms through which retinoids act to modulate cellular differentiation and proliferation.
– AGADIR and CHOMIENNE review the use of retinoids for treating diseases of cell proliferation and differentiation of hematopoietic cells.
– XU and LOTAN provide a general overview of retinoid use and actions in different types of cancer.

Section D grants a comprehensive review of the role of retinoids in directing normal development and their toxic effects in bringing about abnormal development:

– PACKER and WOLGEMUTH review the essential role that retinoids play in spermatogenesis.
– LU, THALLER and EICHELE focus on the actions of retinoids in modulating genes that direct morphogenesis in the embryo.
– MADEN provides a detailed overview of the actions of retinoids in directing neural development.
– ZILE reviews the effects of dietary retinoid deficiency or excess on normal embryologic development.
– NAU and ELMAZAR consider the teratogenetic actions of both natural and synthetic retinoids and the possible mechanisms that underlie their synergistic interactions.

Section E takes up the actions of retinoids in skin and their pharmacologic use in clinical dermatology:

– RANDOLPH and SIEGENTHALER present a review of the basic mechanisms responsible for retinoid actions in skin.
– SAURAT and colleagues consider different approaches for the delivery of retinoids for use in the treatment of dermatologic disorders.
– SHROOT and colleagues focus on the present and potential future use of receptor-specific agonists and antagonists in treating skin diseases.

The final section of this volume, Sect. F, provides reviews of two important physiologic actions of retinoids:

– SAARI offers an update of current research concerning retinoids in the visual process.
– HAYES and colleagues review the essential role of retinoids in the immune response.

Although we have tried to broadly cover the modern field of retinoid research, there are still many areas left unreviewed by this volume. This is due not to the relative significance we ascribe to these areas, but to the need to limit the size of the volume. We have chosen to cover research areas that are more highly investigated and have limited coverage of more specialized research interests. It seems likely, however, that many important physiologic actions of retinoids will ultimately be identified through investigation of specialized research areas.

In the concluding paragraph to Thomas Moore's monograph Vitamin A published in 1957, he muses: "In the lives of scientists, however, it is usually their privilege to learn secrets that were never known to their fathers, and their fate to be denied the knowledge which will in time enlighten their sons." Just as this was true for Moore and his understanding of vitamin A in 1957, we must acknowledge that this fate also awaits the current community of retinoid researchers. Many secrets of retinoid biology and the clinical use of retinoids remain to be uncovered by future researchers. We hope that this volume will serve as a useful milestone to aid future research along this path.

HEINZ NAU
WILLIAM S. BLANER

List of Contributors

AGADIR, A., The Burnham Institute, La Jolla Cancer Research Center, 10901 N. Torrey Pines Road, La Jolla, CA 92037, USA

BEARD, R.L., Retinoid Research, ALLERGAN, 2525 Dupont Drive, P.O. Box 19534, Irvine, CA 92623-9534, USA – e-mail: beard_vichard@allergan.com

BISSONNETTE, R.P., Department of Retinoid Research, Ligand Pharmaceuticals, 10255 Science Center Drive, San Diego, CA 92121, USA

BLANER, W.S., Department of Medicine and Institute of Human Nutrition, College of Physicians and Surgeons, Columbia University, 630 W. 168th Street, New York, NY 10032, USA – e-mail: wsb2@columbia.edn

BRAHAM S., Galderma Research and Development, 5 Cedarbrook Drive, Suite 1, Cranbury, NJ 08512, USA

BUCK, J., Department of Pharmacology, Cornell University Medical College, 1300 York Avenue, New York, NY 10021, USA – FAX: (212)746-8835

CESARIO, R., Department of Retinoid Research, Ligand Pharmaceuticals, 10255 Science Center Drive, San Diego, CA 92121, USA

CHANDRARATNA, R.A.S., Retinoid Research, ALLERGAN, 2525 Dupont Drive, P.O. Box 19534, Irvine, CA 92623-9534, USA – FAX: (714)246-6207

CHOMIENNE, C., Laboratoire de Biologie Cellulaire Hématopoiétique, Hôpital Saint-Louis, 1 Avenue Claude Vellefaux, F-75010 Paris, France – e-mail: lbdn@chu-stlouis.fr

CROMBIE, D., Department of Retinoid Research, Ligand Pharmaceuticals, 10255 Science Center Drive, San Diego, CA 92121, USA

DAVID F.C., Gibson, ACADIA Pharmaceutical, 3911 Sorrento Valley Blvd, San Diego, CA 92121, USA

DIDIERJEAN, L., Clinique de Dermatologie, Hôpital Cantonal de Genève, 24 rue Micheli-du-Crest, CH-1211 Genève 14, SWITZERLAND

DMITROVSKY, E., Department of Pharmacology, Dartmouth Medical School, 7650 Remsen, Hanover, NH 03755-3835, USA – e-mail: Pharmacology.and.Toxicology@Dartmouth.edn

EICHELE, G., Department of Biochemistry, Baylor College of Medicine, One Baylor Plaza, Houston, TX 77030, USA – e-mail: eichele@bcm.tmc.edn and Max-Planck-Institute for Experimental Endocrinology, Feodor-Lynen-Strasse 7, D-30625 Hannover, Germany

ELMAZAR, M.M.A., College of Pharmacy, Mansource University Egypt, presently at King Saud University, College of Pharmacy, Riyadh 1145, SAUDI ARABIA

GAMBLE, M.V., Department of Medicine and Institute of Human Nutrition, College of Physicians and Surgeons, Columbia University, 630 W. 168th Street, New York, NY 10032, USA

GIBSON, D.F.C., Galderma Research Inc., 10835 Altman Row, Suite 250, San Diego, CA 92121, USA

GOMEZ, F.E., Department of Biochemistry, University of Wisconsin-Madison, 33 Babcock Drive, Madison, WI 53706, USA

HARVAT, B., Cell Biology Section, Laboratory of Pulmonary Pathobiology, National Institute of Environmental Health Sciences, National Institutes of Health, Research Triangle Park, NC 27709, USA

HAYES, C.E., Department of Biochemistry, University of Wisconsin-Madison, 33 Babcock Drive, Madison, WI 53706, USA – FAX: (608)262-3453

HEYMAN, R.A., Department of Retinoid Research, Ligand Pharmaceuticals, 10255 Science Center Drive, San Diego, CA 92121, USA – e-mail: rheyman@ligand.com

HOAG, K.A., Department of Biochemistry, University of Wisconsin-Madison, 33 Babcock Drive, Madison, WI 53706, USA

JETTEN, A.M., Cell Biology Section, Laboratory of Pulmonary Pathobiology, National Institute of Environmental Health Sciences,

National Institutes of Health, Research Triangle Park, NC 27709, USA – e-mail: jetten@niehs.nih.gov

LOTAN, R., Departments of Thoracic/Head & Neck Medical Oncology-108, The University of Texas M.D. Anderson Cancer Center, 1515 Holcombe Blvd., Houston, TX 77030, USA

LU, H-C., Verna and Marrs McLean Department of Biochemistry, Baylor College of Medicine, One Baylor Plaza, Houston, TX 77030, USA

LU, X., Department of Clinical Cancer Prevention-236, The University of Texas M.D. Anderson Cancer Center, 1515 Holcombe Boulevard, Houston, TX 77030, USA – FAX: (713)794-0209

LU, X-P., Galderma Research Inc., 10835 Altman Row, Suite 250, San Diego, CA 92121, USA

MADEN, M., Developmental Biology Research Centre, King's College London, 26-29 Drury Lane, London WC2B 5RL, UNITED KINGDOM – e-mail: m.maden@kcl.ac.uk

NASHOLD, F.E., Department of Biochemistry, University of Wisconsin-Madison, 420 Henry Mall, Madison, WI 53706, USA

NASON-BURCHENAL, K., Laboratory of Molecular Medicine, Department of Medicine, Molecular Pharmacology and Therapeutics Program, Memorial Sloan-Kettering Cancer Center, 1275 York Avenue, New York, NY 10021, USA

NAU, H., Department of Food Toxicology, School of Veterinary Medicine Hannover, Bischofholer Damm 15, D-30173 Hannover, Germany – e-mail: hnan@lebensmittel.tiho-hannover.de

NOY, N., Division of Nutritional Sciences, Savage Hall, Cornell University, Ithaca, NY 14853, USA – e-mail: nn14@corwell.edn

PACKER, A.I., Department of Genetics and Development, Columbia University, College of Physicinas and Surgeons, 630 West 168th Street, New York, NY 10032, USA

PFAHL, M., MAXIA Pharmaceuticals Inc., 10835 Altman Row, San Diego, CA 92121, USA – e-mail: pfahl@maxia.com

Piantedosi, R., Department of Medicine,
College of Physicians and Surgeons, Columbia University,
630 W. 168th Street, New York, NY 10032, USA

Piedrafita, F.J., Sidney Kimmel Cancer Center, 10835 Altman Row,
San Diego, CA 92121, USA

Randolph, R.K., Scientific and Regulatory Affairs, Vitamin Shoppe
Industris, Inc., 4800 Westside Avenne, North Bergen, NJ 07047, USA –
e-mail: krandolph@vitaminshoppe.com

Roegner, K., Department of Retinoid Research, Ligand Pharmaceuticals,
10255 Science Center Drive, San Diego, CA 92121, USA

Saari, J.C., University of Washington School of Medicine,
Department of Ophthalmology, Mail Stop RJ-10, Seattle,
WA 98195-6485, USA – e-mail: jsaari@u.washington.edn

Saurat, J.H., Clinique de Dermatologie, Hôpital Cantonal de Genève,
24 rue Micheli-du-Crest, CH-1211 Genève 14, SWITZERLAND –
e-mail: saurat@cmu.unige.ch

Schulman, I.G., Department of Retinoid Research, Ligand Pharmaceuticals,
10255 Science Center Drive, San Diego, CA 92121, USA

Shao, G., Department of Retinoid Research, Ligand Pharmaceuticals,
10255 Science Center Drive, San Diego, CA 92121, USA

Shroot, B., Galderma Research Inc., 10835 Altman Row, Suite 250,
San Diego, CA 92121, USA – FAX: (619)453-5343

Siegenthaler, G., Hôpital Cantonal Universitaire de Genève,
Clinique de Dermatologie, 24, Rue Micheli-du-Crest, CH-1211 Genève
14, SWITZERLAND – FAX: (22)3729470

Sorg, O., Clinique de Dermatologie, Hôpital Cantonal de Genève,
24 rue Micheli-du-Crest, CH-1211 Genève 14, SWITZERLAND

Sykes, A., Institute of Human Nutrition, College of Physicians and Surgeons,
Columbia University, 630 W. 168th Street, New York, NY 10032, USA

Thaller, C., Verna and Marrs McLean Department of Biochemistry,
Baylor College of Medicine, One Baylor Plaza, Houston, TX 77030, USA

Vakiani, E., Department of Pharmacology,
Cornell University Medical College, 1300 York Avenue, New York,
NY 10021, USA

VOGEL, S., Department of Medicine, College of Physicians and Surgeons, Columbia University, 630 W. 168th Street, New York, NY 10032, USA

WOLGEMUTH, D.J., Department of Genetics & Development, Obstetrics and Gynecology, The Center for Reproductive Sciences, Herbert Irving Comprehensive Cancer Center, Columbia University, College of Physicians and Surgeons, New York, NY 10032, USA – e-mail: djw3@columbia.edn

XIAN-PING LU, Galderma Research and Development, 5 Cedarbrook Drive, Suite 1, Cranbury, NJ 08512, USA

XU, X.-C. Department of Clinical Cancer Prevention-236, The University of Texas M.D. Anderson Cancer Center, 1515 Holcombe Boulevard, Houston, TX 77030, USA – FAX: (713)794–0205

ZILE, M.H., Department of Food Science and Human Nutrition, G.M. Trout Building, Michigan State University, East Lansing, MI 48824-1224, USA – e-mail: zile@pilot.msu.edn

Contents

Section A: Metabolism, Pharmocokinetics and Action of Retinoids

CHAPTER 1

Physical-Chemical Properties and Action of Retinoids

N. Noy. With 6 Figures ... 3

A. Introduction ... 3
B. Behavior of Retinoids in an Aqueous Environment ... 3
I. Aqueous Solubility ... 3
II. Self-Association of Retinoids in Water ... 4
III. Lability of Retinoids in Water ... 6
C. Interactions of Retinoids with Membranes ... 6
I. Effects of Membrane Lipid Composition on Membrane-Retinoid Interactions ... 7
II. Effects of the Radius of Curvature of Membranes on Membrane-Retinoid Interactions ... 8
III. Ionization Behavior of Retinoic Acid in Membranes ... 10
D. Interactions of Retinoids with Proteins ... 10
I. Binding Affinities ... 11
1. Ligand Binding Affinities of Retinoid-Binding Proteins ... 12
2. Ligand Binding Affinities of Retinoid Nuclear Receptors ... 14
II. Kinetics of the Interactions of Retinoids with Proteins ... 17
1. Rates of Dissociation of Retinoids from Retinoid-Binding Proteins ... 17
E. Some Implications for Retinoid Transport Processes ... 18
I. Uptake of Retinol by Target Cells ... 18
II. Targeting of Retinoids by IRBP ... 21
F. Concluding Remarks ... 24
References ... 24

CHAPTER 2

Biosynthesis, Absorption, Metabolism and Transport of Retinoids

S. Vogel, M.V. Gamble, and W.S. Blaner. With 4 Figures 31

A. Introduction 31
B. Intestinal Absorption of Retinol and Biosynthesis of Retinol from Carotenoids 32
I. Uptake and Luminal Metabolism of Dietary Preformed Vitamin A (Retinyl Esters, Retinol, and Retinoic Acid) 33
II. Uptake and Metabolism of Dietary Provitamin A Carotenoids 34
III. Intramucosal Metabolism 36
1. Retinoic Acid Formation from Retinal Derived from Carotenoids 36
2. Role of Cellular-Retinol Binding Protein, Type II 36
3. Retinol Esterification and the Packaging of Dietary Retinoid into Chylomicrons 37
C. Circulating Chylomicrons and Chylomicron Remnant Formation 38
I. Hepatic Clearance of Chylomicron Remnants 39
II. Extrahepatic Tissues and Chylomicron Remnant Retinoid 42
D. Retinoid Processing and Metabolism within the Liver 44
I. Uptake and Processing of the Chylomicron Remnant Retinyl Ester by Hepatocytes 45
II. Retinoid Metabolism and Storage in the Hepatocyte 46
1. Cellular Retinol-Binding Protein, Type I 49
2. Hepatic LRAT Activity 50
III. Retinoid Transfer Between Hepatocytes and Stellate Cells 51
IV. Retinoid Metabolism and Storage in Hepatic Stellate Cells 52
E. Extrahepatic Processing and Metabolism of Retinol and Retinyl Ester 54
F. Retinoid Delivery to Target Tissues 55
I. Retinol Bound to RBP 56
1. Cellular Synthesis and Secretion of Retinol-RBP and Its Regulation 58
2. The RBP Gene and Its Regulation 63
3. Structural and Physiochemical Properties of RBP, TTR, and the RBP-TTR Complex 63
4. Retinol-RBP Turnover from the Circulation 66
5. RBP Receptor 67
II. Plasma all-*trans*-Retinoic Acid 77
III. Plasma 13-cis-Retinoic Acid 80

IV. Lipoprotein Bound Retinyl Ester 81
V. Glucuronides of Retinol and Retinoic Acid 82
VI. Retro-Metabolites of Retinol 83
G. Summary 84
References 84

CHAPTER 3

***Retro*-Retinoids: Metabolism and Action**
E. VAKIANI and J. BUCK. With 9 Figures 97

A. Introduction 97
B. Vitamin A and the Immune System 98
C. General Properties of *Retro*-Retinoids 99
D. Cellular Effects of *Retro*-Retinoids 101
I. Growth of B Lymphocytes 101
II. Activation of T Lymphocytes 104
III. Activation of Fibroblasts 104
IV. Cell Growth vs. Cell Differentiation: Retinol Metabolism in HL-60 Cells 106
V. Intracellular Signaling 106
1. Anhydroretinol-Induced Cell Death 106
2. Evidence for a Transmembrane or Cytoplasmic Receptor 108
E. *Retro*-Retinoid Metabolism 108
I. 14-Hydroxy-*Retro*-Retinol 109
II. 13,14-Dihydroxy-Retinol 109
III. Biosynthesis of Anhydroretinol 110
F. Conclusion 112
References 113

CHAPTER 4

Retinoic Acid Synthesis and Metabolism
W.S. BLANER, R. PIANTEDOSI, A. SYKES, and S. VOGEL. With 2 Figures 117

A. Introduction 117
B. Cellular and Extracellular Retinoid Binding Proteins 118
I. Cellular Retinol Binding Proteins 118
II. Cellular Retinoic Acid Binding Proteins 118
III. Other Retinoic Acid Binding Proteins 121
C. Retinoic Acid Synthesis 121
I. Oxidation of Retinol 123
1. Role of Cytosolic Dehydrogenases 123
2. Role of Microsomal Dehydrogenases 125

II. Oxidation of Retinal ... 130
III. Formation of 9-*cis*-Retinoic Acid ... 133
IV. Direct Synthesis of Retinoic Acid from Carotenoids ... 137
D. Oxidative Metabolism of Retinoic Acid ... 137
I. Role of the Cytochrome P450 System in Retinoic Acid Oxidation ... 138
II. Other Metabolism of Retinoic Acid ... 142
E. Summary and Future Needs ... 143
References ... 144

Section B: Binding Proteins and Nuclear Receptors

CHAPTER 5

Nuclear Retinoid Receptors and Mechanisms of Action
F.J. Piedrafita and M. Pfahl. With 5 Figures ... 153

A. Introduction ... 153
B. The Nuclear Retinoid Receptors and Their Cousins ... 154
C. The RAREs: Further Specification of the Ligand Response ... 156
D. Interaction of Retinoid Receptors with Coactivators, Corepressors, and Basal Transcription Factors ... 158
E. Differentiation–Proliferation Switches: Interaction with Other Signal Transducers ... 164
F. CBP/p300, a Cointegrator of Multiple Signaling Pathways ... 165
G. A Role for Histone Acetylation in Receptor-Mediated Transactivation? ... 167
H. Apoptosis Induction by Special Retinoids: A Novel Pathway? ... 169
I. Future Prospects ... 172
References ... 174

CHAPTER 6

RAR-Selective Ligands: Receptor Subtype and Function Selectivity
R.L. Beard and R.A.S. Chandraratna. With 7 Figures ... 185

A. Introduction ... 185
B. Structural Differences Between RAR-Selective and RXR-Selective Ligands ... 186
I. The Effect of Bridging Groups on Conformationally Restricted Retinoids ... 186
II. The α-Methyl Effect ... 187
III. 9-*Cis*-Locked vs 9-*Trans*-Locked Retinoids ... 188
C. RAR Subtype Selective Agonists ... 190
I. RARα Selective Agonists ... 192
II. RARβ Selective Agonists ... 194

III. RARγ Selective Agonists . . . 195
D. RAR Antagonists, Neutral Antagonists, and Inverse Agonists . . . 197
I. RAR Antagonists . . . 197
II. RAR Inverse Agonists and Neutral Antagonists . . . 201
E. Pharmacology of RAR Selective Agonists and Antagonists . . . 203
I. RARα Selective Agonists . . . 204
II. RARβ Selective Agonists . . . 205
III. RARγ Selective Agonists . . . 205
IV. RAR Antagonists and Inverse Agonists . . . 206
References . . . 208

CHAPTER 7

RXR-Specific Agonists and Modulators: A New Retinoid Pharmacology
I.G. SCHULMAN, D. CROMBIE, R.P. BISSONNETTE, R. CESARIO, K. ROEGNER, G. SHAO, and R.A. HEYMAN. With 11 Figures . . . 215

A. Introduction . . . 215
B. Ligand-Dependent Activation of Transcription By RXR . . . 218
I. Activation of Permissive Heterodimers . . . 218
II. RXR Agonists Activate RXR-RAR Heterodimers When The RAR Binding Pocket Is Occupied . . . 221
III. The Phantom Ligand Effect . . . 222
C. RXR Pharmacology . . . 226
I. Sensitization of Diabetic Mice to Insulin by RXR Agonists . . . 226
II. Chemoprevention/Chemotherapy of Carcinogen-Induced Breast Cancer . . . 230
D. Conclusions and Future Directions . . . 231
References . . . 232

Section C: Differentiation, Proliferation, and Cancer

CHAPTER 8

Growth Control by Retinoids: Regulation of Cell Cycle Progression and Apoptosis
B. HARVAT and A.M. JETTEN. With 3 Figures . . . 239

A. Introduction . . . 239
B. Mechanisms of Retinoid Action . . . 240
I. Retinoid Receptor-Dependent Mechanisms . . . 240
II. Retinoid Receptor-Mediated Control of Cell Proliferation . . . 241

III. Receptor-Independent Regulation of Cell Proliferation ... 242
C. Control of Cell Cycle Progression by Retinoids ... 244
I. Overview of Cell Cycle Regulation ... 244
II. Cell Cycle Regulatory Targets of Retinoid Action ... 246
1. Retinoids Can Affect the Rb Pathway at Several Levels ... 246
2. Regulation of c-myc Expression ... 248
3. Modulation of AP-1-Mediated Growth Signals ... 249
4. Effects on Miscellaneous Cell Cycle Proteins ... 250
D. Regulation of Apoptosis ... 250
I. Apoptotic Mechanisms ... 250
II. Regulation of Apoptosis by Retinoids ... 251
1. Receptor-Mediated Induction of Apoptosis ... 252
2. Receptor-Mediated Inhibition of Apoptosis ... 254
3. Retinoid Receptor Independent Mechanisms ... 255
a) Mechanisms of 4-HPR Action in Apoptosis ... 256
b) Mechanisms of AHPN Action ... 257
E. Summary ... 259
References ... 259

CHAPTER 9

Retinoids and Differentiation of Normal and Malignant Hematopoietic Cells
A. AGADIR and C. CHOMIENNE. With 4 Figures ... 277

A. Introduction ... 277
B. Retinoids and Hematopoiesis ... 280
I. Effect of Retinoic Acid on Cellular Proliferation, Differentiation, and Apoptosis of Hematopoietic Leukemic Cells ... 280
1. Effect on Myeloid Leukemic Cell Differentiation ... 280
2. Effect on Acute Promyelocytic Leukemic Cells ... 281
3. Effect on Induced Myeloid Leukemic Cell Apoptosis ... 282
II. Retinoic Acid as a Novel Myeloid Differentiation Factor ... 282
III. Retinoic Acid Control of Myeloproliferative Growth ... 283
C. Retinoic Acid Signaling Pathways in Hematopoietic Cells ... 283
I. Differential Expression of Nuclear Retinoic Acid Receptors and Retinoid-Binding Proteins ... 283
II. Alteration of RARα in Hematopoietic Malignancies ... 284
1. Molecular Characteristics of APL ... 284
2. Alterations of RARα in Other Malignancies ... 285

D. Retinoic Acid as a Therapeutic Agent in Hematopoietic Malignancies ... 287
I. Acute Promyelocytic Leukemia ... 287
1. Retinoid Resistance in APL ... 288
2. Retinoic Acid Syndrome and Other Side Effects of RA Therapy ... 289
3. Mechanisms of retinoic acid induced differentiation in APL patients ... 290
II. Retinoids as Therapeutic Alternative to Myeloproliferative Disorders ... 290
E. Summary and Perspectives ... 291
References ... 291

CHAPTER 10

The Retinoids: Cancer Therapy and Prevention Mechanisms
K. NASON-BURCHENAL and E. DMITROVSKY. With 1 Figure ... 301

I. Summary ... 301
II. Retinoid Clinical Activities ... 301
III. Mechanisms of Retinoid Action ... 303
IV. Retinoids and Cancer Chemoprevention ... 306
V. In Vitro Models for Retinoid Activity in Cancer Therapy and Prevention ... 308
A. The Multipotent NT2/D1 Human Embryonal Carcinoma Line ... 308
B. Acute Promyelocytic Leukemia In Vitro Models ... 310
C. An In Vitro Lung Cancer Prevention Model ... 313
VI. Summary and Future Directions ... 314
References ... 315

CHAPTER 11

Aberrant Expression and Function of Retinoid Receptors in Cancer
X.-C. XU and R. LOTAN ... 323

A. Introduction ... 323
B. Aberrant nuclear retinoid receptor expression and function in cultured cancer cell lines ... 325
I. Expression of Nuclear Retinoid Receptors in Lung Cancer Cell Lines ... 325
II. Expression of Nuclear Retinoid Receptors in Head and Neck Squamous Cell Carcinoma Cell Lines ... 325
III. Expression of Nuclear Retinoid Receptors in Esophageal Cancer Cell Lines ... 326

IV. Expression of Nuclear Retinoid Receptors in Breast Cancer Cell Lines ... 327
V. Expression of Nuclear Retinoid Receptors in Normal Skin and Skin Cancer Cell Lines ... 328
VI. Expression of Nuclear Retinoid Receptors in Cervical Cancer Cell Lines ... 328
VII. Expression of Nuclear Retinoic Acid Receptors in Embryonal Carcinoma Cell Lines ... 329
VIII. Expression of Nuclear Retinoid Receptors in Myeloid Leukemia ... 329
C. Aberrant Nuclear Retinoid Receptor Expression in Specimens from Normal, Premalignant, and Malignant Tissues ... 330
I. Expression of Aberrant RARα in Acute Promyelocytic Leukemia ... 330
II. Expression of Nuclear Retinoid Receptors in Adjacent Tissues and Head and Neck Squamous Cell Carcinoma ... 331
III. Expression of Nuclear Retinoic Acid Receptors in Hamster Cheek-Pouch Mucosa During 7,12-Dimethylbenz[a]anthracene-Induced Carcinogenesis ... 332
IV. Expression of Nuclear Retinoid Receptors in Adjacent Bronchial Epithelium and Non-small Cell Lung Cancer ... 333
V. Expression of Nuclear Retinoid Receptors in Adjacent Normal, Premalignant, and Malignant Breast Tissues ... 333
VI. Expression of Nuclear Retinoid Receptors in Normal Skin and Skin Cancers ... 335
VII. Expression of the Nuclear Retinoid Receptors in Normal Uterine Cervix and Cervical Cancer ... 336
D. Mechanisms of Altered Receptor Expression and Function In Vitro and In Vivo ... 336
References ... 338

Section D: Development and Teratogenesis

CHAPTER 12

Genetic and Molecular Approaches to Understanding the Role of Retinoids in Mammalian Spermatogenesis
A.I. Packer and D.J. Wolgemuth. With 1 Figure ... 347

A. Introduction ... 347
I. Historical Perspective of the Role of Vitamin A in Spermatogenesis ... 347
II. Focus of This Review ... 347
B. Serum and Cellular Retinoid Binding Proteins ... 348
I. Sites of Synthesis and Action ... 348

1. Retinol Binding Protein and Transthyretin 348
2. Cellular Retinol and Retinoic Acid Binding Proteins ... 349
II. Potential Functions During Spermatogenesis 350
C. Receptors 351
I. Retinoic Acid and Retinoid X Receptors and Their Ligands 351
II. Regulation of Transcription by Ligand-Activated Retinoid Receptors 351
III. RAR and RXR Expression in the Testis 352
D. The Vitamin A Deficient Testis 353
E. Genetic Analysis of Retinoid Function 354
I. Overview of Mouse Mutants 354
II. Mutations in Retinoid Receptors and Binding Proteins with no Testicular Phenotype 355
III. Mutations in Retinoid Receptors that Result in Male Sterility 356
1. RARs 356
2. RXRs 357
F. Where Do We Go From Here? 358
I. Assessment of Lineage Specificity of Receptor Function in the Testis 358
II. Identification of In Vivo Cofactors 359
III. Potential Targets of Retinoid Action During Spermatogenesis 360
1. Mutations Affecting Spermatogenesis 360
2. Genes Whose Testicular Expression Is Affected by Retinoids 361
3. Other Retinoid-Responsive Genes 362
References 363

CHAPTER 13

The Role of Retinoids in Vertebrate Limb Morphogenesis: Integration of Retinoid- and Cytokine-Mediated Signal Transduction

H.-C. Lu, C. Thaller, and G. Eichele. With 2 Figures 369

A. Introduction 369
B. An Overview of Retinoid Signal Transduction in the Vertebrate Embryo 370
I. Retinoid Requirement in Development 371
II. Retinoic Acid Synthesizing Enzymes and Degrading Enzymes in Embryos 371
III. RARs and RXRs Mediate Retinoic Acid Signal Transduction In Vivo and Are Required for Development 373

IV. Retinoid Target Genes 375
C. Retinoids in Vertebrate Limb Development 378
I. Embryology of the Vertebrate Limb 380
II. The Zone of Polarizing Activity 381
III. Retinoids and the Generation of the ZPA 383
IV. Retinoids in Limb Differentiation 386
D. Retinoids in Limb Regeneration 387
References 389

CHAPTER 14

Retinoids in Neural Development
M. MADEN. With 2 Figures 399

A. The Induction of Neuronal Differentiation in Culture 399
I. Embryonal Carcinoma Cells and Neuroblastoma Cells 399
II. Dissociated or Explanted Neuronal Cells 402
III. Neural Crest Cells 403
B. The Effects of Excess Retinoids on the CNS 404
I. Effect 1: Posteriorisation 405
II. Effect 2: Loss of Anterior Hindbrain 407
III. Effect 3: Transformation of Anterior Hindbrain 410
C. The Effects of RA on Associated CNS Structures 411
D. The Effects of a Deficiency of Retinoids on the CNS 412
I. Deprivation of Retinoids 412
II. Inhibition of RA Synthesis 414
E. Endogenous RA in the CNS 415
I. HPLC Data 415
II. Transgenic Embryos 416
III. Reporter Cells 417
IV. RA Synthesising Enzymes 419
V. Which Cells in the Nervous System Synthesise RA? 421
F. Binding Proteins and Receptors in the Developing CNS 421
I. CRBP I 421
II. CRABP I 422
III. CRABP II 425
IV. RARs and RXRs 426
V. Knockouts of Binding Proteins and Receptors 427
VI. Disruption of Function 428
VII. Receptor Selective Agonists 429
G. Conclusions 430
References 430

CHAPTER 15

Avian Embryo as Model for Retinoid Function in Early Development
M.H. ZILE. With 1 Figure 443

A. Introduction ... 443
B. Teratogenic Effects of Vitamin A Deficiency and Excess ... 444
C. Mammalian Models ... 446
D. Avian Embryo "Retinoid Ligand Knockout" Model ... 447
E. Cardiogenesis ... 450
F. Left-Right Asymmetry ... 451
G. Vasculogenesis ... 452
H. Central Nervous System ... 453
I. Ethanol-Induced Retinoid Depletion ... 454
J. Retinoid Metabolism ... 455
K. Future Directions ... 456
References ... 457

CHAPTER 16

Retinoid Receptors, Their Ligands, and Teratogenesis: Synergy and Specificity of Effects
H. NAU and M.M.A. ELMAZAR. With 6 Figures ... 465

A. Introduction ... 465
B. Ligands of the Retinoid Receptors ... 467
  I. Endogenous Ligands of the Retinoid Receptors ... 467
  II. Retinoid Receptor Ligands in the Embryo ... 468
C. Retinoid Receptors and Binding Proteins in the Embryo ... 471
  I. Retinoid Binding Proteins in the Embryo ... 471
  II. Retinoid Receptors in the Embryo ... 471
D. Experimental Models for the Study of the Significance of Retinoid Receptors in Teratogenesis ... 472
  I. "Loss of Function" Approach: Studies with Null Mutant Mice ... 472
  II. "Gain of Function" Approach: Administration of Selective Retinoid Receptor Ligands ... 473
E. Synergistic Teratogenic Action Following Combined Administration of RAR and RXR Ligands ... 476
F. Molecular Pathways of Vitamin A and Retinoid Teratogenesis ... 478
G. Conclusions ... 479
References ... 480

Section E: Skin

CHAPTER 17

Vitamin A Homeostasis in Human Epidermis: Native Retinoid Composition and Metabolism
R.K. RANDOLPH and G. SIEGENTHALER. With 4 Figures ... 491

A. Introduction ... 491
B. Total Vitamin A Content of the Epidermis ... 492
C. Vitamin A Metabolites in the Epidermis ... 493
D. Retinoid Composition of Epidermal Compartments ... 497
E. Retinoid Composition of Cultured Human Keratinocytes ... 498
F. Extracellular Retinoid Transport ... 499
I. Retinol ... 499
II. Retinoic Acid ... 500
G. Retinoid Metabolism in Epidermis ... 502
I. Retinoid-Binding Proteins ... 502
1. Cellular Retinol-Binding Protein ... 502
2. Cellular Retinoic Acid Binding Proteins ... 504
II. Vitamin A Storage, Retinyl Ester Synthesis and Hydrolysis ... 507
III. Retinoic Acid Synthesis ... 509
IV. Retinoic Acid Metabolism ... 511
V. 3,4-Didehydroretinol Synthesis and Metabolism ... 513
H. Summary and Working Model for Retinoid Metabolism in Epidermis ... 514
I. Future Prospects ... 515
References ... 515

CHAPTER 18

New Concepts for Delivery of Topical Retinoid Activity to Human Skin
J.-H. SAURAT, O. SORG, and L. DIDIERJEAN. With 2 Figures ... 521

A. Introduction ... 521
B. The "Proligand-Nonligand" Concept ... 524
C. All-*trans*-Retinol ... 525
D. All-*trans*-Retinyl Esters ... 527
E. All-*trans*-Retinal ... 530
F. All-*trans*-Retinoyl-β-Glucuronide ... 533
G. Conclusions and Perspectives ... 534
References ... 535

CHAPTER 19

Retinoid Receptor-Selective Agonists and Their Action in Skin
B. SHROOT, D.F.C. GIBSON, and X.-P. LU. With 4 Figures ... 539

A. The Molecular Aspects of Retinoid Signaling ... 539
I. The Role of Retinoid Receptors ... 539
II. RAR Receptor-Selective Retinoids ... 541
III. RXR Receptor Selective Retinoids ... 542
B. Vitamin A and the Skin ... 543

I. Proliferation and Differentiation Within the Epidermis 543
II. Vitamin A and the Skin: The Historical Aspects 543
III. RARs and RXRs and the Development of the Skin 544
IV. Retinoid Effects upon Keratinocyte Proliferation and Differentiation 545
V. Modulation of Inflammation and the Immune Response in Skin by RA 546
VI. Retinoid Receptors and Skin Function 547
VII. The Development of Synthetic Retinoids and Their Action in Skin 548
1. Adapalene (CD271) 549
2. Tazarotene (AGN 190168) 550
C. Apoptosis: An Emerging Role for Retinoids 551
D. Future Prospects 552
References 552

Section F: Special Effects

CHAPTER 20

Retinoids in Mammalian Vision

J.C. SAARI. With 5 Figures 563

A. Introduction 563
I. Overview of the Visual Process 563
II. Anatomy 564
III. Nourishment of the Outer Retina 566
IV. Overview of the Visual Cycle 567
B. Structure and Function of Visual Pigments 567
I. Interaction of 11-*cis*-Retinal and Opsin 567
II. Cone Visual Pigments 569
III. Identification of the Activated Photoproducts 570
C. Visual Pigment Regeneration 571
I. All-*trans*-Retinol Dehydrogenase 573
II. Movement of Retinoids Between RPE and Photoreceptors 574
III. Regeneration in Rods Compared to Cones 575
IV. Müller Cells 575
V. Control of the Visual Cycle 576
VI. Communication Between Neural Retina and RPE 577
D. Cellular and Extracellular Retinoid-Binding Proteins 578
I. Cellular Retinal-Binding Protein 578
II. Interphotoreceptor Retinoid-Binding Protein 579
III. Cellular Retinol-Binding Protein 580
IV. Retinol-Binding Protein 580

E. Diseases Associated with Retinoid Metabolism or Function in the Retina 581
I. Congenital Stationary Night Blindness 581
II. Autosomal Recessive Retinitis Pigmentosa 582
III. Retinoids and Lipofuscin 582
F. Summary 583
References 583

CHAPTER 21

Retinoids and Immunity
C.E. HAYES, F.E. NASHOLD, F.E. GOMEZ, and K.A. HOAG. With 2 Figures 589

A. Introduction 589
B. Innate Immunity 590
I. Natural Killer Cells 590
II. Phagocytic Cells 591
C. Adaptive Immunity 591
I. Antibody-Mediated Immunity 591
II. Cell-Mediated Immunity 594
D. Immunity to Infections 595
I. Measles Infection 595
II. Human Immunodeficiency Virus Type 1 Infection 596
III. Parasitic Infections 597
E. Autoimmunity 597
I. Rheumatoid Arthritis 597
II. Multiple Sclerosis 598
F. Lymphocyte Turnover 599
G. Vitamin A Metabolism 599
I. Active Metabolites for Immune Function 599
II. Metabolism During Infection or Inflammation 600
H. Mechanisms of Vitamin A Action in the Immune System 601
I. Summary and Future Research Directions 603
References 603

Subject Index 611

Section A
Metabolism, Pharmocokinetics and Action of Retinoids

CHAPTER 1
Physical-Chemical Properties and Action of Retinoids

N. Noy

A. Introduction

This chapter considers how the physical-chemical properties of retinoids and the characteristics of their interactions with the various environments in which they are distributed in vivo affect their biological functions. The discussion focuses on current knowledge on the equilibrium and kinetic parameters that govern the behavior of retinoids within aqueous phases, biological membranes, and binding sites of proteins. The possible implications of this information for the molecular mechanisms underlying some aspects of retinoid biology are then considered.

This chapter is not intended to be all-inclusive and centers mainly on the retinoid forms shown in Fig. 1. As with other retinoids, these compounds are comprised of three distinct structural domains: a β-ionone ring, a polyunsaturated chain, and a polar end-group. The polar end-group can exist at several oxidation states varying from the reduced retinol, to the higher oxidation states in retinal and retinoic acid. In recent years a wide array of synthetic analogs of retinoids have been developed. The β-ionone ring has been systematically replaced by multiple hydrophobic groups, the isoprenoid chain has been derivatized to a variety of cyclic and aromatic rings, and the polar end group has been modified extensively. However, active synthetic analogs, similarly to naturally occurring retinoids, retain an amphipathic nature typified by a hydrophobic moiety and a polar terminus.

B. Behavior of Retinoids in an Aqueous Environment

I. Aqueous Solubility

As a consequence of the presence of a large hydrophobic moiety in retinoids, these compounds are poorly soluble in water. Quantitative data on the aqueous solubilities of retinoids are scarce. In one study the solubility limits of four synthetic retinoids, etretinate, motretinide, fenretinide, and *N*-ethyl retinamide were measured to be about 25 n*M*, or about three orders of magnitude higher than the theoretically predicted solubility, estimated to be 0.01 n*M* (Li et al. 1996). It was suggested that the large discrepancy between the predicted and the observed values stem from self-aggregation of the

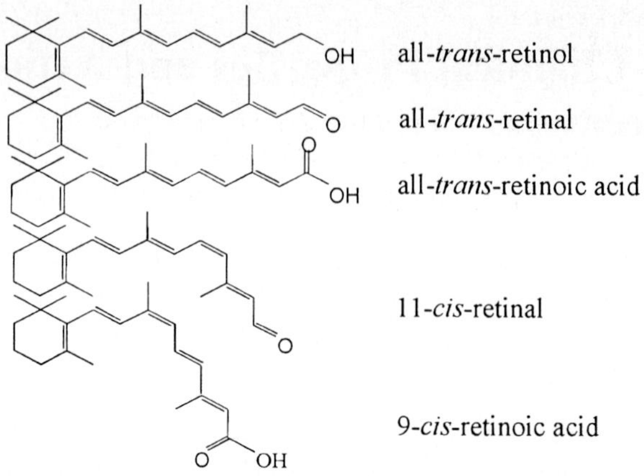

all-*trans*-retinol

all-*trans*-retinal

all-*trans*-retinoic acid

11-*cis*-retinal

9-*cis*-retinoic acid

Fig. 1. Structures of a few retinoids

retinoids in the aqueous environment (see Sect. B.II). Another study used a spectrophotometric method relying on the differential absorption spectra of retinoids in water vs. that in ethanol to measure the aqueous solubilities of three naturally occurring retinoids and found that the solubility limits of retinol, retinal, and retinoic acid in water are 60, 110, and 210 n*M*, respectively (Szuts and Harosi 1991). It should be noted, however, that interpretation of the data concerning retinoic acid may be complicated because while the absorption spectrum of retinoic acid is indeed sensitive to the polarity of the solvent, the spectrum also changes dramatically in response to variations in pH and in the ionization state of retinoic acid (see Sect. B.II).

II. Self-Association of Retinoids in Water

Self-aggregation of retinoids can have significant implications for the biological functions of these compounds. For example, a reasonable explanation for the observation that the solubility of retinoids in water is higher than predicted (Li et al. 1996) is that the observed solubility reflects self-association into micelles rather than the monomeric retinoid solubility. Thus micelle formation may allow higher than expected aqueous concentrations of retinoids to be maintained, and these concentrations may be high enough to support diffusional fluxes of free retinoids that are sufficient for short-distance transport in cells or to allow metabolism of free retinol. Another action of retinoids that may be affected by their self-association properties relates to the often-raised notion that at least some of the toxic effects of vitamin A are due to surface active, "membranolytic," properties. Consideration of the chemical structure of many retinoids suggests that their amphipathic nature result in "detergent-like" characteristics, and several studies have shown that the presence of retinoids affect various aspects of membrane structure and function (DINGLE and LUCY 1965; STILLWELL and RICKETTS 1980; STILLWELL et al. 1982; DEBOECK and ZIDOVETZKI 1988; WASSALL et al. 1988). As with other detergents, retinoids may disrupt membranes both by intercalating into the lipids, thereby changing the characteristics of the bilayer, and by drawing lipids out of the bilayers into retinoid-lipid mixed micelles, leading to dissolution of the membranes.

Few studies have addressed the self-association properties of retinoids. It has been reported that the fluorescence polarization of $2\,\mu M$ retinol in cyclohexane and in an aqueous buffer is 0.04 and 0.35, respectively (RADDA and SMITH 1970). These observations indicate that the rotational volume of retinol in the aqueous buffer is significantly larger than the rotational volume in the apolar solvent and imply that, while retinol is monomeric in cyclohexane, it is associated with much larger structures, i.e., micelles, in water. These data provide an upper limit of $2\,\mu M$ for the critical micellar concentration (CMC) of retinol.

Unlike retinol, retinoic acid (RA) may carry an actual net charge. This feature may result in a stronger detergentlike characteristics of this retinoid and perhaps provide an explanation for the reported differences between the effects of RA vs. other retinoids on membranes (STILLWELL et al. 1982; WASSALL et al. 1988). These considerations raise the question of whether the carboxyl group of RA is protonated or negatively charged at physiological pH, and whether the self-association characteristics of RA affect the compound's ionization state. These questions were examined in a study that utilized the large spectral shift of the absorption maximum of RA upon protonation of the carboxyl end group to monitor the ionization behavior of RA in water (NOY 1992a). The concentration dependence of the pK of RA indicated that the CMC of this retinoid is lower than $1\,\mu M$. The CMC of RA is thus similar to the reported upper limit of the CMC for retinol (RADDA and SMITH 1970), but more than three orders of magnitude lower than reported CMC values for bile acids (SMALL 1971) and tenfold lower than the CMC of the long-chain fatty acid palmitate (CISTOLA et al. 1988). These observations indicate that retinoids self-associate into micelles at substantially lower concentrations than other small amphipathic compounds.

Additionally it has been reported that while the pK of monomeric RA in an aqueous medium is lower than 6.1, the transition from a monomeric to a micellar state results in an increase in pK to 8.5, i.e., micelle formation stabilizes the protonated form of RA (NOY 1992a). Thus at physiological pH and at concentrations of RA that are higher than the CMC (i.e., in the micromolar range) a large fraction of RA is protonated, while at concentrations of RA that are lower than the CMC a predominant fraction of RA is ionized. As the aqueous solubility of the anionic RA is significantly higher than the solubility of the protonated form, these observations suggest that the concentrations of free retinoids in aqueous spaces in vivo are indeed on the order of $20–50\,\mu M$.

III. Lability of Retinoids in Water

The conjugated double-bonds of the isoprenoid chain render retinoids susceptible to photodegradation, isomerization, and oxidation. Consequently, naturally occurring retinoids are extremely labile. Retinol is more labile than either retinal or retinoic acid (Crouch et al. 1992), and retinoid isomers that are energetically unfavorable, especially the 11-*cis* configuration, are more susceptible than other isomers that constitute major fractions at equilibrium, such as all-*trans*-retinoids (Urbach and Rando 1994). These properties raise the question of how the structural integrity of retinoids is maintained in vivo. The answer lies in part with the observation that most retinoids in vivo are associated with retinoid-binding proteins. Many of these proteins bind their ligands in hydrophobic pockets, thereby effectively shielding them from the aqueous environment. Binding to proteins is thus expected to protect retinoids from nonspecific oxidation and may also stabilize their isomerization state. Two studies have examined the ability of a retinoid-binding protein that can bind several isomeric and chemical forms of retinoids, the interphotoreceptor retinoid-binding protein (IRBP), to protect some of its ligands (Ho et al. 1989; Crouch et al. 1992). It was found that placing all-*trans*-, 9-*cis*-, or 11-*cis*-retinol in an aqueous buffer lead to both isomerization and oxidation of these ligands. In contrast, in the presence of IRBP, retinol maintained its integrity for up to 1 h. Another study showed that trp^{19} of β-lactoglobulin serves to protect the structural integrity of bound ligand although this residue did not directly contribute to the formation of the protein-retinol complex (Katakura et al. 1994). These observations exemplify that binding proteins can protect retinoids from the deleterious effects of an aqueous environment.

The life-times of free retinoids in cytosol and in extracellular spaces and the mechanisms by which the integrity of retinoids in these pools is maintained are unknown. However, it has been reported that both degradation and isomerization of retinoids in water can be prevented by addition of biological antioxidants such as tocopherol (Jones et al. 1980; Crouch et al. 1992), suggesting the possibility that under at least some physiological conditions free retinoids can remain stable in cytosol.

Hence, both the concentrations and the stability of retinoids in cytosol may be sufficiently high to allow the free retinoids pools in cells to play important roles in retinoid transport and metabolism. One example that supports this notion is the report that retinol dehydratase, an enzyme that catalyzes the conversion of retinol to anhydroretinol, utilizes free retinol as its substrate (Grun et al. 1996).

C. Interactions of Retinoids with Membranes

Being small hydrophobic compounds, retinoids readily associate with the lipid moieties of biological membranes. For example, the partition constant (expressed as ratio of mole fractions) of retinol between rat liver plasma membranes and an aqueous buffer was reported to be 3.6×10^5 (Noy and Xu 1990b). The high solubility of retinoids in membranes may affect various processes including the inter- and intracellular transport of retinoids, and retinoid metabolism. Theoretically, retinoids distribute in vivo between aqueous phases, membranes and corresponding binding proteins. The concentrations of a particular retinoid in the different phases are determined by the relative local concentrations of the binding proteins and the membranes and by their relative affinities for the retinoid. It is thus important to consider the parameters that may influence the affinity of biological membranes for retinoids.

I. Effects of Membrane Lipid Composition on Membrane-Retinoid Interactions

The association of retinoids with membranes is stabilized by various types of retinoid-lipid interactions. Retinoids intercalate into membranes with the β-ionone ring and the isoprenoid chain packed alongside the fatty acyl chains of membrane lipids, while the polar end group of retinoids is located at the surface, most likely hydrogen bonded to the phospholipid headgroups (for discussion see Noy and Xu 1990a). Variation in the lipid composition of the membrane can thus be expected to affect the energetics of the retinoid-lipid interactions. Quantitative information on this issue is scarce, but it has been reported that the affinity of membranes for retinol indeed varies between various membranes, and that these variations stem from differences in the lipid composition rather than the protein moiety of membranes (Table 1). It may be inferred from these limited data that the membrane solubility of retinol is determined by the exact composition of the acyl chains of membrane lipids. This conclusion suggests that differential lipid compositions of intracellular membranes may serve as a determining factor that directs retinoids to particular subcellular organelles.

Table 1. Effect of lipid composition on solubility of retinol in membranes (from Noy and Xu 1990a)

Lipid composition	Partition constant[a]
Egg yolk phosphatidylcholine (PC)	4.8
1-Oleoyl-2-palmitoyl-PC	4.4
1-Palmitoyl-2-oleoyl-PC	4.3
Dimyristoyl-PC	3.0
Dioleoyl-PC	2.2
Dioleoyl PC + 15 mol% cholesterol	2.0
Total lipids of rat liver plasma membranes	1.2

[a] The partition constants (K_p) of retinol between unilamellar vesicles comprised of the denoted lipids and rat liver plasma membrane fractions are shown; K_p reflects the relative solubility of retinol in the different membranes.

II. Effects of the Radius of Curvature of Membranes on Membrane-Retinoid Interactions

Another property of biological membranes that was reported to significantly affect their interactions with retinoids is the radius of curvature of membranes. It has been shown that membranes with smaller radii of curvature display a lower affinity of retinol as compare with membranes with larger radii, and that the observed decrease in affinity stems to a large degree from the faster rate by which retinol dissociates from smaller vs. large membranes into an aqueous medium (Noy et al. 1995). These findings (Fig. 2) indicate that while the rate of dissociation of retinol from model membranes with radii larger than 0.5 μm is fairly constant with a $t_{1/2}$ of about 7 s, this rate is significantly facilitated as the radius of curvature of the membrane decreases below 0.4 μm, and displays a $t_{1/2}$ of less than 1.5 s for membranes with a radius of about 0.025 μm.

The dramatic effect of the radius of curvature on the interactions of retinol with membranes raises the possibility that the geometry of a cell plays a role in determining the rates and directions by which retinoids enter or leave cells. This possibility is especially intriguing in regard to cells that are involved in retinoid uptake or secretion and that contain microvilli. One striking example is retinal pigment epithelium cells (RPE), polarized cells that are the point of entry of vitamin A (in the form of all-*trans*-retinol) into the eye. The RPE plays several critical roles in supporting the visual function, including the conversion of all-*trans*-retinol into 11-*cis*-retinal, the retinoid that serves as the chromophore for the visual pigment rhodopsin. Following its synthesis, 11-*cis*-retinal is secreted from RPE cells into the extracellular space known as interphotoreceptor matrix (IPM), transverses the IPM, and is taken up by pho-

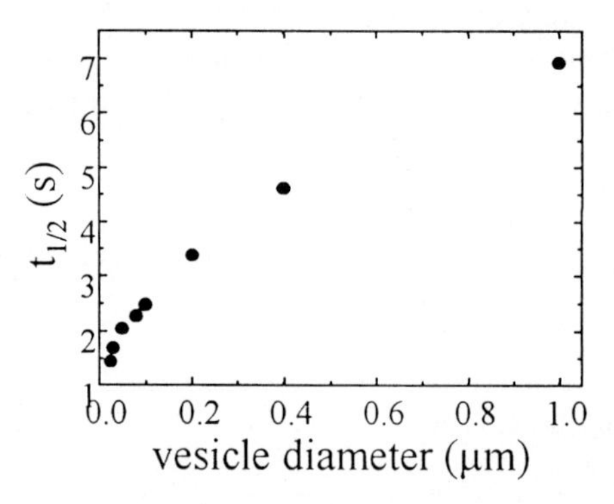

Fig. 2. Effect of radius of curvature on rate of dissociation of retinol from membranes. Retinol was incorporated into vesicles of dioleoylphosphatidylcholine of the designated diameters. These donor vesicles were mixed with acceptor vesicles which contained the fluorescent lipid probe NBD-DPPE. Movement of retinol from donor vesicles was followed by monitoring the fluorescence energy transfer between retinol and NBD-DPPE which increased as retinol arrived at the acceptor vesicles. $t_{1/2}$ was calculated by fitting the data to a single first order reaction. (Data from Noy et al. 1995)

toreceptor cells where it serves to regenerate rhodopsin (for review, see SAARI 1994).

A study aimed at examining the factors that may regulate the process by which 11-*cis*-retinal is secreted from RPE observed that the retinoid leaves cultured pigment epithelium cells preferentially via the cells' apical surface (CARLSON and BOK 1992). To account for these observations it was suggested that the secretion process is mediated by a protein receptor that is present in the apical, but not the basal, membrane of RPE (see also Sect. E.II).

An alternative explanation for these findings can be put forward based on the observations in Fig. 2 and by considering the architecture of RPE. The basal surface of the RPE cell contains extensive infoldings that protrude into the cytoplasm. On the other hand, the apical surface contains multiple processes that project into the IPM. The thickness of both the basal infoldings and the apical microvilli, i.e., the radii of curvature of the respective membranes, are on the order of 0.05 μm while the diameter of the RPE cell is in the 10- to 20-μm range (IWASAKI et al. 1992). As discussed above, the rates by which retinoids leave membranes with small radii of curvature are significantly faster than their rates of dissociation from membrane areas with large radii of curvature. Thus, it can be expected that retinoids enter RPE cells at much larger fluxes through the tips of the basal surface infoldings than through other membrane areas, and that they leave the cells preferentially through the tips of the apical surface processes. It seems then that the characteristics of the interactions of retinoids with membranes, combined with the geometrical properties of the RPE cells may serve to determine fluxes as well as directions by which retinoid enter into and exit from these cells (Fig. 3).

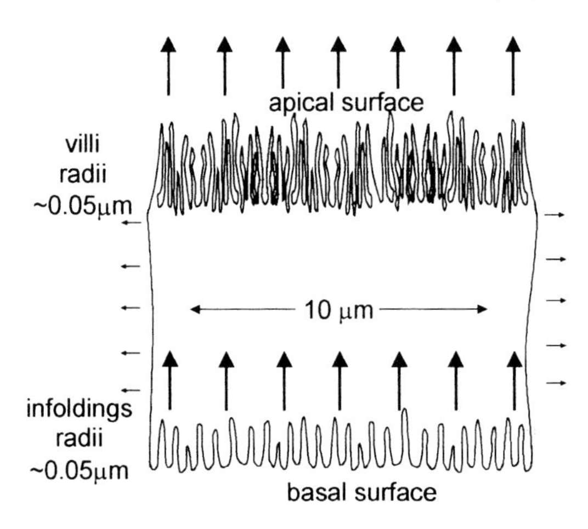

Fig. 3. The geometry of retinal pigment epithelium cells. RPE cells contain basal surface infoldings and apical surface processes with a typical thickness of 0.05 μm (IWASAKI et al. 1992). Based on the data in Fig. 2, it is suggested that, due to the small radii of curvature of the plasma membranes at these locations, retinoid fluxes into RPE are highest at the tips of the basal infoldings, while fluxes of retinoids exiting RPE are highest at the tips of the apical processes (see text)

III. Ionization Behavior of Retinoic Acid in Membranes

The high solubility of retinoids in lipids suggests that, unlike compounds that are freely soluble in water such as sugars and amino acids, crossing a biological membranes may not be a rate limiting step for the processes by which retinoids enter into of exit from cells. However, RA can be deprotonated and thus carry a net negative charge. Deprotonation does not obstruct the association of RA with membranes through its hydrophobic moiety but may significantly diminish the rate by which RA can transverse, or "flip-flop," across membranes. For example, studies with another class of carboxylic acids, long-chain fatty acids, have shown that while $t_{1/2}$ for "flip-flop" of the ionized species across lipid bilayers is about 700 s (Gutknecht 1988), $t_{1/2}$ for translocation of the protonated forms is on the order of milliseconds or faster (Daniels et al. 1985, Kamp et al. 1995). Hence an important question for understanding the intracellular distribution of RA is whether it is protonated or ionized when incorporated into membranes.

It has been reported regarding this issue (Noy 1992b) that, similarly to the behavior of other carboxylic acids (Cabral et al. 1986, Doody et al. 1980), the protonated form of RA is stabilized upon incorporation into lipid bilayers and membranes. More specifically it was reported (Noy 1992b) that the pK_a of RA, which is lower than 6.1 for monomeric RA in water (see Sect. B.II), is shifted to values in the range of 6.8–7.1 upon incorporation into lipid bilayers comprised of neutral lipids, and that it is not affected by variations in either the acyl chains or the headgroup of lipids within neutral (zwitterionic) bilayers. In contrast, it was observed that addition of acidic lipids into lipid bilayers results in further stabilization of the carboxylic group of RA and in a shift of the pK_a by an additional pH unit. pKa values for RA within synthetic bilayers containing acidic lipids were found to be in the range of 8.0–8.4 and were very similar to pK_a values within natural biological membranes such as erythrocyte membranes and rat liver plasma membrane fractions. As most biological membranes carry some negatively charged lipids, these observations indicate that RA in membranes is predominantly protonated, i.e., electrically neutral, at physiological pH and suggest that this compound can readily transverse biological membranes.

D. Interactions of Retinoids with Proteins

A major fraction of retinoids in vivo is associated with specialized retinoid-binding proteins. Such proteins have been identified in extracellular spaces: retinol-binding protein (RBP) in plasma, IRBP in the interphotoreceptor matrix in the eye, and β-lactoglobulin in milks of many mammals. It is generally believed that these proteins serve to increase the concentrations of their hydrophobic ligands in aqueous spaces thereby allowing higher fluxes of the ligands to reach target sites. The molecular mechanisms by which these proteins are loaded with their ligands at their source, i.e., at the sites of storage or synthesis, and by which they unload the ligands at the correct target sites are not completely clear at the present time (see also Sect. E). In addition to extracellular retinoid-binding proteins, proteins that bind specific retinoids have been identified in the cytosol of many cells. These proteins are named after the main retinoid with which they associate in vivo: cellular retinol-binding proteins (CRBP I and II) which, in addition to all-*trans*-retinol, also bind all-*trans*-retinal (reviewed in: Ong et al. 1994); cellular retinoic acid-binding proteins (CRABP I and II, Ong et al. 1994) which associate mainly with the all-*trans*-isomer of retinoic acid; and cellular retinal-binding protein (CRALBP), which is present in RPE, retina and some areas in the brain and specifically binds the 11-*cis*-isomers of retinal and retinol (Saari 1994, Saari et al. 1997). In general, it is believed that cellular retinoid-binding proteins have a dual function: (a) they act to transport their poorly soluble ligands across aqueous spaces between various intracellular sites, and (b) they regulate the metabolic fate of retinoids by modulating the activities of enzymes that catalyze metabolic transformations of their ligands. Another class of proteins that specifically bind retinoids are the members of the nuclear hormone receptor superfamily known as retinoid receptors. Two types and several isoforms of these proteins, the retinoic acid receptors (RARs) and the retinoid X receptors (RXRs) have been identified. RARs can bind and are activated by all-*trans*-RA and by 9-*cis*-RA, while RXRs bind the 9-*cis* isomer exclusively. These proteins serve as ligand-responsive transcription factors. They associate as dimers with particular recognition sequences (response elements) in the promoter regions of target genes and, upon binding of their ligands, either activate or repress transcription (reviewed in Mangelsdorf et al. 1994; Chambon 1996; Chap. 5, this volume; see also Sect. D.I.2).

Here, some aspects of the physical-chemical properties of the interactions of retinoids with their binding proteins are discussed, focusing on the potential implications of these properties for the function of the proteins and for retinoid action.

I. Binding Affinities

Much of the currently available information on equilibrium dissociation constants (K_d) that characterize the interactions of retinoid-binding proteins with their ligands has been obtained by fluorescence titrations, a method that relies on changes in the fluorescence properties of either the ligand or the protein upon formation of a protein-ligand complex. This method, which was first applied to a retinoid-binding protein by Cogan et al. (1976), does not require physical separation of free from bound ligand and is therefore especially useful in studying retinoids because these ligands are hydrophobic and have a tendency to adhere to matrixes used to separate different species in standard binding assays. Fluorescence titrations yield the K_d for protein-ligand interactions as well as the number of active binding sites. The later parameter can be measured with high reliability. However, accurate estimates of K_d by this method cannot be obtained when the K_d of the protein-retinoid complex under investigation is significantly lower than the protein concentration used in reaction mixtures. In these cases, stoichiometric rather than equilibrium binding is observed, and measured K_d values reflect upper limits rather than actual values (for further discussion see: Norris and Li 1998). Comparison between K_d values obtained by fluorescence titrations and values measured by other methods suggests that, under current practices, fluorescence titrations yield reliable values for protein-retinoids complexes with K_d values on the order of 20 nM or higher, but only report upper limits for protein-retinoid complexes with lower K_d values. For example, K_d for the complex of RXRα with 9-*cis*-RA was found by fluorescence titrations to be 15 nM (Kersten et al. 1996), in agreement with the value obtained by a charcoal binding ssay (Allenby et al. 1993). However, fluorescence titration measurement of the K_d for RARα also yielded a value of 15 nM (Kersten et al. 1996), although it is well established that RAR display a much higher binding affinity for RA with K_d in the subnanomolar range (Allenby et al. 1993, Yang et al. 1991).

1. Ligand Binding Affinities of Retinoid-Binding Proteins

Available information on equilibrium dissociation constants of retinoid-binding proteins is shown in Table 2. In general, it seems that the affinities that govern the interactions of extracellular retinoid-binding proteins with their ligands are weaker than those stabilizing complexes of intracellular retinoid-binding proteins with their main ligands. It is interesting to consider whether these observations may reflect that extracellular and intracellular retinoid-binding proteins play different roles in retinoid biology, and whether fulfillment of these roles requires different retinoid-binding affinities. Extracellular retinoid-binding proteins are believed to function as transport vehicles, allowing their poorly soluble ligands to move through the extracellular aqueous spaces in which they are located to target cells. Both RBP and IRBP remain in the extracellular space, blood and interphotoreceptor matrix, respectively, while their ligands move into the target cells in the form of free retinoids. It is thus reasonable to propose that high fluxes of retinoids into the target cells require that the binding proteins have a moderate, rather than a very high binding affinity for their ligand. This property allows the presence of sufficiently high concentrations of free retinoids (also see Sect. E).

In contrast with extracellular proteins, intracellular retinoid binding proteins are believed to function not only as retinoid carriers but also as regulators of retinoid metabolism and thus of retinoid action. It has been shown that metabolic transformations of free retinoids can be catalyzed by multiple enzymes, some of which display very broad substrate specificities in vitro (see Ong 1994; Napoli et al. 1995; Duester 1996). The factors that determine the substrate selectivities of these enzymes in vivo, and thus the identities of the enzymes that catalyze retinoid metabolism, are not completely clear at

Table 2. Equilibrium dissociation constants of retinoid-binding proteins

Protein	Source	Ligand	K_d (n*M*)	Comments
Extracellular				
RBP	Human	All-*trans*-retinol	190[a]	K_d values are ca. twofold lower in the presence of TTR
	Bovine		70[b]	
	Rat		20[b]	
IRBP	Bovine retina	All-*trans*-retinol	Site 1: 80 Site 2: ~150	Two retinoid binding sites. K_d measured in the absence of long chain fatty acids[d]
		11-*cis*-retinal	Site 1: 80 Site 2: ~150	
Cellular				
CRBPI	Rat liver	All-*trans*-retinol	<10[e]	Values for retinol are upper limits only
	Recombinant (rat)	All-*trans*-retinal	50[e]	
CRBPII	Rat small intestine	All-*trans*-retinol	<10[e]	$K_{d\ retinol}$ CRBPI < $K_{d\ retinol}$ CRBPII[f]
	Recombinant (rat)	All-*trans*-retinal	90[e]	
CRABPI	Recombinant (human)	All-*trans*-ra	0.06[g]	
CRABPII	Recombinant (human)	All-*trans*-ra	0.13[g]	
CRALBP	Bovine retina	11-*cis*-retinal	~15[h]	

[a] Cogan et al. (1976); fluorescence titrations.
[b] Noy and Xu (1990c); fluorescence titrations verified by measurements of the rate constants for association (k_{on}) and dissociation (k_{off}) of the retinol-RBP complex and utilizing the relationship: $K_d = (k_{off})/(k_{on})$.
[c] Noy and Blaner (1991); fluorescence titrations.
[d] Chen et al. (1996).
[e] Ong and Chytil (1978), MacDonald and Ong (1987), Levin et al. (1988); fluorescence titrations.
[f] Li et al. (1991).
[g] Dong et al. (1999); based on measured kinetic parameters and calculated by: $K_d = (k_{off})/(k_{on})$.
[h] Noy and Saari, unpublished data.

present, but it has been shown that all three types of intracellular retinoid-binding proteins (CRBPs, CRABPs, and CRALBP) affect the activities of enzymes that catalyze retinoid metabolism (Ong 1994; Napoli et al. 1995; Saari et al. 1994). It has consequently been suggested that the metabolic fates of retinoids in vivo are regulated by binding proteins, acting either by limiting the accessibility of their ligands to some enzymes while allowing catalysis by others, i.e., by "channeling" their ligands to the correct metabolic pathways, or by directly regulating the activities of enzymes. It thus seems that specificity and proper regulation of retinoid metabolism require that a predominant fraction of intracellular retinoids be bound to the corresponding binding proteins, and that intracellular concentrations of free retinoids be minimized. The high binding affinities characterizing the association of retinoids with cellular binding proteins satisfy these conditions.

2. Ligand Binding Affinities of Retinoid Nuclear Receptors

The equilibrium dissociation constants of RXR and RAR with their presumed physiological ligands, 9-*cis*-RA and all-*trans*-RA, respectively, are listed in Table 3. Notably, RARs bind their ligands with a considerably stronger affinities than do RXRs. The most detailed current knowledge on the structures of the ligand-binding domains of retinoid receptors is based on the reported crystal structures of liganded RARγ And unliganded RXRα (Renaud et al. 1995; Bourguet et al. 1995), and thus precise comparison between the modes of interactions of retinoids with the two proteins awaits the solution of the crystal structure of liganded RXR. However, it has been proposed, based on the known structures and on homology modeling of other nuclear receptors, that the core of the protein fold of the ligand binding domains of both retinoid receptors are stabilized by a region of 20 conserved amino acids, and that the ligand binding pockets of the two proteins are very similar and involve predominantly hydrophobic residues (Wurtz et al. 1996). Nonetheless, the very large difference in ligand-binding affinities of RXR and RAR points at a significant difference between the manner by which the two classes of receptors interact with their ligands. The differences between the

Table 3. Equilibrium dissociation constants of retinoid nuclear receptors

Protein	Ligand	K_d (nM)
RARα, RARβ, RARγ	All-*trans*-RA	0.2–0.7[a]
	9-*cis*-RA	0.2–0.7[a]
RXRα, RXRβ, RXRγ	9-*cis*-RA	Monomer/dimer: 15–20[a,b]
RXRα		Tetramer: ~150[c]

[a] Allenby et al. (1993), Yang et al. (1991); charcoal binding assays.
[b] Kersten et al. (1996); fluorescence titrations.
[c] Kersten et al. (1995a); estimated from the difference in partitioning of 9-*cis*-RA between RXRα dimers or tetramers and phospholipids vesicles.

equilibrium dissociation constants of RAR and RXR also suggests that ligand-binding by these proteins have different functional outcomes.

While considering this issue it is important to note that several features of RXR distinguish it from other nuclear hormone receptors. For example, while RXR can bind to cognate DNA and regulate transcription as a homodimer, other hormone nuclear receptors, including RAR, form homodimers with low affinities and bind to DNA only weakly. Instead, these receptors strongly interact with RXR and exert their transcriptional activities mainly via the resulting heterodimeric complexes (YU et al. 1991; LEID et al. 1992; ZHANG et al. 1992; HALLENBECK et al. 1992; DURAND 1992). Thus, RXR serves as a common partner for several hormone nuclear receptors. Another difference between RXR and RAR relates to their ligand-dependent interactions with transcriptional intermediary factors. It has been reported that ligand-binding stabilizes the association of several nuclear receptors, including RAR and RXR, with transcriptional coactivators (CHAMBON 1996; HORWITZ et al. 1996; GLASS et al. 1997). It has also been shown that an important step in ligand-induced activation of RAR is the induction of dissociation of an accessory protein that is bound to the unliganded receptor and serves as a transcriptional corepressor. In contrast with RAR, RXR interacts with the corepressor only weakly and in a ligand-independent fashion, suggesting that ligand-dependent activation of RXR is exerted by a different mechanism (HORWITZ et al. 1996; GLASS et al. 1997; WOLFFE 1997; CHEN and EVANS 1995; HÖRLEIN et al. 1995; SANDE and PRIVALSKY 1996; SEOL et al. 1996).

Another remarkable feature of RXR is that it self-associates into homotetramers with a high affinity, such that the tetrameric species predominates at concentrations as low as 60–70 n*M* (KERSTEN et al. 1995a). The oligomeric behavior of RXR seems to be unique among nuclear hormone receptors, which as a general rule form dimers but not higher order oligomers. One indication that tetramer formation functions to regulate the receptor's transcriptional activity is that ligand binding by RXR results in rapid dissociation of tetramers into dimers and monomers (KERSTEN et al. 1995b,c; CHEN et al. 1998). Interestingly, as seen in Table 3, the ligand-binding affinity of RXR tetramers is about one-tenth that of dimers and monomers. It has also been reported that ligand-induced dissociation of RXR tetramers is characterized by a pronounced positive cooperativity (Fig. 4). Taken together, these observations imply that the positive cooperativity in tetramer dissociation stems from positive cooperativity in ligand-binding by RXR, i.e., that unliganded tetrameric RXR exists in a low-affinity state, and that binding of a ligand to one subunit within tetramers leads to a conformational change that is communicated to the remaining subunits resulting both in an increase in their ligand-binding affinity and in a disruption of protein-protein interactions that stabilize the tetramer (Fig. 5).

Cooperative behavior in the response of an oligomeric protein to its ligand can be an important regulatory feature. A classic example is this of another tetrameric protein, hemoglobin, which, due to the cooperativity characterizing

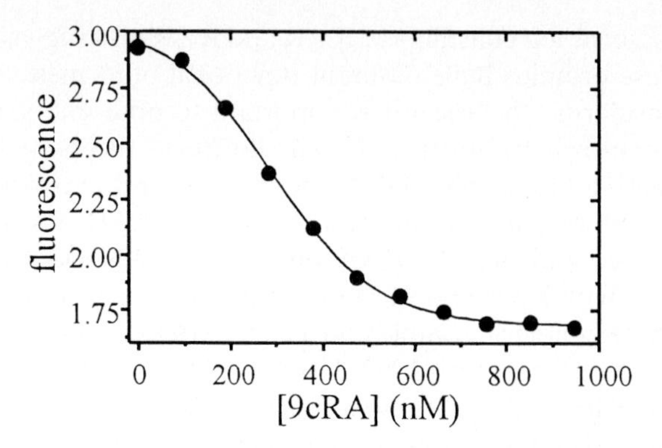

Fig. 4. Positive cooperativity in ligand-induced dissociation of RXR tetramers. RXR lacking the N-terminal A/B domain (RXRΔAB) was covalently labeled with the fluorescent probe fluorescein, and the fluorescence of the labeled protein was monitored upon titration with 9-*cis*-RA. The decrease in fluorescence reflects the change in the environment of the probe resulting from ligand-binding and, mainly, from dissociation of RXR tetramers. (Data from Kersten et al. 1998)

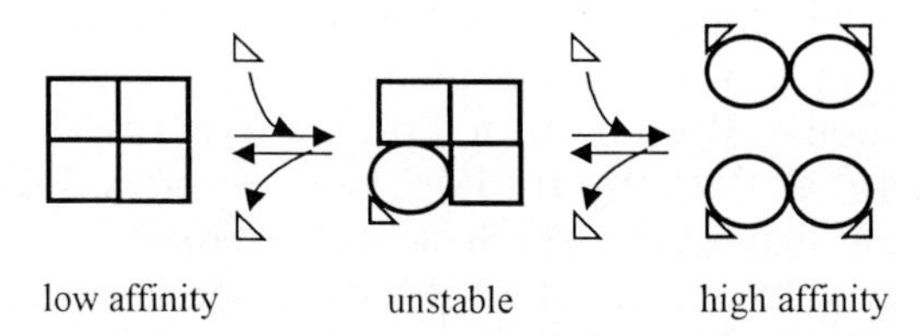

Fig. 5. A model for the mechanism underlying the positive cooperativity in ligand-induced dissociation of RXR tetramers. Nonliganded RXR is tetrameric and exists in a low-affinity state towards its ligand. Binding of ligand to one subunit within a tetramer leads to a conformational change that is communicated to the remaining subunits, resulting both in an increase in their ligand-binding affinity and in a disruption of protein-protein interactions that stabilize the tetramer. *Squares*, low ligand-binding affinity state of the protein; *circles*, high ligand-binding affinity state of the protein; *triangles*, RXR ligand

its interactions with molecular oxygen, is highly sensitive to variations in oxygen pressure, for example, between arterial and venous blood. This feature is critical for the function of hemoglobin as an oxygen carrier (see Streyer 1981). In order to understand how the cooperativity in the interactions of RXR with its ligand supports the receptor's function, the mechanisms by which ligand-binding modulates the activity of RXR need to be considered. It has been proposed in regard to this issue that RXR tetramers are transcriptionally inactive, and that an important function of the ligand is to release active species, dimers and monomers, from the silent tetrameric pool. This notion was supported by the finding that the region of RXR that mediates tetramer formation is critical for the receptor's activity (Kersten et al. 1997) and by the observations that a mutation that impairs the ability of RXR tetramers to dissociate upon ligand binding results in a receptor that cannot be activated by its ligand, while a RXR mutant that is unable to form tetramers displays substantial transcriptional activity even in the absence of ligand (KERSTEN et al. 1998). Hence the positive cooperativity in ligand-binding and in ligand-induced tetramer dissociation can be understood to be a mechanism that ensures that RXR subunits are safely sequestered in an inactive tetrameric pool that does not release active RXR species unless the ligand concentration exceeds a particular threshold.

Induction of tetramer dissociation is clearly not the sole outcome of ligand-binding by RXR (for discussion see KERSTEN et al. 1998). However, this particular function seems to be unique to RXR and may be the function that underlies the remarkable difference between the ligand-binding affinities of RARs and RXRs (Table 3). RAR binds its ligand with a very high affinity and thus can respond to low ligand concentrations by profoundly affecting the rates of transcription of target genes. In contrast, the first step in the RXR response is not a direct transcriptional response. Instead, this step, which requires significantly higher ligand concentrations, comprises the release of subunits from receptor tetramers, making them available for subsequent steps leading to transcriptional regulation.

II. Kinetics of the Interactions of Retinoids with Proteins

Equilibrium dissociation constants reflect the binding affinities characterizing the interactions of retinoids with proteins. K_d values of retinoid-protein complexes thus provide information on the distribution of retinoids in various cellular phases, and on the concentrations of retinoids that are required for their activities via particular proteins. However, processes in vivo may be poised far from the equilibrium. Hence, as demonstrated by the examples in Sect. E, detailed knowledge on the kinetics by which retinoids interact with the various compartments in which they are distributed in cells is important for understanding the factors that govern movement of retinoids between these compartments and the mechanisms of action of retinoid-binding proteins.

1. Rates of Dissociation of Retinoids from Retinoid-Binding Proteins

Current information on the rate constants of the dissociation of retinoid-protein complexes are listed in Table 4. With one exception, retinoids dissociate from their binding proteins at rates in the range of several minutes. The only exception to this general pattern is the rate by which ligands dissociate from IRBP, a reaction that proceeds with a $t_{1/2}$ of about 5 s. The fast rate by which retinoids dissociate from IRBP is likely to relate to the physiological role of this protein. IRBP is believed to mediate transport of its ligands across the aqueous space of the interphotoreceptor matrix, a transport process that serves to regenerate the visual pigment rhodopsin (PEPPERBERG et al. 1993; SAARI 1994, and see Sect. E.II). Quantitatively, the eye is the major user of

Table 4. Rate constants of dissociation of retinoids from retinoid-binding proteins

Protein	Ligand	K_{off} (s^{-1})	$t_{1/2}$ (s)	Comments
RBP	All-*trans*-retinol	0.0019[a]	365	$t_{1/2}$ (bovine) RBP in the presence of TTR, 690 s[c]
		0.003[b]	231	
IRBP	All-*trans*-retinol	0.108[d]	6.4	Mean for the two retinoid-binding sites in the absence of fatty acids
	11-*cis*-retinal	0.138[d]	5.0	
CRBPI	All-*trans*-retinol	0.0095[b]	73	
CRBPII	All-*trans*-retinol	ND		
CRABPI	All-*trans*-ra	0.0036[e]	192	
CRABPII	All-*trans*-ra	0.0062[e]	112	
CRALBP	11-*cis*-retinal	0.0015[f]	449	
	11-*cis*-retinol	0.0057[f]	120	

[a] Noy and Xu (1990c); bovine RBP.
[b] Noy and Blaner (1991); rat RBP, rat CRBP.
[c] Noy et al. (1992); bovine RBP + bovine TTR.
[d] Chen et al. (1996); bovine IRBP.
[e] Ruuska et al.; manuscript in preparation, recombinant human CRABP.
[f] Noy and Saari, unpublished data.

retinoids in the body, and clearly the retinoid fluxes that are required for sustaining high levels of functional rhodopsin under light conditions in the eye are much larger than fluxes required to sustain retinol metabolism in other tissues. It seems then that the unique biological need of the eye is adequately met by a retinoid-binding protein that can supply/release retinoids at a uniquely rapid rate.

E. Some Implications for Retinoid Transport Processes

Examination of the physical-chemical properties of the interactions of retinoids with the various cellular phases in which they distribute in vivo can provide important insights into the molecular mechanisms underlying the mechanisms of action of retinoid-binding proteins. This section considers the implications of current knowledge on the interactions of retinoids with RBP, with IRBP, and with biological membranes for the mechanisms by which the two proteins deliver their ligands to target cells.

I. Uptake of Retinol by Target Cells

Vitamin A is stored mainly in the liver and is delivered to target cells in the form of all-*trans*-retinol. Transport of retinol in plasma is mediated by a tertiary complex where retinol is bound to RBP, which in turn is complexed with transthyretin. In most tissues (though perhaps not all, see MATARESE and LODISH 1993) retinol enters target cells unaccompanied by the binding proteins, i.e., in the form of free retinol (for reviews see SOPRANO and BLANER 1994; see also Chap. 2, this volume).

Beyond these generally accepted observations, the mechanism by which retinol is taken up by cells is a controversial issue, and two alternative views have been presented. On one hand it has been proposed that a receptor embedded in the plasma membranes of target cells mediates transfer of retinol from RBP, across the membrane and into cytosol, or perhaps directly to CRBP (HELLER 1975; RASK and PETERSON 1976; RASK et al. 1980; BAVIK et al. 1991, 1993, 1995; SIVAPRASADARAO et al. 1994; SUNDARAM et al. 1998). The function of the RBP receptor has been assigned to a 63-kDa protein (p63) originally identified in retinal pigment epithelial cells (BAVIK et al. 1991, 1993) and subsequently suggested to also mediate uptake of retinol by cultured keratinocytes (BAVIK et al. 1995) and to be involved in transport of retinol to rat embryos (SIVAPRASADARAO et al. 1994; WARD et al. 1997). However, it has been pointed out in regard to this suggestion, that in order to fulfill its proposed function, p63 is expected to be embedded in the plasma membranes of target cells where it can interact with the RBP moiety of circulating retinol-transporting complex (NICOLLETI et al. 1995), and that, in conflict with this expectation, the primary sequence of p63 lacks regions of hydrophobic residues that would allow it to integrate into a membrane, and immunolocalization studies show that it is localized in the RPE cytoplasm (HAMEL et al. 1993; NICOLLETI et al. 1995). These observations question the ability of p63 to act in direct transfer of retinol from RBP into cells, and thus the exact role of this protein in retinol transport remains ambiguous. Nevertheless, it has recently been reported that transfer of retinol from RBP to CRBP in vitro is facilitated by about 2.5-fold in the presence of placental membranes, which was taken as functional evidence for the presence of a receptor for RBP in these membranes (SUNDARAM et al. 1998).

Alternatively, it has suggested been that movement of retinol from RBP in blood into target cells occurs spontaneously and simply follows the concentration gradients of free retinol (FEX and JOHANNESSON 1987, 1988; CREEK et al. 1989; NOY and XU 1990c; NOY and BLANER 1991; HODAM and CREEK 1998). It was noted that because of the hydrophobic nature of retinol, cellular membranes may not constitute a barrier for it. In addition, it was pointed out that whether uptake of retinol by cells is a facilitated process is a kinetic rather than an equilibrium issue, i.e., that a mediating mechanism such as a receptor would be required for enhancing retinoid fluxes only if the spontaneous rate of any of the steps that constitute the overall uptake process is too slow to support cellular needs for retinol. In order to examine this issue, the spontaneous/unfacilitated rates of the individual reactions that comprise transport of retinol from TTR-RBP, across membranes, to CRBP were measured.

The first step in this process involves the dissociation of retinol from the binding protein complex (reaction 1):

$$\text{TTR-RBP-retinol} \underset{}{\overset{k_{off}RBP}{\rightleftharpoons}} \text{TTR + RBP + retinol} \quad (1)$$

Subsequent steps include association of retinol with the outer leaflet of the plasma membrane of the target cell (or some component of this membrane) as depicted by reaction 2, and transversal (flip-flop) of the membrane to its cytosolic face (reaction 3).

$$\text{retinol + (membrane)}_{out} \overset{k_{on}mem}{\rightleftharpoons} \text{(membrane)}_{out}\text{-retinol} \quad (2)$$

$$\text{(membrane)}_{out}\text{-retinol} \overset{k_{flip\text{-}flop}}{\rightleftharpoons} \text{(membrane)}_{in}\text{-retinol} \quad (3)$$

The final steps in the uptake process are the dissociation of retinol from the plasma membrane into cytosol (reaction 4), and association with CRBP, where the predominant fraction of intracellular retinol is found (reaction 5):

$$\text{(membrane)}_{in}\text{-retinol} \overset{k_{off}mem}{\rightleftharpoons} \text{(membrane)}_{in} + \text{retinol} \quad (4)$$

$$\text{CRBP + retinol} \overset{k_{on}CRBP}{\rightleftharpoons} \text{CRBP-retinol} \quad (5)$$

Available information on the rates of reactions 1–5 and the retinol fluxes that can be accommodated by these reactions are shown in Table 5. Clearly the slowest reaction in the entire process is the dissociation of retinol from the TTR-RBP complex in blood.

As the overall rate of a multiple step process is determined by the slowest step, it can be concluded that the maximal rate of uptake of retinol by cells which can be accommodated without the aid of "a facilitator" corresponds to the rate by which retinol dissociates from its carrier protein in blood. Examination of the literature reveals that reported rates of uptake of retinol by target cells are two to three orders of magnitude slower than the fluxes that can be provided by the spontaneous dissociation of retinol from RBP (compare, e.g., rates reported by Hodam et al. 1991; Ward et al. 1997 with the rate of reaction [1]). The large magnitude of this difference strongly suggests, despite the various assumptions that were made in the course of the calculations (see Table 5), that uptake of retinol is not limited by any of the events described by reactions 1–5. Thus the rate of uptake of retinol by cells is likely to be limited by down-stream reactions such as the intracellular diffusion of CRBP-retinol to metabolism sites, or subsequent metabolic steps.

Overall these considerations suggest that, in most cells, the pools of retinol in blood and in cytosol are near equilibrium, and thus that retinol is selectively taken up only by cells in which a retinol concentration gradient directed inwards can be created. Such a gradient can be generated in cells which actively utilize retinol, or in cells in which CRBP is up-regulated, thereby decreasing the cytosolic concentration of free retinol. It seems then, that CRBP levels not only indicate the cellular need for retinol but also

Table 5. Rates by which reactions comprising the process of movement of retinol into cells can proceed in the absence of a facilitating mechanism

Reaction	k	Maximal spontaneous flux (v; nmol/g tissue per min)
Dissociation from RBP	$0.002\,s^{-1}$	0.24[a]
dissociation from TTR-RBP	$0.001\,s^{-1}$	0.12[a]
Association with a membrane	10^9–$10^{10}\,M^{-1}\,s^{-1}$	Fast[b]
"Flip-flop" across membrane	$>140\,s^{-1}$	>29,400[c]
Dissociation from a membrane	$\sim 0.1\,s^{-1}$	21[d]
Association with CRBP	$>1\times10^6\,M^{-1}\,s^{-1}$	>0.6–1.8[e]

[a] Values of k, see Table 5. Assuming $[\text{TTR-RBP-retinol}]_{blood} = 2\,\mu M$, and that 1 g tissue is associated with 2 nmol TTR-RBP-retinol (i.e., 1 ml blood); $v = k[\text{TTR-RBP-retinol}]_{blood}$.
[b] A diffusion-controlled reaction (Connors 1990).
[c] Value of k from Noy and Xu (1990a). The value for $[\text{retinol}]_{pm}$ was calculated from the equilibrium distribution constant of retinol between TTR-RBP and liver plasma membranes (PM), which is reported to be 141 (ratio of mole fractions; Noy and Xu 1990b), assuming that 1 g tissue is associated with 2 nmol of TTR-RBP-retinol (i.e., 1 ml blood) and contains 1 μmol PM lipids (Cook 1958). An additional assumption is that retinol is equally distributed between the outer and in the inner leaflets of the PM. Hence, $[\text{retinol}]_{pm/out} \sim 3.5$ nmol/g tissue; $v = k[\text{retinol}]_{pm/out}$.
[d] Value of k from Noy et al. (1995). $v = k[\text{retinol}]_{pm/in}$ where $[\text{retinol}]_{pm/in}$ was calculated as detailed in footnote c, above.
[e] Lower limit of k calculated from the upper limit of the K_d (Table 3) and the k_{off} (Table 5) for the CRBPI-retinol complex using the relationship $K_d = k_{off}/k_{on}$. The intracellular concentration of CRBP was taken to be 1 μM (Harrison et al. 1987). The range of the interacellular concentration of free retinol was calculated assuming that the actual K_d of the CRBPI-retinol complex is between 0.1 nM and 1 nM, resulting in $[\text{retinol}]_{cytosol} = 10$–$30$ nM; $v = k[\text{CRBP}]_{cytosol}\,[\text{retinol}]_{cytosol}$.

actively satisfy this need by participating in and by regulating retinol uptake into cells.

II. Targeting of Retinoids by IRBP

The visual chromophore, 11-*cis*-retinal, is generated from all-*trans*-retinol in retinal pigment epithelium cells in the eye. This retinoid is then transported to photoreceptor cells where it serves to regenerate the visual pigment rhodopsin. Exposure to light leads to isomerization of rhodopsin-bound 11-*cis*-retinal to the all-*trans*-form, which is then hydrolyzed from the protein and reduced to all-*trans*-retinol. All-*trans* retinol is transferred back to RPE for reisomerization and oxidation (for review see Saari 1994). Thus, continuous shuttling of retinoids between photoreceptors and pigment epithelium cells across the aqueous compartment that separates them, the IPM, is a critical part of the visual cycle. It is generally believed that this transport process is mediated IRBP, a 140-kDa retinoid and fatty acid binding glycoprotein which is the major soluble protein component of the IPM (reviewed by Pepperberg et al. 1993). IRBP contains two binding sites for retinoids and three or four binding sites for long chain fatty acids (Fong et al. 1984; Saari et al. 1985; Bazan et al. 1985; Chen et al. 1993; Chen et al. 1996). The two retinoid binding sites can bind several chemical and isomeric forms of retinoids but, in accordance to the proposed role of the protein, display the highest affinity towards 11-*cis*-retinal and all-*trans*-retinol (Chen and Noy 1994).

An important question in understanding the visual cycle thus relates to the mechanisms by which IRBP targets specific retinoids to the particular cells in the eye that are their sites of action and metabolism. It has been postulated in regard to this question that IRBP can selectively withdraw 11-*cis*-retinal from pigment epithelium cells, perhaps by binding to receptors in the plasma membranes of these cells. It has also been hypothesized that IRBP is recognized by receptors in the plasma membranes of photoreceptor cells, and that binding to these receptors induces the release of 11-*cis*-retinal from the protein (Flannery et al. 1990; Carlson and Bok 1992; Okajima et al. 1994; Jones et al. 1989; Okajima et al. 1990). Alternatively, it has been proposed that IRBP functions as a reservoir for retinoids in the IPM and that it simply binds and releases retinoids in response to changes in local concentrations of particular species. According to this hypothesis, "targeting" is accomplished passively and follows retinoid concentration gradients that are formed by metabolic and light-generated activities in specific cells (for discussion see Saari 1994).

Recently it has been suggested that the key for understanding how IRBP targets 11-*cis*-retinal to the photoreceptors while transporting all-*trans*-retinol to pigment epithelium lies with the interrelationships between the interactions of the protein with its two classes of ligands, retinoids, and long-chain fatty acids. Several lines of observations led to this view: (a) Photoreceptors and pigment epithelium cells differ in regard to their content of long-chain fatty acids; photoreceptors are enriched in the polyunsaturated fatty acid docosahexaenoic acid (DHA, 22:6, n-3), pigment epithelium cells contain a large fraction of saturated and monounsaturated fatty acids (Chen et al. 1992; Chen et al. 1996). (b) The fatty acid content of IRBP isolated from retina seems to reflect an intermediate between the levels of fatty acids in photoreceptors and in pigment epithelium (Chen et al. 1996). (c) In the absence of fatty acids as well as in the presence of saturated fatty acids, the two retinoid-binding sites of IRBP display similar affinities for both 11-*cis*-retinal and all-*trans*-retinol, and the rates of dissociation of the two retinoids from the two sites cannot be distinguished (Table 4). However, addition of DHA results in a dramatic decrease in the affinity of IRBP's retinoid-binding site 2 for 11-*cis*-retinal, and in an acceleration of the rate of dissociation of 11-*cis*-retinal from this site such that $t_{1/2}$ is decreased from 5 s (Table 4) to about 0.7 s (Chen et al. 1996). Strikingly, the effect is remarkably specific and binding of DHA has little effect on the interactions of retinoids with IRBP site 1, or on the interactions of site 2 with all-*trans*-retinol.

Based on these findings, a model for understanding how IRBP may target specific retinoids to particular cells in the eye was proposed (Fig. 6). Accord-

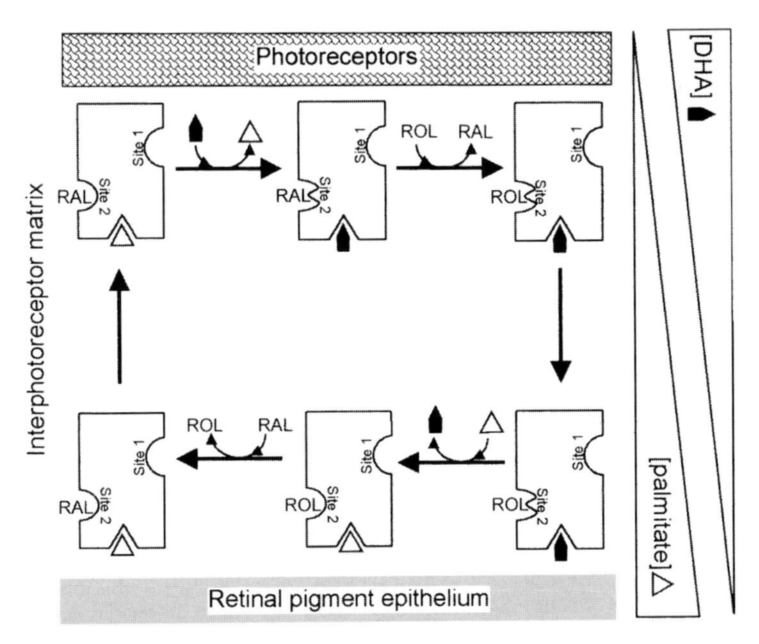

Fig. 6. A model for targeting of retinoids by IRBP. The model is based on the findings that binding of DHA to IRBP specifically switches the protein's retinoid binding site 2 from a state of a high affinity for 11-*cis*-retinal to a state of a low affinity for this ligand, and on the observed concentration gradient of DHA across the IPM. It is proposed that because the region of IPM close to RPE is poor in polyunsaturated fatty acids, IRBP possesses a high affinity for 11-*cis*-retinal in this location. Movement of IRBP, loaded with 11-*cis*-retinal, to the vicinity of photoreceptor cells exposes the protein to a high concentration of DHA, and binding of DHA facilitates the release of 11-*cis*-retinal at this location (see text for further details)

ing to this model, when IRBP is located near the RPE, its fatty acid binding site is occupied with fatty acids that are enriched in these cells, such as palmitate. Under these conditions IRBP adopts a conformation that confers a high affinity for 11-*cis*-retinal on retinoid-binding site 2 (bottom left corner of the scheme in Fig. 6). When IRBP moves across the IPM to the vicinity of the photoreceptors, its fatty acid content is readjusted according to the fatty acids prevalent in these cells to contain a large fraction of DHA. Binding of DHA switches the protein to a conformation in which site 2 has a very low affinity for 11-*cis*-retinal, resulting in rapid release of this ligand. Site 2 then becomes occupied with all-*trans*-retinol, the affinity for which remains high in the presence of DHA. The process is reversed when IRBP moves back to the proximity of the pigment epithelium. The overall outcomes of the cycle are that 11-*cis*-retinal moves to photoreceptors, all-*trans*-retinol is transported to pigment epithelium, and the fatty acids transfer down their concentration gradients. It should be noted that the affinity of IRBP's retinoid-binding site 1 for both all-*trans*-retinol and 11-*cis*-retinal is only marginally affected by fatty acids. This site may thus function simply as a buffer that nonspecifically allows elevated concentrations of retinoids in the IPM.

F. Concluding Remarks

This chapter is based on the basic premise that the physical-chemical nature of retinoids and of the factors that determine their interactions with various biological compartments with which they associate in vivo have important implications for their biological activities. Although some of the models and hypotheses presented above are better substantiated while others are of a more speculative nature at this point, the data cited demonstrate that examination of quantitative information on the characteristics of the interactions of retinoids with aqueous media, cellular membranes and specific proteins can indeed provide new insights and lead to alternative points of view and plausible models for understanding various issues in retinoid biology.

Acknowledgements. Work in the author's laboratory is supported by grants from the National Institute of Health (CA68150; EY09296) and the United State Department of Agriculture (89-34115-4498).

References

Allenby G, Bocquel MT, Saunders M, Kazmer S, Speck J, Rosenberger M, Lovey A, Kastner P, Grippo JF, Chambon P, Levin AA (1993) Retinoic acid receptors and retinoid x receptors interactions with endogenous retinoic acids. Proc Natl Acad Sci USA 90:30–34

Bavik CO, Eriksson U, Allen RA, Peterson PA (1991) Identification and partial characterization of a retinal pigment epithelial membrane receptor for plasma retinol-binding protein. J Biol Chem 266:14978–14985

Bavik CO, Levy F, Hellman U, Wernstedt C, Eriksson U (1993) The retinal pigment epithelial membrane receptor for plasma retinol-binding protein, isolation and cDNA cloning of the 63-KD protein. J Biol Chem 268:20540–20546

Bavik CO, Peterson PA, Eriksson U (1995) Retinol-binding protein mediates uptake of retinol to cultured human keratinocytes. Exp Cell Res 216:358–362

Bazan NG, Reddy TS, Redmond TM, Wiggert B, Chader GJ (1985) Endogenous fatty acids are covalently and noncovalently bound to interphotoreceptor retinoid binding protein in monkey retina. J Biol Chem 260:13677–13680

Bourguet W, Ruff M, Chambon P, Gronemeyer H, Moras D (1995) Crystal structure of the ligand-binding domain of the human nuclear receptor RXR-alpha. Nature 375:377–383

Cabral DJ, Hamilton JA, Small DM (1986) The ionization behavior of bile acids in different aqueous environments. J Lipid Res 27:334–343

Carlson A, Bok D (1992) Promotion of the release of 11-*cis*-retinal from cultured retinal pigment epithelium by interphotoreceptor retinoid binding protein Biochemistry 31:9056–9062

Chambon P (1996) A decade of molecular biology of retinoic acid receptors. FASEB J 10:940–954

Chen JD, Evans RM (1995) A transcriptional co-repressor that interacts with nuclear hormone receptors. Nature 377:454–457

Chen H, Wiegand RE, Koutz CA, Anderson RE (1992) Docosahexaenoic acid increases in frog retinal pigment epithelium following rod photoreceptor shedding, Exp Eye Res 55:93–100
Chen Y, Noy N (1994) Retinoid specificity of interphotoreceptor retinoid-binding protein. Biochemistry 33:10658–10665
Chen Y, Saari JC, Noy N (1993) Studies on the interactions of all-*trans*-retinol and long-chain fatty acids with interphotoreceptor retinoid-binding protein. Biochemistry 32:11311–11318
Chen Y, Houghton LA, Brenna JT, Noy N (1996) Docosahexaenoic acid modulates the interactions of the interphotoreceptor matrix retinoid-binding protein with 11-*cis*-retinal. J Biol Chem 271:20507–20515
Chen Z-P, Iyer J, Bourguet W, Held P, Mioskowski C, Lebeau L, Noy N, Chambon P, Gronemeyer H (1998) Ligand and DNA-induced dissociation of RXR tetramers. J Mol Biol 275:55–65
Cistola D P, Hamilton JA, Jackson D, Small DA (1988) Ionization and phase behavior of fatty acids in water application of the gibbs phase rule Biochemistry 27:1881–1888
Cogan U, Kopelman M, Mokady S, Shinitzky M (1976) Binding affinities of retinol and related compounds to retinol-binding proteins. Eur J Biochemistry 65:71–78
Connors KA (1990) Chemical kinetics, the study of reaction rates in solution. VCH, New York
Cook RP (1958) In: Cook RP (ed) Cholesterol: chemistry, biochemistry, and pathology. Academic, New York, p 145
Creek KE, Silverman-Jones CS, De Luca LM (1989) Comparison of the uptake and metabolism of retinol delivered to primary mouse keratinocytes either free or bound to rat serum retinol-binding protein. J Invest Dermatol 92:283–289
Crouch RK, Hazard ES, Lind T, Wiggert B, Chader G, Corson DW (1992) Interphotoreceptor retinoid-binding protein and α-tocopherol preserve the isomeric and oxidation state of retinol. Photochem Photobiol 56:251–255
Daniels C, Noy N, Zakim D (1985) Rates of hydration of fatty acids bound to unilamellar vesicles of phosphatidylcholine or to albumin. Biochemistry 24:3286–3292
DeBoeck H, Zidovetzki R (1988) NMR study of the interactions of retinoids with phospholipid bilayers. Biochim Biophys Acta 946:244–252
Dingle JT, Lucy JA (1965) Vitamin A, carotenoids and cell function, Biol Rev 40:422–461
Dong D, RvvsKa S, Levinthal D, Noy N (1999) Distinct roles for CRABP-I and II in regulating retinoic acid signaling (submitted)
Doody MC, Pownall HJ, Kao YJ, Smith LC (1980) Mechanism and kinetics of transfer of a fluorescent fatty acid between single-walled phosphatidylcholine vesicles. Biochemistry 19:108–116
Duester G (1996) Involvement of alcohol dehydrogenase, short-chain dehydrogenase reductase, aldehyde dehydrogenase, and cytochrome P450 in the control of retinoid signaling by activation of retinoic acid synthesis. Biochemistry 35:12221–12227
Durand B, Saunders M, Leroy P, Leid M, Chambon P (1992) All-*trans* and 9-*cis* retinoic acid induction of CRABPII transcription is mediated by RAR-RXR heterodimers bound to DR1 and DR2 repeated motifs. Cell 71:73–85
Fex G, Johannesson G (1987) Studies of the spontaneous transfer of retinol from the retinol retinol-binding protein complex to unilamellar liposomes. Biochim Biophys Acta 901:255–264
Fex G, Johannesson G (1988) Retinol transfer across and between phospholipid bilayer membranes. Biochim Biophys Acta 944:249–255
Flannery JG, O'Day W, Pfeffer BA, Horwitz J, Bok D (1990) Uptake processing and release of retinoids by cultured human retinal pigment epithelium Exp Eye Res 51:717–728

Fong S L, Liou GI, Landers RA, Alvarez RA, Bridges CD (1984) Purification and characterization of a retinol-binding glycoprotein synthesized and secreted by bovine neural retina. J Biol Chem 259:6534–6542
Glass CK, Rose DW, Rosenfeld MG (1997) Nuclear receptor coactivators. Curr Opin Cell Biol 9:222–232
Grun F, Noy N, Hammerling U, Buck J (1996) Purification, cloning, and bacterial expression of retinol dehydratase from spodoptera frugiperda. J Biol Chem 271:16135–16138
Gutknecht J (1988) Proton conductance caused by long-chain fatty acids in phospholipid bilayer membranes, J Membr Biol 106:83–93
Hallenbeck PL, Marks MS, Lippoldt RE, Ozato K, Nikodem VM (1992) Heterodimerization of thyroid hormone (TH) receptor with H-2RIIBP (RXR-beta) enhances DNA binding and TH-dependent transcriptional activation. Proc Natl Acad Sci USA 89:5572–5576
Hamel CP, Tsilou E, Pfeffer BA, Hooks JJ, Detrick B, Redmond TM (1993) Molecular cloning and expression of RPE65, a novel retinal pigment epithelium-specific microsomal protein that is post-transcriptionally regulated in vitro. J Biol Chem 268:15751–15757
Harrison EH, Blaner WS, Goodman DS, Ross AC (1987) Subcellular localization of retinoids retinoid-binding proteins and acyl coenzyme a retinol acyltransferase in rat liver. J Lipid Res 28:973–981
Heller J (1975) The interactions of plasma retinol binding protein with its receptor: specific binding of bovine and human RBP to pigment epithelium cells from bovine eyes. J Biol Chem 250:3613–3619
Ho M-TP, Massey JB, Pownall HJ, Anderson RE, Hollyfield JG (1989) Mechanism of vitamin A movement between rod outer segments, interphotoreceptor retinoid-binding protein and liposomes. J Biol Chem 264:928–935
Hodam JR, Hilaire PS, Creek KE (1991) Comparison of the rate of uptake and biologic effects of retinol added to human keratinocytes either directly to the culture medium or bound to serum retinol-binding protein. J Invest Dermatol 97:298–304
Hodam JR, Creek KE (1998) Comparison of the metabolism of retinol delivered to human keratinocytes either bound to serum retinol-binding protein or added directly to the culture medium. Exp Cell Res 238:257–264
Horwitz KB, Jackson TA, Bain DL, Richer JK, Takimoto GS, Tung L (1996) Nuclear receptor coactivators and corepressors. Mol Endocrinol 10:1167–1177
Iwasaki M, Rayborn ME, Tawara A, Hollyfield JG (1992) Proteoglycans in the mouse interphotoreceptor matrix. V. Distribution at the apical surface of the pigment epithelium before and after separation Exp Eye Res 54:415–432
Jones GJ, Crouch RK, Wiggert B, Cornwall MC, Chader GJ (1989) Retinoid requirements for recovery of sensitivity after visual pigment bleaching in isolated photoreceptors. Proc Natl Acad Sci USA 86:9606–9610
Kamp F, Hamilton JA (1992) PH gradients across phospholipid membranes caused by fast flip-flop of un-ionized fatty acids. Proc Natl Acad Sci USA 89:11367–11370
Kamp F, Zakim D, Zhang F, Noy N, Hamilton JA (1995) Fatty acid flip-flop in phospholipid bilayers is extremely fast. Biochemistry 34:11928–11937
Katakura Y, Totsuka M, Ametani A, Kaminogawa S (1994) Tryptophan-19 of beta-lactoglobulin, the only residue completely conserved in the lipocalin superfamily, is not essential for binding retinol, but relevant to stabilizing bound retinol and maintaining its structure. Biochim Biophys Acta 1207:58–67
Kersten S, Pan L, Chambon P, Gronemeyer H, Noy N (1995a) Role of ligand in retinoid signaling: 9-*cis*-retinoic acid modulates the oligomeric state of the retinoid X receptor. Biochemistry 34:13717–13721
Kersten S, Pan L, Noy N (1995b) On the role of ligand in retinoid signaling. Positive cooperativity in the interactions of 9-*cis*-retinoic acid with tetramers of the retinoid X receptor. Biochemistry 34:14263–14269

Kersten S, Kelleher D, Chambon P, Gronemeyer H, Noy N (1995c) The retinoid X receptor forms tetramers in solution. Proc Natl Acad Sci USA 92:8645–8649

Kersten S, Dawson MI, Lewis BA, Noy N (1996) Individual subunits of heterodimers comprised of the retinoic acid- and retinoid X receptors interact with their ligands independently. Biochemistry 35:3816–3824

Kersten S, Reczek PR, Noy N (1997) The tetramerization region of the retinoid X receptor is important for transcriptional activation by the receptor. J Biol Chem 272:29759–29768

Kersten S, Dong D, Lee W-Y, Reczek PR, Noy N (1998) Auto-silencing by the retinoid X receptor. J Mol Biol 284:21–32

Leid M, Kastner P, Chambon P (1992) Trends Biochem Sci 17:427–433

Levin MS, Locke B, Yang NC, Li E, Gordon JI (1988) Comparison of the ligand binding properties of two homologous rat apocellular retinol-binding proteins expressed in E. coli. J Biol Chem 263:17715–17723

Li E, Qian S, Winter NS, d'Avignon A, Levin MS, Gordon JL (1991) Fluorine nuclear magnetic resonance analysis of the ligand binding properties of two homologous rat cellular retinol-binding proteins expressed in E. coli. J Biol Chem 266:3622–3629

Li CY, Zimmerman CL, Wiedmann TS (1996) Solubilization of retinoids by bile salt-phospholipid aggregates, Pharm Res 13:907–913

MacDonald PN, Ong DE (1987) Binding specificities of cellular retinol-binding protein and cellular retinol-binding protein type II. J Biol Chem 262:10550–10556

Mangelsdorf DJ, Umesono K, Evans MR (1994) In: The Retinoids, Biology, Chemistry, and Medicine pp 319–350, eds. Sporn MB, Roberts AB, Goodman DS, Raven, New York

Matarese V, Lodish HF (1993) Specific uptake of retinol-binding protein by variant F9 cell lines. J Biol Chem 268:18859–18865

Napoli J L, Boerman MHEM, Chai X Zhai Y, Fiorella PD (1995) Enzymes and binding proteins affecting retinoic acid concentrations. J Steroid Biochem Mol Biol 53:497–502

Nicoletti A, Wong DJ, Kawase K, Gibson LH, Yang-Feng T, Richards JE, Thompson DA (1995) Molecular characterization of the human gene encoding an abundant 61 kDa protein specific to the retinal pigment epithelium. Hum Mol Gen 4:641–649

Norris AW, Li E (1998) Fluorometric titration of the CRABPs, In: Methods in Molecular Biology, vol. 89, Retinoid Protocols, ed Redfern CPF, Humana, Totowa, New Jersey

Noy N, and Xu Z-J (1990a) Kinetic parameters of the interactions of retinol with lipid bilayers. Biochemistry 29:3883–3888

Noy N, and Xu Z-J (1990b) Thermodynamic parameters of the binding of retinol to binding proteins and to membranes. Biochemistry 29:3888–3892

Noy N, and Xu Z-J (1990c) Interactions of retinol with binding proteins: Implications for the mechanism of uptake by cells. Biochemistry 29:3888–3892

Noy N, Blaner WS (1991) Interactions of retinol with binding proteins: studies with rat cellular retinol-binding protein and with rat retinol-binding protein. Biochemistry 30:6380–6386

Noy N (1992a) The ionization behavior of retinoic acid in aqueous environments and bound to serum albumin. Biochim Biophys Acta 1106:151–158

Noy N (1992b) The ionization behavior of retinoic acid in lipid bilayers and in membranes. Biochim Biophys Acta 1106:159–164

Noy N, Slosberg E, Scarlata S (1992) Interactions of retinol with serum binding proteins: studies with retinol-binding protein, transthyretin and serum albumin. Biochemistry 31:11118–11124

Noy N, Kelleher DJ, Scotto AW (1995) Interactions of retinol with lipid bilayers: studies with vesicles of different radii. J Lipid Res 36:375–382

Okajima T-LL, Pepperberg DR, Ripps H, Wiggert B, Chader GJ (1990) Interphotoreceptor retinoid-binding protein promotes rhodopsin regeneration in toad photoreceptors. Proc Natl Acad Sci USA 87:6907–6911
Okajima T-IL, Wiggert B, Chader GJ, Pepperberg DR (1994) Retinoid processing in retinal pigment epithelium of toad (Bufo marinus) J Biol Chem 269:21983–21989
Ong DE, Newcomer ME, Chytil F (1994) In: The Retinoids, Biology, Chemistry, and Medicine pp 283–318, Eds: Sporn MB, Roberts AB, Goodman DS, Raven, New York
Pepperberg DR, Okajima T-IL, Wiggert B, Ripps H, Crouch RK, Chader GJ (1993) Interphotoreceptor retinoid-binding protein: molecular biology and physiological role in the visual cycle of rhodopsin. Mol Neurobiol 7:61–85
Radda GK, Smith DS (1970) retinol: a fluorescent probe for membrane lipids. FEBS Lett 9:287–289
Rask L, Peterson P (1976) In vitro uptake of vitamin A from the retinol binding plasma protein to mucosal epithelial cells from monkey's small intestines. J Biol Chem 251:6360–6366
Rask L, Gijer C, Bill A, Peterson P (1980) Vitamin A supply to the cornea. Exp Eye Res 31:201–211
Renaud J P, Rochel N, Ruff M, Vivat V, Chambon P, Gronemeyer H, Moras D (1995) Crystal structure of the RAR-gamma ligand-binding domain bound to all-*trans*-retinoic acid. Nature 378:681–689
Saari JC, Teller DC, Crabb JW, Bredberg L (1985) Properties of an interphotoreceptor retinoid-binding protein from bovine retina. J Biol Chem 260:195–201
Saari JC, Bredberg DL, Noy N (1994) Control of substrate flow at a branch in the visual cycle Biochemistry 33:3106–3112
Saari JC (1994) In: The Retinoids, Biology, Chemistry, and Medicine pp 351–386, Eds: Sporn MB, Roberts AB, Goodman DS, Raven, New York
Saari JC, Huang J, Possin DE, Fariss RN, Leonard J, Garwin GG, Crabb JW, Milam AH (1997) Cellular retinal-binding protein is expressed by oligodendrocytes in optic nerve and brain, Glia 21:259–268
Sande S, Privalsky ML (1996) Identification of TRACs (T-3 receptor-associating cofactors), a family of cofactors that associate with, and modulate the activity of, nuclear hormone receptors. Mol Endocrinol 10:813–825
Seol W, Mahon MJ, Lee YK, Moore DD (1996) Two receptor interacting domains in the nuclear hormone receptor corepressor RIP13/N-CoR. Mol Endocrinol 10:1646–1655
Sivaprasadarao A Boudjelal M, Findlay JBC (1994) Solubilization and purification of the retinol-binding protein receptor from human placental membranes. Biochem J 302:245–251
Small DM (1971) In: The Bile Acids. Vol 1, pp 249–355 Eds. Nair PP, Kritchevsky S, Plenum
Soprano, Blaner (1994) In: The Retinoids, Biology, Chemistry, and Medicine pp 257–282, Eds: Sporn MB, Roberts AB, Goodman DS, Raven, New York
Stillwell W, Ricketts M (1980) Effect of *trans*-retinol on the permeability of egg lecithin liposomes. Biochem Biophys Res Commun 97:148–153
Stillwell W, Ricketts M, Hudson H, Nahmias S (1982) Effect of retinol and retinoic acid on permeability, electrical resistance and phase transition of lipid bilayers. Biochim Biophys Acta 668:653–659
Stryer L (1981) Biochemistry, 2nd edn, chap 4, Freeman, New York
Sundaram M, Sivaprasadarao A, DeSousa MM, Findlay JBC (1998) The transfer of retinol from serum retinol-binding protein to cellular retinol binding protein is mediated by a membrane receptor. J Biol Chem 273:3336–3342
Szuts EZ, Harosi IH (1991) Solubility of retinoids in water. Arch Biochem Biophys 287:297–304
Tschanz C, Noy N (1997) Binding of retinol in both retinoid-binding sites of IRBP is stabilized mainly by hydrophobic interactions. J Biol Chem 272:30201–30207

Urbach J, Rando RR (1994) Isomerization of all-*trans*-retinoic acid to 9-*cis*-retinoic acid. Biochem J 299:459–465
Ward SJ, Chambon P, Ong DE, Bavik CA (1997) A retinol-binding protein receptor-mediated mechanism for uptake of vitamin A to postimplantation rat embryos. Biol Reprod 57:751–755
Wassall SR, Phelps TM, Albrecht MR, Langsford CA, Stillwell W (1988) ESR study of the interactions of retinoids with a phospholipid model membrane. Biochim Biophys Acta 939:393–402
Wolffe AP (1997) Histones, nucleosomes and the role of chromatin structure in transcriptional control. Biochem Soc Trans 25:354–358
Wurtz JM, Bourguet W, Renaud JP, Vivat V, Chambon P, Moras D, Gronemeyer H (1996) A canonical structure for the ligand-binding domain of nuclear receptors, Nature Struct Biol 3:87–94
Yang N, Schule R, Mangelsdorf DJ, Evans RM (1991) Characterization of DNA binding and retinoic acid binding properties of retinoic acid receptor. Proc Natl Acad Sci USA 88:3559–3563
Yu V, Delsert C, Andersen B, Holloway J, Devary OV, Naeaer AM, Kim SY, Boutin JM, Glass CK, Rosenfeld MG (1991) RXR-beta a coregulator that enhances binding of retinoic acid, thyroid hormone and vitamin D receptors to their cognate response elements. Cell 67:1251–1266
Zhang XK, Hoffmann B, Tran PB-V, Graupner G, Pfahl M (1992) Retinoid X receptor is an auxiliary protein for thyroid hormone and retinoic acid receptors. Nature 355:441–446

CHAPTER 2

Biosynthesis, Absorption, Metabolism and Transport of Retinoids

S. VOGEL, M.V. GAMBLE, and W.S. BLANER

A. Introduction

All retinoids (vitamin A and its analogs) in the body originate in the diet either as preformed vitamin A, usually from animal food sources, or as provitamin A carotenoids, usually from plant food sources or as supplements or additives to processed foods. Both preformed vitamin A and provitamin A undergo metabolism within the intestine (BLANER and OLSON 1994), including a series of metabolic conversions, extracellularly in the lumen of the intestine and intracellularly in the intestinal mucosa, which result in the preponderance of the dietary vitamin A being converted to retinol (vitamin A alcohol). The retinol, along with other dietary lipids in the intestinal mucosa, is packaged as retinyl ester in nascent chylomicrons. The chylomicrons are secreted into the lymphatic system, and approximately 75% of chylomicron retinoid is eventually taken up as part of the chylomicron remnants by the liver, where the majority of the body's retinoid reserves are stored. Some retinoid is delivered to extrahepatic tissues where this retinoid is either stored as retinyl ester, released back into the circulation for delivery to the liver and other tissues, metabolized to active retinoid forms such as retinoic acid or metabolized to catabolic forms destined for excretion from the body.

The predominant retinoid present in the fasting circulation is the retinoic acid precursor, retinol. Since retinol is not soluble in aqueous environments, all circulating retinol is bound to retinol-binding protein (RBP). Other retinoids are also present in blood, albeit at much lower concentrations than retinol. Both all-*trans*- and 13-*cis*-retinoic acid can be found in fasting plasma of both animals and humans. Some retinyl ester is present as a component of lipoproteins, especially in very low density lipoprotein (VLDL) and low density lipoprotein (LDL). Glucuronides of both retinol and retinoic acid can be found in the circulation along with low levels of retro-metabolites of retinol.

This chapter focuses primarily on recent advances in our understanding of the uptake and metabolism of dietary retinoid and provitamin A carotenoids, retinoid storage and metabolism in the liver and the delivery of retinoid to target tissues. Thus this review is concerned primarily with the metabolism and transport of the precursor retinol and its storage form, retinyl ester. Later chapters review the formation and catabolism of active retinoid forms such as retinoic acid and retro-retinoids from retinol.

B. Intestinal Absorption of Retinol and Biosynthesis of Retinol from Carotenoids

The body acquires retinoid both as preformed vitamin A and as provitamin A carotenoid. Within the small intestine the first processes that are important for dietary retinoid absorption take place within the lumen of the intestine. Other important metabolic events, specifically the hydrolysis of dietary retinyl ester, take place at the brush border of the enterocyte. However, most of the metabolic processing important to the absorption of both preformed vitamin A and provitamin A carotentoids takes place intracellularly within the enterocyte. Information about these processes is summarized below. To assist the reader, a diagram summarizing these processes is provided in Fig. 1.

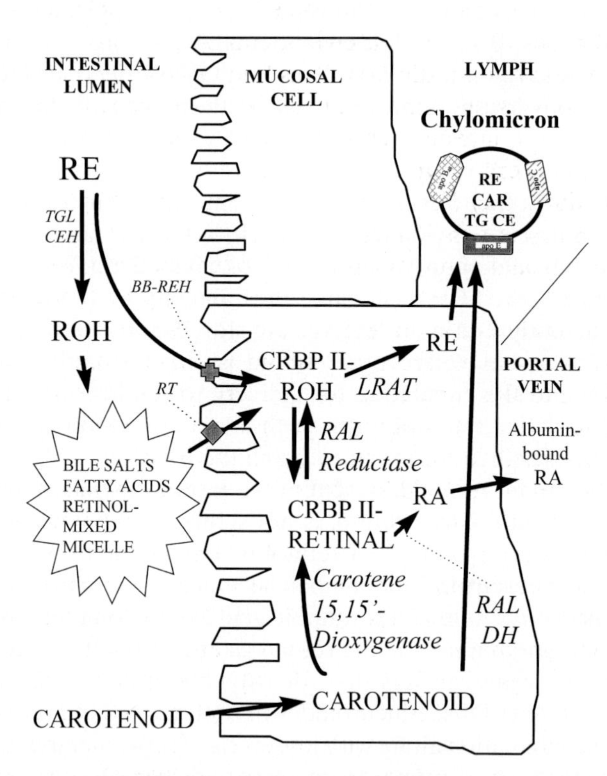

Fig. 1. Schematic summary for the intestinal absorption and metabolism of preformed vitamin A and of provitamin A carotenoids. Details of these processes are described in the text. *apoB*$_{48}$, Apolipoprotein B$_{48}$; *apoC*, apolipoprotein C; *apoE*, apolipoprotein E; *BB-REH*, brush-border retinyl ester hydrolase; *CAR*, carotenoids; *CEH*, cholesteryl ester hydrolase; *CRBP II*, cellular retinol-binding protein, type II; *LRAT*, lecithin:retinol acyltransferase; *RA*, retinoic acid; *RAL*, retinal; *RAL DH*, retinal dehydrogenase; *RE*, retinyl esters; *RT*, retinol transporter; *TG*, triacylglycerol; *TGL*, triglyceride lipase

I. Uptake and Luminal Metabolism of Dietary Preformed Vitamin A (Retinyl Esters, Retinol, and Retinoic Acid)

Preformed dietary vitamin A consists primarily of retinyl esters, retinol, and very small amounts of retinoic acid. Extracellularly within the lumen of the small intestine, the dietary retinyl esters are hydrolyzed to retinol. The retinol formed through retinyl ester hydrolysis and the retinol arriving as such from the diet is taken up by the mucosal cells. It has long been known that pancreatic lipases, such as triglyceride lipase and cholesteryl ester hydrolase (also known as carboxy ester lipase and bile salt dependent retinyl ester hydrolase), are able to hydrolyze retinyl esters in vitro, and it has been assumed that these enzymes are responsible for the luminal hydrolysis of dietary retinyl ester (GOODMAN and BLANER 1984; BLANER and OLSON 1994). More recent studies by Ong and colleagues have demonstrated that a retinyl ester hydrolase activity intrinsic to the brush border membrane of the small intestine plays a central role in the hydrolysis of dietary retinyl ester in the rat (RIGTRUP and ONG 1992) and in humans (RIGTRUP et al. 1994). This brush border retinyl ester hydrolase from the rat small intestine mucosa is probably identical to the brush border phospholipase B characterized earlier for this tissue (RIGTRUP et al. 1994). The intrinsic brush border retinyl ester hydrolase activity is stimulated by both trihydroxy and dihydroxy bile salts, and preferentially hydrolyzes long-chain retinyl esters such as retinyl palmitate (RIGTRUP and ONG 1992). It appears that this activity makes a quantitatively greater contribution to the hydrolysis of dietary retinyl ester in the rat than the contribution made by pancreatic enzymes (RIGTRUP and ONG 1992; RIGTRUP et al. 1994). Thus the brush border retinyl ester hydrolase likely plays a key physiological role in the hydrolysis of dietary retinyl ester. This conclusion is strengthened by recent studies of induced mutant mice that totally lack pancreatic cholesteryl ester hydrolase. Studies of this mouse model indicate that both the rate and the amount of vitamin A taken up by the small intestine from bolus doses of retinol, retinyl acetate, and retinyl palmitate do not differ between wild-type and cholesteryl ester hydrolase deficient mice (WENG et al. 1999; VAN BENNEKUM et al. 1999a).

Retinol formed through the hydrolysis of dietary retinyl esters and the dietary retinol arriving as such in the diet are taken up by the mucosal cells. The uptake process requires emulsification of the retinol by bile salts and free fatty acids (GOODMAN and BLANER 1984). Recent findings in the suckling rat indicate that there is a specific retinol transport protein within the brush border of the enterocyte, which facilitates retinol uptake (DEW and ONG 1994). The brush border retinol transporter transports both all-*trans*-retinol and 3-dehydroretinol specifically and at similar rates. Specific uptake of these two retinoids was inhibited by treatment of brush border preparations with *N*-ethylmaleimide. Both 13-*cis*- and 9-*cis*-retinol were taken up by the brush boarder preparations; however, unlike uptake of all-*trans*-retinol, this uptake could not be competed away using excessive quantities of the unlabeled retinol isomer and hence was considered nonspecific. Interestingly, although the retinol transporter showed specificity for only all-*trans*-retinol and 3-dehydroretinol, once internalized, both of these retinoids and the nonspecifically absorbed 13-*cis*- and 9-*cis*-retinols are quickly converted to retinyl esters by the actions of LRAT within the enterocyte (DEW and ONG 1994).

Dietary retinoic acid is absorbed by the mucosal cells as such and is delivered to the portal circulation where it is taken into the body bound to albumin (BLANER and OLSON 1994).

II. Uptake and Metabolism of Dietary Provitamin A Carotenoids

Provitamin A carotenoids such as β-carotene, α-carotene, and β-cryptoxanthin are absorbed intact by the mucosal cells of the proximal small intestine (OLSON 1989; VAN VLIET 1996). As with other neutral lipids in the diet, carotenoids are very insoluble in aqueous environments and they must be emulsified with bile salts and free fatty acids within the lumen of the intestine for uptake into the mucosa (GOODMAN and BLANER 1984; OLSON 1989, 1990).

Within the mucosal cells β-carotene is cleaved through the action of carotene 15,15'-dioxygenase (also known as the carotene cleavage enzyme) to retinal (vitamin A aldehyde) (GOODMAN and HUANG 1965; OLSON and HAYAISHI 1965). The dioxygenase is known to be cytosolic and influenced by bile salts and retinoid levels in the diet (VAN VLIET et al. 1996). In the rat it is located primarily in mature jejunal enterocytes (DUSZKA et al. 1996). Although this enzyme was first identified in the mid-1960s, it has not been purified to homogeneity, nor has its cDNA or genomic clone been identified. The properties of this enzyme and others that may be important for the conversion of provitamin A carotenoids such as β-carotene to retinal have been the subject of some disagreement in the literature (OLSON 1989; BLANER and OLSON 1994). This disagreement centers on whether the β-carotene cleavage reaction is a specific oxidation reaction which takes place at the central double bond yielding retinal as the only product (GOODMAN and HUANG 1965; OLSON and HAYAISHI 1965; OLSON 1989; DUSZKA et al. 1996; NAGAO et al. 1996), or whether some relatively nonspecific eccentric cleavage of any double bound of the carotenoid, yielding two of several different β-apo-carotenals, can occur (WANG 1994; WANG et al. 1994, 1996). It has been proposed from work carried out in chickens, rats, and ferrets that the intestine and liver are able to shorten β-apo-carotenals through a β-oxidation-like mechanism (sequential removal of 2 carbon units) to retinal and retinoic acid (WANG et al. 1996). In general, the recent literature increasingly suggests that it is likely that the central cleavage pathway is the predominant, if not sole route through which provitamin A carotenoids are converted to retinoids. A proposed catalytic mechanism for carotene cleavage by carotene 15,15'-dioxygenase is shown in Fig. 2.

Both provitamin A and nonprovitamin A carotenoids can be absorbed intact from the diet. Thus, carotenoids are present throughout the body. However, there are marked species differences in carotenoid absorption

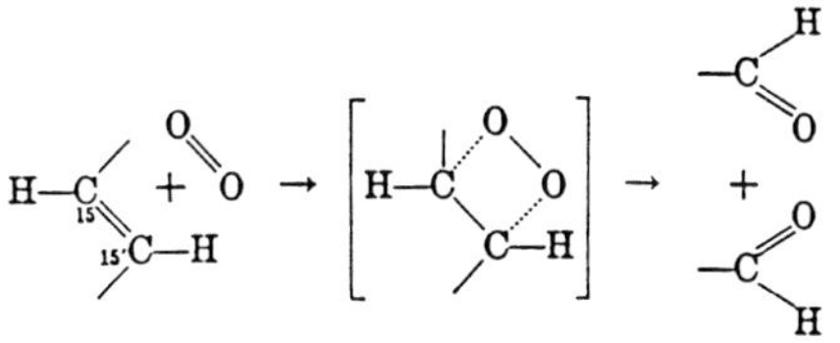

Fig. 2. Proposed reaction mechanism for the oxidative cleavage of β-carotene (or other provitamin A carotenoids) catalyzed by carotene 15,15′-dioxygenase. This mechanism for molecular O_2 addition across the central 15–15' carbon-carbon double bond was proposed originally by GOODMAN and HUANG (1965) and OLSON and HAYAISHI (1965)

and/or metabolism; various species show markedly different characteristics and properties with respect to carotenoid absorption and metabolism (MATSUNO 1991; VAN VLIET 1996). In humans the majority of absorbed β-carotene (60%–70%) is converted to retinal and the remainder is absorbed intact and deposited in adipose tissue (OLSON 1989, 1990). In addition, humans, unlike most nonprimate species, transport β-carotene primarily in the LDL fraction (VAN VLIET 1996). The most commonly used animal models to study β-carotene uptake and metabolism include the ferret, cow, rodents, rabbit, pig, and chicken. The ferret is reported to be a poor converter of provitamin A carotenoids to vitamin A and is known to absorb large amounts of cleavage products (reportedly including β-apo-carotenals) (LEDERMAN et al. 1998) through the portal vein, unlike in humans where this route of absorption is a minor pathway (VAN VLIET 1996). Unlike humans, both ferrets and cows have relatively high fasting retinyl ester levels in blood (RIBAYA-MERCADO et al. 1994; LEDERMAN et al. 1998). Rodents and chickens are very efficient carotenoid converters, but they will absorb intact carotenoids only when carotenoids are provided at high doses (VAN VLIET 1996). The pig, in general, is a good model for digestion and absorption of carotenoids but there are reports that this species is not a very efficient converter of β-carotene to vitamin A (SCHWEIGERT et al. 1995). Moreover, the pig seems to store most of the absorbed β-carotene in lung and not adipose tissue and liver, two major sites of β-carotene deposition in humans (SCHWEIGERT et al. 1995). Rabbits similarly do not deposit (store) carotenoids in fat (VAN VLIET 1996).

Since the liver, adipose tissue, lung, and other tissues possess carotene 15,15′-dioxygenase activity (OLSON and HAYASHI 1965; GOODMAN and BLANER 1984; OLSON 1990) and also a separate activity able to convert carotenoids directly to retinoic acid not requiring retinal as an intermediate (NAPOLI and RACE 1988) and relatively high levels of carotenoids, the uptake of dietary cartenoids intact from the gastrointestinal tract may be an important route through which some tissues acquire retinoid. Recently this intriguing possibility has been explored experimentally. THATCHER et al. (1998) investigated whether β-carotene present (stored) within tissues of the Mongolian gerbil was used as a source for retinoid synthesis when the gerbils were placed on a retinoid-deficient diet. Specifically, THATCHER et al. (1998), using different nutritional protocols, observed that loss of β-carotene from serum and tissues was not influenced by the presence of retinoid in the diet. These investigators concluded from their studies that β-carotene stored in tissues is not conserved to meet later need for retinoid. Thus, it would appear that previous feeding with β-carotene does not serve to maintain the retinoid status of the organism.

III. Intramucosal Metabolism

1. Retinoic Acid Formation from Retinal Derived from Carotenoids

A very small percentage of the retinal formed from the cleavage of the dietary β-carotene can be oxidized to retinoic acid and taken into the circulation bound to albumin (GOODMAN and BLANER 1984; BLANER and OLSON 1994). A large percentage of the retinoic acid of dietary origin appears to be removed from the circulation by tissues (BLANER and OLSON 1994). It is clear from the literature that dietary intake of either preformed vitamin A and/or of provitamin A carotenoids can give rise to increased circulating levels of all-*trans*- and 13-*cis*-retinoic acid (FOLMAN et al. 1989; TANG et al. 1995). For rabbits maintained for several weeks on a β-carotene-supplemented diet a markedly higher level of plasma retinoic acid is observed, as compared to rabbits maintained on a control diet (FOLMAN et al. 1989). Thus, it would appear that dietary β-carotene intake (and presumably dietary intake of other provitamin A carotenoids) contributes directly to the concentration of retinoic acid present in the circulation.

2. Role of Cellular-Retinol Binding Protein, Type II

Although biochemical details regarding the conversion of β-carotene to vitamin A remain elusive, the metabolism of retinal following its formation is well understood. Upon formation, retinal is immediately reduced by a microsomal enzyme, retinal reductase, to retinol (GOODMAN and BLANER 1984; BLANER and OLSON 1994). Intestinal retinal reductase is known to require the participation of a second cytosolic protein, cellular retinol-binding protein, type II (CRBP II) (ONG et al. 1994a). CRBP II, which binds both retinal and retinol, plays a central role in the processing of dietary retinoid within the mucosal cell (ONG et al. 1994a). In the adult rat CRBP II is localized exclusively in the enterocytes of the small intestine (ONG et al. 1994a) and is thought to act in only dietary vitamin A uptake and metabolism. CRBP II is a member of a family of highly related proteins that also includes the intracellular fatty acid-binding proteins and the intracellular-retinoid binding proteins as members. CRBP II shows the highest degree of sequence homology with cellular retinol-binding protein type I (CRBP I) but is immunologically distinct from the much more widely distributed CRBP I (ONG et al. 1994a). Retinal, when bound to CRBP II, is the preferred substrate for reduction to retinol by intestinal retinal reductase. This carotenoid-derived retinol formed through oxidation of CRBP II bound retinal is metabolically indistinguishable from retinol arriving in the diet as preformed retinoid and hence, undergoes the same metabolic events described below for preformed retinoid.

TAKASE and colleagues (1998; SURUGA et al. 1995) have reported that CRBP II expression is markedly influenced by dietary fatty acid intake. These authors observed that when rats were maintained on a fat-free diet containing a sufficient amount of vitamin A, expression of CRBP II mRNA in rat jejunum is reduced to 50% of that in control fed rats. This depression in CRBP II mRNA expression was accompanied by a decrease in expression of peroxisomal proliferator-activated receptor α (PPARα). When rats receiving a vitamin A deficient diet were orally administered corn oil, within 6h CRBP II mRNA levels increased by approximately threefold. Administration of 9-*cis*-retinoic acid alone (without a source of long-chain fatty acids) or together with long-chain triglycerides brought about no further increase in jejunum CRBP II mRNA levels over that which was observed upon administration of corn oil alone. Although these investigators did not show a direct effect of fatty acids on the rate of CRBP II gene transcription through a PPAR response element, they have provided interesting data that suggests dietary fat intake directly influences the small intestine's ability to take up and process dietary retinol.

3. Retinol Esterification and the Packaging of Dietary Retinoid into Chylomicrons

Within the intestinal mucosa all retinol, regardless of dietary origin, is reesterified with long-chain fatty acids (primarily as the esters of palmitic, stearic, oleic, and linoleic acids) through the action of the enzyme lecithin:retinol acyltransferase (LRAT). The newly synthesized retinyl esters are packaged along with other dietary lipids into nascent chylomicrons and secreted into the lymphatic system for uptake into the circulation. Intestinal LRAT uses retinol bound to CRBP II as a substrate for the esterification reaction (ONG et al. 1994a). Thus, CRBP II plays a central role in directing or channeling dietary retinol to nascent chylomicrons for uptake into the body. Retinal formed by carotene cleavage and dietary retinol (arriving as retinol or formed through retinyl ester hydrolysis) are metabolically channeled, through binding to CRBP II, towards retinyl ester formation and packaging in chylomicrons.

LRAT has been proposed to play a key role in directing the metabolism of retinol throughout the body. This enzyme catalyzes the transfer of the *sn-1* fatty acid from the membrane-associated phosphatidyl choline to all-*trans*-retinol and has been reported to be present in the liver, eye, small intestine and testes (BLANER and OLSON 1994). It is not presently clear whether the LRAT species present in each of these tissues are distinct gene products but it has been established that hepatic but not intestinal LRAT is actively regulated by vitamin A nutritional status (RANDOLPH and ROSS 1991b). The biochemical characteristics of LRAT have been summarized previously (BLANER and OLSON 1994). Most important among these is that LRAT utilizes retinol bound to CRBP II (in the intestine) or to CRBP I (for other tissues) as its preferred substrate (ONG et al. 1994a). The apparent K_m values of rat intestinal and hepatic LRAT for the retinol-binding protein complexes are in the low micromolar range, a concentration that is physiological (BLANER and OLSON 1994). Properties and roles of hepatic LRAT are discussed below in Sect. D.II.

Together with triglycerides, cholesterol esters, phospholipids, and other dietary lipids, retinyl esters are incorporated into the apolipoprotein B_{48} ($apoB_{48}$) containing chylomicrons and secreted into the lymphatic system. Presently no information is available concerning the biochemical processes through which the newly synthesized retinyl ester is packaged by the mucosal cells into nascent chylomicrons.

C. Circulating Chylomicrons and Chylomicron Remnant Formation

Once in the general circulation, chylomicrons interact with lipoprotein lipase (LPL) bound to the luminal surface of the vascular endothelium and rapid lipolysis of the triglyceride occurs. LPL catalyzed hydrolysis gives rise to free fatty acids and a smaller lipoprotein particle termed the chylomicron remnant. The free fatty acids generated from triglyceride hydrolysis are taken up by extrahepatic tissues, including muscle and adipose tissue where they are ultimately used as substrates for energy metabolism (COOPER 1997). The chylomicron remnants acquire apolipoprotein E (apoE) either in the plasma or in the space of Disse, where apoE can accumulate after its secretion from hepatocytes (WILLIAMS et al. 1985; HAMILTON et al. 1990). The acquisition of apoE by the remnant particles is essential for their clearance by the liver (COOPER 1997). The importance of apoE in chylomicron remnant clearance is underscored by apoE-deficient mice, which clear postprandial cholesterol very slowly (ISHIBASHI et al. 1994, 1996; MORTIMER et al. 1995), and which have very high circulating fasting levels of total cholesterol and retinyl esters (ISHIBASHI et al. 1994, 1996; MORTIMER et al. 1995). Data accumulated from studies carried out over the past 35 years consistently show that approximately 75% of chylomicron retinyl ester is removed from the circulation by the liver and the remaining 25% by extrahepatic tissues including skeletal muscle, adipose tissue, heart, spleen, and kidney (GOODMAN et al. 1965)

In general, dietary retinoid uptake and its clearance from the postprandial circulation is not very different for rats, mice, rabbits, or humans. However, there is one major exception to this generalization, which can become important in human disease states that bring about impaired (decreased rates of) clearance of postprandial lipids. Specifically, some species including rodents, totally lack plasma cholesteryl ester transfer protein (CETP) (TALL 1995). CETP is able to catalyze the transfer within the circulation of retinyl esters from triglyceride rich lipoproteins such as chylomicrons to low density lipoprotein (LDL) and high density lipoprotein (HDL) (TALL 1995). Thus, for both humans and rabbits, species which express CETP, some postprandial retinyl ester is transferred to these lipoprotein fractions if postprandial lipid clearance is impaired either as the result of disease or experimental intervention (GOODMAN and BLANER 1984; BLANER and OLSON 1994). The retinyl ester transferred to LDL or HDL would then be processed as a component of these lipoproteins. Since for most healthy individuals the amount of postprandial retinyl ester transferred by CETP to LDL and/or HDL is very small, the significance of this transfer process in normal retinoid transport and metabolism has been discounted; however, it should be recognized that this process may be significant in some pathological states.

I. Hepatic Clearance of Chylomicron Remnants

It has long been accepted that apoE is required for uptake of chylomicron remnants by the liver, but the exact mechanisms involved in this process are still unsettled. Several distinct cell surface receptors that are able to bind apoE-containing lipoproteins have been proposed to be involved in the receptor-mediated uptake of chylomicron remnants by hepatocytes. Among those are the LDL receptor (LDL-R), the LDL receptor related protein (LRP) and the lipolysis-stimulated receptor (LSR) (COOPER 1997). In addition, heparan sulfate proteoglycans (HSPG), located on the surface of hepatocytes, have been proposed to be responsible for the initial interaction of remnants with the cells of the liver (COOPER 1997). Data suggesting roles for LPL and hepatic lipase in the hepatic uptake process have also been reported (BEISIEGEL et al. 1991; SHAFI et al. 1994). The processes/events thought to be important for the uptake of chylomicron remnants by hepatocytes are summarized in Fig. 3.

Gene targeting of the LDL-R provides evidence for the involvement of the LDL-R in chylomicron remnant clearance by hepatocytes. Investigations using a mouse strain generated to lack totally the LDL receptor (LDL-$R^{-/-}$) (ISHIBASHI et al. 1994, 1996; MORTIMER et al. 1995; HERZ et al. 1995) indicate that while the rate of plasma clearance of chylomicron remnants in LDL-$R^{-/-}$ mice was unaffected by LDL-R deficiency, endocytosis of the remnant by the hepatocyte was delayed. Thus, the rate of clearance of chylomicron remnant cholesteryl ester from the circulation was comparable to that which was observed in wild type. Moreover, LDL-$R^{-/-}$ mice showed no significant difference in accumulation of apoB_{48} in plasma (MORTIMER et al. 1995; HERZ et al. 1995). However, uptake of chymomicron remnant cholesteryl ester into the livers of LDL-$R^{-/-}$ mice was significantly slower than in wild-type mice (MORTIMER et al. 1995; HERZ et al. 1995). Using confocal microscopy and a fluorescently labeled cholesteryl ester derivative, fluorescent cholesteryl ester, was observed to accumulate in the extracellular sinusoidal space (Space of

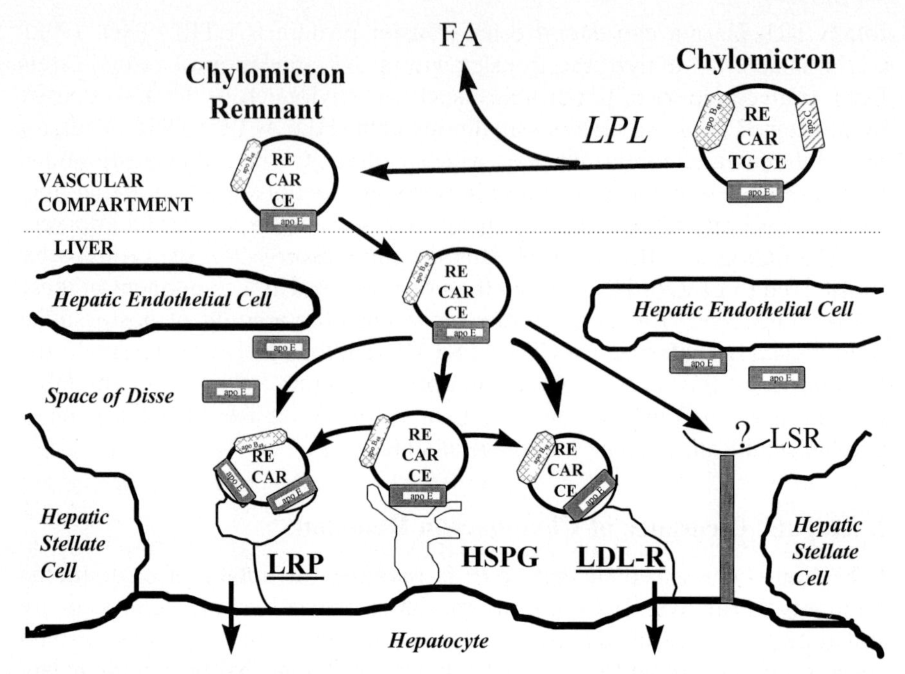

Fig. 3. Summary of the processes/events important for the receptor mediated uptake of chylomicron remnants by hepatocytes. Details of these processes/events are described in the text. *apoB$_{48}$*, Apolipoprotein B$_{48}$; *apoC*, apolipoprotein C; *apoE*, apolipoprotein E; *CAR*, carotenoids; *CE*, cholesteryl ester; *FA*, free fatty acid; *HSPG*, heparan sulfate proteoglycans; *LDL-R*, low-density lipoprotein receptor; *LPL*, lipoprotein lipase; *LRP*, lipoprotein-related protein; *LSR*, lipolysis stimulated receptor; *RE*, retinyl ester; *TG*, triacylglycerol

Disse), and internalization of the particles by the hepatocytes was delayed in LDL-R$^{-/-}$ mice (MORTIMER et al. 1995). Similarly, accumulation of unesterified cholesterol in the livers of LDL-R$^{-/-}$ mice occurred significantly later than in wild-type mice, suggesting delayed uptake and hydrolysis of chylomicron remnant cholesteryl ester (HERZ et al. 1995). In double-mutant mice totally lacking both LDL-R and apoE, chylomicron remnants accumulated in the circulation to levels comparable to those observed in apoE-deficient mice carrying the wild-type LDL-R alleles (ISHIBASHI et al. 1996). When these data are considered together, it seems that the LDL-R contributes to hepatic uptake of chylomicrons remnants; however, the data suggest that chylomicron remnants are also taken up by the liver through alternative apoE-dependent processes which are distinct from the LDL-R pathway, and which are able to compensate for the absence of the LDL-R pathway.

One such alternative apoE-dependent process may involve LRP, a 600-kDa, Ca^{2+}-dependent multifunctional receptor protein that is a member of the LDL receptor protein family. LRP is identical to the receptor reported to be the activated receptor form of the α_2-macroglobulin receptor and therefore also is known as α_2-macroglobulin receptor/LRP. LRP is known to interact with multiple ligands that are present in the circulation (COOPER 1997). These include apoE, activated α_2-macroglobulin, LPL, lactoferrin, RAP and various serum proteases. The involvement of LRP in chylomicron remnant metabolism is supported by studies using the protein RAP. RAP is a 39-kDa protein, that binds to LRP in the presence of Ca^{2+} and upon binding suppresses binding of other ligands to LRP (COOPER 1997). For mice totally deficient in RAP ($RAP^{-/-}$), a phenotype of reduced expression of LRP in liver and brain, no difference in the rate chylomicron remnant clearance was observed; however, for double mutants, mice lacking both the LDL-R and the RAP protein, chylomicron remnants (as assessed by $apoB_{48}$ concentrations) accumulate in the circulation to high levels (WILLNOW et al. 1995). Overexpression of RAP in mice leads to accumulation of $apoB_{48}$ and apoE containing particles in the circulation of $LDL\text{-}R^{-/-}$ mice (WILLNOW et al. 1994). Other studies in isolated hepatocytes and in perfused rat liver have implicated involvement of LPL with LRP-mediated clearance of chylomicron remnants. These in vitro studies indicate that LPL stimulates binding of triglyceride-rich particles such as chylomicrons to LRP (BEISIEGEL et al. 1991). The enhancement of binding appears to be independent of the lipolytic activity of LPL and is based solely on the structural properties of LPL (BEISIEGEL et al. 1991; SKOTTOVA et al. 1995). Other cell culture studies (KRAPP et al. 1996; DE FARIA et al. 1996) have suggested that hepatic lipase can also facilitate LRP action through binding of its C-terminal to LRP in a manner which enhances chylomicron remnant uptake by cultured hepatocytes. In summary, it appears that LRP, as with LDL-R, is not essentially needed for clearance of chylomicron remnants by the liver. As with LDL-R, LRP contributes to hepatic clearance of chylomicron remnants in a manner that can be compensated for through the actions of other receptors.

A third receptor proposed to be present on hepatocytes, LSR, is distinct from the LDL-R and LRP and may also contribute to uptake of chylomicron remnants by hepatocyte (BIHAIN and YEN 1992; YEN et al. 1994; MANN et al. 1995). The role of this protein in chylomicron remnant clearance has been studied in both human and rat cells. LSR activity seems to depend on two distinct but interacting membrane proteins (BIHAIN and YEN 1992; YEN et al. 1994). Free fatty acids are thought to activate the LSR by causing a conformational shift that reveals cryptic binding sides for apoE and apoB. LSR binds to lipoproteins, preferentially triglyceride-rich particles, and to RAP at high concentrations. Unlike lipoprotein particle binding to the LDL-R and LRP, binding of ligands to LSR is Ca^{2+} independent (YEN et al. 1994). The contribution of LSR to chylomicron remnant clearance in vivo remains to be established.

While internalization of chylomicron remnants by hepatocytes is mediated by cell surface receptors, initial binding of the remnant to the surface of the hepatocyte may involve the actions of cell surface heparan sulfate proteoglycans (HSPG). HSPGs are situated on cell surfaces and are especially abundant in the Space of Disse. Studies employing hepatocytes in culture have demonstrated that apoE containing particles bind to HSPGs in a manner that is dependent on apoE concentration (Ji et al. 1994). When cultured hepatocytes were treated with heparinase, an enzyme which cleaves cell surface heparan sulfate and thus destroys the binding capacity of HSPGs for apoE (Ji et al. 1995), remnant binding and uptake by hepatocytes were both abolished (Ji et al. 1994, 1995). Using a different approach to investigate the role of hepatocyte HSPGs in remnant clearance, Mortimer et al. (1997) transfected HepG2 cells with antisense oligodeoxynucleotides (ODNs) for syndecan, a core protein of hepatic HSPGs. Antisense administration significantly suppressed HepG2 HSPG expression and reduced remnant uptake by 50%–70% compared to HepG2 cultures transfected with sense ODNs (Mortimer et al. 1997). Data supporting a role for HSPGs in facilitating hepatocyte removal of chylomicron remnants also has been obtained through investigations of mice. Following intravenous administration of heparinase to mice, the rate of removal of apoE-rich remnant particles by the liver was greatly reduced for both wild-type and LDL-$R^{-/-}$ mice (Ji et al. 1995; Mortimer et al. 1995). Taken together, these cell culture and animal model data provide support for the notion that HSPGs located on the surface of hepatocytes help facilitate chylomicron remnant clearance by the liver. It seems reasonable to speculate that remnant binding to hepatocyte HSPGs helps increase the resident time of apoE-containing remnant particles at the cell surface, thus allowing sufficient time for the remnant to bind to receptors such as LDL-R and LRP.

In summary, apoE plays an essential role in the uptake of chylomicron remnants by the liver. It seems likely that initially the chylomicron remnants are sequestered in close proximity to the hepatocytes through the binding to HSPGs in the space of Disse. Subsequently, the remnant is able to interact with a cell surface receptor that is then able to bring about or catalyze endocytosis of the remnant particle. Both the LDL-R and LRP appear to be important cell surface receptors for the uptake of chylomicron remnants. De Faria et al. (1996), employing RAP or antibodies against LDL-R have estimated that approximately 55% of chylomicron remnants are cleared through the LDL-R mediated pathway and approximately 25% through the LRP-mediated pathway. Other studies by Ishibashi et al. (1996) estimate that approximately 75% of chylomicron remnants are cleared through the LDL-R pathway. Although the contribution which the LSR makes towards chylomicron remnant clearance by the liver has not been convincingly established, it is possible that this receptor too will be found to contribute substantially to hepatic chylomicron remnant uptake.

II. Extrahepatic Tissues and Chylomicron Remnant Retinoid

Early work by Goodman et al. (1965) indicated that approximately 25% of postprandial retinyl ester is taken up by extrahepatic tissues. As shown in Table 1, following intravenous injection of doses of rat mesenteric chylomi-

Table 1. Recovery and tissue distribution of [^{14}C]cpm in rat tissues at various times after intravenous injection of rat chylomicrons containing newly absorbed [^{14}C]retinyl esters[a]

	A	B	C	D	E	F	G
Period after injection (h)	0.28	1.0	3.0	7.75	23.5	72	143
Total tissue recovery (% of dose)	92.0	7.88	79.2	80.7	63.6	59.8	55.7
Tissue (% distribution of recovered ^{14}C)							
Liver	67.81	76.3	67.35	72.39	66.85	72.51	78.73
Fat	2.58	9.33	3.39	6.73	9.57	8.13	8.96
Kidneys	0.42	1.14	3.57	4.04	9.28	12.16	4.20
Adrenals	0.053	0.19	0.13	0.13	0.17	0.12	0.26
Small Intestine	0.68	1.90	2.74	2.20	1.62	0.71	0.76
Spleen	0.20	0.83	0.18	0.16	0.09	0.04	0.04
Skeletal muscle	3.60	6.55	7.36	4.74	6.19	4.09	5.13
Heart ventricle	1.29	0.26	0.19	0.07	0.08	0.04	0.03
Lungs	0.66	0.51	0.86	0.55	2.52	1.24	0.92
Plasma	20.0	2.83	3.75	2.07	1.87	0.94	0.95
Red cells	0.10	0.15	0.07	0.06	0.04	0.01	0.02
Rest of carcass	2.6	0	10.4	6.9	1.7	0	0

[a] This table was adapted from GOODMAN et al. (1965). Rat A was given 0.95 μCi and rats B–G were given 1.58 μCi ^{14}C-labeled chylomicrons. Tissue were analyzed for [^{14}C]retinyl esters various times after injection as indicated in the table.

crons containing [^{14}C]retinyl ester, these investigators demonstrated substantial portion of the ^{14}C-label in skeletal muscle, depot fat, heart, spleen, and kidney. Although later investigators have reported similar findings (VAN BENNEKUM et al. 1998a,b), until recently neither the physiological significance nor the biochemical basis for the observation that some chylomicron or chylomicron remnant retinyl ester is taken up by extrahepatic tissues has been the subject of systematic investigation. Recent investigations suggest that LPL plays an important role in facilitating uptake of postprandial vitamin A by extrahepatic tissues.

Studies exploring the ability of LPL to hydrolyze chylomicron retinyl esters were first reported in 1994 (BLANER et al. 1994). These investigations showed that after most of the chylomicron triglyceride has been hydrolyzed, chylomicron retinyl esters become substrates for LPL-catalyzed hydrolysis. The hydrolysis of chylomicron retinyl ester was found to enhance uptake of the chylomicron vitamin A by cultures of murine BFC-1β adipocytes (BLANER et al. 1994). This observation led to the hypothesis that LPL plays an important role in vivo in facilitating clearance of dietary retinoid by extrahepatic tissues (BLANER et al. 1994). This hypothesis was tested in studies using three strains of mice having different patterns of expression of mouse and human LPL; wild-type mice, LPL-null mice overexpressing human LPL in skeletal muscle, and wild-type mice overexpressing human LPL in skeletal muscle (VAN BENNEKUM et al. 1999b). Mice were injected intravenously with bolus doses of rat mesenteric chylomicrons containing [^{3}H]retinyl esters and after 20 min clearance, tissues were removed to assess uptake of the ^{3}H label. Mice over-expressing human LPL in skeletal muscle took up approximately twice as much chylomicron vitamin A as wild-type mice. For the LPL-null mice over-expressing human LPL in skeletal muscle and which do not express LPL in adipose tissue, much less ^{3}H label was found associated with fat depots than was observed for mice from wild-type backgrounds. Other studies carried out in rats injected with bolus doses of [^{3}H]retinyl ester containing chylomicrons and for which plasma levels of LPL were manipulated through feeding, fasting, or fasting with simultaneous injection of a dose of heparin indicated that increased or decreased LPL expression also directly influences the amount of ^{3}H-labeled retinoid taken up by skeletal muscle, adipose tissue, and heart in these rats (VAN BENNEKUM et al. 1999b). The data from each of these studies in mice and rats were consistent in that the level of expression of LPL activity in skeletal muscle, adipose tissue, and heart is directly correlated with the amount of chylomicron vitamin A taken up by the tissues. Thus, it seems that the amount of LPL activity in skeletal muscle, adipose tissue and heart is a key determinant of the amount of postprandial vitamin A taken up by these tissues. While these data provide a biochemical basis for the observation that some extrahepatic tissues are able to take up retinoid from chylomicrons or their remnants, it should be noted that the dietary retinoid taken up by these extrahepatic tissues must equilibrate with body pools since the steady state total retinol levels for the tissues are much less than would be expected if the tissue retained retinoid commensurate with what it takes up from the diet. Retinol recycling is discussed below in the section on retinol/RBP plasma transport.

Other important findings regarding the clearance of chylomicron retinyl esters have come from studies of chylomicron clearance in rabbits (COOPER 1997). MAHLEY and HUSSAIN (1991) reported that the bone marrow of marmosets and rabbits takes up substantial amounts of chylomicrons. Chylomicron uptake by rabbit bone marrow was 50%–100% of that of the liver. However, unlike hepatic uptake of chylomicrons and their remnants where LPL plays an important role in facilitating uptake, rabbit bone marrow takes up chylomicron retinyl ester even in the total absence of LPL activity (HUSSAIN et al. 1997). Thus, bone marrow may well play a dynamic role in the overall metabolism of postprandial vitamin A.

D. Retinoid Processing and Metabolism within the Liver

The liver is the major organ in the body for the storage and metabolism of retinoid. The preponderance of the body's retinoid reserves are stored as retinyl esters in hepatic stellate cells (also called Ito cells, fat-storing cells, lipocytes or pericytes), although other tissues including adipose tissue, lung, and kidney contain significant retinoid stores and/or are active in retinoid

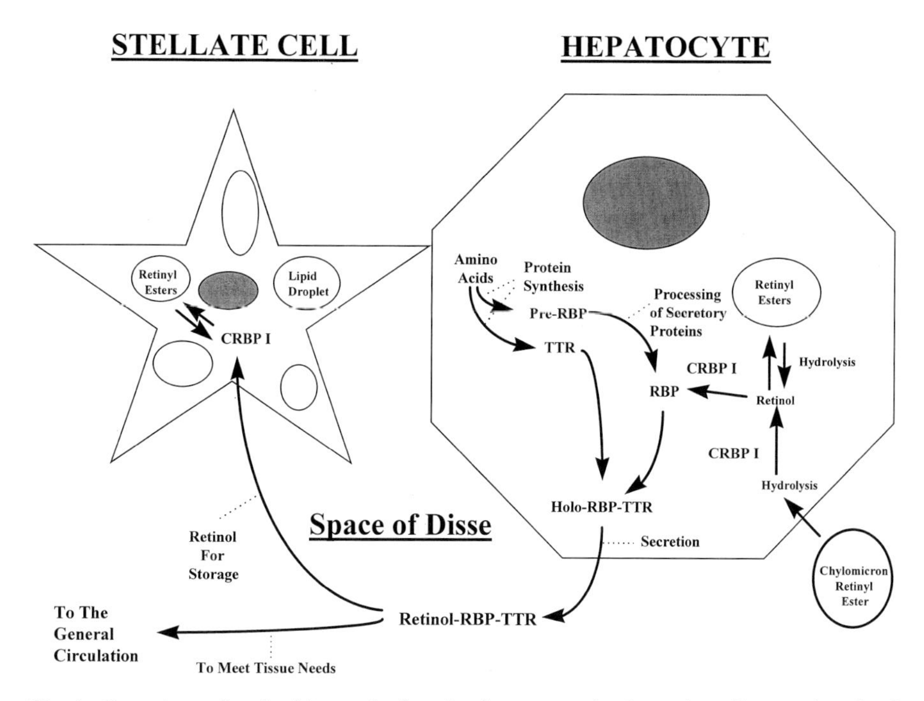

Fig. 4. Overview of retinoid metabolism in the two major hepatic cell types involved in retinoid metabolism, the hepatocytes and the stellate cells. Details of these processes are described in the text. *CRBP I*, Cellular retinol-binding protein, type I; *RBP*, retinol-binding protein; *TTR*, transthyretin

metabolism (Blaner and Olson 1994). The processes of retinoid storage and metabolism in the liver are complex and many details regarding these processes and their regulation are still not fully understood (Blaner 1994). Hepatic retinoid metabolism is mediated by enzymes such as LRAT and retinyl ester hydrolase (REH), which esterify retinol with long-chain fatty acids and which hydrolyze retinyl esters to retinol, and by CRBP I (Blaner and Olson 1994; Ong et al. 1994a). To meet tissue needs for retinoids, retinol is secreted from hepatocytes (hepatic parenchymal cells) bound to its specific plasma transport protein, retinol-binding protein (RBP) (Soprano and Blaner 1994). As with hepatic retinoid storage and metabolism, the factors and processes that regulate retinol mobilization from the liver are not fully understood. An overview cartoon summarizing hepatic retinoid metabolism is provided in Fig. 4.

I. Uptake and Processing of the Chylomicron Remnant Retinyl Ester by Hepatocytes

Chylomicron remnants are thought to be internalized solely by hepatocytes (Cooper 1997). Other hepatic cell types do not appear to participate in the uptake process. Confirming earlier work, Mortimer et al. (1995) using confocal microscopy showed that fluorescent-labeled cholesteryl ester derivatives arriving in chylomicron remnants accumulate solely in hepatocytes and not in other hepatic cell types (Mortimer et al. 1995). It is believed that very shortly after endocytosis by hepatocytes the chylomicron remnant retinyl esters are hydrolyzed to retinol (Blaner et al. 1987; Blomhoff et al. 1988; Harrison et al. 1995) Work by Harrison et al. (1995) showed that the newly endocytosed retinyl esters colocalize with a retinyl ester hydrolase in early endosomes. It is thought that chylomicron remnant retinyl esters are hydrolyzed to retinol at this stage of endocytosis before the remnant particle has reached the lysosome (Harrison et al. 1995). At this point in the uptake process the metabolism of dietary vitamin A diverges from the metabolism of dietary cholesterol, and these dietary lipids no longer share common transport pathways. The retinol formed from retinyl ester hydrolysis is thought to bind apo-CRBP I, which is found in relatively high concentrations in hepatocytes (Blaner et al. 1985). One fate of the CRBP I-bound retinol is for it to be delivered to newly synthesized RBP for secretion into the circulation. The amount of newly endocytosed retinol that is secreted from the liver depends on the vitamin A nutritional status of the animal; more of the newly arrived retinol is secreted from hepatocytes of animals experiencing poor vitamin A nutritional status (Batres and Olson 1987). An alternative fate for newly absorbed retinol is storage as retinyl ester in hepatocytes or in stellate cells following transfer from the hepatocyte. The mechanism through which retinol is transferred from hepatocytes to hepatic stellate cells has not been fully established. The specific events important to hepatic vitamin A storage and metabolism that occur in hepatocytes and in stellate cells is summarized below, separately for each cell type.

II. Retinoid Metabolism and Storage in the Hepatocyte

The hepatocyte is an important cellular site of retinoid metabolism and storage. These liver cells have relatively high concentrations of total retinol (retinol + retinyl ester), RBP, CRBP I, LRAT, bile salt stimulated and bile salt independent retinyl ester hydrolases, retinol dehydrogenases, and retinal dehydrogenases which catalyze the formation of retinoic acid from retinol and enzymes which catalyze the oxidative and/or catabolic metabolism of retinoic acid. Table 2 provides a summary of rat hepatocyte levels of retinoids, retinoid-binding proteins, and some of the enzymes involved in hepatic retinoid metabolism. The hepatocyte plays a central role in retinoid metabolism and is probably one of the few cell types in the body that is active in all aspects of retinoid metabolism including retinoid storage and mobilization, retinoid activation to retinoic acid and retinoic acid catabolism and excretion. This section focuses on the metabolism of retinol and does not specifically consider retinoic acid formation or its oxidative metabolism.

The total retinol concentration of hepatocytes isolated from rats fed a control diet was reported to be 5.2 ± 2.4 nmol retinol/10^6 cells (or 2.6 nmol

Table 2. Distribution of total retinol (retinol + retinyl ester; ROH), retinol-binding protein (RBP), cellular retinol-binding protein, type I (CRBP I), cellular retinoic acid-binding protein, type I (CRABP I), bile salt dependent-retinyl ester hydrolase (BSD-REH), bile salt independent-retinyl ester hydrolase (BSI-REH), and lecithin:retinol acyltransferase (LRAT) in isolated rat liver hepatocytes and stellate cells

Parameter	Hepatocytes	Stellate cells	Reference
Total retinol	5.2 ± 2.4 nmol ROH/10^6 cells	293 ± 107 nmol ROH/10^6 cells	Blaner et al. 1985
	2.6 ± 1.2 nmol ROH/mg protein	1558 ± 569 nmol ROH/mg protein	
RBP	138 ± 67.9 ng RBP/10^6 cells	7.4 ± 3.5 ng RBP/10^6 cells	Blaner et al. 1985
	71.4 ± 35.1 ng RBP/mg protein	39.0 ± 19.3 ng RBP/mg protein	
CRBP I	470 ± 238 ng CRBP-I/10^6 cells	236 ± 89 ng CRBP-I/10^6 cells	Blaner et al. 1985
	243 ± 124 ng CRBP-I/mg protein	1256 ± 477 ng CRBP-I/mg protein	
CRABP I	5.6 ± 3.9 ng CRABP-I/10^6 cells	8.7 ± 5.7 ng CRABP-I/10^6 cells	Blaner et al. 1985
	2.9 ± 3.3 ng CRABP-I/mg protein	44 ± 30 ng CRABP-I/mg protein	
BSD-REH	826 ± 47 pmol FFA/10^6 cells per h	1152 ± 144 pmol FFA/10^6 cells per h	Blaner et al. 1985
	427 ± 24 pmol FFAmg protein per h	6129 ± 768 pmol FFA/mg protein per h	
BSI-REH	Not determined	10.98 ± 1.08 nmol FFA/mg protein per h	Friedman et al. 1993
LRAT	158 ± 53 pmol retinyl ester/mg microsomal protein per min	383 ± 54 pmol retinyl ester/microsomal protein per min	Blaner et al. 1990

retinol/mg cellular protein; BLANER et al. 1985). Approximately 95%–99% of this total retinol is present as retinyl esters, primarily as the esters of palmitic, stearic, oleic and linoleic acids (BLANER et al. 1985). It has been estimated that approximately 5%–30% of the total retinol present in the livers of rats maintained on a control diet is present in hepatocytes (BLOMHOFF et al. 1985; BLANER et al. 1985; HENDRIKS et al. 1985; BATRES and OLSON 1987), with the remainder in hepatic stellate cells. The relative abundance of hepatocyte total retinol is thought to be inversely related to hepatic stores (BATRES and OLSON 1987). Thus, as hepatic vitamin A stores decline, the abundance of vitamin A in hepatocytes increases relative to hepatic stellate cells. Conversely, as hepatic retinol stores increase, relative hepatocyte levels decrease. Since the hepatocyte is the site of RBP synthesis in the liver (SOPRANO and BLANER 1994), this observation could be taken to suggest that hepatocyte total retinol levels are defended in order to maintain circulating retinol-RBP levels.

Chylomicron remnant retinyl esters are thought to be hydrolyzed through the actions of a bile salt independent retinyl ester hydrolase that is present in the hepatocyte plasma membrane and that shows optimal activity in the neutral pH range (GAD and HARRISON 1991; SUN et al. 1997). Recent studies indicate that this hydrolase is identical to rat liver carboxylesterase ES-2 (SUN et al. 1997). Purified rat liver ES-2 shows a substrate preference for retinyl palmitate over triolein and does not catalyze the hydrolysis of cholesteryl oleate (SUN et al. 1997). With retinyl palmitate as substrate, the enzyme shows apparent saturation kinetics with a half-maximal activity achieved at substrate concentrations (K_m) of approximately $70\,\mu M$.

Generally it is now accepted that rat liver carboxylesterase ES-2 is important for catalyzing the hydrolysis of dietary retinyl esters as they enter the hepatocyte (SUN et al. 1997). This enzyme was identified originally in the literature as the bile salt independent retinyl ester hydrolase as opposed to the bile salt stimulated retinyl ester hydrolase which was first characterized for rat liver in the late-1970s. Until the early 1990s it was believed that the bile salt stimulated retinyl ester hydrolase is the key if not sole enzyme responsible for hydolysis of retinyl ester in the liver (GOODMAN and BLANER 1984). In the late 1980s work on the rat bile salt stimulated retinyl ester hydrolase demonstrated that this enzyme is identical to cholesteryl ester hydrolase, a lipid hydrolase that is expressed in both liver and pancreas (HARRISON 1993). Although rat hepatocytes possess both bile salt stimulated and bile salt independent retinyl ester hydrolases, it is now understood that the bile salt stimulated enzyme is not required for hepatic retinyl ester hydrolysis. This conclusion is reached from study of cholesteryl ester hydrolase-deficient mice. Livers of these mice, which totally lack bile salt stimulated retinyl ester hydrolase activity, both take up chylomicron remnant retinyl esters and mobilize retinol-RBP identically to wild-type mice (VAN BENNEKUM et al. 1999a). These data provide convincing evidence that the bile salt stimulated retinyl ester hydrolase is not absolutely required for either dietary retinoid uptake or retinoid mobilization from the liver. At most, the physiological actions of this enzyme in the liver are redundant with those of other enzymes

Whether the bile salt independent retinyl ester hydrolase/carboxylesterase ES-2 plays a unique or essential role in retinyl ester hydrolysis within hepatic cells remains to be demonstrated. However, this possibility seems unlikely since several other enzymes that are proposed to be important for hepatic retinyl ester hydrolysis have also been described in the literature. One such retinyl ester hydrolase is the lysosomal (acid) retinyl ester hydrolase that has been described for rat liver (MERCIER et al. 1994). This enzymatic activity was reported by MERCIER et al. (1994) to be distinct from the lysosomal cholesteryl ester hydrolase and to catalyze the hydrolysis of retinyl palmitate with an apparent K_m of 315 μM. Although the cellular distribution of this enzyme in the liver was not reported by MERCIER et al. (1994), later reports from this group of investigators (AZAÏS-BRESCO et al. 1995) have provided compelling in vitro evidence that the lysosomal retinyl ester hydrolase acts to hydrolyze retinyl esters present in stellate cell lipid droplets. Thus, the lysosomal retinyl ester hydrolase may play an important role in mobilizing retinyl ester stores from stellate cell lipid droplets.

MATA et al. (1996) characterized a "*cis*" retinyl ester hydrolase activity in bovine liver microsomes. Since the activity of this enzyme for 11-*cis*-retinyl palmitate could be competed with triolein, MATA et al. (1996) suggested that the activity may be a property of one of the "broad-specificity" hydrolases present in liver microsomes.

Others have reported the characterization of a porcine retinyl ester hydrolase (SCHINDLER et al. 1998) which shows a broad pH optimum ranging from 7.0 to 9.2. This hepatic retinyl ester hydrolase was shown to be distinct from bile salt-activated lipases and cholesterol esterases although enzymatic activity can be stimulated in vitro by addition of CHAPS and several nonionic detergents to the assay mixture. Based on its structural, immunological, and catalytic features, SCHINDLER et al. (1998) proposed that this retinyl ester hydrolase is a nonspecific carboxylesterase isoform. These investigators indicated that although many of the properties of this retinyl ester hydrolase resemble those of nonspecific carboxylesterase ES-4, the molecular size of this hydrolase is much greater than that of ES-4, and consequently that this pig liver retinyl ester hydrolase is not likely ES-4.

1. Cellular Retinol-Binding Protein, Type I

CRBP I concentrations in hepatocytes are relatively high (470 ± 238 ng CRBP I/10^6 cells or 243 ± 124 ng CRBP I/mg cellular protein), and it has been estimated that approximately 90% of the CRBP I present in the liver is localized in hepatocytes (BLANER et al. 1985). Based on measures made in rat liver and testis that show CRBP I is present at concentrations in excess of those of retinol, it is believed that all retinol present within these tissues is bound to

CRBP I (HARRISON et al. 1987; BISHOP and GRISWOLD 1987). It is generally accepted that CRBP I plays an important role in retinol metabolism (ONG et al. 1994a). Holo-CRBP I has been reported to deliver retinol to LRAT for esterification (ONG et al. 1994a) and to provide retinol to retinol dehydrogenases which catalyzes its oxidation, the first of two oxidation steps needed for formation of retinoic acid. Both LRAT and several retinol-CRBP I utilizing retinol dehydrogenases (ONG et al. 1994a; NAPOLI 1996) are known to be present in hepatocytes. In addition, apo-CRBP I has been reported to enhance hepatic bile salt independent retinyl ester hydrolase activity (ONG et al. 1994a; NAPOLI 1996). It also has been postulated that holo-CRBP I delivers retinol to newly synthesized RBP for secretion from the liver into the circulation. Finally, based on the binding characteristics of retinol to CRBP I, it has been proposed that CRBP I plays an important role in facilitating retinol uptake by cells from the circulating retinol-RBP complex. Thus, CRBP I serves as an intracellular transporter of retinol, linking, and facilitating the processes of retinol uptake, metabolism, and mobilization within the hepatocyte.

2. Hepatic LRAT Activity

Hepatic LRAT is present in both hepatocytes and stellate cells, but its specific activity is highest in the stellate cells (BLANER et al. 1990; MATSUURA et al. 1997). However, because of the large size and the large number of hepatocytes present in the liver, approximately 90% of hepatic LRAT activity is present in hepatocytes (BLANER et al. 1990; MATSUURA et al. 1997). Retinol bound to CRBP I is the preferred substrate for hepatic LRAT, with an apparent K_m for holo-CRBP I in the physiological micromolar range (ONG et al. 1994a). To date LRAT has not been purified from any tissue or cell, and its cDNA and gene have not been cloned. Employing radiation inactivation analyses, ROSS and KEMPER (1993) estimated that hepatic LRAT has a molecular size of approximately 52 ± 10 kDa.

Studies by RANDOLPH and ROSS (1991b) and MATSUURA and ROSS (1993) in rats indicated that hepatic but not intestinal LRAT activity is regulated by retinoid nutritional status. These investigators demonstrated that hepatic LRAT activity falls progressively during retinoid depletion and within 8 h after repletion of retinoid deficient rats with retinol, hepatic LRAT activity starts to rise and reaches control levels within 24 h after the start of repletion. Repletion with retinoic acid also increased hepatic LRAT activity in retinoid deficient rats, but this effect was abolished when hepatocytes were pretreated with inhibitors of gene transcription and protein synthesis (RANDOLPH and ROSS 1991b). Later studies by SHIMADA et al. (1997) demonstrated that feeding of retinoid-deficient rats with synthetic retinoic acid receptor (RAR)-specific agonists but not retinoid X receptor (RXR)-specific agonists induced a higher level of hepatic LRAT activity than was observed for vehicle treated retinoid-deficient rat livers. Moreover, SHIMADA et al. (1997) demonstrated that administration of receptor-specific synthetic retinoids with strong RARα binding activity is directly correlated with the ability of the retinoid to induce hepatic LRAT activity. Taken together, these data suggest that hepatic LRAT expression is regulated by retinoic acid, probably through the actions of RARα. This provides a convenient regulatory mechanism through which retinol, in times of sufficient intake, can be converted to retinyl ester for storage or, in times of insufficient intake, can remain in the unesterified form to serve as a substrate for retinoic acid synthesis.

III. Retinoid Transfer Between Hepatocytes and Stellate Cells

For retinoid homeostasis to be maintained, interactions between hepatocytes and stellate cells must be tightly regulated. However, little is known about how hepatocytes and stellate cells communicate. The processes and cellular events that are involved in the transfer of vitamin A between the two cell types also have not been elucidated. BLANER and OLSON (1994) reviewed possible mechanisms that have been proposed to explain the movement of retinoid between hepatocytes and stellate cells. It is well established that chylomicron remnant retinyl ester must be hydrolyzed before it is transferred from hepatocytes to stellate cells; however, the specific roles which RBP, CRBP I, and/or intercellular contacts between hepatocytes and stellate cells have in the transfer process have not been definitively established. It is thought by most investigators that retinoid transfer involves the actions of RBP and bulk movement through the extracellular space. However, the possibility that movement of free retinol, retinol-CRBP I or retinyl ester occurs via intercellular contacts between hepatocytes and stellate cells has not been definitively ruled out.

Since it is generally assumed that the former process is most likely, the main focus of recent research in this area has been on whether RBP is essentially involved in intercellular transfer and, if so, how. Early studies supporting the participation of RBP in the uptake process demonstrated that RBP is taken up by rat liver stellate cells both in vivo and in vitro (BLOMHOFF et al. 1988; SENOO et al. 1990; MALABA et al. 1993, 1995). This was taken to indicate that RBP is important for the delivery of retinol from hepatocytes to stellate cells. Additional support for this hypothesis was provided by studies carried out in a perfused rat liver model, which demonstrated that transfer of newly absorbed retinol between hepatocytes and stellate cells is effectively blocked when the animals are perfused with antibodies against RBP (BLOMHOFF et al. 1988; SENOO et al. 1990). Recent studies that provide data supporting this possibility consist of coculture experiments investigating the transfer of RBP between human HepG2 hepatoma cells, which secrete RBP, and cultured rat hepatic stellate cells. After coculture for 18h the uptake of human RBP by the cocultured stellate cell was assessed by indirect immunolabeling of the human RBP. Human RBP was identified on the surface, in coated pits, and in vesicles in the stellate cells, thus suggesting movement of the hepatoctye-secreted human RBP to the stellate cells. The amount of human RBP associated with the stellate cells is significantly reduced when antibodies against human RBP are added to the culture medium (BLOMHOFF et al. 1988). TROEN et al. (1994) provided data that the intracellular pool of vitamin A within stellate cells influence the amount of retinol that is esterified when retinol-RBP is added to the cell medium compared to addition as free retinol (TROEN et al. 1994). Other recent studies which do not support a need for RBP in intercellular transfer indicated that while stellate cells may take up RBP and retinol bound to RBP, the uptake of retinol into stellate cells does not depend on RBP (MATSUURA et al. 1993a,b). MATSUURA et al. (1993a,b) employed microscopic autoradiography and fluorescent cell sorting analysis to study retinol uptake by cultures of rat stellate cells from the retinol-RBP complex. These investigators observed that uptake of retinol from the media is not facilitated by RBP, as there was no difference in the intracellular retinol concentration in the stellate cells when retinol was added to the culture medium alone or bound to RBP. Thus, studies exploring the mechanism of transfer of retinol from hepatocytes to stellate cells do not conclusively agree that RBP plays an essential role in the process.

IV. Retinoid Metabolism and Storage in Hepatic Stellate Cells

In the vitamin A sufficient adult rat 70%–95% of hepatic vitamin A is stored as retinyl ester in lipid droplets of stellate cells (BLOMHOFF et al. 1984, 1985; BLANER et al. 1985; HENDRIKS et al. 1985; BATRES and OLSON 1987). Hepatic stellate cells are nonparenchymal cells located perisinusoidally in the space of Disse, in recesses between parenchymal cells (GEERTS et al. 1994). They comprise about 5%–8% of total rat liver cells and 1% of the total liver mass (GEERTS et al. 1994). Approximately 99% of the retinoid present in stellate cells is present as retinyl ester in the lipid droplets located in the cytoplasm of the stellate cells. These lipid droplets are the characteristic morphological feature of these cells. By electron microscopic analysis, the lipid droplets present in rat stellate cells were found to be larger (up to 8 μm in diameter) than those found in rat parenchymal cells (up to 2.5 μm in diameter) (GEERTS et al. 1994). The size and number of the stellate cell lipid droplets is markedly influenced by dietary retinoid intake (WAKE 1980) and intraportal injection of retinol (WAKE 1980).

Stellate cell lipid droplets are formed through vacuolization of cisternae of the rough endoplasmic reticulum and exist in membrane bound and non-membrane bound forms (WAKE 1980). The lipid composition of stellate cell lipid droplets has been investigated by a number of laboratories, and there is general agreement regarding the composition of these droplets (YAMADA et al. 1987; HENDRIKS et al. 1987; YUMOTO et al. 1988). One study reported that the lipid composition of isolated lipid droplets from vitamin A sufficient rats consists of approximately 42% retinyl ester, 28% triglyceride, 13% total cholesterol (free plus ester), and 4% phospholipid (YAMADA et al. 1987). This work also indicated that the retinyl esters in lipid droplets consist of approximately 70% retinyl palmitate, 15% retinyl stearate, 8% retinyl oleate, 4% retinyl linoleate, and lesser percentages of other long-chain retinyl esters. A later study showed that the lipid composition of the stellate cell lipid droplets in rats is markedly affected by dietary retinol intake, but not by dietary triglyceride (calorie) intake (MORIWAKI et al. 1988). The effects of dietary fat and retinoid intake on the lipid composition of stellate cell lipid droplets is shown in Table 3.

The mechanism by which retinoids regulate and maintain the lipid composition of the lipid droplets in stellate cells is not understood. However, it has been suggested that retinoic acid is important in this process (YUMOTO et al. 1989). Because retinoic acid has a sparing effect on hepatic retinol levels, intracellular retinoic acid may well influence hepatic retinol secretion (SHANKAR and DE LUCA 1988; JONES et al. 1994). Although the liver contains some retinoic acid (KURLANDSKY et al. 1995), its cellular distribution has not been determined. Because retinoic acid receptors α, β, and γ (RARα, RARβ, and RARγ) and retinoid X receptor α (RXRα) are all expressed in stellate cells (WEINER et al. 1992; FRIEDMAN et al. 1993), retinoids might well play a role in regulating the state of differentiation of and regulate metabolism within stellate cells.

Stellate cells are highly enriched in CRBP I and the enzymes that are able to hydrolyze (both bile salt-dependent and bile salt-independent retinyl ester hydrolase activities) and to synthesize retinyl esters (LRAT) (BLANER and OLSON 1994). In addition, the stellate cells contain an intracellular binding protein for retinoic acid, cellular retinoic acid-binding protein, type I (CRABP I) (BLANER et al. 1985). However, stellate cells contain very little if any RBP (SOPRANO and BLANER 1994). Levels of these enzymes and of retinoid-binding proteins reported to be present in isolated rat hepatocytes and hepatic stellate cells are provided in Table 2.

In vivo, rat liver stellate cells exhibit a dual phenotype, that is, a quiescent phenotype in normal healthy liver and an activated phenotype in chronically diseased liver (GEERTS et al. 1994). The quiescent stellate cell phenotype is characterized by lipid droplets rich in retinoid, a low cell proliferative rate and low levels of collagen synthesis. In contrast, the activated or myoblastlike phenotype, is distinguished by loss of retinoid-containing lipid droplets,

Table 3. Effect of dietary fat and retinoid intake on the relative lipid distribution for isolated stellate cell lipid droplets (from MORIWAKI et al. 1988)

Diet group	Diet content		Relative distribution (% of total lipid)					
	Retinol/kg	Percentage calories as fat	Retinoid	Triglyceride	Cholesteryl ester	Cholesterol	Free fatty acid	Phospholipid
Control	8.4 μmol	20.5[a]	40	32	15	5	2	6
Low retinol	2.1 μmol	20.5	13	40	27	10	4	6
High retinol	84 μmol	20.5	65	24	7	1	2	1
Low fat	8.4 μmol	5	36	33	18	5	2	6
High fat	8.4 μmol	45	35	33	18	5	3	6

[a] Triglyceride.

increased cell proliferation and increased synthesis of collagen. The activated stellate cell form predominates in liver fibrosis (FRIEDMAN 1993). Interestingly, the activated stellate cell phenotype is observed in livers of rats experiencing vitamin A toxicity. Although these observations suggest possible linkage between stellate cell storage of retinoids and liver disease, the pathophysiological mechanisms underlying these observations are not yet fully understood. It has been established that retinoic acid exacerbates rat liver fibrosis by inducing activation of latent transforming growth factor-β (TGF-β) (OKUNO et al. 1997), the major cytokine implicated in the pathogenesis of liver fibrosis and cirrhosis (FRIEDMAN 1993). It is possible that this action of retinoic acid on TGF-β activity is an important process linking vitamin A toxicity and hepatic fibrosis.

In culture, stellate cells isolated from healthy control rats rapidly lose their in vivo quiescent phenotype and become activated. One characteristic of this activated phenotype observed in cultured rat stellate cells is the rapid loss of the capability of these cells to store retinyl ester (TROEN et al. 1994). Freshly isolated rat stellate cells contain about 144 nmol retinol/mg cellular protein; this concentration declines to 32.9 nmol/mg after 2 days in culture (TROEN et al. 1994) and continues to decline until no retinol or retinyl ester is present in the cells. Thus, it is possible that the many in vitro studies utilizing primary cultures of rat stellate cells may not be relevant for understanding normal hepatic retinoid physiology, since these cells have an activated rather than a quiescent phenotype.

E. Extrahepatic Processing and Metabolism of Retinol and Retinyl Ester

Extrahepatic tissues play significant and important roles in the storage and mobilization of retinol. This was first suggested by kinetic modeling studies of retinol transport and storage in the rat (LEWIS et al. 1981, 1990; GREEN et al. 1985, 1987). LEWIS et al. (1981) initially suggested, for both vitamin A sufficient and deficient rats, that retinol is extensively recycled among the liver, plasma, interstitial fluid, and peripheral tissues. According to the multicompartmental models proposed in these studies, extrahepatic tissues of rats with normal plasma retinol levels but with very low total liver stores (<0.35 μmol vitamin A) should contain as much as 44% of whole body vitamin A. The ability of retinol to recycle from the periphery to the liver is also supported by observations by SOPRANO et al. (1986) who demonstrated that a variety of tissues, including the kidney, lungs, heart, spleen, skeletal muscle, adipose tissue, eyes and testis express RBP. Thus, these tissues may well play a role in the recycling of retinol back to the liver. Finally, retinol and retinyl esters as well as enzymes that are able to esterify retinol and to hydrolyze retinyl esters are found in most peripheral tissues. Moreover, many rat tissues, including lung, kidney, and small intestine contain cells which morphologically and structurally resemble hepatic stellate cells (OKABE et al. 1984; NAGY et al. 1997). These extrahepatic stellate-like cells contain lipid droplets which resemble those of hepatic stellate cells, and both the size and number of these droplets is reported to increase in response to administration of excess dietary vitamin A (OKABE et al. 1984; NAGY et al. 1997). Taken together, these data are consistent and support strongly the notion that extrahepatic tissues play an important role in retinol storage and metabolism.

Many tissues are able to take up and store retinol. Retinol storage and metabolism in the eye (see Chap. 20), adipose tissue, testis, kidney, lung, and bone marrow has been relatively well characterized in the literature (BLANER and OLSON 1994). In addition, GERLACH et al. (1989) has reported that trachea, intestine, and spleen of vitamin A sufficient rats contain retinyl esters. In the guinea pig BIESALSKI (1990) has reported that the kidney, lung, testes, epididymis, vas deferens, trachea, nasal mucosa, tongue, and inner ear contain significant concentrations of retinyl esters. It seems likely that other tissues will also be found to store retinoid. These local stores may be used to meet both local tissue demands as well as total body needs. Whatever the fate of these local reserves, gaining an understanding of the metabolism and storage of retinoid in extrahepatic tissues will aid in formulating the total body dynamics of vitamin A. Considering the possible importance of local stores for providing retinoid needed to prevent local deficiencies and to maintain the good health of the tissues, this area of research holds much future promise for understanding fully retinoid storage and metabolism.

F. Retinoid Delivery to Target Tissues

In the fasted state, the predominant retinoid form in the circulation is retinol, bound to RBP. RBP bound retinol accounts for greater than 95% of all retinoid present in the fasting circulation. However, the fasting circulation also contains low levels of all-*trans*- and 13-*cis*-retinoic acid, 14-hydroxy, and other retro-metabolites of retinol, glucuronides of both retinoic acid and retinol and some retinyl ester bound to lipoproteins. In the postprandial circulation the concentration of chylomicron and chylomicron remnant retinyl ester can greatly exceed the concentration of plasma retinol. In healthy individuals the postprandial retinyl ester response usually subsides 6–8h after consumption of dietary retinoid. Since in a state of vitamin A sufficiency the liver tightly regulates plasma retinol-RBP levels, dietary intake of retinoid does not result in changes in circulating retinol-RBP levels. However, immediately following a retinoid and/or provitamin A carotenoid containing meal, circulating levels of the retinoic acid isomers and the glucuronides of both retinoic acid and retinol are affected (BARUA and OLSON 1986; FOLMAN et al. 1989; ECKHOFF et al. 1991; TANG et al. 1995; BARUA et al. 1997). Specific details regarding the transport and significance of each of the different retinoid forms present in the circulation are described below.

I. Retinol Bound to RBP

Transport of retinol from retinoid stores in the liver to target tissues is accomplished exclusively by means of its specific plasma transport protein, RBP (GOODMAN 1984a; SOPRANO and BLANER 1994). Although most RBP in the body is synthesized in the liver, other tissues including adipose tissue, kidney, lungs, heart, skeletal muscle, spleen, eyes, and testis are known to express RBP (SOPRANO and BLANER 1994). RBP has been extensively studied in humans and rats. RBP is a single polypeptide chain comprising 183 amino acids and with a molecular weight of about 21,000. RBP has one binding site for one molecule of all-*trans*-retinol. In the plasma, RBP circulates as a 1:1 molar complex with another plasma protein, transthyretin (TTR; formerly called prealbumin). Studies of TTR-deficient mice have demonstrated that the formation of the RBP-TTR complex reduces the glomerular filtration and renal catabolism of RBP. Normal levels of plasma retinol and RBP in healthy well-nourished Western populations range between approximately 2 and $3\,\mu M$ (GOODMAN 1984a; SOPRANO and BLANER 1994). Slightly lower plasma levels of retinol and RBP are present in the circulatory systems of well nourished rats, mice, and rabbits. Since essentially all retinol in the circulation is bound to RBP, plasma retinol and RBP levels are highly correlated in well nourished humans as well as retinoid sufficient rats and mice (GOODMAN 1984a,b).

It is well established that circulating levels of retinol-RBP remain very constant except in response to extremes in vitamin A, protein, calorie, and zinc nutriture, or hormonal factors or stress or as a consequence of some disease states (GOODMAN 1984a; SOPRANO and BLANER 1994). The physiological process responsible for maintaining and regulating retinol-RBP levels in the circulation are beginning to be characterized (see below). Plasma retinol-RBP levels can be maintained through regulation of RBP synthesis and/or secretion from the hepatocyte, through regulation of retinol-RBP plasma clearance and catabolism or through a combination of these two mechanisms. RBP synthesis and secretion from hepatocytes is a regulated process (GOODMAN 1984a; SOPRANO and BLANER 1994). The available data also indicate that hormonal and physiological factors probably play roles in regulating the efflux of RBP from hepatocytes. At present we do not fully understand which biochemical signals or processes are important for retaining RBP within hepatocytes or, conversely, for allowing the secretion of RBP. Similarly, there is very little information regarding the catabolism of RBP or its regulation. It is believed that the kidney plays an important role in RBP catabolism since humans and rats with chronic renal failure show elevated plasma RBP levels (GOODMAN 1984a; JACONI et al. 1995). However, at present, it is not understood how the kidney or other organs influence or regulate RBP turnover.

Investigations of plasma retinol homeostasis in the rat have been carried out by several groups (SUNDARESAN 1977; KEILSON et al. 1979; UNDERWOOD et al. 1979; LEWIS et al. 1981; GREEN et al. 1985, 1987; GERLACH and ZILE 1990, 1991; LEWIS et al. 1990). Each of these investigations concludes that a feedback control mechanism regulates mobilization and/or release of retinol-RBP from hepatic stores. It has been proposed based on the data reported in these studies that in response to peripheral tissue needs for retinoid, a signal is sent from the periphery to the liver to regulate retinol-RBP release. Although this feedback signal has not been identified, it has been suggested that circulating retinoic acid levels (SUNDARESAN 1977; KEILSON et al. 1979; UNDERWOOD et al. 1979; LEWIS et al. 1981), apo-RBP levels (SUNDARESAN 1977; GREEN et al. 1985, 1987; LEWIS et al. 1990), or a modified form of circulating RBP (KEILSON et al. 1979; UNDERWOOD et al. 1979) may serve as such a signal.

Even though data have been lacking, there have been several suggestions in the literature that the magnitude of retinoid uptake from the diet may play some role in influencing or possibly regulating plasma retinol-RBP levels (KEILSON et al. 1979; LEWIS et al. 1990). Among other possibilities, UNDERWOOD et al. (1979) suggested that the rate of release of retinol from retinyl ester stores in the liver may be an important site of action of a signal for regulating plasma retinol levels. GREEN and colleagues (1985, 1987; LEWIS et al. 1990) suggested that when RBP synthesis is not compromised, the rate-limiting factor for liver retinol secretion is the appropriate positional availability of unesterified retinol near the intracellular site of RBP synthesis. GREEN and colleagues (1985, 1987; LEWIS et al. 1990) further made the suggestion that an acute influx of a reasonable load of diet-derived retinoid increases hepatocyte retinol-RBP secretion up to some saturation point at which RBP becomes rate limiting.

It is established that the liver is the major tissue site of synthesis of RBP in the body. The major and, in the view of many investigators, possibly the sole cellular site of hepatic RBP synthesis is the parenchymal cell (SOPRANO and BLANER 1994). Initial immunocytochemical localization studies of RBP within the liver indicated that RBP distribution is restricted to the parenchymal cells (KATO et al. 1984). When purified liver cell preparations consisting of highly purified fractions of parenchymal cells, Kupffer cells, endothelial cells, and stellate cells were examined by specific and sensitive radioimmunoassay procedures for RBP, only the purified parenchymal cell preparations were found to contain RBP (BLANER et al. 1985). Later studies extended this work by demonstrating, using Northern blot analysis, that RBP mRNA was present only in isolated parenchymal cells (YAMADA et al. 1987). Subsequent investigations also have confirmed that RBP mRNA is localized solely in parenchymal cells and not stellate cells (WEINER et al. 1992; FRIEDMAN et al. 1993). A high-resolution immunoelectron microscopic study has also provided a detailed description of the distribution of RBP in the liver; consistent with earlier reports, this study was not able to demonstrate immunoreactive RBP in any hepatic cell type other than parenchymal cells (SUHARA et al. 1990). However, some data that suggest that RBP can be present in stellate cells are available in the literature (SENOO et al. 1990). When a very large dose of human RBP (consisting of approximately twice the amount of RBP present in the entire circulation of a rat) was injected into the circulation of rats, a small amount of human RBP was detected by immunohistochemical techniques in the stellate cells (SENOO et al. 1990). Considering the currently available literature on the distribution or RBP within liver cells, it seems that the great majority, if not all, of RBP of hepatic origin is synthesized by parenchymal cells.

1. Cellular Synthesis and Secretion of Retinol-RBP and Its Regulation

RBP synthesis and secretion from the liver are highly regulated processes that are still not fully understood. The factors and processes that regulate RBP secretion are localized primarily to the endoplasmic reticulum (ER) (GOODMAN 1984a; SOPRANO and BLANER 1994; BLANER 1994). It seems that retinol availability within the cell is the most critical factor for regulating the secretion of RBP (GOODMAN 1984a; SOPRANO and BLANER 1994). In addition, the available data indicate that other hormonal and physiological factors probably play roles in regulating the efflux of RBP from cells (GOODMAN 1984a; SOPRANO and BLANER 1994). It has long been known, for instance, that when retinoic acid is provided chronically in the diet to rats, plasma retinol-RBP levels decline by 25%–50% (SHANKAR and DE LUCA 1988; BARUA et al. 1997). At present we do not understand completely which biochemical signals or processes are important for regulating the transcription of RBP, for retaining RBP within cells or conversely, for allowing the secretion of RBP. It is not presently known, for instance, whether the information needed to bring about RBP secretion is inherent in the primary sequence of RBP or if this information resides in some still undescribed chaperone molecule that is responsible for regulating RBP secretion. Nevertheless, over the past 5 years there has been significant progress in understanding the secretory pathway for RBP.

In vitamin A deficiency, RBP protein levels in the liver are three- to tenfold higher than those observed in livers of animals receiving sufficient dietary retinoid (GOODMAN 1984a; SOPRANO et al. 1986; SOPRANO and BLANER 1994). Hepatic RBP mRNA levels show no differences between the two nutritional states (SOPRANO and BLANER 1994). Thus, it is clear that retinol-deficiency specifically inhibits the secretion of RBP from the liver. Several biochemical studies have shown that retinol deficiency largely prevents the movement of newly synthesized RBP from the ER to the Golgi complex (SOPRANO and BLANER 1994). RBP concentrations in ER fractions isolated from retinol-deficient rat livers were reported to be substantially elevated over levels in similar fractions isolated from normal rat livers (SOPRANO and BLANER 1994). In cultured hepatocytes prepared from retinol-deficient rats, pulse-labeled RBP was found to accumulate in the ER and not to be secreted into the cell medium (DIXON and GOODMAN 1987). Interestingly, the transit time of RBP through the ER of cultured hepatocytes prepared from normal rats is shorter than that of albumin and transferrin (FRIES et al. 1984). Studies investigating the subcellular localization of RBP in normal, retinol-deficient, and retinol-repleted retinol-deficient rats using electron microscopic techniques have provided some information regarding the RBP secretory pathway (SUHARA et al. 1990). In the normal liver parenchymal, cell RBP was found to be localized in the synthetic and secretory structures including ER, Golgi complex, and secretory vesicles. The distribution of RBP within parenchymal cells changed markedly with retinol depletion. A heavy accumulation of RBP in the ER accompanied by a marked decrease in RBP-positive Golgi complex and secretory vesicles was observed in the parenchymal cells of retinol-deficient rats. After repletion of the vitamin A deficient rats with retinol, the RBP present in the ER appeared to move rapidly through the Golgi complex and the secretory vesicles to the surface of the cell.

One factor that is important for regulating the release of newly synthesized RBP from the ER appears to be TTR. It was originally proposed, based on the observation that plasma RBP levels drop in vitamin A deficiency whereas those of TTR do not, that RBP must complex with TTR in the plasma after both proteins are independently secreted from the hepatocyte (NAVAB et al. 1977). However, recent studies by several independent research groups have suggested that the RBP-TTR complex is formed within the ER of hepatocytes prior to secretion. MELHUS et al. (1991) were able to trap RBP in HepG2 cells by cotransfecting cells with a cDNA for RBP together with a cDNA for a mutant TTR form generated to contain the ER retention signal KDEL at its C-terminus. The secretory protein purpurin which shares a strong sequence homology with RBP (MELHUS et al. 1991) and which was cotransfected with the mutant TTR cDNA to serve as a control, was secreted normally. The RNLL sequence located in the C-terminus of RBP, while resembling the KDEL ER retention sequence, has been shown by NATARAJAN et al. (1996) not to be responsible for intracellular retention of RBP. NATARAJAN et al. (1996) generated a mutant RBP cDNA lacking the nucleotides encoding these four C-terminal amino acids and demonstrated through transient transfections of COS cells with both wild-type and mutant RBP cDNAs that both RBP species are secreted to the same extent, both in the presence and absence of retinol.

Direct evidence supporting the formation of the RBP-TTR complex within the ER has been provided by BELLOVINO et al. (1996). These authors were able to demonstrate the presence of the RBP-TTR complex within metabolically labeled HepG2 cells after lysing them through mild detergent treatment followed by immunoprecipitation and SDS-PAGE. BELLOVINO et al. (1996) further confirmed through Western blot analysis and chemical cross-linking of the RBP-TTR complex that RBP and TTR bind prior to secretion from HepG2 cells. Moreover, the RBP-TTR bind could be found in the microsomal fraction following treatment of the HepG2 cells with brefeldin A, an agent which blocks the exit of newly synthesized proteins from the ER. Interestingly, the RBP-TTR complex could be co-precipitated with antiserum to calnexin. Since calnexin is an ER integral membrane chaperone protein, this suggests that calnexin may also be involved in RBP (and possibly TTR) retention in the ER. This later experiment was carried out in the presence of dithiothreotol (DTT) to minimize folding of RBP and TTR, and this may have facilitated the detection of the interaction of RBP with calnexin. Interestingly, calnexin is known to associate transiently with incompletely folded glycoproteins (BELLOVINO et al. 1996). The potential role of other chaperone proteins and possibly ubiquitination pathways (KLAUSNER and SITIA 1990) for targeting RBP towards degradation or secretion is a topic worthy of further investigation. In this respect RBP synthesis and secretion may mimic the situation established for hepatic apolipoprotein B (DAVIS et al. 1990), another lipid-transporting secretory protein which, as with RBP, is both synthesized in excess of the amount secreted and requires the presence of its ligand (triglyceride) for maximal secretion.

Studies of TTR knockout mice (EPISKOPOU et al. 1993; WEI et al. 1995) also suggest that TTR plays an important role in regulating RBP synthesis and/or secretion from the hepatocyte. While phenotypically normal, TTR-deficient mice have plasma levels of RBP and retinol which are 5% and 6% of wild-type levels, respectively. While hepatic total retinol levels are the same for wild-type and TTR-deficient mice, hepatic RBP levels are 60% higher for TTR-deficient mice than wild type. This suggests there may be a partial blockage of RBP secretion from TTR-deficient hepatocytes. When RBP secretion from primary cultures of wild-type and TTR-deficient hepatocytes was studied, the rates of RBP release into the medium was determined to be the same for hepatocytes obtained from the two strains of mice. RBP mRNA levels were also the same for the hepatocyte cultures from the two mouse strains. However, RBP protein levels were approximately 60% higher in TTR-deficient hepatocytes than in the wild-type hepatocytes. Thus, it appears that in the absence of TTR, hepatocytes synthesize more RBP in order to maintain the same rate of RBP secretion as wild-type hepatocytes. Taken together these observations are consistent with the hypothesis that TTR is needed, but not absolutely required, to facilitate RBP secretion from hepatocytes.

Interestingly, very low plasma levels of RBP have been identified in patients who have a specific point mutation in their TTR gene. This point mutation results in a single amino acid substitution, isoleucine to serine, at position 84 in the primary sequence of TTR (BERNI et al. 1994; WAITS et al. 1995). The interactions of human RBP with recombinant wild-type and Ser-84 human TTR were investigated by monitoring the fluorescence anisotropy of RBP-bound retinol upon addition of wild-type and mutant TTR species. The Ser-84 TTR was found to have very little affinity for interaction with RBP. Hence it appears that RPB interacts poorly with TTR in patients which carry this mutation in the TTR gene. This observation in humans seems to resemble the situation which is observed for TTR-deficient mice. Thus, the presence of mutant Ser-84 TTR could bring about increased renal filtration of RBP and its irreversible loss from the circulation and also bring about impaired secretion of RBP from the liver.

Although TTR seems to play a role in facilitating RBP secretion from hepatocytes, this may not be the case for all tissues. In addition, TTR secretion may not be influenced by RBP presence in the cell. ONG et al. (1994b) investigated the synthesis and secretion of RBP and TTR from cultured bovine RPE cells and found that approximately 50-fold more TTR tetramer is synthesized and released by the RPE cells than RBP. Similar results were obtained with cultured human fetal RPE cells. This is unlike hepatocytes where the rate of TTR secretion exceeds that of RBP by only two- to three-fold (DIXON and GOODMAN 1987). As is the case for TTR-deficient mice, it is clear from transfected HeLa cells that there is not an absolute requirement that RBP associate with TTR prior to its secretion (MELHUS et al. 1991), nor is such an association required for secretion of RBP from primary cultures of rat Sertoli cells (DAVIS and ONG 1992). Cultured BFC-1β adipocytes that do not synthesize TTR are also able to synthesize and secrete RBP (ZOVICH et al. 1992). Finally, RITTER et al. (1995) found that fenretinide (a synthetic retinoid which binds to RBP, but prevents RBP from binding to TTR) induces secretion of RBP from vitamin A deficient rat liver, a finding that is inconsistent with an absolute need for TTR binding to RBP prior to RBP secretion from the liver.

In studies of the biochemistry of RBP secretion, TOSETTI et al. (1992) have demonstrated that exogeneous addition of millimolar concentrations of β-mercaptoethanol can induce RBP secretion from retinol-depleted HepG2 hepatocytes, suggesting a possible role for sulfhydryl groups in the retention of RBP in the ER. KAJI and LODISH (1993a) have identified disulfide-bonded RBP intermediates in the ER of HepG2 cells. Using DTT, a strong reducing agent which allows continued protein synthesis but prevents synthesized proteins from forming disulfide bonds, these authors (KAJI and LODISH 1993b) identified two RBP folding intermediates that precede the formation of mature disulfide-bonded RBP. These intermediate forms of RBP were termed compact I and II. Compact I and II comigrate with mature RBP upon native PAGE, but they can be distinguished by their different sensitivity to DTT-induced unfolding. Moreover, compact I and II intermediates can be stabilized against DTT-induced unfolding to different degrees, by both retinol and retinoic acid. This was taken by KAJI and LODISH (1993a,b) to suggest that RBP folding in the ER is altered in the presence of a bound ligand. Since RBP synthesized in the presence of retinol but not retinoic acid (which also is able to bind RBP; MALPELI et al. 1996; see below) is rapidly secreted, these investigators suggested that proper folding of RBP such that it is stabilized to DTT reduction is necessary but not sufficient for allowing the exit of RBP from the ER. The stabilization of RBP folding in the ER by retinoids is specific, since lipophilic reagents such as oleic acid and cholesterol had no effect on the stability of RBP to DTT unfolding. Furthermore, neither of these lipophilic reagents enhanced RBP secretion, but in fact inhibited it. Another study (KAJI and LODISH 1993b) demonstrated that the sensitivity of RBP to DTT reduction is specific to intracellular RBP, and that this sensitivity results through

RBP interaction with intracellular factors including protein disulfide isomerase and other ATP-dependent factors in the ER rather than through inherent differences in the intra- and extracellular forms of RBP itself. While these findings are consistent with a role for disulfide bond formation and subsequent proper peptide folding being requisite for RBP secretion, the secretion of other disulfide bond containing proteins from the ER, for example, IgM assembly intermediates and free joining (J) chains, is enhanced in the presence of reducing agents (ALBERINI et al. 1990). The secretory pathway for proteins which do not require disulfide bond formation are not altered in the presence of DTT (LODISH and KONG 1993). Thus, RBP seems to require disulfide bond formation for its secretion, whereas some disulfide bond-containing secretory proteins are more efficiently secreted when disulfide bond formation is blocked.

SMITH and colleagues (1998), analyzing subcellular fractions prepared from the livers of control and vitamin A depleted rats collected at time points ranging from 2 to 20 min after i.v. injection of [^{3}H]retinol, have provided insight as to where complex formation between retinol and nascent RBP takes place. These investigators demonstrated that the holo-RBP complex can form in the rough ER of the liver. Secretion of the complex could be inferred from the presence of the [^{3}H]retinol-RBP complex in the smooth ER, Golgi fraction and the blood at successively later times. Thus, it appears that retinol is incorporated into RBP either during its synthesis in the rough ER or immediately after synthesis. This finding is consistent with the data of KAJI and LODISH (1993a,b) described above.

Other work by RITTER and SMITH (1996) has examined the effects of *N*-(4-hydroxyphenyl)retinamide (fenretinide; HPR), a synthetic retinoid which has chemopreventive activity in animal models, and colchicine, which delays secretion of proteins that are vectorially processed through the Golgi complex by inhibiting microtubule formation, on RBP synthesis and secretion. Retinol treatment of vitamin A deficient rats produced a 12-fold increase in plasma RBP, whereas colchicine treatment prior to retinol repletion allowed only a 5.5-fold increase in plasma RBP. Colchicine treatment alone resulted in accumulation of RBP primarily in the rough ER, but also in the Golgi (at levels significantly less than that seen in the presence of retinol). RITTER and SMITH (1996) took this to suggest that there is continued progression of a small amount of apo-RBP through the smooth ER even though sufficient ligand retinol is unavailable (RITTER and SMITH 1996). While treatment of retinoid-deficient rats with HPR permitted RBP to progress out of the rough ER, albeit at a significantly lower rate than upon retinol treatment, a concurrent rise in plasma RBP was not observed. RITTER and SMITH (1996) attributed the lack of rise in plasma RBP level to the low affinity that the HPR-RBP complex has for TTR and suggested that the HPR-RBP is either rapidly filtered by the kidney and/or rapidly diffuses into the interstitial fluid. TTR was also released from the rough ER in response to retinol treatment under conditions of vitamin A deficiency but was not released in response to HPR treatment. Thus, in addition to providing insight into the actions of HPR on RBP secretion from the liver, these findings by RITTER and SMITH (1996) are also consistent with the hypothesis that the formation of the holo-RBP:TTR complex occurs within the rough ER.

2. The RBP Gene and Its Regulation

The gene for RBP has been the subject of recent investigations by Colantouni and colleagues (MOUREY et al. 1994; PANARIELLO et al. 1996). These researchers have shown that the human gene for RBP can be upregulated in HepG2 hepatocytes upon exposure to either retinol or retinoic acid. This upregulation was both time- and dose-dependent, and an increase in RBP mRNA levels was accompanied by an increase in RBP protein. The region in the promoter of the human RBP gene responsible for this upregulation was identified using a combination of band-shift assays and site-directed mutagenesis (PANARIELLO et al. 1996). Retinoid responsiveness in the RBP gene was attributed to two degenerate retinoic acid response elements separated by 30 nucleotides in the promoter of the gene. Included in this element is a GC-rich Sp1 consensus-like sequence. The entire element was shown to be required in order to confer retinoic acid inducibility to a heterologous promoter. In addition to retinoids, Sp1, or a related protein, was shown to play an important role in regulating transcription of the RBP gene (PANARIELLO et al. 1996).

3. Structural and Physiochemical Properties of RBP, TTR, and the RBP-TTR Complex

In the last several years significant advances have been made towards understanding the structural and biochemical properties of RBP and TTR responsible for, and important to, their interaction. Most of this new information has come from X-ray crystalographic studies (NEWCOMER et al. 1984; COWAN et al. 1990; ZANOTTI et al. 1993a,b 1994; MONACO et al. 1994, 1995).

The X-ray crystal structure of the RBP-TTR complex has now been determined to 3.1 Å resolution (COWAN et al. 1990) and the structures for apo- and holo-RBP forms binding various ligands (retinol, all-*trans*-retinoic acid, fenretinamide and axerophthene) have been determined to a resolution as fine as 1.8–1.9 Å (NEWCOMER et al. 1984; ZANOTTI et al. 1993a,b). RBP is comprised of eight strands of antiparallel β sheets and a short α-helical segment. One end of the β-barrel is open and provides an entrance to a hydrophobic binding cavity, whereas the other end of the barrel is closed by the N-terminus of the RBP molecule. Retinol binds to RBP with a hand-in-glove-like fit, with the cyclohexene ring of retinol buried innermost within the β-barrel. The hydroxyl group of retinol points toward the outside of the β-barrel and is positioned near the entrance loops of the barrel (MALPELI et al. 1996; NEWCOMER 1995). Structural differences between the apo- and holo- forms are found in a short loop at the entrance to the β-barrel, in the region of amino acids 34–37 (ZANOTTI et al. 1993a,b). The positioning of the retinol within the binding cavity of RBP has functional consequences towards the ligand-binding specificity of RBP. Since tight complementarity exists between the cyclohexyl ring and its binding pocket within the RBP molecule, RBP shows a strong ligand binding preference towards ligands having a cyclohexyl ring or a similar structure at one end of the molecule. RBP shows less specificity towards the functional group present at the polar tail of the molecule. The hydroxyl group constituting the polar tail of retinol is positioned near the surface of the protein in a relatively unconstrained position. Consequently, in vitro RBP accommodates retinoid ligands other than retinol, for example, all-*trans*-retinoic acid or -retinal or the 13-*cis*-isomers of retinol, retinal or retinoic acid (NEWCOMER 1995).

TTR consists of a tetramer made up of four identical subunits of 127 amino acids each. The monomers are arranged into two four-stranded β sheets. Two monomers form a stable dimer and two dimers form a tetramer. Three-dimensional structures for wild-type and several mutant human TTRs and for TTRs from various species have been reported in the literature (MONACO et al. 1995). Tetrameric TTR has a channel of approximately 10 Å diameter which contains two thyroxine-binding sites (BLAKE and OATLEY 1977; BLANEY et al. 1982). Interestingly, retinoic acid and other molecules (e.g., products of thyroxine deiodination, pharmacological agents such as penicillin, salicylates, steroids, flavenoids, and environmental contaminants such as hydroxylated polychlorinated biphenyls, dibenzodioxins and dibenzofurans) can bind to this channel within TTR with relatively high affinity (SMITH et al. 1994; ZANOTTI et al. 1995). Aside from the binding of thyroxine, the physiological significance of the binding of these other bioactive molecules to tetrameric TTR is not understood.

Determination of the crystal structure of the RBP-TTR complex at the 3.1 Å resolution was reported by MONACO et al. (1994, 1995). For these studies crystals were prepared from complexes containing human TTR and chicken RBP, and the structure of the complex solved using the molecular replacement method. This crystallographic analysis revealed a complex containing one TTR tetramer plus two RBP molecules. Amino acid residues involved in the RBP:TTR contact region are listed in Table 4 (MONACO et al. 1995). These occur in regions of the RBP and TTR molecules that are conserved across species, as might be expected considering the species cross-reactivity observed for RBP and TTR binding. The X-ray structure demonstrates that the retinol (C-15) hydroxyl group participates in the contacts with TTR. This is in agreement with the observation obtained from studies of RBP and TTR interaction in solution which demonstrated that complex formation stabilizes the binding of retinol to RBP (MALPELI et al. 1996), and substitution of the retinol ligand impairs interaction between RBP and TTR (MALPELI et al. 1996; see below). MONACO et al. (1994, 1995) was able to confirm the proposed interaction of Lys and Trp residues in the RBP molecule with TTR (see Table 4) by demonstrating that lysine residues 89 and 99 and tryptophan residues 67 and 91 of RBP contact TTR. These investigators further confirmed that Leu-35 plays a

Table 4. Significant amino acid contacts in the RBP:TTR complex as reported by MONACO et al. (1995)

RBP residue	Distance (Å)	TTR residue[a]
Leu-35	3.41	B-Gly-83
Trp-67	4.20	C-Val-120
Trp-67	3.45	C-Ile-84
Lys-89	2.73	A-Asp-99
Lys-89	4.54	A-Asp-99
Trp-91	4.07	A-Ser-100
Ser-95	3.03	B-Tyr-114
Phe-96	2.69	B-Ser-85
Phe-96	3.38	B-Tyr-114
Phe-96	3.37	B-Ile-84
Phe-96	3.44	C-Arg-21
Leu-97	2.80	B-Ser-85
Lys-99	2.70	A-Asp-99
Lys-99	3.87	B-Ser-85
Retinol (OH)	3.15	B-Gly-83

[a] Chains A and B refer to the two subunits of one of the TTR dimers; chain C is a TTR monomer from the other dimer.

role in the holo- to apo- transition of RBP resulting in lessened affinity of RBP for TTR.

For a more detailed summary of the X-ray crystallographic studies on RBP and other retinoid-binding proteins, the reader is referred to a review by NEWCOMER (1995).

Using fluorescence polarization techniques, MALPENI et al. (1996) studied the interaction of RBP with TTR and how various retinoids influence this interaction. These investigators determined the dissociation constants for the interaction of human TTR with human RBP, either in the apo-form or binding retinoids which differ chemically in the functional groups present at the terminal C-15 position. Dissociation constants for the interaction of TTR for apo-RBP and several holo-RBPs were: all-*trans*-retinol-RBP, K_d approx. $0.2\,\mu M$; apo-RBP, K_d approx. $1.2\,\mu M$, all-*trans*-retinoic acid-RBP, K_d approx. $0.8\,\mu M$; and all-*trans*-retinyl methyl ether-RBP, K_d approx. $6\,\mu M$. HPR-RBP exhibited only negligible affinity for TTR. These authors deduced from their data that the retinoid moiety affected RBP-TTR recognition in a manner which was correlated with the steric effects exerted by the functional group at the C-15 position of the retinoid. The larger more bulky functional groups at the C-15 position lessened the affinity of the ligand-RBP complex for TTR. This result, as well as those obtained by others (NOY 1992a,b; ZANOTTI et al. 1994), is consistent with information obtained from the X-ray crystalographic studies of retinol binding to RBP (NEWCOMER et al. 1984; ZANOTTI et al. 1993a,b) and of retinol-RBP binding to TTR (MONACO et al. 1994, 1995). In other studies, ZANOTTI et al. (1994) likewise found weaker interactions between axerophthene-RBP (axerophthene has the same chemical structure as all-*trans*-retinol but lacks the hydroxyl group) and TTR. This finding is consistent with the observation from X-ray crystalographic studies that the hydroxy group of retinol stabilizes RBP-TTR interactions. These authors also found more drastic steric hindrance for HPR-RBP and retinoic acid–RBP that affected interaction of the liganded-RBP with TTR, causing conformational changes in the loop region of RBP at the entrance of its β-barrel and lessening affinity for TTR.

4. Retinol-RBP Turnover from the Circulation

Several studies investigating either the turnover of retinol and/or RBP from the circulation of humans or animal models or of the biochemical parameters regulating this plasma turnover have been published in the recent literature. Detailed treatments of the older literature in this area can be obtained from the reviews of GOODMAN (1984a) and SOPRANO and BLANER (1994).

In one recent study SIVAKUMAR et al. (1994) reported investigations of the fractional rates of synthesis (FRS) of RBP and TTR in infant pigs infused with $[^2H_3]$leucine for 6h. The mean FRS of TTR (1.97% ± 0.13% pools/h) and of RBP (3.89 ± 0.40% pools/h) were significantly different. Thus, it appears that the two proteins involved in the transport of retinol in the circulation turnover from the plasma at very different rates. This raises an interesting question regarding how the body distinguishes between RBP and TTR, proteins that interact tightly in the circulation, but that undergo different rates of turnover from the circulation.

Some insight into this question has been provided by JACONI et al. (1995) who explored the mechanisms through which the kidney recognizes RBP destined for proteolysis and removal from the circulation. These investigators, employing mass spectrometric analysis of the intact RBP, concluded that at least three different forms of RBP can be present in the human circulation. Each of these RBP forms differ solely by the number of leucine residues present at the C-terminal of the molecule. Native human RBP consists of 183 amino acids and contains two C-terminal leucine residues. JACONI et al. (1995) identified in normal human plasma a second RBP species (termed RBP_1) consisting of 182 amino acids and containing only one C-terminal leucine. In plasma obtained from patients experiencing chronic renal failure, a third RBP species (termed RBP_2) lacking both C-terminal leucines was identified. Only native RBP and RBP_1 could be detected in plasma from healthy individuals whereas all three species were found to be present in plasma obtained from chronic renal failure patients. Based on their data, JACONI et al. (1995) hypothesized that removal of the terminal leucine residues from RBP by a carboxypeptidase provides a signal that causes the RBP molecule to undergo clearance from the circulation in the kidney. This interesting hypothesis has not been experimentally tested in animal models.

Using stably labeled $[^{13}C]$retinol, VON REINERSDORFF et al. (1996) investigated the plasma kinetics of a single oral dose of 105 μmol retinyl palmitate in healthy male volunteers. The total retinyl ester concentration in the plasma peaked 4.45 h after administration of the oral dose. Over the 7 days of this study the incorporation of the labeled retinol into the plasma retinol-RBP pool was characterized by fast distribution kinetics and is consistent with the concept that the liver follows the principle of "last in–first out" in maintaining retinoid homeostasis. Thus, data from this study suggest that newly absorbed chylomicron remnant retinyl ester serves as a preferred source for retinol to be incorporated into nascent RBP within hepatocytes.

Isotope dilution methodologies utilizing oral doses of either stably or radioactively labeled retinol to estimate liver retinoid reserves have been described in the recent literature. One such approach was reported by ADAMS and GREEN (1994) and employs an oral dose of [^{3}H]retinol to estimate the mass of retinoid present in livers of rats. ADAMS and GREEN (1994) were able to develop and validate a biexponential equation describing the relationship at each sampling time between the fraction of the dose remaining in the plasma and the size of the liver retinoid pool. This equation proved useful for distinguishing animals with deficient, marginal, adequate and high retinoid nutritional status. HASKELL et al. (1998) described a similar retinol dilution approach, using an oral dose of [$^{2}H_4$]retinyl acetate, for estimating total body stores of retinoid in healthy adults living in the United States and Bangladesh and in a healthy child living in the United States. In these human studies the size of hepatic retinoid reserves did not appear to affect the time needed for the test dose to equilibrate with endogeneous body stores for the range of hepatic stores studied. The hepatic retinoid concentrations estimated using this retinol dilution technique were similar to values obtained through hepatic biopsy reported previously for populations in the United States and Bangladesh. The use of this relatively noninvasive method employing stably labeled retinol to estimate hepatic or total body retinol pools in humans may prove useful for obtaining understanding of the effects of retinoid deficiency in human populations.

5. RBP Receptor

The mechanism through which retinol is taken up from the circulation by cells has not been conclusively established. Almost all circulating retinol is bound to RBP, and nearly all circulating retinol-RBP is complexed to TTR. The potential role(s) of RBP and/or TTR in the cellular uptake of retinol is a matter of considerable current debate and speculation. Resolution of this issue ill be essential for gaining a complete understanding of retinol metabolism.

Existence of a plasma membrane receptor for RBP was first postulated and investigated by HELLER (1975), BOK and HELLER (1976) and independently by RASK and PETERSEN (1976) in the mid-1970s. It was felt that the presence of a receptor for RBP on the cell surface would provide a mechanism through which cells could regulate the amount of retinol entering the cell. Thus, expression of such a receptor on the cell surface would allow the cell to take up needed retinol from the retinol-RBP complex and the lack of expression of a receptor for RBP on the cell surface would allow cells that do not require retinol to exclude it. Through modulation of RBP receptor expression the cell could modulate retinol uptake. In the mid-1970s it was still not recognized that isomers of retinoic acid are the primary active forms of vitamin A (GOODMAN 1984b) and also that the circulation contains low levels of retinoic acid and retinoic acid metabolites (see Sects. F.II, F.III. below), in addition to the more abundant retinol-RBP (DE LEENHEER et al. 1982). Moreover, in the mid-1970s it was not yet recognized that nearly all tissues, if not nearly all cell types, have some need for retinoids (GOODMAN 1984b). We now often see the word "pleiotropic" used when describing the effects of retinoids, this was not true in the mid-1970s when the actions of retinoids were felt to be limited to only a relatively selective number of tissues within the body. Many studies investigating the receptor mediated uptake of retinol from RBP have been carried out since the mid-1970s. At present only one putative cell surface receptor for RBP, located almost solely in the retinal pigment epithelium and termed p63 (also termed RPE65 by some investigators (NICOLETTI et al. 1995; TSILOU et al. 1997; GU et al. 1997)), has been cloned or purified to homogeneity (BÅVIK et al. 1993). As outlined below, there also are substantial data in the recent literature that p63 is not a cell surface receptor for RBP (NICOLETTI et al. 1995; TSILOU et al. 1997; GU et al. 1997).

Although many investigators have reported the identification and characterization of cell surface receptors for RBP that facilitate retinol uptake by target cells and/or confer cell type specificity to the uptake process, it is not clear from what is known about the physical chemistry of retinol binding to RBP and to membranes that a cell surface receptor is required in order for retinol to be taken up by cells. Free retinol rapidly "flip-flops" across phospholipid bilayer membranes in a manner that is not dependent on any protein components (FEX and JOHANNESSON 1987, 1988; HO et al. 1989; NOY and BLANER 1991; BLANER 1994) and with a rate constant that is faster than those reported for RBP- or RBP-TTR-bound retinol (NOY and XU 1990; SIVAPRASADARAO and FINDLAY 1988a). Furthermore, the dissociation of retinol from retinol-RBP and retinol-RBP-TTR complexes is very rapid and is not rate limiting for retinol uptake from cells (NOY and XU 1990). Based on these considerations, it is not clear how spontaneous uptake of retinol by cells could be prevented if one were to accept the existence and necessity of an RBP receptor. Some studies have suggested that contrary to facilitating retinol uptake by cells, RBP or RBP-TTR may retard cellular uptake of retinol from the extracellular space (CREEK et al. 1989; RANDOLPH and ROSS 1991a). This hypothesis is consistent with the faster rate constants observed for uptake of free versus protein-bound retinol (NOY and XU 1990). Proponents arguing in favor of the nonreceptor-mediated uptake of retinol by cells have proposed a model whereby the level of apo-cellular retinol-binding protein, type I (apo-CRBP I) dictates the amount of retinol that can be taken up by cells. According to this model, any process that generates apo-CRBP I (e.g., metabolism or esterification of retinol or new CRBP I synthesis) enhances retinol uptake by cells (Noy and Blaner 1991). Nevertheless, the recent literature is replete with reports concerning the nature of the cell surface receptor for RBP and exploring the receptor-mediated uptake of retinol by cells. A summary of recent reports in this area is provided in the text below and also in tabular form in Table 5.

In 1991 Båvik et al. reported the identification of a RPE membrane receptor for RBP, termed p63 (based on its apparent molecular weight of 63 kDa), by means of a membrane binding assay. RBP binding sites were found primarily in microsomes prepared from bovine RPE fractions, but also in liver and kidney microsomes, but not in microsomes prepared from lung or muscle. These authors later went on to generate a monoclonal antibody, A52, to p63, and found that p63 expression is limited to RPE cells, that p63 occurs in multiple forms (some of which are unable to bind RBP), and that by immunohistochemical staining p63 is primarily found associated with intracellular membranes (Båvik et al. 1992). When Båvik et al. isolated and cloned p63 (Båvik et al. 1993), hydropathy analysis revealed that p63 has no N-terminal signal sequence or transmembane spanning region, suggesting that p63 is not an integral membrane protein. Northern blot analysis revealed that p63 is present only in the RPE (and not in liver, kidney, testis, adrenal gland, lung or brain, even after long exposures of the Northern blots). Southern blot analysis suggested complex organization of the p63 gene. The coding region of p63 is 1599-bp, and the deduced amino acid sequence consists of 533 amino acids. Different p63 cDNA clones were found to have heterogeneous 5′ UTRs. When the protein was translated in vitro in the presence of human microsomes, only small amounts of p63 associated with microsomes and Båvik et al. (1993) postulated that microsomal association might occur posttranslationally.

Later these investigators reported raising a second and inhibitory monoclonal antibody to p63, named P142 (Båvik et al. 1995). This monoclonal antibody to p63 reportedly inhibits binding of radiolabeled RBP to RPE membranes in a concentration-dependent manner. Inclusion of P142 in the medium of cultured keratinocytes partially protects them from inhibition of terminal differentiation. Terminal differentiation of cultured keratinocytes is known to be inhibited by retinoids. Moreover, incubation of keratinocytes with increasing concentrations of retinol-RBP suppressed in a concentration-dependent manner expression of the terminal differentiation marker keratin K_1 in a manner which was at least ten-fold more efficient than addition of free retinol (not bound to RBP). Thus, the ability of P142 to protect cultured keratinocytes from inhibition of terminal differentiation upon exposure to increasing concentrations of retinol-RBP was taken by these investigators to suggest that P142 interferes with retinol uptake by the cultured keratinocytes. Surprisingly, P142 did not detect a keratinocyte protein upon western analysis, and no p63 mRNA expression was detected upon northern blot analysis of keratinocyte total RNA (although p63 was detected in control human RPE total RNA). Although the data regarding the ability of P142 to block

Table 5. Partial summary of recent reports focusing on the characterization of cell surface receptors for RBP

Reference	Tissue or cell source	K_d of receptor for RBP	Requirement for TTR	Is RBP internalized?	Comments
SIVAPRASADARAO and FINDLAY (1988a,b)	Human placental brush border membrane vesicles	High affinity $3 \pm 2.7\,nM$ Low affinity $5 \pm 5\,nM$	TTR inhibited binding	No	Heterogeneity in RBP; high- and low-affinity binding forms of the RBP receptor; trypsin-, heat-, and thiol group-specific reagent sensitive
SIVAPRASADARAO and FINDLAY (1988b)	Human placental brush border membrane vesicles	Apparent K_m $116 \pm 13\,nM$ V_{max} 6.9 ± 0.5 pmol/mg per min	TTR decreased rate of [^{3}H]ROH: RBP uptake	No	Velocity of [^{3}H]ROH:RBP uptake was rapid, time and temperature dependent
SIVAPRASADARAO and FINDLAY (1993)	Human placental brush border membrane vesicles and recombinant RBP	$50 \pm 20\,nM$	No	NR[a]	Serum RBP loses its ability to bind to the RBP receptor upon storage/handling
SIVAPRASADARAO et al. (1994)	Human placental brush border membrane vesicles	$600\,nM$ in octyl glucoside $270\,nM$ in Nonidet P-40	TTR increased nonspecific binding	NR	Studies of detergent solubilized RBP receptor from human placental membranes
SIVAPRASADARAO and FINDLAY (1994)	Human placental membranes	NR	TTR and RBP receptor binding sites are overlapping; RBP cannot bind RBP receptor while bound to TTR	NR	Site-directed mutagenesis of RBP to determine residues important in RBP binding to its receptor, to TTR and to retinol

SENOO et al. (1990)	Parenchymal and stellate cells from rat liver	NR	NR	Yes	RBP uptake by parenchymal and stellate cells, studied by immunocytochemistry at the electron microscopy level; RBP localized first on cell surface, later in vesicles near the cell surface, then in larger vesicles deeper in the cytoplasm
BÅVIK et al. (1991)	Microsome fractions from bovine retinal pigment epithelial (RPE) cells, liver and kidney	31–72 n*M*	No	NR	Membrane binding assays using ^{125}I-RBP; identification of RBP receptor termed p63
BÅVIK et al. (1992)	Bovine RPE cells	NR	NR	NR	Generated monoclonal antibodies to p63; p63 expression is limited to RPE, primarily intracellular membranes
BÅVIK et al. (1993)	Bovine RPE cells	NR	NR	NR	Isolated and cloned p63; no N-terminal signal sequence or transmembrane region; northern blot analysis indicates expression only in RPE cells
BÅVIK et al. (1995)	Cultured human keratinocytes	NR	NR	NR	
MALABA et al. (1993)	Rat and rabbit primary hepatocyte cultures	NR	NR	Yes	Iodination method used to iodinate RBP affects receptor-mediated endocytosis of RBP
MALABA et al. (1995)	Rat hepatocyte and nonparenchymal preparations	NR	NR	Yes	Endocytosis via caveolae, not clatherin-coated pits

Table 5. *Continued*

Reference	Tissue or cell source	K_d of receptor for RBP	Requirement for TTR	Is RBP internalized?	Comments
MATARESE and LODISH (1993)	F9 embryonal carcinoma cells	200–300 n*M*	NR	Yes	Demonstrated RBP receptor activity is heritable
MELHUS et al. (1995)	Bovine RPE cell membranes	NR	No	NR	Generated monoclonal antibodies to RBP and tested their ability to interfere binding to bovine RPE membrane fractions
SMELAND et al. (1995)	Placenta, RPE cells, bone marrow, kidney, intestine, spleen, liver, lung, choroid plexus, pineal gland, immature testes, undifferentiated keratinocytes	Multiple K_ds reported, each dependent on tissue or cell source of the membrane	NR	NR	Cell-free binding assays, measured binding of ^{125}I-RBP to membrane preparations
SUNDARAM et al. (1997)	Permeabilized human placental membrane	NR	NR	Not applicable	Role of CRBP I in facilitating retinol uptake explored
WARD et al. (1997)	Postimplantation rat embryos and visceral yolk sac endoderm studied by immunohistochemistry	ROH uptake shows $K_m = 0.4\,\mu M$ and $V_{max} = 2 \times 10^{-18}$ mol/µg protein per s	NR	NR	Study the mechanism of transport of maternal retinol-RBP to the embryo; antibody to p63 reduced [^{3}H]ROH-RBP uptake by the visceral yolk sac endoderm by 21.3%

[a]NR = not reported

retinol-RBP inhibition of keratinocyte differentiation are convincing, when these observations are considered together, they are somewhat difficult to reconcile.

P142 also was employed to immunoprecipitate a 63-kDa protein from yolk sac endoderm of rat embryos by WARD et al. (1997). Thus, p63 is expressed in rat embryonic tissues in addition to adult bovine retinal pigment epithelial cells. Addition of P142 to cultured rat conceptuses was found to reduce [^{3}H]retinol uptake from [^{3}H]retinol-RBP by 21.3% but uptake was not affected by the presence of the same concentration of control antiserum. Based on these data, WARD et al. (1997) proposed that initial embryonic uptake of RBP-carried maternal retinol is mediated by p63 in the visceral yolk sack endoderm of the rat. Based on these data and data regarding the presence of CRBP I in the visceral yolk sac endoderm, these authors propose an interesting model for the transcytosis of retinol from the maternal circulation to that of the developing embryo (WARD et al. 1997).

Several reports have appeared in the recent literature that provide strong arguments that p63 is not a cell surface receptor for RBP (HAMEL et al. 1993; NICOLETTI et al. 1995; TSILOU et al. 1997; GU et al. 1997). The data provided in these reports indicate that p63 is localized on intracellular membranes, that p63 is not an integral membrane protein, and that p63 lacks any transmembrane domains or signal peptides within the p63 molecule. Together these data suggest that p63 does not function as an RBP receptor. NICOLETTI et al. (1995) reported cloning the cDNA for a human protein that is 98.7% identical in its amino acid sequence (525 of 533 amino acids identical) to the bovine protein cloned by BÅVIK et al. (1991). NICOLETTI et al., based on the intracellular localization of this protein within the human RPE and its lack of a membrane spanning domain, concluded that this protein could not be a cell surface receptor for RBP. Studies by HAMEL et al. (1993) investigating the intracellular distribution of p63 within bovine RPE cells agreed with those of NICOLETTI et al. (1995) in that p63 protein was localized immunocytochemically to intracellular membranes (specifically to the smooth endoplasmic reticulum) and not to the basolateral surface as would be expected of an RBP receptor. These investigators also concluded that p63 could not be a cell surface receptor for RBP. TSILOU et al. (1997) demonstrated that p63 readily associates with phospholipids and suggested that the association between p63 and phospholipids may be related to the still undefined function of p63. Interestingly, work by GU et al. (1997) has identified the gene for p63 to be the site for mutations which cause autosomal recessive childhood-onset severe retinal dystrophy (GU et al. 1997).

In the late 1980s, SIVAPRASADARAO and FINDLAY (1988a,b) reported studies of the binding of ^{125}I-labeled RBP to a putative RBP receptor in human placental brush-border membranes. These investigators indicated that there are high- and low-affinity binding forms of a plasma membrane receptor for RBP in human placenta. TTR was found to inhibit the binding of RBP to the membrane preparations. Binding of RBP to this RBP receptor was trypsin-, heat-

and thiol-group-specific-reagent sensitive. Binding was taken to be specific for RBP since [^{3}H]retinol-RBP uptake was reduced by addition of native or apo-RBP but not by albumin. When vesicles generated from placental brush-border membranes were treated with p-chloromercuribenzene sulfonate (which inhibited ^{125}I-RBP binding) the uptake of retinol from RBP also was inhibited, whereas the uptake of free retinol was not. High-pressure gel-filtration chromatography indicated that the [^{3}H]retinol was associated with a membrane component with an apparent molecular weight of 125 kDa.

Later work by SIVAPRASADARAO and FINDLAY (1993) employing recombinant human RBP demonstrated that recombinant RBP had a B_{max} for placental brush-border membranes which is about six- to eight-fold higher than RBP purified from serum. This suggested that a significant portion of purified serum RBP is incapable of interacting with this cell surface RBP receptor. While the RBP receptor showed high- and low-affinity forms for serum RBP with K_d values equal to 3 and 80 nM, respectively, the recombinant protein had only one K_d value of 50 nM. When the putative RBP receptor was solubilized from human placental membranes (SIVAPRASADARAO and FINDLAY 1993), the purified material was found to contain a major species of 63 kDa, believed to represent the RBP receptor, and a minor species of 55 kDa. The 63 kDa RBP receptor could not be identical to p63 identified by BÅVIK et al. (1991), since the protein characterized by SIVAPRASADARAO and FINDLAY (1993) is an integral membrane protein, whereas p63 is not. The minor 55-kDa protein was postulated to be a contaminant or a proteolysed form of the 63 kDa species, but data were not presented to rule out the possibility that this protein is a subunit of the receptor protein. These authors also conducted structure-function studies using recombinant RBP and site directed mutagenesis (SIVAPRASADARAO and FINDLAY 1994) to identify amino acid residues responsible for the binding of RBP to both TTR and the RBP receptor. These studies suggested that the sites on RBP important for TTR and cell surface receptor binding are over-lapping but not identical, since RBP when bound to TTR cannot simultaneously bind to the RBP receptor. This finding explains the earlier observation made by these investigators that TTR inhibits the binding of RBP to placental brush border membranes (SIVAPRASADARAO and FINDLAY 1988b).

The effects of detergent solubilization on the human placenta RBP receptor were also explored by SIVAPRASADARAO et al. (1994). These studies, which were carried out using several detergents including octyl-β-glucoside, Nonidet P-40 and CHAPS, led these investigators to several conclusions. First, the RBP receptor requires a nonionic detergent for stability, suggesting that the receptor is embedded in the lipid bilayer. Relatively high detergent concentrations were required for maintaining optimal binding of RBP to the RBP receptor. This was taken to indicate the need for a lipid environment for a functioning RBP receptor. Finally, an apparent increase in the number of available binding sites observed upon increasing detergent concentrations was taken to suggest that aggregation of the receptor occurs at low detergent concentrations. It was concluded that RBP binding may involve hydrophobic interactions since the receptor is unable to bind RBP at low ionic strength, a finding that is consistent with a previous site-directed-mutagenesis study which identified the hydrophobic residues of the loops of RBP which bind to the receptor (SIVAPRASADARAO and FINDLAY 1994).

Findlay and colleagues also studied retinol transfer to recombinant CRBP I from [^{3}H]retinol-RBP upon incubation in the presence or absence of human placental membranes (SUNDARAM et al. 1997). Maximal [^{3}H]retinol uptake was achieved in the presence of plasma membranes from human placenta, and retinol transfer to CRBP I was reduced when RBP receptor-deficient erythrocyte membranes were added to the incubation mixture. Heat-treatment of the placental membrane proteins, treatment of the membranes with *p*-chloromercuribenzene sulfonate which inhibits RBP-receptor interaction (see above) and competition with unlabeled RBP all lessened CRBP I uptake of [^{3}H]retinol from the [^{3}H]retinol-RBP complex. Retinol transfer was judged to be specific since other proteins which bind retinol (β-lactoglobulin and serum albumin) failed to substitute for either RBP or apo-CRBP I in the transfer assay. The authors proposed that their data suggest that the RBP receptor, through specific interactions with these retinol-binding proteins, participates in the process which mediates the transfer of retinol from the extracellular RBP to the intracellular CRBP.

SENOO et al. (1990) studied RBP uptake by primary rat liver cells by electron microscopy. These investigators found that 10 min after injection of large doses of human ^{125}I-RBP into the circulatory systems of rats, the labeled RBP was detected on the cell surface and in vesicles near the surface of rat parenchymal cells, primarily in association with the membrane of the vesicles. Two hours after injection RBP was also found in larger vesicles deeper in the cytoplasm of the parenchymal cells. When compared to the distribution of asialo-ovosomucoid, a protein known to be internalized by parenchymal cells through receptor-mediated endocytosis, the two proteins were seldom localized in the same small vesicles, but larger vesicles contained both ligands.

MALABA et al. (1993) studied the effect of four different methods for labeling RBP with ^{125}I on RBP binding, uptake, release and degradation by isolated rat liver parenchymal cells. These studies were designed to investigate the mechanism through which liver cells takes up the retinol-RBP complex. One important finding reported by MALABA et al. (1993) is that the method used for ^{125}I labeling of purified RBP influences RBP receptor binding and the amount of the labeled RBP taken up by the liver cells. Other studies by MALABA et al. (1995) explored the uptake of RBP by parenchymal and stellate cells in vivo and in vitro. Data obtained from these studies suggest that RBP is taken up by potocytosis, a mechanism of uptake whereby small molecules are first concentrated in specialized pits called caveoli on the cell surface. The RBP concentrated in caveoli was internalized through invagination followed by recycling of the membrane components back to the plasma membrane. Immunoelectron microscopy experiments demonstrated ^{125}I-labeled human RBP injected into the femoral vein of rats appeared after 10 min in small vesicles close to the plasma membrane. This localization was interpreted by MALABA et al. (1995) to suggest that RBP is taken up by a receptor-mediated process.

The possible role played by TTR in facilitating retinol uptake by primary cultures of rat hepatocytes and nonparenchymal cells was studied by YAMAMOTO et al. (1997). These investigators reported that retinol is taken up from the retinol-RBP complex in a manner that suggests the presence of a membrane receptor on the surfaces of the hepatic cells. Similar to the findings of FINDLAY and colleagues for their work employing human placental membranes, YAMAMOTO et al. (1997) concluded that uptake of [^{3}H]retinol by hepatic cells is inhibited by addition of an excess of free TTR to the uptake assays. This was the case when retinol was added bound to the RBP-TTR complex or bound only to RBP. YAMAMOTO et al. (1997) suggest that TTR may play a broad physiological role in regulating hepatic retinoid storage and processing by influencing retinol uptake by hepatocytes and nonparenchymal cells.

Data supportive of the hypothesis that retinol is taken up by cells in a receptor-dependent fashion have also been provided by MATARESE and LODISH (1993). These authors demonstrated that retinoic acid-differentiated F9 embryonal carcinoma cells associate with fluorescein-derivatized or iodinated RBP in a time- and concentration-dependent manner at half-maximal concentrations which are well below physiological serum RBP levels. Competition by simultaneous incubation with excess unlabeled RBP indicated that this association was specific and saturable with an apparent dissociation constant of 200–300 n*M*. When visualized by fluorescence microscopy, the labeled RBP appeared to have been incorporated into vesicles distributed throughout the cytoplasm, rather than cell-surface associated vesicles. Indeed, when the amount of cell surface bound versus internalized protein was quantified using ^{125}I-labeled RBP, over 80% of the cell-associated RBP was found to be internalized, and less than 20% was bound to the cell surface. Specific binding of the RBP to the cell membranes could not be measured and this was interpreted to be due to rapid dissociation of RBP from cell surface receptors. By random subcloning, natural variant F9 cell lines which either possessed or lacked the ability to internalize RBP were identified. Since the ability to take up RBP was a heritable trait, this was taken as support that RBP uptake was mediated by expression of a cell surface RBP receptor.

MELHUS et al. (1995) explored the ability of monoclonal antibodies directed against human RBP to interfere with RBP binding to its putative receptor. Two monoclonal antibodies against human RBP that could block RBP binding to its receptor were identified. These antibodies recognized conserved epitopes within human RBP since the antibodies also recognized rat RBP. This conservation of epitopes was taken as supportive of conservation of a structure-function motif within the RBP molecule which these authors suggested was possibly a RBP receptor binding domain. Since these antibodies also were capable of coimmunoprecipitating RBP and TTR, these investigators suggested that the TTR-binding sites involve separate regions of RBP than the membrane receptor binding sites. This is in contrast to findings of SIVAPRASADARAO and FINDLAY (1988b) and YAMAMOTO et al. (1997) who reported that for their RBP receptor studies, high concentrations of TTR block binding of RBP to the RBP receptor. One of the monoclonal antibodies generated by MELHUS et al. (1995), L48, showed immunoreactivity to a peptide corresponding to amino acid residues 60–70 which correspond to one of the entrance loops of the retinol binding site within RBP.

SMELAND et al. (1995) studied the tissue distribution of the RBP receptor by incubating membrane vesicles prepared from various rabbit tissues with ^{125}I-labeled RBP. These authors found high specific binding activity in placenta, retinal pigment epithelial cells, bone marrow and kidneys. Specific binding activity was also found in small intestine, spleen, liver, and to a lesser extent in lung and testis – all from mature rabbit. When the tissue sites of uptake of ^{125}I-labeled RBP after its intravenous injection into rats was investigated, the highest specific activity for uptake was found in kidney, but bone marrow, spleen, liver and small intestine also accumulated significant amounts of radioactivity. Closer examination of renal uptake in the rabbit showed that the cortex took up significantly more ^{125}I-labeled cpm/g wet weight than the medulla and was responsible for almost all of the specific binding activity observed in kidney. High levels of specific binding were also observed in the choroid plexus and the pineal gland, but not the brain. Binding activity was two to three times higher in testis from immature rabbits than in mature rabbits, and higher in membrane vesicles isolated from Sertoli cells than from whole testis. The binding activity observed in undifferentiated human keratinocytes was found to be the highest of all the tissues and cells presented. Based on their results, SMELAND et al. (1995) concluded that the level of retinol-binding protein receptor expression varies considerably between cell types. This final result is contrary to data reported earlier by CREEK et al. (1989) regarding the lack of specific RBP binding to mouse keratinocytes.

In summary, over the past decade there has been extensive research activity focused on identifying and characterizing a cell surface receptor for RBP. The great majority of published studies report convincing data regarding the existence and nature of the membrane receptor for RBP. Yet to date there are no reports describing the purification or cloning of a protein that is unequivocally accepted to be a cell surface receptor for RBP. Moreover, when these recent reports characterizing a cell surface RBP receptor are considered together, there is no consensus regarding the properties of such a receptor. Thus, although there has been considerable effort aimed at identifying and characterizing a membrane receptor for RBP, unequivocal information on such a receptor (or receptors) is still missing.

II. Plasma all-*trans*-Retinoic Acid

A small percentage of dietary retinoid is converted to retinoic acid in the intestine (or may arrive as such in the diet) and is absorbed via the portal system as retinoic acid bound to albumin (OLSON 1990; BLANER and OLSON 1994). In plasma, retinoic acid circulates bound to albumin (BLANER and OLSON 1994). The fasting plasma level of all-*trans*-retinoic acid is very low and is in the range of 4–14 nmol/l in humans (about 0.2–0.7% of plasma retinol levels; DE LEENHEER et al. 1982; ECKHOFF and NAU 1990) and 7.3–9 nmol/l in rats (CULLUM and ZILE 1985; NAPOLI et al. 1985). It is not known whether the retinoic acid present in the circulation arises solely from the diet (i.e., is of intestinal origin), or whether some arises through export of retinoic acid from tissues which synthesize it from retinol. The possibility that the kidney is a site of synthesis and export of retinoic acid has been raised in the literature (BHAT et al. 1988a,b); however at present no data are available to support the possibility that some tissues (except for the intestine after dietary intake of carotenoids or retinoids) are able to provide retinoic acid to the circulation for delivery to other tissues.

Retinoic acid can be taken up efficiently by cells. All-*trans*-retinoic acid, although fully ionized in free solution at pH 7.4, is uncharged when within a lipid environment (NOY 1992a,b). In the uncharged state all-*trans*-retinoic acid moves rapidly between the outer and inner leaflets of the plasma membrane. Thus, all-*trans*-retinoic acid is able to traverse cellular membranes and rapidly enters the cell.

Although based on this work by NOY (1992a,b) a cell surface receptor for all-*trans*-retinoic acid would not be needed to bring about internalization of circulating retinoic acid, studies by KANG et al. (1998) have demonstrated that all-*trans*-retinoic acid binds the mannose-6-phosphate/insulin-like growth factor II receptor present on plasma membranes of rat cardiac myocytes. Scatchard analysis of binding data revealed that the mannose-6-phosphate/insulin-like growth factor II receptor displays a single class of high-affinity binding sites for all-*trans*-retinoic acid with a K_d of $2.5 \pm 0.3\,nM$. It appears from the work of KANG et al. (1998) that the binding of all-*trans*-retinoic acid to this membrane receptor is important for modulating the activity of cellular signal transduction pathways and cellular activities and not for facilitating uptake of all-*trans*-retinoic acid from the circulation. However, this work does raise an interesting question for understanding retinoic acid metabolism. Specifically, are there membrane factors that function to regulate cellular retinoic acid uptake and availability.

This question has been examined by HODAM and CREEK (1996) who investigated the uptake and metabolism of all-*trans*-[^{3}H]retinoic acid by human foreskin keratinocytes. For these studies, retinoic acid was added to the keratinocyte culture medium either bound to serum albumin or added directly to the serum-free culture medium. These investigations showed that retinoic acid added directly to the culture medium is rapidly taken up and rapidly converted to polar compounds that are subsequently excreted back into the culture medium. In contrast, retinoic acid added to the culture medium bound to albumin is taken up slowly by the human keratinocytes and metabolized slowly to more polar metabolites. These investigators concluded that the binding of retinoic acid to albumin both protects retinoic acid from conversion to polar metabolites and provides for its controlled delivery from the aqueous extracellular environment to the cell surface.

Recently HAYAISHI and colleagues (EGUCHI et al. 1997; TANAKA et al. 1997) using polarization of fluorescence approaches have demonstrated that the lipocalin known as β-trace (also identified as prostaglandin D synthase) binds with relatively high-affinity all-*trans*- and 9-*cis*-retinoic acid but not all-*trans*-retinol. Homogeneously purified recombinant rat β-trace was shown to bind each of the retinoic acid isomers with a K_d of 80 n*M* (TANAKA et al. 1997). Since β-trace is a very abundant component of the cerebrospinal fluid (CSF), these investigators have proposed that β-trace may play a role in the transport of retinoic acid in the CSF (EGUCHI et al. 1997). Outside of the central nervous system, β-trace is expressed in reproductive tissues and in heart (EGUCHI et al. 1997). Although this work by Hayaishi and colleagues (EGUCHI et al. 1997; TANAKA et al. 1997) raises some interesting possibilities regarding the extracellular transport of retinoic acid within the central nervous system, the physiological relevance of this hypothesis remains to be established. At present there is no information regarding whether retinoic acid isomers are present in CSF.

KURLANDSKY et al. (1995) explored the contribution which plasma all-*trans*-retinoic acid makes to tissue pools of all-*trans*-retinoic acid in chow fed male rats. Using steady state tracer kinetic approaches, these authors determined how much of the all-*trans*-retinoic acid in a tissue was derived from the circulation. For the liver and brain, greater than 75% of the all-*trans*-retinoic acid present in these tissues was derived from the circulation (88.4% for brain and 78.2% for liver). The seminal vesicles, epididymis, kidney, epididymal fat, perinephric fat, spleen, and lungs derived, respectively, 23.1%, 9.6%, 33.4%, 30.2.%, 24.5%, 19.0%, and 26.7% of their tissue all-*trans*-retinoic acid pools from the circulation. Only 2.3% and 4.8%, respectively, of the retinoic acid present in the pancreas and eyes was contributed by the circulation. For all rats studied, the testes did not take up any (<1%) all-*trans*-retinoic acid from the circulation. KURLANDSKY et al. (1995) also reported a fractional catabolic rate for all-*trans*-retinoic acid in plasma of 30.4 plasma pools/hr and an absolute catabolic rate for all-*trans*-retinoic acid of 640 pmol/h. These rates are very rapid compared to those of the only other naturally occurring form of vitamin A studied under normal physiological conditions, all-*trans*-retinol (LEWIS et al. 1990). Very little 9-*cis*- or 13-*cis*-retinoic acid could be detected in any of these tissues. These data demonstrate that plasma all-*trans*-retinoic acid is a significant source of all-*trans*-retinoic acid for some, but not all, tissues; and, by inference, these studies suggest that tissues have different capacities for the conversion of retinol to retinoic acid.

Since plasma and tissues contain only very low levels of all-*trans*-retinoic acid, it was necessary to measure these levels for pooled plasma samples taken from adult male C57Bl mice. These values are provided below in Table 6. As with rat tissue all-*trans*-retinoic acid levels, levels in mice are approximately two to three orders of magnitude less than tissue total retinol levels.

Table 6. Concentrations of all-*trans*-retinoic acid for tissue and plasma pools obtained from 3 month-old male wild-type C57BL mice (R. PIANTEDOSI ZOTT and W.S. BLANER, unpublished observations)

Tissue	All-*trans*-retinoic acid (pmol/g tissue)
Liver	15.0 ± 8.3
Kidney	20.3 ± 15.8
Lung	18.3 ± 8.3
Testis	16.0 ± 6.0
Skeletal muscle	5.0 ± 2.7
Spleen	15.6 ± 6.0
Plasma	6.0 ± 2.0[a]

Values are mean ± standard deviation for four pools of tissue or plasma each obtained from four mice. Each tissue or plasma pool was constructed so that tissues and plasma from the same four mice constitute a pool. Thus, tissues and plasma from the same four mice were pooled to constitute a pool of liver, kidney, lung, testis, skeletal muscle (gastrocnemius), spleen, and plasma.

[a] The concentration of all-*trans*-retinoic acid in plasma is given as picomoles per milliliter.

Studies in rabbits indicate that the β-carotene content of the diet influences serum levels of all-*trans*-retinoic acid (FOLMAN et al. 1989). From studies in which female rabbits were given for 9 weeks graded doses of β-carotene in their diets it was concluded that β-carotene intake was associated with higher serum concentrations of all-*trans*-retinoic acid. Thus, for the rabbit not only is all-*trans*-retinoic acid present in the circulation, the levels of this active form of vitamin A are influenced by dietary β-carotene intake.

III. Plasma 13-*cis*-Retinoic Acid

CULLUM and ZILE (1985) reported that 13-*cis*-retinoic acid is an endogenous retinoid present in the intestinal mucosa, intestinal muscle, and plasma of vitamin A sufficient rats. When a dose of all-*trans*-retinoic acid was administered by intrajugular injection into vitamin A depleted rats, 13-*cis*-retinoic acid appeared in the plasma and small intestine within 2 min after dosing. The endogeneous plasma concentrations of all-*trans*- and 13-*cis*-retinoic acid in vitamin A sufficient rats were reported by CULLUM and ZILE to be 9.7 and 3.0 nmol/l, respectively. CULLUM and ZILE (1985) concluded that 13-*cis*-retinoic acid is a naturally occurring metabolite of all-*trans*-retinoic acid. NAPOLI et al. (1985) similarly demonstrated that 13-*cis*-retinoic acid is a naturally occurring form of retinoic acid in rat plasma. BHAT and JETTEN (1987) demonstrated that cultures of rabbit tracheal epithelial cells can convert all-*trans*-retinoic acid to 13-*cis*-retinoic acid, and TANG and RUSSELL (1990, 1991) showed that 13-*cis*-retinoic acid is an endogenous component in human serum. Fasting serum levels of all-*trans*- and 13-*cis*-retinoic acid determined in 26 human volunteers ranged from 3.7 to 6.3 nmol/l and from 3.7 to 7.2 nmol/l, respectively (TANG and RUSSELL 1990), a level which is essentially the same as observed for the rat. TANG and RUSSELL (1991) later showed that plasma levels of all-*trans*- and 13-*cis*-retinoic acid levels rose 1.3- and 1.9-fold, respectively, over fasting levels, in human subjects who had received a physiological dose of retinyl palmitate. Similarly, ECKHOFF et al. (1991) demonstrated that 13-*cis*-retinoic acid is present as an endogeneous component of human plasma and that administration of an oral dose of retinyl palmitate elevated plasma levels of 13-*cis*-retinoic acid.

IV. Lipoprotein Bound Retinyl Ester

Studies of fasting human plasma have conclusively demonstrated that low levels of retinyl ester can be found in the VLDL and LDL fractions (SCHINDLER and KLOPP 1986; KRASINSKI et al. 1990). SCHINDLER and KLOPP (1986) reported that in humans approximately 70% of circulating retinyl palmitate was associated with the VLDL fraction and the remainder with the LDL fraction. No esterified retinol was found by SCHINDLER and KLOPP in the high-density lipoprotein (HDL) fraction or in the soluble fraction (d > 1.21 g/ml). It is not presently understood how retinyl esters come to be present in the VLDL and LDL fractions. It is possible in humans that this occurs through the actions of cholesteryl ester transfer protein (CETP), which is known to be able to transfer retinyl ester between triglyceride rich chylomicrons and other lipoprotein fractions (GOODMAN and BLANER 1984; KRANSINSKI et al. 1990; BLANER and OLSON 1994; TALL 1995; see Sect. C. above). Alternatively, human hepatocytes may package and secrete some retinyl ester in nascent VLDL. If this later possibility proves to be correct, this would suggest that some vitamin A can be mobilized from the liver through an RBP-independent pathway. Although either or both of these alternative hypothesizes may be valid, neither possibility has been established experimentally in humans.

Early studies in normal and cholesterol-fed rabbits conclusively demonstrated that rabbit liver does not secrete retinyl ester in VLDL (THOMPSON et al. 1983). For this species it was felt that all of the retinyl ester present in the circulation ultimately arises from transfer of chylomicron retinyl ester to other lipoprotein fractions. However, this is not the case for all animal species. Fasting blood of some species, especially ferrets and dogs, contains very substantial levels of retinyl ester in VLDL, LDL and HDL (WILSON et al. 1987; SCHWEIGERT 1988; RIBAYA-MERCADO et al. 1994; LEDERMAN et al. 1998). For both dogs and ferrets, plasma retinyl ester is the major form of vitamin A in the fasting circulation. Studies with dogs have suggested that lipoprotein bound retinyl ester in the fasting circulation arises from hepatic secretion of retinyl ester in VLDL (WILSON et al. 1987). Recent studies of mice indicate that low levels of retinyl ester is present in VLDL, LDL, and HDL lipoprotein fractions in the circulatory systems of animals fasted 18h (BLANER et al. 1998). Although it was not determined experimentally whether the retinyl ester present in fasting mouse serum was secreted into the circulation bound to VLDL, the lack of CETP in mice and the relative long duration of the fasting period could be taken to suggest that some retinyl ester is secreted bound to VLDL from the livers of mice.

Overall, different species show very distinct patterns of distribution of retinyl ester in fasting blood. The level of retinyl ester present in fasting rodent and rabbit blood closely resembles that which is present in fasting human blood. However, since rodents do not express CETP and since rabbits do not secrete retinyl ester in VLDL, one must be cautious when trying to extrapolate from these species to humans.

V. Glucuronides of Retinol and Retinoic Acid

When all-*trans*-retinoic acid is orally administered to rats, all-*trans*-retinoyl β-glucuronide is secreted into the bile in significant amounts (DUNAGIN et al. 1966; SWANSON et al. 1981; SKARE et al. 1982; FROLIK 1984). Retinoyl β-glucuronide is thought to be synthesized from retinoic acid and uridine diphosphoglucuronic acid in the liver, intestine, kidney, and other tissues (BARUA and OLSON 1986; BARUA 1997). Recently a specific rat liver microsomal UDP-glucouronosyltransferase which uses all-*trans*-[^{3}H]retinoic acid as a substrate has been identified (LITTLE and RADOMINSKA 1997). This enzyme was identified through photoincorporation of both [^{32}P]5N_3UDP-glucuronic acid and all-*trans*-[^{3}H]retinoic acid into the 52-kDa protein. A partial purification of this enzyme through UDP-hexanolamine-Sepharose affinity chromatography has been reported (LITTLE and RADOMINSKA 1997). Based on the data presented by these investigators, it seems likely that this enzyme (and possibly others resembling it) plays a physiologically relevant role in hepatic synthesis of retinoyl β-glucuronides.

Retinoyl β-glucuronides can be hydrolyzed to retinoic acid by various organelles of liver, kidney and intestine (KAUL and OLSON 1998). Plasma retinoic acid levels have long been known to increase after administration of a dose of retinoyl β-glucuronide and it is thought that this increase arises from hydrolysis of retinoyl β-glucuronides. In vitro rates of hydrolysis of the retinoyl β-glucuronide were found by KAUL and OLSON (1998) to be higher in organelles obtained from tissues of vitamin A deficient rats than from vitamin A sufficient rat. Based on their data, KAUL and OLSON (1998) proposed that the vitamin A deficiency has an enhancing effect on tissue β-glucuronidases which brings about an increased rate of retinoic acid formation from retinoyl β-glucuronides. These investigators did not address specific mechanisms which might account for this increase in β-glucuruonidase activity.

In an attempt to understand how retinoid β-glucuronides act within cells, SANI et al. (1992) investigated whether retinoyl β-glucuronides interacted with either cellular retinoic acid-binding protein type I (CRABP I) or RARα,

RARβ or RARγ. These investigators reported that retinoyl β-glucuronide did not interact well with either CRABP I or the RARs. However, when retinoyl β-glucuronide was allowed to incubate for 24 or 48 h with cytosol from chick skin, detectable amounts of retinoic acid was generated and this retinoic acid interacted with the RARs. Based on this observation SANI et al. (1992) concluded that the biological activity of retinoyl β-glucuronide arises from its conversion to retinoic acid.

Retinol is also conjugated with glucuronic acid in vivo and in vitro to form retinyl β-glucuronide (BARUA 1997). As with retinoyl β-glucuronide, retinyl β-glucuronide is also present in human plasma at a mean concentration of 6.8 nmol/l (BARUA et al. 1989). Retinyl β-glucuronide is hydrolyzed in vivo to retinol and is not directly converted to retinoic acid (BARUA 1997).

Approximately one-third of the retinoid β-glucuronides that is excreted in the bile of rats is recycled back to the liver, thereby forming an enterohepatic circulation (ZACHMAN et al. 1966; SWANSON et al. 1981). Very little retinol and retinyl ester are involved in the enterohepatic circulation, however, and the water-soluble retinyl β-glucuronide and retinoyl β-glucuronide are presumably reabsorbed by the portal, rather than by the lymphatic, route (BARUA 1997). Nonetheless, disruption of this enterohepatic circulation by intestinal disorders may well have adverse nutritional sequelae.

VI. Retro-Metabolites of Retinol

In 1990 BUCK et al. (1990) showed that human lymphoblastoid cells in culture are dependent for growth on a constant supply of retinol. In the absence of retinol, these cells were found to perish within days. Retinoic acid was unable to prevent cell death. By employing sensitive trace-labeling techniques, BUCK et al. (1991a) later demonstrated that retinoic acid and 3,4-didehydroretinoic acid could not be detected as metabolites of retinol in activated B lymphocytes. Nevertheless, B lymphocytes formed several other metabolites of retinol, which were able to sustain B cell growth in the absence of an external source of retinol (BUCK et al. 1991a). BUCK et al. (1991b) then identified one of these active metabolites of retinol as optically active 14-hydroxy-4,14-retro-retinol. The 14-hydroxy-4,14-retro-retinol was isolated, by reverse phase HPLC, from cells of the lymphoblastoid line 5/2 which were grown in the presence of [^{3}H]retinol-retinol-binding protein complex. The chemical structure and properties of 14-hydroxy-4,14-retro-retinol were determined by use of high-resolution electron ionization mass spectrometry, proton nuclear magnetic resonance, and measurements of its UV spectrum and circular dichroism. This retinol metabolite was active in sustaining the growth of cells of the lymphoblastoid line 5/2 (and of T cell lines) in the absence of retinol. In confirmation of their initial studies, retinoic acid was not active in these growth assays. BUCK et al. (1991b) demonstrated that 14-hydroxy-4,14-retro-retinol is a direct biosynthetic product of retinol in 5/2 lymphoblastoid cells. Thus, in addition to the retinoic acid pathway, a second pathway of retinol metabolism is mediating the actions of retinoids in cellular growth and differentiation, at least in specific cells. This pathway is discussed in detail in Chap. 3 dealing with retro-retinoids.

G. Summary

The first reports describing the nuclear receptors for retinoic acid were published over a decade ago. Since then there has been substantial growth in our understanding of retinoid uptake, transport, and metabolism. Thus, we now are starting to have a more full understanding of the complexity of these processes. It is clear that there is much redundancy or plasticity in how organisms transport and metabolize retinoids. For instance, we now understand that there are multiple retinoid forms present in the circulation. Moreover, it is starting to be understood that under specific physiological or pathological conditions the presence of some of these circulating retinoids may be of importance to specific tissues and cells. We also are starting to understand that there are species differences in how retinoids are transported and metabolized. However, there are still many major questions regarding retinoid uptake, transport, and metabolism that remain to be resolved. One major and controversial question concerns how retinoids enter and leave cells and if and how these processes are regulated. Another still poorly understood question centers on how tissues and cells regulate intracellular retinoid concentrations. One would expect that these and other significant questions regarding retinoid physiology will be addressed in future research to be carried out in this area.

Acknowledgements. The work from the authors' laboratory mentioned in this review was supported by NIH grant DK 52444.

References

Adams WR, Green MH (1994) Prediction of liver vitamin A in rats by an oral isotope dilution technique. J Nutr 124:1265–1270

Alberini CM, Bet P, Milstein C, Sitia R (1990) Secretion of immunoglobulin M assembly intermediates in the presence of reducing agents. Nature 347:485–487

Azaïs-Braesco V, Dodeman I, Delpal S, Alexandre-Gouabau M-C, Partier A, Borel P, Grolier P (1995) Vitamin A contained in the lipid droplets of rat liver stellate cells is substrate for acid retinyl ester hydrolase. Biochim Biophys Acta 1259:271–276

Båvik CO, Eriksson U, Allen RA, Peterson PA (1991) Identification and partial characterization of a retinal pigment epithelial membrane receptor for plasma retinol-binding protein. J Biol Chem 266:14978–14985

Båvik CO, Busch C, Eriksson U (1992) Characterization of a plasma retinol-binding protein membrane receptor expressed in the retinal pigment epithelium. J Biol Chem 267:23035–23042

Båvik CO, Levy F, Hellman U, Wernstedt C, Eriksson U (1993) The retinal pigment epithelial membrane receptor for plasma retinol-binding protein. Isolation and cDNA cloning of the 63-kDa protein. J Biol Chem 268:20540–20546

Båvik CO, Peterson A, Eriksson U (1995) Retinol-binding protein mediates uptake of retinol to cultured human keratinocytes. Exp Cell Res 216:358–362

Batres RO, Olson JA (1987) A marginal vitamin A status alters the distribution of vitamin A among parenchymal and stellate cells in rat liver. J Nutr 117:874–879

Barua AB, Olson JA (1986) Retinoyl-β-glucuronide: An endogenous compound of human blood. Am J Clin Nutr 43:481–485

Barua AB, Batres RO, Olson JA (1989) Characterization of retinyl beta-glucuronide in human blood. Am J Clin Nutr 50:370–374

Barua AB, Duitsman PK, Kostic D, Baura M, Olson JA (1997) Reduction of serum retinol levels following a single oral dose of all-*trans* retinoic acid in humans, Int J Vit Nutr Res 67:423–426

Barua AB (1997) Retinoyl beta-glucuronide: a biological active form of vitamin A. Nutr Rev 55:259–267

Beisiegel U, Weber W, Bengtsson-Olivecrona G (1991) Lipoprotein lipase enhances binding of chylomicrons to low density lipoprotein receptor-related protein, Proc Natl Acad Sci USA 88:8342–8346

Bellovino D, Morimoto T, Tosetti F, Gaetani S (1996) Retinol binding protein and transthyretin are secreted as a complex formed in the endoplasmic reticulum in HepG2 human hepatocarcinoma cells. Exp Cell Res 222:77–83

Berni R, Malpeli G, Folli C, Murrell JR, Liepnieks JJ, Benson MD (1994) The Ile-84–>Ser amino acid substitution in transthyretin interferes with the interaction with plasma retinol-binding protein. J Biol Chem 269:23395–23398

Bhat PV, Jetten AM (1987) Metabolism of all-*trans*-retinol and all-*trans*-retinoic acid in rabbit tracheal epithelial cells in culture. Biochim Biophys Acta 922:18–27

Bhat PV, Poissant L, Lacroix A (1988a) Properties of retinal-oxidizing enzyme activity in rat kidney. Biochim Biophys Acta 967:211–217

Bhat PV, Poissant L, Falardeau P, Lacroix A (1988b) Enzymatic oxidation of all-*trans*-retinal to retinoic acid in rat tissues. Biochem Cell Biol 66:735–740

Biesalski HK (1990) Separation of retinyl esters and their geometric isomers by isocratic adsorption high-performance liquid chromatography. Methods Enzymol 189:181–189

Bihain BE, Yen FT (1992) Free fatty acids activate a high-affinity saturable pathway for degradation of low-density lipoproteins in fibroblasts from a subject homozygous for familial hypercholesterolemia. Biochemistry 31:4628–4636

Bishop PD, Griswold MD (1987) Uptake and metabolism of retinol in cultured Sertoli cells: evidence for a kinetic model. Biochemistry 26:7511–7518

Blake CCF, Oatley SJ (1977) Protein-DNA and protein-hormone interactions in prealbumin: a model of the thyroid hormone nuclear receptor. Nature 268:115–120

Blaner WS, Hendriks HFJ, Brouwer A, de Leeuw AM, Knook DL, Goodman DS (1985) Retinoids, retinoid-binding proteins, and retinyl palmitate hydrolase distributions in different types of rat liver cells. J Lipid Res 26:1241–1251

Blaner WS, Dixon JL, Moriwaki H, Martino RA, Stein O, Stein Y, Goodman DS (1987) Studies on the in vivo transfer of retinoid from parenchymal to stellate cells in the rat liver. Eur J Biochem 164:301–307

Blaner WS, van Bennekum AM, Brouwer A, Hendriks HFJ (1990) Distributions of lecithin: retinol acyltransferase activity in different types of rat liver cells and subcellular fractions. FEBS Lett 274:89–92

Blaner WS, Obunike JC, Kurlandsky SB, Al-Haideri M, Piantedosi R, Deckelbaum RJ, Goldberg IJ (1994) Lipoprotein lipase hydrolysis of retinyl ester. J Biol Chem 269:16559–16565

Blaner WS, Olson JA (1994) Retinol and retinoic acid metabolism. In: Sporn MB, Roberts AB, Goodman DS (eds) The retinoids: biology, chemistry, and medicine 2nd edn. Raven, New York, pp 229–256

Blaner WS (1994) Retinoid (vitamin A) metabolism and the liver. In: Arias IM, Boyer JL, Fausto N, Jakoby WB, Schachter D, Shafritz DA (eds) The liver: biology and pathobiology 3rd edn. Raven, New York, pp 529–542

Blaney JM, Jorgensen EC, Connolly ML, Ferrin TE, Largridge R, Oatley SJ, Burridge JM, Blake CC (1982) Computer graphics in drug design: molecular modeling of thyroid hormone-prealbumin interactions. J Med Chem 25:785–790

Blomhoff R, Holte K, Naess L, Berg T (1984) Newly administered [^{3}H]retinol is transferred from hepatocytes to stellate cells in liver for storage. Exp Cell Res 150: 186–193

Blomhoff R, Rasmussen M, Nilsson A, Norum KR, Berg T, Blaner WS, Kato M, Mertz JR, Goodman DS, Eriksson U, Peterson PA (1985) Hepatic retinol metabolism: distribution of retinoids, enzymes and binding proteins in isolated rat liver cells. J Biol Chem 260:13560–13565

Blomhoff R, Berg T, Norum KR (1988) Transfer of retinol from parenchymal to stellate cells in liver is mediated by retinol-binding protein. Proc Natl Acad Sci USA 85:3455–3458

Bok D, Heller J (1976) Transport of retinol from blood to retina: Autoradiographic study of pigment epithelial cell surface receptor for plasma retinol-binding protein, Exp Eye Res 22:395–402

Buck J, Ritter G, Dannecker L, Katta V, Cohen SL, Chait BT, Hämmerling U (1990) Retinol is essential for growth of activated human B cells. J Exp Med 171:1613–1624

Buck J, Myc A, Garbe A,Cathomas G (1991a) Differences in the action and metabolism between retinol and retinoic acid in B lymphocytes. J Cell Biol 115:851–859

Buck J, Derguini F, Levi E, Nakanishi K, Hämmerling UL (1991b) Science 254:1654–1656

Cowan SW, Newcomer ME, Jones TA (1990) Crystallographic refinement of human serum retinol binding protein at 2.0 Å resolution. Proteins 8:44–61

Creek KE, Silverman-Jones CS, DeLuca LM (1989) Comparison of the uptake and metabolism of retinol delivered to primary mouse keratinocytes either free or bound to rat serum retinol-binding protein. J Invest Dermatol 92:283–289

Cooper AD (1997) Hepatic uptake of chylomicron remnants. J Lipid Res 38:2173–2192

Cullum ME, Zile MH (1985) Metabolism of all-*trans*-retinoic acid and all-*trans*-retinyl acetate: demonstration of common physiological metabolites in rat small intestinal mucosa and circulation. J Biol Chem 260:10590–10596

De Leenheer AP, Lambert WE, Claeys I (1982) All-*trans*-retinoic acid: Measurement of reference values in human serum by high performance liquid chromatography. J Lipid Res 23:1362–1367

Davis JT, Ong DE (1992) Synthesis and secretion of retinol-binding protein by culture rat Sertoli cells. Biol Reprod 47:528–533

Davis RA, Thrift RN, Wu CC, Howell KE (1990) Apolipoprotein B is both integrated into and translocated across the endoplasmic reticulum membrane. J Biol Chem 265:10005–10011

De Faria E, Fong LG, Komaromy M, Cooper AD (1996) Relative roles of the LDL receptor, the LDL receptor-like protein, and hepatic lipase in chylomicron removal by the liver. J Lipid Res 37:197–209

Dew SE, Ong DE (1994) Specificity of the retinol transporter of the rat small intestine brush border. Biochemistry 33:12340–12345

Dixon JL, Goodman DS (1987) Studies on the metabolism of retinol-binding protein by primary hepatocytes from retinol-deficient rats. J Cell Physiol 130:14–20

Dunagin PE Jr, Zachman RD, Olson JA (1966) The identification of metabolites of retinol and retinoic acid in rat bile. Biochim Biophys Acta 124:71–85

Duszka C, Grolier P, Azim E-M, Alexandre-Gouabau M-C, Borel P, Azaïs-Braesco V (1996) Rat intestinal β-carotene dioxygenase activity is located primarily in the cytosol of mature jejunal enterocytes. J Nutr 126:2550–2556

Eckhoff C, Nau H (1990) Identification and quantitation of all-*trans*- and 13-*cis*-retinoic acid and 13-*cis*-4-oxoretinoic acid in human plasma. J Lipid Res 31:1445–1454

Eckhoff C, Collins MD, Nau H (1991) Human plasma all-*trans*- 13-*cis*- and 13-*cis*-4-oxoretinoic acid profiles during subchronic vitamin A supplementation: comparison to retinol and retinyl ester plasma levels. J Nutr 121:1016–1025
Eguchi Y, Eguchi N, Oda H, Seiki K, Kijima Y, Matsuura Y, Urade Y, Hayaishi O (1997) Expression of lipocalin-type prostaglandin D synthase (β-trace) in human heart and its accumulation in the coronary circulation of angina patients. Proc Natl Acad Sci USA 94:14689–14694
Episkopou V, Maeda S, Nishiguchi S, Shimada K, Gaitanaris GA, Gottesman ME, Robertson EJ (1993) Disruption of the transthyretin gene results in mice with depressed levels of plasma retinol and thyroid hormone. Proc Natl Acad Sci USA 90:2375–2379
Fex G, Johannesson G (1987) Studies on the spontaneous transfer of retinol from the retinol: retinol-binding protein complex to unilamellar liposomes, Biochem Biophys Acta 901:255–264
Fex G, Johannesson G (1988) Retinol transfer across and between phospholipid bilayer membranes. Biochim Biophys Acta 944:249–255
Folman Y, Russell RM, Tang GW, Wolf DG (1989) Rabbits fed on beta-carotene have higher serum levels of all-*trans* retinoic acid than those receiving no beta carotene. Br J Nutr 62:195–201
Friedman SL (1993) The cellular basis of hepatic fibrosis. N Engl J Med 331:1286–1292
Friedman SL, Wei S, Blaner WS (1993) Retinol released by activated rat hepatic lipocytes: regulation by Kupffer cell-conditioned medium and PDGF. Am J Physiol 264:G947–G953
Fries E, Gustafsson L, Peterson PA (1984) Four secretory proteins synthesized by hepatocytes are transported from endoplasmic reticulum to Golgi complex at different rates. EMBO J 3:147–152
Frolik CA (1984) Metabolism of retinoids. In: Spron MB, Roberts AB, Goodman DS (eds) The retinoids, vol 2, Academic Press, Orlando, pp 177-208
Gad MZ, Harrison EH (1991) Neutral and acid retinyl ester hydrolases associated with rat liver microsomes: relationships to microsomal cholesteryl ester hydrolases. J Lipid Res 32:684–693
Geerts A, Bleser PD, Hautekeete ML, Niki T, Wisse E (1994) Fat-storing (Ito) cell biology. In: Arias IM, Boyer JL, Fausto N, Jakoby WB, Schachter D, Shafritz DA (eds) The liver: biology and pathobiology 3rd edn. Raven, New York, pp 819–838
Gerlach T, Biesalski HK, Weiser H, Haeussermann B, Baessler KH (1989) Vitamin A in parenteral nutrition: uptake and distribution of retinyl esters after intravenous application. Am J Clin Nutr 50:1029–1038
Gerlach TH, Zile MH (1990) Upregulation of serum retinol in experimental acute renal failure. FASEB J 4:2511–2517
Gerlach TH, Zile MH (1991) Metabolism and secretion of retinol transport complex in acute renal failure. J Lipid Res 32:515–520
Goodman DS, Huang HS (1965) Biosynthesis of vitamin A with rat intestinal enzymes. Science 149:879–880
Goodman DS, Huang HS, Shiratori T (1965) Tissue distribution and metabolism of newly absorbed vitamin A in the rat. J Lipid Res 6:390–396
Goodman DS (1984a) Plasma retinol-binding protein. In: Sporn MB, Roberts AB, Goodman DS (eds) The retinoids, vol 2. Academic Press, Orlando, pp 41–87
Goodman DS (1984b) Vitamin A and retinoids in health and disease. N Engl J Med 310:1023–1031
Goodman DS, Blaner WS (1984) Biosynthesis, absorption, and hepatic metabolism of retinol In: Sporn MB, Roberts AB, Goodman DS (eds) The retinoids, vol 2, Academic Press, Orlando, pp 1–40
Green MH, Uhl L, Green JB (1985) A multicompartmental model of vitamin A kinetics in rats with marginal liver vitamin A stores. J Lipid Res 26:806–818
Green MH, Green JB, Lewis KC (1987) Variation in retinol utilization rate with vitamin A status in the rat. J Nutr 117:694–703

Gu S-M, Thompson DA, Srisailapathy Srikumari CR, Lorenz B, Finckh U, Nicoletti A, Murthy KR, Rathmann M, Kumaramanickavel G, Denton MJ, Gal A (1997) Mutations in RPE65 cause autosomal recessive childhood-onset severe retinal dystrophy. Nat Genet 17:194–197

Hamel CP, Tsilou E, Pfeffer BA, Hooks JJ, Detrick B, Redmond TM (1993) Molecular cloning and expression of RPE65, a novel retinal pigment epithelium-specific microsomal protein that is posttranscriptionally regulated in vitro. J Biol Chem 268:15751–15757

Hamilton RL, Wong JS, Guo LSS, Krisans S, Havel RJ (1990) Apolipoprotein E localization in rat hepatocytes by immunogold labeling of cryothin sections. J Lipid Res 31:1589–1603

Harrison EH, Blaner WS, Goodman DS, Ross AC (1987) Subcellular localization of retinoids, retinoid-binding proteins, and acyl-CoA: retinol acyltransferase in rat liver. J Lipid Res 28:973–981

Harrison EH (1993) Enzymes catalyzing the hydrolyses of retinyl esters. Biochim Biophys Acta 1170:99–108

Harrison EH, Gad MZ, Ross AC (1995) Hepatic uptake and metabolism of chylomicron retinyl esters: probable role of plasma membrane/endosomal retinyl ester hydrolases. J Lipid Res 36:1498–1506

Haskell MJ, Islam MA, Handelman GJ, Peerson JM, Jones AD, Wahed MA, Mahalanabis D, Brown KH (1998) Plasma kinetics of an oral dose of [$^{2}H_{4}$]retinyl acetate in human subjects with estimated low or high total body stores of vitamin A. Am J Clin Nutr 68:90–95

Heller J (1975) Interactions of plasma retinol-binding protein to its receptor: specific binding of bovine and human retinol-binding protein to pigment epithelial cells from bovine eyes. J Biol Chem 250:3613–3619

Hendriks HFJ, Verhoofstad WAMM, Brouwer A, de Leeuw AM, Knook DL (1985) Perisinusoidal fat-storing cells are the main vitamin A storage sites in rat liver. Exp Cell Res 160:138–149

Hendriks HFJ, Brekelmans PJAM, Buytenhek R, Brouwer A, de Leeuw AM, Knook DL (1987) Liver parenchymal cells differ from the fat-storing cells in their lipid composition. Lipids 22:266–273

Hendriks HFJ, Blaner WS, Wennekers HM, Piantedosi R, Brouwer A, de Leeuw AM, Goodman DS, Knook DL (1988) Distribution of retinoids, retinoid-binding proteins, and related parameters in different types of liver cells isolated from young and old rats. Eur J Biochem 171:237–244

Herz J, Qiu SQ, Oesterle A, DeSilva HV, Shafi S, Havel RJ (1995) Initial hepatic removal of chylomicron remnants is unaffected but endocytosis is delayed in mice lacking the low density lipoprotein receptor. Proc Natl Acad Sci USA 92:4611–4615

Ho MTP, Pownall HJ, Hollyfield JG (1989) Spontaneous transfer of retinoic acid, retinyl acetate, and retinyl palmitate between single unilamellar vesicles. J Biol Chem 264:17759–17763

Hodam JR, Creek KE (1996) Uptake and metabolism of [^{3}H]retinoic acid delivered to human foreskin keratinocytes either bound to serum albumin or added directly to the culture medium. Biochim Biophys Acta 311:102–110

Hodam JR, Creek KE (1998) Comparison of the metabolism of retinol delivered to human keratinocytes either bound to serum retinol-binding protein or added directly to the culture medium. Exp Cell Res 238:257–264

Hussain MM, Goldberg IJ, Weisgraber KH, Mahley RW, Innerarity TL (1997) Uptake of chylomicrons by the liver, but not by the bone marrow, is modulated by lipoprotein lipase activity. Arterioscler Thromb Vasc Biol 17:1407–1413

Ishibashi S, Herz J, Maeda N, Goldstein JL, Brown MS (1994) The two-receptor model of lipoprotein clearance: tests of the hypothesis in "knockout" mice lacking the low density lipoprotein receptor, apolipoprotein E, or both proteins. Proc Natl Acad Sci USA 91:4431–4435

Ishibashi S, Perrey S, Chen Z, Osuga J, Shimada M, Ohashi K, Harada K, Yazaki Y, Yamada N (1996) Role of the low density lipoprotein (LDL) receptor pathway in the metabolism of chylomicrons. J Biol Chem 37:22422–22427

Jaconi S, Rose K, Hughes GJ, Saurat J-H, Siegenthaler G (1995) Characterization of two post-translationally processed forms of human serum retinol-binding protein: altered ratios in chronic renal failure. J Lipid Res 36:1247–1253

Ji ZS, Fazio S, Lee YL, Mahley RW (1994) Secretion-capture role for apolipoprotein E in remnant lipoprotein metabolism involving cell surface heparan sulfate proteoglycans. J Biol Chem 269:2764–2772

Ji ZS, Sanan DA, Mahley RW (1995) Intravenous heparinase inhibits remnant lipoprotein clearance from the plasma and uptake by the liver: in vivo role of heparan sulfate proteoglycans. J Lipid Res 36:583–592

Jones CS, Sly L, Chen LC, Ben T, Brugh-Collins M, Lichti U, De Luca LM (1994) Retinol and beta-carotene concentrations in skin, papillomas and carcinomas, liver, and serum of mice fed retinoic acid or beta-carotene to suppress skin tumor formation, Nutr. Cancer 21:83–93

Kang JX, Li Y, Leaf A (1998) Mannose-6-phosphate/insulin-like growth factor-II receptor is a receptor for retinoic acid. Proc Natl Acad Sci USA 95:13671–13676

Kaji EH, Lodish HF (1993a) Unfolding of newly made retinol-binding protein by dithiothreitol Sensitivity to retinoids. J Biol Chem 268:22188–22194

Kaji EH, Lodish HF (1993b) In vitro unfolding of retinol-binding protein by dithiothreitol endoplasmic reticulum-associated factors. J Biol Chem 268:22195–22202

Kato M, Kato K, Goodman DS (1984) Immunocytochemical studies on the localization of plasma and of cellular retinol-binding proteins and of transthyretin (prealbumin) in rat liver and kidney. J Cell Biol 98:1696–1704

Kaul S, Olaon JA (1998) Effect of vitamin A deficiency on the hydrolysis of beta-glucuronide to retinoic acid by rat tissue organelles in vitro, Int J Vit Nutr Res 68:232–236

Keilson B, Underwood BA, Loerch JD (1979) Effects of retinoic acid on the mobilization of vitamin A from the liver in rats. J Nutr 109:787–795

Klausner RD, Sitia R (1990) Protein degradation in the endoplasmic reticulum. Cell 62:611–614

Krapp A, Ahle S, Kersting S, Hua Y, Kneser K, Nielsen M, Gliemann J, Beisiegel U (1996) Hepatic lipase mediates the uptake of chylomicrons and β-VLDL into cells via the LDL receptor-related protein (LRP). J Lipid Res 37:926–936

Krasinski SD, Cohn JS, Russell RM, Schaefer EJ (1990) Postprandial plasma vitamin A metabolism in humans: a reassessment of the use of plasma retinyl esters as markers for intestinally derived chylomicrons and their remnants. Metabolism 39:357–365

Kurlandsky SB, Gamble MV, Ramakrishnan R, Blaner WS (1995) Plasma delivery of retinoic acid to tissues in the rat. J Biol Chem 270:17850–17857

Lederman JD, Overton KM, Hofmann NE, Moore BJ, Thornton J, Erdman JW (1998) Ferrets (Mustela putoius furo) inefficiently convert beta-carotene to vitamin A. J Nutr 128:271–279

Lewis KC, Green MH, Underwood BA (1981) Vitamin A turnover in rats as influenced by vitamin A status. J Nutr 111:1135–1144

Lewis KC, Green MH, Green JB, Zech LA (1990) Retinol metabolism in rats with low vitamin A status; a compartmental model. J Lipid Res 31:1535–1548

Little JM, Radominska A (1997) Application of photoaffinity labeling with [11,12–^{3}H]all-*trans*-retinoic acid to characterization of rat liver microsomal UDP-glucuronosyltransferase(s) with activity toward retinoic acid. Biochem Biophys Res Commun 230:497–500

Lodish HF, Kong N (1993) The secretory pathway is normal in dithiothreitol-treated cells, but disulfide-bonded proteins are reduced and reversibly retained in the endoplasmic reticulum. J Biol Chem 268:20598–20605

Mahley RW, Hussain MM (1991) Chylomicron and chylomicron remnant catabolism, Curr Opin Lipid 2:170–176
Malaba L, Kindberg GM, Norum KR, Berg T, Blomhoff R (1993) Receptor-mediated endocytosis of retinol-binding protein by liver parenchymal cells: interference by radioactive iodination. Biochem J 291:187–191
Malaba L, Smeland S, Senoo H, Norum KR, Berg T, Blomhoff R, Kindberg GM (1995) Retinol-binding protein and asialo-orosomucoid are taken up by different pathways in liver cells. J Biol Chem 270:15686–15692
Malpeli G, Folli C, Berni R (1996) Retinoid binding to retinol-binding protein and the interference with the interaction with transthyretin. Biochim Biophys Acta 1294: 48–54
Mann CJ, Khallou J, Chevreuil O, Trousard AA, Guermani LM, Launay K, Delplanque B, Yen FT, Bihain BE (1995) Mechanism of activation and functional significance of the lipolysis-stimulated receptor. Evidence for a role as chylomicron remnant receptor. Biochemistry 34:10421–10431
Mata NL, Mata JR, Tsin ATC (1996) Comparison of retinyl ester hydrolase activities in bovine liver and retinal pigment epithelium. J Lipid Res 37:1947–1952
Matarese V, Lodish HF (1993) Specific uptake of retinol-binding protein by variant F9 cell lines. J Biol Chem 268:18859–18865
Matsuno T (1991) Xanthophylls as precursors of retinoids, Pure Appl Chem 63:81–88
Matsuura T, Nagamori S, Hasumura S, Sujine H, Niiya M, Shimizu K (1993a) Regulation of vitamin A transport into cultured stellate cells of rat liver: studies by anchored cell analysis and sorting system. Exp Cell Res 209:33–37
Matsuura T, Nagamori S, Hasumura S, Sujine H, Shimizu K, Niiya M, Hirosawa K (1993b) Retinol transport in cultured stellate cells of rat liver: studies by light and electron microscope autoradiography. Exp Cell Res 206:111–118
Matsuura T, Ross AC (1993) Regulation of hepatic lecithin: retinol acyltransferase activity by retinoic acid. Arch Biochem Biophys 301:221–227
Matsuura T, Gad MZ, Harrison EH, Ross AC (1997) Lecithin: retinol acyltransferase and retinyl ester hydrolase activities are differentially regulated by retinoids and have distinct distributions between hepatocytes and nonparenchymal cell fractions of rat liver. J Nutr 127:218–224
Melhus H, Nilsson T, Peterson PA, Rask L (1991) Retinol-binding protein and transthyretin expressed in HeLa cells form a complex in the endoplasmic reticulum in both the absence and the presence of retinol. Exp Cell Res 197:119–124
Melhus H, Båvik CO, Rask L, Peterson PA, Eriksson U (1995) Epitope mapping of a monoclonal antibody that blocks the binding of retinol-binding protein to its receptor, Biochem Biophys Res Comm 210:105–112
Mercier M, Forget A, Grolier P, Azaïs-Braesco V (1994) Hydrolysis of retinyl esters in rat liver. Description of a lysosomal activity. Biochim Biophys Acta 1212:176–182
Monaco HL, Mancia F, Rizzi M, Coda A (1994) Crystallization of the macromolecular complex transthyretin-retinol-binding protein. J Mol Biol 244:110–113
Monaco HL, Rizzi M, Coda A (1995) Structure of a complex of two plasma proteins: transthyretin and retinol-binding protein. Science 268:1039–1041
Moriwaki H, Blaner WS, Piantedosi R, Goodman DS (1988) Effects of dietary retinoid and triglyceride on the lipid composition of rat liver stellate cells and stellate cell lipid droplets. J Lipid Res 29:1523–1534
Mortimer BC, Beveridge DJ, Martins IJ, Redgrave TG (1995) Intracellular localization and metabolism of chylomicron remnants in the livers of low density lipoprotein receptor-deficient mice and apoE-deficient mice. J Biol Chem 270:28767–28776
Mortimer BC, Martins I, Zeng BJ, Redgrave TG (1997) Use of gene manipulated models to study the physiology of lipid transport, Clin Exp Phar Phys 24:281–285
Mourey MS, Quadro L, Panariello L, Colantuoni V (1994) Retinoids regulate expression of the retinol-binding protein gene in hepatoma cells in culture. J Cell Physiol 160:596–602

Nagao A, During A, Hoshino C, Terao J, Olson JA (1996) Stoichiometric conversion of all *trans* β-carotene to retinal by pig intestinal extract. Arch Biochem Biophys 328:57–63
Nagy NE, Holven KB, Roos N, Senoo H, Kojima N, Norum KR, Blomhoff R (1997) Storage of vitamin A in extrahepatic stellate cells in normal rats. J Lipid Res 38:645–658
Napoli JL, Pramanik BC, Williams JB, Dawson MI, Hobbs PD (1985) Quantification of retinoic acid by gas-liquid chromatography-mass spectrometry: total versus all-*trans*-retinoic acid in human plasma. J Lipid Res 26:387–392
Napoli JL, Race KR (1988) Biogenesis of retinoic acid from β-carotene. J Biol Chem 263:17372–17377
Napoli JL (1996) Retinoic acid biosynthesis and metabolism. FASEB J 10:993–1001
Natarajan V, Holven KB, Reppe S, Blomhoff R, Moskaug JO (1996) The C-terminal RNLL sequence of the plasma retinol-binding protein is not responsible for its intracellular retention. Biochem Biophys Res Commun 221:374–379
Navab M, Smith JE, Goodman DS (1977) Rat plasma prealbumin. Metabolis studies on effects of vitamin A status and on tissue distribution. J Biol Chem 252:5107–5114
Newcomer ME, Jones TA, Aqvist J, Sundelin J, Eriksson U, Rask L, Peterson PA (1984) The three dimensional structure of retinol-binding protein. EMBO J 3:1451–1454
Newcomer ME (1995) Retinoid-binding proteins: structural determinants important for function. FASEB J 9:229–239
Nicoletti A, Wong DJ, Kawase K, Gibson LH, Yang-Feng TL, Richards JE, Thompson DA (1995) Molecular characterization of the human gene encoding an abundant 61 kDa protein specific to the retinal pigment epithelium. Hum Mol Genet 14:641–649
Noy N, Xu ZJ (1990) Interactions of retinol with binding proteins: implications for the mechanism of uptake by cells. Biochemistry 29:3878–3883
Noy N, Blaner WS (1991) Interactions of retinol with binding proteins: Studies with rat cellular retinol-binding protein and with rat retinol-binding protein. Biochemistry 30:6380–6386
Noy N (1992a) The ionization behavior of retinoic acid in aqueous environments and bound to serum albumin. Biochim Biophys Acta 1106:152–158
Noy N (1992b) The ionization behavior of retinoic acid in lipid bilayers and in membranes. Biochim Biophys Acta 1106:159–164
Noy N, Slosberg E, Scarlata S (1992) Interactions of retinol with binding proteins: studies with retinol-binding protein and with transthyretin. Biochemistry 31: 11118–11124
Olson JA, Hayaishi O (1965) The enzymatic cleavage of β-carotene into vitamin A by soluble enzymes of rat liver and intestine. Proc Natl Acad Sci USA 54:1364–1370
Olson JA (1989) Provitamin A function of carotenoids: the conversion of β-carotene into vitamin A. J Nutr 119:105–108
Olson JA (1990) Vitamin A. In: Machlin LJ (ed) Handbook of vitamins 2nd edn. Dekker, New York, pp 1–57
Okabe T, Yorifuji H, Yamada E, Takaku F (1984) Isolation and characterization of vitamin-A-storing lung cells. Exp Cell Res 154:125–135
Okuno M, Moriwaki H, Imai S, Muto Y, Kawanda N, Suzuki Y, Kojima S (1997) Retinoids exacerbate rat liver fibrosis by inducing the activation of latent TGF-β in liver stellate cells. Hepatology 26:913–921
Ong DE, Newcomer ME, Chytil F (1994a) Cellular retinoid-binding proteins. In: Sporn MB, Roberts AB, Goodman DS (eds) The retinoids, biology, chemistry, and medicine 2nd edn. Raven, New York, pp 283–317
Ong DE, Davis JT, O'Day WT, Bok D (1994b) Synthesis and secretion of retinol-binding protein and transthyretin by cultured retinal pigment epithelium. Biochemistry 33:1835–1842

Panariello L, Quadro L, Trematerra S, Colantuoni V (1996) Identification of a novel retinoic acid response element in the promoter region of the retinol-binding protein gene. J Biol Chem 271:25524–25532

Randolph RK, Ross AC (1991a) Regulation of retinol uptake and esterification in MCF-7 and HepG2 cells by exogenous fatty acids. J Lipid Res 32:809–820

Randolph RK, Ross AC (1991b) Vitamin A status regulates hepatic lecithin: retinol acyltransferase activity in rats. J Biol Chem 266:16453–16457

Rask L, Peterson PA (1976) In vitro uptake of vitamin A from the retinol-binding plasma protein to mucosal epithelial cells from the monkey's small intestine, J Biol Chem 251:6360–6366

Ribaya-Mercado JD, Blanco MC, Fox JG, Russell RM (1994) High concentrations of vitamin A esters circulate primarily as retinyl stearate and are stored primarily as retinyl palmitate in ferret tissues, J Am Coll Nut 13:83–86

Rigtrup KM, Ong DE (1992) A retinyl ester hydrolase activity intrinsic to the brush border membrane of rat small intestine. Biochem 31:2920–2926

Rigtrup KM, McEven LR, Said HM, Ong DE (1994) Retinyl ester hydrolytic activity associated with human intestinal brush border membranes. Am J Clin Nutr 60:111–116

Ritter SJ, Green MH, Adams WR, Kelley SK, Schaffer EM, Smith JE (1995) N-(4-hydroxyphenyl)retinamide (Fenretinide) and nephrectomy alter normal plasma retinol-binding protein metabolism, J Nutr Biochem 6:689–696

Ritter SJ, Smith JE (1996) Retinol-binding protein secretion from the liver of N-(4-hydroxyphenyl) retinamide-treated rats. Biochim Biophys Acta 1290:157–164

Ross AC, Kempner A (1993) Radiation inactivation analysis of acyl-CoA: retinol acyltransferase and lecthin: retinol acyltransferase in rat liver. J Lipid Res 34:1201–1207

Sani BP, Barua AB, Hill DL, Shih TW, Olson JA (1992) Retinoyl beta-glucuronide: lack of binding to receptor proteins of retinoic acid as related to biological activity. Biochem Pharmacol 43:919–922

Schindler R, Mentlein R, Feldheim W (1998) Purification and characterization of retinyl ester hydrolase as a member of the non-specific carboxylesterase supergene family. Eur J Biochem 251:863–873

Schindler R, Klopp A (1986) Transport of esterified retinol in fasting human blood, Int J Vit Nutr Res 56:21–27

Schweigert FJ (1988) Insensitivity of dogs to the effects of nonspecific bound vitamin A in plasma, Int J Vit Nutr Res 58:23–25

Schweigert FJ, Rosival I, Rambeck WA, Gropp J (1995) Plasma transport and tissue distribution of $[^{14}C]\beta$-carotene and $[^{3}H]$retinol administered orally to pigs, Int J Vit Nutr Res 65:95–100

Senoo H, Stang E, Nilsson A, Kindberg GM, Berg T, Roos N, Norum KR, Blomhoff R (1990) Internalization of retinol-binding protein in parenchymal and stellate cells of rat liver. J Lipid Res 31:1229–1239

Shafi S, Brady SE, Bensadoun A, Havel RJ (1994) Role of hepatic lipase in the uptake and processing of chylomicron remnants in rat liver. J Lipid Res 35:709–720

Shankar S, De Luca LM (1988) Retinoic acid supplementation of a vitamin A-deficient diet inhibits retinoid loss from hamster liver and serum pools. J Nutr 118:675–680

Shimada T, Ross AC, Muccio DD, Brouillette WJ, Shealy YF (1997) Regulation of hepatic lecithin: retinol acyltransferase activity by retinoic acid receptor-selective retinoids. Arch Biochem Biophys 344:220–227

Sivakumar B, Jahoor F, Burrin DG, Frazer EM, Reeds PJ (1994) Fractional synthesis rates of retinol-binding protein, transthyretin, and a new peptide measured by stable isotope techniques in neonatal pigs. J Biol Chem 269:26196–26200

Sivaprasadarao A, Findlay JB (1988a) The mechanism of uptake of retinol by plasma-membrane vesicles. Biochem J 255:571–579

Sivaprasadarao A, Findlay JB (1988b) The interaction of retinol-binding protein with its plasma-membrane receptor. Biochem J 255:561–569

Sivaprasadarao A, Findlay JB (1993) Expression of functional human retinol-binding protein in Escherichia coli using a secretion vector. Biochem J 296:209–215
Sivaprasadarao A, Boudjelal M, Findlay JB (1994) Solubilization and purification of the retinol-binding protein receptor from human placental membranes. Biochem J 302:245–251
Sivaprasadarao A, Findlay JB (1994) Structure-function studies on human retinol-binding protein using site-directed mutagenesis. Biochem J 300:437–442
Skare KL, Schnoes HK, DeLuca HF, (1982) Biliary metabolites of all-*trans*-retinoic acid in the rat: isolation and identification of a novel polar metabolite. Biochemistry 21:3308–3317
Skottova N, Savonen R, Lookene A, Hultin M, Olivecrona G (1995) Lipoprotein lipase enhances removal of chylomicron remnants by the perfused rat liver. J Lipid Res 36:1334–1344
Smeland S, Bjerknes T, Malaba L, Eskild W, Norum KR, Blomhoff R (1995) Tissue distribution of the receptor for plasma retinol-binding protein. Biochem J 305:419–424
Smith JE, DeMoor LM, Handler CE, Green EL, Ritter SJ (1998) The complex between retinol and retinol-binding protein is formed in the rough microsomes of liver following repletion of vitamin A-depleted rats. Biochim Biophys Acta 1380:10–20
Smith TJ, Davis FB, Deziel MR, Davis PJ, Ramsden DB, Schoenl M (1994) Retinoic acid inhibition of thyroxine binding to human transthyretin. Biochim Biophys Acta 1199:76–89
Soprano DR, Soprano KJ, Goodman DS (1986) Retinol-binding protein messenger RNA levels in the liver and in extrahepatic tissue of the rat. J Lipid Res 27:166–172
Soprano DR, Blaner WS (1994) Plasma retinol-binding protein, In: Sporn MB, Roberts AB, Goodman DS (eds) The retinoids, biology, chemistry, and medicine 2nd edn. Raven, New York, pp 257–282
Suhara A, Kato M, Kanai M (1990) Ultrastructural localization of plasma retinol-binding protein in rat liver. J Lipid Res 31:1669–1681
Sun G, Alexson SEH, Harrison EH (1997) Purification and characterization of a neutral, bile salt-independent retinyl ester hydrolase from rat liver microsomes. J Biol Chem 272:24488–24493
Sundaram M, Sivaprasadarao A, DeSousa M, Findlay JB (1997) The transfer of retinol from serum retinol-binding protein to cellular retinol-binding protein is mediated by a membrane receptor. J Biol Chem 273:3336–3342
Sundaresan PR (1977) Rate of metabolism of retinol in retinoic acid-maintained rats after a single dose of radioactive retinol. J Nutr 107:70–78
Suruga K, Suzuki R, Goda T, Takase S (1995) Unsaturated fatty acids regulate gene expression of cellular retinol-binding protein, type II in rat jujunum. J Nutr 125:2039–2044
Swanson BM, Frolik CA, Zaharevitz DW, Roller PP, Sporn MB (1981) Dose-dependent kinetics of all-*trans*-retinoic acid in rats. Biochem Pharmacol 30:107–113
Takase S, Tanaka K, Suruga K, Kitagawa M, Igarashi M, Goda T (1998) Dietary fatty acids are possible key determinants of cellular retinol-binding protein II gene expression, Am J Physiol (Gastrointest Liver Physiol) 274:G626–G632
Tall A (1995) Plasma lipid transfer proteins. Annu Rev Biochemistry 64:235–257
Tanaka T, Urade Y, Kimura H, Eguchi N, Nishikawa A, Hayaishi O (1997) Lipocalin-type prostaglandin D synthase (β-trace) is a newly recognized type of retinoid transporter. J Biol Chem 272:15789–15796
Tang G, Russell RM (1990) 13-*cis* Retinoic acid is an endogenous compound in human serum. J Lipid Res 30:175–182
Tang G, Russell RM (1991) Formation of all-*trans*-retinoic acid and 13-*cis*-retinoic acid from all-*trans*-retinyl palmitate in humans, J Nutr Biochem 2:2100–2213
Tang G, Shiau A, Russell RM, Mobarhan S (1995) Serum retinoic acid levels in patients with resected benign and malignant colonic neoplasias on beta-carotene supplementation, Nutr. Cancer 23:291–298

Thatcher AJ, Lee CM, Erdman JW Jr (1998) Tissue stores of β-carotene are not conserved by later use as a source of vitamin A during compromised vitamin A status in Mongolian gerbils (Meriones unguiculatus). J Nutr 128:1179–1185
Thompson HK, Hughes LB, Zilversmit DB (1983) Lack of secretion of retinyl ester by livers of normal and cholesterol-fed rabbits. J Nutr 113:1995–2001
Tosetti F, Ferrari N, Pfeffer U, Brigati C, Vidali G (1992) Regulation of plasma retinol binding protein secretion in human HepG2 cells. Exp Cell Res 200:467–472
Troen G, Nilsson A, Norum KR, Blomhoff R (1994) Characterization of liver stellate cell retinyl ester storage. Biochem J 300:793–798
Tsilou E, Hamel CP, Yu S, Redmond TM (1997) RPE65, the major retinal pigment epithelium microsomal membrane protein, associates with phospholipid liposomes. Arch Biochem Biophys 346:21–27
Underwood BA, Loerch JD, Lewis KC (1979) Effects of dietary vitamin A deficiency, retinoic acid and protein quantity and quality on serially obtained plasma and liver levels of vitamin A in rats. J Nutr 109:796–806
Van Bennekum AM, Li L, Piantedosi R, Shamir R, Vogel S, Fisher EA, Blaner WS, Harrison EH (1999a) Carboxyl ester lipase overexpression in rat hepatoma cells and CEL-deficiency in mice have no impact on hepatic uptake and metabolism of chylomicron-retinyl ester. Biochemistry 38:4150–4156
Van Bennekum AM, Kako Y, Weinstock PH, Harrison EH, Deckelbaum RJ, Goldberg IJ, Blaner WS (1999b) Lipoprotein lipase expression level influences tissue clearance of chylomicron retinyl ester, J Lipid Res 40:565–574
Van Vliet T (1996) Absorption of β-carotene and other carotenoids in humans and animal models. Eur J Clin Nutr 50:S32–S37
Van Vliet T, van Vlissingen MF, van Schaik F, van den Berg H (1996) β-Carotene absorption and cleavage in rats is affected by the vitamin A concentration of the diet. J Nutr 126:499–508
Von Reinersdorff D, Bush E, Liberato DJ (1996) Plasma kinetics of vitamin A in humans after a single oral dose of [8,9,19–^{13}C]retinyl palmitate. J Lipid Res 37:1875–1885
Waits RP, Yamada T, Uemichi T, Benson MD (1995) Low plasma concentrations of retinol-binding protein in individuals with mutations affecting position 84 of the transthyretin molecule. Clin Chem 41:1288–1291
Wake K (1980) Perisinusoidal stellate cells (fat-storing cells, interstitial cells, lipocytes), their related structure in and around liver sinusoids, and vitamin A-storing cells in extrahepatic organs. Int Rev Cytol 66:303–353
Ward SJ, Chambon P, Ong DE, Båvik C (1997) A retinol-binding protein receptor-mediated mechanism for uptake of vitamin A to postimplantation rat embryos. Biol Reprod 57:751–755
Wang X-D (1994) Absorption and metabolism of β-carotene. J Am Coll Nutr 13:314–325
Wang X-D, Krinsky NI, Benotti PN, Russell RM (1994) Biosynthesis of 9-*cis*-retinoic acid from 9-*cis*-β-carotene in human intestinal mucosa in vitro. Arch Biochem Biophys 313:150–155
Wang X-D, Russell RM, Liu C, Stickel F, Smith DE, Krinsky NI (1996) β-Oxidation in rabbit liver in vitro and in the perfused ferret liver contributes to retinoic acid biosynthesis from β-apocarotenoic acids. J Biol Chem 271:26490–26498
Weng W, Li L, van Bennekum AM, Potter SH, Blaner WS, Breslow JL, Fisher EA (1999) Pancreatic carboxyl ester lipase knockout mice have partially decreased cholesteryl ester absorption and normal retinyl ester absorption: evidence of alternative enzyme(s) hydrolyzing dietary cholesteryl ester and retinyl ester. Biochemistryx 38:4143–4150
Wei S, Episkopou V, Piantedosi R, Maeda S, Shimada K, Gottesman ME, Blaner WS (1995) Studies on the metabolism of retinol and retinol-binding protein in transthyretin-deficient mice produced by homologous recombination. J Biol Chem 270:866–870

Weiner FR, Blaner WS, Czaja MJ, Shah A, Geerts A (1992) Ito cell expression of a nuclear retinoic acid receptor. Hepatology 15:336–342
Williams DL, Dawson PA, Newman TC, Rudel LL (1985) Apolipoprotein E synthesis in peripheral tissues of nonhuman primates. J Biol Chem 260:2444–2451
Willnow TE, Armstrong SA, Hammer RE, Herz J (1995) Functional expression of low density lipoprotein receptor-related protein is controlled by receptor-associated protein in vivo. Proc Natl Acad Sci USA 92:4537–4541
Willnow TE, Sheng Z, Ishibashi S, Herz J (1994) Inhibition of hepatic chylomicron remnant uptake by gene transfer of a receptor antagonist. Science 264:1471–1474
Wilson DE, Hejazi J, Elstad NL, Chan IF, Gleeson JM, Iverius PH (1987) Novel aspects of vitamin A metabolism in the dog: distribution of lipoprotein retinyl esters in vitamin A-deprived and cholesterol-fed animals. Biochim Biophys Acta 922:247–258
Yamada M, Blaner WS, Soprano DR, Dixon JL, Kjeldbye HM, Goodman DS (1987) Biochemical characteristics of isolated rat liver stellate cells. Hepatology 7:1224–1229
Yamamoto Y, Yoshizawa T, Kamio S, Aoki O, Kawamata Y, Masushinge S, Kato S (1997) Interactions of transthyretin (TTR) and retinol-binding protein (RBP) in the uptake of retinol by primary rat hepatocytes. Exp Cell Res 234:373–378
Yen FT, Mann CJ, Guermani LM, Hannouche NF, Hubert N, Hornick CA, Bordeau VN, Agnani G, Bihain BE (1994) Identification of lipolysis stimulated receptor that is distinct from the LDL receptor and the LDL receptor-related protein. Biochemistry 33:1172–1180
Yumoto S, Ueno K, Mori S, Takebayashi N, Handa S (1988) Morphological and biochemical analyses of lipid granules isolated from fat-storing cells in rat liver, Biomed Res 2:147–160
Yumoto S, Yumoto K, Yamamoto M (1989) Effects of retinoic acid on Ito cells (fat-storing cells) in vitamin A-deficient rats. In: Wisse E, Knook DL, Decker K (eds) Cells of the hepatic sinusoid, vol 2. Kupffer Cell Foundation, Rijswijk, p 33
Zachman RD, Dunagin PE Jr, Olson JA (1966) Formation and enterohepatic circulation of metabolites of retinol and retinoic acid in bile duct-cannulated rats. J Lipid Res 7:3–9
Zanotti G, Ottonello S, Berni R, Monaco HL (1993a) Crystal structure of liganded and unliganded forms of bovine plasma retinol-binding protein. J Biol Chem 268:10728–10738
Zanotti G, Ottonello S, Berni R, Monaco HL (1993b) Crystal structure of the trigonal form of human plasma retinol-binding protein at 2.5 Å resolution. J Mol Biol 230:613–624
Zanotti G, Marcello M, Malpeli G, Folli C, Sartori G, Berni R (1994) Crystallographic studies on complexes between retinoids and plasma retinol-binding protein. J Biol Chem 269:29613–29620
Zanotti G, D'Acunto MR, Malpeli G, Folli C, Berni R (1995) Crystal structure of the transthyretin-retinoic-acid complex. Eur J Biochem 234:563–569
Zovich DC, Orologa A, Okuno M, Wong Yen Kong L, Talmage D, Piantedosi R, Goodman DS, Blaner WS (1992) Differentiation-dependent expression of retinoid-binding proteins in BFC-1β adipocytes. J Biol Chem 267:13884–13889

CHAPTER 3

Retro-Retinoids: Metabolism and Action

E. VAKIANI and J. BUCK

A. Introduction

In a classic histological study published in 1925 WOLBACH and HOWE described the widespread pathology associated with vitamin A deficiency in rats. The reported abnormalities included stunted growth, blindness, keratinization of the epithelial linings of the respiratory and alimentary tracts, defective development of testis and atrophy of the central and peripheral lymphoid organs. Subsequent studies with vitamin A deficient animals have confirmed these findings and have established a critical role for retinol in the normal development and maintenance of most tissues.

The range and complexity of the pathology of vitamin A deficiency suggests that retinol can regulate cellular function in multiple ways. Retinol is not known to have biological activity itself, rather it serves as the parent compound for the biosynthesis of active metabolites. During the past few decades several of these bioactive metabolites have been characterized (Fig. 1). The pioneering work of G. WALD, for which he was awarded the 1967 Nobel Prize in Physiology, identified 11-*cis* retinal (RAL) as the essential chromophore for vision. In photoreceptor cells 11-*cis* RAL is bound to G protein coupled transmembrane spanning receptors, the opsins, and isomerizes to all-*trans* RAL in response to light (reviewed in WALD 1968).

Retinoic acids (RAs) are important for cellular differentiation (reviewed in LOVE and GUDAS 1994) and morphogenesis (reviewed in HOFMANN and EICHELE 1994). Extensive studies into their mechanism of action have led to the characterization of the nuclear retinoic acid receptors RARs and RXRs. These receptors are members of the steroid/thyroid receptor superfamily and regulate transcription of target genes upon binding of RAs (reviewed in MANGELSDORF et al. 1994). Recently, two nonacid derivatives of retinol, 4-oxo-retinol (ACHKAR et al. 1996) and 4-oxo-retinal (BLUMBERG et al. 1996), were also shown to transactivate RARs. Outside the visual system, RAs appear to mediate many, but not all, of the actions of vitamin A. In vitamin A deficient animals, for example, RA is inferior to retinol in reversing defects in spermatogenesis (HOWELL et al. 1963) or the immune system (DAVIS and SELL 1983).

The fact that retinol can act in a manner independent of RA (outside vision) has been known for many years, yet there has been little progress in

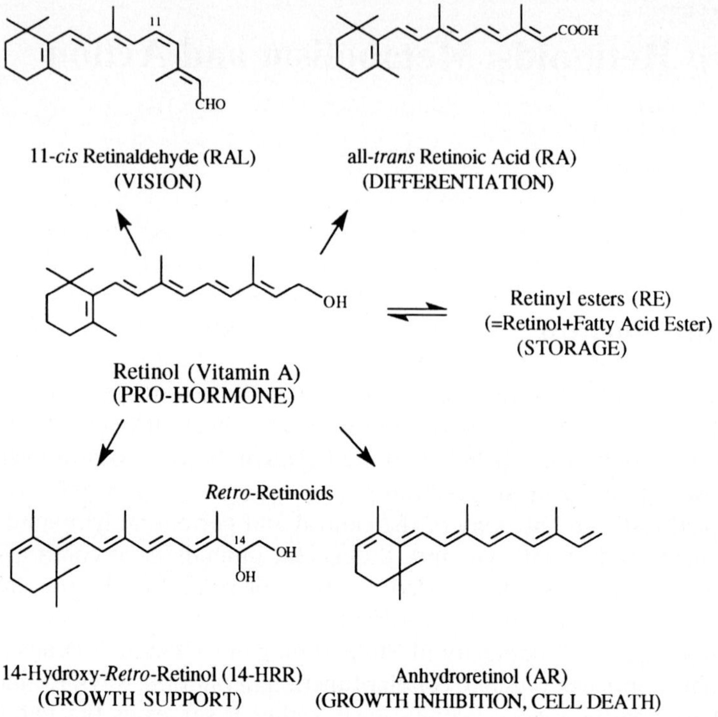

11-*cis* Retinaldehyde (RAL)
(VISION)

all-*trans* Retinoic Acid (RA)
(DIFFERENTIATION)

Retinol (Vitamin A)
(PRO-HORMONE)

Retinyl esters (RE)
(=Retinol+Fatty Acid Ester)
(STORAGE)

Retro-Retinoids

14-Hydroxy-*Retro*-Retinol (14-HRR)
(GROWTH SUPPORT)

Anhydroretinol (AR)
(GROWTH INHIBITION, CELL DEATH)

Fig. 1. Structures of several retinol metabolites

understanding the mechanisms underlying such RA-independent effects. A significant advance in this area was the characterization over the last few years of new bioactive retinoid metabolites, namely 14-hydroxy-*retro*-retinol (14-HRR) and anhydro-*retro*-retinol (also referred to as anhydroretinol; AR) (Fig. 1). These two *retro*-retinoids were initially identified for their putative role in lymphocyte physiology, and subsequent studies revealed that they may have a more general role in cell growth and proliferation. While their mechanism of action remains to be elucidated, present evidence suggests that they act through novel pathway(s). This chapter discusses what is presently known about the biosynthesis and actions of these recently identified *retro*-retinoid metabolites and how research in this area can contribute to a better understanding of the actions of vitamin A.

B. Vitamin A and the Immune System

The link between vitamin A deficiency and impaired immune function was suggested by the early studies of WOLBACH and HOWE (1925) when vitamin A deficiency was shown to result in the underdevelopment of lymphoid organs.

Their work was followed by a large number of experiments using various animal models that showed vitamin A deficiency compromises the ability to mount both cell-mediated and humoral immune responses. Among the dysfunctions identified were decreased lymphocyte proliferation and antibody production as well as reduced activity of cytotoxic T cells and natural killer cells (reviewed in Ross and Hämmerling 1994).

The work of A. Sommer, recipient of the 1997 Lasker Award, focused attention on the increased mortality, especially among children, associated with vitamin A deficiency. Vitamin A deficiency is an important health concern in poor countries where a large part of the population is malnourished. When Sommer et al. (1983) studied children in rural areas of Indonesia, they found that children suffering from even mild vitamin A deficiency were more likely to die from infections. Additional field and hospital-based studies (reviewed in Underwood 1994) confirmed their findings and revealed an increased severity of infection in vitamin A deficient children, even though the susceptibility to infection appeared unchanged.

Despite the strong evidence for a direct involvement of retinol in the immune response, the molecular mechanisms underlying its actions have remained largely unknown. A number of in vivo and in vitro studies (reviewed in Ross and Hämmerling 1994) have found that all-*trans* RA can stimulate lymphocyte proliferation in response to a variety of activators. However, the physiological role of all-*trans* RA in the regulation of lymphocyte function is not yet clear. Moreover, experiments using vitamin A deficient animals have indicated that at least some of the effects of retinol on the immune system are independent of RA.

In 1990 retinol was shown for the first time to be essential for the survival and growth of activated human B lymphocytes grown in culture (Buck et al. 1990). The impetus for this discovery was the observation that normal and malignant human B lymphoblasts died when transferred into serum-free medium with the rate of cell death dependent on cell density. Efforts to identify the growth factor(s) present in serum that can prevent cell death resulted in the isolation of all-*trans* retinol bound to its carrier protein, retinol-binding protein (RBP). Addition of all-*trans* retinol at concentrations normally present in human serum was sufficient to prevent the death of B lymphoblastoid cells. In contrast, all-*trans* RA was not able to sustain cell growth in this assay. This finding triggered a number of studies into the mechanism by which retinol regulates the growth of activated B cells, resulting in the identification of new retinol metabolites, most notably 14-HRR (Buck et al. 1991b) and AR (Buck et al. 1993).

C. General Properties of *Retro*-Retinoids

The term "*retro*-" is used to describe the retinoid skeleton of these compounds, which is characterized by a shift in the conjugated double bond system resulting in the presence of a double bond between carbons 6 and 7 (Fig. 2a). In the *β*-series retinoids (e.g., retinol, retinal, retinoic acid) there is rotation about the 6–7 carbon bond which allows the side chain to be out of the plane of the ring. In contrast, the side chain of *retro*-retinoids is forced in the same plane as the cyclohexene ring, and the double bond on the ring becomes part of the conjugated double bond system. Hence, *retro*-retinoids have a different three-dimensional structure from the *β*-series retinoids, and their ultraviolet spectrum exhibits a distinct and characteristic vibronic fine structure (Fig. 2b). It is interesting to note that a coplanar conformation between the ring and the side chain has also been shown for retinal when it is bound to protein in bacteriorhodopsin (SCHRECKENBACH et al. 1977) and for retinol when it is bound to the cellular retinol-binding protein I (CRBP I) (ONG and CHYTIL 1978).

Retro-retinoids have been known for a long time to synthetic chemists who studied them for their interesting physical properties (see, e.g., OROSHNIK et al. 1952). AR was the first *retro*-retinoid to be described in nature. It was identified in 1939 in fish liver oils (EMBREE 1939), even though it was

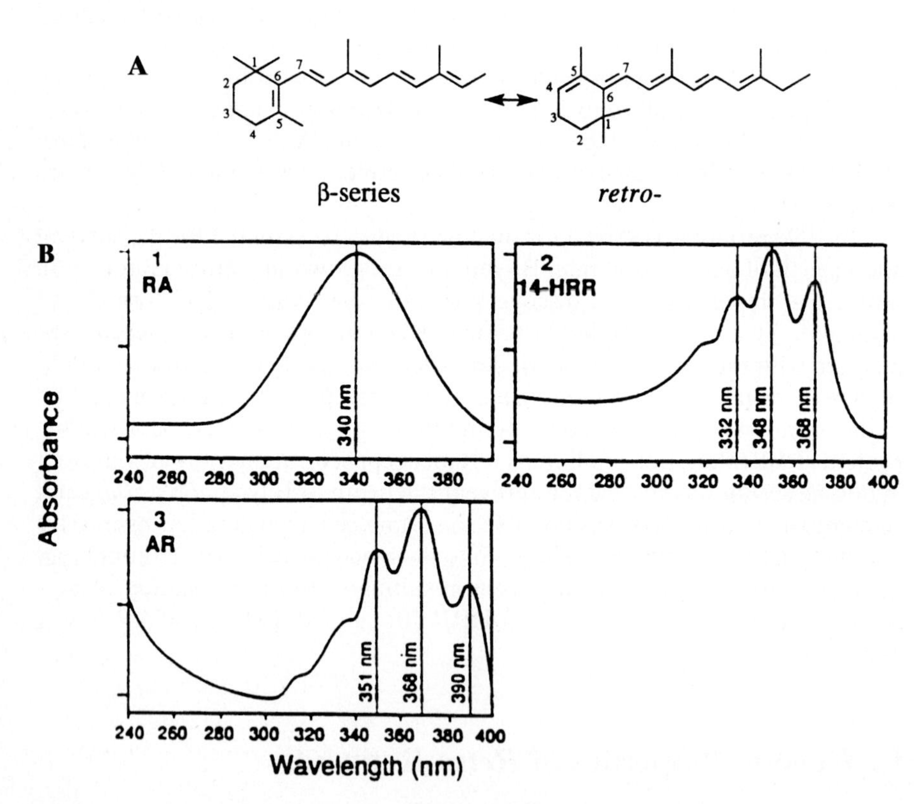

Fig. 2. A Structure of the retinoid skeleton in β-series compounds compared to *retro* compounds. **B** Ultraviolet absorption spectra of RA, 14-HRR and AR

mistakenly referred to as "cyclized vitamin A" until 1943 (SHANTZ et al. 1943). Until recently it was believed to be an inert retinoid metabolite, because it failed to prevent the vitamin A deficiency syndrome in rats maintained on a vitamin A deficient diet (VARMA et al. 1965). Interest in the *retro*-retinoids was renewed after the characterization of 14-HRR (BUCK et al. 1991b), which was the first described bioactive *retro*-retinoid.

14-HRR and AR can bind to retinol-binding protein (RBP) in the plasma and to CRBP I but do not bind to cellular retinoic acid-binding protein I (CRABP I) (BUCK et al. 1993). The proposed functions for RBP and CRBP include transporting and sequestering retinol and preventing nonspecific hydrophobic interactions (reviewed in SOPRANO and BLANER 1994; ONG et al. 1994). It will be interesting to investigate the physiological roles of these proteins in *retro*-retinoid transport and metabolism and to see whether they serve similar functions for 14-HRR and AR as they do for retinol.

Both AR and 14-HRR can be chemically synthesized, which is important to ensure their availability for bioassays. AR is easily synthesized by acid catalyzed dehydration of retinol (EDISBURY et al. 1932; DERGUINI et al. 1994a), while the organic synthesis of 14-HRR involves several steps (DERGUINI et al. 1994b; COREY et al. 1995). Synthetic AR and 14-HRR have the same chemical and biological profiles as the native compounds (DERGUINI et al. 1994a 1994b).

D. Cellular Effects of *Retro*-Retinoids

I. Growth of B Lymphocytes

The finding that known retinol metabolites, in particular all-*trans* RA, cannot account for the growth-supporting effect of retinol on activated B cells suggested that unidentified metabolites mediate this action. To explore this hypothesis BUCK et al. (1991b) analyzed the metabolic products of retinol in B lymphoblastoid cells by reversed-phase high-pressure liquid chromatography (HPLC) techniques and found 14-HRR to be a major metabolite. Specifically, when these cells were grown in medium containing 10% fetal calf serum (FCS) (corresponding to approximately $1–2 \times 10^{-7} M$ retinol concentration), the cellular levels of 14-HRR were found to be in the range of $1–3 \times 10^{-8} M$. In contrast, there was no detectable RA production. When 14-HRR was added to activated B cells in serum-free medium, it prevented cell death and was sufficient to replace retinol. The potency of 14-HRR was 10- to 30-fold higher than that of retinol, provided 14-HRR was replenished periodically to compensate for its short cellular half-life (4h for 14-HRR compared to 24h for retinol; Fig. 3). Increased potency and short-half life are both characteristics of second messenger molecules and are consistent with the putative role of 14-HRR as the intracellular mediator of the retinol effect.

In addition to 14-HRR, B lymphoblastoid cells also produced small amounts of AR. When this compound was added to cultured B cells, it failed

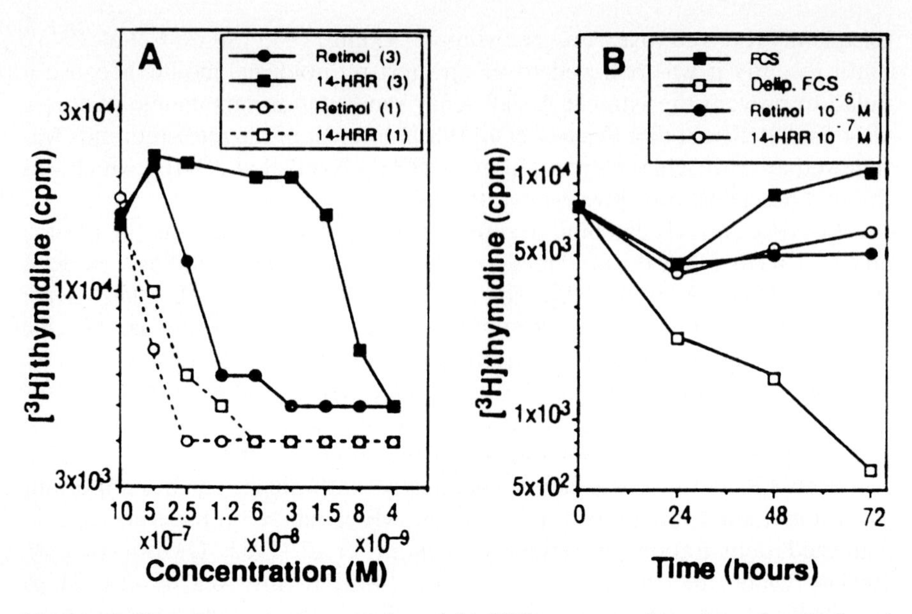

Fig. 3A,B. Growth-supporting activity of 14-HRR compared to that of retinol and serum. **A** Dependency of dose response on repeated 14-HRR application. **B** Growth kinetics of lymphoblastoid cells. Lymphoblastoid 5/2 cells grown in RPMI/5% FCS were taken from their exponential growth phase, washed twice, and seeded at 30,000 cells/ml in RPMI medium containing in delipidated bovine albumin (1.2 mg/ml), 10^{-6} *M* linoleic acid, bovine insulin (5 µg/ml), and transferrin (5 µg/ml; **A**); in 5% FCS, 5% delipidated FCS with or without 10^{-6} *M* retinol or 10^{-7} *M* 14-HRR (**B**). The assay was performed in 96-well microtiter plates in a final volume of 200 µl per well. The cells were cultured for 72 h, and we determined cell growth by labeling for the last 16 h (**A**) or 8 h (**B**) with [^{3}H]thymidine (0.8 µCi per well) at the indicated time points. Retinoids at the indicated concentration were added either once at the initiation of culture (**A**, *broken lines*) or every 24 h thereafter as well (**A** and **B**, *solid lines*). Data represent the mean of triplicate measurements and SDs were <16%. (Reprinted with permission from BUCK et al. 1991b; copyright 1991 American Association for the Advancement of Science)

to sustain cell growth; on the contrary, it accelerated the cell death seen in the absence of retinol or 14-HRR. The growth-inhibitory effect of AR was reversed by the addition of retinol or 14-HRR, and dose-response studies revealed that AR behaved as a competitive inhibitor of retinol and 14-HRR (Fig. 4) (BUCK et al. 1993).

These findings suggested for the first time the existence of a retinol-dependent but RA-independent pathway that can control cell survival. Given the structural similarity between 14-HRR and AR and the observed competition between them, it seems likely the two compounds act on a common receptor molecule that controls a signaling cascade important for cell growth. According to this model, if both *retro*-retinoids are present in a cell, the balance between them determines cell survival versus death.

A

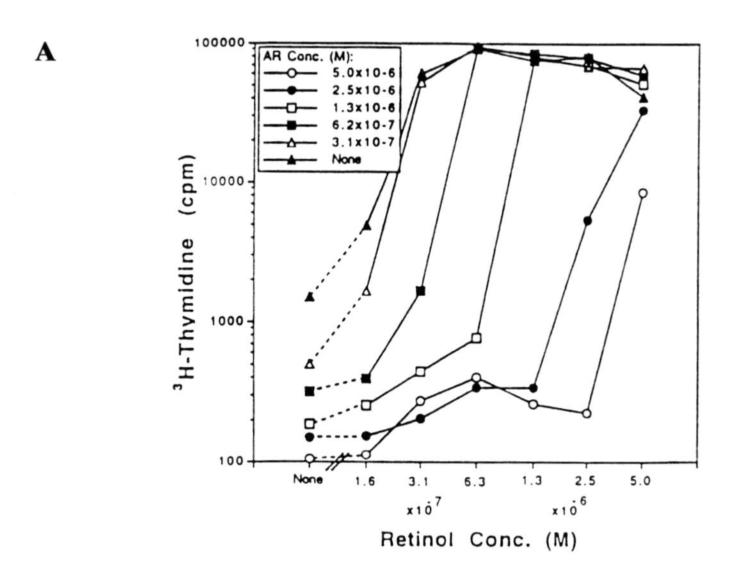

B

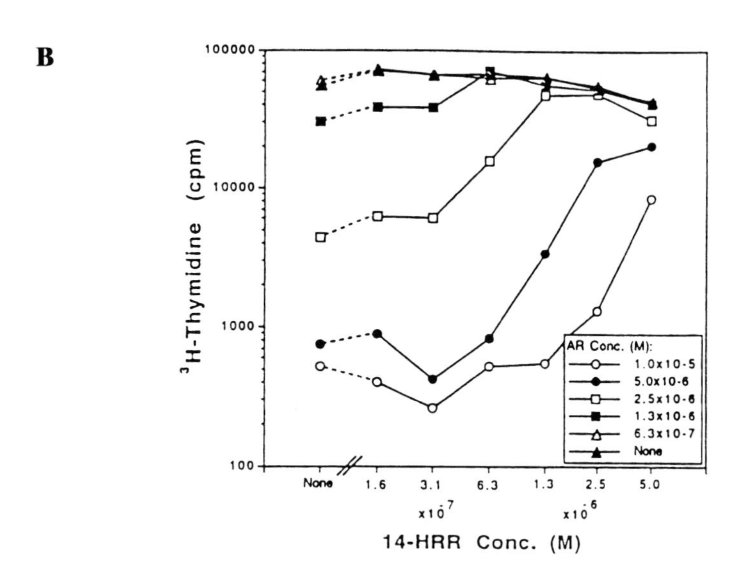

Fig. 4A,B. AR as reversible inhibitor of B lymphocyte proliferation. AR was produced from all-*trans* retinol by hydrochloric acid catalyzed dehydration as described (Edisbury et al. 1932), and the all-*trans* isomer was purified to homogeneity by reversed-phase high-pressure liquid chromatography (HPLC). Human B lymphoblastoid cells (e.g., cell line 5/2) were cultured in medium RPMI 1640 with 7% FCS. On the day of the assay, cells were washed and transferred to serum-free medium, RPMI 1640 supplemented with ITLB medium as described (Garbe et al. 1992). Test reagents at indicated concentrations were added to triplicate cultures of 5×10^3 cells/well in 96-well plates (final volume of 200 μl/well) and cell proliferation tested 1(**B**) or 3 days later (**A**) by [^{3}H]thymidine incorporation into DNA. The agonists used were all-*trans* retinol (**A**) and 14-HRR (**B**). Serum contains $1–2 \times 10^{-6}$ *M* all-*trans* retinol. (Modified with permission from Buck et al. 1993; copyright Rockefeller University Press)

II. Activation of T Lymphocytes

Retro-retinoids have also been shown to affect the function of T lymphocytes in similar ways. In serum-free medium, murine thymus and lymph node T cells failed to proliferate when activated either by mitogens or by cross-linking antibodies to the T cell receptor (GARBE et al. 1992). Addition of retinol or 14-HRR was sufficient to support proliferation of activated T cells, and the response was similar to that seen when thymocytes were activated in serum-containing medium. The subsets of T cells proliferating with the assistance of retinoid cofactor were $CD4^+$ and $CD8^+$ thymic cells and $CD4^+$ peripheral cells, while $CD8^+$ peripheral cells required interleukin 2 in addition to retinol or 14-HRR. All-*trans* RA could not replace retinol in this assay, which again indicated that this effect was RA-independent. In addition, interleukins were unable to overcome the requirement for retinol (GARBE et al. 1992). AR behaved as a competitive inhibitor of retinol and 14-HRR and could prevent the proliferation seen upon thymocyte activation in a dose-dependent manner (BUCK et al. 1993).

III. Activation of Fibroblasts

The identification of 14-HRR and AR in a number of cell types outside the immune system suggested that their role is not limited to lymphocyte physiology. A study of retinol and its *retro*-retinoid derivatives in mouse 3T3 fibroblasts supports this hypothesis. Murine 3T3 cells grown in culture can be induced to proliferate by a number of growth factors, most notably platelet-derived growth factor (PDGF). When they are transferred to serum-free or low serum medium (starved) they arrest in a quiescent, nondividing state (HOLLEY and KIERNAN 1968). Cells starved for 1 day can be activated by PDGF; however, cells that have been starved for longer fail to do so and can only be activated by the addition of serum (BROOKS et al. 1990). Retinol, or 14-HRR, in combination with PDGF, was shown in our laboratory to be able to replace serum and result in full activation (CHEN et al. 1997). Other lipophilic molecules including all-*trans* RA were unable to substitute for retinol or 14-HRR (Fig. 5). AR inhibited 3T3 cell activation (Fig. 6) and the inhibition could be reversed by retinol and 14-HRR.

Murine 3T3 cells, as with most cells, have intracellular retinyl esters (RE) which represent the storage form of retinol and can serve as a retinoid source when no retinol is present. Cells become critically dependent on exogenous retinol, when the intracellular RE pools are depleted. In the case of 3T3 cells, this occurs after 1–2 days of starvation (Chen and Buck, unpublished data) which would explain why cells that are starved longer require exogenous retinol.

Additional experiments revealed that PDGF and retinol had distinct roles in 3T3 cell activation (CHEN et al. 1997). PDGF appeared to control entry into the cell cycle, and retinol had an effect on the total number of activated cells.

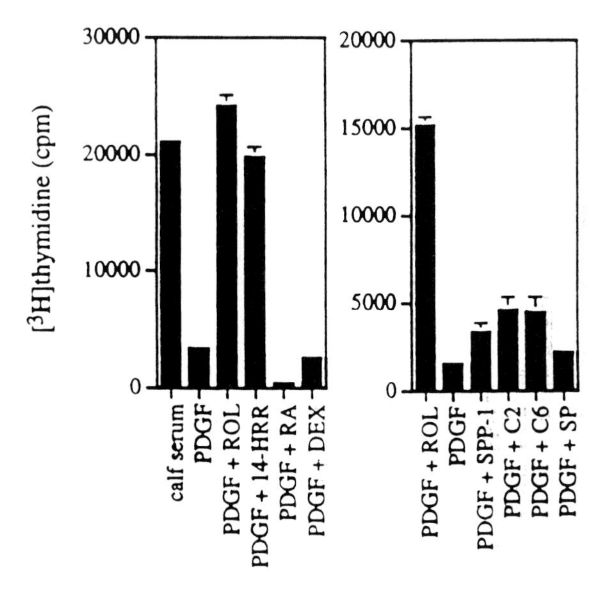

Fig. 5. Effect of several lipophilic molecules on PDGF activation of 2-day serum-starved 3T3 cells: 50 ng/ml PDGF, 2 µ*M* retinol (*ROL*), 4 µ*M* 14-HRR, 4 µ*M* retinoic acid (*RA*), 4 µ*M* dexamethasone (*DEX*), 10 µ*M* ceramide C-2 (*C2*), 10 µ*M* ceramide C-6 (*C6*), and 5 µ*M* sphingosine (*SP*). Data of activation assays represent the mean and SDs of triplicate measurements. (Reprinted with permision from CHEN et al. 1997; copyright 1997 National Academy of Sciences)

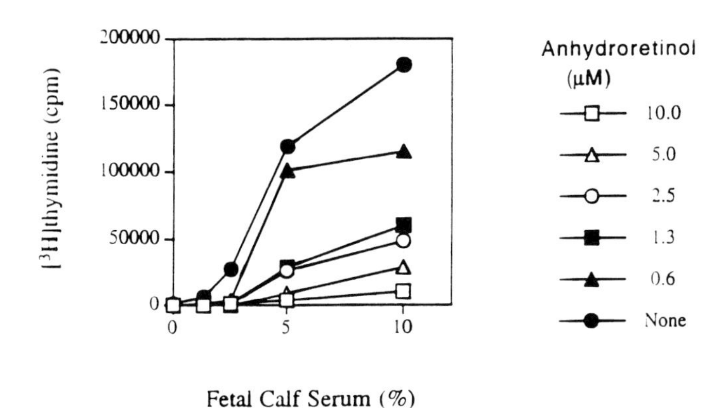

Fig. 6. Anhydroretinol blocks the activation of 3T3 cells by serum. Two-day starved cells were activated with serum in the presence and absence of various concentrations of anhydroretinol. Data of activation assay (mean and SDs of triplicate measurements) were less than 12%

To assess cell viability Chen at al. used WST-1, a dye that indicates mitochondrial activity, and found that addition of retinol to serum-starved cells resulted in an increased percentage of viable cells. They concluded that while PDGF initiates the cell cycle, retinol and 14-HRR ensure cell cycle progression by preventing cell death.

IV. Cell Growth vs. Cell Differentiation: Retinol Metabolism in HL-60 Cells

A large number of studies have focused on all-*trans* RA and its effects on the differentiation of various cell types including embryonic stem cells, keratinocytes, carcinomas, and leukemia cells (reviewed in Sect. C, this volume). The relationship, if any, between the cellular effects of RA and the effects of the newly characterized retinoid metabolites was studied using the HL-60 cell line. HL-60 cells are derived from a patient with acute promyelocytic leukemia and have the potential to differentiate to mature monocytes, granulocytes and osteoclasts. All-*trans* RA induces growth arrest and terminal differentiation to cells with a granulocyte phenotype (Breitman et al. 1980).

A number of retinol metabolites are present in HL-60 cells including all-*trans* RA, 14-HRR, and AR. In order to show that these retinoids have distinct specific effects on HL-60 cells, each one of these metabolites was applied exogenously (Eppinger et al. 1993). HL-60 cells were dependent on vitamin A for growth under serum-free conditions, and 14-HRR was able to substitute for retinol and to support cell proliferation. AR inhibited proliferation and behaved as a competitive inhibitor of 14-HRR. Supranormal levels of vitamin A led to cellular RA production and, as previously shown, to growth arrest and granulocyte differentiation. In contrast to RA, AR could not induce differentiation, nor could it inhibit RA-induced differentiation.

HL-60 cells have a large intracellular RE pool which allows them to withstand prolonged starvation (1 week) before dying. Analysis of endogenous retinoids revealed that the levels of RE decreased over time, when cells were transferred to serum-free medium, while the levels of 14-HRR remained relatively constant until the collapse of the culture (Fig. 7) (Eppinger et al. 1993). This led to the hypothesis that during starvation the RE pools are used up for the production of 14-HRR which is critical for cell survival. Cell death follows the depletion of endogenous RE and retinol pools from which 14-HRR is synthesized.

V. Intracellular Signaling

1. Anhydroretinol-Induced Cell Death

AR-induced cell death was morphologically indistinguishable from the cell death caused by retinol deficiency and was cell cycle-independent, i.e., there was no preferential cell death in any specific phase of the cell cycle (Buck et

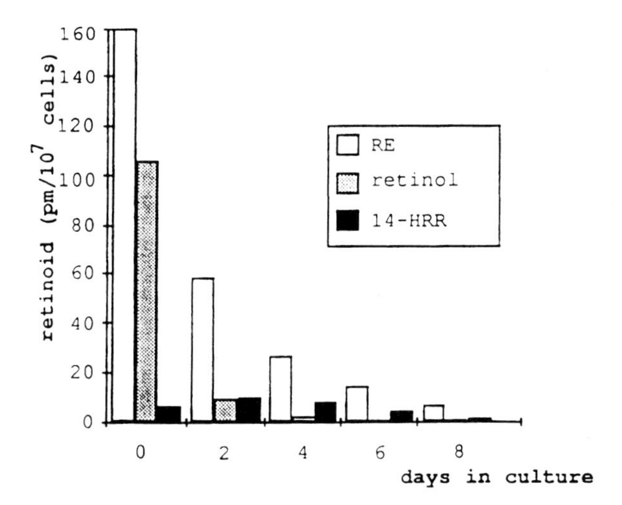

Fig. 7. Cellular retinoids of HL-60 over the course of culture in serum-free ITLB medium. Intracellular retinoids of HL-60 cells were labeled by exposing cells to a mixture of cold and [^{3}H]retinol overnight. Cells were then washed and reseeded in ITLB medium. Lipid extracts of cells of culture aliquots were HPLC analyzed immediately after reseeding and after 2, 4, 6, and 8 days of culture under retinol-free conditions. Shown are the amounts of labeled retinol derivatives recovered from cells of culture aliquots calculated from the specific activity of precursor [^{3}H]retinol, 14-HRR, retinol, and RE over time. (Reproduced with permission from EPPINGER et al. 1993; copyright Rockefeller University Press)

al. 1991a). The acceleration of cell death by AR was dose dependent, inducing within 4–8 h widespread surface blebbing, ballooning, and osmotic bursting of cells. The cell cytoplasm disintegrated and the vacuoles fused, while the nuclei, with their nucleoli, remained intact for prolonged periods of time (O'CONNELL et al. 1996). These observations were true for all the cell types tested so far, including B lymphoblastoid cells (BUCK et al. 1991a), T cells (O'CONNELL et al. 1996) and NIH 3T3 cells (Chen and Buck, unpublished data).

The finding that retinol and 14-HRR can rescue cells from death following AR treatment argues against generalized AR toxicity. AR-induced cell death appears to be a regulated type of cell death, but its characteristics are distinct from those of classical apoptosis. In ERLD cells, a mouse thymoma cell line which is very AR sensitive, AR treatment did not result in any DNA fragmentation or chromatin condensation both of which are hallmarks of classical apoptosis. There was, however, DNA damage as detected by the terminal deoxynucleotidyl transferase (TUNEL) assay. Microscopic inspection using the vital dye trypan blue showed that cell swelling and rupture occurred simultaneously with or prior to DNA damage (O'CONNELL et al. 1996).

Several important regulators of classical apoptotic death, including Bcl-2 and caspase inhibitors (reviewed in WHITE 1996), were tested for their ability to inhibit AR-induced cell death. The Bcl-2 protein inhibits several forms of apoptosis, though its mechanism of action is not yet clear. Caspases on the other hand, are proteinases activated in response to a number of cell death signals and are believed to mediate apoptosis. Bcl-2 overexpression in 3T3 cells had no effect on AR-induced cell death. Similarly, peptide inhibitors of the Ced-3-like family and the ICE family of caspases did not prevent cell death following AR treatment (Chen and Buck, unpublished data).

2. Evidence for a Transmembrane or Cytoplasmic Receptor

14-HRR and AR were initially thought to act similarly to all-*trans* RA, i.e., bind nuclear receptors to affect transcription of target genes. However, physiological concentrations of *retro*-retinoids were unable to transactivate not only the RARs and RXRs but also a number of other orphan nuclear receptors. In addition, AR-induced death was inhibited by neither actinomycin D, which inhibits transcription, nor cycloheximide, which inhibits protein translation, indicating that macromolecular synthesis is not a requirement for cell death (O'CONNELL et al. 1996).

Several pharmacological blockers of signal transduction including staurosporin (an inhibitor of protein C), methyl 2,5-dihydroxycinnamate, lavendustin A and genistein (inhibitors of tyrosine kinases) were also found to have no effect on AR-induced cell death. However, herbimycin A, an inhibitor of *src*-like tyrosine kinases, did prevent cell death (O'CONNELL et al. 1996). This finding supports the hypothesis that AR does not cause generalized cytotoxicity, but that it results in cell death by acting on specific molecular targets. It also suggests that its signaling pathway includes at least one tyrosine kinase, although it is possible that herbimycin A exerts its inhibitory effect via an as yet unidentified mechanism.

Retro-retinoids are relatively lipophilic compounds that can readily cross membrane barriers. The studies performed to date suggest that they may act on a membrane or a cytoplasmic receptor to activate (14-HRR) or inhibit (AR) a growth-stimulating signaling pathway. The components of this pathway are the subject of present research and their identification is likely to reveal new mechanisms of vitamin A action.

E. *Retro*-Retinoid Metabolism

Biosynthesis of a large number of retinoid metabolites requires many metabolizing enzymes catalyzing a variety of chemical reactions. Despite the fact that such enzyme activities have long been known, progress in the purification of proteins and in the cloning of their genes has been remarkably slow (reviewed in NAPOLI 1996). Lecithin-retinol acyltransferase, for example, a microsomal enzyme responsible for the esterification of retinol, has been studied extensively for its biochemistry, but cloning of a cDNA was only recently reported (RUIZ et al. 1999). Given the interest in all-*trans* RA most of the attention has focused on the biochemical pathway of RA synthesis (reviewed in NAPOLI 1996). Genes have been isolated for retinol dehydrogenases, which are responsible for the oxidation of retinol to retinal (CHAI et al. 1995a 1995b 1996; KEDISHVILLI et al. 1997; SIMON et al. 1994), as well as for retinal dehydrogenases, which can oxidize retinal to retinoic acid (CHEN et al. 1994; PENZES et al. 1997; WANG et al. 1996). Studies of some of these enzymes have led investigators to favor a model of retinol metabolism where the preferred substrate is holo-CRBP, i.e., retinol bound to cellular retinol binding protein (HERR and ONG 1992; POSCH et al. 1991).

Characterization of 14-HRR and AR, the novel retinol metabolites, provides yet another example where vitamin A is a pro-hormone. The fact that various retinol metabolites can act through distinct pathways to produce their specific cellular effects implies that vitamin A metabolism must be a critical point of regulation for the actions of vitamin A. Therefore to understand the pleiotropic effects of retinol it is important to characterize the retinol metabolizing enzymes that are part of the 14-HRR and AR biosynthetic pathways and to study their mechanism of action, their modes of regulation and the patterns of their spatial and temporal expression.

I. 14-Hydroxy-*Retro*-Retinol

14-HRR was originally purified from HeLa cells and has been subsequently found in a large number of cell types from insects to mammals including human B lymphocytes, fibroblasts, leukemia cells and *Drosophila* Schneider cells (BUCK et al. 1991b). 14-HRR is also present as a physiological retinol derivative in human plasma (ARNOLD et al. 1996) and in several tissues of mice, rats, and rabbits (TZIMAS et al. 1996a,b). When pregnant mice were fed teratogenic amounts of retinol, their embryos were shown to contain relatively high levels of 14-HRR (TZIMAS et al. 1996a). This is intriguing because all of the teratogenic effects of retinol have been attributed to RA. The elevated concentration of 14-HRR suggests it may account for some of the teratogenic potential of vitamin A (TZIMAS et al. 1996a).

Very little is known about the biochemical pathway(s) of 14-HRR production. All tested cell lines convert radiolabeled retinol to 14-HRR in a reaction that appears to be enzymatic. Phorbol esters, which activate protein kinase C, were found to increase 14-HRR production two- to fourfold (DERGUINI et al. 1995). In addition, in B and T lymphoblastoid cells, disulfiram, which modifies protein function by binding to thiol groups, inhibited the synthesis of 14-HRR in a dose-dependent manner (Buck, unpublished data).

II. 13,14-Dihydroxy-Retinol

13,14-Dihydroxy-retinol (DHR) (Fig. 8) is another recently identified β-series retinoid which was originally thought to be an intermediate in the biosynthesis of 14-HRR. It was found in the same cell lines as 14-HRR, and similar to

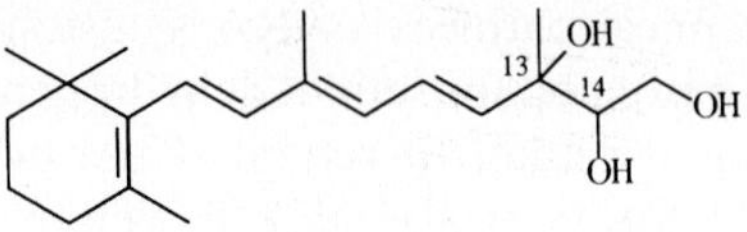

Fig. 8. Structure of 13,14-dihydroxy-retinol

14-HRR its production in lymphoblastoid cells was inhibited by disulfiram (Buck, unpublished data). Bioactivity studies revealed the effects of 13,14-DHR resembled those of 14-HRR (DERGUINI et al. 1995). It could replace retinol to support growth of promyelocytic HL-60 cells and activate T cells, and its effects were antagonized by AR in a dose-dependent manner. In addition, 13,14-DHR can be readily converted to 14-HRR by simple mild acid treatment.

In order to explore the hypothesis that 13,14-DHR is a precursor of 14-HRR, B lymphoblastoid cells were incubated with radiolabeled 13,14-DHR and the metabolic products were analyzed by HPLC (DERGUINI et al. 1995). Contrary to expectations, cells failed to convert 13,14-DHR to 14-HRR. At the same time they were able to convert cold retinol to 14-HRR, which indicated that the enzymatic pathways were intact. HeLa cells and T lymphocytes also failed to produce 14-HRR from cold 13,14-DHR. These experiments led to the conclusion that 13,14-DHR is a distinct end point of retinol metabolism and not a biosynthetic intermediate in the conversion of retinol to 14-HRR (DERGUINI et al. 1995).

III. Biosynthesis of Anhydroretinol

AR was initially purified from *Drosophila* S2M3 cells and has been identified in other insect and mammalian cell lines, such as *Spodoptera frugiperda* cells, mouse fibroblasts and human lymphoblastoid cells (BUCK et al. 1993). AR was found in significant amounts in the stellate cells (also referred to as lipocytes or Ito cells) of mammalian liver, and it was also detected in homogenized mouse lung tissue (Buck, unpublished data). With the exception of liver stellate cells, the amounts of AR present in mammalian cells were lower than the micromolar concentrations of AR needed to induce cell death in the bioactivity assays. However, in vitro experiments may require higher than physiological amounts of AR given the fact that AR is a very lipophilic compound. In addition, the production of AR can be tightly regulated within cells allowing for only very transient increases in intracellular AR that are difficult to detect.

Enzymatic production of cellular AR was suggested for the first time in 1979 when it was shown that spontaneously transformed mouse fibroblasts can metabolize radiolabeled retinol to AR and that this dehydratase activity was destroyed by boiling (BHAT et al. 1979). Almost 20 years later it was discovered that AR is the major retinol metabolite in a moth cell line, *Spodoptera frugiperda* Sf-21. This allowed the purification of a 41-kDa protein with retinol dehydratase activity (GRUN et al. 1996). The gene for the insect retinol dehydratase was cloned, and recombinant protein was expressed in *E. Coli* and shown to catalyze the conversion of retinol to AR. Both the native and the recombinant enzyme have very high affinity for retinol with the K_m and K_d values for retinol being in the low nanomolar range. This implies that retinol dehydratase can utilize retinol at its physiological intracellular concentrations. Moreover, retinol dehydratase seems to prefer as its substrate-free retinol over protein-bound retinol. This is an important finding because it is contrary to what has been reported previously for other retinol metabolizing enzymes.

The role that retinol dehydratase and AR play in Sf-21 cells, and in insect cells in general, is not known. Interestingly, the *Spodoptera frugiperda* cell line Sf-9, which is a subclone of Sf-21, did not show any AR production (GRUN et al. 1996). Northern blot analysis showed that retinol dehydratase mRNA was not present in Sf-9 cells, accounting for the absence of enzymatic activity (Grün and Buck, unpublished data). This raises the possibility that the expression of insect retinol dehydratase can be transcriptionally regulated.

Retinol dehydratase is homologous to the family of cytosolic sulfotransferases, and it represents the first insect homologue of this growing family of enzymes. A number of mammalian and plant sulfotransferases have been cloned and studied (reviewed in WEINSHILBOUM et al. 1997; FALANY 1997; VARIN et al. 1997). They catalyze bisubstrate reactions where a sulfonate group is transferred from 3′-phosphoadenosine-5′-phosphosulfate (PAPS) to the target substrate. Accordingly, retinol dehydratase uses PAPS as a cosubstrate (with a K_m value in the low micromolar range) (GRUN et al. 1996). The known sulfotransferases have distinct but overlapping substrate specificities and can bind a variety of structurally different compounds. We have also found that retinol dehydratase can bind and sulfate a number of compounds such as *p*-nitrophenol, dopamine, and estradiol, albeit with lower affinity than it binds retinol (Luz and Buck, unpublished data). Very recently the structure of the mouse estrogen sulfotransferase was solved providing important information about the topology of this class of enzymes (KAKUTA et al. 1997). Our laboratory has also obtained crystals of the insect retinol dehydratase with PAPS and with its nonphysiological substrate *p*-nitrophenol (Luz and Buck, unpublished data).

Since retinol dehydratase appears to be a sulfotransferase, it is reasonable to assume that AR production proceeds via retinyl sulfate (Fig. 9), although the existence of such an intermediate remains to be proven. Moreover, it is likely that mammalian retinol dehydratase is also a sulfotransferase. For this reason we have tested several known rat sulfotransferases but none were able to metabolize retinol to AR (Vakiani and Buck, unpublished data). Sulfation (also referred to as sulfonation) has been known for a long time to be an important phase II conjugation reaction in mammals that facilitates the excretion of a large number of endogenous compounds and xenobiotics. Recently it was shown to serve additional important functions, and deregulated

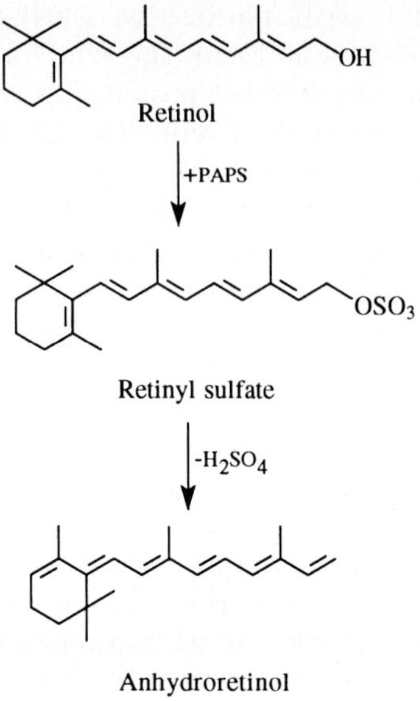

Fig. 9. Putative mechanism of anhydroretinol production

sulfotransferase expression was found to be associated with pathological states (reviewed in RUNGE-MORRIS 1997). Sulfation can reversibly inactivate bioactive compounds such as steroids, thyroid hormones and monoamine neurotransmitters, and thus it plays a role in regulating their biological activity. It can also serve to activate compounds, as in the case of minoxidil, where minoxidil sulfate is the active form. Moreover, a number of xenobiotics are converted by sulfotransferases to potent electrophiles that can mutagenize DNA (reviewed in GLATT 1997). Understanding the mechanism of AR production in mammalian cells may identify a role for sulfation in retinoid metabolism.

F. Conclusion

14-HRR and AR are the first members of the *retro*-retinoid family of compounds to be characterized with respect to their biological activity, and they constitute the first known naturally occurring agonist/antagonist pair of retinoids. They are present in a large number of cell types from insects to mammals, and their conservation over a long evolutionary period suggests they serve important cellular functions. The studies with mammalian cells implicate these two molecules in the regulation of cell survival and growth in a positive (14-HRR) or negative (AR) way. The cellular effects appear to be specific and to be mediated by a signaling cascade independent of RA. However, many questions remain to be answered. When and how are 14-HRR and AR produced, and how is their synthesis regulated? Is there a receptor molecule that both molecules can bind to and how does this receptor regulate cell death and cell survival? Much more needs to be learned about the metabolism and functions of the *retro*-retinoid metabolites before we can have a clear understanding of their physiological role, but research in this field is bound to expand our understanding of the mechanisms of vitamin A action.

Acknowledgements We thank Dr Lorraine J. Gudas and Dr Lonny Levin for their helpful suggestions. The work reported here from our laboratory was supported by grants from the American Cancer Society and the National Institutes of Health. J.B. is a Pew Scholar in the Biomedical Sciences.

References

Achkar CC, Derguini F, Blumberg B, Langston A, Levin AA, Speck J, Evans RM, Bolado J, Nakanishi K, Buck JB, Gudas LJ (1996) 4-Oxoretinol, a new natural ligand and transactivator of the retinoic acid receptors. Proc Natl Acad Sci USA 93:4879–4884

Arnold T, Tzimas G, Wittfoht W, Plonait S, Nau H (1996) Identification of 9-*cis*-retinoic acid, 9,13-di-*cis*-retinoic acid, and 14-hydroxy-4,14-*retro*-retinol in human plasma after liver consumption. Life Sci 59:PL169–177

Bhat PG, De Luca LM, Adamo S, Akalovsky I, Silverman-Jones CS, Peck GL (1979) Retinoid metabolism in spontaneously transformed mouse fibroblasts (Balb/c 3T12-3): enzymatic conversion of retinol to anhydroretinol. J Lipid Res 20:357–362

Blumberg B, Bolado J, Derguini F, Craig AG, Moreno TA, Chakravarti D, Heyman RA, Buck JB, Evans RM (1996) Novel retinoic acid receptor ligands in Xenopus embryos. Proc Natl Acad Sci USA 93:4873–4878

Breitman TR, Selonick SE, Collins SJ (1980) Induction of differentiation of the human promyelocytic leukemia cell line (HL-60) by retinoic acid. Proc Natl Acad Sci USA 77:2936–2940

Brooks RF, Howard M, Leake DS, Riddle PN (1990) Failure of platelet-derived growth factor plus insulin to stimulate sustained proliferation of Swiss 3T3 cells. Requirement for hydrocortisone, prostaglandin E1, lipoproteins, fibronectin and an unidentified component derived from serum. J Cell Sci 97:71–78

Buck J, Ritter G, Dannecker L, Katta V, Cohen LS, Chait BT, Hämmerling U (1990) Retinol is essential for growth of activated human B cells. J Exp Med 171:1613–1623

Buck J, Myc A, Garbe A, Cathomas G (1991a) Differences in the Action and Metabolism between Retinol and Retinoic Acid in B Lymphocytes. J Cell Biol 115:851–859

Buck J, Derguini F, Levi E, Nakanishi K, Hämmerling U (1991b) Intracellular Signaling by 14-Hydroxy-4,14-*Retro*-Retinol. Science 254:1654–1656

Buck J, Grün F, Derguini F, Chen Y, Kimura S, Noy N, Hämmerling U (1993) Anhydroretinol: A Naturally Occurring Inhibitor of Lymphocyte Physiology. J Exp Med 178:675–680

Chai X, Boerman MHEM, Zhai Y, Napoli JL (1995a) Cloning of a cDNA for liver microsomal NADP-dependent retinol dehydrogenase: A tissue-specific, short-chain alcohol dehydrogenase. J Biol Chem 270:3900–3904

Chai X, Zhai Y, Popescu G, Napoli JL (1995b) Cloning of a cDNA for a second retinol dehydrogenase, type II: expression of its mRNA relative to type I. J Biol Chem 270:28408–28412

Chai X, Zhai Y, Napoli JL (1996) Cloning of a rat cDNA encoding retinol dehydrogenase isozyme type III. Gene 169:219–222

Chen M, Achkar C, Gudas LJ (1994) Enzymatic Conversion of Retinal to Retinoic Acid by Cloned Murine Cytosolic and Mitochondrial Aldehyde Dehydrogenases. Mol Pharmacol 46:88–96

Chen Y, Derguini F, Buck J (1997) Vitamin A in serum is a survival factor for fibroblasts. Proc Natl Acad Sci USA 94:10205–10208

Corey EJ, Noe MC, Guzman-Perez A (1995) Catalytic Enantioselective Synthesis of (14R)-14-Hydroxy-4,14-*retro*-retinol from Retinyl Acetate. Tetrahedron Letters 36:4171–4174

Davis CY, Sell JL (1983) Effect of all-*trans* Retinol and Retinoic Acid Nurtiture on the Immune System of Chicks. J Nutr 113:1914–1919

Derguini F, Nakanishi K, Buck J, Hämmerling U, Grün F (1994a) Spectroscopic Studies of Anhydroretinol, an Endogenous Mammalian and Insect *retro*-Retinoid. Angew Chem Int Ed Engl 33:1837–1839

Derguini F, Nakanishi K, Hämmerling U, Buck J (1994b) Intracellular Signaling Activity of Synthetic (14R)-, (14S)-, and (14RS)-14-Hydroxy-4,14-*retro*-retinol. Biochemistry 33:623–628

Derguini F, Nakanishi K, Hämmerling U, Chua R, Eppinger T, Levi E, Buck J (1995) 13,14-Dihydroxy-retinol, a New Bioactive Retinol Metabolite. J Biol Chem 270:18875–18880

Edisbury JR, Gillam AM, Heilbron IM, Morton RA (1932) Absorption Spectra of Substances derived from Vitamin A. Biochem J 26:1164–1173

Embree ND (1939) The occurrence of cyclized vitamin A in fish liver oils. J Biol Chem 128:187–198

Eppinger TM, Buck J, Hämmerling U (1993) Growth Control or Terminal Differentiation: Endogenous Production and Differential Activities of Vitamin A Metabolites in HL-60 Cells. J Exp Med 178:1995–2005

Falany CN (1997) Enzymology of human cytosolic sulfotransferases. FASEB 11:206–216

Garbe A, Buck J, Hämmerling U (1992) Retinoids Are Important Cofactors in T Cell Activation. J Exp Med 176:109–117

Glatt H (1997) Bioactivation of mutagens via sulfation. FASEB 11:314–321

Grün F, Noy N, Hämmerling U, Buck J (1996) Purification, Cloning, and Bacterial Expression of Retinol Dehydratase from *Spodoptera frugiperda*. J Biol Chem 271:16135–16138

Herr FM, Ong DE (1992) Differential interaction of lecithin-retinol acyltransferase with cellular retinol-binding proteins. Biochemistry 31:6748–6755

Hofmann C, Eichele G (1994) Retinoids in Development. In: Sporn MB, Roberts AB, Goodman DS (eds) THE RETINOIDS: Biology, Chemistry and Medicine. 2nd edn. Raven, New York, p 387

Holley RW, Kiernan JA (1968) “Contact inhibition” of cell division in 3T3 cells. Proc Natl Acad Sci USA 60:300–304

Howell JM, Thompson JN, Pitt GA (1963) Histology of the lesions produced in the reproductive tract of animals fed a diet deficient in vitamin A alcohol but containing vitamin A acid. J Reprod Fertl 5:159–178

Kakuta Y, Pedersen LG, Carter CW, Negishi M, Pedersen LC (1997) Crystal structure of estrogen sulphotransferase. Nat Struct Biol 4(11):904–908

Kedishvilli NY, Gough WH, Chernoff EAG, Hurley TD, Stone CL, Bowman KD, Popov KM, Bosron WF, Li TK (1997) cDNA Sequence and Catalytic properties of a Chick Embryo Alcohol Dehydrogenase that Oxidizes Retinol and $3\beta,5\alpha$-Hydroxysteroids. J Biol Chem 272:7494–7500

Love JM, Gudas LJ (1994) Vitamin A, Differentiation and Cancer. Current Opin Cell Biol 6:825–831

MacDonald PN, Ong DE (1988) Evidence for a lecithin-retinol acyltransferase activity in the rat small intestine. J Biol Chem 263:12478–12482

Mangelsdorf DJ, Umesono K, Evans RM (1994) The Retinoid Receptors. In: Sporn MB, Roberts AB, Goodman DS (eds) THE RETINOIDS: Biology, Chemistry and Medicine. 2nd edn. Raven, New York, p 319

Napoli JI (1996) Biochemical Pathways of Retinoid Transport, Metabolism, and Signal Transduction. Clin Immunol Immunopathol 80:S52–62
O'Connell MJ, Chua R, Hoyos B, Buck J, Chen Y, Derguini F, Hämmerling U (1996) *Retro*-Retinoids in Regulated Cell Growth and Death. J Exp Med 184:549–555
Ong DE, Chytil F (1978) Cellular Retinol-binding Protein from Rat Liver. J Biol Chem 253:828–832
Ong DE, Newcomer ME, Chytil F (1994) Cellular Retinoid-Binding Proteins. In: Sporn MB, Roberts AB, Goodman DS (eds) THE RETINOIDS: Biology, Chemistry and Medicine. 2nd edn. Raven, New York, p 283
Oroshnik O, Karmas G, Mebane AD (1952) Synthesis of Polyenes. I. Retrovitamin A Methyl Ether. Spectral Relationships of the β-Ionylidene and Retroionylidene Series. J Am Chem Soc 74:295–304
Penzes P, Wang X, Sperkova Z, Napoli JL (1997) Cloning of a rat retinal dehydrogenase isozyme type I and its expression in E. Coli. Gene 191:167–172
Posch KC, Boerman MHEM, Burns RD, Napoli JL (1991) Holo-cellular retinol binding protein as a substrate for microsomal retinal synthesis. Biochemistry 30:6224–6230
Ross AC, Hämmerling UG (1994) Retinoids and the Immune System. In: Sporn MB, Roberts AB, Goodman DS (eds) THE RETINOIDS: Biology, Chemistry and Medicine. 2nd edn. Raven, New York, p 521
Ruiz A, Winston A, Lim Y, Gilbert BA, Rando RR, Bok D (1999) Molecular and Biochemical Characterization of Lecithin Retinol Acyltransferase. J Biol Chem 274:3834–3841
Runge-Morris MA (1997) Regulation of expression of the rodent cytosolic sulfotransferases. FASEB 11:109–117
Schreckenbach T, Walckhoff B, Oesterhelt D (1977) Studies on the Retinal-Protein Interaction in Bacteriorhodopsin. Eur J Biochemistry 76:499–511
Shantz EM, Cawley JD, Embree ND (1943) Anhydro ("Cyclized") Vitamin A. J Am Chem Soc 65:901–906
Simon A, Hellman U, Wernstedt C, Eriksson U (1994) The Retinal Pigment Epithelial-specific 11-*cis* Retinol Dehydrogenase Belongs to the Family of Short Chain Alcohol Dehydrogenases. J Biol Chem 270:1107–1112
Sommer A, Tarwotjo I, Hussaini G, Susanto D (1983) Increased mortality in children with mild vitamin A deficiency. Lancet 327:1169–1173
Soprano DR, Blaner WS (1994) Plasma Retinol-Binding Protein. In: Sporn MB, Roberts AB, Goodman DS (eds) THE RETINOIDS: Biology, Chemistry and Medicine. 2nd edn. Raven, New York, p 257
Tzimas G, Collins MD, Nau H (1996a) Identification of 14-hydroxy-4,14-*retro*-retinol as an in vivo metabolite of vitamin A. Biochim Biophys Acta 1301:1–6
Tzimas G, Collins MD, Burgin H, Hummler H, Nau H (1996b) Embryotoxic doses of vitamin A to rabbits result in low plasma but high embryonic concentrations of all-*trans*-retinoic acid: risk of vitamin A exposure in humans. J Nutr 126:2159–2171
Underwood BA (1994) Vitamin A in Human Nutrition: Public Health Considerations. In: Sporn MB, Roberts AB, Goodman DS (eds) THE RETINOIDS: Biology, Chemistry and Medicine. 2nd edn. Raven, New York, p 21
Varin L, Marsolais F, Richard M, Rouleau M (1997) Biochemistry and molecular biology of plant sulfotransferases. FASEB 11:517–525
Varma TNR, Erdody P, Murray TK (1965) The utilization of Anhydrovitamin A by the Vitamin A Deficient Rat. Biochim Biophys Acta 104:71–77
Wald G (1968) Molecular basis of visual excitation. Science 162:230–239
Wang X, Penzes P, Napoli JL (1996) Cloning of a cDNA encoding an aldehyde dehydrogenase and its expression in Escheria Coli: recognition of retinal as substrate and role in retinoic acid synthesis. J Biol Chem 271:16288–16293
Weinshilboum RM, Otterness DM, Aksoy IA, Wood TC, Her C, Raftogianis RB (1997) Sulfotransferase molecular biology: cDNAs and genes. FASEB 11:3–14
White E (1996) Life, death, and the pursuit of apoptosis. Genes & Development 10:1–15
Wolbach SB, Howe PR (1925) Tissue Changes following deprivation of fat-soluble A vitamin. J Exp Med 42:753–777

CHAPTER 4

Retinoic Acid Synthesis and Metabolism

W.S. BLANER, R. PIANTEDOSI, A. SYKES, and S. VOGEL

A. Introduction

It is well known from the early literature that liver alcohol dehydrogenase catalyzes the oxidation of retinol to retinal (BLANER and OLSON 1994). Similarly, it was established from early work that several members of the aldehyde dehydrogenase family of enzymes are able to catalyze irreversibly the oxidation of retinal to retinoic acid (BLANER and OLSON 1994). Thus, it has long been understood that retinoic acid synthesis from its precursor retinol involves a two step oxidation that resembles the oxidation of ethanol to acetic acid.

Investigations carried out in the 1960s and 1970s provided insights into products that are formed as the result of oxidative and/or conjugative metabolism of retinoic acid (FROLIK 1984; BLANER and OLSON 1994). Most of these early investigations were aimed at identifying retinoic acid metabolites that are formed by tissues in vivo or by tissue extracts in vitro (FROLIK 1984). This work was seminal for identifying important retinoic acid metabolites such as 4-hydroxy-retinoic acid, 4-oxo-retinoic acid and retinoyl-β-glucuronide (FROLIK 1984; BLANER and OLSON 1994). Taken together, this early literature provides a sound framework for our present understanding of retinoic acid synthesis and its subsequent metabolism.

In the decade of the 1990s there have been many substantial advances towards understanding the biochemical processes that are important for retinoic acid synthesis and for its metabolism. Among these is the understanding that intracellular binding proteins for both retinol and retinoic acid play important roles in regulating/facilitating the oxidation of retinol to retinoic acid and also the oxidative metabolism of retinoic acid. Many previously unknown enzymes that catalyze either retinol or retinal oxidation have been identified. These are now the subject of active study aimed at understanding their roles in the synthesis of retinoic acid. A retinoic acid inducible cytochrome P450 isoform that catalyzes the oxidative metabolism of retinoic acid has been identified and is now being characterized. Although these recent advances greatly extend our understanding of retinoic acid synthesis and metabolism, there is much important information that is still missing.

This chapter focuses primarily on recent literature exploring the roles of retinoid-binding proteins in retinoic acid formation and metabolism, the enzymes responsible for retinoic acid synthesis, and the enzymes or enzyme systems responsible for the oxidative metabolism of retinoic acid. The older literature in this area is well summarized by FROLIK (1984) and BLANER and OLSON (1994).

B. Cellular and Extracellular Retinoid Binding Proteins

As outlined in Chap. 1 of this volume, retinoids lack appreciable solubility in water and consequently must be bound to proteins if they are to be present in the aqueous environment within the cell. Many retinoid-specific binding proteins have been identified in higher animals (ONG et al. 1994). Some of these are found solely intracellularly, whereas others are found only extracellularly. Some of these bind solely retinol, others bind only retinoic acid and others can bind retinol, retinal, or retinoic acid. A partial listing of some of the better-characterized retinoid-binding proteins and some of their properties is provided in Table 1.

I. Cellular Retinol Binding Proteins

In addition to being involved in retinyl ester formation and retinyl ester hydrolysis (see Chap. 2 for more details), cellular retinol-binding protein, type I (CRBP I) has been proposed to play a critical role in regulating the oxidation of retinol and retinal (BOERMAN and NAPOLI 1995; CHAI et al. 1995; NAPOLI 1996; CHAI and NAPOLI 1996). As detailed in Sect. C.I, several microsomal enzymes, termed retinol dehydrogenase types I, II, and III [RoDH(I), RoDH(II) and RoDH(III)], have been identified in rat tissues and shown to have substrate preferences for retinol bound to CRBP I, as opposed to unbound retinol (POSCH et al. 1991; BOERMAN and NAPOLI 1995; CHAI et al. 1995; NAPOLI 1996; CHAI and NAPOLI 1996; GOUGH et al. 1998). These microsomal retinol dehydrogenases catalyze the $NADP^+$-dependent oxidation of retinol to retinal. Moreover, retinal dehydrogenases that are able to oxidize retinal bound to CRBP I have also been identified for the rat and mouse (POSCH et al. 1992; EL AKAWI and NAPOLI 1994; NAPOLI 1996; YAMAMOTO et al. 1998). This suggests that retinal-CRBP I formed through oxidation of retinol-CRBP I by the microsomal retinol dehydrogenases can be directly oxidized to retinoic acid through the actions of retinal dehydrogenases (NAPOLI 1996). Thus, CRBP I may well play a role in channeling the metabolic activation of all-*trans*-retinol to all-*trans*-retinoic acid. More details regarding these enzymes and processes are provided in Sect. C.I and C.II.

II. Cellular Retinoic Acid Binding Proteins

A direct role for cellular retinoic acid-binding protein, type I (CRABP I) in the oxidative metabolism of retinoic acid has been proposed by FIORELLA and NAPOLI (1991). When all-*trans*-retinoic acid is bound to CRABP I, microsomal

Table 1. Partial listing of intracellular and extracellular retinoid binding proteins and their binding characteristics

Protein	Family[a]	Ligand	K_d	References
Intracellular binding proteins				
CRBP I	FABP	All-*trans*-retinol	16 n*M*	Ong and Chytil 1978
			10 n*M*	Levin et al. 1988
			13 n*M*	Noy and Blaner 1991
		13-*Cis*-retinol	40 n*M*	MacDonald and Ong 1987
		All-*trans*-retinal	Weak binding	Li et al. 1991
		All-*trans*-retinoic acid	No binding	Levin et al. 1988
CRBP II	FABP	All-*trans*-retinol	10 n*M*	MacDonald and Ong 1987
		All-*trans*-retinal	10 n*M*	MacDonald and Ong 1987
		All-*trans*-retinoic acid	No binding	Cheng et al. 1991
		Palmitic acid	No binding	Cheng et al. 1991
CRABP I	FABP	All-*trans*-retinoic acid	4–10 n*M*	Fiorella and Napoli 1991; Ong and Chytill 1978
		13-*Cis*-retinoic acid	156 n*M*	Fiorella and Napoli 1991
		All-*trans*-retinal	No binding	Fiorella and Napoli 1991
		All-*trans*-retinol	No binding	Fiorella and Napoli 1991
		4-Hydroxyretinoic acid	17 n*M*	Fiorella and Napoli 1991
		4-Oxoretinoic acid	20 n*M*	Fiorella and Napoli 1991
		4-Oxo-13-*cis*-retinoic acid	183 n*M*	Fiorella and Napoli 1991
		16-Hydroxy-4-oxo-retinoic acid	20 n*M*	Fiorella and Napoli 1991
		18-Hydroxy-retinoic acid	22 n*M*	Fiorella and Napoli 1991
		3,4-Didehydro-retinoic acid	6 n*M*	Fiorella and Napoli 1991
CRABP II	FABP	All-*trans*-retinoic acid	64 n*M*	Bailey and Siu 1988
		All-*trans*-retinol	No binding	Bailey and Siu 1988
		All-*trans*-retinal	No binding	Bailey and Siu 1988
		13-*Cis*-retinoic acid	No binding	Bailey and Siu 1988
CRalBP	Opsin	11-*Cis*-retinal	90 n*M*	Livrea and Tesoriere 1990
		11-*Cis*-retinol	NR	Saari and Bredberg 1988
		9-*Cis*-retinol	NR	Saari and Bredberg 1988
		All-*trans*-retinol	No binding	Saari and Bredberg 1988
		All-*trans*-retinal	No binding	Saari and Bredberg 1988
		11-*Cis*-retinyl palmitate	No binding	Livrea et al. 1987

Table 1. *Continued*

Protein	Family[a]	Ligand	K_d	References
Extracellular binding proteins				
RBP	Lipocalin	All-*trans*-retinol	20 n*M*	Noy and Blaner 1991
IRBP	Lipocalin	All-*trans*-retinol	1 μ*M*	Fong et al. 1984
		All-*trans*-retinoic acid	NR	Fong et al. 1994
		Cholesterol	NR	Fong et al. 1994
		α-Tocopherol	>1 μ*M*	Alvarez et al. 1987
		Stearic acid	0.36 μ*M*	Putilina et al. 1993
		Palmitic acid	0.35 μ*M*	Putilina et al. 1993
β-TRACE	Lipocalin	All-*trans*-retinoic acid	80 n*M*	Tanaka et al. 1997
		9-*Cis*-retinoic acid	80 n*M*	Tanaka et al. 1997
		All-*trans*-retinal	70 n*M*	Tanaka et al. 1997
		13-*Cis*-retinal	70 n*M*	Tanaka et al. 1997
		All-*trans*-retinol	No binding	Tanaka et al. 1997
E-RABP	Lipocalin	All-*trans*-retinoic acid	–	Newcomer and Ong 1990

CRBP I and CRBP II, cellular retinol-binding protein, types I and II, respectively; CRABP I and CRABP II, cellular retinoic acid-binding protein, types I and II, respectively; CRalBP, cellular retinal-binding protein; RBP, retinol-binding protein; IRBP, interphotoreceptor retinol-binding protein; E-RABP, epididymal retinoic acid-binding protein; FABP, fatty acid-binding protein family; NR, binding of the ligand was reported but no dissociation constant given.

[a] Retinoid-binding proteins are identified as belonging to one of three different protein families. These are listed for each protein under this heading.

enzymes of rat testes catalyze the conversion of retinoic acid to 3,4-didehydro-, 4-hydroxy-, 4-oxo-, 16-hydroxy-4-oxo-, and 18-hydroxy-retinoic acids. Ketoconazole inhibited oxidation by testis microsomes of both free and CRABP I bound retinoic acid suggesting that cytochrome P450 isozymes are involved in this metabolism. Based on these data, it appears that CRABP I may well play a direct role in the oxidative metabolism of all-*trans*-retinoic acid. Furthermore, the binding of all-*trans*-retinoic acid to CRABP I may provide a mechanism for discriminating metabolically between all-*trans*- and 13-*cis*-and 9-*cis*-retinoids, which do not bind CRABP I well.

III. Other Retinoic Acid Binding Proteins

At least two of the extracellular retinoic acid-binding proteins, epididymal retinoic acid-binding protein (E-RABP) (Newcomer and Ong 1990) and β-trace (Eguchi et al. 1997; Tanaka et al. 1997), are known to bind all-*trans*-retinoic acid with very high affinity (see Table 1). Both of these proteins (as well as serum retinol-binding protein) are members of the lipocalin family of extracellular lipid-binding proteins (Flower 1996). The high binding affinities of E-RABP and β-trace for retinoic acid suggest that these proteins may be involved physiologically in retinoic acid transport and/or metabolism. Recently published work indicates that the targeted disruption of the mouse β-trace gene results in the lack of tactile pain (allodynia) for β-trace deficient mice; however, it is still unclear whether this phenotype is in some way linked to retinoic acid transport and/or metabolism (Eguchi et al. 1999). Thus at present there is only very limited experimental evidence supporting the possibility that either ERABP or β-trace is essentially involved in retinoic acid transport or metabolism. If we are to understand better the physiological roles of these proteins in retinoic acid transport and/or metabolism in the living organism, further study of the proteins and their in vivo actions is required.

C. Retinoic Acid Synthesis

It has long been known that the relatively nonspecific zinc-dependent alcohol dehydrogenase of liver can catalyze the oxidation of retinol to retinal (Blaner and Olson 1994; Duester 1996: Napoli 1996), and that aldehyde oxidase can convert retinal to retinoic acid (Blaner and Olson 1994; Duester 1996; Napoli 1996). Because retinoic acid in tiny amounts shows such potent physiological actions, however, one must question the role of abundant enzyme systems, which possess relatively broad substrate specificities (e.g., alcohol dehydrogenase, aldehyde dehydrogenase, and aldehyde oxidase), in forming retinoic acid. This question is now a focus of considerable research and debate that are described below in Sect. C.I.

There have been tremendous gains made in the past decade for identifying enzymes, many of which had been previously unknown that are able to catalyze either retinol or retinal oxidation, essential steps that are needed for forming retinoic acid. Nevertheless, the physiological importance of each of these enzymes for catalyzing the formation of all-*trans*-retinoic acid from all-*trans*-retinol has not been unequivocally established. This is because most of the work identifying enzymes that are able to catalyze retinol or retinal oxidation has been carried out in vitro and there is only limited work linking this biochemical data with physiological processes within the living organism.

A scheme summarizing the metabolic pathway thought to be responsible for the formation of all-*trans*- and 9-*cis*-retinoic acid from all-*trans*-retinol is provided for the reader in Fig. 1.

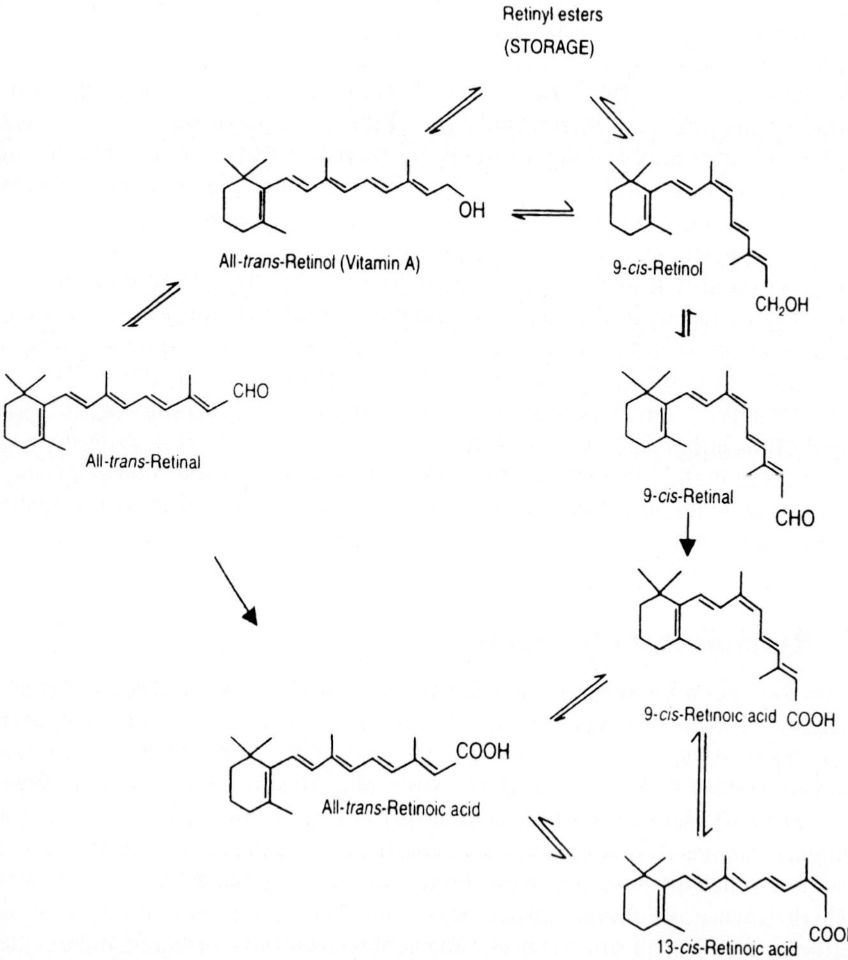

Fig. 1. Metabolic pathway for the formation of all-*trans*-, 13-*cis*- and 9-*cis*-retinoic acid from all-*trans*-retinol. The enzymes and factors involved in these metabolic conversions are described in the text

I. Oxidation of Retinol

The enzymes that catalyze retinol oxidation to retinal, the first of the two oxidative steps needed for retinoic acid formation, have not been unequivocally established. Enzymes from two distinct enzyme families have been proposed as being important for catalyzing this reaction. Members of the family of cytosolic medium-chain alcohol dehydrogenases (ADHs) have been demonstrated to catalyze retinol oxidation in vitro (BOLEDA et al. 1993; YANG et al. 1994; DUESTER 1996; NAPOLI 1996). Moreover, a very significant body of circumstantial evidence has accumulated that supports the idea that some of these enzymes are importantly involved in retinoic acid formation in vivo. Members of a second family of enzymes, the short-chain alcohol dehydrogenase (SCAD) family, have also been proposed as being physiologically relevant for catalyzing retinol oxidation (DUESTER 1996; NAPOLI 1996). [The SCAD family is also known in the literature as the short-chain dehydrogenase/reductase (SCDR) family (JÖRNVALL et al. 1995).] The members of this enzyme family that catalyze retinol oxidation are present in cells and tissues at relatively low concentrations and are associated with membrane fractions. Several of the SCADs which oxidize retinol prefer as a substrate, over unbound retinol, retinol bound to CRBP I (or for the case of the 11-*cis*-retinol dehydrogenase present in the eye, cellular retinal-binding protein (CRalBP), see Chap. 20 of this volume for more details) (DUESTER 1996; NAPOLI 1996). This is unlike the cytosolic ADHs, which require unbound retinol as their substrate (DUESTER 1996; NAPOLI 1996). From presently available data, it is not possible to conclude whether only some or all of the enzymes described in the literature as being important for retinoic acid formation are indeed physiologically essential. The discussion below is divided so as to focus separately on possible roles of the cytosolic medium-chain ADHs and the membrane bound SCADs in catalyzing retinol oxidation.

1. Role of Cytosolic Dehydrogenases

In vitro, all-*trans*-retinol is oxidized to all-*trans*-retinoic acid by multiple enzymes present in the cytosol or in microsomes (DUESTER 1996; NAPOLI 1996). The cytosolic activity has been linked to the medium-chain ADHs that have 40 kDa subunits (BOLEDA et al. 1993). Medium-chain ADHs are encoded in humans by nine different genes. These ADH isoenzymes have been grouped into six classes based on their catalytic properties and primary structures (JÖRNVALL and HÖÖG 1995). All members of this enzyme family are dimeric zinc metalloenzymes that require NAD^+ for catalyzing the oxidation of a variety of primary, secondary and a number of cyclic alcohols (JÖRNVALL and HÖÖG 1995). Several human and rat ADH isozymes (classes I, II, and IV) oxidize retinol to retinal in vitro (DUESTER 1996; NAPOLI 1996). The class IV ADH human isozyme (ADH4) is the most catalytically efficient isozyme for retinol oxidation (YANG et al. 1994) and consequently it has been proposed that ADH4 contributes importantly to retinoic acid biosynthesis in vivo.

The expression pattern of ADH4 appears to support this hypothesis. Ang et al. (1996a,b) have demonstrated in the developing mouse embryo that the pattern of expression of ADH4 overlaps both temporally and spatially the pattern of retinoic acid distribution within the embryo, thereby providing circumstantial in vivo evidence associating ADH4 with retinoic acid presence. Investigations of class I alcohol dehydrogenase (ADH1) and ADH4 expression in the developing mouse embryo were made employing both transgenic mice expressing lacZ under control of the mouse ADH4 promoter and immunocytochemical detection of ADH1 and ADH4. These studies demonstrate that both of these ADH isoforms are present in embryonic day 11.5 adrenal blastomas (Haselbeck and Duester 1998a). The presence of both ADH1 and ADH4 during the earliest stages of adrenal gland development, combined with the earlier observation of high levels of retinoic acid in the embryonic adrenal gland, has led these investigators to suggest that one of the earliest functions of ADH1 and ADH4 is to provide an embryonic endocrine source of retinoic acid for growth and development (Haselbeck et al. 1997; Haselbeck and Duester 1998a). Further investigations employing the ADH4-lacZ transgenic mice showed that ADH4 expression was highly localized in the brain and craniofacial region of the mouse embryo as early as embryonic days 8.5–9.5, during neurulation (Haselbeck and Duester 1998b). At day 8.5 ADH4-lacZ expression was noted in several dispersed regions throughout the head and by embryonic day 9.5 expression was evident in regions that corresponded to the otic vesicles and migrating neural crest cells, particularly the mesencephalic, trigeminal, facial and olfactory neural crest (Haselbeck and Duester 1998b). Since it is well established that retinoic acid and the retinoic acid nuclear receptors are important for directing this stage of mouse embryogenesis, especially for central nervous system and craniofacial development, Haselbeck and Duester (1998b) take their observations to be strong circumstantial evidence that ADH4 is importantly involved in retinoic acid synthesis in the mouse embryo.

Work by Parés and colleagues (Boleda et al. 1993) has shown that ADH4 purified from rat stomach is able to catalyze the oxidation of both ω-hydroxy fatty alcohols and free retinol. Later work from this group has demonstrated that both all-*trans*-retinol and 9-*cis*-retinol but not 13-*cis*-retinol are good substrates for human ADH4 (Allali-Hassani et al. 1998). The apparent K_m values of purified human gastric ADH4 for all-*trans*-retinol and 9-*cis*-retinol were reported to be, respectively, 15 ± 4 and $36 \pm 5\,\mu M$. The apparent K_m values for the reverse (reduction) reactions were very similar to those determined for the two retinol isomers (Allali-Hassani et al. 1998). In addition, both all-*trans*-retinoic acid and 13-*cis*-retinoic acid were found to be potent competitive inhibitors of ADH4 catalyzed retinol oxidation with K_i values in the range of $3–10\,\mu M$.

Investigations by Kedishvili et al. (1998) employing purified recombinant human ADH4 and purified recombinant rat CRBP I have demonstrated that retinol bound to CRBP I cannot be channeled to the active site of ADH4.

These investigators concluded that the contribution of ADH isozymes to retinoic acid biosynthesis depends on the amount of free retinol present in a cell. Since most retinol within a cell is thought to be bound to CRBP I and since ADH1 and ADH4 catalyze the oxidation of free retinol and not retinol bound to CRBP I, this raises a question as to the physiological significance of ADH1 and ADH4 for forming retinoic acid (DUESTER 1996; NAPOLI 1996). Alternatively, if ADH1 and ADH4 do play a physiological role in retinoic acid synthesis, this would suggest that CRBP I does not play a direct role to facilitate this process.

2. Role of Microsomal Dehydrogenases

Other reports in the recent literature indicate that oxidation of all-*trans*-retinol to all-*trans*-retinal is catalyzed by microsomal enzymes that can use all-*trans*-retinol bound to CRBP I as a substrate. These microsomal enzymes are members of the SCAD family of enzymes. To date, over 60 members of the SCAD enzyme family have been identified (BAKER 1994; JÖRNVALL et al. 1995; BAKER 1998). Members of the SCAD family consist of peptides of approximately 28–32 kDa, and do not require zinc for activity. Many of the members of this enzyme family are reported to catalyze oxidations/reductions of hydroxysteroids or prostaglandins (BAKER 1994; JÖRNVALL et al. 1995; BAKER 1998). Thus, the SCAD family of enzymes appear to be importantly involved in the metabolism of several potent classes of highly bioactive molecules.

Napoli and colleagues have cloned and characterized three microsomal retinol dehydrogenases from rat liver [termed retinol dehydrogenase, type I (RoDH(I)], type II [RoDH(II)] and type III [RoDH(III))] (POSCH et al. 1991; BOERMAN and NAPOLI 1995; CHAI et al. 1995; CHAI and NAPOLI 1996). Each of these dehydrogenases recognizes all-*trans*-retinol bound to CRBP I as substrate. Since most retinol within cells is bound to CRBP I, this substrate specificity suggests that these enzymes are physiologically relevant for retinol oxidation. Sequence analysis indicates that these enzymes are 82% identical to each other and highly similar to other members of the SCAD family. Each requires $NADP^+$ as an electron acceptor and is expressed most prominently in liver. RoDH(I) is the best studied isoform and is also reported to be present in kidney, brain, lung and testis but at levels which are less than 1% of that measured in liver. RoDH(II) is also expressed in kidney, brain, lung and testis at levels which are 25, 8, 4, and 3%, respectively, of that observed in liver (CHAI et al. 1996), whereas RoDH(III) is expressed solely in liver (CHAI and NAPOLI 1996). Interestingly, RoDH(I) does not use 9-*cis*-retinol as substrate (POSCH et al. 1991; BOERMAN and NAPOLI 1995;). The substrate specificities of RoDH(II) and RoDH(III) for different retinol isomers have not been reported (CHAI et al. 1995; CHAI and NAPOLI 1996). Based on the properties of these three enzymes it seems likely that they are physiologically involved in the oxidation of all-*trans*-retinol to all-*trans*-retinal, the first step of retinoic acid formation.

In situ hybridization studies by ZHAI et al. (1997) have demonstrated that CRBP I mRNA and mRNA species for RoDH(I) and RoDH(II) are coexpressed in adult rat hepatocytes and in the proximal tubules of the rat renal cortex. In the rat testis, CRBP I and RoDH(I) and RoDH(II) also were coexpressed in Sertoli cells with weaker coexpression in spermatogonia and primary spermatocytes (ZHAI et al. 1997). Since CRBP I and RoDH(I) and/or RoDH(II) expression occur in the same cellular loci in vivo, ZHAI et al. (1997) proposed that these data support the hypothesis that holo-CRBP I serves as a substrate for RoDH isozyme catalyzed retinoic acid synthesis.

HAESELEER et al. (1998) have reported the cloning and characterization of another SCAD that is able to catalyze oxidation of all-*trans*-retinol and reduction of all-*trans*-retinal. This enzyme, termed retinol short-chain dehydrogenase/reductase (retSDR1), was cloned from human, bovine and mouse retinal tissue cDNA libraries (HAESELEER et al. 1998). Human, bovine and murine retSDR1 show approximately 35% amino acid similarity to rat RoDH(I), RoDH(II) and RoDH(III). Human retSDR1 was found by northern blot analysis to be expressed in heart, liver, kidney, pancreas and retina. Recombinant retSDR1 catalyzes reduction of all-*trans*-retinal but not 11-*cis*-retinal suggesting that retSDR1 shows specificity for all-*trans*-retinoids. It was proposed by HAESELEER et al. (1998) that retSDR1 plays an important role in the visual cycle acting to reduce the bleached visual pigment, all-*trans*-retinal, to all-*trans*-retinol.

Rat RoDH(II) has been purified and characterized by IMAOKA et al. (1998) as a binding protein that is associated with cytochrome P4502D1 (CYP2D1) in the liver. These investigators were initially interested in understanding the identity of a protein that copurified with CYP2D1 through a series of column chromatography steps. Upon isolation and cDNA cloning of the protein, IMAOKA et al. (1998) recognized that that they had purified the same protein described by CHAI et al. (1995). IMAOKA et al. (1998) have demonstrated that recombinant RoDH(II) binds tightly to CYP2D1 even in the presence of 1% sodium cholate. Since CYP2D1 contributes to steroid metabolism and can hydroxylate testosterone, estrogen and cortisol, IMAOKA et al. (1998) suggested that the binding of RoDH(II) to CYP2D1 may be important for the metabolism of diverse bioactive substances such as retinoids and steroids.

Similar linkages between the metabolism of retinoids and steroids had been suggested a bit earlier by BISWAS and RUSSELL (1997) who studied 17β- and 3α-hydroxysteroid dehydrogenases cloned from rat and human prostate. BISWAS and RUSSELL (1997) recognized that the human prostate hydroxysteroid dehydrogenases shared a high degree of primary sequence homology with rat RoDH(I). This observation led these investigators to hypothesize that the microsomal retinol dehydrogenases might use hydroxysteroids as substrates. For recombinant rat RoDH(I) and recombinant protein generated from the newly cloned human homologue of RoDH(I), BISWAS and RUSSELL (1997) demonstrated that both rat and human RoDH(I) catalyze the oxidation of 5α-androstan-3,17-diol to dihydrotestosterone. The apparent K_m values of both human and rat retinol dehydrogenases for 5α-androstan-3,17-diol were reported to be $0.1\,\mu M$ [as compared to approximately $2\,\mu M$ for retinol-CRBP I for rat RoDH(I) (BOERMAN and NAPOLI 1995)]. Thus, these investigations have raised the possibility that microsomal retinol dehydrogenases may play important roles both in the generation of retinoic acid and in the generation of active steroids.

A human liver SCAD with all-*trans*-retinol dehydrogenase activity has been cloned and characterized by GOUGH et al. (1998). This NAD^+-dependent enzyme comprises 317 amino acids and exhibits strong primary sequence similarity with rat RoDH(I), RoDH(II) and RoDH(III). This human microsomal protein, designated RoDH-4 by GOUGH et al. (1998), is expressed highly in the adult liver and weakly in the lung but not in heart, brain, placenta, skeletal muscle or pancreas. Expression of RoDH-4 in human fetal liver and lung but not fetal brain and kidney was also observed (GOUGH et al. 1998). The apparent K_m values of RoDH-4 for 3α-adiol, androsterone and dihydrotestosterone were reported to be well under $1\,\mu M$, apparent K_m values that are similar to those of rat RoDH(I) for these hydroxysteroids (BISWAS and RUSSELL 1997). GOUGH et al. (1998) further report that all-*trans*-retinol is a much better substrate for RoDH-4 than 13-*cis*-retinol; however, other kinetic details concerning retinol oxidation were not provided. Interestingly, however, all-*trans*-retinol, 13-*cis*-retinol and all-*trans*-retinol bound to CRBP I were relatively potent inhibitors of RoDH-4 catalyzed androsterone oxidation with respective K_i values of 5.8, 3.5, and $3.6\,\mu M$. Apo-CRBP I had no effect on androsterone oxidation (GOUGH et al. 1998). Thus, as BISWAS and RUSSELL (1997), GOUGH et al. (1998) reported data suggesting that some of the enzymes involved in the metabolism of steroids and retinoids are shared.

This suggestion has also been made by Napoli and colleagues (CHAI et al. 1997; SU et al. 1998) who have recently identified two previously unknown members of the SCAD enzyme family that can use both sterols and retinols as substrates. These enzymes, *cis*-retinol/3α-hydroxysteroid dehydrogenases (CRADs), catalyze the oxidation of *cis*-retinols and are described in more detail below in Sect. C.III. The existence of the CRADs adds further weight to the suggestion that steroid and retinoid metabolism intersect at key multifunctional enzymes. This raises a very intriguing opportunity for identifying linkages between these two very potent families of bioactive molecules.

Table 2 provides a summary of SCADs that are reported in the literature to catalyze oxidation/reduction in both retinols and hydroxysterols. This table provides the relative binding affinities of these enzymes for both retinoid and steroid substrates.

It is worthy to note that members of the ADH family of enzymes also can catalyze both retinol and hydroxysteroid oxidation (KEDISHVILI et al. 1997). Two cDNAs encoding for ADHs, one a class III ADH and the other a previously unknown ADH, were cloned by RT-PCR from chick embryo limb bud and heart RNA (KEDISHVILI et al. 1997). The previously unknown ADH cDNA clone exhibited 67 and 68% sequence identity with chicken class I and III

Table 2. Partial listing of short-chain dehydrogenases discussed in the text and able to catalyze oxidation/reduction of retinoids and steroids

Enzyme	Species	Retinoid substrate	K_m or $K_{0.5}$[a] (μM)	Steroid substrate	K_m or $K_{0.5}$ (μM)
RoDH[b]	Rat	All-*trans*-retinol	1.6	–	–
RoDH(I)[c]	Rat	All-*trans*-retinol	0.9	3α-Adiol	0.1[d]
				Androsterone	0.1[d]
				Dihydrotestosterone	NR[d]
				Testosterone	2.8[d]
RoDH(II)[e]	Rat	All-*trans*-retinol	2	–	–
RoDH(III)[f]	Rat	All-*trans*-retinol	NR	–	–
9cRDH[g]	Human	9-*Cis*-retinol	NR	–	–
		11-*Cis*-retinol	NR	–	–
CRAD1[h]	Mouse	9-*Cis*-retinol	5.4[b]	3α-Adiol	0.2
		11-*Cis*-retinol	7[b]	Androsterone	0.1
				Testosterone	>35
CRAD2[i]	Mouse	11-*Cis*-retinol	5[b]	3α-Adiol	2.2[a]
		All-*trans*-retinol	6.9[b]	Androsterone	0.6[a]
		9-*Cis*-retinol	>28[b]	Testosterone	1.9[a]
				Dihydrotestosterone	25[a]
RoDH(I)[d]	Human			3α-Adiol	0.1
				Androsterone	0.1
				dihydrotestosterone	0.1
				Testosterone	3.3
				Estradiol	NR
RoDH-4[j]	Human	All-*trans*-retinol	NR	3α-Adiol	0.20
		13-*Cis*-retinol	NR	Androsterone	0.14
				Dihydrotestosterone	0.80

retSDR1[k]	Human	All-*trans*-retinal	NR		
	Mouse			–	–
	Bovine			–	–
11cRDH[m]	Mouse	9-*Cis*-retinol	NR	–	–
		11-*Cis*-retinol	NR	–	–
		13-*Cis*-retinol	NR	–	–
		9-*Cis*-retinal	NR	–	–
		11-*Cis*-retinal	NR	–	–
		13-*Cis*-retinal	NR	–	–

RoDH, retinol dehydrogenase; RoDH (I), (II), (III), retinol dehydrogenases types I, II, and III, respectively; CRAD1 and CRAD2, *cis*-retinol/androgen dehydrogenases types I and II, respectively; 9cRDH, 9-*cis*-retinol dehydrogenase; RoDH-4, retinol dehydrogenase type 4; ret SDR1, short-chain dehydrogenase/reductase; 11cRDH, 11-*cis*-retinol dehydrogenase; NR, reported to be a substrate for the enzyme but no K_m or $K_{0.5}$ value given.

[a] K_m for the substrate is listed except for allosteric enzymes where a $K_{0.5}$ value for a substrate is reported. Where $K_{0.5}$ values are provided, they are denoted with the superscript "a" whereas values without a superscript represent K_m values.

[b] Posch et al. (1991).

[c] Chai et al. (1995).

[d] Biswas and Russell (1997).

[e] Chai et al. (1996).

[f] Chai et al. (1996).

[g] Mertz et al. (1997).

[h] Chai et al. (1997).

[i] Su et al. (1998).

[j] Gough et al. (1998).

[k] Haeseleer et al. (1998).

[l] Driessen et al. (1998).

ADHs, respectively, and had lower identity with mammalian class II and IV ADH isozymes. Expression of this cDNA in *Escherichia coli* yielded an active ADH species that was stereospecific for the $3\beta,5\alpha$-hydroxysteroids versus $3\beta,5\beta$-hydroxysteroids (KEDISHVILI et al. 1996). Moreover, this cytosolic enzyme catalyzed retinol oxidation with an apparent K_m of 56 μM for all-*trans*-retinol, as compared to a K_m of 31 μM for epiandrosterone (KEDISHVILI et al. 1997). Thus, as with members of the SCAD family, ADH enzyme family members also can catalyze reactions involving both steroids and retinoids.

In summary, a relatively large number of different enzymes have been proposed to be importantly involved in catalyzing retinol oxidation, the first oxidative step needed for retinoic acid synthesis. Considering the very diverse and sometimes contradictory characteristics of these enzymes, it seems possible that all of these enzymes do not function in vivo in the synthesis of retinoic acid from retinol. Alternatively, considering the very diverse actions of retinoic acid, multiple cell type- or tissue-specific pathways for retinoic acid synthesis may exist. If this were the case, one might expect many different enzyme species to be involved in retinoic acid formation. At present, it is not possible to distinguish between these alternatives.

II. Oxidation of Retinal

LEE et al. (1991) explored the ability of the 13 aldehyde dehydrogenases that had been identified at the time of their studies to be present in mouse tissues to catalyze the oxidation of all-*trans*-retinal to all-*trans*-retinoic acid. Three of the six aldehyde dehydrogenases present in mouse liver cytosol were able to catalyze this oxidation. One of these, AHD-2, was estimated by LEE et al. (1991) to catalyze about 95% of retinal oxidation to retinoic acid in the liver. The apparent K_m of AHD-2 for all-*trans*-retinal was determined to be 0.7 μM. None of the aldehyde dehydrogenases present in the particulate fractions of mouse liver were able to catalyze significant retinal oxidation (LEE et al. 1991). Based on these data, it was concluded that the enzymes responsible for retinoic acid formation from retinal are cytosolic, NAD^+-linked, substrate nonspecific dehydrogenases (LEE et al. 1991).

Other investigators have demonstrated that cytosol preparations from rat kidney, testis, and lung can catalyze the oxidation of retinal to retinoic acid (BHAT et al. 1988a; BHAT et al. 1988b; HUPERT et al. 1991). HUPERT et al. (1991) focused on the enzymatic activity present in rat liver cytosol responsible for the formation of retinoic acid from retinal. The enzymatic activity was observed to be linear with respect to protein concentration (0–2.4 mg/ml) and time (0–30 min), showed a broad pH maximum of 7.7–9.7, and had an apparent K_m of 0.25 mM for all-*trans*-retinal.

Bhat and colleagues focused on the purification and characterization of the cytosolic retinal dehydrogenase present in rat kidney (BHAT et al. 1988a,b; LABRECQUE et al. 1993, 1995). Purification of the enzyme was obtained through a combination of anion exchange, dye-binding affinity and chromatofocusing chromatography steps. The enzyme was found to have a subunit molecular weight of 53,000 (LABRECQUE et al. 1993). The purified enzyme is NAD^+-dependent and catalyzes the oxidation of both all-*trans*- and 9-*cis*-retinal to the corresponding retinoic acid isomer (LABRECQUE et al. 1993, 1995). The rat kidney cytosolic aldehyde dehydrogenase displayed typical Michaelis-Menten kinetics towards all-*trans*-retinal with an apparent K_m of 8–10 μM whereas the apparent K_m for 9-*cis*-retinal is 5.7 μM. The cDNA for rat kidney aldehyde dehydrogenase was subsequently cloned by BHAT et al. (1995) and its amino acid sequence was found to be very similar to those of other cytosolic aldehyde dehydrogenases cloned from rat, mouse, and human livers (BHAT et al. 1995). RT-PCR analysis indicated that the enzyme is also strongly expressed in rat lung, testis, intestine, stomach and trachea.

Initially, this cDNA clone was used to define the expression pattern of this aldehyde dehydrogenase in fetal and adult rat kidney (BHAT et al. 1998). This group of investigators subsequently focused on the expression of this cytosolic aldehyde dehydrogenase in the stomach and small intestine of rats during postnatal development and in vitamin A deficiency (BHAT 1998). Expression of the enzyme was reported to be high in the small intestine 2 days prior to birth, whereas for this age expression was not detectable in the stomach. Aldehyde dehydrogenase expression in the intestine was found to decrease progressively postnatally. After birth however, expression of the dehydrogenase in the stomach increased and reached its highest level at postnatal day 42. Vitamin A deficiency was found to upregulate enzyme expression in the stomach and small intestine and administration of retinoids was found to downregulate expression of this dehydrogenase for these tissues (BHAT 1998).

Two distinct aldehyde dehydrogenases ($ALDH_1$ and $ALDH_2$) able to catalyze retinal oxidation have been purified through a combination of gel exclusion and anion exchange chromatography from bovine kidney (BHAT et al. 1996). Both $ALDH_1$ and $ALDH_2$ have relatively low apparent K_m values for all-*trans*-retinal, 6.4 and 9.1 μM, respectively; however, based on their data these investigators suggested that $ALDH_1$ is likely the primary enzyme responsible for oxidizing retinal to retinoic acid in the bovine kidney (BHAT et al. 1996).

The kinetic properties of a human liver cytosolic aldehyde dehydrogenase ($ALDH_1$) for retinal isomers have been explored (BHAT and SAMAHA 1999). Human $ALDH_1$ shares 87% amino acid identity with the rat kidney retinal dehydrogenase previously described by BHAT and colleagues (1995). This human aldehyde dehydrogenase isozyme catalyzes the oxidation of all-*trans*-, 9-*cis*- and 13-*cis*-retinal with apparent K_m values of 2.2, 5.5 and 4.6 μM respectively. Thus, each of these retinal isomers is a relatively good substrate for this enzyme. Interestingly, all-*trans*-retinol was found to be a potent uncompetitive inhibitor of retinal oxidation. Unlike human $ALDH_1$, all-*trans*-retinol inhibits the rat kidney cytosol retinal dehydrogenase in a competitive manner (BHAT et al. 1995); thus, these investigators suggested that human $ALDH_1$ and rat kidney aldehyde dehydrogenases may have different actions in retinal metabolism (BHAT and SAMAHA 1999).

A retinal dehydrogenase [designated RalDH(I)] has been purified from rat liver cytosol and characterized by Napoli and colleagues (POSCH et al. 1992; EL AKAWI and NAPOLI 1994). This isozyme is reported to be the quantitatively major aldehyde dehydrogenase isoform in rat liver, kidney and testis (POSCH et al. 1992). Purified rat liver RalDH(I) is reported to display allosteric kinetics with a $K_{0.5}$ towards free all-*trans*-retinal of approximately $0.8\,\mu M$ and a Hill coefficient of 1.5. All-*trans*-retinal concentrations greater than $6\,\mu M$ were reported to be inhibitory to RalDH(I). RalDH(I) is reported to also recognize all-*trans*-retinol bound to CRBP I as a substrate. RalDH(I) exhibits allosteric kinetics towards retinal-CRBP I with a $K_{0.5}$ of $0.13\,\mu M$ and a Hill coefficient of 1.75 (POSCH et al. 1992).

Later work by EL AKAWI and NAPOLI (1994) demonstrated that RalDH(I) also catalyzes the oxidation of both all-*trans*- and 9-*cis*-retinal in an NAD^+-dependent manner but 13-*cis*-retinal was found not to be an effective substrate. The partially purified rat liver enzyme displayed allosteric kinetics for 9-*cis*-retinal with a $K_{0.5}$ of $5.2\,\mu M$ and with a Hill coefficient of 1.4. Oxidation of retinal was not inhibited by either all-*trans*- or 9-*cis* retinoic acid, nor by holo-CRABP I. EL AKAWI and NAPOLI (1994), based on these data, suggested that RalDH(I) may serve as a common enzyme in the conversion of all-*trans*- and 9-*cis*-retinal into their acids.

WANG et al. (1996) report the cloning of another aldehyde dehydrogenase from rat testis, termed RalDH(II) that is distinct from the RalDH(I) isoform characterized earlier by NAPOLI and colleagues (POSCH et al. 1992; EL AKAWI and NAPOLI 1994). At the amino acid level, rat testis RalDH(II) is 85% identical to rat RalDH(I), 85% identical to mouse AHD-2, 87% identical to human $ALDH_1$ and 87% identical to bovine retina retinal dehydrogenase (WANG et al. 1996). Recombinant RalDH(II) expressed in *E. coli* recognizes as substrate both unbound all-*trans*-retinal and all-*trans*-retinal in the presence of CRBP I (WANG et al. 1996). In addition, RalDH(II) utilizes as a substrate all-*trans*-retinal generated in situ by the action of rat liver microsomal retinol dehydrogenase(s) from holo-CRBP I.

An aldehyde dehydrogenase present at high levels in the basal forebrain of mice was studied by MCCAFFERY and DRÄGER (1994) for its ability to catalyze the formation of retinoic acid from all-*trans*-retinal. The enzyme was found in axons and terminals of a subpopulation of dopaminergic neurons of the mesostriatal and mesolimbic system. Based on the reported properties of this enzyme, it appears that this aldehyde dehydrogenase present in nervous tissues is identical to mouse AHD-2 (LEE et al. 1991). From these studies, MCCAFFERY and DRÄGER (1994) also identified other aldehyde dehydrogenase isoforms able to catalyze all-*trans*-retinal oxidation in the ventral part of the embryonic mouse retina. In later work, ZHAO et al. (1996) report the characterization of one of these aldehyde dehydrogenases present in the ventral part of the embryonic mouse retina. This enzyme, termed by these investigators

RALDH-2, is distinct from murine AHD-2 and is expressed very early in mouse embryonic development both in mesoderm and neuroectoderm (ZHAO et al. 1996). Expression of this enzyme in mouse embryos receiving a teratogenic dose of all-*trans*-retinoic acid at embryonic day 8.5 results in a down-regulation of RALDH-2 expression (NIEDERREITHER et al. 1997). RALDH-2 was demonstrated by RT-PCR to be expressed in adult mouse testis but not liver (ZHAO et al. 1996).

YAMAMOTO et al. (1998) report that CRBP I facilitates all-*trans*-retinoic acid synthesis from all-*trans*-retinal for cytosol fractions from adult mouse liver, mouse early postnatal retina, embryonic meninges and postnatal choroid plexus. In addition, CRBP I enhanced retinoic acid synthesis by RALDH-2 was seen in the choroid plexus when the enzyme was assayed in isolation from other cytoplasmic aldehyde dehydrogenases by zymographic assay (YAMAMOTO et al. 1998). For this zymographic assay, the ratio of CRBP I to retinal was found to be crucial, since excess apo-CRBP I was found to be inhibitory to the reaction. These investigators also reported that CRABP I facilitated uptake and concentration of all-*trans*-retinoic acid by the postnatal day 1 mouse cerebellum after i.p. injection of 0.3 μmol all-*trans*-retinoic acid but this retinoic acid was almost entirely lost from the cerebellum within 8 h of injection (YAMAMOTO et al. 1998).

ROBERTS et al. (1992) have reported that P450 isoforms 1A2 and 3A6 from rabbit liver microsomes are able to oxidize retinal to retinoic acid. To date, no work in the rabbit exploring the roles of cytosolic retinal dehydrogenases in retinoic acid has been reported.

Based on the data summarized above, it would appear that several aldehyde dehydrogenases present in the cytosol of rat, mouse, bovine and human tissues can catalyze the oxidation of retinal to retinoic acid. Unlike the situation described in Sect. C.I for enzymes able to catalyze retinol oxidation, the candidate enzymes proposed to be important in catalyzing the final step of retinoic acid formation are far more limited in number. Consequently it seems reasonable that some if not all of the aldehyde dehydrogenases mentioned above act physiologically in the synthesis of retinoic acid.

III. Formation of 9-*cis*-Retinoic Acid

It is clear that 9-*cis*-retinoic acid acting through RARs and RXRs is an essential retinoid form that is required for regulating retinoid responsive gene activity. It is similarly clear that any factor that influences 9-*cis*-retinoic acid availability to, or within a cell, will have a broad impact on retinoic acid signaling pathways and on cellular responses. There is presently only very limited information available regarding how 9-*cis*-retinoid isomers are formed. This is in contrast to our understanding of how 11-*cis*-retinoid isomers are formed, where it is now well established for the visual process that the isomerization of all-*trans*-retinoids to 11-*cis*-retinoids is catalyzed by a specific enzyme. In the visual cycle the isomerization takes place at the level of the retinols and not the retinaldehydes (see Chap. 20, this volume). Since the first reports in 1992 that 9-*cis*-retinoic acid is a ligand for the RXRs, several studies have explored possible pathways for 9-*cis*-retinoic acid formation. Three possible pathways for 9-*cis*-retinoic acid formation have been proposed and investigated. These are pictured in Fig. 2. The proposed pathways consist of: (a) isomerization of all-*trans*-retinoic acid to 9-*cis*-retinoic acid, probably through nonenzymatic processes; (b) enzymatic oxidation of 9-*cis*-retinol to 9-*cis*-retinoic acid through a pathway similar to the oxidation of all-*trans*-retinol to all-*trans*-retinoic acid; and (c) cleavage of 9-*cis*-β-carotene either directly to 9-*cis*-retinoic acid or to 9-*cis*-retinol or 9-*cis*-retinal that is directly oxidized to 9-*cis*-retinoic acid. Although each of these possible pathways seems reasonable and the in vitro data supporting them are individually convincing, it is still unclear from the literature whether any or all of these possibilities are important for 9-*cis*-retinoic acid formation in vivo.

URBACH and RANDO (1994) have reported that membranes prepared from bovine liver catalyze nonenzymatically the isomerization of all-*trans*-retinoic acid to 9-*cis*-retinoic acid. This isomerization was shown to depend on free

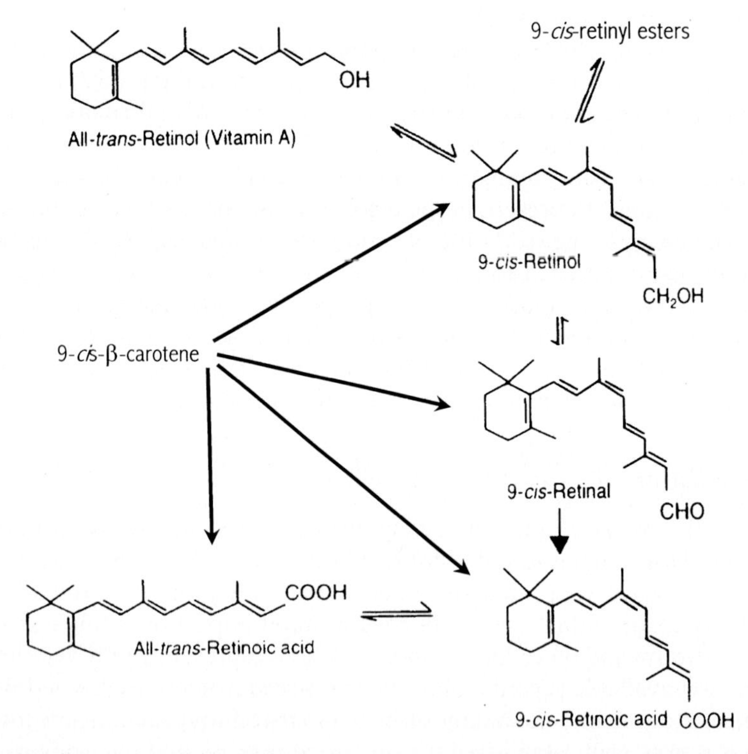

Fig. 2. Possible biosynthetic pathways for the formation of 9-*cis*-retinoic acid within cells and tissues. The formation of 9-*cis*-retinoic acid from all-*trans*-retinoic acid, 9-*cis*-β-carotene and 9-*cis*-retinol is discussed in the text

sulfhydryl groups present in the microsomes and not to involve the participation of an enzyme (URBACH and RANDO 1994). Later studies by SHIH et al. (1997) show that the interconversion of 9-*cis*-retinoic acid, all-*trans*-retinoic acid, 13-*cis*-retinoic acid, and 9,13-di-*cis*-retinoic acid may be catalyzed by sulfhydryl compounds of low molecular weight (including L-cysteine methyl ester, glutathione and *N*-acetyl-L-cysteine), as well as by proteins containing sulfhydryl groups (apoferritin, native microsomes and boiled microsomes). Data from these studies are convincing that free sulfhydryl groups can bring about the nonenzymatic formation of 9-*cis*-retinoic acid from all-*trans*-retinoic acid.

Krinsky, Russell, and colleagues have reported that 9-*cis*-β-carotene serves as a precursor for 9-*cis*-retinoic acid in vivo in the ferret (HEBUTERNE et al. 1995). Intestinal mucosa concentrations of 9-*cis*-retinoic acid for ferrets undergoing perfusion with 9-*cis*-β-carotene are markedly greater than those of ferrets undergoing perfusion with either the control micellular solution used to solubilize the carotenoid or a micellular solution of all-*trans*-β-carotene. Moreover, portal blood concentrations of 9-*cis*-retinoic acid were much greater after intestinal perfusion with 9-*cis*-β-carotene than either portal blood levels before perfusion or those of ferrets perfused with the control micellular solution or all-*trans*-β-carotene (HEBUTERNE et al. 1995). This same result was also observed in the ferret for hepatic 9-*cis*-retinoic acid levels following perfusion with 9-*cis*-β-carotene (HEBUTERNE et al. 1995). However, the significance of this pathway for providing 9-*cis*-retinoic acid has been questioned since the rate of cleavage of 9-*cis*-β-carotene is only 6%–7% of that of all-*trans*-β-carotene (NAGAO and OLSON 1994). Furthermore, since rats maintained on carotenoid free diets display normal health, the conversion of 9-*cis*-β-carotene to 9-*cis*-retinoic acid cannot be an essential pathway for formation of this retinoic acid isomer.

The possibility that 9-*cis*-retinoic acid may be formed through the two step oxidation of 9-*cis*-retinol to 9-*cis*-retinal followed by oxidation of 9-*cis*-retinal to 9-*cis*-retinoic acid was first suggested by studies of retinal oxidation by LABRECQUE et al. (1993). These investigators (LABRECQUE et al. 1993; LABRECQUE et al. 1995) exploring the retinal dehydrogenase present in rat kidney reported that this enzyme is able to catalyze the oxidation of 9-*cis*-retinal to 9-*cis*-retinoic acid. This was taken by LABRECQUE et al. (1993) to suggest that a pathway starting with 9-*cis*-retinol may be important for 9-*cis*-retinoic acid formation. To further substantiate this possibility, these investigators have demonstrated the presence of 9-*cis*-retinol in rat kidney at a level that was approximately 10% of that of all-*trans*-retinol (LABRECQUE et al. 1993, 1995).

An NAD^+-dependent retinol dehydrogenase which specifically oxidizes 9-*cis*-retinol but not all-*trans*-retinol has been cloned from a human mammary tissue cDNA library and characterized upon its expression in CHO cells (MERTZ et al. 1997). This enzyme, which is a member of the SCAD enzyme family, is expressed in adult human mammary tissue, kidney, liver, and testis.

The mouse homologue of this enzyme has a very similar tissue distribution and shares many similar properties with the human enzyme (GAMBLE et al. 1999). It was proposed by MERTZ et al. (1997) that this enzyme may play an important role in the synthesis of 9-*cis*-retinoic acid. Interestingly, this enzyme, termed 9-*cis*-retinol dehydrogenase (9cRDH), is expressed in kidney, a tissues which has been reported to contain significant quantities of 9-*cis*-retinol (LABRECQUE et al. 1993, 1995).

Investigations by DRIESSEN et al. (1998) indicate that 9cRDH (MERTZ et al. 1997) is identical to the 11-*cis*-retinol dehydrogenase (11cRDH) present in the eye (see Chap. 20, this volume, for more details regarding this enzyme). Although the vision literature had indicated that 11cRDH protein and 11cRDH mRNA are found only in the retinal pigment epithelium (SIMON et al. 1995), and that 11cRDH does not catalyze 9-*cis*-retinol oxidation (SUZUKI et al. 1993), this literature appears to be incorrect (DRIESSEN et al. 1998). Based on presently available data, it now appears that 9cRDH and 11cRDH are the same enzyme. It seems that this enzyme that does not catalyze all-*trans*-retinol oxidation should be renamed *cis*-retinol dehydrogenase.

A second *cis*-retinol dehydrogenase, also a member of the SCAD family, has been described by NAPOLI and colleagues (CHAI et al. 1997). The cDNA for this mouse liver enzyme, termed the *cis*-retinol/3α-hydroxy sterol short-chain dehydrogenase (CRAD1), encodes an enzyme comprising 317 amino acids that recognizes 9-*cis*- and 11-*cis*-retinol, 5α-androstan-3α,17β-diol and 5α-androstan-3α-ol-17-one as substrates. Based on apparent K_m values for these substrates the *cis*-retinol dehydrogenase has a greater affinity for sterol than retinol substrates. This mouse enzyme, which is most similar to mouse RoDH(I) and RoDH(II) isozymes (86% and 91%, respectively), uses NAD^+ as its preferred cofactor. This enzyme is expressed in liver, kidney, small intestine, heart, retinal pigment epithelium, brain, spleen, testis, and lung. It has been proposed by CHAI et al. (1997) that this multifunctional enzyme is important for generating both 9-*cis*-retinoic acid and bioactive androgens. It is possible that this enzyme and others that resemble it provide a linkage between the actions of retinoids and androgens.

This suggestion is further supported by more recent work from Napoli's laboratory (SU et al. 1998) reporting the cloning of a second *cis*-retinol/3α-hydroxysteroid short-chain dehydrogenase isozyme from a mouse embryonic cDNA library. This enzyme, termed CRAD2, encodes a 316 amino acid short chain alcohol dehydrogenase that shares 87% amino acid identity with CRAD1. This enzyme recognizes as substrates both androgens and retinols and exhibits cooperative kinetics towards 3α-adiol and testosterone but Michaelis-Menten kinetics towards androsterone, all-*trans*-, 11-*cis*- and 9-*cis* retinols. Reported $K_{0.5}$ values for 3α-adiol, testosterone, androsterone, 11-*cis*-, all-*trans*- and 9-*cis*-retinol were, respectively, 2.2, 1.9, 0.6, 5, 6 and greater than 28 μM. CRAD2 mRNA was highly expressed in mouse liver and at much lower levels in lung, eye, kidney and brain (SU et al. 1998). It was suggested by SU et al. (1998) that the exact physiological function of CRAD2 (as well as that of CRAD1) may depend on the loci of expression, substrate availability, hormonal influences and other as yet unappreciated factors.

At present we only have limited understanding of how 9-*cis*-retinoic acid is formed or of how concentrations of 9-*cis*-retinoic acid are regulated within tissues and cells. As outlined above, it is possible that the 9-*cis*-isomer of retinoic acid is formed directly from all-*trans*-retinoic acid. Alternatively, it is possible that a pathway independent of all-*trans*-retinoic acid formation exists within at least some cells and tissues. If this later possibility is proven to be correct, this would raise many interesting new questions regarding the regulation of all-*trans*- and 9-*cis*-retinoic acid formation and tissue concentrations and of how the two isomers of retinoic acid interact to regulate gene expression.

IV. Direct Synthesis of Retinoic Acid from Carotenoids

Napoli and Race (1988) reported that cytosol preparations from rat tissues could catalyze the direct formation of retinoic acid from β-carotene. The rate of retinoic acid synthesis from $10\,\mu M$ β-carotene was found to range from 120–224 pmol/ protein per hour for intestinal cytosol, and from 334 to 488 pmol/h/mg protein for cytosols prepared from kidney, lung, testes, and liver. Retinol that was generated during β-carotene metabolism was determined, by isotope dilution, not to be the major substrate for retinoic acid synthesis. Retinal was not detected by these investigators as a free intermediate in this process. Thus, Napoli and Race (1988) suggested that retinal might be tightly bound by the enzyme, or β-carotene might be oxidized to a 15,15'-enediol before dioxygenase cleavage, by analogy to the conversion of catechol to *cis*, *cis*-muconic acid. Napoli and Race (1988) also proposed other mechanisms for producing retinoic acid from β-carotene without first yielding retinal, but as yet no intermediates have been characterized.

Wang et al. (1991) reported that homogenates of liver, lung, kidney, and fat from monkey, ferret, and rat were all able to form retinoic acid with a pH optimum of 7.0 upon incubation with $2\,\mu M$ β-carotene. It was suggested that some retinoic acid formation may occur through pathways that do not have retinal as an intermediate (Wang et al. 1991). Later studies by Wang et al. (1992) explored retinoic acid formation from β-carotene in human intestinal mucosa homogenates in vitro. Because citral, which inhibits the conversion of retinal to retinoic acid, did not affect retinoic acid formation from β-carotene or β-apocarotenals, Wang et al. (1992) concluded that retinoic acid is formed through a biochemical process that does not involve retinal as an intermediate. Biochemical details regarding this process are still missing.

D. Oxidative Metabolism of Retinoic Acid

Metabolites of all-*trans*-retinoic acid that have been reported to be generated in vivo include 13-*cis*-retinoic acid, 9-*cis*-retinoic acid, retinoyl-β-glucuronide,

5,6-epoxy-retinoic acid, 4-hydroxy-retinoic acid, 4-oxo-retinoic acid, and 3,4-didehydro-retinoic acid (FROLIK 1984; BLANER and OLSON 1994). Some of these metabolites are active in mediating retinoic acid function, whereas others are likely catabolic products. The enzymatic processes responsible for generating these retinoic acid metabolites are now starting to be identified and characterized. It appears that members of the cytochrome P450 family play key roles in the formation of many of these metabolites.

I. Role of the Cytochrome P450 System in Retinoic Acid Oxidation

It has long been known that the cytochrome P450 system is active in metabolizing retinoic acid. ROBERTS et al. (1979) reported that the formation of polar metabolites of retinoic acid was catalyzed by an activity present in the 100,000 *g* pellet of rat intestine and liver homogenates. This activity required NADPH and oxygen, and was strongly inhibited by carbon monoxide. In addition, the activity was reported to be markedly induced by retinoids, but only to a minor extent by phenobarbital or 3-methylcholanthrene. ROBERTS et al. (1979) concluded that the rat intestinal and liver activities responsible for the formation of polar metabolites of retinoic acid are members of a class of mixed function oxidases containing the cytochrome P450s.

Later investigations (ROBERTS et al. 1992) demonstrated that many rabbit liver cytochrome P450 isoforms, including 2A4, 1A2, 2E1, 2E2, 2C3, 2G1, and 3A6, can catalyze the 4-hydroxylation of retinoic acid. These rabbit liver cytochromes also catalyzed the 4-hydroxylation of both retinol and retinal, but not the conversion of 4-hydroxy-retinoids to the corresponding 4-oxo-retinoids.

VAN WAUWE et al. (1992) showed that oral administration of a dose (40 mg/kg body weight) of liarazole, a 1-substituted imidazole derivative which inhibits cytochrome P450 activity, enhances the endogenous plasma concentrations of retinoic acid from less than 1.7 n*M* to 10–15 n*M*. Based on these data, VAN WAUWE et al. (1992) concluded that the cytochrome P450 system plays an important physiological role in retinoic acid metabolism and homeostasis.

More recently, work carried out on human skin by DUELL et al. (1996) demonstrated that application of all-*trans*-retinoic acid produced a marked increase in retinoic acid 4-hydroxylase activity. Application of 0.1% all-*trans*-, 0.1% 9-*cis*-, and 0.1% 13-*cis*-retinoic acid to human skin for 2 days resulted in induction of only an all-*trans*-retinoic acid 4-hydroxylase activity. The inducible 4-hydroxylase activity was present in microsomes and in vitro this activity was able to catalyze 4-hydroxylation of all-*trans*-retinoic acid but not of 9-*cis*- or 13-*cis*-retinoic acid (DUELL et al. 1996). Neither all-*trans*-retinol nor all-*trans*-retinal were able to compete with all-*trans*-retinoic acid as a substrate for the 4-hydroxylase activity.

HAN and CHOI (1996) reported that when human breast cancer T47-D cells were treated with all-*trans*-retinoic acid, both all-*trans*-retinoic acid 4-

hydroxylase and all-*trans*-retinoic acid 18-hydroxylase activities were induced in microsomes obtained from T47-D cells. This induction occurred in a time- and dose-dependent manner and appeared to be regulated at the transcriptional level (HAN and CHOI 1996). The retinoic acid-inducible 4- and 18-hydroxylase activities present in microsomes from T47-D cells showed high specificity for all-*trans*-retinoic acid, with respective apparent K_m values of 99 nM and 65 nM for all-*trans*-retinoic acid. In addition, HAN and CHOI (1996) demonstrated that these inducible microsomal 4- and 18-hydroxylase activities could be inhibited by treatment with liarazole. Based on these and other data, HAN and CHOI (1996) suggested that retinoic acid-inducible 4- and 18-hydroxylase activities present in T47-D human mammary carcinoma cells represented a novel P450 isozyme not previously described in the literature.

A previously unknown cytochrome P450 isoform, CYP26, that is able to metabolize retinoic acid was cloned independently by several groups (FUJII et al. 1997; RAY et al. 1997; WHITE et al. 1996, 1997). One cDNA clone (originally termed P450RA) was obtained from a subtraction library prepared from retinoic acid-treated and untreated murine P19 embryonic carcinoma cells (FUJII et al. 1997). FUJII et al. (1997) reported that when expressed, the murine P450RA cDNA catalyzed the oxidation of all-*trans*-retinoic acid to 5,8-epoxy-all-*trans*-retinoic acid. Both 13-*cis*- and 9-*cis*-retinoic acid were found to serve as substrates for P450RA. P450RA was expressed in a stage- and region-specific fashion during mouse development but expression did not appear to be inducible upon exposure of mouse embryos to excess retinoic acid. In the adult mouse P450RA was expressed only in liver (FUJII et al. 1997). Based on these data, FUJII et al. (1997) proposed that P450RA is important in the adult for regulating retinoic acid concentrations in blood (FUJII et al. 1997).

WHITE et al. (1996) reported the isolation and characterization of a cDNA for CYP26 (that they originally referred to as P450RAI). P450RAI was found to be expressed during gastrulation of the zebrafish. This cDNA was obtained utilizing mRNA differential display in a search for zebrafish genes responding to exogenous all-*trans*-retinoic acid exposure. P450RAI was found to be expressed normally during gastrulation and in a defined pattern in epithelial cells of the regenerating caudal fin in response to administration of exogenous all-*trans*-retinoic acid. When the cDNA for P450RAI was expressed in COS-1 cells, all-*trans*-retinoic acid was rapidly metabolized to more polar metabolites. Two of these metabolites were identified as 4-oxo-all-*trans*-retinoic acid and 4-hydroxy-all-*trans*-retinoic acid. Thus, P450RAI, which is induced upon retinoic acid exposure, catalyzes the oxidative metabolism of retinoic acid.

Later, these investigators (WHITE et al. 1997) reported cloning the human homologue of the zebrafish P450RAI. As with the zebrafish enzyme, human P450RAI was able to catalyze hydroxylation of all-*trans*-retinoic acid to its 4-hydroxy-, 4-oxo- and 18-hydroxy-forms (WHITE et al. 1997b). Moreover, P450RAI mRNA expression was highly induced by retinoic acid treatment of human LC-T, MCF-7, HB4 and HepG2 cells. Human P450RAI was found to share 68% amino acid identity with zebrafish P450RAI (WHITE et al. 1997).

This group further reported the cloning of the mouse homologue for P450RAI (ABU-ABED et al. 1998). Mouse P450RAI catalyzes metabolism of retinoic acid into 4-hydroxy-retinoic acid, 4-oxo-retinoic acid and 18-hydroxy-retinoic acid (ABU-ABED et al. 1998). At the amino acid level, mouse P450RAI was found to be identical to that of P450RA, as reported by FUJII et al. (1997). However, the metabolic products described by FUJII et al. (1997) differ from those observed for mouse P450RAI. ABU-ABED et al. (1998) suggested that the basis for this discrepancy may arise from technical difficulties experienced during the extraction and characterization of metabolites by FUJII et al. (1997). ABU-ABED et al. (1998) also observed for wild-type F9 cells and derivatives lacking retinoic acid receptors (RARs) and/or retinoid X receptors (RXRs), a direct relationship between the level of retinoic acid metabolic activity and retinoic acid-induced P450RAI mRNA. Finally, these investigators provided data suggesting that the RARγ and RXRα mediate the effects of retinoic acid for inducing the expression of the P450RAI gene (ABU-ABED et al. 1998).

Simultaneously, RAY et al. (1997) reported the cloning of CYP26 (that is identical to P450RA and P450RAI) from a mouse embryonic stem cell library. Expression of CYP26 was reported to be limited in the adult mouse to liver and brain by RAY et al. (1997). CYP26 was detectable in mouse embryos as early as embryonic day 8.5. Moreover, CYP26 mRNA was reported by RAY et al. (1997) to be upregulated during the retinoic acid-induced neural differentiation of mouse embryonic stem cells in vitro. These investigators further reported the RT-PCR cloning of human CYP26 from liver mRNA. For human CYP26, 195 of 200 amino acids were found to be identical to those of mouse CYP26 (RAY et al. 1997). In the adult human, CYP26 is expressed in liver, brain and the placenta (RAY et al. 1997).

Other interesting recent work on CYP26 has been carried out by SONNEVELD et al. (1998). These investigators demonstrated by Northern blot analysis that CYP26 is induced within 1 h upon all-*trans*-retinoic acid treatment in retinoic acid-sensitive T47-D human breast carcinoma cells but not in retinoic acid-resistant MDA-MB-231 human breast cancer cells or HCT 116 human colon cancer cells (SONNEVELD et al. 1998). Stable transfection of RARα and RARγ and to a lesser extent RARβ into HCT 116 cells showed that CYP26 induction by retinoic acid is dependent on these retinoic acid nuclear receptors. Contrary to the findings of FUJII et al. (1997), SONNEVELD et al. (1998) reported that retinoic acid-induced CYP26 is highly specific for the hydroxylation of all-*trans*-retinoic acid and does not recognize either 13-*cis*- or 9-*cis*-retinoic acid. These investigators also confirm that human CYP26 is highly sensitive to inhibition by liarazole.

Induction of oxidative metabolism of all-*trans*-retinoic acid in MCF-7 human mammary carcinoma cells upon treatment of the cells with all-*trans*-retinoic acid was reported by KREKELS et al. (1997). These workers indicate that MCF-7 cells show low constitutive levels of activities able to catalyze all-*trans*-retinoic acid oxidation. However, concentration-dependent induction of an enzyme species able to catalyze all-*trans*-retinoic acid oxidation was obtained upon preincubation of the cells with all-*trans*-retinoic acid (10^{-9}–10^{-6} *M*). Induction was detected within 1h of treatment. The apparent K_m of the induced activity for all-*trans*-retinoic acid was determined to be 0.33 μM Induction of all-*trans*-retinoic acid oxidation also was observed upon treatment of MCF-7 cells with 13-*cis*- and 9-*cis*-retinoic acid and several keto-forms of retinoic acid. Induction of all-*trans*-retinoic acid oxidation was inhibited by actinomycin D, and liarozole was found to inhibit all-*trans*-retinoic acid oxidation after induction of the oxidizing activity. Furthermore, all-*trans*-retinoic acid treatment did not increase CRABP I mRNA levels in the MCF-7 cells. Although KREKELS et al. (1997) did not specifically examine CYP26 induction in MCF-7 mammary carcinoma cells, it seems likely that these investigators were exploring CYP26 actions or those of a closely related isoform in MCF-7 cells.

KIM et al. (1998) examined eight different human squamous cell carcinoma cell lines to determine their capacity to induce P450-mediated oxidation of all-*trans*-retinoic acid. Among eight different cell lines examined, all-*trans*-retinoic acid was found to induce its own oxidative metabolism in four of these cell lines (KIM et al. 1998). Induction was blocked when retinoic acid was added to the cells along with either actinomycin D or cyclohexamide. The inducible metabolism was inhibited by the addition of ketoconazole to the cells, suggesting involvement of a P450 isozyme in this process. KIM et al. (1998) further reported that retinoic acid-induced metabolism was first detectable within 4h after retinoic acid treatment and that 13-*cis*- and 9-*cis*-retinoic acid and all-*trans*-retinal also were effective inducers of this P450 mediated oxidative metabolism. As was the case for the studies of KREKELS et al. (1997) recounted above, it seems likely that KIM et al. (1998) have been exploring in their studies the actions of CYP26 or a closely related isoform in human squamous carcinoma cell lines.

A possible linkage between the cytochrome P450 catalyzed oxidation of all-*trans*-retinoic acid and human disease has been proposed by RIGAS et al. (1996). These investigators had initially observed marked intersubject differences in the pharmacokinetics of all-*trans*-retinoic acid in patients with non-small cell lung cancer (RIGAS et al. 1993). These initial studies further indicated that patients who exhibit rapid plasma clearance of all-*trans*-retinoic acid also have significantly lower endogeneous plasma levels of all-*trans*-retinoic acid compared to subjects that clear all-*trans*-retinoic acid from the circulation more slowly (RIGAS et al. 1993). Since administration of ketoconazole attenuated the high rate of all-*trans*-retinoic acid plasma clearance, RIGAS et al. (1993) concluded that rapid all-*trans*-retinoic acid plasma clearance was probably catalyzed by a cytochrome P450-dependent oxidative pathway. Based on their initial observations, RIGAS et al. (1996) carried out a study examining the association between this metabolic phenotype and the risk of lung cancer in 85 subjects. The area under the plasma concentration X time curve (AUC) was calculated after a single oral dose of all-*trans*-retinoic acid (45 mg/m^3). The mean AUC for patients who had either squamous or large cell carcinomas was significantly lower than that of patients with adenocarcinomas ($p = 0.0001$) or control subjects ($p = 0.01$). Individuals with an AUC of less than 250 ng/ml per hour had a greater likelihood of having squamous or large cell carcinoma (odds ratio = 5.93). Based on their data, RIGAS et al. (1996) concluded that rapid oxidation (clearance) of all-*trans*-retinoic acid is linked to an increased risk of squamous or large cell cancers of the lung. It is unclear from these studies whether the all-*trans*-retinoic acid rapid clearance phenotype is a cause or an effect of the squamous or large cell lung cancer. Nevertheless, this work does link the constitutive rapid plasma clearance, and putative P450-catalyzed oxidation, of all-*trans*-retinoic acid to an increased risk of developing squamous or large cell lung cancer (RIGAS et al. 1996).

It is now generally accepted that CYP26 plays an important physiological role in catalyzing the oxidative metabolism of retinoic acid. It remains unclear, however, whether other cytochrome P450 isoforms also have physiological roles in this metabolism. This possibility seems likely considering the diverse actions of retinoic acid. However, this will require experimental confirmation, but undoubtedly searches for such P450 isoforms are presently underway. Based on the available data, it seems likely that the oxidative metabolism of retinoic acid plays a very important role in regulating plasma and tissue levels of retinoic acid. Thus, complete understanding of this metabolism will likely be of central importance for understanding how retinoids act in maintaining normal health and how retinoids can be used clinically in the treatment of disease.

II. Other Metabolism of Retinoic Acid

Conjugated forms of retinoic acid, especially retinoyl-β-glucuronide, are well characterized metabolites of retinoic acid. As described in Chap. 2 of this volume, when all-*trans*-retinoic acid is orally administered to rats, all-*trans*-retinoyl-β-glucuronide is secreted into the bile in significant amounts (DUNAGIN et al. 1966; SWANSON et al. 1981; SKARE et al. 1982; FROLIK 1984).

Retinoyl-β-glucuronide can be synthesized from retinoic acid and uridine diphosphoglucuronic acid by extracts from rat liver, intestine, kidney and other tissues (BARUA and OLSON 1986; BARUA 1997). GENCHI et al. (1996) demonstrated that the relative rates of retinoid-β-glucuronide formation from non-induced rat liver microsomes were: 9-*cis*-retinoic acid > 13-*cis*-retinoic acid > all-*trans*-4-oxo-retinoic acid > all-*trans*-retinoic acid > 13-*cis*-retinol > 9-*cis*-retinol > all-*trans*-retinol. Pretreatment of rats with 3-methylcholanthrene or clofibrate markedly increased retinoyl-β-glucruonide formation, whereas phenobarbital pretreatment had little effect. GENCHI et al. (1996) concluded that the rates of glucuronidation of retinoids depend both on the isomeric state and chemical structure of the retinoids. These investigators also suggested from their studies that various UDP-glucruonosyltransferases may well act on different geometric isomers of retinoic acid (GENCHI et al. 1996).

A specific UDP-glucouronosyltransferase which uses all-*trans*-[^{3}H]retinoic acid as a substrate has been identified in rat liver microsome preparations (LITTLE and RADOMINSKA 1997). This enzyme was identified through photoincorporation of both [^{32}P]5N_3UDP-glucuronic acid and all-*trans*-[^{3}H]retinoic acid into the 52 kDa microsomal protein. Partial purification of this enzyme through UDP-hexanolamine-Sepharose affinity chromatography has been has been obtained (LITTLE and RADOMINSKA 1997) and it appears that this enzyme (and possibly others resembling it) probably plays a physiologically relevant role in hepatic synthesis of retinoyl-β-glucuronides. In other studies, RADOMINSKA et al. (1997) have demonstrated that recombinant rat UDP:glucuronosyltransferase 1.1 (UGT1.1) catalyzes the glucuronidation of all-*trans*-retinoic acid. Rat UGT1.1 displays an apparent K_m of 59.1 ± 5.4 μM towards all-*trans*-retinoic acid. RADOMINSKA et al. (1997) further report that microsomes prepared from livers of Gunn rats, a strain or rat that lacks UGT1.1, still have significant all-*trans*-retinoic acid glucuronidating activity, suggesting that other hepatic UDP:glucuronosyltransferase isoforms also contribute to retinoic acid glucuronidation.

Although retinoyl-β-glucuronides can be excreted into the bile, they also can be hydrolyzed to retinoic acid by enzymes present in rat liver, kidney and intestine (KAUL and OLSON 1998). Plasma retinoic acid levels have long been known to increase after administration of a dose of retinoyl-β-glucuronide and it is thought that this increase arises from hydrolysis of administered retinoyl-β-glucuronide. In vitro rates of hydrolysis of the retinoyl-β-glucuronide were found by KAUL and OLSON (1998) to be higher in tissue preparations from vitamin A-deficient rats than from control vitamin A-sufficient rats. Based on their data, KAUL and OLSON (1998) proposed that in vitamin A-deficiency a regulatory mechanism may exist that enhances the actions of tissue β-glucuronidases to bring about an increased rate of retinoic acid formation from retinoyl-β-glucuronides.

E. Summary and Future Needs

Our understanding of the biochemistry of retinoic acid synthesis and its subsequent metabolism has expanded greatly in the past decade. Nevertheless, we still have only very fragmentary knowledge of these processes. There is presently no information available regarding how cells or tissues regulate retinoic acid formation from retinol. It is unclear whether such regulation involves transcriptional, translational or posttranslational processes or whether allosteric factors or factors that regulate substrate availability are important for regulating retinoic acid synthesis. The findings of BISWAS and RUSSELL (1997), CHAI et al. (1997), SU et al. (1998), and GOUGH et al. (1998) that some enzymes initially identified as retinol dehydrogenases also possess hydroxysteroid dehydrogenase activity suggest metabolic overlaps between these two families of potent bioactive molecules. This raises many interesting new possibilities for how steroids and retinoids can interact to regulate physiological processes within the body. The demonstration that retinoic acid is able to induce expression of a previously unidentified cytochrome P450 form that catalyzes retinoic acid oxidation raises the question of whether still other unknown cytochrome P450 forms able to catalyze the oxidative metabolism of retinoic acid remain to be identified.

It is clear that we are starting to gain insight into some of the biochemical processes that are important for the synthesis and degradation of transcriptionally active retinoids. However, there is almost no information available regarding how retinoic acid synthesis and metabolism are regulated in the context of the intact living organism. We presently have no insight into whether retinoic acid synthesis and degradation are regulated solely at the cellular or tissue levels or whether these processes are regulated in a concerted manner throughout the body. Although it would seem reasonable that retinoic acid synthesis and degradation would be subject to regulation at the level of the whole body, this remains to be demonstrated experimentally. If we are to gain full understanding of retinoic acid synthesis and metabolism, and more importantly of retinoic acid actions within the body, it will be necessary to acquire additional information that links the new and evolving understanding of the biochemistry of retinoic acid formation and metabolism to the physiological context of the whole organism.

Acknowledgements. The work from the authors' laboratory mentioned in this review was supported by NIH grant DK 52444.

References

Abu-Abed SS, Beckett BR, Chiba H, Chithalen JV, Jones G, Metzger D, Chambon P, Petkovich M (1998) Mouse P450RAI (CYP26) expression and retinoic acid-inducible retinoic acid metabolism in F9 cells are regulated by retinoic acid receptor γ and retinoid X receptor α. J Biol Chem 278:2409–2415

Allali-Hassani A, Peralba JM, Martras S, Farrés J, Parés X (1998) Retinoids, ω-hydroxyfatty acids and cytotoxic aldehydes as physiological substrates, and H_2-receptor antagonists as pharmacological inhibitors, of human class IV alcohol dehydrogenase. FEBS Lett 426:362–366

Alvarez RA, Liou GI, Fong S-L, Bridges CDB (1987) Levels of α- and γ-tocopherol in human eyes: evaluation of the possible role of IRBP in intraocular α-tocopherol transport. Am J Clin Nutr 46:481–487

Ang HL, Deltour L, Zgombic-Knight M, Wagner MA, Duester G (1996a) Expression patterns of class I and class IV alcohol dehydrogenase genes in developing epithelia suggest a role for alcohol dehydrogenase in local retinoic acid synthesis. Alcohol Clin Exp Res 20:1050–1064

Ang HL, Deltour L, Hayamizu TF, Zgombic-Knight M, Duester G (1996b) Retinoic acid synthesis in mouse embryos during gastrulation and crainofacial development linked to class IV alcohol dehydrogenase gene expression. J Biol Chem 271:9526–9534

Bailey JS, Siu C-H (1988) Purification and partial characterization of a novel binding protein for retinoic acid from neonatal rat. J Biol Chem 263:9326–9332
Baker ME (1994) Sequence analysis of steroid- and prostaglandin-metabolizing enzymes: application to understanding catalysis. Steroids 59:248–258
Baker ME (1998) Evolution of mammalian 11β- and 17β-hydroxysteroid dehydrogenases-type 2 and retinol dehydrogenases from ancestors in *Caenorhabditis elegans* and evidence for horizontal transfer of a eukaryote dehydrogenase to *E. coli*. J Steroid Biochem Mol Biol 66:355–363
Barua AB, Olson JA (1986) Retinoyl-β-glucuronide: An endogenous compound of human blood. Am J Clin Nutr 43:481–485
Barua AB (1997) Retinoyl beta-glucuronide: a biological active form of vitamin A. Nutr Rev 55:259–267
Bhat PV, Poissant L, Lacroix A (1988a) Properties of retinal-oxidizing enzyme activity in rat kidney. Biochim Biophys Acta 967:211–217
Bhat PV, Poissant L, Falardeau P, Lacroix A (1988b) Enzymatic oxidation of all-*trans*-retinal to retinoic acid in rat tissues. Biochem Cell Biol 66:735–740
Bhat PV, Labrecque J, Boutin J-M, Lacroix A, Yoshida A (1995) Cloning of a cDNA encoding rat aldehyde dehydrogenase with high activity for retinal oxidation. Gene 166:303–306
Bhat PV, Poissant L, Wang XL (1996) Purification and partial characterization of bovine kidney aldehyde dehydrogenase able to oxidize retinal to retinoic acid. Biochem Cell Biol 74:695–700
Bhat PV (1998) Retinal dehydrogenase gene expression in stomach and small intestine of rats during postnatal development and in vitamin A deficiency. FEBS Lett 426:260–262
Bhat PV, Marcinkiewicz M, Li Y, Mader S (1998) Changing patterns of renal retinal dehydrogenase expression parallel nephron development in the rat. J Histochem Cytochem 46:1025–1032
Bhat PV, Samaha H (1999) Kinetic properties of the human liver cytosolic aldehyde dehydrogenase for retinal isomers. Biochem Pharmacol 57:195–197
Biswas MG. Russell DW (1997) Expression cloning and characterization of oxidative 17β- and 3α-hydroxysteroid dehydrogenases from rat and human prostate. J Biol Chem 272:15959–15966
Blaner WS, Olson JA (1994) Retinol and retinoic acid metabolism. In: MB Sporn, AB Roberts, DS Goodman (eds) The retinoids: biology, chemistry, and medicine, 2nd edn. Raven, New York, pp 229–256,
Boerman MHEM, Napoli JL (1995) Characterization of a microsomal retinol dehydrogenase: a short-chain alcohol dehydrogenase with integral and peripheral membrane forms that interacts with holo-CRBP (type I). Biochemistry 34:7027–7037
Boleda MD, Saubi N, Farrés J, Parés X (1993) Physiological substrates for rat alcohol dehydrogenase classes: aldehydes of lipid peroxidation, ω-hydroxy-fatty acids, and retinoids. Arch Biochem Biophys 307:85–90
Chai X, Zhai Y, Popescu G, Napoli JL (1995) Cloning of a cDNA for a second retinol dehydrogenase type II. J Biol Chem 270:28408–28412
Chai X, Napoli JL (1996) Cloning of a rat cDNA encoding retinol dehydrogenase isozyme type III. Gene 169:219–212
Chai X, Zhai Y, Napoli JL (1997) cDNA cloning and characterization of a *cis*-retinol/3α-hydroxysterol short-chain dehydrogenase. J Biol Chem 272:33125–33131
Cheng L, Qian S-J, Rothschild C, d'Avignon A, Lefkowith JB, Gordon JI, Li E (1991) Alteration of the binding specificity of cellular retinol-binding protein II by site-directed mutagenesis. J Biol Chem 266:24404–24412
Driessen CAGG, Winkens HJ, Kuhlmann ED, Janssen APM, van Vugt AF, Janssen JJM (1998) The visual cycle retinol dehydrogenase: possible involvement in the 9-*cis* retinoic acid biosynthetic pathway. FEBS Lett 428:135–140

Duell EA, Kang S, Woorhees JJ (1996) Retinoic acid isomers applied to human skin in vivo each induce a 4-hydroxylase that inactivates only *trans* retinoic acid. J Invest Dermatol 106:316–320

Duester G (1996) Involvement of alcohol dehydrogenase, short-chain dehydrogenase/reductase, aldehyde dehydrogenase, and cytochrome P450 in the control of retinoid signaling by activation of retinoic acid synthesis. Biochemistry 35:12221–12227

Dunagin PE, Jr, Zachman RD, Olson JA (1966) The identification of metabolites of retinol and retinoic acid in rat bile. Biochim Biophys Acta 124:71–85

Eguchi Y, Eguchi N, Oda H, Seiki K, Kijima Y, Matsuura Y, Urade Y, Hayaishi O (1997) Expression of lipocalin-type prostaglandin D synthase (β-trace) in human heart and its accumulation in the coronary circulation of angina patients. Proc Natl Acad Sci USA 94:14689–14694

Eguchi N, Minami T, Shirafuji N, Kanaoka Y, Tanaka T, Nagata A, Yoshida N, Urade Y, Ito S, Hayaishi O (1999) Lack of tactile pain (allodynia) in lipocalin-type prostaglandin D synthase-deficient mice. Proc Natl Acad Sci USA 96:726–730

El Akawi Z, Napoli J (1994) Rat liver cytosolic retinal dehydrogenase: comparison of 13-*cis*-, 9-*cis*-, and all-*trans*-retinal as substrates and effects of cellular retinoid-binding proteins and retinoic acid on activity. Biochemistry 33:1938–1943

Fiorella PD, Napoli JL (1991) Expression of cellular retinoic acid binding protein (CRABP) in *Escherichia coli*. J Biol Chem 266:16572–16579

Flower DR (1996) The lipocalin protein family: structure and function. Biochem J 15:1–14

Frolik CA (1984) Metabolism of retinoids. In: The Retinoids, pp 177–208, Academic, Orlando

Fujii H, Sato T, Kaneko S, Gotoh O, Fujii-Kuriyama Y, Osawa K, Kato S, Hamada H (1997) Metabolic inactivation of retinoic acid by a novel P450 differentially expressed in developing mouse embryos. EMBO J 15:4163–4173

Gamble MV, Shang E, Piantedosi R, Mertz JR, Wolgemuth DJ, Blaner, WS (1999) 9-*cis*-retinol dehydrogenase: biochemical properties, tissue expression and gene structure (submitted)

Genchi G, Wang W, Barua A, Bidlack WR, Olson JA (1996) Formation of β-glucuronides and β-galacturonides of various retinoids catalyzed by induced and noninduced microsomal UDP-glucruonosyltransferases of rat liver. Biochim Biophys Acta 1289:284–290

Gough WH, VanOoteghem S, Stint T, Kedishvili NY (1998) cDNA cloning and characterization of a new human microsomal NAD^+-dependent dehydrogenase that oxidizes all-*trans*-retinol and 3α-hydroxysteroids. J Biol Chem 273:19778–197785

Haeseleer F, Huang J, Lebioda L, Saari JC, Palczewski K (1998) Molecular characterization of a novel short-chain dehydrogenase/reductase that reduces all-*trans*-retinal. J Biol Chem 273:21790–21799

Han IS and Choi J-H (1996) Highly specific cytochrome P450-like enzymes for all-*trans*-retinoic acid in T47D human breast cancer cells. J Clin Endocrinol Metab 81:2069–2075

Haselbeck RJ, Ang HL, Deltour L, Duester G (1997) Retinoic acid and alcohol/retinol dehydrogenase in the mouse adrenal gland: a potential endocrine source of retinoic acid during development. Endocrinology 138:3035–3041

Haselbeck RJ, Duester G (1998a) ADH1 and ADH4 alcohol/retinol dehydrogenases in the developing adrenal blastema provide evidence for embryonic reitnoid endocrine function. Dev Dyn 213:114–120

Haselbeck RJ, Duester G (1998b) ADH4-lacZ transgenic mouse reveals alcohol dehydrogenase localization in embryonic midbrain/hindbrain, otic vesicles, and mesencephalic, trigeminal, facial and olfactory neural crest. Alcohol Clin Exp Res 22:1607–1613

Hubuterne X, Wang XD, Johnson EJ, Krinsky NI, Russell RM (1995) Intestinal absorption and metabolism of 9-*cis*-β-carotene in vivo: biosynthesis of 9-*cis*-retinoic acid. J Lipid Res 36:1264–1273

Hupert J, Mobarhan S, Layden TJ, Papa VM, Lucchesi DJ (1991) In vitro formation of retinoic acid from retinal in rat liver. Biochem Cell Biol 69:509–514

Imaoka S, Wan J, Chow T, Hiroi T, Eyanagi R, Shigematsu H, Funae Y (1998) Cloning and characterization of the CYP2D1-binding protein, retinol dehydrogenase. Arch Biochem Biophys 353:331–336

Jörnvall H, Höög JO (1995) Nomenclature of alcohol dehydrogenases. Alcohol Alcohol 30:153–161

Jörnvall H, Persson B, Krook M, Atrian S, Gonzalez-Duarte R, Jeffery J, Ghosh D (1995) Short-chain dehydrogenases/reductases (SDR). Biochemistry 34:6003–6013

Kaul S, Olson JA (1998) Effect of vitamin A deficiency on the hydrolysis of beta-glucuronide to retinoic acid by rat tissue organelles in vitro. Int J Vit Nutr Res 68:232–236

Kedishvili NY, Gough WH, Chernoff EAG, Hurley TD, Stone CL, Bowman KD, Popov KM, Bosron WF, Li T-K (1997) cDNA sequence and catalytic properties of a chick embryo alcohol dehydrogenase that oxidizes retinol and 3β,5α-hydroxysteroids. J Biol Chem 272:7494–7500

Kedishvili NY, Gough WH, Davis WI, Parsons S, Li TK, Bosron WF (1998) Effect of cellular retinol-binding protein on retinol oxidation in human class IV retinol/alcohol dehydrogenase and inhibition by ethanol. Biochem Biophys Res Commun 249:191–196

Kim SY, Han IS, Yu HK, Lee HR, Chung JW, Choi, JH, Kim SH, Byun Y, Carey TE, Lee KS (1998) The induction of P450-mediated oxidation of all-trans-retinoic acid by retinoids in head and neck squamous cell carcinoma cell lines. Metabolism 47:955–958

Krekels MDWG, Verhoeven A, van Dun J, Van Hove C, Dillen L, Coene M-C, Wouters W (1997) Induction of the oxidative catabolism of retinoic acid in MCF-7 cells. Br J Cancer 75:1098–1104

Labrecque J, Bhat PV, Lacroix A (1993) Purification and partial characterization of a rat kidney aldehyde dehydrogenase that oxidizes retinal to retinoic acid. Biochem Cell Biol 71:85–89

Labrecque J, Dumas F, Lacroix A, Bhat PV (1995) A novel isoenzyme of aldehyde dehydrogenase specifically involved in the biosynthesis of 9-*cis*- and all-trans-retinoic acid. Biochem J 305:681–684

Lee M-O, Manthey CL, Sladek NE (1991) Identification of mouse liver aldehyde dehydrogenases that catalyze the oxidation of retinal to retinoic acid. Biochem Pharm 42:1279–1285

Levin MS, Locke B, Yang NC, Li E, Gordon JI (1988) Comparison of the ligand binding properties of two homologous rat apocellular retinol-binding proteins expressed in *Escherichia coli*. J Biol Chem 263:17715–17723

Li E, Qian SJ, Winter NS, d'Avignon A, Levin MS, Gardon JI (1991) Flavorite nuclear magnetic resonance analysis of the ligand binding properties of two homologous rat cellular retinal-binding proteins expressed in Escherichia coli. J Biol Chem 266:3622–3629

Little JM, Radominska A (1997) Application of photoaffinity labeling with [11,12-^{3}H]all-*trans*-retinoic acid to characterization of rat liver microsomal UDP-glucuronosyltransferase(s) with activity toward retinoic acid. Biochem Biophys Res Commun 230:497–500

Livrea MA, Bongierno A, Tesoriere L, Nicotran C, Bone A (1987) Binding of 11-*cis*-retinal to the partially purified cellular retinal binding protein from barine retinal pigment epithelium. Experientia 43:582–586

Livrea MA, Tesoriere L (1990) Binding of 11-*cis*-retinal to cellular retinal-binding protein from pigment epithelium. Methods Enzy nd 189:369–373

MacDonald PN, Ong DE (1987) Binding specificities of cellular retinol-binding protein and cellular retinol-binding protein, type II. J Biol Chem 262:10550–10556
McCaffery P, Dräger UC (1994) High levels of retinoic acid-generating dehydrogenase in the meso-telencephalic dopamine system. Proc Natl Acad Sci USA 91:7772–7776
Mertz JR, Shang E, Piantedosi R, Wei S, Wolgemuth DJ, Blaner WS (1997) Identification and characterization of a stereospecific human enzyme that catalyzes 9-*cis*-retinol oxidation. J Biol Chem 272:11744–11749
Nagao A, Olson JA (1994) Enzymatic formation of 9-*cis*, 13-*cis* and all-*trans* retinals from isomers of β-carotene. FASEB J 8:968–973
Napoli JL, Race KR (1988) Biogenesis of retinoic acid from β-carotene. J Biol Chem 263:17372–17377
Napoli JL (1996) Retinoic acid biosynthesis and metabolism. FASEB J 10:993–1001
Newcomer ME, Ong DE (1990) Purification and crystallization of a retinoic acid-binding protein from rat epididymis. J Biol Chem 265:12876–12879
Niederreither K, McCaffery P, Dräger UC, Chambon P, Dollé P (1997) Restricted expression and retinoic acid-induced downregulation of the retinal dehydrogenase type 2 (RALDH-2) gene during mouse development. Mech Dev 62:67–78
Noy N, Blaner WS (1991) Interactions of retinol with binding proteins: studies with rat cellular reitnol-binding protein and with rat retinol-binding protein. Biochemistry 30:6380–6386
Ong DE, Chytil F (1978) Cellular retinol-binding protein from rat liver. J Biol Chem 253:828–832
Ong DE, Newcomer ME, Chytil F (1994) Cellular retinoid-binding proteins, In: Sporn MB, Roberts AB, Goodman DS (eds) The retinoids, biology, chemistry, and medicine, 2nd edn. Raven, New York, pp 283–312
Posch KC, Boerman MHEM, Burns RD, Napoli JL (1991) Holo-cellular retinol-binding protein as a substrate for microsomal retinal synthesis. Biochemistry 30:6224–6230
Posch KC, Burns RD, Napoli JL (1992) Biosynthesis of all-*trans*-retinoic acid from retinal. Recognition of retinal bound to cellular retinol binding protein (type I) as substrate by a purified cytosolic dehydrogenase. J Biol Chem 267:19676–19682
Putilina T, Sittenfeld D, Chader GJ, Wiggert B (1993) Study of a fatty acid binding site of interphotoreceptor retinoid-binding protein using fluorescent fatty acids. Biochemistry 32:3797–3803
Radominska A, Little JM, Lehman PA, Samokyszyn V, Rios GR, King CD, Green MD, Tephly TR (1997) Glucuronidation of retinoids by rat recombinant UDP:glucuronosyltransferase 1.1 (bilirubin UGT). Drug Met. Dispos. 25:889–892
Ray WJ, Bain G, Yao M, Gottlieb DI (1997) CYP26, a novel mammalian cytochrome P450, is induced by retinoic acid and defines a new family. J Biol Chem 272:18702–18708
Rigas, JR, Francis PA, Muindi JRF, Kris MG, Huselton C, DeGrazia F, Orazem JP, Young CW, Warrell RP, Jr (1993) Constitutive variability in the pharmacokinetics of the natural retinoid, all-*trans*-retinoic acid, and its modulation by ketoconazole. J Natl Cancer Inst 85:1921–1926
Rigas JR, Miller VA, Zhang Z-F, Klimstra DS, Tong WP, Kris MG, Warrell RP, Jr (1996) Metabolic phenotypes of retinoic acid and the risk of lung cancer. Cancer Res 56:2692–2696
Roberts AB, Nichols MD, Newton DL, Sporn MB (1979) In vitro metabolism of retinoic acid in hamster intestine and liver. J Biol Chem 254:6296–6302
Roberts ES, Vaz ADN, Coon MJ (1992) Role of isozymes of rabbit microsomal cytochrome P-450 in the metabolism of retinoic acid, retinol, and retinal. Mol Pharmacol 41:427–433
Saari JC, Bredberg DL (1988) Purification of cellular retinal-binding protein from barine retina and retinal pigment epithelium Exp Eye Res 46:569–578
Shih T-W, Lin T-H, Shealy YF, Hill DL (1997) Nonenzymatic isomerization of 9-cis-retinoic acid catalyzed by sulfhydryl compounds. Drug Metabol Dispos 25:27–32

Simon A, Hellman U, Wernstedt C, Eriksson U (1995) The retinal pigment epithelial-specific 11-cis retinol dehydrogenase belongs to the family of short chain alcohol dehydrogenases. J Biol Chem 270:1107–1112

Skare KL, Schnoes HK, DeLuca HF, (1982) Biliary metabolites of all-*trans*-retinoic acid in the rat: isolation and identification of a novel polar metabolite. Biochemistry 21:3308–3317

Sonneveld E, van den Brink CE, van der Leede BM, Schulkes RK, Petkovich M, van der Burg B, van der Saag PT (1998) Human retinoic acid (RA) 4-hydroxylase (CYP26) is highly specific for all-*trans*-retinoic acid and can be induced through retinoic acid receptors in human breast and colon carcinoma cells. Cell Growth Differ 9:629–637

Su J, Chai X, Kahn B, Napoli JL (1998) cDNA cloning, tissue distribution, and substrate characteristics of a *cis*-retinol/3α-hydroxysterol short-chain dehydrogenase isozyme. J Biol Chem 273:17910–17916

Suzuki Y, Ishiguro S-I, Tamai M (1993) Identification and immunohistochemistry of retinol dehydrogenase from bovine retinal pigment epithelium. Biochim Biophys Acta 1163:201–208

Swanson BM, Frolik CA, Zaharevitz DW, Roller PP, Sporn MB (1981) Dose-dependent kinetics of all-*trans*-retinoic acid in rats. Biochem Pharmacol 30:107–113

Tanaka T, Urade Y, Kimura H, Eguchi N, Nishikawa A, Hayaishi O (1997) Lipocalin-type prostaglandin D synthase (β-trace) is a newly recognized type of retinoid transporter. J Biol Chem 272:15789–15796

Urbach J, Rando RR (1994) Isomerization of all-*trans*-retinoic acid to 9-*cis*-retinoic acid. Biochem J 299:459–465

van Wauwe J, van Nyen G, Coene M-C, Stoppie P, Cools W, Goossens J, Borghgraef P, Janssen PAJ (1992) Liarazole, an inhibitor of retinoic acid metabolism, exerts retinoid-mimetic effects in vivo. J Pharmacol Exp Ther 261:773–779

Wang X-D, Tang G-W, Fox JG, Krinsky NI, Russell RM (1991) Enzymatic conversion of β-carotene into β-apo-carotenals and retinoids by human, monkey, ferret, and rat tissues. Arch Biochem Biophys 285:8–16

Wang X-D, Krinsky NI, Tang G, Russell RM (1992) Retinoic acid can be produced from excentric cleavage of β-carotene in human intestinal mucosa. Arch Biochem Biophys 293:298–304

Wang X, Penzes P, Napoli JL (1996) Cloning of a cDNA encoding an aldehyde dehydrogenase and its expression in *Escherichia coli*. J Biol Chem 271:16288–16293

White JA, Guo Y-D, Baetz K, Beckett-Jones B, Bonasoro J, Hsu KE, Dilworth FJ, Jones G, Petkovich M (1996) Identification of the retinoic acid-inducible all-*trans*-retinoic acid 4-hydroxylase. J Biol Chem 271:29922–29927

White JA, Beckett-Jones B, Guo Y-D, Dilworth FJ, Bonasoro J, Jones G, Petkovich M (1997) cDNA cloning of human retinoic acid-metabolizing enzyme (hP450RAI) identifies a novel family of cytochromes P450 (CYP26). J Biol Chem 272:18538–18541

Yamamoto M, Dräger UC, Ong DE, McCaffery P (1998) Retinoid-binding proteins in the cerebellum and choroid plexus and their relationship to regionalized retinoic acid synthesis and degradation. Eur J Biochem 257:344–350

Yang ZN, Davis GJ, Hurley TD, Stone CL, Li TK, Bosron WF (1994) Catalytic efficiency of human alcohol dehydrogenases for retinol oxidation and retinal reduction. Alcohol Clin Exp Res 18:587–591

Zhai Y, Higgins D, Napoli JL (1997) Coexpression of the mRNAs encoding retinol dehydrogenase isozymes and cellular retinol-binding protein. J Cell Physiol 173:36–43

Zhao D, McCaffery P, Ivins KJ, Neve RL, Hogan P, Chin WW, Dräger UC (1996) Molecular identification of a major retinoic-acid-synthesizing enzyme, a retinal-specific dehydrogenase. Eur J Biochem 240:15–22

Section B
Binding Proteins and Nuclear Receptors

CHAPTER 5

Nuclear Retinoid Receptors and Mechanisms of Action

F. JAVIER PIEDRAFITA and M. PFAHL

A. Introduction

Fat-soluble small molecule hormones and vitamins have long attracted the curiosity of molecular biologists, since they appeared to regulate gene expression in eukaryotic cells via similar mechanisms as certain signal molecules that had been studied extensively in prokaryotic systems. Vitamin A and its natural and synthetic analogues and derivatives (the retinoids) were of particular interest because of a variety of reasons. Starting a century ago it became more and more obvious that vitamin A played a very central role in the regulation and timing of many important biological processes, including development, differentiation, morphogenesis, growth, metabolism and homeostasis. Because of their central biological roles it appeared possible that retinoids serve as therapeutic agents, increasing the desire and need to understand their molecular mechanism of action. This became particularly important when synthetic vitamin A derivatives were sought with fewer undesirable side effects. Thus not surprisingly, as soon as the molecular biology technologies became available, a rapid progress was made in the deciphering of molecular signal transduction pathways of small fat soluble hormones and vitamins including the retinoids.

Consistent with their pleiotropic biological effects, the retinoid signaling pathways proved to be particularly complex. At the same time, however, it became apparent that many of the small molecule hormones and vitamins functioned similar to prokaryotic signal molecules, by directly regulating the activity of a class of DNA-binding proteins now well known as the nuclear receptor superfamily. While these nuclear receptors are the executive signal transducers, responding to a number of natural vitamin A derivatives – the most important of which are all-*trans* retinoic acid (tRA) and 9-*cis* retinoic acid (RA) – it has also become apparent that vitamin A signal control involves many levels, including: uptake, storage, transport, activation, modification and breakdown, and the interaction with cellular retinol and cellular retinoic acid proteins – whose roles are not yet very well understood. This chapter focuses on the final step in the retinoid signal transduction, the nuclear retinoid receptors and their mechanism of action. Since much of this has been reviewed in great detail before, we concentrate particularly on the more recent observations while more briefly summarizing previously reviewed mechanisms.

B. The Nuclear Retinoid Receptors and Their Cousins

Six bonafide retinoid receptors were discovered which fall into two classes, the retinoic acid receptors (RARs) α, β, and γ (PETKOVICH et al. 1987; GIGUERE et al. 1987, 1990; BENBROOK et al. 1988; BRAND et al. 1988; KRUST et al. 1989) and the retinoid X receptors (RXR) α, β, and γ (HAMADA et al. 1989; MANGELSDORF et al. 1990, 1992; YU et al. 1991; LEID et al. 1992) (Fig. 1). Each subtype is encoded by a specific gene from which usually multiple isoforms can be generated involving differential splicing and multiple promoters (LEHMANN et al. 1991a; LEROY et al. 1992). The receptor subtypes and isoforms are expressed in a developmental and tissue specific manner, suggesting that each one of them may have specific tasks in the regulation of developmental and cell type or tissue specific biological processes. The two classes of receptors, the RARs and RXRs are quite distinct from each other. While, as with all nuclear receptors, they contain the typical domains including a DNA- and a ligand-binding domain (DBD and LBD), the two classes of receptors appear to have evolved independently because of their low amino acid homologies in their LBD (for a detailed review of the receptor building plans and their phylogenetic relationships see PFAHL et al. 1994). Not surprisingly, the ligand specificities of the receptors differ. RARs bind tRA and 9-*cis* RA while RXRs bind only 9-*cis* RA. In addition, so-called RXR selective retinoids – a class of syn-

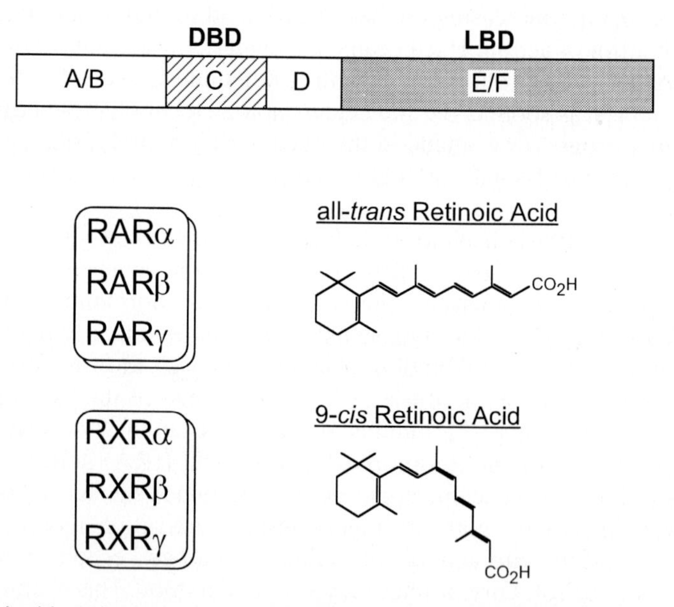

Fig. 1. Retinoid nuclear receptors. Schematic domain structure of a nuclear hormone receptor. The position of the DBD and the LBD is shown. Both RAR and RXR comprise subfamilies that include three independent receptor subtypes (α, β, and γ) encoded by different genes. The chemical structure of the natural retinoids, tRA and 9-*cis*-RA are also indicated

thetic retinoids that only bind RXRs with high affinity – have been discovered (Lehmann et al. 1992), while natural high-affinity ligands selective for RXR have not been found so far.

Consistent with their apparent independent evolution, the roles of RARs and RXRs are also very distinct. The RARs appear to be the major vitamin A signal transducers. To convert a retinoid signal into transcriptional activation of a gene, the RARs function however, as heterodimers with RXRs. Only the RXR-RAR heterodimers were found to bind effectively to their DNA recognition sequences, the retinoic acid response elements (RARE) (Zhang et al. 1992a; Kliewer et al. 1992a; Marks et al. 1992; Bugge et al. 1992). In vivo gene knockout studies in mice (Kastner et al. 1997) as well as in cell lines (Chiba et al. 1997) have now clearly confirmed the essential role of the RXR-RAR heterodimer for retinoid signal transduction. Activation of the heterodimer is controlled by the RARs since the RXRα-RARα heterodimer was found to respond only to RAR ligands but not to RXR selective ligands (Kurokawa et al. 1994). Once the RAR has bound a ligand, RXR ligands can lead to further, optimal activation of the heterodimer (Minucci et al. 1997). From this it appears that the RXRs function only as enhancers of retinoid signals. However, we have observed that the strong dominance observed for RARα over RXRα (Kurokawa et al. 1994) was not seen with RARβ and RARγ(La Vista-Picard et al. 1996). Both RARβ- and RARγ-containing heterodimers could be at least partially activated by RXR selective ligands in the absence of RAR ligands (La Vista-Picard et al. 1996). However, since it is unknown at this point whether natural high-affinity RXR selective ligands exist, the RAR dominance discussed above may be of limited significance for normal retinoid signal transduction, i.e., tRA activates only RARs, while 9-*cis*-RA would simultaneously activate the RAR and RXR components in the heterodimer. In contrast, the RAR dominance over RXR may turn out to be very important in some retinoid therapies, approaches where RXR selective retinoids can be used without activating the general retinoid response or where 9-*cis*-RA is inactive (see below).

Simultaneous to the discovery that RARs required RXRs for efficient DNA binding and transactivation it was also found that RXR was an essential cofactor for several other nonsteroidal nuclear receptors, including the thyroid hormone receptors (TRs) (Zhang et al. 1992a; Kliewer et al. 1992a; Marks et al. 1992; Bugge et al. 1992), and the vitamin D receptor (VDR) (Kliewer et al. 1992a), where RXRs appear to function mostly as silent – not ligand activated – partners. A clearly central role for RXRs was revealed when it was found to function as coreceptor for a still increasing group of nuclear receptors that included novel ligand binding receptors as well as orphan receptors (Kliewer et al. 1992b 1995; Forman et al. 1995a; Devchand et al. 1996; Lala et al. 1997; Apfel et al. 1994; Bes et al. 1994), many of which may not interact with high-affinity ligands (Escriva et al. 1997).

In most of these latter cases the RXRs appear to function as ligand responsive receptors. That is, when RXR heterodimerizes with PPARγ, LXR, or FXR, these heterodimers can be activated by either RXR selective ligands, or by the partner's ligand: fatty acids and thiazolidinediones (PPARγ) (FORMAN et al. 1995b; KLIEWER et al. 1995), oxysterols (LXR/RLD-1) (JANOWSKI et al. 1996), and farnesoids (FXR) (FORMAN et al. 1995a; ZAVACKI et al. 1997). RXR has also been shown to form heterodimers with the orphan receptor nur77/NGFB-1, which binds to DR-5 elements and is activated by RXR ligands (PERLMANN and JANSSON 1995). Furthermore, RXR can bind and activate DNA as a homodimer, which is induced in the presence of its natural ligand, 9-*cis*-RA (ZHANG et al. 1992b). It should be pointed out here that since RXR homodimers bind to some of the same response elements as some of the heterodimers it is often difficult to determine whether at high RXR ligand concentration the increased activation observed results from the RXR molecule in the heterodimer or from RXR homodimers.

Nevertheless, it is apparent that the RXRs have a broad role as coreceptors thereby allowing cross-talk between certain retinoid signals and other hormonal and vitamin signals. This is likely to occur not only through the RXR containing heterodimers but also through competition between various RXR partners for RXR, when the RXR molecules are in limited supply. In the presence of high concentrations of RXR ligands, competition between RXR homodimers and heterodimers may also occur (LEHMANN et al. 1993). Thus retinoid receptors are part of a complex network that allows interactions and cross-talk between a large number of nuclear receptors and their ligands.

C. The RAREs: Further Specification of the Ligand Response

The nuclear receptors are transcription factors that modulate transcription through binding to specific DNA sequences, named hormone response elements (HREs) or RAREs in the case of the retinoid receptors. HREs and/or RAREs are usually located in the promoter region of responsive genes. The RAREs normally consist of two hexameric half sites with the consensus sequence AGGTCA, which can be arranged as direct repeats (DR), palindromes (Pal), or inverted palindromes (IP), separated by a spacer with variable number of nucleotides (Fig. 2) (GLASS 1994). The differential arrangement of the half sites and the distance between them are critical determinants of the specificity and mode of binding of the receptors. RARs can bind as heterodimers with RXR to direct repeats separated by 1, 2, or 5 nucleotides (UMESONO et al. 1991; HUSMANN et al. 1992; SMITH et al. 1991; MANGELSDORF et al. 1991; DURAND et al. 1992; PIEDRAFITA et al. 1996; HOFFMANN et al. 1990; SUCOV et al. 1990; DE THÉ et al. 1990). RAR-RXR heterodimers can also bind to palindromes with no spacer (GLASS et al. 1988; UMESONO et al. 1988) or with a nine-nucleotide spacer (LEE et al. 1994), as well as to inverted palindromes with 6 or 8 bases in the spacer (BANIAHMAD et al. 1990; TINI et al.

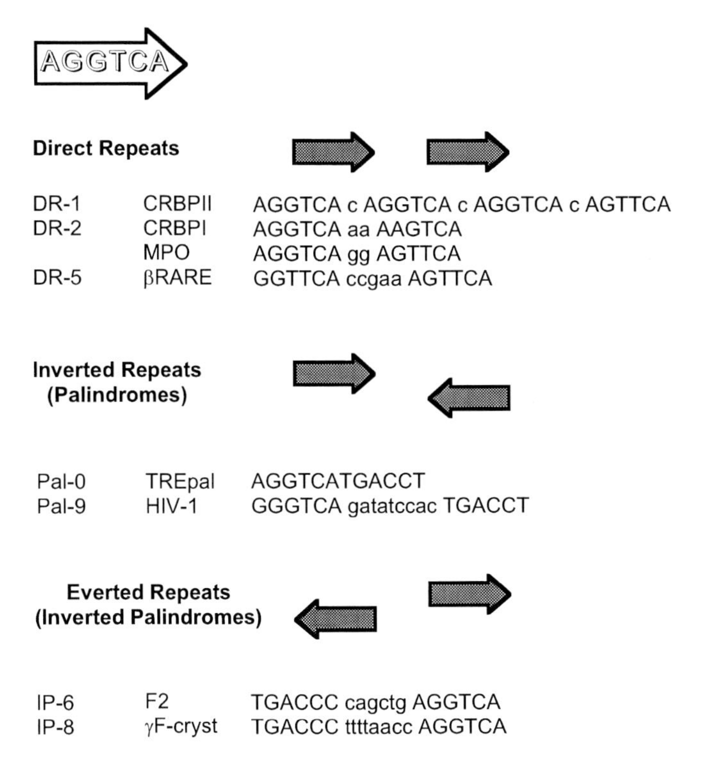

Fig. 2. Retinoid response elements. The nucleotide sequence of some retinoid response elements is shown. *Arrow*, half site binding site and consensus sequence; *upper-case letters*, binding motif; *lower-case letters*, spacer between half sites. Retinoid response elements are classified according to the arrangement of the half site motif in direct repeats, palindromes, or inverted palindromes. CRBPI and CRBPII are found in the promoter region of the cellular retinol binding protein I and II, respectively (Husmann et al. 1992; Smith et al. 1991; Mangelsdorf et al. 1991; Durand et al. 1992). MPO is a DR-2 found in an Alu element located 5' the myeloperoxidase gene (Piedrafita et al. 1996), and βRARE is a DR-5 located near the TATA-box of the RARβ 2 gene (Hoffmann et al. 1990; Sucov et al. 1990; de Thé et al. 1990). A palindromic element with no spacer derived from the growth hormone sequence (TREpal) binds RARs, as well as RXRs and TRs (Glass et al. 1988; Umesono et al. 1988). Another palindromic retinoid response element with a 9-bp spacer is found in the HIV promoter (Lee et al. 1994). Two inverted palindromes with 6- and 8-bp spacers are found in the chicken lysozyme silencer (F2) and in the γF-crystallin gene, respectively (Baniahmad et al. 1990; Tini et al. 1993)

1993). RXR homodimers can bind to and activate palindromic elements (Zhang et al. 1992b; Lee et al. 1994) and DR-1 elements (Zhang et al. 1992b).

DNA binding of most of the nuclear receptors – with the exception of some of the steroid receptors – is ligand independent. In fact, it was observed early on that TR and RARs can have dual functional roles operating as transcriptional repressors in the absence of ligands and as transcriptional activators in the presence of ligand (GRAUPNER et al. 1989; DAMM et al. 1989). This was later confirmed by the discovery of the cofactors necessary for repression and activation functions (see below).

Interestingly, some of the RAREs are also bound by other receptors, including orphan receptors. For the orphan receptor COUP-TF and its subtypes, it was first recognized that this receptor can bind as a homodimer to several RAREs with high affinity, and thereby block access of RAR-RXR to DNA, resulting in repression of the retinoid response (TRAN et al. 1992). Thus, temporal and spacial coexpression patterns of COUP-TF and certain other orphan receptors such as TOR and Tak-1 (ORTIZ et al. 1995; HIROSE et al. 1995) can lead to a restriction of the retinoid response to such genes that carry RAREs recognized only by the retinoid receptor heterodimers but not by the orphan receptors expressed. For example, COUP-TF is expressed at high levels at certain stages of brain development and is therefore likely to strongly repress certain genes that in other tissues are retinoid responsive (LU et al. 1994). One explanation for the observed diversity in the RAREs is therefore that this diversity allows for tissue- and time-specific retinoid responses of certain genes.

In addition, a detailed analysis of the ligand responsiveness of various RAR-RXR heterodimers when complexed with different response elements revealed that it is in fact the RAR-RXR heterodimer-RARE complex that determines the affinity and thereby the response to a retinoid, and not the receptor subtype and/or isoforms alone. That is, the same RARα-RXRα heterodimer can show different ligand responsiveness on the various RAREs. Thus, the different RARE structures allow for a further diversification of the retinoid response, which in the natural setting may allow selective or preferential activation of certain receptor-RARE complexes by specific retinoic acid derivatives and metabolites. For retinoid drug development, it may allow the design of "gene selective" retinoids (LA VISTA-PICARD et al. 1996).

D. Interaction of Retinoid Receptors with Coactivators, Corepressors, and Basal Transcription Factors

During recent years important details have been elucidated on how the nuclear receptors can mediate repression and transcriptional activation by interacting with coactivators, corepressors, and basal transcription factors. This relatively new material is reviewed here in some detail (for a summary the reader is referred to Figs. 3, 4). As noted above, RARs and TRs can inhibit basal transcription in the absence of ligand and activate transcription in a ligand-dependent manner. The resolution of the three-dimensional structures of the unliganded RXRα LBD (BOURGUET et al. 1995) and the RARγ LBD bound to tRA (RENAUD et al. 1995) suggested that significant conformational changes happen upon ligand binding. One of the most striking conclusions from these crystallographic studies is that the amphipatic α-helix H12, which

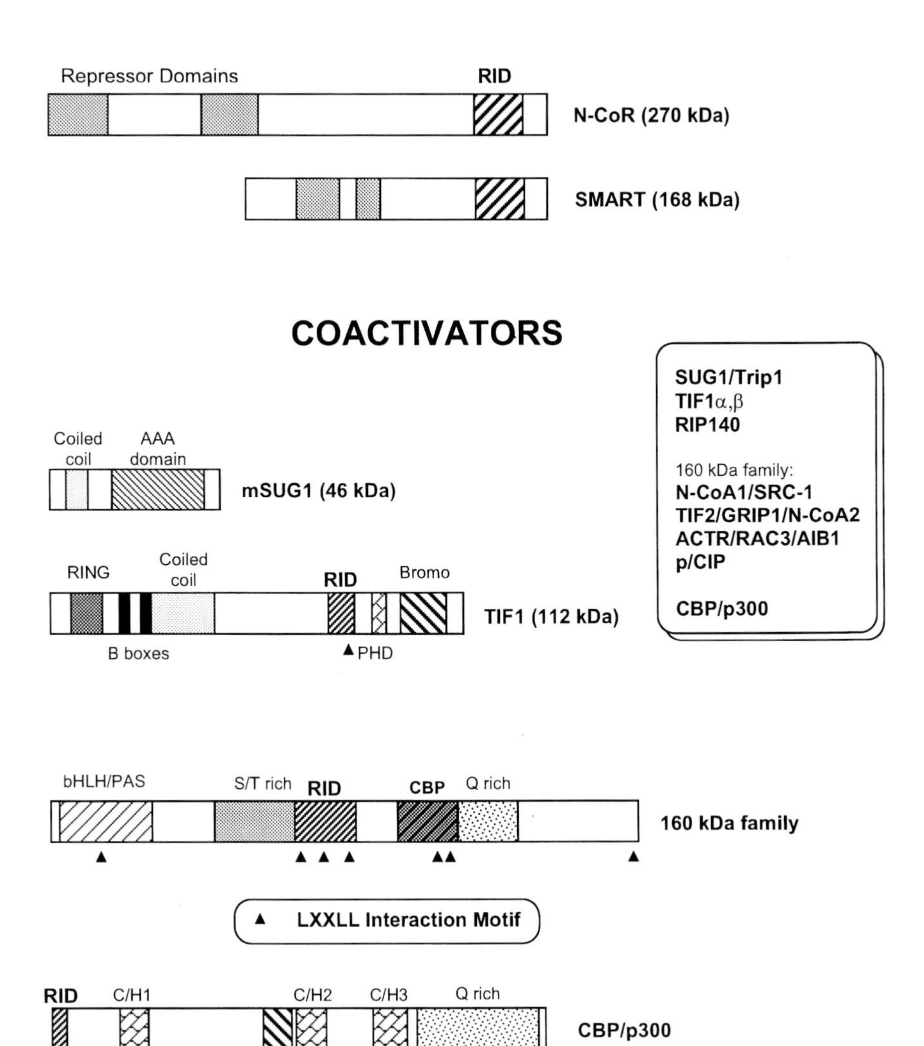

Fig. 3. Structure of corepressors and coactivators of nuclear hormone receptors. The receptor interaction domain is indicated, as well as two independent repressor domains present in N-CoR and SMRT (*shaded*). Various proteins with different structural characteristics can serve as coactivators for the nuclear hormone receptors. The largest group, the p160 family, comprises several different proteins which share similar structural characteristics (see text). *Closed triangle*, position of the helical LXXLL receptor interaction motif found in several coactivators; *open triangles*, position of interaction domains with other proteins, found also in the cointegrator CBP/p300

contains the associated factor (AF)-2 activation domain, protrudes beyond the core of the RXRα apo-LBD (unliganded LBD), but it is compactly folded along the core of the RARγ holo-LBD (ligand bound LBD). A comparable folding of the H12 has been observed in the TRα holo-LBD (Wagner et al. 1995). This predicted ligand-induced conformational change indicates the possibility of novel interactions between the retinoid receptors and other regulatory proteins, such as components of the general transcription machinery and other putative cofactors. In fact, ligand-dependent interactions of nuclear receptors with several basal transcription factors have been reported, suggesting a simple model of how liganded receptors could recruit the TATA box binding protein (TBP) and other TBP-associated factors (TAFs), thereby promoting the assembly of the preinitiation complex and allowing gene transcription by RNA pol II. However, the fact that various receptors can interfere with the transcriptional activity of each other suggested the existence of limiting factors, now called coactivators, required for receptor activity. Similarly, the observation that RAR or v-erbA can reverse the transcriptional repression by unliganded TR, indicated that other cofactors, named corepressors, are required for repression.

It is generally assumed that binding of TBP is the rate-limiting step in transcription (reviewed in Tjian and Maniatis 1994; Buratowski 1994), and that the role of transactivators such as the retinoid receptors could be to interact with TBP and/or other TAFs and increase the rate of formation of a preinitiation complex. Association between TBP and nuclear receptors has been indeed reported, for instance the ligand-dependent interaction of RXRα with TBP in yeast (Schulman et al. 1995). TBP can also interact with ER in a ligand independent manner (Sadovsky et al. 1995). In addition to TBP, nuclear receptors also interact with several TAFs. For example, $TAF_{II}30$ has been shown to interact with ER (Jacq et al. 1994), while $TAF_{II}110$ can interact with RXR and TR in a ligand-dependent manner (Schulman et al. 1995). In addition, RAR as well as TR, VDR, and other receptors, have been reported to interact with TFIIB (Blanco et al. 1995; Baniahmad et al. 1993; Malik and Karathanasis 1996). Most interesting is the recent observation that TFIIH can interact with RARα, and this interaction permits phosphorylation of Ser-77 of the receptor by cdk7, a component of TFIIH, which activates RAR-mediated transcription (Rochette-Egy et al. 1997).

The existence of putative corepressors or silencing factors necessary for the ligand-independent repression by nuclear receptors was suggested by several studies (Tong et al. 1996; Baniahmad et al. 1995; Casanova et al. 1994). Two proteins, nuclear receptor corepressor (N-CoR) (Horlein et al. 1995) and silencing mediator for retinoid and thyroid hormone receptors (SMRT) (Chen and Evans 1995), have been isolated and shown to interact with unliganded RAR and TR, but not with ligand bound receptors.

In solution, both N-CoR and SMRT strongly interact with RAR and TR in the absence of ligand, but not with VDR and only weakly with RXR, although this interaction is not affected by the ligand. The corepressors can form a ternary complex with a RAR/RXR heterodimer bound to a DR-5-

RARE, and this complex dissociates in the presence of RAR ligands but not when RXR selective retinoids are present (KUROKAWA et al. 1995). The receptor interaction domain of N-CoR has been mapped to a small region near the carboxyl end of the molecule (HORLEIN et al. 1995), and this domain shows high homology with a similar region in SMRT. On the nuclear receptor molecule, the hinge region of both TR and RAR contains the corepressor interaction domain, or CoR box. The CoR box is highly conserved in both RARs and TRs, and mutations within this domain have been shown to affect both binding to corepressors and repression activity of the receptor mutants (HORLEIN et al. 1995), confirming that interaction with the corepressor is essential for the transrepression activity of unliganded receptors. In addition to the receptor interaction domain, both corepressors contain two independent repressor domains in the amino terminal half of the molecule (Fig. 3). The gene repression activity of orphan receptors COUP-TFs and Rev-erbA is also mediated by N-CoR and SMRT (ZAMIR et al. 1996; SHIBATA et al. 1997).

Interaction of nuclear receptors with a corepressor in solution is not always correlated with repression activity, and binding to DNA is an important determinant for receptor-corepressor interactions. Recent evidence suggests that the hinge and the LBD regions of the receptor acquire a particular conformation upon binding to a selective HRE which allows for specific interaction with a corepressor. The configuration of a RAR/RXR heterodimer on a DR-1 or a DR-5-RARE is different, in that both receptors bind with opposite polarity depending on the spacer (KUROKAWA et al. 1994). This would explain why retinoids disrupt the interaction of N-CoR with a DR-5 bound RAR/RXR heterodimer, but not when RAR/RXR is bound to a DR-1-RARE. This provides a molecular argument for the lack of transactivation by RAR/RXR heterodimers on this particular RARE (KUROKAWA et al. 1995). N-CoR, but not SMRT, can interact with Rev-erbA when bound to DR-2 elements as a dimer, but not when it binds as a monomer, correlating with the transrepression activities of this orphan receptor (ZAMIR et al. 1997). Similarly, corepressors can bind to TR homodimers but not to monomers on DNA, in agreement with the strong repression activity of the TR homodimers (ZAMIR et al. 1997; PIEDRAFITA et al. 1995). In addition, both N-CoR and SMRT can interact with PPAR in solution, but not when the receptor is bound to DNA, correlating with the observation that PPAR cannot repress transcription on a DR-1-PPARE (ZAMIR et al. 1997). Therefore binding of the receptors to specific DNA elements promotes a particular conformational change in the LBD that allows binding (repression) and ligand-mediated dissociation (release of repression) of the corepressors.

Biochemical studies have identified two protein species of 140 (RIP140) and 160 kDa (ERAP160) which are able to interact with transcriptionally active ER in an AF-2 dependent manner but do not interact with transactivation-defective ER mutants or antagonist-bound receptor, suggesting that these proteins can mediate ER transcription (CAVAILLES et al. 1994; HALACHMI et al. 1994). Such putative coactivators, p140 and p160, can also interact with the RAR/RXR heterodimer bound to an active DR-5-RARE, in a ligand dependent manner (Kurokawa et al. 1995). Using GR bound to a GRE as a probe to detect protein-protein interactions, three GR-interacting proteins (GRIPs) of 170, 120, and 95 kDa were identified. A GRIP-170 enriched fraction was reported to stimulate GR-dependent transcription in vitro (Eggert et al. 1995). In another study, up to nine different TR associated proteins (TRAPs) were isolated and shown to interact with TR in cells growing in the presence, but not in the absence of T3 (Fondell et al. 1996).

Several coactivators have been now isolated and identified, that interact with nuclear receptors in a ligand-dependent manner and mediate their transcriptional activity (Fig. 3). Trip1 was isolated as a thyroid hormone receptor interacting protein by the yeast two-hybrid system, and shown to be very similar to the yeast transcriptional mediator Sug1 (Lee et al. 1995b; Swaffield et al. 1992, 1995). The murine homolog, mSUG1, has been recently cloned and shown to interact both in yeast and in vitro with RAR, TR, VDR, and ER, but not with RXR. This interaction is dependent on the presence of ligand and the AF-2 (vom Baur et al. 1996). Trip1/SUG1 contains a coiled coil region in its aminoterminus and an AAA domain at the carboxyl end, which is conserved in a family of ATPases with various cellular activities (Confalonieri and Duguet 1995). A receptor interaction domain is also present within the AAA region. It has recently been shown that ySug1 is a component of the 26S proteasome (Rubin et al. 1996), and it might therefore affect transcription indirectly by modulating the rate of nuclear receptors degradation.

Transcription intermediary factor 1 (TIF1) was isolated using the RXR LBD as a bait. It can interact in a ligand and AF-2 dependent manner with the LBD of RXR, RAR, VDR, PR, and ER, but not with TR, both in yeast and in vitro (Le Douarin et al. 1995; vom Baur et al. 1996). TIF1 contains several structural characteristics common for transcriptional activators: a RING finger followed by two B boxes with zinc finger-like structures, and a coiled-coil domain are found in the amino terminus. In addition, there is a bromodomain in the carboxyterminus, which has also been observed in several other transcriptional regulatory proteins. Unexpectedly, TIF1 inhibits ligand-dependent transactivation by RXR, RAR, and ER when overexpressed, probably by sequestration of a limiting factor required for nuclear receptor activity (Le Douarin et al. 1995). Recently, a TIF1 related protein, named TIF1β, has been cloned which shares very similar structural features with TIF1, now called TIF1α (Le Douarin et al. 1996). However, TIF1β does not contain a typical receptor interaction domain, and it is unknown whether TIF1β interacts with nuclear receptors.

RIP140 has been cloned from a human cDNA library, encoding a protein of 1158 amino acids with a predicted molecular mass of 127 kDa (Cavailles et al. 1995). Antibodies raised against RIP140 recognized the p140 identified in previous biochemical studies. No structural features have been revealed from its amino acid sequence that foretell its function, except for a serine/threonine-rich domain in the central part of the molecule. RIP140 can interact with the ER LBD only in the presence of agonists, but not in the presence of antiestrogens. That RIP140 is a coactivator for ER is suggested by the observation that transcriptionally defective ER mutants no longer interact with RIP140. Two independent receptor interaction domains have been found in RIP140, one located in the amino terminus and another near the carboxyl end of the molecule. In addition to ER, RIP140 can interact in a hormone-dependent manner with TR and RAR. Furthermore, RXR can interact with the amino terminal interaction domain, but not with the carboxyl end site (L'Horset et al. 1996). In contrast to SUG1 and TIF1, RIP140 can function as a transcriptional activator when fused to a heterologous DBD. However, similar to TIF1 and SUG1, overexpression of RIP140 inhibits ER-mediated transcriptional activation (Cavailles et al. 1995).

Thus, neither SUG1, TIF1, nor RIP140 can efficiently stimulate hormone receptor-mediated transcriptional activation, and their role as true coactivators of the nuclear receptors is still unclear. A different group of proteins have been identified which can stimulate hormone dependent transcription by nuclear receptors. Most of these proteins belong to the same family of coactivators and they are represented by the p160 protein identified in biochemical studies (see above), and includes SRC-1/N-CoA$_1$, TIF2/GRIP1/N-CoA$_2$, and ACTR/RAC$_3$/pCIP.

Several p160 family members have been isolated and identified in two-hybrid system screens using the LBD of different nuclear receptors (SRC-1, TIF2/GRIP1, ACTR/RAC3; Oñate et al. 1995; Voegel et al. 1996; Hong et al. 1996, 1997; Li et al. 1997; Chen et al. 1997) or using CREB-Binding Protein (CBP) (SRC-1/N-CoA$_1$, N-CoA$_2$, p/CIP; Torchia et al. 1997; Kamei et al. 1996; Yao et al. 1996) as a bait. In addition, one of these genes, AIB1 (also named ACTR or RAC3), was discovered by a search for amplified and overexpressed genes in human breast and ovarian cancer (Anzick et al. 1997). They all share similar structural features: a basic helix-loop-helix (bHLH) and a PAS domain, involved in protein dimerization, are present in the amino terminus. A serine/threonine rich domain, containing the receptor interaction domain, is found in the central portion of the molecule, and the carboxiterminus contains a Q-rich region (see Fig. 3). Coexpression of any of the isolated p160 family members significantly stimulates the ligand dependent transactivation by retinoid receptors, as well as all steroid receptors, TR, and VDR. Importantly, these coactivators can overcome the squelching effect observed when two different nuclear receptors are cotransfected, indicating that the coactivators are expressed in limiting amounts and are required for receptor activity. An autonomous activation domain has been reported in TIF2/GRIP1 and ACTR/RAC3 (Voegel et al. 1996; Hong et al. 1997; Chen et al. 1997; Li et al. 1997). The agonist-dependent interaction between these coactivators and the receptors has been confirmed in vitro and in yeast, and requires an intact AF-2 domain. Coactivators of the p160 family also interact with another protein, CBP, which is described below. p/CIP was isolated as a p300/CBP-interacting protein, and it was found to be necessary together with CBP for the transcriptional activation by retinoid receptors, as well as TR, ER, and PR (Torchia et al. 1997). In addition, it has been recently shown that ACTR, SRC-1, and certain nuclear receptors (PR), can interact with P/CAF, a p300/CBP-Associated Factor with histone acetyltransferase activity (see below; Chen et al. 1997; Jenster et al. 1997).

The receptor interaction domain is located in the central region of the molecule and a second such domain has been identified in SRC-1, which can bind PR but not other receptors (Oñate et al. 1995). Comparing the sequence of the receptor interaction domain from several coactivators, an α-helical domain with a consensus core LXXLL sequence – where the conserved leucine residues form a hydrophobic surface – has been identified and shown to be required for functional interactions between coactivators and nuclear receptors (Heery et al. 1997; Torchia et al. 1997). This sequence was first noted in $TIF1_{\alpha}$, where a ten amino acids peptide (726–735: I*LTSLL*LNSS) was shown to be necessary and sufficient to functionally interact with the receptors (Le Douarin et al. 1996). Nine of these sequences have been found in RIP140, and all of them are able to interact in a ligand and AF-2 dependent manner with ER and RAR, although with variable efficiency. Mutation of either one of the conserved leucines abolishes interaction of the peptides with the receptors (Heery et al. 1997; Le Douarin et al. 1996; Torchia et al. 1997). SRC-1 and TIF2 contain three such conserved motifs within the receptor interaction domain, which can strongly interact with ER and RAR. Other LXXLL motifs found in the aminoterminus, but outside the receptor interaction domain, do not interact with ER or RAR. p/CIP also contains several LXXLL motifs, three within the receptor interaction domain and two more within the CBP interaction domain (Torchia et al. 1997).

E. Differentiation–Proliferation Switches: Interaction with Other Signal Transducers

Retinoids not only induce transcriptional activation of specific genes, but also can specifically repress gene transcription. Although for some nuclear receptors "negative" response elements have been reported, inhibition of gene transcription or transrepression by retinoids involves a different mechanism and in general does not require direct retinoid receptor-DNA interaction. Thus, no positive or negative RAREs are involved. Interestingly, transrepression could be linked to the antiproliferative activity of retinoids, since transrepression affects in the majority of cases genes that are activated by the transcription factor AP-1 (Yang-Yen et al. 1991; Salbert et al. 1993; Schüle et al. 1991; Li et al. 1996). Although ligand bound RARs and RXRs can downregulate AP-1 activity, the reverse – repression of the retinoid signal – occurs when AP-1 or its components Jun and Fos are in excess. Therefore, a mechanism exists where the two types of transcription factors, the retinoid receptors and the AP-1 can control each others' activities (for a more detailed review of this mechanism see Pfahl 1993). The transcriptional regulator AP-1 mediates signals from growth factors, oncogenes and tumor promoters and usually signals cell proliferation, whereas the retinoid receptors often signal differentiation. Thus, the retinoid receptor–AP-1 interaction represents a molecular switch between differentiation and proliferation.

The exact details of this mechanism still remain to be elucidated. From earlier studies it appears that the receptors can directly interact with the AP-1 components cJun and cFos (YANG-YEN et al. 1991). However, this interaction seemed to be relatively weak and the existence of a third factor required for stabilizing the retinoid receptor–AP-1 complex, was hypothesized (PFAHL 1993). From more recent studies, this third factor appears to be CBP/p300, which has also been suggested as a target for retinoid receptor AP-1 competition (KAMEI et al. 1996; CHAKRAVARTI et al. 1996). The more general role of the transcription factor CBP/p300 in retinoid signal mediation is discussed below. AP-1 is not the only transcription factor that can be inhibited by the ligand bound retinoid receptors and some other transcriptional regulators have been reported to be inhibited by the ligand activated receptors (PFAHL 1993; PFITZNER et al. 1995). Importantly, it was found that the ligand requirements for transcriptional activation and transrepression of RARs are different (FANJUL et al. 1994). Of particular interest are retinoid molecules that show no transcriptional activation activity but are potent AP-1 inhibitors (FANJUL et al. 1994).

Recently we have observed that the sensitivity of an AP-1 complex to retinoid inhibition can change significantly depending on whether AP-1 is acting on its own or whether it is part of a complex with other proteins. We found that in the case of the endothelin I gene, which is activated by an AP-1/GATA-2 complex, only tRA and an RXR selective retinoid were efficient inhibitors, and that most antiAP-1 selective retinoids were not (HSU and PFAHL 1998). Thus, certain genes activated by an AP-1 complex appear to show more stringent retinoid response requirements. This suggests that retinoids that selectively repress some but not all AP-1 activated genes can be obtained.

The observed differences in retinoid structures that are able to induce retinoid receptor mediated transcriptional activation and transrepression of AP-1 or AP-1/GATA-2 complexes suggest that the ligand bound receptors assume different configurations in transcriptional activation and transrepression. As discussed above, different transactivation configurations are additionally induced through the interaction of the heterodimer with structurally distinct response elements and transrepression configuration can depend on the context of the AP-1 complex.

F. CBP/p300, a Cointegrator of Multiple Signaling Pathways

Several lines of evidence point to a role of CBP/p300 as a coactivator of nuclear receptors. CBP was identified as a phosphoCREB-binding protein, and is required for transcriptional activation by CREB as well as AP-1 (ARIAS et al. 1994; CHRIVIA et al. 1993; BANNISTER et al. 1995; BANNISTER and KOUZARIDES 1995). Another protein, p300, was independently isolated as an E1A-associated protein and found to be highly homologous with CBP, in fact almost identical in the functional domains (ARANY et al. 1995). Since the two proteins are functionally interchangeable, they are usually referred to as CBP/p300 (LUNDBLAD et al. 1995). They are large proteins with over 2400 amino acids and a well-defined structure. CBP/p300 is characterized by the presence of three zinc fingers (C/H) spread along the molecule, a bromodomain in the center region, and a Q-rich domain at the carboxyl terminus of the protein (Fig. 3). Several regions along the protein have been shown to be required for the interaction of CBP/p300 with nuclear receptors and other transcription factors such as CREB and AP-1, STAT, myb, myoD, p45/NF-E2, p53 and basal transcription factors such as TBP and TFIIB (BANNISTER et al. 1995; BANNISTER and KOUZARIDES 1995; AVANTAGGIATI et al. 1997; HORVAY et al. 1997; DAI et al. 1996; ECKNER et al. 1996; CHENG et al. 1997; LEE et al. 1995a; BHATTACHARYA et al. 1996; ARIAS et al. 1994; CHRIVIA et al. 1993). As mentioned above, CBP/p300 also interacts with all the known coactivators of the p160 family, but not with p140 (TORCHIA et al. 1997; HANSTEIN et al. 1996; SMITH et al. 1996; KAMEI et al. 1996).

Interaction of CBP/p300 with the LBD of RAR, RXR, TR, and ER has been demonstrated in vitro and in vivo experiments. This interaction is functional, since cotransfection of CBP/p300 potentiates hormone-dependent transcriptional activation by nuclear receptors, and also relieves inhibition of AP-1 by RAR or GR (KAMEI et al. 1996; CHAKRAVARTI et al. 1996). As with other receptor coactivators, an intact AF-2 domain and the presence of the nuclear receptor ligand are required for this interaction to occur. The receptor interaction domain is located within the first 100 aminoterminal amino acids, which contain one LXXLL helical motif similar to that found in the receptor interaction domain of the p160 coactivators. A second LXXLL motif has been found in the first zinc finger (C/H1) of CBP and p300, which interacts weakly with RXR but not with TR or RAR (CHAKRAVARTI et al. 1996). A third LXXLL motif in CBP/p300, located at the carboxyl terminus within the Q-rich region, is necessary for interaction with the coactivators SRC-1 and p/CIP (TORCHIA et al. 1997). The fact that the receptor interaction domain and the SRC-1 interacting region are not overlapping but rather distant, suggests that both nuclear receptors and coactivators can bind simultaneously to CBP forming a trimeric complex. In support of this, coexpression of both SRC-1 and CBP or p/300 synergistically potentiates the transcriptional activation by ER or PR (SMITH et al. 1996), and both CBP and p/CIP are necessary for nuclear receptor activity (TORCHIA et al. 1997).

Since CBP/p300 can interact with and is required for the activity of several transcription factors, it has been proposed that CBP/p300 is an integrator of multiple signaling pathways. CBP/p300 is expressed in very limited amounts, and that CBP is important for normal cellular function has been recently demonstrated by the discovery that the loss of one functional copy of CBP causes the Rubistein-Taybi syndrome (PETRIJ et al. 1995). Furthermore, a CBP fusion product with MOZ, a putative acetylase homolog of yeast SAS2, has been associated with some types of acute myeloid leukemias (BORROW et al. 1996). Competition between several transcription factors for CBP/p300 binding may account for the observed negative interference between them, for instance, the well-known inhibition by retinoid receptors of several transcription factors, such as AP-1 and CREB (see above). In the case of AP-1, however, the mechanism of inhibition by retinoid receptors is not simply due to squelching of CBP/p300, since both AP-1 and retinoid receptors can bind simultaneously to CBP (KAMEI et al. 1996). In this case we propose that CBP/p300 stabilizes the retinoid receptor/AP-1 interaction that constitutes the molecular switch between differentiation and proliferation. When CBP/p300 is present at elevated concentrations, for instance when CBP/p300 is overexpressed, nuclear receptors such as RARs, RXRs and GR can bind to separate CBP/p300 molecules. This would preclude direct nuclear receptor–AP-1 contact and interaction and thereby eliminate AP-1 inhibition and cross-talk between the proliferation/differentiation pathways.

G. A Role for Histone Acetylation in Receptor-Mediated Transactivation?

For many years it has been proposed that chromatin structure plays a critical role in gene expression, and that histone acetylation is involved in chromatin remodeling (see reviews GRUNSTEIN 1997; WADE et al. 1997). Histone acetylation has been indeed shown to affect transcription in vivo (DURRIN et al. 1991), and therefore that histone acetyltransferases (HATs) could be involved in gene regulation. Recent cloning of HATs has indeed identified several transcription factors or coactivators with HAT activity: GCN5, CBP/p300, and $hTAF_{II}250$ (BROWNELL et al. 1996; OGRYZKO et al. 1996; MIZZEN et al. 1996). P/CAF, which shows high homology with GCN5, has also HAT activity (YANG-YEN et al. 1996). Interestingly, it was recently shown that two of the nuclear receptor coactivators of the p160 family, ACTR and SRC-1, have intrinsic HAT activity (CHEN et al. 1997; JENSTER et al. 1997), and it will be interesting to determine whether additional members of the family have this enzyme activity. P/CAF was isolated as a protein associated with CBP/p300 (YANG et al. 1996), and it can also interact with p160 coactivators (ACTR, SRC-1) and with certain nuclear receptors (PR) (CHEN et al. 1997; JENSTER et al. 1997). Thus, a multimeric complex with HAT activity (containing P/CAF, CBP/p300, and p160 coactivators) is required for transcriptional activation by nuclear hormone receptors, providing strong evidence for a role of histone acetylation in gene expression.

Histone deacetylation has, in addition, been associated with transcriptional repression, probably by returning nucleosomes to a tightly repressive

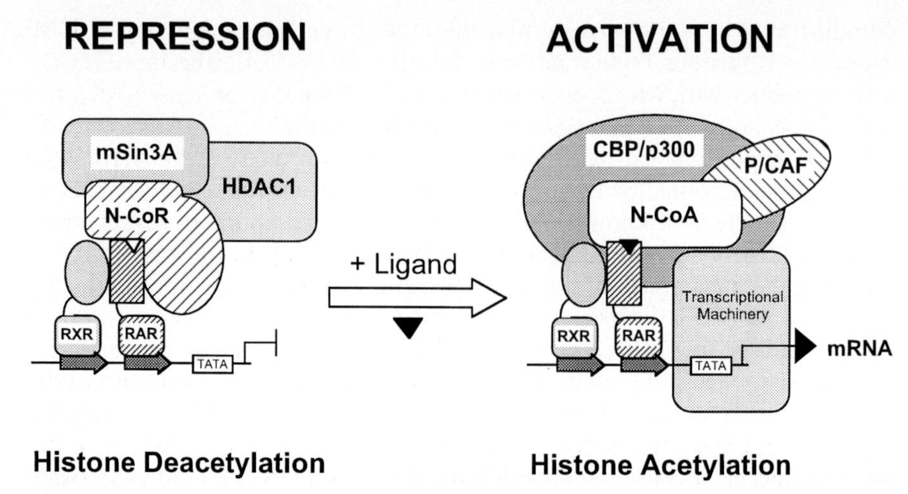

Fig. 4. A role for histone acetylation in retinoid receptor transcriptional activation. A model for retinoid receptor action: a RXR-RAR heterodimer bound to a DR-5-RARE recruits a trimeric complex composed by a corepressor (N-CoR), Sin3, and a histone deacetylase, which maintains the DNA in a tightly closed repressive conformation. Binding of the retinoid disassociates the corepressor complex, allowing binding of a coactivator (*N-CoA*) together with a cointegrator (*CBP*) and a CBP-associated factor (*P/CAF*), all of them showing histone acetyltransferase activity. Acetylation of histones opens the chromatin structure, allowing binding of the transcriptional machinery and gene expression

conformation (PAZIN and KADONAGA 1997). Several histone deacetylases have been recently isolated (AYER et al. 1995; SCHREIBER-AGUS et al. 1995; TAUNTON et al. 1996; RUNDLETT et al. 1996; VIDAL and GABER 1991), which can form a trimeric complex with N-CoR/SMRT and mSin3 (a corepressor for Mad; HEINZEL et al. 1997; ALLAND et al. 1997; NAGY et al. 1997). Microinjection of antibodies against each of the components of this complex prevents transcriptional repression, indicating that all of them are required for the repressor activity of nuclear receptors (HEINZEL et al. 1997). The importance of histone deacetylation in repression by unliganded nuclear receptors was further supported by the use of histone deacetylase inhibitors (HEINZEL et al. 1997; NAGY et al. 1997).

Thus a role for histone deacetylation/acetylation can be proposed for the regulation of transcription by nuclear receptors (Fig. 4). In this model, binding of unliganded RXR-RAR heterodimer to a DR-5 HRE recruits a corepressor complex with histone deacetylase activity, promoting nucleosome assembly and therefore potentiating transcriptional repression. N-CoR/SMRT serves as a bridge to link the histone deacetylase activity to the DNA bound receptors. In the presence of ligand, N-CoR/SMRT dissociates from the RXR-RAR heterodimer, resulting in displacement of the mSin3/HDAC complex from the DNA. The liganded receptor heterodimer now recruits a coactivator complex with HAT activity, formed by N-CoA, CBP, and probably P/CAF. This complex, which contains histone acetyltransferase activity, acetylates the core histones which alters the nucleosome structure, facilitating the entrance and assembly of the transcriptional machinery and allowing for transcription by RNA pol II.

H. Apoptosis Induction by Special Retinoids: A Novel Pathway?

Apoptosis is a type of programmed cellular suicide which plays important roles in development and tissue homeostasis. Apoptosis can be morphologically distinguished from other forms of cell death and is characterized by cell shrinkage, membrane blebbing, chromatin condensation, and cell fragmentation into apoptotic bodies which are phagocytosed by macrophages or surrounding cells (STELLER 1995). Apoptosis can be induced by a large variety of stimuli and occurs through multiple distinct pathways. Several gene products have been isolated and identified in recent years that can positively or negatively regulate apoptosis (WHITE 1996). An important breakthrough was the identification in *C. elegans* of Ced3 and Ced4 as essential for programmed cell death during development, while Ced9 was found to be sufficient and necessary to prevent apoptosis (YUAN and HORVITZ 1990; YUAN et al. 1993; HENGARTNER et al. 1992). Ced9 is homologue to mammalian Bcl-2, a protooncogene which blocks apoptosis when overexpressed (HENGARTNER and HORVITZ 1994; VAUX et al. 1992). On the other hand, Ced3 was found to be homologous to the interleukin-1_{β} converting enzyme, a cysteine protease that belongs to the family of caspases (YUAN et al. 1993). A large body of evidence now points to an essential role for caspases and proteolysis in apoptosis (MARTIN and GREEN 1995). The mammalian homologue of Ced4 has been recently identified as apoptotic protease activating factor-1 (Apaf1) (ZOU et al. 1997). Apaf1 was first isolated as a factor required for the activation of caspase 3 in vitro (LIU et al. 1996a).

A dysfunction of apoptosis contributes to the pathogenesis of several human diseases, including cancer, and the use of agents capable of inducing apoptosis in malignant cells could have significant implications for the chemoprevention and treatment of cancer. Retinoids are promising agents for the prevention and treatment of cancer, and the recent observation that some retinoids induce apoptosis in malignant cells has raised great interest. Natural and synthetic retinoids show antiproliferative activity against a large number of cancer cells. However, tRA and, more efficiently 9-*cis*-RA, were first shown to inhibit activation-induced TCR/CD3-mediated apoptosis in T cell hybridomas, while neither retinoid had any effect on the viability of those cells (IWATA et al. 1992; YANG et al. 1993, 1995). This protective effect of the natural retinoids required both RARs and RXRs and occurred by repressing the expression of Fas ligand upon T cell activation (BISSONETTE et al. 1995). Interestingly, activation-induced apoptosis also required a nuclear receptor, the orphan nur77, which is up-regulated upon stimulation (WORONICZ et al. 1994; LIU et al. 1994).

On the other hand, retinoids have been found to show some apoptotic activity in a large variety of cancer cell lines. tRA can induce apoptosis in RA-sensitive breast cancer cell lines, by a mechanism presumably involving induction of RARβ by RXR-RARα heterodimers (LIU et al. 1996b). Interestingly, activation of RXR-nur77 heterodimers by RXR selective retinoids is believed to be involved in the induction of RARβ and apoptosis in RA-resistant breast cancer cell lines (WU et al. 1997). tRA and 9-*cis*-RA can also induce apoptosis in undifferentiated cells, such as F9 and P19 (CHEN et al. 1996; OKAZAWA et al. 1996). Apoptosis is induced in lung cancer cells (NCI-H460), as well as in nontumorigenic epithelial cells (SPOC-1), by tRA and RARα-selective retinoids, but not by RXR-selective compounds, and this apoptosis is prevented in the presence of retinoid antagonists and is correlated with the induction of transglutaminase II (ZHANG et al. 1995). In addition, acyclic retinoid – a synthetic retinoid – but not the natural compounds tRA or 9-*cis*-RA, were found to induce apoptosis in a human hepatoma cell line (HUH-7) in serum-free medium, and this effect could be inhibited by TGFα or EGF (NAKAMURA et al. 1995, 1996). Activation of RXR has been shown to be required for retinoid-induced apoptosis in leukemia cells (HL60), since 9-*cis*-RA alone or combinations of RXR-selective with RAR-selective retinoids, but not either one of the selective compounds alone, could induce cell death (NAGY et al. 1995; BOEHM et al. 1995). Thus, tRA and 9-*cis*-RA can induce apoptosis under certain circumstances in some cancer cell lines, but in general the apoptosis inducing activity of these retinoids is rather weak and thus may not be sufficiently effective for cancer treatment. In addition, these natural compounds induce many other effects in the animal that make them undesirable as chemotherapeutic agents. However, novel synthetic retinoids with particular selectivities have been recently identified, which can function as potent inducers of apoptosis in malignant cells.

One of such retinoids is fenretinide [4-(*N*-hydroxyphenyl)retinamide, HPR]. HPR was found to be effective against prostate, breast, and ovarian cancers in animal models and low-dose regiment is currently being evaluated in several clinical trials. At these doses HPR may be promising for cancer prevention because of its low toxicity and high antiproliferative activity. However, for cancer treatment, these low doses are unlikely to be effective. Although its mechanism of action is still unknown, it has recently been shown that high doses of HPR effectively induce apoptosis in leukemia cells (PIEDRAFITA and PFAHL 1997; DELIA et al. 1993), cervical carcinoma (ORIDATE et al. 1995), neuroblastoma (PONZONI et al. 1995), prostate (HSIEH et al. 1995), small-cell lung (KALEMKERIAN et al. 1995), and breast cancer (FANJUL et al. 1996; PELLEGRINI et al. 1995) cells. While Bcl-2 was found to prevent tRA-induced apoptosis in

P19 cells, the proto-oncogene did not inhibit HPR-induced apoptosis in leukemia cells, although it delayed its onset (DELIA et al. 1995; OKAZAWA et al. 1996), suggesting that the natural and the synthetic retinoid induce cell death by different mechanisms, although certain cell type specific effects may also have to be considered. In fact, HPR differentially affected the expression of p53 and c-myc in a cell type selective manner (DELIA et al. 1995; KALEMKERIAN et al. 1995). HPR-induced apoptosis required the synthesis of mRNA and protein in some cell lines, and some inhibition was observed by antioxidants. Interestingly, HPR activates preferentially RARγ and RARβ, while showing antiAP-1 activity with all three RARs as well as RXR (FANJUL et al. 1996). The observation that HPR can induce apoptosis in RA-resistant HL-60 cells, however, has led to the proposal that HPR-induced apoptosis is independent of the retinoid receptors (DELIA et al. 1993), which remains to be proven.

Another potent apoptosis inducer is the synthetic retinoid CD437 which has been reported to induce apoptosis in breast cancer cells (SHAO et al. 1995), melanoma (SCHADENDORF et al. 1996), leukemia cells (PIEDRAFITA and PFAHL 1997), as well as in mouse thymocytes (SZONDY et al. 1997). As with HPR, CD437 is also selective for RARγ/RARβ (BERNARD et al. 1992), and induces apoptosis in a p53-independent manner (SHAO et al. 1995, and our unpublished observations). However, when we analyzed CD 437 and HPR against more than 50 different cell types, we observed rather indiscriminate induction of apoptosis in almost all cell types be these two compounds. In addition, we were unable to satisfactorily separate the antitumor activity from general toxicity of these compounds in animal models (our unpublished observation). Interestingly, a novel series of synthetic retinoids structurally related to CD437, has been recently found to induce apoptosis more selectively. One of them, MX335, was found to be effective against non-small-cell lung cancer cells and inhibit the growth of solid tumors in an animal model (LU et al. 1997). How retinoids in general, and particularly these RARγ-selective retinoids, induce apoptosis is still unknown. Importantly, a role for the retinoid receptors in this process remains to be established. While one study supports that CD437 can induce apoptosis independently of retinoid receptors, since it is equally effective in RA-resistant leukemia cells (SHAO et al. 1995), another study suggest an RARγ-mediated process, since only RARγ-selective retinoids could induce apoptosis in thymocytes (SZONDY et al. 1997). Activation of AP-1 has been observed during CD437-induced apoptosis in melanoma and leukemia cells (SCHADENDORF et al. 1996, and our unpublished data), with different effects on Bcl-2 depending on the cell line analyzed (SCHADENDORF et al. 1996; SHAO et al. 1995).

We have recently analyzed the antiproliferative activity of several selective retinoids in different leukemia cell lines, and found that only RARγ-selective compounds, including CD437 and HPR, exhibited a strong antiproliferation profile against most of the cell lines analyzed. This antiproliferative activity is correlated with the induction of apoptosis (PIEDRAFITA and PFAHL 1997 and unpublished results). The structurally related retinoids, CD437 and

MX2870–1, were far more potent than HPR as apoptosis inducers in Jurkat cells. Interestingly, retinoid-induced apoptosis required the activation of caspases, since specific caspase inhibitors could effectively block the effect of CD437 (PIEDRAFITA and PFAHL 1997). In contrast to previous observations in mouse thymocytes (SZONDY et al. 1997), CD437-induced apoptosis was found to be independent of transcription (PIEDRAFITA and PFAHL 1997). According to our results, it appears that CD437 and other related retinoids such as MX335 and MX2870-1 can induce apoptosis by a novel mechanism of action which does not involve transactivation. In agreement with this is our observation that retinoid antagonists can also induce apoptosis in leukemia cells as well as in breast cancer cells, in support for a novel mechanism of action which is independent of transcriptional activation (FANJUL et al. 1998). The observation that transcription/translation is required in some cell types such as mouse thymocytes but not in Jurkat cells suggests a cell type specific mechanism, or that the protein(s) mediating the apoptotic signal is(are) not constitutively present in such cells as thymocytes. Thus, RARγ-selective apoptosis-inducing retinoids, which show transactivation activity with RARγ/RARβ, can bind RARγ and induce a particular conformational change in the receptor that allows specific protein-protein interactions with some components of the apoptotic machinery. However, although the universal selectivity for RARγ observed with apoptotic retinoids suggest RARγ as the apoptosis signal mediator, other as yet unknown proteins cannot be excluded at this time (see Fig. 5). The apoptotic machinery is located in the cell cytoplasm. The observation that the apoptosis inducing retinoids were found not to bind to the cytosolic retinoic acid binding proteins (CRABPs) (M. Pfahl, unpublished) may therefore be of importance. Thus, we hypothesize a direct mechanism of activation of the apoptotic machinery in the cytoplasm by certain retinoid receptor complexes. In such a model tRA and other retinoids would be much less effective as direct triggers of apoptosis, since they would be bound with high affinity by CRABPs present in the cytosol. In addition, they may induce a different configuration in the receptor. Other more indirect pathways of apoptosis induction by tRA and 9-*cis*-RA that involve induction of transcriptional cascades are also most likely to exist in some cell types.

I. Future Prospects

The deciphering of most of the retinoid signaling mechanisms for gene regulation clearly underscores the exceptional potential of these molecules for the treatment of various types of diseases. In general the multitude and complexity of the retinoid signaling pathways discovered are well consistent with the known broad biological activities of these compounds. Perhaps the most important discoveries for the practical and optimal exploitation of retinoids

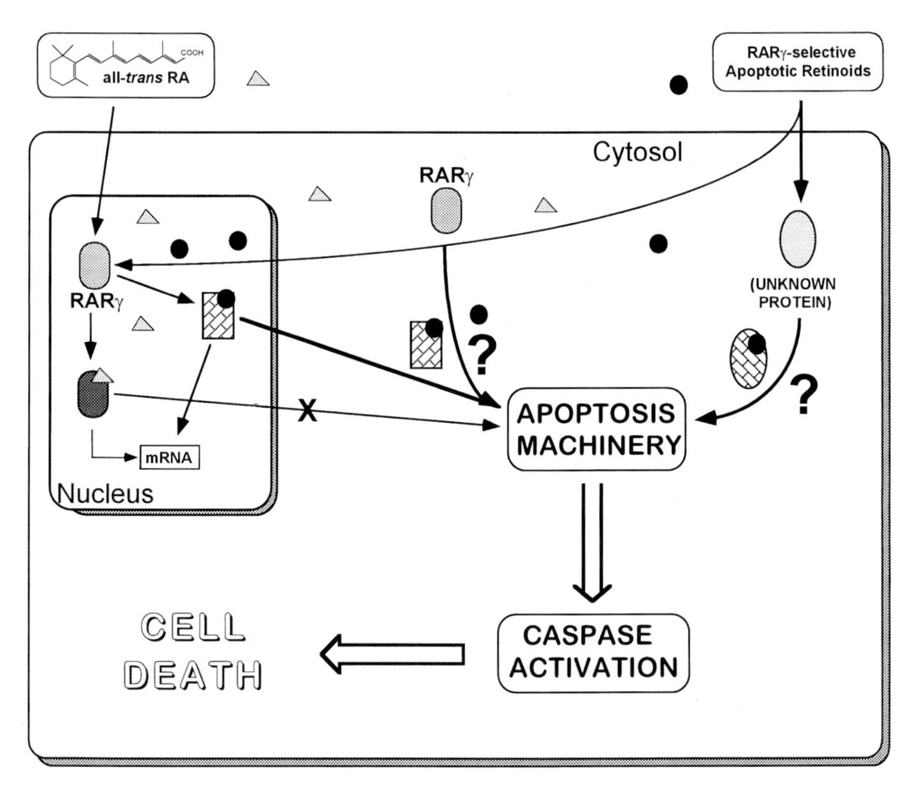

Fig. 5. Apoptosis induced by RARγ-selective retinoids. A novel mechanism of action for selective apoptotic retinoids is proposed, in which transcriptional activation is not required. This model proposes that protein-protein interactions between the retinoid bound receptor (RARγ or another, yet unidentified, protein) interacts with a component of the apoptotic machinery and triggers caspase activation, leading to cell death. Other retinoids (*tRA*), which cannot induce a particular conformational change in the retinoid receptor or cannot bind efficiently to RARγ (or the unidentified protein), function through a transcriptional activation pathway but do not directly induce apoptosis

in therapeutic applications, were the findings that synthetic retinoids could be found that only activate selected or limited portions of the complex retinoid signaling mechanism (Lehmann et al. 1991b 1992; Graupner et al. 1991; Fanjul et al. 1994; Lu et al. 1997). Although it has not yet been fully resolved how to correlate the molecular selectivity type of a retinoid with effective treatments of specific diseases, first examples for the effectiveness of novel selective retinoids are on hand. For dermatological (acne vulgares) applications, the RARβ/RARγ-selective retinoid adapalene (CD 271) has been shown to be as effective as tRA, but without most of the side effects. Adapalene (Differin) has now been approved for acne treatment in many countries, including the United States. For the treatment of cancers, even such unresponsive types as non-small-cell lung carcinoma, the recently described apoptosis-inducing synthetic retinoid MX355 appears to hold particular promise (LU et al. 1997). A comparable strong apoptosis inducing activity has never been discovered with natural retinoids and related synthetic broad activity compounds. RXR-selective retinoids have also shown promise in the treatment of several human cancers. It is important to note that for instance LGD1069 can be effective in the absence of significant side effects (MILLER et al. 1997). However, RXR-selective ligands show even more promise for the treatment of type II diabetes (MUKHERJEE et al. 1997; Lernhardt, Fanjul, Al-Shamma and M.P., unpublished). The new potential for retinoids has now been recognized by several specialized biotechnology companies and additional modern highly selective retinoids are scheduled for clinical evaluation. The molecular deciphering of a vitamin signaling system has thus already led to significant new therapeutics and drug candidates and holds more promise for the future.

References

Alland L, Muhle R, Hou H, Jr, Potes J, Chin L, Schreiber-Agus N, DePinho R (1997) Role for N-CoR and histone deacetylase in Sin3-mediated transcriptional repression. Nature 387:49–55

Anzick SL, Kononen J, Walker RL, Azorsa DO, Tanner MM, Guan X-Y, Sauter G, Kallioniemi O-P, Trent JM, Meltzer PS (1997) AIB1, a steroid receptor coactivator amplified in breast and ovarian cancer. Science 277:965–968

Apfel R, Benbrook D, Lernhardt E, Ortiz MA, Pfahl M (1994) A novel orphan receptor with a unique ligand binding domain and its interaction with the retinoid/thyroid hormone receptor subfamily. Mol Cell Biol 14:7025–7035

Arany Z, Newsome D, Oldread E, Livingston DM, Eckner R (1995) A family of transcriptional adaptor proteins targeted by the E1 a oncoprotein. Nature 374:81–84

Arias J, Alberts AS, Brindle P, Claret FX, Smeal T, Karin M, Feramisco J, Montminy M (1994) Activation of cAMP and mitogen responsive genes relies on a common nuclear factor. Nature 370:226–229

Avantaggiati ML, Ogryzko V, Gardner K, Giordano A, Levine AS, Kelly K (1997) Recruitment of p300/CBP in p53-dependent signal pathways. Cell 89:1175–1184

Ayer DE, Lawrence QA, Eisenman RN (1995) Mad-Max transcriptional repression is mediated by ternari complex formation with mammalian homologs of yeast repressor Sin3. Cell 80:767–776

Baes M, Gulick T, Choi H-S, Martinoli MG, Simha A, Moore DD (1994) A new orphan member of the nuclear hormone receptor superfamily that interacts with a subset of retinoic acid response elements. Mol Cell Biol 14:1544–1552

Baniahmad A, Ha I, Reinberg D, Tsai S, Tsai M-J, O'Malley B (1993) Interaction of human thyroid hormone receptor b with transcription factor TFIIB may mediate target gene derepression and activation by thyroid hormone. Proc Natl Acad Sci USA 90:8832–8836

Baniahmad A, Leng X, Burris TP, Tsai SY, Tsai M-J, O'Malley BW (1995) The $\tau 4$ activation domain of the thyroid hormone receptor is required for release of a putative corepressor(s) necessary for transcriptional silencing. Mol Cell Biol 15:76–86

Baniahmad A, Steiner CH, Köhne AC, Renkawitz R (1990) Modular structure of a chicken lysozym silencer: Involvement of an unusual thyroid hormone receptor binding site. Cell 61:505–514

Bannister AJ, Kouzarides T (1995) CBP-induced stimulation of c-Fos activity is abrogated by E1 a. EMBO J 14:4758–4762

Bannister AJ, Oehler T, Wilhelm D, Angel P, Kouzarides T (1995) Stimulation of c-Jun activity by CBP: c-Jun residues Ser63/73 are required for CBP induced stimulation in vivo and CBP binding in vitro. Oncogene 11:2509–2514

Benbrook D, Lernhardt W, Pfahl M (1988) A new retinoic acid receptor identified from a hepatocellular carcinoma. Nature 333:669–672

Bernard BA, Bernardon J-M, Delescluse C, Martin B, Lenoir M-C, maignan J, Charpentier B, Pilgrim WR, Reichert U, Shroot B (1992) Identification of synthetic retinoids with selectivity for human nuclear retinoic acid receptor γ. Biochem Biophys Res Commun 186:977–983

Bhattacharya S, Eckner R, Grossman S, Oldread E, Arany Z, D'Andrea A, Livingston DM (1996) Cooperation of Stat2 and p300/CBP in signalling induced by interferon-α. Nature 383:344–347

Bissonette RP, Brunner T, Lazarchick SB, Yoo NJ, Boehm MF, Green DR, Heyman RA (1995) 9-*cis* retinoic acid inhibition of activation-induced apoptosis is mediated via regulation of Fas ligand and requires retinoic acid receptor and retinoid X receptor activation. Mol Cell Biol 15:5576–5585

Blanco JCG, Wang I-M, Tsai SY, Tsai M-J, O'Malley BW, Jurutka PW, Haussler MR, Ozato K (1995) Transcription factor TFIIB and the vitamin D receptor cooperatively activate ligand-dependent transcription. Proc Natl Acad Sci USA 92:1535–1539

Boehm MF, Zhang L, Zhi L, McClurg MR, Berger E, Wagoner M, Mais DE, Suto CM, Davies PJ A, Heyman RA, Nadzan AM (1995) Design and synthesis of potent retinoid X receptor selective ligands that induce apoptosis in leukemia cells. J Med Chem 38:3146–3155

Borrow J, Stanton VPJ, Andressen JM, Becher R, Behm FG, Chaganti RSK, Civin CL, Disteche C, Dube I, Frischauf AM, et al (1996) The translocation t(8;16)(p11;p13) of acute myeloid leukaemia fuses a putative acetyltransferase to the CREB-binding protein. Nat Genet 14:33–41

Bourguet W, Ruff M, Chambon P, Gronemeyer H, Moras D (1995) Crystal structure of the ligand-binding domain of the human nuclear receptor RXRα. Nature 375:377–382

Brand N, Petkovich M, Krust A, de Thé H, Marchio A, Tiollais P, Dejean D (1988) Identification of a second human retinoic acid receptor. Nature (London) 332:850–853

Brownell JE, Zhou J, Ranalli T, Kobayashi R, Edmonson DG, Roth SY, Allis CD (1996) Tetrahymena histone acetyltransferase A: a homolog to yeast Gcn5p linking histone acetylation to gene activation. Cell 84:843–851

Bugge TH, Pohl J, Lonnoy O, Stunnenberg HG (1992) RXRa, a promiscuous partner of retinoic acid and thyroid hormone receptors. EMBO J 11:1409–1418

Buratowski S (1994) The basics of basal transcription by RNA polymerase II. Cell 77:1–3

Casanova J, Helmer E, Selmi-Ruby S, Qi J-S, Au-Fliegner M, Desai-Yajnik V, Koudinova N, Yarm F, Raaka BM, Samuels HH (1994) Functional evidence for ligand-dependent dissociation of thyroid hormone and retinoic acid receptors from an inhibitory cellular factor. Mol Cell Biol 14:5756–5765

Cavailles V, Dauvois S, Danielian PS, Parker MG (1994) Interaction of proteins with transcriptionally active estrogen receptors. Proc Natl Acad Sci USA 91: 10009–10013

Cavailles V, Dauvois S, L'Horset F, Lopez G, Hoare S, Kushner PJ, Parker MG (1995) Nuclear factor RIP140 modulates transcriptional activation by the estrogen receptor. EMBO J 14:3741–3751

Chakravarti D, LaMorte VJ, Nelson MC, Nakajima T, Schulman IG, Juguilon H, Montminy M, Evans RM (1996) Role of CBP/p300 in nuclear receptor signalling. Nature 383:99–103

Chen H, Lin RJ, Schiltz RL, Chakravarti D, Nash A, Nagy L, Privalsky ML, Nakatani Y, Evans RM (1997) Nuclear receptor coactivator ACTR is a novel histone acetyl-

transferase and forms a multimeric activation complex with P/CAF and CBP/p300. Cell 90:569–580
Chen J-Y, Clifford J, Zusi C, Starrett J, Tortolani D, Ostrowski J, Reczek PR, Chambon P, Gronemeyer H (1996) Two distinct actions of retinoid-receptor ligands. Nature 382:819–822
Chen JD, Evans RM (1995) A transcriptional corepressor that interacts with nuclear hormone receptors. Nature 377:454–457
Cheng X, Reginato MJ, Andrews NC, Lazar MA (1997) The transcriptional integrator CREB-binding protein mediates positive cross talk between nuclear hormone receptors and the hematopoietic bZip p45/NF-E2. Mol Cell Biol 17:1407–1416
Chiba H, Clifford J, Metzger D, Chambon P (1997) Distinct retinoid X receptor-retinoic acid receptor heterodimers are differentially involved in the control of expression of retinoid target genes in F9 embryonal carcinoma cells. Mol Cell Biol 17:3013–3020
Chrivia JC, Kwok RPS, Lamb, N, Hagiwara M, Montminy MR, Goodman RH (1993) Phosphorylated CREB binds specifically to the nuclear protein CBP. Nature 365:855–859
Confalonieri F, Duguet M (1995) A 200-aminoacid ATPase module in search of a basic function. BioEssays 17:639–650
Dai P, Akimaru H, Tanaka Y, Hou D-X, Yasukawa T, Kanei-Ishii C, Takahashi T, Ishii S (1996) CBP as a transcriptional coactivator of c-Myb. Genes Dev 10:528–540
Damm K, Thompson CC, Evans RM (1989) Protein encoded by v-erbA functions as a thyroid-hormone receptor antagonist. Nature (London) 339:593–597
de Thé H, Vivanco-Ruiz MM, Tiollais P, Stunnenberg H, Dejean A (1990) Identification of a retinoic acid response element in the retinoic acid receptor gene. Nature 343:177–180
Delia D, Aiello A, Formelli F, Fontanella E, Costa A, Miyashita T, Reed JC, Pierotti MA (1995) Regulation of Apoptosis Induced by the Retinoid N-(4-Hydroxyphenyl) Retinamide and Effect of Deregulated bcl-2. Blood 85:359–367
Delia D, Aiello A, Lombardi L, Pellici PG, Grignani F, Formelli F, Menard S, Costa A, Veronesi U, Pierotti MA (1993) N-(4-hydroxyphenyl) retinamide induces apoptosis of malignant hemopoietic cell lines including those unresponsive to retinoic acid. Cancer Res 53:6036–6041
Devchand PR, Keller H, Peters JM, Vazquez M, Gonzalez FJ, Wahli W (1996) The PPARα-leukotriene B4 pathway to inflammation control. Nature 384:39–43
Durand B, Saunders M, Leroy P, Leid M, Chambon P (1992) All-*trans* and 9-*cis* retinoic acid induction of CRABPII transcription is mediated by RAR-RXR heterodimers bound to DR1 and DR2 repeated motifs. Cell 71:73–85
Durrin LK, Mann RK, Kayne PS, Grunstein M (1991) Yeast histone H4 N terminal sequence is required for promoter activation in vivo. Cell 65:1023–1031
Eckner R, Yao T-P, Oldread E, Livingston DM (1996) Interaction and functional collaboration of p300/CBP and bHLH proteins in muscle and B-cell differentiation. Genes Dev 10:2478–2490
Eggert M, Mows CC, Tripier D, Arnold R, Michel J, Nickel J, Schmidt S, Beato M, Renkawitz R (1995) A fraction enriched in a novel glucocorticoid receptor-interacting protein stimulates receptor-dependent transcription in vitro. J Biol Chem 270:30755–30759
Escriva H, Safi R, Hanni C, Langlois M-C, Saumitou-Laprade P, Stehelin D, Capron A, Pierce R, Laudet V (1997) Ligand binding was acquired during evolution of nuclear receptors. Proc Natl Acad Sci USA 94:6803–6808
Fanjul A, Hobbs PD, Jong L, Cameron JF, Harlev E, Graupner G, Lu X-P, Dawson MI, Pfahl M (1994) A novel class of retinoids with selective anti-AP-1 activity exhibit anti-proliferative activity. Nature 372:107–111
Fanjul AN, Delia D, Pierotti MA, Rideout D, Qiu J, Pfahl M (1996) 4-hydroxyphenyl retinamide is a highly selective activator of retinoid receptors. J Biol Chem 271:22441–22446

Fanjul AN, Piedrafita FJ, Al-Shamma H, Pfahl M (1998) Apoptosis induction and potent anti-ER negative breast cancer activity *in vivo* by a retinoid antagonist. Cancer Res 58:4607–4610

Fondell JD, Ge H, Roeder RG (1996) Ligand induction of a transcriptionally active thyroid hormone receptor coactivator complex. Proc Natl Acad Sci USA 93:8329–8333

Forman BM, Goode E, Chen J, Oro AE, Bradley DJ, Perlmann T, Noonan DJ, Burka LT, McMorris T, Lamph WW, Evans RM, Weinberger C (1995a) Identification of a nuclear receptor that is activated by farnesol metabolites. Cell 81:687–693

Forman BM, Tontonoz P, Chen J, Brun RP, Spiegelman BM, Evans RM (1995b) 15-deoxy-Δ12,14-prostaglandin J2 is a ligand for the adipocyte determination factor PPARγ. Cell 83:803–812

Giguere V, Ong ES, Seigi P, Evans RM (1987) Identification of a receptor for the morphogen retinoic acid. Nature 330:624–629

Giguere V, Shago M, Zirngibl R, Tate P, Rossant J, Varmuza S (1990) Identification of a novel isoform of the retinoic acid receptor γ expressed in the mouse embryo. Mol Cell Biol 10:2335–2340

Glass CK (1994) Differential recognition of target genes by nuclear receptor monomers, dimers, and heterodimers. Endocr Rev 15:391–407

Glass CK, Holloway JM, Devary OV, Rosenfeld MG (1988) The thyroid hormone receptor binds with opposite transcriptional effects to a common sequence motif in thyroid hormone and estrogen response elements. Cell 54:313–323

Graupner G, Malle G, Maignan J, Lang G, Prunieras Mand Pfahl M (1991) 6′-Substituted naphthalene-2-carboxylic acid analogs, a new class of retinoic acid receptor subtype-specific ligand. Biochem Biophys Res Commun 179:1554–1561

Graupner G, Wills KN, Tzukerman M, Zhang X-K, Pfahl M (1989) Dual regulatory role for thyroid-hormone receptors allows control of retinoic-acid receptor activity. Nature 340:653–656

Grunstein M (1997) Histone acetylation in chromatin structure and transcription. Nature 389:349–352

Halachmi S, Marde E, Martin G, MacKay H, Abbondanza C, Brown M (1994) Estrogen receptor-associated proteins: possible mediators of hormone-induced transcription. Science 264:1455–1458

Hamada K, Gleason SL, Levi B-Z, Hirschfeld S, Apella E, Ozato K (1989) H-2RIIBP, a member of the nuclear hormone receptor superfamily that binds to both the regulatory element of major histocompatibility class I genes and the estrogen response element. Proc Natl Acad Sci USA 86:8289–8293

Hanstein B, Eckner R, DiRenzo J, Halachmi S, Liu H, Searcy B, Kurokawa R, Brown M (1996) p300 is a component of an estrogen receptor coactivator complex. Proc Natl Acad Sci USA 93:11540–11545

Heery DM, Kalkhoven E, Hoare S, Parker MG (1997) A signature motif in transcriptional coactivators mediates binding to nuclear receptors. Nature 387:733–736

Heinzel T, Lavinsky RM, Mullen T-M, Soderstrom M, Laherty CD, Torchia J, Yang W-M, Brard G, Ngo SD, Davie JR, Seto E, Eisenman RN, Rose DW, Glass CK, Rosenfeld MG (1997) A complex containing N-CoR, mSin3 and histone deacetylase mediates transcriptional repression. Nature 387:43–48

Hengartner MO, Horvitz HR (1994) C. elegans cell survival gene ced-9 encodes a functional homolog of the mammalian proto-oncogene bcl-2. Cell 76:665–676

Hengartner MO, Ellis RE, Horvitz HR (1992) Caenorhabditis elegans gene ced-9 protects cells from programmed cell death. Nature 356:494–499

Hirose T, Apfel R, Pfahl M, Jetten AM (1995) The orphan receptor Tak1 acts as a competitive repressor or RAR/RXR-mediated signaling pathways. Biochem Biophys Res Commun 211:83–91

Hoffmann B, Lehmann JM, Zhang X-K, Hermann T, Graupner G, Pfahl M (1990) A retinoic acid receptor specific element controls the retinoic acid receptor-β promoter. Mol Endocrinol 4:1734–1743

Hong H, Kohli K, Garabedian MJ, Stallcup MR (1997) GRIP1, a transcriptional coactivator for the AF-2 transactivation domain of steroid, thyroid, retinoid, and vitamin D receptors. Mol Cell Biol 17:2735–2744

Hong H, Kohli K, Trivedi A, Johnson DL, Stallcup MR (1996) GRIP1, a novel mouse protein that serves as a transcriptional coactivator in yeast for the hormone binding domains of steroid receptors. Proc Natl Acad Sci USA 93:4948–4952

Horlein AJ, Naar AM, Heinzel T, Torchia J, Gloss B, Kurokawa R, Ryan A, Kamei Y, Soderstrom M, Glass CK, Rosenfeld MG (1995) Ligand-independent repression by the thyroid hormone receptor mediated by a nuclear receptor corepressor. Nature 377:397–404

Horvay AE, Xu L, Korzus E, Brard G, Kalafus D, Mullen T-M, Rose DW, Rosenfeld MG, Glass CK (1997) Nuclear integration of JAK/STAT and Ras/AP-1 signaling by CBP and p300. Proc Natl Acad Sci USA 94:1074–1079

Hsieh T-c, Ng C, Wu JM (1995) The synthetic retinoid N-(4-hydroxyphenyl) retinamide (4-HPR) exerts antiproliferative and apoptosis-inducing effects in the androgen-independent human prostatic JCA-1 cells. Biochem Mol Biol Int 37:499–506

Hsu J-Y, Pfahl M (1998) ET-1 expression and growth inhibition of prostate cancer cells: a retinoid target with novel specificity. Cancer Res 58:4817–4822

Husmann M, Hoffmann B, Stump DG, Chytil F, Pfahl M (1992) A retinoic acid response element from the rat CRBPI promoter is activated by an RAR/RXR heterodimer. Biochem Biophys Res Commun 187:1558–1564

Iwata M, Mukai M, Nakai Y, Iseki R (1992) Retinoic acids inhibit activation-induced apoptosis in T cell hybridomas and thymocytes. J Immunol 149:3302–3308

Jacq X, Brou C, Lutz Y, Davidson I, Chambon P, Tora L (1994) Human TAFII30 is present in a distinct TFIID complex and is required for transcriptional activation by the estrogen receptor. Cell 79:107–117

Janowski BA, Willy PJ, Devi TR, Falck JR, Mangelsdorf DJ (1996) An oxysterol signalling pathway mediated by the nuclear receptor LXRα. Nature 383:728–731

Jenster G, Spencer TE, Burcin MM, Tsai SY, Tsai M-J, O'Malley BW (1997) Steroid receptor induction of gene transcription: a two-step model. Proc Natl Acad Sci USA 94:7879–7884

Kalemkerian GP, Slusher R, Ramalingam S, Gadgeel S, Mabry M (1995) Growth inhibition and induction of apoptosis by fenretinide in small-cell lung cancer cell lines. J Natl Cancer. Inst 87:1674–1680

Kamei Y, Xu L, Heinzel T, Torchia J, Kurokawa R, Gloss B, Lin S-C, Heyman RA, Rose DW, Glass CK, Rosenfeld MG (1996) A CBP integrator complex mediates transcriptional activation and AP-1 inhibition by nuclear receptors. Cell 85:403–414

Kastner P, Mark M, Ghyselinck N, Krezel W, Dupe V, Grondona JM, Chambon P (1997) Genetic evidence that the retinoid signal is transduced by heterodimeric RXR/RAR functional units during mouse development. Development 124:313–326

Kliewer SA, Lenhard JM, Willson TM, Patel I, Morris DC, Lehmann JM (1995) A prostaglandin J2 metabolite binds peroxisome proliferator-activated receptor γ and promotes adipocyte differentiation. Cell 83:813–819

Kliewer SA, Umesono K, Mangelsdorf DJ, Evans RM (1992a) Retinoid X receptor interacts with nuclear receptors in retinoic acid, thyroid hormone and vitamin D3 signaling. Nature 355:446–449

Kliewer SA, Umesono K, Noonan DJ, Heyman RA, Evans RM (1992b) Convergence of 9-*cis* retinoic acid and peroxisome proliferator signaling pathways through heterodimer formation of their receptors. Nature 358:771–774

Krust A, Kastner PH, Petkovich M, Zelent A, Chambon P (1989) A third human retinoic acid receptor, hRAR-γ. Proc Natl Acad Sci USA 86:5310–5314

Kurokawa R, DiRenzo J, Boehm M, Sugarman J, Gloss B, Rosenfeld M, Heyman RA, Glass CK (1994) Regulation of retinoid signalling by receptor polarity and allosteric control of ligand binding. Nature 371:528–531

Kurokawa R, Soderstrom M, Horlein A, Halachmi S, Brown M, Rosenfeld MG, Glass CK (1995) Polarity-specific activities of retinoic acid receptors determined by a corepressor. Nature 377:451–454

L'Horset F, Dauvois S, Heery DM, Cavailles V, Parker MG (1996) RIP140 interacts with multiple nuclear receptors by means of two distinct sites. Mol Cell Biol 16:6029–6036

La Vista-Picard N, Hobbs PD, Pfahl M, Dawson MI (1996) The receptor-DNA complex determines the retinoid response: a mechanism for the diversification of the ligand signal. Mol Cell Biol 16:4137–4146

Lala DS, Syka PM, Lazarchik SB, Mangelsdorf DJ, Parker KL, Heyman RA (1997) Activation of the orphan nuclear receptor steroidogenic factor 1 by oxysterols. Proc Natl Acad Sci USA 94:4895–4900

Le Douarin B, Nielsen AL, Garnier J-M, Ichinose H, Jeanmougin F, Losson R, Chambon P (1996) A possible involvement of TIF1α and TIF1β in the epigenetic control of transcription by nuclear receptors. EMBO J 15:6701–6715

Le Douarin B, Zechel C, Garnier J-M, Lutz Y, Tora L, Pierrat B, Heery D, Gronemeyer H, Chambon P, Losson R (1995) The N-terminal part of TIF1, a putative mediator of the ligand-dependent activation function (AF-2) of nuclear receptors, is fused to B-raf in the oncogenic protein T18. EMBO J 14:2020–2033

Lee M-O, Hobbs PD, Zhang X-K, Dawson MI, Pfahl M (1994) A new retinoid antagonist inhibits the HIV-1 promoter. Proc Natl Acad Sci USA 91:5632–5636

Lee J-S, Galvin KM, See RH, Eckner R, Livingston DM, Moran E, Shi Y (1995a) Relief of YY1 transcriptional repression by adenovirus E1 a is mediated by E1A-associated protein p300. Genes Dev 9:1188–1198

Lee JW, Ryan F, Swaffield JC, Johnston SA, Moore DD (1995b) Interaction of thyroid-hormone receptor with a conserved transcriptional mediator. Nature 374:91–94

Lehmann JM, Hoffmann B, Pfahl M (1991a) Genomic organization of the retinoic acid receptor γ gene. Nucleic Acids Res 19:573–578

Lehmann JM, Dawson MI, Hobbs PD, Husmann M, Pfahl M (1991b) Identification of retinoids with nuclear receptor subtype-selective activities. Cancer Res 51: 4804–4809

Lehmann JM, Jong L, Fanjul A, Cameron JF, Lu XP, Haefner P, Dawson MI, Pfahl M (1992) Retinoids selective for retinoid X receptor response pathways. Science 258 1944–1946

Lehmann JM, Zhang X-K, Graupner G, Lee M-O, Hermann T, Hoffmann B, Pfahl M (1993) Formation of retinoid X receptor homodimers leads to repression of T3 response: hormonal crosstalk by ligand-induced squelching. Mol Cell Biol 13(12):7698–7707

Leid M, Kastner P, Lyons R, Nakshatri H, Saunders M, Zacharewski T, Chen J-Y, Staub A, Garnier J-M, Mader S, Chambon P (1992) Purification, cloning, and RXR identity of the HeLa cell factor with which RAR or TR heterodimerizes to bind target sequences efficiently. Cell 68:377–395

Leroy P, Krust A, Kastner P, Mendelsohn C, Zelent A, Chambon P (1992) Retinoic acid receptors. In: Morris-Kay G (ed) Retinoids in normal development and teratogenesis. Oxford University Press, New York, pp 2–25

Li H, Gomes PJ, Chen JD (1997) RAC3, a steroid/nuclear receptor-associated coactivator that is related to SRC-1 and TIF2. Proc Natl Acad Sci USA 94:8479–8484

Li J-J, Dong Z, Dawson MI, Colburn NH (1996) Inhibition of tumor promoter-induced transformation by retinoids that transrepress AP-1 without transactivating retinoic acid response element. Cancer Res 56:483–489

Liu X, Kim CN, Yang J, Jemmerson R, Wang X (1996a) Induction of apoptotic program in cell-free extracts: requirement for dATP and cytochrome c. Cell 86:147–157

Liu Y, Lee M-O, Wang H-G, Li Y, Hashimoto Y, Klaus M, Reed JC, Zhang X-K (1996b) Retinoic acid receptor β mediates the growth-inhibitory effect of retinoic

acid by promoting apoptosis in human breast cancer cells. Mol Cell Biol 16:1138–1149
Liu Z-G, Smith SW, McLaughlin KA, Schwartz LM, Osborne BA (1994) Apoptotic signals delivered through the T-cell hybrid require the immediate-early gene nur77. Nature 367:281–284
Lu XP, Salbert G, Pfahl M (1994) An evolutionary conserved COUP-TF binding element in a neural specific gene and COUP-TF expression patterns support a major role for COUP-TF in neural development. Mol Endocrinol 8:1774–1788
Lu XP, Fanjul AN, Picard N, Pfahl M, Rungta D, Nared-Hood K, Carter B, Piedrafita FJ, Tang S, Fabbrizio E (1997) Novel retinoid-related molecules as apoptosis inducers and effective inhibitors of human lung cancer cells in vivo. Nat Med 3:686–690
Lundblad JR, Kwok RPS, Laurance ME, Harter ML, Goodman RH (1995) Adenoviral E1A-associated protein p300 as a functional homologue of the transcriptional coactivator CBP. Nature 374:85–88
Malik S, Karathanasis SK (1996) TFIIB-directed transcriptional activation by the orphan nuclear receptor hepatocyte nuclear factor 4. Mol Cell Biol 16:1824–1831
Mangelsdorf DJ, Borgmeyer U, Heyman RA, Zhan JY, Ong ES, Oro AE, Kakizuka A, Evans RM (1992) Characterization of three RXR genes that mediate the action of 9-*cis* RA retinoic acid. Genes Dev 6:329–344
Mangelsdorf DJ, Ong ES, Dyck JA, Evans RM (1990) Nuclear receptor that identifies a novel retinoic acid response pathway. Nature 345:224–229
Mangelsdorf DJ, Umesono K, Kliewer S, Borgmeyer U, Ong ES, Evans RM (1991) A direct repeat in the cellular retinol-binding protein type II gene confers differential regulation by RXR and RAR. Cell 66:555–561
Marks MS, Hallenbeck PL, Nagata T, Segars JH, Appella E, Nikodem VM, Ozato K (1992) H-2RIIBP (RXRb) heterodimerization provides a mechanism for combinatorial diversity in the regulation of retinoic acid and thyroid hormone responsive genes. EMBO J 11:1419–1435
Martin SJ, Green DR (1995) Protease activation during apoptosis: death by a thousand cuts? Cell 82:349–352
Miller VA, Benedetti FM, Rigas JR, Verret AL, Pfister DG, Strauss D, Kris MG, Crisp M, Heyman R, Loewen GR, Truglia JA, Warrell RP Jr (1997) Initial clinical trial of a selective retinoid X receptor ligand, LGD1069. J Clin Oncol 15:790–795
Minucci S, Leid M, Toyama R, Saint-Jeannet J-P, Peterson VJ, Horn V, Ishmael JE, Bhattacharyya N, Dey A, Dawid IB, Ozato K (1997) Retinoid X receptor (RXR) within the RXR-retinoic acid receptor heterodimer binds its ligand and enhances retinoid-dependent gene expression. Mol Cell Biol 17:644–655
Mizzen CA, Yang X-J, Kokubo T, Brownell JE, Bannister, AJ, Owen-Hughes T, Workman J, Wang L, Berger SL, Kouzarides T, Nakatani Y, Allis CD (1996) The $TAF_{II}250$ subunit of TFIID has histone acetyltransferase activity. Cell 87:1261–1270
Mukherjee R, Davies PJ A, Crombie DL, Bischoff ED, Cesario RM, Jow L, Hamann LG, Boehm MF, Mondon CE, Nadzan AM, Paterniti JR Jr, Heyman R A (1997) Sensitization of diabetic and obese mice to insulin by retinoid X receptor agonists. Nature 386:407–410
Nagy L, Kao H-Y, Chakravarti D, Lin RJ, Hassig CA, Ayer DE, Schreiber SL, Evans RM (1997) Nuclear receptor repression mediated by a complex containing SMRT, mSin3 a, and histone deacetylase. Cell 89:373–380
Nagy L, Thomazy VA, Shipley GL, Fesus L, Lamph W, Heyman RA, Chandraratna RAS, Davies PJ A (1995) Activation of retinoid X receptors induces apoptosis in HL-60 cell lines. Mol Cell Biol 15:3540–3551
Nakamura N, Shidoji Y, Moriwaki H, Muto Y (1996) Apoptosis in human hepatoma cell line induced by 4,5-didehydro geranylgeranoic acid (acyclic retinoid) via down-regulation of transforming growth factor-α. Biochem Biophys Res Commun 219:100–104

Nakamura N, Shidoji Y, Yamada Y, Hatekayama H, Moriwaki H, Muto Y (1995) Induction of apoptosis by acyclic retinoid in the human hepatoma-derived cell line HUH-7. Biochem Biophys Res Commun 207:382–388

Ogryzko VV, Schiltz RL, Russanova V, Howard BH, Nakatani Y (1996) The transcriptional coactivators p300 and CBP are histone acetyltransferases. Cell 87: 953–959

Okazawa H, Shimizu J, Kamei M, Imafuku I, Hamada H, Kanazawa I (1996) Bcl-2 inhibits retinoic acid-induced apoptosis during the neural differentiation of embryonal stem cells. J Cell Biol 132:955–968

Oñate S, Tsai S, Tsai M-J, O'Malley BW (1995) Sequence and characterization of a coactivator for the steroid hormone receptor superfamily. Science 270:1354–1357

Oridate N, Lotan D, Mitchell MF, Hong WK, Lotan R (1995) Induction of apoptosis by retinoids in human cervical carcinoma cell lines. Int J Oncol 7:433–441

Ortiz MA, Piedrafita FJ, Pfahl M, Maki R (1995) TOR: a new orphan receptor expressed in the thymus that can modulate retinoid and thryroid hormone signals. Mol Endocrinol 9:1679–1691

Pazin MJ, Kadonaga JT (1997) What's up and down with histone deacetylation and transcription? Cell 89:325–328

Pellegrini R, Mariotti A, Tagliabue E, Bressan R, Bunone G, Coradini D, Della Valle G, Formelli F, Cleris L, Radice P, Pierotti MA, Colnaghi MI, Menard S (1995) Modulation of markers associated with tumor aggressiveness in human breast cancer cell lines by N-(4-hydroxyphenyl) retinamide. Cell Growth Differ. 6:863–869

Perlmann T, Jansson L (1995) A novel pathway for vitamin A signaling mediated by RXR heterodimerization with NGFI-B and NURR1. Genes Dev 9:769–782

Petkovich M, Brand NJ, Krust A, Chambon P (1987) Human retinoic acid receptor belongs to the family of nuclear receptors. Nature 330:444–450

Petrij F, Giles RH, Dauwerse HG, Saris JJ, Hennekam RCM, Masuno M, Tommerup N, van Ommen, G-J B, Goodman RH, Peters, DJ M, Breunig MH (1995) Rubinstein-Taybi syndrome caused by mutations in the transcriptional coactivator CBP. Nature 376:348–351

Pfahl M (1993) Nuclear receptor/AP-1 interaction. Endocr Rev 14(5):651–658

Pfahl M, Apfel, R, Bendik I, Fanjul A, Graupner G, Lee M-O, La-Vista N, Lu X-P, Piedrafita FJ, Ortiz MA, Salbert G, Zhang X-K (1994) Nuclear retinoid receptors and their mechanism of action. In: Litwac G (ed) Vitamins and hormones. Academic, San Diego, pp 327–392

Pfitzner E, Becker P, Rolke P, Schüle R (1995) Functional antagonism between the retinoic acid receptor and the viral transactivator BZLF1 is mediated by protein-protein interactions. Proc Natl Acad Sci USA 92:12265–12269

Piedrafita FJ, Pfahl M (1997) Retinoid-induced apoptosis and Sp1 cleavage occur independently of transcription and require caspase activation. Mol Cell Biol 17:6348–6358

Piedrafita FJ, Bendik I, Ortiz MA, Pfahl M (1995) Thyroid hormone receptor homodimers can function as ligand-sensitive repressors. Mol Endocrinol 9:563–578

Piedrafita FJ, Molander RB, Vansant G, Orlova EA, Pfahl M, Reynolds WF (1996) An Alu element in the myeloperoxidase promoter contains a composite SP1-thyroid hormone-retinoic acid response element. J Biol Chem 271:14412–14420

Ponzoni M, Bocca P, Chiesa V, Decensi A, Pistoia V, Raffaghello L, Rozzo C, Montaldo PG (1995) Differential effects of N-(4-hydroxyphenyl) retinamide and retinoic acid on neuroblastoma cells: apoptosis versus differentiation. Cancer Res 55: 853–861

Renaud J-P, Rochel N, Ruff M, Vivat V, Chambon P, Gronemeyer H, Moras D (1995) Crystal structure of the RARg ligand-binding domain bound to all-*trans* retinoic acid. Nature 378:381–689

Rochette-Egly C, Adam S, Rossignol M, Egly J-M, Chambon P (1997) Stimulation of RARα activation function AF-1 through binding to the general transcription factor TFIIH and phosphorylation by CDK7. Cell 90:97–107

Rubin DM, Coux O, Wefes I, Hengartner C, Young RA, Goldberg AL, Finley D (1996) Identification of the gal4 suppressor Gug1 as a subunit of the yeast 26 s proteasome. Nature 379:655–657
Rundlett SE, Carmen AA, Kobayashi R, Bavykin S, Turner BM, Grunstein M (1996) HDA1 and RPD3 are members of distinct yeast histone deacetylase complexes that regulate silencing and transcription. Proc Natl Acad Sci USA 93:14503–14508
Sadovsky Y, Webb P, Lopez G, Baxter JD, Fitzpatrick PM, Gizang-Ginsberg E, Cavailles V, Parker MG, Kushner PJ (1995) Transcriptional activators differ in their responses to overexpression of TATA-box-binding protein. Mol Cell Biol 15:1554–1563
Salbert G, Fanjul A, Piedrafita FJ, Lu X-P, Kim S-J, Tran P, Pfahl M (1993) RARs and RXRα down-regulate the TGF-β1 promoter by antagonizing AP-1 acitivity. Mol Endocrinol 7:1347–1356
Schadendorf D, Kern MA, Artuc M, Pahl HL, Rosenbach T, Fichtner I, Nurnberg W, Stuting S, Stebut Ev, Worm M, Makki A, Jurgovsky K, Kolde G, Henz BM (1996) Treatment of melanoma cells with the synthetic retinoid CD437 induces apoptosis via activation of AP-1 in vitro, and causes growth inhibition in xenografts in vivo. J Cell Biol 135:1889–1898
Schreiber-Agus N, Chin L, Chen K, Torres R, Rao G, Guida P, Skoultchi AI, DePinho RA (1995) An amino-terminal domain of Mxi1 mediates anti-myc oncogenic activity and interacts with a homolog of the yeast transcriptional repressor Sin3. Cell 80:777–786
Schüle R, Rangarajan P, Yang N, Kliewer S, Ransone LJ, Bolado J, Verma IM, Evans RM (1991) Retinoic acid is a negative regulator of AP-1-responsive genes. Proc Natl Acad Sci USA 88:6092–6096
Schulman IG, Chakravarti D, Juguilon H, Romo A, Evans RM (1995) Interactions between the retinoid X receptor and a conserved region of the TATA-binding protein mediate hormone-dependent transactivation. Proc Natl Acad Sci USA 92:8288–8292
Shao Z-M, Dawson MI, Li XS, Rishi AK, Sheikh MS, Han Q-X, Ordonez V, Shroot B, Fontana JA (1995) p53 independent G_0/G_1 arrest and apoptosis induced by a novel retinoid in human breast cancer cells. Oncogene 11:493–504
Shibata H, Nawaz Z, Tsai SY, O'Malley BW, Tsai M-J (1997) Gene silencing by chicken ovalbumin upstream promoter-transcription factor I (COUP-TF I) is mediated by transcriptional corepressors, Nuclear receptor corepressor (N-CoR) and silencing mediator for retinoic acid receptor and thyroid hormone receptor (SMRT). Mol Endocrinol 11:714–724
Smith CL, Oñate S, Tsai M-J, O'Malley BW (1996) CREB binding protein acts synergistically with steroid receptor coactivator-1 to enhance steroid receptor-dependent transcription. Proc Natl Acad Sci USA 93:8884–8888
Smith WC, Nakshatri H, Leropy P, Rees J, Chambon P (1991) A retinoic acid response element is present in the mouse cellular retinol binding protein I (mCRBPI) promoter. EMBO J 10:2223–2230
Steller H (1995) Mechanisms and genes of cellular suicide. Science 267:1445–1449
Sucov HM, Murakami KK, Evans RM (1990) Characterization of an autoregulated response element in the mouse retinoic acid receptor type β gene. Proc Natl Acad Sci USA 87:5392–5396
Swaffield JC, Bromberg JF, Johnston SA (1992) Alterations in a yeast protein resembling HIV Tat-binding protein relieve requirement for an acidic activation domain in GAL4. Nature 357:698–700
Swaffield JC, Melcher K, Johnston SA (1995) A highly conserved ATPase protein as a mediator between acidic activation domains and the TATA-binding protein. Nature 374:88–91
Szondy Z, Reichert U, Bernardon J-M, Michel S, Toth R, Ancian P, Ajzner E, Fesus L (1997) Induction of apoptosis by retinoids and retinoic acid receptor γ-selective compounds in mouse thymocytes through a novel apoptosis pathway. Mol Pharmacol 51:972–982

Taunton J, Hassig CA, Schreiberg SL (1996) A mammalian histone deacetylase related to the yeast transcriptional regulator Rpd3p. Science 272:408–411
Tini M, Otulakowski G, Breitman ML, Tsui L-C, Giguere V (1993) An everted repeat mediates retinoic acid induction of the γF-crystallin gene: evidence of a direct role for retinoids in lens development. Genes Dev 7:295–307
Tjian R, Maniatis T (1994) Transcriptional activation: a complex puzzle with few easy pieces. Cell 77:5–8
Tong G-X, Jeyakumar M, Tanen MR, Bagchi MK (1996) Transcriptional silencing by unliganded thyroid hormone receptor β requires a soluble corepressor that interacts with the ligand-binding domain of the receptor. Mol Cell Biol 16 1909–1920
Torchia, J, Rose DW, Inostroza J, Kamei Y, Westin S, Glass CK, Rosenfeld MG (1997) The transcriptional coactivator p/CIP binds CBP and mediates nuclear receptor function. Nature 387:677–684
Tran P, Zhang X-K, Salbert G, Hermann T, Lehmann J, Pfahl M (1992) COUP orphan receptors are negative regulators of retinoic acid response pathways. Mol Cell Biol 12:4666–4676
Umesono K, Giguere V, Glass CK, Rosenfeld MG, Evans RM (1988) Retinoic acid and thyroid hormone induce gene expression through a common response element. Nature 336:262–265
Umesono K, Murakami KK, Thompson CC, Evans RM (1991) Direct repeats as selective response elements for the thyroid hormone, retinoic acid and vitamin D3 receptors. Cell 65:1255–1266
Vaux DL, Weissman IL, Kim SK (1992) Prevention of programmed cell death in Caenorhabditis elegans by human bcl-2. Science 258 1955–1957
Vidal M, Gaber RF (1991) RPD3 encodes a second factor required to achieve maximum positive and negative transcriptional states in Sacharomyces cerevisiae. Mol Cell Biol 11:6317–6327
Voegel JJ, Heine MJS, Zechel C, Chambon P, Gronemeyer H (1996) TIF2, a 160 kDa transcriptional mediator for the ligand-dependent activation function AF-2 of nuclear receptors. EMBO J 15:3667–3675
vom Baur E, Zechel C, Heery D, Heine MJS, Garnier JM, Vivat V, Le Douarin B, Gronemeyer H, Chambon P, Losson R (1996) Differential ligand-dependent interactions between the AF-2 activating domain of nuclear receptors and the putative transcriptional intermediary factors mSUG1 and TIF1. EMBO J 15:110–124
Wade PA, Pruss D, Wolffe AP (1997) Histone acetylation: chromatin in action. Trends Biochem Sci 22:128–132
Wagner RL, Apriletti JW, McGrath ME, West BL, Baxter JD, Fletterick RJ (1995) A structural role for hormone in the thyroid hormone receptor. Nature 378:690–697
White E (1996) Life, death, and the pursuit of apoptosis. Genes Dev 10:1–15
Woronicz JD, Calnan B, Ngo V, Winoto A (1994) Requirement for the orphan steroid receptor Nur77 in apoptosis of T-cell hybridomas. Nature 367:277–281
Wu, Q, Dawson MI, Zheng Y, Hobbs PD, Agadir A, Jong L, Li Y, Liu R, Lin B, Zhang X-k (1997) Inhibition of *trans*-retinoic acid-resistant human breast cancer cell growth by retinoid X receptor-selective retinoids. Mol Cell Biol 17:6598–6608
Yang Y, Minucci S, Ozato K, Heyman RA, Ashwell JD (1995) Efficient inhibition of activation-induced Fas ligand up- regulation and T cell apoptosis by retinoids requires occupancy of both retinoid X receptors and retinoic acid receptors. J Biol Chem 270:18672–18677
Yang X-J, Ogryzko VV, Nishikawa J, Howard BH (1996) A p300/CBP-associated factor that competes with the adenoviral oncoprotein E1 a. Nature 382:319–324
Yang Y, Vacchio MS, Ashwell JD (1993) 9-*cis*-retinoic acid inhibits activation-driven T-cell apoptosis: implications for retinoic X receptor involvement in thymocyte development. Proc Natl Acad Sci USA 90:6170–6714
Yang-Yen H-F, Zhang X-K, Graupner G, Tzukerman M, Sakamoto B, Karin M, Pfahl M (1991) Antagonism between retinoic acid receptors and AP-1: implication for tumor promotion and inflammation. New Biol 3:1206–1219

Yao T-P, Ku G, Zhou N, Scully R, Livingston DM (1996) The nuclear hormone receptor coactivator SRC-1 is a specific target of p300. Proc Natl Acad Sci USA 93:10626–10631

Yu VC, Delsert C, Andersen B, Holloway JM, Devary OV, Naar AM, Kim SY, Boutin J-M, Glass CK, Rosenfeld MG (1991) RXRβ: a coregulator that enhances binding of retinoic acid, thyroid hormone, and vitamin D receptors to their cognate response elements. Cell 67:1251–1266

Yuan J, Horvitz HR (1990) Genetic mosaic analysis of ced-3 and ced-4, two genes that control programmed cell death in the nematode C. elegans. Dev Biol 138:33–41

Yuan J, Shaham S, Ledoux S, Ellis HM, Horvitz HR (1993) The C. elegans cell death gene ced-3 encodes a protein similar to mammalian interleukin-1b-converting enzyme. Cell 75:641–652

Zamir I, Harding HP, Atkins GB, Horlein A, Glass CK, Rosenfeld MG, Lazar MA (1996) A nuclear hormone receptor corepressor mediates transcriptional silencing by receptors with distinct repression domains. Mol Cell Biol 16:5458–5465

Zamir I, Zhang J, Lazar MA (1997) Stoichiometric and steric principles governing repression by nuclear hormone receptors. Genes Dev 11:835–846

Zavacki AM, Lehmann JM, Seol W, Willson TM, Kliewer SA, Moore DD (1997) Activation of the orphan receptor RIP14 by retinoids. Proc Natl Acad Sci USA 94:7909–7914

Zhang L-X, Mills KJ, Dawson MI, Collins SJ, Jetten AM (1995) Evidence for the involvement of retinoic acid receptor RARα-dependent signaling pathway in the induction of tissue transglutaminase and apoptosis by retinoids. J Biol Chem 270:6022–6029

Zhang X-K, Hoffmann B, Tran P, Graupner G, Pfahl M (1992a) Retinoid X receptor is an auxiliary protein for thyroid hormone and retinoic acid receptors. Nature 355:441–446

Zhang X-K, Lehmann J, Hoffmann B, Dawson MI, Cameron J, Graupner G, Hermann T, Pfahl M (1992b) Homodimer formation of retinoid X receptor induced by 9-*cis* retinoic acid. Nature 358:587–591

Zou H, Henzel WJ, Liu X, Lutschg A, Wang X (1997) Apaf-1, a human protein homologous to C. elegans CED-4, participates in cytochrome c-dependent activation of caspase-3. Cell 90:405–413

CHAPTER 6

RAR-Selective Ligands: Receptor Subtype and Function Selectivity

R.L. BEARD and R.A.S. CHANDRARATNA

A. Introduction

There are two primary reasons for developing retinoic acid receptor (RAR) subtype-selective agonists and antagonists. First, it would be desirable to decrease the toxic side effects associated with retinoid therapy. Retinoids are known to mediate many different biological processes, including cellular differentiation and proliferation, epithelial homeostasis, immunocompetence, and embryogenesis (SPORN et al. 1994; MORRIS-KAY 1992). Currently, retinoids are useful for the treatment of a limited number human diseases, including skin disorders such as acne and psoriasis, and for the treatment of some cancers. The use of retinoids in other therapeutic areas is limited because they cause a wide variety of toxic side effects such as teratogenicity and mucocutaneous toxicity. Thus, the full therapeutic potential of retinoids may only be realized if new compounds with vastly improved therapeutic indices can be developed. The probability of finding retinoids with better therapeutic indices has increased dramatically in the past few years because our understanding of the molecular mechanisms of retinoid action has progressed rapidly since the discovery of the six known retinoid receptors, RARα, RARβ, RARγ, and RXRα, RXRβ, RXRγ. It is known that each of the RAR subtypes has a specific tissue distribution pattern (REES 1992). For example, RARγ is the most abundant retinoid receptor in the skin (ELDER et al. 1991; RECZEK et al. 1995), whereas RARβ is found primarily in the heart, lung and spleen. RARα is expressed at low levels in many tissue types. It follows that retinoids that activate a particular receptor subtype may elicit biological responses in specific tissues. Such compounds are likely to be associated with fewer toxic side effects and have greatly improved therapeutic indices over nonspecific retinoids in the treatment of retinoid responsive diseases.

The second major reason for developing RAR-subtype specific retinoids is for their use in determining the physiological role of each RAR subtype. In this case it is important to distinguish between retinoids that are RAR subtype-selective from those that are RAR subtype-specific. In this chapter, we refer to compounds that are subtype-selective as those that have a clear preference in terms of receptor binding affinity or transactivation activity for one or two of the three RAR subtypes. In contrast, compounds that are termed RAR subtype-specific have true pharmacological specificity, i.e., greater than

1000-fold selectivity in receptor binding or transactivation. It is easy to show that the biology associated with compounds that are RAR-selective and compounds that are RAR-specific are very different. This is so because even low levels of receptor activity may elicit a pronounced biological response due to factors such as pharmacokinetic effects and uneven tissue distribution of the receptors. Thus, RAR subtype-specific compounds are necessary for elucidating the biology associated with a particular receptor subtype.

This chapter begins with a brief discussion of important structural differences between RAR selective and RXR selective compounds, followed by a review of the structural characteristics of known RAR subtype-selective and subtype-specific retinoids. We then turn to the subject of RAR antagonists and their relationships to RAR inverse agonists. The final section points out some important relationships between a compound's receptor subtype or function selectivity and any specific biological responses that it causes.

B. Structural Differences Between RAR-Selective and RXR-Selective Ligands

Prior to the identification of 9-*cis*-RA as the putative hormone for the RXRs (LEVIN et al. 1992; HEYMAN et al. 1992), virtually every synthetic retinoid reported was selective for the RARs. In 1992, the first series of retinoids that selectively activated RXR-RXR homodimers over RAR-RXR heterodimers in receptor cotransfection assays was described (LEHMANN et al. 1992). Shortly thereafter, a synthesis of [^{3}H]-9-*cis*-RA was reported (BOEHM et al. 1994a), which led to the development of competitive binding assays for each of the RXR subtypes. These high-throughput receptor assays allowed retinoid medicinal chemistry to progress rapidly, and structural differences between RAR-selective and RXR-selective ligands quickly emerged.

I. The Effect of Bridging Groups on Conformationally Restricted Retinoids

One successful approach to designing more potent and selective retinoids is to decrease the rotational degrees of freedom accessible to retinoic acid (RA) by confining the polyene side chain in a cyclic system consisting of two or more aromatic rings. A typical example of this approach is illustrated by (*E*)-4-[2-(5,6,7,8-tetrahydro-5,5,8,8-tetramethyl-2-naphthalenyl)propen-1-yl]benzoic acid (TTNPB, 1), in which the cyclohexene ring and the C7-C8 double bond of RA are confined in a bicyclic tetrahydronapthalene ring system and the C11-C15 dienoic acid side chain is replaced by a benzoic acid group. In TTNPB, the two ring systems are connected by an *trans*-alkene bridge, which corresponds to the 9-*trans* double bond position of all-*trans*-RA (ATRA). Thus, it is not surprising that TTNPB is a potent, RAR specific compound. However, many other types of linking groups have been used, and it is the nature of the linking group that often determines the selectivity of a compound for a particular receptor subtype. As one might expect, compounds bridged by two atoms which can adopt a *trans*-relationship are generally RAR selective. However, stilbene analogs linked by a *cis*-alkene display only weak receptor binding and transactivational activity and are not selective (Jong et al. 1993). A more effective isostere for the 9-*cis* double bond of 9-*cis*-RA in conformationally restricted systems is a one-atom linking group, and benzoic acid derivatives of this type tend to be RXR selective. For example, the RXR-selective compounds described by Lehmann et al. are characterized by a sp^3-hybridized methylene group linking the tetralin ring system to the benzoic acid ring. Similarly, the use of a geminally disubstituted cyclopropyl ring to bridge the tetralin ring to a benzoic or nicotinic acid moiety led to potent and selective RXR analogs (Boehm et al. 1995). Other tetralin-substituted benzoic acids bridged by sp^2-hybridized carbons (Boehm et al. 1994b, 1995; Dawson et al. 1995a), or by heteroatoms (Beard et al. 1996), are also RXR-selective.

II. The α-Methyl Effect

The data in Tables 1 and 2 illustrate the "α-methyl effect" associated with the stilbene structure (Beard et al. 1994). As noted above, stilbene based arotinoids, such as TTNPB (**1**), are potent RAR specific agonists. However, introduction of a substituent α- to the double bond on the tetrahydronaphthalene

Table 1. Receptor assay data for analogs of TTNPB (**1**): receptor cotransfection assay data (from Beard et al. 1994)

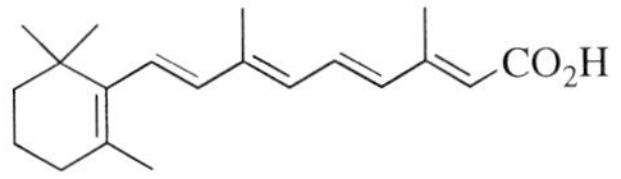

CO_2H ATRA

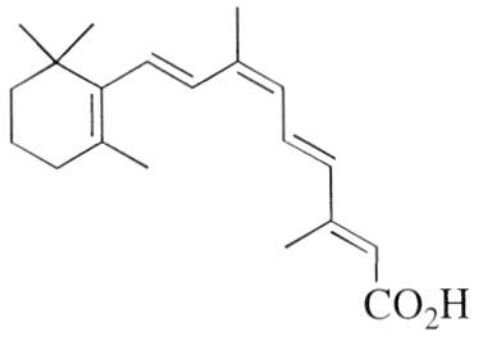

CO_2H 9-cis-RA

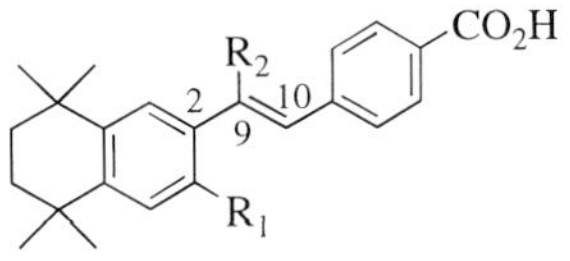

R_2 2 10 9 CO_2H R_1

	Stilbene substitution		EC_{50} (n*M*)			
	R_1	R_2	RARα	RARβ	RARγ	RXRα
ATRA	–	–	5.0	1.5	0.5	NA
9-*cis*-RA	–	–	102	3.3	6.0	13
1	H	Me	21	4.0	2.4	NA
2	Me	Me	4580	74	152	385
3	Cl	Me	>1000	21	77	275
4	Br	Me	989	21	91	298
5	Me	H	15	0.4	1.4	NA

NA, not active (efficacy <20% that of ATRA or 9-*cis*-RA at a dose of 10^{-5} *M*).

Table 2. Receptor assay data for analogs of TTNPB (**1**) (see Table 1): competitive receptor binding assay data (from BOEHM et al. 1994a)

	Stilbene substitution		K_d (nM)					
	R_1	R_2	RARα	RARβ	RARγ	RXRα	RXRβ	RXRγ
ATRA			15	13	18	>1000	>1000	350
9-*cis*-RA			7	7	17	32	12	4
1	H	Me	36	3	26	NA	NA	NA
2	Me	Me	638	1169	645	32	>1000	100
3	Cl	Me	640	465	482	ND	ND	ND
4	Br	Me	778	508	524	ND	ND	ND
5	Me	H	22	14	56	ND	ND	ND

NA, Not active (K_d > 10000 nM); ND, not determined.
[a] K_d values (mean of triplicate determinations) were determined via competition of [^{3}H]ATRA (5 nM) binding with unlabeled test retinoid at baculovirus expressed RARs as described (CHENG and PRUSOFF 1973).

ring, as in (*E*)-4-[2-(5,6,7,8-tetrahydro-3,5,5,8,8-pentamethyl-2-naphthalenyl) propen-1-yl]benzoic acid (3-Me-TTNPB, **2**), results in a decrease in activity at the RARs and concomitant increase in activity at the RXRs. The 3-chloro (**3**) and 3-bromo (**4**) analogs have very similar receptor transactivation and binding properties as 3-Me-TTNPB (**2**), discounting the possibility that an oxidative metabolite of the 3-methyl substituent may be the true RXR-selective agent. The concomitant presence of the C9 methyl group is also necessary for RXR activity, as is illustrated by the pure RAR activity of the 9-desmethyl analog, **5**. Thus, the "α-methyl effect" is probably due to a conformational change in the molecule that allows it to interact more favorably with the RXRs and less favorably with the RARs.

The α-methyl effect is not limited to stilbene systems. A similar effect is also known to occur in benzophenone based retinoids (BOEHM et al. 1994a). For example, the desmethyl analog, **7**, has pan-agonist properties (i.e., it is capable of activating both RARs and RXRs) while the α-methyl analog, **8**, is much more selective for the RXRs (Fig. 1). It is likely that this α-methyl substituent will be useful in conferring RXR selectivity in other novel structural classes as well.

III. 9-*Cis*-Locked vs 9-*Trans*-Locked Retinoids

The C9-C10 double bond of RA may be locked in either a *cis*- or *trans*- confirmation by using a cyclopropyl group as an alkene bioisostere. The receptor activities for some of these analogs is presented in Table 3 (VULIGONDA et al. 1996). It is interesting that the 9-*cis*-locked cyclopropyl compound **9a** (R_1 = H) is highly specific for the RXRs over the RARs since 9-*cis*-RA itself binds to and activates both RARs and RXRs. Evidently, the active conformations

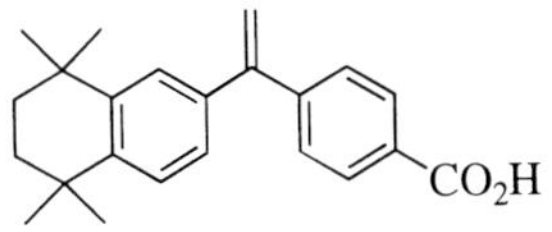

7 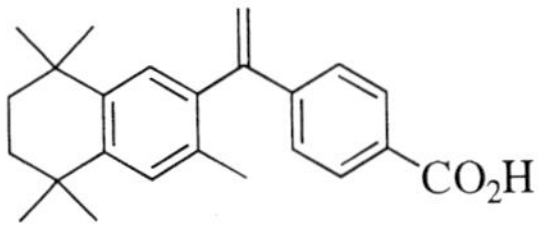 8

Fig. 1. Benzophenone-derived retinoids having an α-methyl effect

Table 3. Cotransfection assay data for cyclopropyl locked analogs of RA (from VULIGONDA et al. 1996)

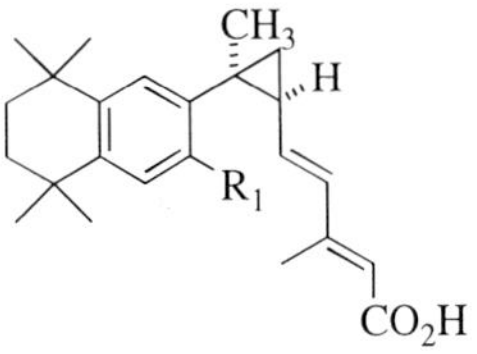

9 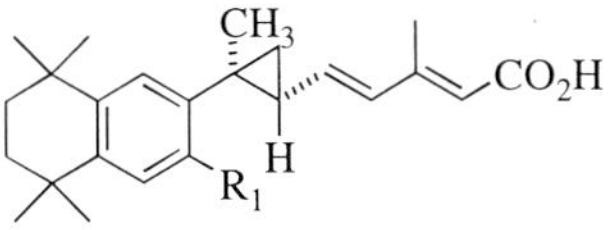 10

	R_1	EC_{50} (n*M*) RARα	RARβ	RARγ	RXRα	RXRβ	RXRγ
ATRA	–	15	13	18	>1000	>1000	350
9-*cis*-RA	–	7	7	17	32	12	4
9a	H	>1000	>1000	>1000	1.5	2.0	1.0
10	H	>1000	340	200	>1000	>1000	>1000
9b	Me	>1000	>1000	>1000	130	100	71

Transactivation assays were performed in CV-1 cells transfected with an expression vector for the indicated retinoid receptor and a luciferase reporter plasmid. A ΔMTV-TREp-Luc reporter was used for the RARs. A CRABPII-tk-Luc reporter was used for RXRα and RXRγ, and a CPRE3-tk-Luc reporter was used for RXRβ. EC_{50} values are reported as the mean of triplicate determinations.

adopted by 9-*cis*-RA in binding to the RARs cannot be accessed by the 9-*cis*-locked cyclopropyl compound, strongly suggesting that 9-*cis*-RA binds to the RARs and RXRs in very distinct conformations. It is also surprising that the 9-*trans*-locked cyclopropyl compound, **10**, which is RAR specific, is about 100-fold less potent than ATRA. Thus, the cyclopropyl ring acts as a good isostere for the C9-C10 double bond of 9-*cis*-RA in terms of the RXR activity of 9-*cis*-RA, but it is a poor isostere for the C9-C10 double bond in terms of retaining the RAR activity of either ATRA or 9-*cis*-RA. Interestingly, the introduction of an α-methyl substituent in the C9-C10 *cis* cyclopropyl-locked series is detrimental to RXR activity as illustrated by the 50-fold or greater decrease in potency observed for the methyl-substituted analog, **9b**, compared to its desmethyl congener. The conformational constraints introduced by the

α-substituent in this series are not favorable for RXR activity unlike in the stilbene and benzophenone based series. As in the stilbene series, the presence of the C9-methyl group is important for RXR activity in the cyclopropyl series since the C9 desmethyl analog has about a 100-fold lower affinity for the RXRs (APFEL et al. 1995).

C. RAR Subtype Selective Agonists

With the identification of the three RAR subtypes, medicinal chemists began designing receptor subtype selective compounds as a means of reducing the side effects associated with RAR agonist therapy, and to elucidate the biology associated with each receptor subtype. Some of the earliest examples of RAR subtype-selective compounds are the amide-linked retinoids, **11** and **12**, which were synthesized by SHUDO and coworkers (KAGECHIKA et al. 1988) before retinoid receptor binding and transactivation assays were established as high-throughput screening assays (Fig. 2). Thus, the RARα selectivity displayed by these compounds went unreported for several years (CRETTAZ et al. 1990). As the retinoid receptor assays were developed, compounds displaying moderate selectivity for each of the RAR subtypes began appearing in the literature. Since then, systematic identification of the structural features necessary for receptor subtype selectivity has led to development of compounds that are truly pharmacologically specific at the RARα and RARβ subtypes. These compounds have been, and continue to be, valuable tools for delineating the pleiotropic biological effects displayed by retinoids. Moreover, receptor selective retinoids are likely to have decreased toxic side effects and be useful as drugs of improved therapeutic index in the treatment of specific retinoid responsive diseases.

In 1991, a series of 6-substituted 2-naphthoic acid retinoids **13** (Fig. 3) were discovered (GRAUPNER et al. 1991) that have gene transcriptional activity at the RARβ and RARγ subtypes but not at RARα. Several laboratories have since described other retinoids that also selectively bind to and activate RARβ and RARγ, examples of which are shown in Fig. 3. As can be seen, compounds of various structural types have this type of receptor activity profile, including highly rigid derivatives such as the benzofuran analog, **14** (DELESCLUSE et al. 1991), and the bisnaphthalene derivative, **15** (LEHMANN et al. 1991), as well as

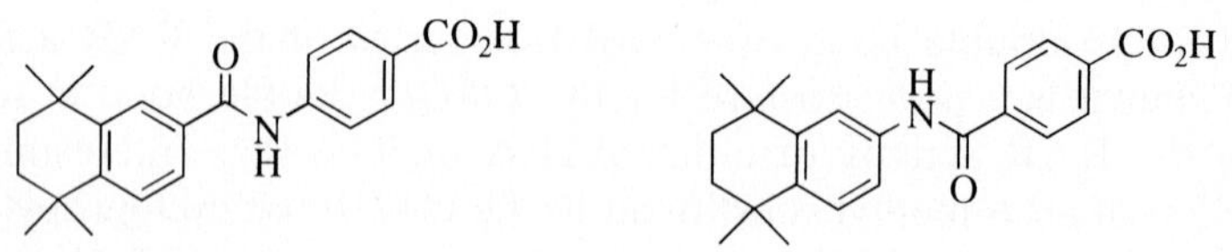

11 **12**

Fig. 2. RARα selective retinoids bearing amide linking groups

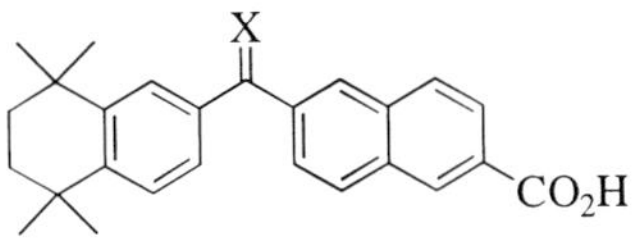

13 ($X=H_2$, O)

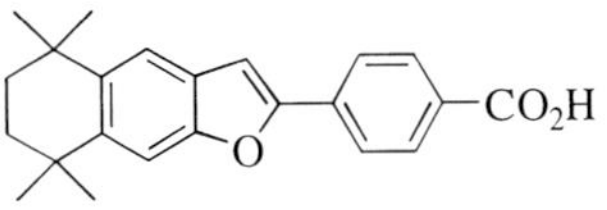

14

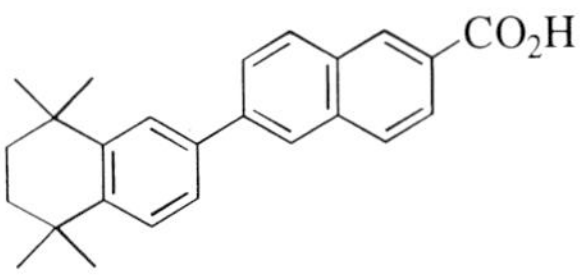

15

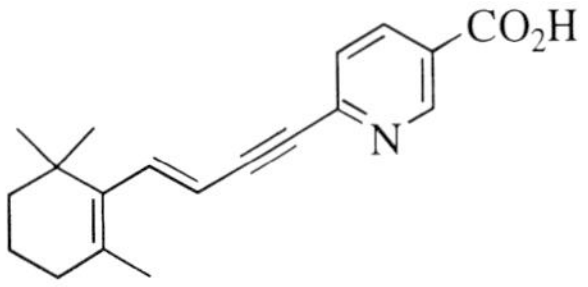

16

Fig. 3. Retinoids having selectivity for the RARβ and RARγ subtypes

the less conformationally restricted compound **16** (CHANDRARATNA et al. 1995). The development of these compounds demonstrated that ligands could be synthesized with differential RAR subtype binding affinities and gene transcriptional activities, and many of these compounds were useful templates for the design of compounds with greater subtype selectivity.

A structural modification of the lipophilic left-hand ring system of bisnaphthalene **15** led to the discovery of retinoids such as **17–22** (Table 4) that are selective for either the RARβ or RARγ subtypes (CHARPENTIER et al. 1995). In these compounds, the 2-naphthoic acid moiety is linked to a phenyl group substituted in the 3-position with a bulky, lipophilic group and in the 4-position with either methoxy or hydroxyl group. Regardless of the substituents (i.e., R_1 or R_2) on the phenyl group, these compounds bind to RARα with only low affinity, having K_i values of greater than 1 μm and thus are selective for RARβ or RARγ. Compounds substituted in the 4-position with a methoxy group, **17–19**, are slightly selective for RARβ, with the C3 *t*-butyl-substituted compound, **17**, having 12-fold higher affinity for RARβ than for RARγ. On the other hand, compounds substituted in the 4-position with a hydroxyl group, **20–22**, are more selective for RARγ, with the highest selectivity being associated with the C3-adamantyl-substituted derivative, **22**, which has a 32-fold higher affinity for RARγ than for RARβ. These compounds were also tested for their ability to induce differentiation of F9 murine teratocarcinoma cells (F9 cells), but their appears to be no correlation between receptor subtype selectivity and the ability of a compound to induce differentiation in these cells.

These compounds demonstrate the feasibility of designing compounds with RAR subtype selectivity, and show that ligands bind to each of the RAR subtypes in distinct conformations. However, because these compounds have activity at two or more RAR subtypes, they are of little use in determining

Table 4. Retinoid analogs with moderate selectivity for the RARβ and RARγ subtypes (from CHARPENTIER et al. 1995)

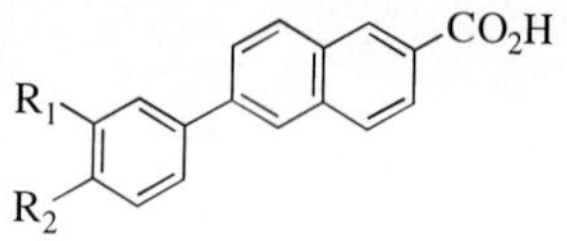

17-22

	Substitution		K_i (nM)			F9 differentiation
	R_1	R_2	RARα	RARβ	RARγ	AC_{50} (nM)
ATRA	–		15	13	18	200
17		OMe	6500	36	426	200
18	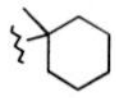	OMe	1120	26	160	150
19	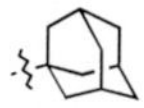	OMe	1100	34	130	37
20		OH	NA	3550	200	NA
21	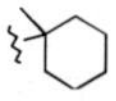	OH	NA	660	121	175
22		OH	6500	2480	77	33

Results are the mean of three separate experiments. NA, Not active.

which receptor subtype may be associated with a particular biological response. In order to elucidate the biological selectivity associated with each of the RAR subtypes, it is necessary to synthesize compounds with true pharmacological specificity. In the past few years there have been a few reports of RAR agonists that are subtype specific, and the biology associated with each of the retinoid subtypes is beginning to emerge.

I. RARα Selective Agonists

The early finding that the amide-linked compound, **11**, was about 30-fold selective for the RARα subtype over RARβ and RARγ was a major advance in the development of RAR subtype selective retinoids. When compared to other arotinoids such as TTNPB (**1**), it is expected that the introduction of a polar, hydrophilic amide linking group would lead to destabilization of ligand binding in the lipophilic pocket of the retinoid receptors and, in fact, this is what is observed for compound **11** at RARβ and RARγ. However, compound **11** binds to RARα with the same affinity as TTNPB (**1**), suggesting that the amide linking group somehow stabilizes the ligand's interaction with RARα.

Table 5. Receptor binding data for RARα selective retinoids (from TENG et al. 1996)

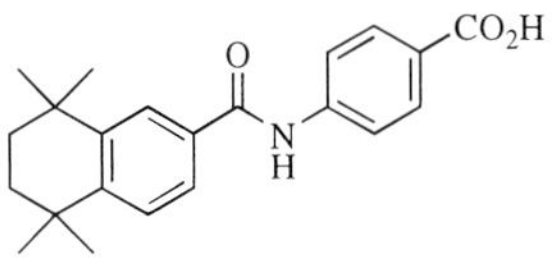

11

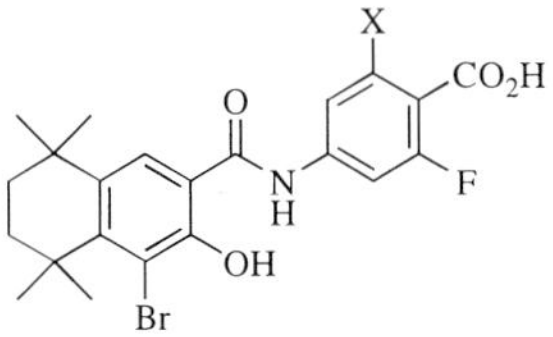

23-24

	Substitution X	K_d (nM) RARα	RARβ	RARγ
ATRA	–	15	13	18
11	–	36	1361	3824
23	H	4.4	3037	>30000
24	F	8.4	17374	>30000

K_d values (mean of triplicate determinations) were determined via competition of [^{3}H]ATRA (5 nM) binding with unlabeled test retinoid at baculovirus expressed RARs as described (CHENG and PRUSOFF 1973). NA, Not active (i.e., K_d > 10000 nM).

Interestingly, from extrapolations of the crystal structure of RARγ (RENAUD et al. 1995), it is known that RARα contains a polar serine residue in helix 3, which is approximately in the same position that is occupied by alanine residues in RARβ and RARγ. Thus, the amide group of **11** may interact favorably with RARα by forming hydrogen bonds with the hydroxyl group of serine, an interaction that is not possible in RARβ or RARγ. Working off this hypothesis, the hydrogen bonding properties of the amide group were varied by placing different substituents on the aromatic rings of **11**. From this research, two new amide-linked compounds, **23** and **24**, were discovered that had much improved selectivity for RARα over RARβ, RARγ (TENG et al. 1996). As depicted in Table 5, compound **23** has approximately 700-fold higher affinity for RARα than for RARβ without measurable binding affinity to RARγ. Moreover, compound **24** shows true pharmacological specificity, binding to RARα with about 2000-fold higher affinity than to RARβ and has no measurable binding affinity to RARγ. In terms of gene transactivation, compound **11** has full agonist activity at all three RAR subtypes, with EC_{50} values of less than 100 nM at RARα and RARβ, and less than 1 μM at RARγ (data not shown). On the other hand, compound **23** has full agonist activity only at RARα, with partial agonist activity at RARβ at a dose of 1 μM, and is inactive at RARγ. Compound **24** transactivates exclusively through RARα, and is completely inactive at RARβ and RARγ (data not shown). With their high degree of selectivity, compounds **23** and **24** will be valuable tools for elucidating the biology associated with the RARα subtype.

II. RARβ-Selective Agonists

In terms of both receptor binding affinity and gene transcriptional activation, only compounds which are specific for the RARα subtype have yet been described. However, we recently reported on a series of compounds that clearly activate gene transcription only through RARβ (JOHNSON et al. 1996). The RAR binding and transcriptional activation properties of these compounds are shown in Table 6. The common structural feature in these compounds is a 2-thienyl dihydronaphthalene ring system, which is connected by acetylene (**25**), amide (**26**), or ester (**27**) linking groups to benzoic acid. Compounds **25** and **27** are similar in that they bind to each of the receptor subtypes but induce gene transcription only through RARβ. Compounds **25** and **27** are also potent RARβ activators, having EC_{50} values of 25 nM and 26 nM, respectively, but have only 30%–50% of the efficacy of ATRA and are therefore partial agonists at RARβ. As is the case with amides in general, compound **26** binds to RARα with about 10-fold higher affinity than to RARβ, and does not bind to RARγ. However, as with compounds **25** and **27**, compound **26** induces transactivation only through RARβ with an EC_{50} of 115 nM and 61% of ATRA efficacy. Thus, the 2-thienyl group at C8 of the dihydronaphthalene ring selects exclusively for RARβ transactivation over the other RAR subtypes. It was also determined that these compounds antagonize the

Table 6. Receptor binding and transactivation data for RARβ selective compounds (from JOHNSON et al. 1996)

S CO2H

25

S O CO2H X

26-27

	Substitution X	K_d (nM)			EC_{50} (nM) (% efficacy)[a]		
		RARα	RARβ	RARγ	RARα	RARβ	RARγ
ATRA	–	15	13	18	459	87	20
25	–	129	20	104	NA	25 (32)	NA
26	NH	94	996	>10 000	NA	115 (61)	NA
27	O	303	189	1 490	NA	26 (47)	NA

K_d values are reported as the mean value of three determinations by competition of [^{3}H]ATRA (5 nM) binding with unlabeled test retinoid using baculovirus expressed RARs. NA, Not active (i.e., $K_d > 10\,000$ nM). EC_{50} values are the mean of at least three experiments performed in triplicate using CV-1 cells cotransfected with the luciferase reporter plasmid MTV-4(R5G)-Luc and an expression vector of the indicated RAR.
[a] Percentage activity at 1 μM relative to ATRA.

RARα and RARγ mediated transcriptional activity of ATRA in a dose-dependent manner that is consistent with each of the compound's K_d values (data not shown). CHEN and coworkers (1995) recently reported on similar series of RARβ agonists with RARα /RARγ antagonist properties, but the structures of these compounds were not disclosed. Thus, direct comparisons with these compounds cannot be made.

III. RARγ-Selective Agonists

Compounds with increasing degrees of selectivity for the RARγ subtype have been developed in the past few years (YU et al. 1996; DAWSON et al. 1995b). The receptor binding and transcriptional activities for these analogs were reported (CHAO et al. 1997) as shown in Table 7. Building on the lead compounds **20–22** described in Table 4, each of these compounds have a polar hydroxyl group placed near the center of the molecule. As with the RARα selective amides, it is expected that placement of a polar hydroxyl group in the lipophilic binding pocket of the RARs would lead to a destabilizing interaction between the ligand and the receptor when compared to more lipophilic retinoids such as ATRA. Thus, it is not surprising that compounds **28–30** have lower binding affinities than ATRA at all three RAR subtypes. However, it is surprising that each of these compounds bind with relatively high affinity to RARγ without binding to RARα or RARβ at doses as high as 1 μM. Thus, it

Table 7. Receptor binding and transactivation data for RARγ selective compounds (from CHAO et al. 1997)

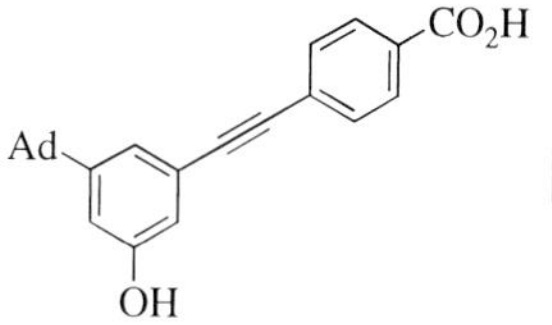

Ad, OH, CO_2H — **28**; OH, CO_2H — **29**; HO, N, CO_2H — **30**

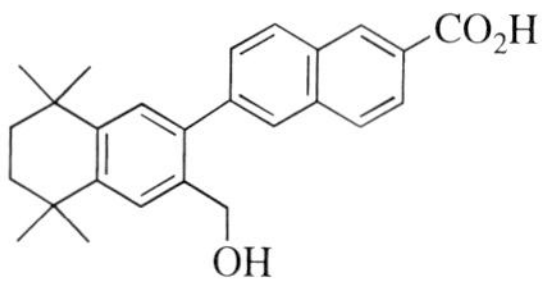

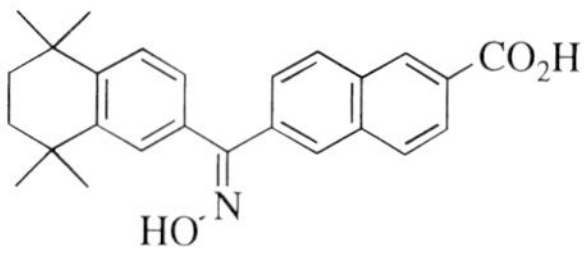

	K_d (nM)[a]			EC_{50} (nM) (% efficacy)[b]		
	RARα	RARβ	RARγ	RARα	RARβ	RARγ
ATRA	2.8	2.1	0.6	45 (100)	5 (86)	2 (87)
28[c]	>1000	>1000	15	>1000 (−1)	2800 (15)	340 (67)
29	>1000	>1000	22	>1000 (8)	>1000 (39)	170 (92)
30	>1000	>1000	20	>1000 (13)	1600 (46)	22 (84)

Numbers in parentheses refer to percentage activity at 1 μM relative to ATRA.
[a] Competitive binding affinity using [11,12]-[$^3H_2$]*trans*-RA.
[b] Determined by RAR transcriptional activation in CV-1 cells using TREpal-*tk*-CAT reporter construct.
[c] 1-Adamantyl (Ad).

is likely that the hydroxyl groups in compounds **28–30** interact favorably with a specific residue in the binding pocket of RARγ, and that this interaction is absent in RARα and RARβ. With regards to receptor transcriptional activation, compound **30** is about 80-fold selective for RARγ, having EC_{50} values of 22 n*M* at RARγ and 1600 n*M* at RARβ, and is inactive at RARα.

A second series of RARγ selective agonists was recently described (RECZEK et al. 1995; SWANN et al. 1996). These compounds, some of which are displayed in Table 8, are characterized by an α-hydroxyacetamide linking group, and they are therefore similar in structure to other RARγ selective compounds. The transactivation ratios (i.e., the relative transactivational activities that are normalized to the EC_{50} value of ATRA) at each of the RARs are shown in Table 8 for compounds **31–34**. In this series, the unsubstituted analog, **31**, is the most potent RARγ agonist, but is somewhat less selective than the other analogs shown, having partial agonist activity at RARβ at the highest dose tested (1 μ*M*). The 3-chloro-substituted compound, **32**, appears to be specific for RARγ with a transactivation ratio of 13 at RARγ, and no activity at RARα or RARβ. The RARγ potency is decreased for the 3-hydroxy and 3-methyl substituted derivatives, **33** and **34**, respectively. Nevertheless, these compounds are important additions to the increasing number of RAR subtype selective retinoids that are now available to help delineate the complex biology associated with retinoids.

Table 8. Receptor transactivation ratios for RARγ selective compounds (from SWANN et al. 1996)

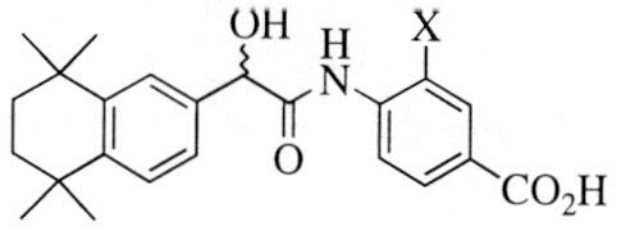

31-34

	X	Transactivation ratio[a]		
		RARα	RARβ	RARγ
ATRA	–	1	1	1
31	H	NA	SA	3.8
32	Cl	NA	NA	13
33	OH	NA	NA	40
34	CH_3	NA	NA	67

NA, Not active; SA, slight activity (<35% maximum transactivation at dose of 1 μ*M*).

[a] The ratio of a test compound's EC_{50} for a given receptor to the EC_{50} for ATRA for the same receptor.

D. RAR Antagonists, Neutral Antagonists, and Inverse Agonists

In the past decade, along with the improvements in the development of RAR subtype selective retinoids, we have also witnessed the development of function selective RAR ligands. Several groups have now reported on RAR ligands that antagonize the effects of ATRA and other RAR agonists both at the receptor level, and in in vitro and in vivo models of retinoid efficacy and toxicity. At the receptor level, compounds have been developed that are pan-RAR antagonists as well as those which are selective for the RARα subtype. In addition, a new class of compounds called RAR inverse agonists (i.e., compounds which are able to *decrease* the basal transcriptional activity of the RARs) have recently been described. This section reviews retinoid antagonists and how they differ from retinoid inverse agonists.

I. RAR Antagonists

The first examples of retinoid antagonists were described by KANEKO and coworkers in 1991. These compounds, **35–36**, have a bulky diadamantyl group in the lipophilic region of the molecule, as shown in Fig. 4. These compounds were tested for their ability to induce differentiation of HL-60 cells to granulocytes (KAGECHIKA et al. 1986, 1988), and found to be completely inactive at concentrations of $10^{-5}\,M$ or less. However, each of these compounds was able to completely antagonize the ATRA-induced HL-60 cell differentiation at concentrations of approximately 100-fold in excess of the concentration of ATRA used. In addition, compounds **35** and **36** were tested in a competitive

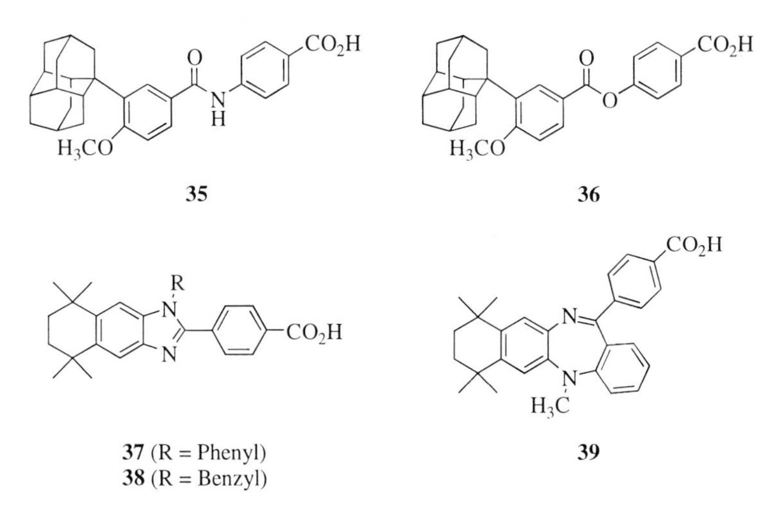

Fig. 4. Retinoid antagonists of ATRA induced HL-60 cell differentiation

binding assay at RARβ and their inhibition constants (K_i values) were estimated to be 6×10^{-7} M and 8×10^{-7} M, respectively.

Another series of compounds displaying retinoid antagonist activity in the HL-60 differentiation assay include the benzimidazole derivatives, **37** and **38**, and the benzodiazapine analog, **39** (Fig. 4). All three of these compounds were unable to induce HL-60 cell differentiation at concentrations below 10^{-6} M. The retinoid agonist, 4-[(5,6,7,8-tetrahydro-4,4,8,8-tetrahydro-2-naphthalenyl) carbamoyl]benzoic acid (**12**), is able to induce the differentiation of approximately 80% of HL-60 cells to mature granulocytes at a concentration of 1.2 nM. At this concentration of **12**, compounds **37**, **38** and **39** were able to reduce the number of differentiated cells to 20%, 28% and 10%, respectively, when applied at a concentration of 110 nM. However, these compounds were found to have only weak binding affinity to RARα and RARβ (binding to RARγ was not reported) in a competitive binding assay.

YOSHIMURA and coworkers (1995) have reported on a series of highly potent retinoid antagonists containing N-substituted pyrole and pyrazole rings. The binding affinity to HL-60 nuclear extracts and antagonist activity against ATRA were determined in the HL-60 cell assay and the data are presented in Table 9 for some of these analogs. Presumably, these HL-60 cells contain primarily RARα receptors and the binding affinities reported in this paper refer to RARα. As can be seen in Table 9, the nature of the R substituent is important in determining a compound's receptor binding affinity and antagonism of ATRA-induced differentiation of HL-60 cells. The binding affinities of these compounds are highly consistent with the ability of the compounds to inhibit ATRA-induced HL-60 cell differentiation. Thus, of the compounds tested, the 3-pyridylmethyl substituted compounds, **42**, **43**, and **45**, are the most potent antagonists, having binding IC_{50} values of 0.34, 0.16 and 0.91 nM, respectively, and are also the most potent inhibitors of ATRA-induce HL-60 cell differentiation, having antagonism IC_{50} values of 4.9, 5.6 and 1.3 nM, respectively.

A few anthracene-derived retinoid antagonists, including compounds **46** and **47** (Fig. 5), have been described in the literature (LEE et al. 1994, 1996). These compounds were unable to induce receptor transcriptional activity at

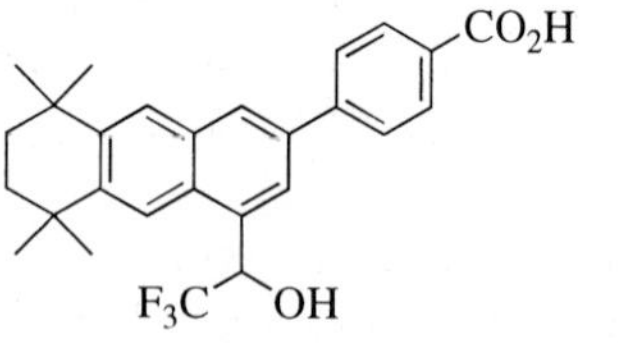

CO_2H

F_3C OH

46 (R = H)
47 (R = Me)

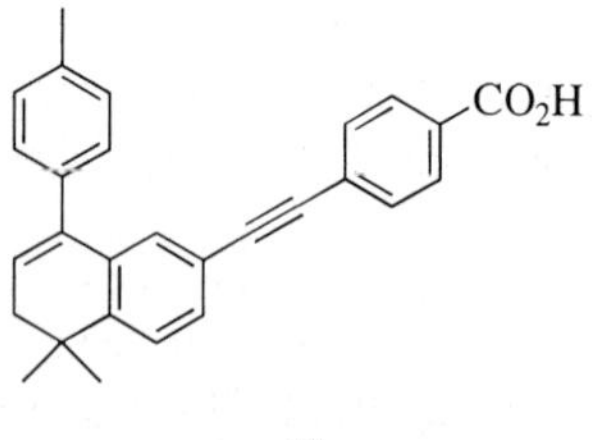

CO_2H

48

Fig. 5. RAR antagonists

Table 9. Binding affinities and differentiation-inducing and differentiation-antagonizing activities of ATRA and compounds **40–45** in HL-60 cells (from YOSHIMURA et al. 1995)

40-43

44-45

	R	X	Binding IC_{50}[a] (n*M*)	HL-60 differentiation $EC_{1/3}$[b] (n*M*)	HL-60 antagonism IC_{50}[c] (n*M*)
ATRA	–		0.63	0.34	
40	Benzyl	CH_2	1.2	NA	210
41	2-Furylmethyl	CH_2	3.4	NA	51
42	3-Pyridylmethyl	CH_2	0.34	NA	4.9
43	3-Pyridylmethyl	S	0.16	NA	5.6
44	Benzyl		1.5	NA	56
45	3-Pyridylmethyl		0.91	NA	1.3

Values are the mean of two experiments. NA, Not active.
[a] Concentration of the test compound inhibiting the binding of [^{3}H]-ATRA to HL-60 cell nuclear extracts. IC_{50} values were obtained from plots of binding affinity at 5.0×10^{-5} [^{3}H]ATRA vs. the logarithmic concentration of the test compound.
[b] $ED_{1/3}$ values were obtained from the percentage of the differentiated cells ($CD11^+$ cells).
[c] Concentration of the test compound inhibiting the differentiation-inducing activity of ATRA (10 n*M*). IC_{50} values were obtained from plots of activity at 10 n*M* ATRA vs. the logarithmic concentration of the test compound.

any of the RARs or at RXRα at concentrations as high as 10^{-5} *M*. The compounds were able to effectively inhibit the transcriptional activity of 10^{-8} *M* ATRA at RAR-RXR heterodimers, but not at RXRα-RXRα homodimers. In mechanistic studies, it was determined that the inhibition displayed by these antagonists was due to ligand competition for the binding sites of the RARs rather than by other possible mechanisms of antagonism, such as interfering with the binding of the receptor to retinoic acid response elements (RAREs) on DNA.

We recently described the synthesis and characterization of a toluyl derivative of the diarylacetylene class of retinoids, compound **48**, shown in Fig. 5 (JOHNSON et al. 1995). This compound bound with very high affinity (K_d = 2–3 n*M*) to all three RAR subtypes and had an approximately 5-fold higher affinity for the RARs than the natural hormone, ATRA. Compound **48** had absolutely no activity in transactivation assays using all three RAR subtypes.

Also, the compound functioned as an effective antagonist of ATRA in these transactivation assays and at equimolar concentrations it inhibited ATRA activity by 85%, 62% and 100% at RARα, RARβ and RARγ, respectively. At a 10-fold excess of compound **48** over ATRA, activity of ATRA at all three RARs was completely eliminated. Clearly, compound **48** is the most potent and effective antagonist of all three RARs that has been described to date. In addition, compound **48** functioned as an effective antagonist of ATRA action in in vitro (AGARWAL et al. 1996) and in vivo (STANDEVEN et al. 1996) models.

Only two groups have reported on receptor subtype-selective retinoid antagonists, both describing compounds that selectively antagonize RARα transactivation. The stilbene based analog **49** (Fig. 6) was reported to bind to RARα with an IC_{50} value of 60 nM, and to RARβ and RARγ with IC_{50} values of 2400 nM and 3300 nM, respectively (APFEL et al. 1992). The compound was also unable to induce transcriptional activity through RARα at concentrations of up to 500 nM. At a given concentration of ATRA, a 2- to 10-fold excess of compound **49** was needed to show inhibition of ATRA-induced transcriptional activity at RARα whereas a 50- to 100-fold excess of **49** was required to antagonize RARβ and RARγ mediated transactivation. Compound **49** was able to completely inhibit ATRA-induced RARα and RARβ mediated transactivational activity at high antagonist to agonist concentration ratios, but was able to only partially inhibit ATRA-induced transactivation by RARγ. Thus, compound **49** acts as a RARα selective antagonist, preferentially antagonizing ATRA-induced transactivation mediated by RARα over RARβ and RARγ.

Recently we described a series of highly potent RARα selective antagonists based on a combination of the structural features which imparted RAR antagonism to compound **48** and those features which resulted in RARα selective binding in compounds such as **23** and **24** (TENG et al. 1997). As shown in Table 10, these compounds bind with high affinity to RARα but do not induce gene transactivation through any of RARs at concentrations as high as 1 μM. These compounds are highly potent and effective antagonists, being able to completely inhibit 10 nM ATRA-induced transcriptional activity mediated by RARα at doses as low as 10 nM The difluorobenzoic acid analogs, **52** and **54**, are potent and selective RARα antagonists, which are able to completely block 10 nM ATRA-induced RARα mediated transactivation at concentrations of 1 μM and 100 nM, respectively. More importantly, compounds **52** and

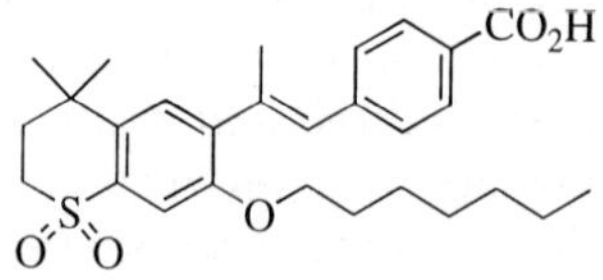

CO_2H

O

S

O O

49

Fig. 6. An RARα-selective antagonist

Table 10. Receptor binding affinity and inhibition of $10^{-8} M$ ATRA-induced transcriptional activity for RARα selective antagonists (from TENG et al. 1997)

	Substitution		K_d (nM)			IC_{50} (nM)[a]		
	X	Y	RARα	RARβ	RARγ	RARα	RARβ	RARγ
ATRA			15	13	18	–	–	–
50	H	H	27	1020	3121	3	300	1000
51	H	F	6	622	863	4.5	180	200
52	F	F	32	2256	>30000	60	NA	NA
53	H	F	2.8	320	7258	<1	50	100
54	F	F	1.5	898	10.18	<1	NA	NA

NA, Not Active.
[a] Antagonism of transactivation induced by ATRA at a dose of $1 \mu M$.

54 are highly specific antagonists which can only partially antagonize 10nM ATRA-induced transcriptional activity mediated by RARβ and RARγ. Thus, for RARβ and RARγ mediated transactivation, 60%–90% of the full ATRA-induced transactivational activity was maintained in the presence of $1 \mu M$ antagonist. Because of their high degree of selectivity, these compounds will have important applications in delineating the biology associated with the RARα subtype, and may themselves be useful as therapeutic agents in diseases associated with the RARα subtype.

II. RAR Inverse Agonists and Neutral Antagonists

The concept of inverse antagonism was first described in the area of G protein coupled receptors where certain compounds were found to bind and decrease the activity of unliganded receptor (DELEAN et al. 1980). In this adrenoreceptor model, the receptors are postulated to exist, in the absence of ligand, in an equilibrium between inactive and spontaneously active forms. An agonist would bind the receptor and shift the equilibrium toward the active form and increase activity while an inverse agonist would shift the equilibrium towards the inactive form and decrease basal activity. A neutral antagonist would bind to the receptor and not change the basal equilibrium. The concept of inverse agonism has not been previously described in the area of nuclear receptor hormones although antagonists of RARs and other steroid receptors have been known for many years. During the course of investigations into the pharmacology and mechanism of action of RAR antagonists related to compound **48**, we noticed clear differences in the biology of these structurally related antagonists (Klein et al. 1996). As shown in Table 11, these compounds, which differ only in the nature of the 4-substituent in the C-1 phenyl ring, were all effective full antagonists of RARα and RARγ. Compounds **55** and **56**, which have smaller (H, F) 4-substituents, were only partial antagonists at RARβ while compounds **48**, **57** and **58**, which had larger (CH_3, CF_3, Cl) 4-substituents, were full antagonists. However, the most striking differences between these two sets of compounds were observed when they were examined in RARγ transactivation assays. An assay was set up using a RARγ-VP16 construct which consists of the full length RARγ fused to the constituitively active transactivation domain of the herpes simplex virus, VP16. RARγ-VP16 gave high levels of basal transcriptional activity in the absence of any added ligand. When the compounds listed in Table 11 were tested in the RARγ-VP16 assay, compounds **48**, **57** and **58** all decreased the basal transcriptional activity in a dose dependent manner and hence functioned as inverse agonists. In contrast, compounds **55** and **56** did not change the high basal level activity and hence functioned as neutral antagonists. In the same assay, typical RAR agonists such as ATRA increased the basal level activity in a dose dependent manner. To further

Table 11. IC_{50} values of C1 substituted acetylenic retinoids (from Klein et al. 1996)

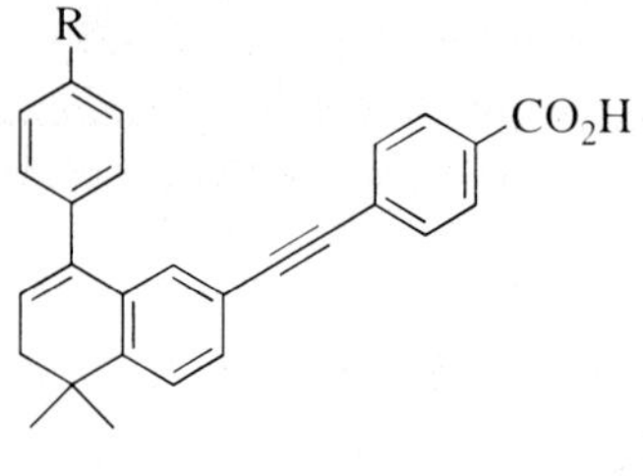

	Substitution R	IC_{50} (nM) RARα	RARβ	RARγ
55	H	87 ± 42	pa	32 ± 11
56	F	42 ± 17	pa	22 ± 12
48	CH_3	9 ± 1	24 ± 3	5 ± 1
57	CF_3	20 ± 9	60 ± 44	14 ± 7
58	Cl	15 ± 5	44 ± 7	8 ± 4

IC_{50} values represent represent the mean ± SE of at least three independent assays, essentially as shown in Fig. 1, performed using antagonism of a 10^{-8} M dose of ATRA at each of the RARs.

emphasize the differences between these compounds, the neutral antagonist compounds, **55** and **56**, were able to antagonize both the gene transcriptional stimulatory effects of agonists such as ATRA and the gene transcriptional inhibitory effects of inverse agonists such as **48**. In an alternative transactivation assay, chimeric ER-RAR receptors containing the DEF domain of RAR fused to the ABC domain of the estrogen receptor were used in conjunction with an estrogen response element containing a reporter construct. The basal reporter activity was increased upon transfection with ER-RARβ and ER-RARγ but not with ER-RARα. Compound **48** repressed the basal activity of ER-RARβ and ER-RARγ in a dose-dependent manner, again demonstrating that it functions as an inverse agonist.

The differences between the neutral antagonists and inverse agonists described above were all obtained in genetically engineered cells transfected with chimeric RARs. However, clear differences in the pharmacology of neutral antagonists and inverse agonists have also been observed in cultured normal human keratinocytes (see below). These differences were observed with respect to regulation of an endogenous gene presumably by mechanisms involving the regulation of the endogenous complement of retinoid receptors, further validating the reclassification of retinoid antagonists into neutral antagonists and inverse agonists. Moreover, the concept of nuclear receptor agonists, inverse agonists and neutral antagonists is in keeping with the recently emerging general model of nuclear receptor transcriptional activation which involves ligand-mediated changes in receptor interaction with coactivator (Glass et al. 1997) and corepressor (Hörlein et al. 1995) proteins. Based on this model, an agonist would stabilize interaction between nuclear receptor and coactivators thereby shifting the equilibrium towards a transcriptionally more active state while an inverse agonist would increase nuclear receptor/corepressor interaction and shift the equilibrium towards a transcriptionally less active state (Chandraratna and Klein 1997).

E. Pharmacology of RAR Selective Agonists and Antagonists

The recent availability of ligands that are highly selective for individual RAR subtypes will be of great benefit in studies directed toward understanding the physiological roles of the subtypes. Studies utilizing RAR subtype selective ligands will complement those that involve targeted disruption of the RARs. A clearer understanding of the physiological roles of the individual RARs in various tissues will result in the identification of new therapeutic applications for receptor and function selective retinoids. It is also expected that such selective retinoids will have fewer toxic side effects and hence be more suitable for human clinical use than the currently available nonspecific retinoids. In this section we review some of the more recent literature suggesting potential therapeutic applications for these new retinoids.

I. RARα-Selective Agonists

The growth and differentiation of certain myeloid cells in vitro have been shown to be regulated by ATRA signaling through RARα (Collins et al. 1990). More recent studies using a dominant-negative RARα construct suggest an important role for RARα in hematopoiesis in the development of myeloid progenitor cells (Tsai et al. 1992). The often fatal disease of acute promyelocytic leukemia (APL) is associated with a specific cytogenetic abnormality involving a mutual translocation between a region of chromosome 17 which codes for RARα and a region of chromosome 15 encoding the promyelocytic myeloid leukemia (PML) gene (Larson et al. 1984). In spite of the fact that APL apparently arises from disrupted RARα signaling, several studies have shown that ATRA, presumably acting through the altered RARα, is highly effective in inducing complete remission in APL patients (Meng-Er et al. 1988; Castaigne et al. 1990; Warrell et al. 1991). However, an important limitation of ATRA therapy of APL is the development of resistance to therapy by a mechanism involving autoinduction of ATRA metabolism (Muindi et al. 1992). Recently, the synthetic RARα-selective agonist **12** (Am-80) has been used successfully to induce complete remission in APL patients who had relapsed on ATRA therapy (Takeshita et al. 1996). These findings suggest that an RARα-specific agonist such as **23** or **24** should be ideal for the treatment of APL and may be associated with fewer side effects such as skin irritation (Standeven et al. 1997). Also, RARα-specific antagonists such as compound **54** will be very useful tools in studying the role of RARα in myeloid differentiation and may even find uses in myeloid disease.

Early studies have established retinoids as effective agents in the chemoprevention of breast cancer development in animal models (Moon et al. 1979). The potential usefulness of RARα-selective agonists in the treatment of ER-positive breast cancer has been suggested by the observations that ER-positive breast cancer cell lines and tumors express higher levels of RARα than the corresponding ER-negative cells and tumors (Roman et al. 1992; Han et al. 1997). Retinoids inhibit the growth of ER-positive breast cancer cells (Fontana 1987) and the sensitivity of retinoid-resistant, ER-negative breast cancer cells to retinoid growth inhibition was restored by transfection of RARα (Sheikh et al. 1994). In addition, the ability of a series of retinoid analogs to inhibit the growth of ER-positive MCF-7 breast cancer cell lines correlated well with their ability to bind to RARα (Dawson et al. 1995). Recently, the highly RARα-selective agonist, compound **23**, was shown to be effective and much more potent than a RARβ/RARγ-selective ligand in inhibiting the growth of ER-positive T-47D breast cancer cells (Fitzgerald et al. 1997). Interestingly, compound **23** also effectively inhibited the growth of ER-negative, SK-BR-3 breast cancer cells which expressed high levels of RARα. Thus, RARα-specific agonists may be useful for the treatment of ER-positive breast tumors and some ER-negative tumors as well. Also, if such

RARα-specific agonists have significantly attenuated side effect profiles, they would be suitable candidates for breast cancer chemoprevention studies.

II. RARβ Selective Agonists

Several lines of evidence suggest that normal RARβ signaling is important for the suppression of certain types of cancers. RARβ was first identified in a hepatocellular carcinoma where the RARβ gene surrounds an integration site of the hepatitis B virus (Benbrook et al. 1988). Presumably, misregulation of RARβ expression as a consequence of hepatitis B virus integration can contribute to hepatocarcinogenesis. Similarly, RARβ is abnormally expressed in about 50% of human lung cancer cell lines and primary tumors (Gebert et al. 1991). Expression of RARβ mRNA is decreased or lost in the premalignant lesions of oral leukoplakia (Lotan et al. 1995) and in head and neck tumors (Xu et al. 1994) suggesting that loss of RARβ signaling is an early event in oral carcinogenesis. Significantly, isotretinoin (13-*cis*-RA) has been used successfully to treat oral leukoplakia (Hong et al. 1986) and RARβ expression is restored concomitant with the clinical response (Lotan et al. 1995). Isotretinoin has also been used to prevent the development of second primary tumors in head and neck cancer patients (Benner et al. 1994). It remains to be determined whether RARβ selective agonists would also be effective in treatment and chemoprevention of oral cancers. The availability of RARβ specific analogs such as compound **25** will greatly facilitate such determination in preclinical models and ultimately in human trials.

III. RARγ Selective Agonists

Retinoids have been widely used in dermatology for the treatment of psoriasis, acne and photodamaged skin (Peck and Digiovanna 1994). Clinical trials have also demonstrated that retinoids can be useful in the treatment of squamous cell (Lippman and Meyskens 1987) and basal cell carcinomas of the skin (Peck et al. 1988). RARγ has been shown to be the most abundant retinoid receptor subtype in adult human skin and amounts to almost 90% of total RAR protein suggesting that RARγ is the most likely mediator of retinoid efficacy in the skin (Fisher et al. 1994). A study utilizing several RARγ selective retinoids has demonstrated a good correlation between RARγ transcriptional activity and topical efficacy and irritation in animal models (Chen et al. 1995). In keeping with these observations, CD 271 (compound **19**), an RARβ/RARγ-selective agonist, has been shown to be effective in the topical treatment of acne. Tazarotene, a potent RARβ/RARγ-selective agonist of the acetylene class of retinoids, has been shown to be an effective monotherapy for the topical treatment of plaque psoriasis (Weinstein et al. 1997). Further clinical studies are required to determine whether these RARβ/RARγ-selective compounds or other RARγ selective compounds are of improved benefit in other dermatological conditions such as photodamaged skin, keratinizing disorders and skin cancers.

IV. RAR Antagonists and Inverse Agonists

Although antagonists of other nuclear receptor hormones such as Tamoxifen, an estrogen receptor antagonist, and RU 486, a progesterone antagonist, have been in human clinical use for some time, there have been no clinical trials conducted with retinoid antagonists. However, the recent availability of more potent and effective antagonists will facilitate studies in preclinical models leading to the identification of potential therapeutic applications. One possible application of a retinoid antagonist is as an antidote to specific toxicities produced by pharmacological doses of a retinoid agonist. The currently marketed retinoids produce a broad spectrum of toxicities which are characteristic of hypervitaminosis A and are produced by RAR activation (STANDEVEN et al. 1996a–c). Skin irritation is an almost universally produced side effect associated with both topical retinoids and systemic retinoids. In a clinically relevant setting, a female mouse model was used to study whether a topical retinoid antagonist can prevent the cutaneous toxicity produced by a systemic retinoid agonist. Topical application of compound **48**, a potent RAR antagonist, completely abrogated the severe cutaneous toxicity produced by the systemic administration of the RAR agonist, TTNPB (STANDEVEN et al. 1996a). These results suggest that a topical RAR antagonist can effectively prevent or treat the mucocutaneous toxicity produced by systemic retinoids such as Accutane. Such mucocutaneous toxicity is manifested only in relatively limited areas of skin and mucous tissue and can be treated with relatively small amounts of a topical RAR antagonist. Thus, the systemic absorption of the RAR antagonist will be expected to be minimal and will not interfere with the therapeutic activity of the systemic retinoid drug. Compound **48** also antagonized the hypertriglyceridemia (STANDEVEN et al. 1996b) and the premature epiphyseal plate closure (STANDEVEN et al. 1996c) produced by systemic RAR agonists. The use of potent synthetic retinoids at high doses in a variety of clinical applications such as chemotherapy is likely to lead to more incidents of acute retinoid intoxication, including life-threatening pancreatitis associated with severe hypertriglyceridemia. The future availability of an RAR antagonist as an antidote to such acute retinoid intoxication would be quite desirable.

Another potential area of clinical applications for RAR antagonists would be in situations where endogenous retinoic acid signaling could maintain or exacerbate a pathogenic condition. Many such pathogenic conditions should exist because retinoids play important physiological roles in several fundamental biological processes and the availability of potent antagonists will greatly facilitate the identification of such processes. Several recent reports suggest that retinoids can play a role in regulating viral replication (GHAZAL et al. 1992; HUAN and SIDDIQUI 1992; ORCHHARD et al. 1993). A retinoic acid response element (RARE) has been identified in the long terminal repeat of the human immunodeficiency virus type 1 (HIV-1) promoter (LEE et al. 1994). This response element can be activated by ATRA and, most importantly, this ATRA activation can be blocked by the synthetic RAR antagonist **46**. Similarly, activation of RARs, but not RXRs, promotes replication of the human cytomegalovirus in embryonal cells (ANGULO et al. 1995). It remains to be determined whether endogenous retinoids actually enhance replication of these viruses during infection. If this is the case, RAR antagonists may have important practical applications in the treatment of specific viral infections.

The above discussion is related to the potential clinical applications of RAR antagonists in blocking the unwanted effects of pharmacologically applied or physiological RAR agonists. However, recent discoveries suggest that nuclear receptors exist in an equilibrium of interactions with other coactivator and corepressor proteins in the absence of any ligand. Thus, even an unliganded RAR can potentially drive transcription from an RARE and, in addition, would effect the activity of other transcription factors by sequestering common coactivator/corepressor proteins. RAR inverse agonists not only can suppress the basal transcriptional activity of the unliganded RAR on an RARE but can change the activity of other factors by selectively releasing coactivators and sequestering corepressors. Thus, the RAR inverse agonists, in contrast to neutral antagonists, will be expected to have biological activities of their own as a consequence of binding to RARs. We have studied the effects of RAR inverse agonists in cultured human keratinocytes with particular emphasis on the regulation of expression of the differentiation marker, MRP-8 (KLEIN et al. 1996). MRP-8 is not expressed in normal human skin but is expressed in psoriatic skin suggesting that it is a marker of the abnormal differentiation patterns of the psoriatic keratinocyte. Tazarotene, a recently approved antipsoriatic topical retinoid, down-regulates MRP-8 levels in psoriatic lesions in conjunction with clinical improvement of psoriasis (NAGPAL et al. 1996). The RAR inverse agonist, compound **48**, down-regulates MRP-8 levels in cultured keratinocytes in a dose-dependent manner in much the same way as the antipsoriatic RAR agonist, tazarotene (KLEIN et al. 1996). Interestingly, simultaneous addition of an RAR agonist and an RAR inverse agonist results in mutual antagonism of their MRP-8 repressive activities suggesting that these two types of ligands regulate MRP-8 by distinct mechanisms. These data suggest that RAR inverse agonists may be useful in the treatment of psoriasis and that their mechanism of action in the disease will be much different from that of an agonist.

It is apparent that we are entering a new era of retinoid therapeutics with the availability of novel receptor subtype and function selective retinoid ligands. RAR subtype selective agonists will not only broaden the scope of retinoid clinical applications but also provide compounds of improved therapeutic indices in the more traditional roles of application. Finally, with the emergence of the concept of inverse agonism in nuclear receptor function, one must conceive of a new spectrum of retinoid action which encompasses both

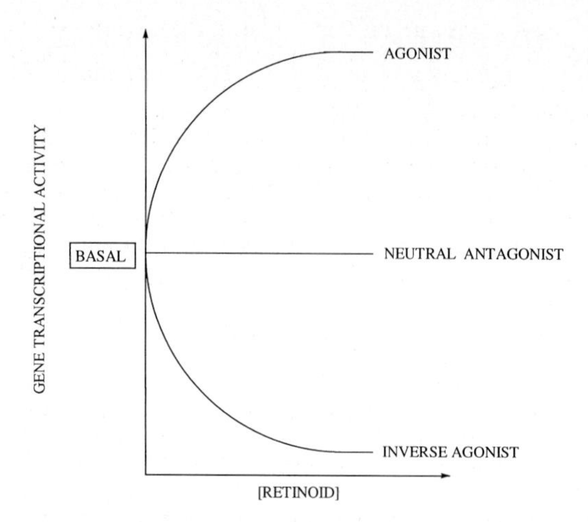

Fig. 7. New spectrum of retinoid action

activation and repression of retinoid target genes (Fig. 7). While the biology associated with retinoid agonists has been quite well investigated, the biology of retinoid inverse agonists is an unexplored and exciting new area for future investigations.

References

Agarwal C, Chandraratna RAS, Johnson AT, Rorke EA, Eckert RL (1996) AGN 193109 is a highly effective antagonist of retinoid action in human ectocervical cells. J Biol Chem 271:12209–12212

Angulo A, Suto C, Boehm MF, Heyman RA, Ghazal P (1995) Retinoid activation of the retinoic acid receptors but not of retinoid X receptors promotes cellular differentiation and replication of human cytomegalovirus in embryonal cells. J Virol 69:3831–3837

Apfel C, Bauer F, Cretaz M, Forni L, Kamber M, Kaufmann F, LeMotte P, Pirson W, Klaus M (1992) A retinoic acid receptor α antagonist selectively counteracts retinoic acid effects. Proc Natl Acad Sci USA 89:7129–7133

Apfel C, Bauer F, Kamber M, Klaus M, Mohr P, Keidel S, LeMotte P (1995) Enhancement of HL-60 differentiation by a novel class of retinoids with selective activity on retinoid x receptor. J Biol Chem 270:30765–30772

Beard RL, Gil DW, Marler DK, Henry E, Colon DF, Gillett SJ, Arefieg T, Breen TS, Krauss H, Davies PJA, Chandraratna RAS (1994) Structural basis for the differential RXR and RAR activity of stilbene retinoid analogs, Bioorg Med Chem Lett 4:1447–1452

Beard RL, Colon DF, Song TK, Davies PJA, Kochhar DM, Chandraratna RAS (1996) Synthesis and structure-activity relationships of retinoid x receptor selective diaryl sulfide analogs of retinoic acid. J Med Chem 39:3556–3563

Benbrook D, Lernhardt E, Pfahl M (1988) A new retinoic acid receptor identified from a hepatocellular carcinoma. Nature 333:669–672

Benner SE, Pajak TF, Lippman SM, Earley C, Hong WK (1994) Prevention of secondary primary tumors with isotretinoin in patients with squamous cell carcinoma of the head and neck. J Natl Cancer Inst 86:140–141

Boehm MF, McClurg MR, Pathirana C, Mangelsdorf D, White SK, Hebert J, Winn D, Goldman ME, Heyman RA (1994a) Synthesis of high specific activity [^{3}H]-9-*cis*-retinoic acid and its application for identifying retinoids with unusual binding properties. J Med Chem 37:408–414

Boehm MF, Zhang L, Badea BA, White SK, Mais DE, Berger E, Suto CM, Goldman ME, Heyman RA (1994b) Synthesis and structure-activity relationships of novel retinoid x receptor-selective retinoids. J Med Chem 37:2930–2941

Boehm MF, Zhang L, Zhi L, McClurg MR, Berger E, Wagoner M, Mais DE, Suto CM, Davies PJA, Heyman RA, Nadzan AM (1995) Design and synthesis of potent retinoid x receptor selective ligands that induce apoptosis in leukemia cells. J Med Chem 38:3146–3155

Castaigne S, Chomienne C, Daniel MT, Ballerini P, Berger R, Fenaux P, Degos L (1990) All-*trans*-retinoic acid as a differentiation therapy for acute promyelocytic leukemia. I. Clinical results. Blood 76:1704–1709

Chandraratna RAS, Gillett SJ, Song TK, Attard J, Vuligonda S, Garst ME, Arefieg T, Gil DW, Wheeler L (1995) Synthesis and pharmacological activity of conformationally restricted retinoid analogs, Bioorg Med Chem Lett 5:523–527

Chandraratna RAS, Klein ES (1997) Retinoid nuclear receptors. In: Lowe NJ, Marks R (eds) Retinoids a clinicians guide, 2nd edn. Dunitz, London, pp 21-32

Charpentier B, Bernardon J-M, Eustache J, Millois C, Martin B, Michel S, Shroot B (1995) Synthesis, structure-activity relationships, and biological activities of ligands binding to retinoic acid receptor subtypes. J Med Chem 38:4993–5006

Chao W, Hobbs PD, Jong L, Zhang X, Zheng Y, Wu Q, Shroot B, Dawson MI (1997) Effects of receptor class- and subtype-selective retinoids and an apoptosis-inducing retinoid on the adherent growth of the NIH: OVCAR-3 ovarian cancer cell line in culture. Cancer Lett 115:1–7

Chen J-Y, Penco S, Ostrowski J, Balaguer P, Pons M, Starrett JE, Reczec P, Chambon P, Gronemeyer H (1995) RAR specific agonist/antagonists which dissociate transactivation and AP1 transrepression inhibit anchorage-independent cell proliferation. EMBO J 14:1187–1197

Chen S, Ostrowski J, Whiting G, Roalsvig T, Hammer L, Currier SJ, Honeyman J, Kwasniewski B, Yu K-L, Sterzycki R, Kim CU, Starrett J, Mansuri M, Reczek PJ (1995) Retinoic acid receptor gamma mediates topical retinoid efficacy and irritation in animal models. J Invest Dermatol 104:779–783

Cheng Y-C, Prusoff WH (1973) The relationship between the inhibitory constant K_I and the concentration of inhibitor which causes 50% inhibition (IC_{50}) of an enzymatic reaction. Biochem Pharmacol 22:3099–3108

Collins SJ, Robertson KA, Mueller L (1990) Retinoic acid-induced granulocytic differentiation of HL-60 myeloid leukemia cells is mediated directly through the retinoic acid receptor (RAR-alpha). Mol Cell Biol 10:2154–2163

Crettaz M, Baron A, Siegenthaler G, Hunziker W (1990) Ligand specificities of recombinant retinoic acid receptors RARα and RARβ. Biochem J 272:391–397

Dawson MI, Jong L, Hobbs PD, Cameron JF, Chao W-R, Pfahl M, Lee M-O, Shroot B, Pfahl M (1995a) Conformational effects on retinoid receptor selectivity. 2. Effects of retinoid bridging group on retinoid x receptor activity and selectivity. J Med Chem 38:3368–3383

Dawson MI, Chao W-R, Pine P, Jong L, Hobbs PD, Rudd CK, Quick TC, Niles RM, Zhang X, Lombardo A, Ely KR, Shroot B, Fontana JA (1995b) Correlation of retinoid binding affinity to retinoic acid receptor α with retinoid inhibition of estrogen receptor-positive MCF-7 mammary carcinoma cells. Cancer Res 55:4446–4451

DeLean A, Stadel JM, Lefkowitz RJ (1980) A ternary complex model explains the agonist-specific binding properties of the adenylate cyclase-coupled beta-adrenergic receptor. J Biol Chem 255:7108–7117

Delescluse C, Cavey MT, Martin B, Bernard BA, Reichert U, Maignan J, Darmon M, Shroot B (1991) Selective high affinity retinoic acid receptor α or β-γ ligands. Mol Pharmacol 40:556–562

Elder JT, Fisher GJ, Zhang Q-Y, Eisen D, Krust A, Kastner P, Chambon P, Voorhees JJ (1991) Retinoic acid receptor gene expression in in human skin J Invest Dermatol 96:425–433
Eyrolles L, Kagechika H, Kawachi E, Fukasawa E, Iijima T, Matsushima Y, Hashimoto Y, Shudo K (1994) Retinobenzoic acids. 6. Retinoid antagonists with a heterocyclic ring. J Med Chem 37:1508–1517
Fisher GJ, Talwar HS, Xiao J-H, Datta SC, Reddy AP, Gaub M-P, Rochette-Egly C, Chambon P, Voorhees JJ (1994) Immunological Identification and functional quantitation of retinoic acid and retinoid X receptor proteins in human skin. J Biol Chem 269:20629–20635
Fitzgerald P, Teng M, Chandraratna RAS, Heyman RA, Allegretto EA (1997) Retinoic acid receptor α expression correlates with retinoid-induced growth inhibition of human breast cancer cells regardless of estrogen receptor status. Cancer Res 57:2642–2650
Fontana JA (1987) Interaction of retinoids and tomoxifen on the inhibition of human mammary carcinoma cell proliferation. Exp Cell Res 55:136–144
Gebert JF, Moghal N, Frangioni JV, Sugarbaker DJ, Neel BG (1991) High frequency of retinoic acid receptor beta abnormalities in human lung cancer. Oncogene 6:1859–1868 [Erratum, (1992) Oncogene 7:821]
Ghazal P, DeMattei C, Giulietti E, Kliewer SA, Umesono K, Evans RM (1992) Retinoic acid receptors initiate induction of the cytomegalovirus enhancer in embryonal cells. Proc Natl Acad Sci USA 89:7630–7634
Glass CK, Rose DW, Rosenfeld MG (1997) Nuclear receptor coactivators. Curr Opin Cell Biol 9:222–232
Graupner G, Malle G, Maignan J, Lang G, Prunieras M, Pfahl M (1991) 6′-Substituted naphthalene-2-carboxylic acid analogs, a new class of retinoic acid receptor subtype-specific ligands. Biochem Biophys Res Commun 179:1554–1561
Han QX, Allegretto EA, Shao ZM, Kute TE, Ordonez J, Aisner SC, Fontana JA (1997) Elevated expression of retinoic acid receptor-α in estrogen receptor-positive breast carcinomas as detected by immunohistochemistry. Diagn Mol Pathol 6:42–48
Heyman RA, Mangelsdorf DJ, Dyck JA, Stein RB, Eichele G, Evans RM, Thaller C (1992) 9-Cis-retinoic acid is a high affinity ligand for the retinoid x receptor. Cell 68:397–406
Hong WK, Endicott J, Itri LM, et al (1986) 13-*cis*-Retinoic acid in the treatment of oral leukoplakia. N Engl J Med 315:1501–1505
Hörlein AJ, Näär A, Heinzel T, Torchia J, Gloss B, Kurokawa R, Ryan A, Kamei Y, Söderström M, Glass CK, Rosenfels MG (1995) Ligand-independent repression by the thyroid hormone receptor mediated by a nuclear receptor co-repressor. Nature 377:397
Huan B, Siddiqui A (1992) Retinoid X receptor RXRα binds to and *trans*-activates the hepatitis B virus enhancer. Proc Natl Acad Sci USA 89:9059–9063
Huang M, Ye Y, Chen S, Chai J, Lu J-X, Zhoa L, Gu L, Wang Z (1988) Use of all-*trans*-retinoic acid in the treatment of acute promyelocytic leukemia. Blood 72:567–572
Johnson AT, Klein ES, Gillett SJ, Wang L, Song TK, Pino ME, Chandraratna RAS (1995) Synthesis and characterization of a highly potent and effective antagonist of retinoic acid receptors. J Med Chem 38:4764–4767
Johnson AT, Klein ES, Wang L, Pino ME, Chandraratna RAS (1996) Identification of retinoic acid receptor β subtype specific agonists. J Med Chem 39:5027–5030
Jong L, Lehmann JM, Hobbs PD, Harlev E, Huffmann JC, Pfahl M, Dawson MI (1993) Conformational effects on retinoid receptor selectivity. 1. Effect of 9-double bond geometry on retinoid x receptor activity. J Med Chem 36:2605–2613
Kagechika H, Kawachi E, Hashimoto Y, Shudo K (1986) Differentiation inducers of human promyelocytic leukemia cells HL-60. Chem Pharm Bull 34:2275–2278

Kagechika H, Kawachi E, Hashimoto Y, Himi T, Shudo K (1988) Retinobenzoic acids. 1. Structure-activity relationships of aromatic amides with retinoidal activity. J Med Chem 31:2182–2192

Kaneko S, Kagechika H, Kawachi E, Hashimoto Y, Shudo K (1991) Retinoid antagonists, Med Chem Res 1:220–225

Klein ES, Pino ME, Johnson AT, Davies PJA, Nagpal S, Thacher SM, Krasinski G, Chandraratna RAS (1996) Identification and functional separation of retinoic acid receptor neutral antagonists and inverse agonists. J Biol Chem 271:22692–22696

Larson RA, Kondo K, Vardiman JW, Butler AE, Golomb HM, Rowley JD (1984) Evidence for a 15:17 translocation in every patient with acute promyelocytic leukemia. Am J Med 76:827–841

Lee M-O, Hobbs PD, Zhang X-K, Dawson MI, Pfahl M (1994) A synthetic retinoid antagonist inhibits the human immunodeficiency virus type 1 promoter. Proc Natl Acad Sci USA 91:5632–5636

Lee M-O, Dawson MI, Picard N, Hobbs PD, Pfahl M (1996) A novel class of retinoid antagonists and their mechanism of action. J Biol Chem 271:11897–11903

Lehmann JM, Dawson MI, Hobbs PD, Husmann M, Pfahl M (1991) Identification of retinoids with nuclear receptor subtype-selective activities. Cancer Res 51:4804–4809

Lehmann JM, Jong L, Fanjul A, Cameron JF, Lu XP, Haefner P, Dawson MI, Pfahl M (1992) Retinoids selective for retinoid x receptor response pathways. Science 258:1944–1946

Levin AA, Sturzenbecker LJ, Kazmer S, Bosakowski T, Huselton C, Allenby G, Speck J, Kratzeisen CL, Rosenberger M, Lovey A, Grippo JF (1992) 9-Cis-retinoic acid stereoisomer binds and activates the nuclear receptor RXRα. Nature 355:359–361

Lippman SM, Meyskens Fl Jr (1987) Treatment of advanced squamous cell carcinoma of the skin with isotretinoin. Ann Intern Med 107:499–501

Lotan R, Xu X-C, Lippman SM, Ro JY, Lee JS, Lee JJ, Hong WK (1995) Suppression of retinoic acid receptor-β in premalignant oral lesions and its up-regulation by isotretinoin. N Engl J Med 332:1405–1410

Meng-Er H, Yu-Chen Y, Shu-Rong C, Jin-Ren C, Jia-Xiang L, Lin Z, Long-Jun G, Zhen-Yi W (1988) Use of all-trans retinoic acid in the treatment of acute promyelocytic leukemia Blood 72 (2):567–572

Moon RC, Thompson HJ, Becci PJ, Grubbs CJ, Gander CJ, Newton Dl, Smith JM, Phillips SL, Henderson WR, Mullen LT, Brown CC, Sporn MB (1979) N-(4-Hydroxyphenyl)retinamide, a new retinoid for the prevention of breast cancer in the rat. Cancer Res 39:1339–1346

Morris-Kay G (ed) (1992) Retinoids in normal development and teratogenesis. Oxford Science Publications, Oxford

Muindi J, Frankel SR, Miller WH, Jakubowski A, Scheinberg DA, Young CW, et al (1992) Continuous treatment with all-*trans*-retinoic acid causes a progressive reduction in plasma drug concentrations: implications for relapse and retinoid resistance in patients with acute promyelocytic leukemia. Blood 79:299–303

Nagpal S, Thacher SM, Patel S, Friant S, Malhotra M, Shafer J, Krasinski G, Asano AT, Teng M, Duvic M, Chandraratna RAS (1996) Negative regulation of two hyperproliferative differentiation markers by a retinoic acid receptor-specific retinoid: insight into the mechanism of retinoid action in psoriasis. Cell Growth Differ 7:1783–1791

Orchard K, Lang G, Harris J, Collins M, Latchman D (1993) A palindromic element in the human immunodeficiency virus long terminal repeat binds retinoic acid receptors and can confer retinoic acid responsiveness on a heterologous promoter. AIDS 6:440–445

Peck G, Digiovanna JJ, Sarnoff DS, Gross EG, Butkus D, Olsen TG, Yoder FW (1988) Treatment and prevention of basal cell carcinoma with oral isotretinoin. J Am Acad Dermatol 19:176–185

Peck G, Digiovanna JJ (1994) Synthetic retinoids in dermatology. In: Sporn MB, Roberts AB, Goodman DS (eds) The retinoids, 2nd edn. Raven, New York, p 631
Reczek PR, Ostrowski J, Yu K-L, Chen S, Hammer L, Roalsvig T, Starrett JE Jr, Driscoll JP, Whiting G, Spinazze PG, Tramposch KM, Mansuri MM (1995) Role of retinoic acid receptor gamma in the rhino mouse and rabbit irritation models of retinoid activity. Skin Pharmacol 8:292–299
Rees J (1992) The molecular biology of retinoic acid receptors: orphan from good family seeks home, Br J Dermatol 126:97–104
Renaud J-P, Rochel N, Ruff M, Vivat V, Chambon P, Gronemeyer H, Moras D (1995) Crystal structure of the RAR-γ ligand binding domain bound to all-*trans* retinoic acid. Nature 378:681–689
Roman SD, Clarke CL, Hall RE, Alexander IE, Sutherland RL (1992) Expression and regulation of retinoic acid receptors in human breast cancer cells. Cancer Res 52:2236–2242
Sheikh MS, Shao Z-M, Li X-S, Dawson M, Jetten AM, Wu S, Conley BA, Garcia M, Rochefort H, Fontana JA (1994) Retinoid-resistant estrogen receptor-negative human breast cancer cells transfected with retinoic acid receptor-α acquire sensitivity to growth inhibition by retinoids. J Biol Chem 34:21440–21447
Sporn MB, Roberts AB, Goodman DS (eds) (1994) The retinoids, vol 2. Academic, Orlando
Standeven AM, Johnson AT, Escobar M, Chandraratna RAS (1996a) Specific antagonist of retinoid toxicity in mice. Toxicol Appl Pharmacol 138:169–175
Standeven AM, Beard RL, Johnson AT, Boehm MF, Escobar M, Heyman RA, Chandraratna RAS (1996b) Retinoid-induced hypertriglyceridemia in rats is mediated by retinoic acid receptors. Fundam Appl Toxicol 33:264–271
Standeven AM, Chandraratna RAS, Mader DR, Johnson AT, Thomazy VA, (1996c) Retinoid-induced epiphyseal plate closure in guinea pigs. Fundam Appl Toxicol 33:91–98
Standeven AM, Teng M, Chandraratna RAS (1997) Lack of involvement of retinoic acid receptor α in retinoid-induced skin irritation in hairless mice. Toxicol Lett 92:231–240
Swann TR, Smith D, Tramposch KM, Zusi FC (1996) RARγ-specific retinobenzoic acid derivatives, European patent application EP 0 747 347 A1
Takeshita A, Shibata Y, Shinjo K, Yanagi M, Tobita T, Ohnishi K, Miyawaki S, Shudo K, Ohno R (1996) Successful treatment of relapse of acute promyelocytic leukemia with a new synthetic retinoid, AM80. Ann Intern Med 124:893–896
Teng M, Duong TT, Klein ES, Pino ME, Chandraratna RAS (1996) Identification of a retinoic acid receptor α subtype specific agonist. J Med Chem 39:3035–3038
Teng M, Duong TT, Johnson AT, Klein ES, Wang L, Khalifa B, Chandraratna RAS (1997) Identification of highly potent retinoic acid receptor α-selective antagonists. J Med Chem 40:2445–2451
Tsai S, Bartelmez S, Heyman R, Damm K, Evans R, Collins SJ (1992) A mutated retinoic acid receptor-α exhibiting dominant-negative activity alters the lineage development of a multipotent hematopoietic cell line. Genes Dev 6:2258–2269
Vershoore M, Langer A, Wolska H, Jablonska S, Czerntelewski J, Schaefer H (1990) Efficacy and safety of topical CD271 alcoholic gels. A new treatment candidate for acne vulgaris. Br J Dermatol 124:368–371
Vuligonda V, Lin Y, Chandraratna RAS (1996) Synthesis of highly potent RXR-specific retinoids: the use of a cyclopropyl group as a double bond isostere, Bioorg Med Chem Lett 6:213–218
Warrell RP, Frankel SR, MillerWH Jr, Scheinberg DA, Itri LM, Hittelman WN, Vyas R, Andreeff M, Tafuri A, Jakubowski A, Gabrilove J, Gordon MS, Dmitrovsky E (1991) Differentiation therapy of acute promyelocytic leukemia with tretinoin (all-*trans*-retinoic acid). N Engl J Med 324:1385–1393
Weinstein GD, Krueger GG, Lowe NJ, Duvic M, Friedman DJ, Jegasothy BV, Jorizzo JL, Shmunes E, Tschen EH, Lew-Kaya DA, Lue JC, Sefton J, Gibson JR,

Chandraratna RAS (1997) Tazarotene gel, a new retinoid, for topical therapy of psoriasis: vehicle-controlled study of safety, efficacy, and duration of action. J Am Acad Dermatol 37:85–92

Xu X-C, Ro JY, Lee JS, Shin DM, Hong WK, Lotan R (1994) Differential expression of nuclear retinoid receptors in normal, premalignant, and malignant head and neck tissues. Cancer Res 54:3580–3587

Yoshimura H, Nagai M, Hibi S, Kikuchi K, Abe S, Hida T, Higashi S, Hishinuma I, Yamanaka T (1995) A novel type of retinoic acid receptor antagonist: synthesis and structure-activity relationships of heterocyclic ring-containing benzoic acid derivatives. J Med Chem 38:3163–3173

Yu K-L, Spinazze P, Ostrowski J, Currier SJ, Pack EJ, Hammer L, Roalsvig T, Honeyman JA, Tortolani DR, Reczek PR, Mansuri MM, Starrett, Jr JE (1996) Retinoic acid receptor β,γ-selective ligands: synthesis and biological activity of 6-substituted 2-naphthoic acid retinoids. J Med Chem 39:2411–2421

CHAPTER 7

RXR-Specific Agonists and Modulators: A New Retinoid Pharmacology

I.G. SCHULMAN, D. CROMBIE, R.P. BISSONNETTE, R. CESARIO, K. ROEGNER, G. SHAO, and R.A. HEYMAN

A. Introduction

Receptors for retinoic acid (RARs and RXRs), thyroid hormone (TRs), vitamin D (VDR), prostanoids (PPARs) and numerous orphan receptors for which ligands have not been identified play vital roles in vertebrate cell growth, development, differentiation, metabolism and homeostasis. These receptors, like most members of the nuclear hormone receptor superfamily, function as hormone-dependent transcription factors that regulate the expression of large gene networks. The important biological activities of nuclear hormone receptors combined with the ability to directly regulate their activity with small molecules has long made this superfamily of transcription factors a target for drug discovery. This review focuses on the retinoid X receptors (RXRs), first characterized as receptors for the vitamin A derivative 9-*cis* retinoic acid (for recent reviews see MANGELSDORF and EVANS 1995; MANGELSDORF et al. 1995; PERLMANN and EVANS 1997). Recent work demonstrating that ligands for RXR may provide effective treatments for several types of cancers as well as for type II diabetes has lead to great interest in this class of nuclear hormone receptors that until recently were thought of as a "silent partner."

The RXRs were first identified as orphan members of the nuclear receptor superfamily based on their homology to the previously characterized RARs. As with the three RAR isoforms (α, β, and γ) there are three RXR isoforms (α, β, and γ) each encoded by a single gene. Interestingly, RXRs weakly activate transcription in response to high concentrations of all-*trans* retinoic acid, a vitamin A derivative known to be a high-affinity ligand for RAR. Unlike the case with RAR, however, direct binding of all-*trans* retinoic acid to RXR could not be demonstrated (MANGELSDORF et al. 1990, 1992; see Table 1). The paradoxical observation that all-*trans* retinoic acid can activate RXR without direct binding prompted a search for other vitamin A metabolites that may be high-affinity ligands for RXR. This successful search led to the identification of 9-*cis* retinoic acid, a stereoisomer of all-*trans* retinoic acid, as a naturally occurring ligand for RXR (HEYMAN et al. 1992; LEVIN et al. 1992). 9-*cis* Retinoic acid binds with high affinity to both the RARs and RXRs and is considered a pan-agonist that activates all six receptors (Tables 1, 2). In contrast, all-*trans* retinoic acid only activates RXRs when administered at con-

Table 1. K_d (nm) and EC_{50} values for all-*trans* retinoic acid, 9-*cis* retinoic acid, LGD1069, LG100268, and LG100754

Compound	RARα	RARβ	RARγ	RXRα	RXRβ	RXRγ	RARαEC_{50}	RXRαEC_{50}
All-*trans* RA	15	17	17	>1000	>1000	>1000	200	>1000
9-*cis* RA	15.2	13.4	14.7	6.7	6.2	9.7	200	50
LGD1069	>1000	>1000	>1000	36	21	29	>1000	28
LG100268	>1000	>1000	>1000	3.2	6.2	9.7	ND[c]	4
LG100754	>1000	>1000	>1000	3.4	10	12.2	300	ND

K_d values were determined by the ability of ligands to displace [^{3}H]9-*cis* retinoic acid from RXR or [^{3}H]all-*trans* retinoic acid from RAR. EC_{50} values were determined for RXRα homodimers and RXRα-RARα heterodimers by transfection assays in CV1 cells using reporters with the appropriate hormone response elements. ND, Not determined. Alone LG100268 does not activate RXR-RAR heterodimers, and LG100754 does not activate RXR homodimers. See text for details.

Table 2. Agonist activity of LGD1069, LG100268, and LG100754 on RXR homodimers and RXR-dependent heterodimers

Compound	RXR-RXR	RXR-RAR	RXR-TR	RXR-VDR	RXR-PPAR	RXR-LXR	RXR-FXR	RXR-NGFIB
All-*trans* RA	+/–	+	–	–	–	–	–	–
9-*cis* RA	+	+	–	–	+	+	+	+
LGD1069	+	–	–	–	+	+	+	+
LG100268	+	–	–	–	+	+	+	+
LG100754	–	+	–	–	+	–	–	–

+, Agonist activity; –, agonist activity is not observed.

centrations high enough that significant isomerization to 9-*cis* retinoic acid occurs.

RXRs, as with most members of the nuclear hormone receptor superfamily, are tripartite in structure comprised of an amino terminal hormone-independent transactivation function, a central DNA binding domain and a carboxy terminal ligand-binding domain (LBD). The LBD is functionally complex and along with ligand binding encodes domains required for dimerization, repression of transcription in the absence of ligand, and ligand-dependent activation of transcription (for review see MANGELSDORF et al. 1995). The recently published crystal structures of the RXR, RAR, TR, and estrogen receptor (ER) LBDs support a long standing hypothesis that binding of ligand induces a significant conformational change in receptors. Upon ligand binding helix 12, the receptor domain encompassing the ligand-dependent activation function (named the τc or AF-2 domain), appears to move almost 90° from a position extended away from the rest of the LBD to a position loosely packed upon the surface (BOURGUET et al. 1995; BRZOZOWSKI et al. 1997; RENAUD et al. 1995; WAGNER et al. 1995). The functional consequence of this conformational change arises from the ability of different *trans*-acting factors to distinctly recognize either nonliganded or liganded receptors. In the absence of ligand many receptors interact with a family of corepressors (The Silencing Mediator of Retinoid and Thyroid receptors [SMRT] and the Nuclear receptor Corepressor [NCoR]) that actively repress transcription. Upon ligand binding, the ligand-dependent conformational change results in the release of corepressors and promotes interactions with positively acting coactivators such as the CREB binding protein (CBP), the steroid receptor coactivators (SRC-1, SRC-2, and SRC-3) and components of the basal transcription machinery (for recent reviews see GLASS et al. 1997; HORWITZ et al. 1996). The essential role of the τc/AF-2 domain in receptor activity is illustrated by the observation that deletion or mutation of this domain produces receptors that bind ligand normally but fail to release corepressors or interact with coactivators (BANIAHMAD et al. 1995; CHAKRAVARTI et al. 1996; CHEN et al. 1997; CHEN and EVANS 1995; HONG et al. 1997; HORLEIN et al. 1995; KAMEI et al. 1996; ONATE et al. 1995; SCHULMAN et al. 1995, 1996; TORCHIA et al. 1997; VOEGEL et al. 1996).

Around the same time that 9-*cis* retinoic acid was identified as a high-affinity ligand for RXR, a second paradox in the nuclear receptor field was also resolved. Although individual nuclear receptors can bind ligand in solution, RARs, TRs, VDR, and peroxisome proliferator activated receptors (PPARs) do not function as monomeric transcription factors. These receptors, as well as many orphan receptors, require an additional factor present in nuclear extracts for high-affinity DNA binding. This additional factor was identified simultaneously in several laboratories as RXR and DNA binding was shown to require a heterodimer of RXR with the partner (for review see MANGELSDORF and EVANS 1995). The identification of a structural and functional homologue of RXR in *Drosophila* (ORO et al. 1990; YAO et al. 1992), further emphasized the critical role of RXR in signal transduction by nuclear hormone receptors. The finding that RXR functions as the common heterodimeric subunit in multiple hormonal signaling pathways suggested the identification of RXR-specific ligands would provide unique tools for deciphering nuclear receptor pharmacology and could have important therapeutic benefits.

Unlike RXR's partners described above, RXR itself also can function as a homodimer that binds and regulates gene expression in response to the natural ligand 9-*cis* retinoic acid. The ability of RXR to bind and respond to its own ligands as well as to participate as a dimeric partner in multiple hormonal signaling pathways presents organisms and pharmacologists with a quandary. How does an organism refrain from simultaneously activating all the signaling pathways RXR participates in? Not surprisingly, nature has developed ways to restrict the broad activity of RXR. Recent studies have defined two classes of RXR-dependent heterodimers, permissive (i.e., activated by RXR-specific ligands as well as ligands binding to the partner) and nonpermissive. In nonpermissive heterodimers (e.g., RAR, TR, VDR) signaling by RXR agonists is inhibited. Two mechanisms for the inhibition of RXR signaling by nonpermissive partners have been discovered (see Sect. B.I). In the absence of its own ligand, nonpermissive partners recruit members of the recently identified class of *trans*-acting corepressors of transcription (SMRT and NCoR). Also, when dimerized with nonpermissive partners several studies have indicated the ability of RXR to bind 9-*cis* retinoic acid and some synthetic ligands is reduced (FORMAN et al. 1995b; KUROKAWA et al. 1994). These data, however, remain controversial (LI et al. 1997; MINUCCI et al. 1997). The inability of RXR to respond to agonists when a component of nonpermissive heterodimers gave rise to the idea that RXR functions simply as a "silent partner."

In contrast, when dimerized with permissive partners [e.g., dimers between RXR and PPAR, the liver X receptor (LXR), the farnosoid X receptor (FXR), and the nerve growth factor induced receptor (NGFIB)] RXR as well as the partner are competent to bind natural hormones or synthetic ligands and regulate gene expression. Indeed, the observation that heterodimers of RXR with PPARγ, an adipocyte-enriched member of the nuclear receptor superfamily, are permissive led to the finding that RXR agonists can function as insulin sensitizers in animal models of type II diabetes (MUKHERJEE et al. 1997). This dichotomy in RXR-dependent heterodimer activity, nature's way to reign in the promiscuity of RXR, suggests it should be possible to target RXR for drug discovery without unwanted side effects resulting from simultaneous activation of multiple hormonal signaling pathways.

This review describes recent work on the synthesis and characterization of two classes of high-affinity RXR ligands (Rexinoids). RXR-specific agonists that activate most if not all permissive heterodimers (BOEHM et al. 1994, 1995; LALA et al. 1996; MUKHERJEE et al. 1997) and a new class of compounds, RXR modulators, that use a novel mechanism to activate only a subset of RXR-dependent heterodimers (CANAN KOCH et al. 1996; LALA et al. 1996; SCHULMAN et al. 1997). This review first highlights recent advances in the molecular mechanisms underlying RXR activity in both permissive and nonpermissive heterodimers. Second it illustrates how these new molecular insights have been exploited in order to better understand nuclear receptor pharmacology and to develop effective RXR-based therapeutics.

B. Ligand-Dependent Activation of Transcription By RXR

I. Activation Of Permissive Heterodimers

To identify RXR agonists, the ability of RXR homodimers to activate transcription of a reporter gene containing a RXR-response element (RXRE) was examined by transfection in CV1 cells. Table 1 describes the binding and activity profiles of two ligands, LGD1069 and LG100268 (Fig. 1), that are highly selective for RXR in solution binding assays and potent in the RXR homodimer cotransfection assay (see Fig. 5A; BOEHM et al. 1994, 1995). To further

LGD1069

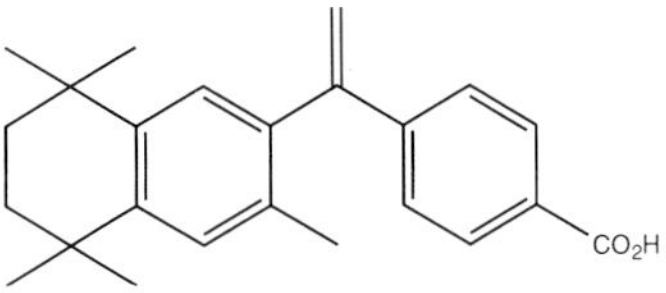

4-[1-(3,5,5,8,8-pentamethyl-5,6,7,8-tetrahydro-2-naphtyl)ethenyl]benzoic acid

LG100268

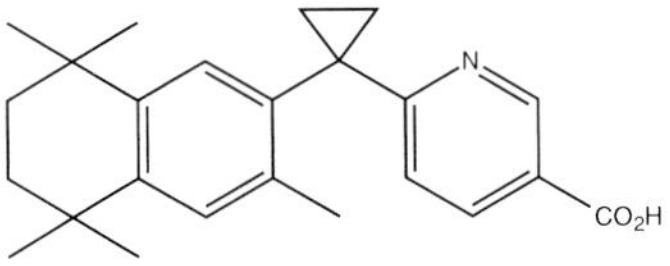

6-[1-(3,5,5,8,8-pentamethyl-5,6,7,8-tetrahydronaphthlen-2-yl)-cyclopropyl]-nicotinic acid

LG100754

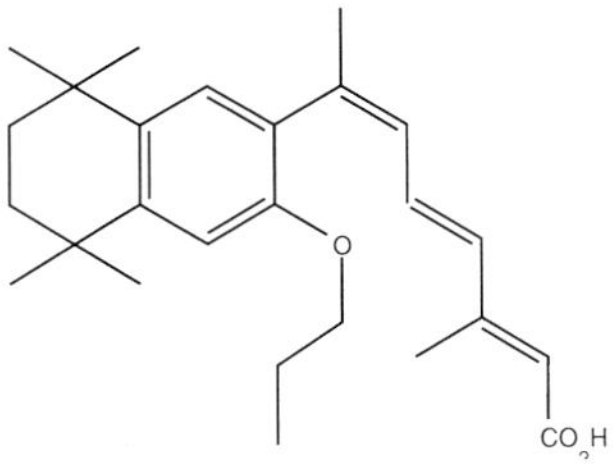

(2E,4E,6Z)-7-(3-N-propoxy-5,6,7,8-tetrahydro-5,5,8,8-tetramethylnaphthalen -2-yl)-3-methylocta-2,4,6-trienoic acid

Fig. 1. Structures of RXR-specific ligands: the RXR-specific agonists LGD1069 and LG100268 and the RXR modulator LG100754. See text for details

characterize LGD1069 and LG100268, their ability to activate a number of heterodimers was examined in cotransfection assays using the appropriate hormone response elements (HREs). The results of these experiments (Table 2) indicate LGD1069 and LG100268 activate RXR-PPAR, RXR-LXR, RXR-FXR, and RXR-NGFIB heterodimers but have little or no activity when assayed with RXR-RAR, RXR TR, and RXR-VDR. Not surprisingly, ligands specific for the various partners only activate the appropriate heterodimers (data not shown). Similar activities are also observed in cell-based systems that examine the activity of endogenous receptors. For instance, LG100268 does not activate RXR-RAR heterodimers or promote the differentiation of NB-4 cells while RAR agonists do activate the dimer and promote differentiation (LALA et al. 1996). On the other hand, LG100268 does promote the differentiation of 3T3-L1 preadipocytes to mature adipocytes, an RXR-PPARγ driven response. Thus the identification and characterization of LGD1069 and LG100268 defines the concept of permissive and nonpermissive RXR-dependent heterodimers at both the molecular and cellular level.

To further characterize the differences between permissive and nonpermissive heterodimers at the molecular level, the ability of permissive and nonpermissive heterodimers to interact with corepressors and coactivators has been examined using biochemical and two-hybrid assays. When RXR homodimers or permissive heterodimers are treated with RXR-specific agonists, interactions between RXR and the coactivators SRC-1, SRC-2, SRC-3, RIP140, CBP, and the basal transcription factor TBP can be detected (Fig. 2B). The agonist-dependent interaction between RXR and these coactivators requires the RXR τc/AF-2 domain. Mutations in the τc/AF-2 domain that eliminate RXR-agonist dependent transcription, eliminate the ability to detect interactions between RXR and coactivators (for recent reviews see GLASS et al. 1997; HORWITZ et al. 1996). Together with the recent receptor LBD crystal structures, the results of the coactivator interaction studies indicate the repositioning of the τc/AF-2 domain upon agonist binding allows receptors to assume a conformation that favors interaction with coactivators (Fig. 2B; for review see SCHWABE 1996).

In the absence of agonists, nonpermissive heterodimers interact strongly with the corepressors SMRT and NCoR. Treatment of these heterodimers with partner-specific agonists results in the dissociation of corepressors and recruitment of coactivators (for recent reviews see GLASS et al. 1997; HORWITZ et al. 1996). On the other hand, treatment with RXR-specific agonists does not release corepressors (Fig. 2C; CHEN et al. 1997; CHEN and EVANS 1995; HORLEIN

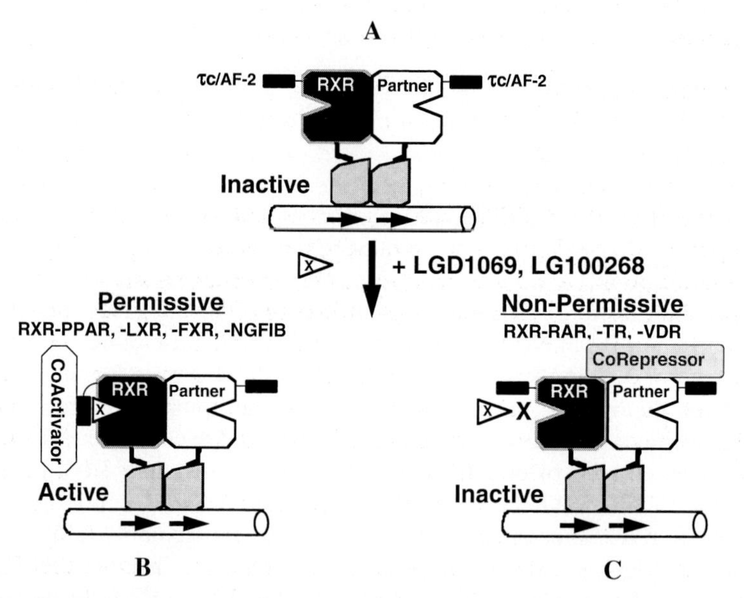

Fig. 2. RXR-dependent heterodimers (**A**) can be subdivided into two classes, permissive (**B**) and nonpermissive heterodimers (**C**). Permissive heterodimers allow both RXR and the partner to bind ligands and undergo a conformational change that includes a repositioning of the ligand-dependent activation domain (τc/AF-2 domain) from an extended location to a position loosely packed on the surface of the LBD core. In contrast, nonpermissive heterodimers inhibit RXR signaling by recruiting corepressors, reducing ligand binding by RXR and by other allosteric effects. See text for details and references

et al. 1995; KUROKAWA et al. 1995; SCHULMAN et al. 1995, 1997; VOEGEL et al. 1996). Interestingly, RXR homodimers and permissive heterodimers interact weakly if at all with corepressors (SCHULMAN et al. 1997; ZAMIR et al. 1997). This difference in corepressor affinity between permissive and nonpermissive heterodimers may be one mechanism nonpermissive partners use to inhibit RXR signaling.

Differences in corepressor affinity, however, cannot be the whole story. In vitro biochemical assays indicate treatment of nonpermissive heterodimers with RXR agonists does not allow recruitment of coactivators, even when corepressors are absent. Although studies have indicated the ability of RXR to bind 9-*cis* retinoic acid and LGD1069 is reduced when dimerized with RAR and TR (FORMAN et al. 1995b; KUROKAWA et al. 1994) inhibition of agonist binding by RXR cannot completely account for the repressive effects of non-permissive partners. For instance a study by VIVAT et al. (1997) has demonstrated that a constitutively active RXR mutant is still repressed when dimerized with RAR in a corepressor-independent manner. The results of molecular studies indicate that when dimerized with nonpermissive partners RXR can be locked in an inactive state (Fig. 2C).

II. RXR Agonists Activate RXR-RAR Heterodimers When The RAR Binding Pocket Is Occupied

As described in Sect. B.I, RXR agonists alone do not activate nonpermissive RXR-RAR heterodimers. Several studies, however, have shown that combining an RXR agonist such as LGD1069 or LG100268 with an RAR agonist has a greater effect than the RAR agonist alone (APFEL et al. 1995; CHEN et al. 1996; FORMAN et al. 1995b; MINUCCI et al. 1997). These observations suggest binding of ligand to the partner unlocks RXR and the nonpermissive nature of the heterodimer is lost (Fig. 3). Interestingly, the ability to unlock the RXR subunit can be observed even if the RAR binding pocket is occupied by an antagonist (CHEN et al. 1996). The ability of RAR antagonists to allow activation by RXR agonists is illustrated by the experiment in Fig. 4. Treatment of NB4 cells with the RAR agonist ALRT1550 results in a cell cycle arrest in G_1 (Fig. 4, column 1). The effect of ALRT1550 can be competitively inhibited by the RAR antagonist LG030282 (Fig. 4, column 2). Alone, neither the antagonist LG030282 (column 3) nor the RXR agonist LG100268 (column 4) induces G_1 arrest. Together, however, the RAR antagonist (LG030282) and RXR agonist (LG100268) are almost as effective as a RAR agonist ALRT1550 (column 5). Thus, simply binding of ligand to the RAR ligand binding pocket is sufficient to unlock RXR, independent of whether RAR is occupied by agonist or antagonist.

The results described above suggest that occupation of the RAR binding pocket results in conformational change throughout the heterodimer that allows RXR to achieve an active conformation and recruit coactivators in a manner similar to that observed for permissive heterodimers (Fig. 3). A pre-

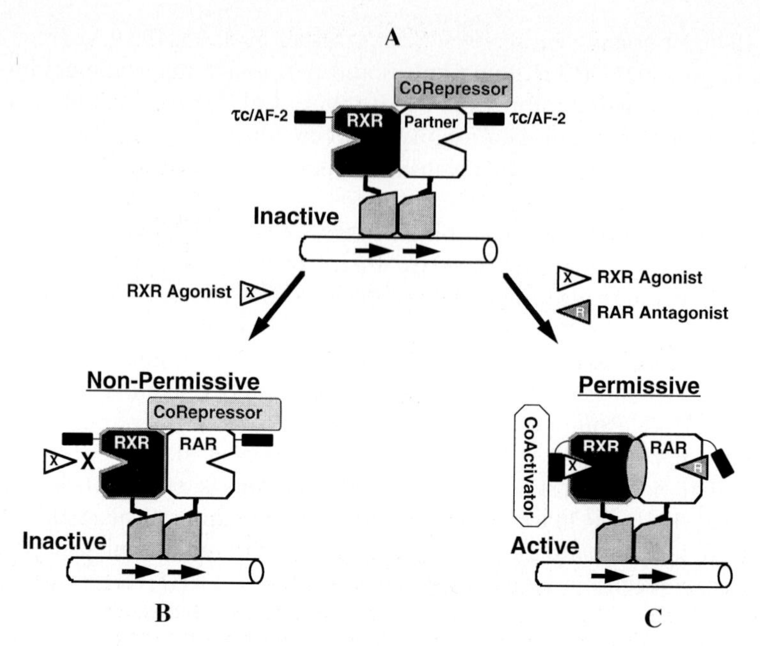

Fig. 3A–C. Occupying the RAR-binding pocket transforms nonpermissive heterodimers to permissive. **A** Nonliganded RXR-RAR heterodimer. **B** Alone RXR agonists such as LGD1069 and LG100268 do not activate RXR-RAR heterodimers (see text and Table 2 for details). **C** When the RAR-binding pocket is occupied by agonists or antagonists, however, RXR-RAR heterodimers are now permissive for activation by RXR agonists. The model suggests that occupation of the RAR-binding pocket results in a conformational change that is propagated through the dimer interface to RXR (*gray oval at dimer interface*). This conformation change allows RXR to respond to ligand in a typical fashion. See the legend to Fig. 1 and the text for details

diction of this model is that activation of transcription by the combination of RXR agonist and RAR antagonists would require the τc/AF-2 domain of RXR, as has been shown for permissive heterodimers (see Sect. B.I). To our knowledge, however, experiments to support this prediction have not been reported. The ability of ligands occupying the RAR binding pocket, either agonists or antagonists, to influence ligand binding and the transcriptional activity of the RXR subunit suggests a mechanism of allosteric control whereby conformational changes initiated by ligand binding to RAR are transmitted to the nonliganded RXR subunit.

III. The Phantom Ligand Effect

Most if not all RXR-specific agonists behave as LGD1069 and LG100268 when profiled on RXR-dependent heterodimers (see Sect. B.I). To determine if RXR-specific ligands can be identified that selectively activate only one or a subset of heterodimers, a large number of synthetic ligands with nanomolar

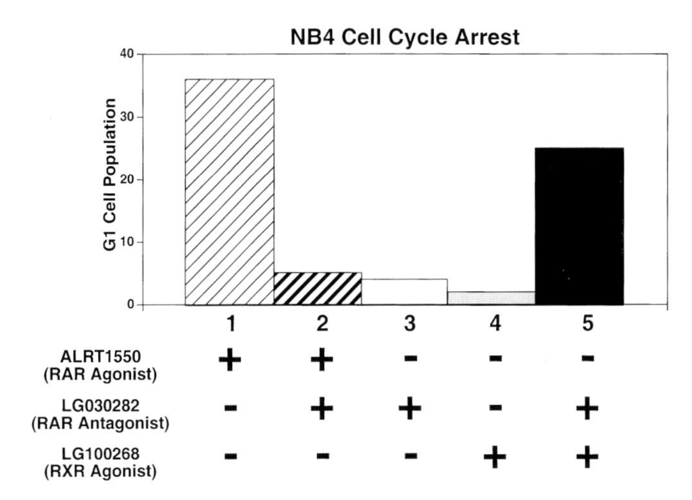

Fig. 4A,B. The combination of an RXR agonist and RAR antagonist arrests NB4 cells in G_1. NB4 cells were treated with ligands for 4 days and cell cycle position was determined by FACS analysis of propidium iodide stained cells. The percentage increase in G_1 cells compared to untreated controls is shown. *Column 1*, 1.0 n*M* ALRT1550 (RAR agonist); *column 2*, 1.0 n*M* ALRT1550 + 1.0 μ*M* LG030282 (RAR antagonist): *column 3*, 1.0 μ*M* LG030282; *column 5*, 10 n*M* LG100268 (RXR agonist); *column 6*, 100 n*M* LG030282 + 10 n*M* LG100268

binding affinity for RXR were examined for activity on a battery of heterodimers. This analysis has identified a class of compounds characterized by LG100754 (Fig. 1). LG100754, as the RXR agonists described in Sect. B.I, binds with high affinity to RXR (Table 1). In contrast to LGD1069 and LG100268, LG100754 does not activate RXR homodimers (Fig. 5A; also see Tables 1, 2). LG100754, however, competitively antagonizes the activation of RXR homodimers by LGD1069 and LG100268 (data not shown). LG100754, as with LGD1069 and LG100268 does activate RXR-PPAR heterodimers (Canan Koch et al. 1996; Lala et al. 1996). Thus, the heterodimer activity profile of LG100754 indicates that activation of RXR homodimers and RXR-PPAR heterodimers via the RXR subunit can be pharmacologically separated (Table 2).

Surprisingly, LG100754 also activates nonpermissive RXR-RAR heterodimers (Fig. 5B; also see Tables 1, 2; Lala et al. 1996; Schulman et al. 1997). The trivial explanation that LG100754 is metabolized by tissue culture cells to compounds that bind and activate RAR has been ruled out by HPLC analysis of ethyl acetate extracts from cells treated with LG100754. The results of the metabolism experiments and in vitro ligand binding experiments clearly indicate that LG100754 activates RXR-RAR heterodimers via binding to the RXR subunit (Schulman et al. 1997). The ability of the RXR-specific ligand LG100754 to activate what were considered to be nonpermissive RXR-RAR

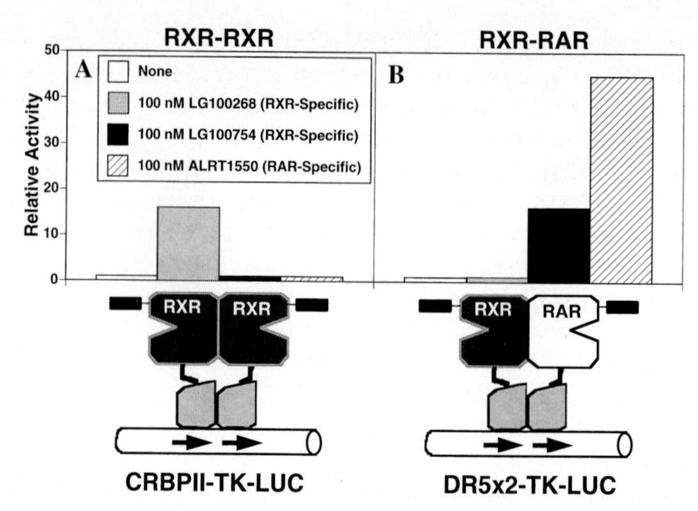

Fig. 5A,B. The RXR-specific ligand LG100754 activates nonpermissive RXR-RAR heterodimers. **A** CV1 cells were transfected with a luciferase (*LUC*) reporter that contains a RXR response element from the cellular retinol-binding protein type II gene (*CRBPII*) fused to the minimal herpes simplex virus thymidine kinase (*TK*) promoter (*CRBPII-TK-LUC*) (MANGELSDORF et al. 1991) and a plasmid expressing human RXR_{α}. After transfection, cells were treated for 36 h in the absence (*open bars*) or presence of 100 n*M* LG100268 (*gray bars*), LG100754 (*black bars*), or ALRT1550 (*striped bars*). Activity relative to the CRBPII-TK-LUC reporter alone is presented. **B** CV1 cell were transfected with a luciferase reporter that contains two copies of a synthetic direct repeat (AGGTCA) spaced by five nucleotides (*DR5×2-TK-LUC*) (UMESONO et al. 1991) and a plasmid expression human RAR_{α}. In this experiment endogenous RXR provides the dimerization partner for transfected RAR. After transfection, cells were treated for 36 h in the absence (*open bars*) or presence of 100 n*M* LG100268 (*gray bars*), LG100754 (*black bars*) or ALRT1550 (*striped bars*). Activity relative to the DR5×2-TK-LUC reporter alone is reported. All luciferase values were corrected for transfection efficiency by cotransfecting an expression plasmid for β-galactosidase

heterodimers indicates the concept of nonpermissiveness is an oversimplification. We suggest it is more appropriate to classify the activities of the individual ligands and not the heterodimer pairs. Therefore we have classified compounds such as LGD1069 and LG100268 as RXR-specific agonists, and ligands with more restricted profiles of activity such as LG100754 as RXR modulators.

Why does LG100754 activate RXR-RAR heterodimers while LGD1069 and LG100268 do not? One simple explanation could be that dimerization with RAR does not restrict the binding of LG100754 to RXR as it does for LGD1069 and LG100268. In this case binding of LG100754 to the RXR subunit of RXR-RAR heterodimers would lead to the activation of transcription mediated by the RXR ligand-dependent activation domain (τc/AF-2 domain), in much the same way that RXR homodimers and permissive heterodimers activate transcription in response to binding LGD1069 or LG100268 (Fig. 2). This simple model, however, is incorrect. Although

LG100754 can bind to RXR when it is dimerized with RAR, activation of transcription by RXR-RAR heterodimers in response to LG100754 is completely independent of the RXR ligand-dependent activation domain (τc/AF-2 domain). Surprisingly, activation of transcription by RXR-RAR heterodimers in response to LG100754 absolutely depends upon the τc/AF-2 domain of the nonliganded RAR subunit (SCHULMAN et al. 1997). As shown in the model depicted in Fig. 6, binding of LG100754 to RXR results in a conformational change throughout the heterodimeric complex that alters the structure of the noncovalently bound RAR subunit. This conformational change results both in the dissociation of corepressors and association of coactivators in a fashion mediated by the τc/AF-2 domain of the nonliganded RAR. Thus, although LG100754 binds to RXR, it is instead RAR that behaves as if it is liganded. We have termed this unique form of allosteric control the phantom ligand effect (SCHULMAN et al. 1997).

How does LG100754 binding to RXR lead to transactivation by RAR? Clearly binding of LG100754 must result in some type of conformational change in the RXR LBD. This hypothesis is supported by the observation that LG100754 promotes an interaction between RXR and *Drosophila* TAF110 in yeast two-hybrid assays. LG100754, however, does not promote interactions between RXR and other coactivators including CBP, SRC-1 and TBP (LALA et al. 1996). Interestingly, while ligand-dependent interactions between RXR and CBP, SRC-1 or TBP require the activity of the RXR τc/AF-2 domain, ligand-dependent interactions between RXR and TAF110 do not

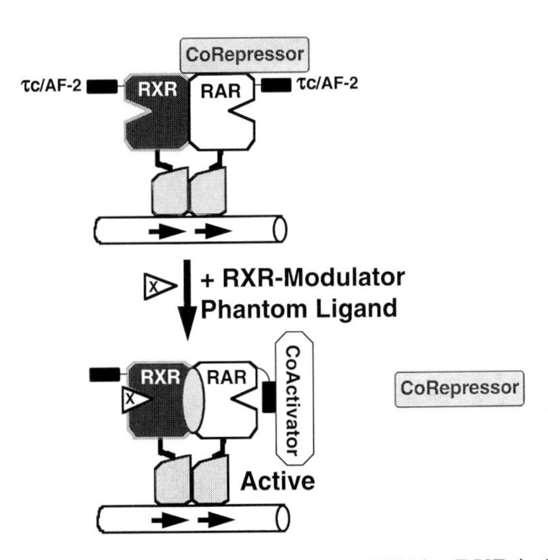

Fig. 6. The phantom ligand effect. Binding of LG100754 by RXR induces a structural change in RXR that is propagated through the dimer interface (*gray oval at dimer interface*) to RAR. This conformational change in RAR allows the release of corepressors and the RAR τc/AF-2 domain to achieve an active conformation. See text for details

(CHAKRAVARTI et al. 1996; KAMEI et al. 1996; ONATE et al. 1995; SCHULMAN et al. 1995). Thus the unique properties of LG100754 appear to arise from combining the qualities of both agonists and antagonists into one molecule.

Binding of ligand by nuclear hormone receptors is thought to initiate a 180° flip in the omega loop connecting helices 1 and 3 of RAR (helices 2 and 3 of RXR) and a 90° angular shift in the τc/AF-2 domain (helix 12; BOURGUET et al. 1995; BRZOZOWSKI et al. 1997; RENAUD et al. 1995; WAGNER et al. 1995). These changes are believed to result in a conformation that promotes association with positive coactivators and/or components of the basal transcription machinery as well as dissociation of *trans*-acting corepressors (for review see SCHWABE 1996). Activation by the RAR τc/AF-2 domain after binding of ligand to RXR suggests that an altered RXR structure triggers a similar change in RAR. It seems inescapable that this change must be propagated through the heterodimerization interface (helices 9 and 10). This in turn has two implications. First, ligand binding not only changes the omega loop and τc/AF-2 domain but also alters the dimer interface. Second, the dimer interfaces are linked such that a change in one (i.e., RXR) results in a consequential linked change in the partner. This change is then transmitted throughout the nonliganded LBD resulting in a conformation that mimics ligand binding even in the absence of bound ligand. Thus it is more appropriate to consider RXR-dependent heterodimers as structural and functional units. Accordingly the DNA-binding, ligand-binding, repression and transactivation properties are products of the complex and are not simply the sum of two independent subunits.

C. RXR Pharmacology

I. Sensitization of Diabetic Mice to Insulin by RXR Agonists

The observation that a certain number of the RXR dependent heterodimers are permissive (i.e., they respond to RXR ligands) raises the question whether RXR selective agonists may replicate the biological activity of ligands for several of the RXR heterodimer pairs. Recent studies have demonstrated that PPARγ is an adipocyte enriched member of the intracellular receptor superfamily and has been shown to be a master regulator of adipocyte differentiation (for review see MANDRUP and LANE 1997). Ligands that activate PPARγ promote adipocyte differentiation and PPARγ has been identified as the molecular target for a class of antidiabetic agents referred to as thiazolidinediones (TZDs). These compounds which function as insulin sensitizers are effective in the treatment of noninsulin dependent diabetes mellitus (NIDDM) (FUJIWARA et al. 1988; SUTER et al. 1992; WILLSON et al. 1996).

We examined the responsiveness of the RXR-PPARγin response to RXR agonists alone, PPARγ ligands such as the TZD BRL49653, or the combination of both agents. This heterodimer is activated by RXR ligands, such as

LG100268 and is also responsive to TZDs (Table 2, Fig. 7). Furthermore, the magnitude of the response to RXR agonists is significantly enhanced in the presence of a fixed concentration of BRL49653 resulting in a synergistic transcriptional response when both ligands are present (Fig. 7). These results demonstrate that RXR-PPARγ heterodimers respond to ligands that bind to either receptor heterodimer and that maximal efficacy requires both ligands.

The high-affinity PPARγ ligands including TZDs and PGJ2 metabolites induce conversion of preadipocytes to adipocytes (FORMAN et al. 1995a; KLIEWER et al. 1995; TONTONOZ et al. 1994). Thus, activation of PPARγ initiates an adipogenic signaling cascade. If RXR ligands can function to activate RXR-PPARγ heterodimers, these compounds might also be expected to promote adipocyte differentiation. To test this possibility, 3T3L1 preadipocytes were treated with LG100268 and differentiation was assayed by monitoring lipid accumulation. LG100268 produces a strong induction of lipid formation indicating that this agent alone induces adipocyte differentiation (Fig. 8). In addition, LG100268 treatment strongly enhances the ability of the TZD BRL49653 to induce 3T3-L1 differentiation. Taken together, the cotransfection assay and the adipocyte differentiation assays provide strong evidence that RXR ligands can activate RXR-PPARγ heterodimers to produce molecular and cellular

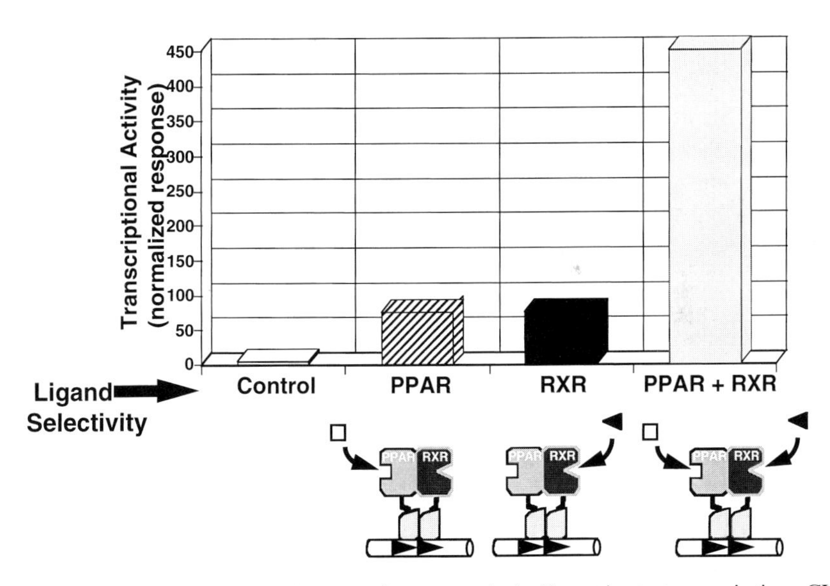

Fig. 7. RXR- and PPARγ-specific agonists synergistically activate transcription. CV1 cells were transfected with a reporter that has three copies of the peroxisome proliferator response element from the acyl CoA oxidase gene upstream of the minimal promoter from the herpes simplex virus thymidine kinase gene fused to luciferase. Plasmids expressing human PPARγ and/or human RXRα were also included in the transfection. After transfection cells were cultured in the absence (*control*) or presence of 1.0 μM of the PPARγ-specific agonist BRL49653 (*PPAR*), 1.0 μM of the RXR-specific agonist LG100268 (*RXR*) or the combination of 1.0 μM BRL49653 and1.0 μM LG100268 (*PPAR + RXR*)

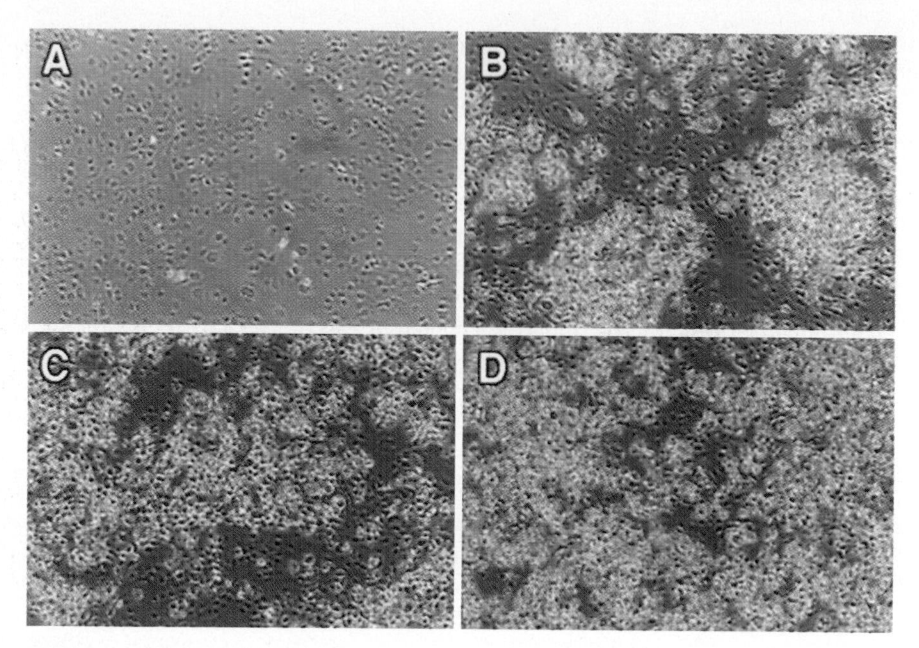

Fig. 8A–D. RXR-specific agonists promote the differentiation of 3T3-L1 cells. 3T3-L1 cells were grown to confluence and then cultured in the absence (**A**) or presence of 1.0 μM BRL49653 (**B**), 1.0 μM LG100268 (**C**), or 1.0 μM BRL49653 + 1.0 μM LG100268 (**D**). After 7 days, cells were fixed and stained with oil red O to visualize lipids

responses similar or identical to activation of the heterodimer via PPARγ ligands.

The ability of TZDs to bind and activate PPARγ is correlated with their ability to reduce hyperglycemia in animal models of NIDDM and obesity (FORMAN et al. 1995a; LEHMANN et al. 1995; WILLSON et al. 1996). Since the antidiabetic activity of TZDs is thought to be via activation of RXR-PPARγ heterodimers, we examined whether RXR ligands might also alter glucose homeostasis in these animal models. In mouse models of NIDDM and obesity, RXR agonists, similar to TZDs, function as insulin sensitizers decreasing hyperglycemia, and hyperinsulinemia (Fig. 9). We have also recently demonstrated that this antidiabetic activity can be further enhanced by combination treatment with PPARγ agonists (MUKHERJEE et al. 1997).

To further explore the effectiveness of RXR agonists as insulin sensitizers we determined whether these ligands can prevent the onset of overt diabetes. We initiated LG100268 treatment to 4 week old db/db mice; at this age these animals are prediabetic and rapidly become hyperglycemic and hypertriglyceridemic. LG100268 (5 mg/kg) prevented this age dependent increase in glucose levels and prevented progression of the disease state (Fig. 10). We also examined whether RXR agonists function as true insulin sensitizers as determined by their ability to enhance the activity of insulin (Fig. 11). LG100268

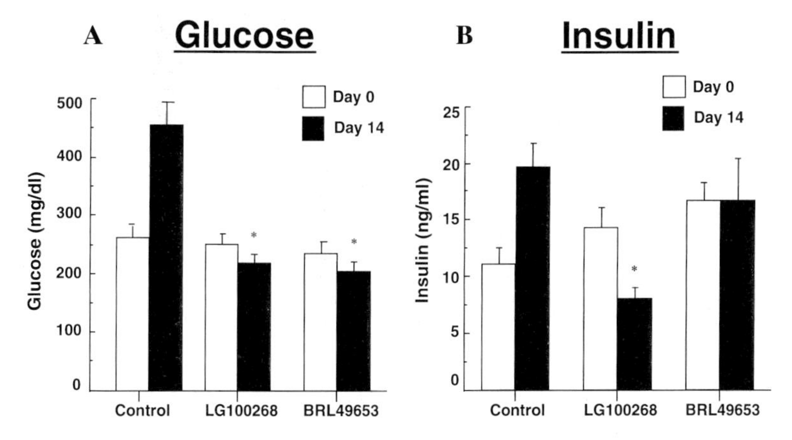

Fig. 9. RXR agonists decrease hyperglycemia and hyperinsulinemia in *ob/ob* mice. Animals ($n = 6$) were initiated on treatment at 8 weeks of age and treated with either vehicle, BRL49653 (1 mg/kg), or LG100268 (20 mg/kg) daily by oral gavage. Immediately before treatment was initiated (*Day 0*) and after 14 days (*Day 14*) glucose and insulin levels was determined as described (MUKHERJEE et al. 1997). Values are mean ± SEM; $*p < 0.05$ day 14 vs control

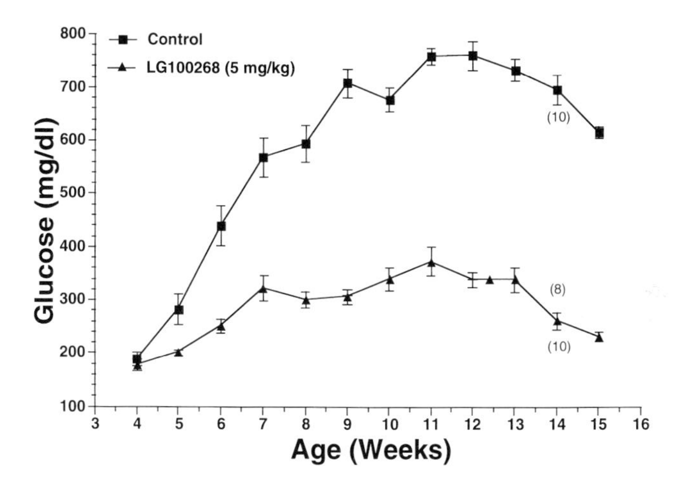

Fig. 10. LG100268 prevents overt hyperglycemia in *db/db* mice. Animals were initiated on treatment at 4 weeks of age and treated with LG100268 (5 mg/kg). Samples were taken weekly 3 h after dosing in the fasted state (3 h), five females and five males per group. When no difference was observed between sexes, the data were combined to give $n = 10$ per group. Values are the mean ± SEM

was coadministered with insulin to highly insulin resistant (15–16 week old) db/db mice. No effect was observed after 2 weeks with single agent therapy of insulin (1 U/mouse) or LG100268 (5 mg/kg). However, a combination of the two dramatically reduced glucose levels (Fig. 11).

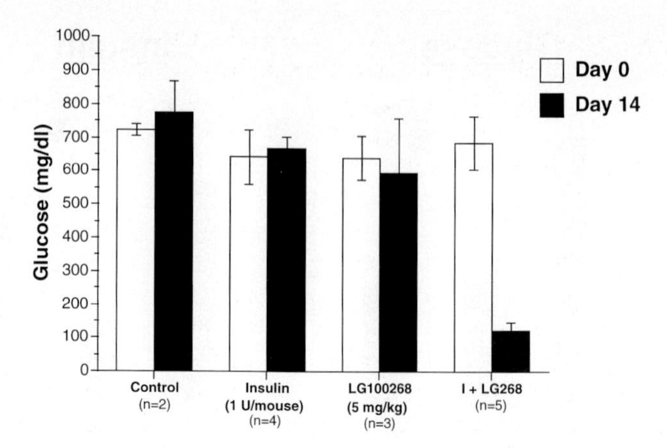

Fig. 11. LG100268 functions as an insulin sensitizer in *db/db* mice. Animals were initiated on treatment with vehicle, insulin (1 U/mouse, s.c.), LG100268 (5 mg/kg, p.o.) or a combination of the two at approximately 15 weeks of age. Samples were taken for analysis 3 h after dosing and following a 3 h fast. n = 2, 4, 3, and 5 for the groups, respectively

These studies and those that we have previously reported demonstrate that RXR agonists function as insulin sensitizers markedly decreasing serum glucose, and insulin in animal models of insulin resistance. The combination of RXR and PPARγ ligands has additive or synergistic effects in stimulating transcription and reducing the degree of hyperglycemia and hypertriglyceridemia in animal models of NIDDM. The combination of RXR agonist with insulin markedly decreases hyperglycemia associated with late stage disease. In the present models of insulin resistance RXR agonists appear to have pharmacological properties comparable to drugs with proven efficacy in humans for treatment of NIDDM, suggesting that these RXR agonists offer a novel approach to the treatment of insulin resistance.

II. Chemoprevention/Chemotherapy of Carcinogen-Induced Breast Cancer

For the past 20 years numerous preclinical and human clinical investigations have examined retinoids such as all-*trans* retinoic acid (ATRA) and 13-*cis* retinoic acid as chemopreventive agents (COSTA et al. 1994; HONG and SPORN 1997; LIPPMAN et al. 1994; LOTAN 1996). However, in many human clinical trials, chronic treatment has been associated with significant toxicities, including mucocutaneous toxicity's, thereby limiting their use for chemoprevention. Anzano and colleagues (ANZANO et al. 1994) demonstrated that compounds such as 9-*cis* retinoic acid that activate both RARs and RXRs are more effective in the chemoprevention of breast cancer than compounds such as all-*trans* retinoic acid that only binds RARs. To examine the contribution that the individual RAR and RXR subtypes play in both the therapeutic potential as well as the side effect profile of these receptor pathways we have recently examined the efficacy of an RXR agonist in the carcinogen induced breast cancer model. LGD1069, a high-affinity ligand for the RXRs, was shown to have efficacy equivalent to tamoxifen as a chemopreventive agent in the *N*-nitroso-*N*-methlyurea (NMU)-induced rat mammary carcinoma model. Furthermore, LGD1069 was very well tolerated during chronic therapy with no classic signs of "retinoid associated" toxicities (GOTTARDIS et al. 1996).

Due to the high efficacy and benign profile of this RXR agonist as a suppressor of carcinogenesis, we have recently initiated a series of additional studies examining its role as a therapeutic agent on established mammary carcinomas. In the rat mammary carcinoma model, NMU was utilized to induce tumors and the tumors were allowed to grow to an established size prior to initiation of treatment. LGD1069-treated animals showed complete regression in 72% of treated tumors and had a reduced tumor load compared to control. In addition, combination of LGD1069 and tamoxifen showed increased efficacy over either agent alone. These data demonstrate that the RXR-selective ligand LGD1069 is a highly efficacious therapeutic agent for mammary carcinoma and enhances the activity of tamoxifen.

There are a number of potential mechanisms for LGD1069's antitumor efficacy, including direct action on tumor cell proliferation signals, paracrine effects, and modulation of differentiation. Nevertheless, these data suggest RXR agonists may be effective in both the chemoprevention and chemotherapy of certain types of breast cancer.

D. Conclusions and Future Directions

As described in Sect. D, the participation of RXR as an obligate heterodimeric partner in multiple hormonal signaling pathways has allowed targeting of this transcription factor for pharmocological intervention in several human diseases. The ability of RXR agonists to activate RXR-PPARγ heterodimers has identified LGD1069 and LG100268 as potential new therapeutics for the treatment of NIDDM, a chronic disease effecting millions each year. Furthermore, recent work has also shown that in animal models of mammary carcinogenesis RXR agonists perform as well as or better than the widely used chemotherapeutic agent tamoxifen. In the mammary carcinogenesis models it is not known whether the therapeutic effects of RXR agonists are mediated via a permissive heterodimer, as observed in the NIDDM models, or if an allosteric mechanism such as those described in Sect. B is responsible. Future studies should shed light on this question.

The observations made with the combination of RXR-specific agonists and RAR antagonists and with the synthetic ligand LG100754 alone indicate a remarkable flexibility in the ability of heterodimeric nuclear receptors to respond to ligands. Regulation of hormonal signaling through RXR suggests two important directions for further study. First, it may be possible to regulate previously independent signaling pathways such as RXR-RAR and RXR-PPAR with a single ligand. Second, it should be possible to identify RXR modulators that activate only a single RXR-dependent heterodimer. Indeed, we have recently identified RXR modulators that selectively activate only RXR-PPAR heterodimers. As expected, at least one of these heterodimer-specific RXR modulators reduces glucose levels in animal models of type II diabetes (data not shown). The ability of ligand binding to one subunit of a heterodimer to influence the conformation and activity of the other subunit indicates RXR-dependent heterodimers should be considered as single functional units. This conclusion also dictates that each heterodimer has unique properties that are not simply the sum of the individual monomeric receptors. The ability to exploit the uniqueness of individual RXR-dependent heterodimers to identify high-affinity heterodimer-specific RXR ligands will allow a unique approach to pharmocological regulation of nuclear receptors with many potential benefits. For instance, RXR-specific agonists do not appear to be as teratogenic as RAR-specific or RAR/RXR pan-agonists. Given the multiple contributions of heterodimeric nuclear hormone receptors to vertebrate development, human malignancies and metabolic diseases, RXR-specific compounds such as LGD1069, LG100268, and LG100754 are likely to provide new insight into how nuclear hormone receptors influence these processes and provide potential new and exciting therapies for human diseases.

Acknowledgements. The authors thank Marcus Boehm, Deepak Lala, Bill Lamph, Eric Bischoff, Marco Gottardis, Bernadette Maddatu and Dale Mais for their contributions in the synthesis and characterization of LGD1069, LG100268, and LG100754.

References

Anzano MA, Byers SW, Smith JM, Peer CW, Mullen LT, Brown CC, Roberts AB, Sporn MB (1994) Prevention of breast cancer with 9-*cis*-retinoic acid as a single agent and in combination with tamoxifen. Cancer Res 54:4614–4617

Apfel CM, Kamber M, Klaus M, Mohr P, Keidel S, LeMotte PK (1995) Enhancement of HL-60 differentiation by a new class of retinoids with selective activity on retinoid X receptor. J Biol Chem 270:30765–30772

Baniahmad A, Leng X, Burris TP, Tsai SY, Tsai, M-J, O'Malley BW (1995) The τ4 activation domain of the Thyroid Hormone Receptor is required for the release of a putative corepressor(s) necessary for transcriptional silencing. Mol Cell Biol 15:76–86

Boehm MF, Zhang L, Badea BA, White SA, Mais DE, Berger E, Suto CM, Goldman ME, Heyman RA (1994) Synthesis and structure-activity relationships of novel retinoid X receptor-selective retinoids. J Med Chem 37:2930–2941

Boehm MF, Zhang L, Zhi L, McClurg MR, Berger E, Wagoner M, Mais DE, Suto CM, Davies PJ A, Heyman RA, Nadzan AM (1995) Design and synthesis of potent retinoid X receptor selective ligands that induce apoptosis in leukemia cells. J Med Chem 38:3146–3155

Bourguet W, Ruff M, Chambon P, Gronemeyer H, Moras D (1995) Crystal structure of the ligand-binding domain of the human nuclear receptor RXR-alpha. Nature 375:377–382

Brzozowski AM, Pike AC W, Dauter Z, Hubbard RE, Bonn T, Engstrom O, Ohman L, Greene GL, Gustafsson J-A, Carlquist M (1997) Molecular basis of agonism and antagonism in the oestrogen receptor. Nature 389:753–758

Canan Koch SS, Dardashti LJ, Hebert JJ, Croston GE, Flatten KS, Heyman RA, Nadzan AM (1996) Identification of the first retinoid X receptor homodimer antagonist. J Med Chem 39:3229–3234

Chakravarti D, LaMorte VJ, Nelson MC, Nakajima T, Schulman IG, Juguilon H, Montminy M, Evans RM (1996) Role of CBP/P300 in nuclear receptor signaling. Nature 383:99–103

Chen H Lin RJ, Schiltz RL, Chakravarti D, Nash A, Nagy L, Privalsky ML, Nakatani Y, Evans RM (1997) Nuclear receptor coactivator ACTR is a novel histone acetyltransferase and forms a multimeric activation complex with P/CAF and CBP/p300. Cell 90:569–580

Chen JD, Evans RM (1995) A transcriptional co-repressor that interacts with nuclear hormone receptors. Nature 377:454–457

Chen JY, Clifford J, Zusi C, Starrett J, Tortolini D, Ostrowski J, Reczek PR, Chambon P, Gronemeyer H (1996) Two distinct actions of retinoid-receptor ligands. Nature 382:819–822

Costa A, Formelli F, Chiesa F, Decensi A, De Palo G, Veronesi U (1994) Prospects of chemoprevention of human cancers with the synthetic retinoid fenretinide. Cancer Res 54:2032S-2037 s

Forman BM, Tontonoz P, Chen J, Brun RP, Speigelman BM, Evans RM (1995a) 15-deoxy-$\Delta^{12,14}$-prostaglandin J_2 is a ligand for the adipocyte determination factor PPARγ. Cell 83:803–812

Forman BM, Umesono K, Chen J, Evans RM (1995b) Unique response pathways are established by allosteric interactions among nuclear hormone receptors. Cell 81:541–550

Fujiwara T, Yoshioka S, Toshioka T, Ushiyama I, Horikoshi H (1988) Characterization of a new oral antidiabetic agent CS-045. Diabetes 37:1549–1558

Glass CK, Rose DW, Rosenfeld MG (1997) Nuclear receptor coactivators. Curr Opin Cell Biol 9:222–232

Gottardis MM, Bischoff ED, Shirley MA, Wagoner MA, Lamph WW, Heyman RA (1996) Chemoprevention of mammary carcinoma by LGD1069 (targretin): an RXR selective ligand. Cancer Res 15:5566–5570

Heyman RA, Mangelsdorf DJ, Dyck JA, Stein RB, Eichele G, Evans RM, Thaller C (1992) 9-Cis retinoic acid is a high affinity ligand for the retinoid X receptor. Cell 68:397–406

Hong H, Kulwant K, Garabedian M, Stallcup MR (1997) GRIP1, a transcriptional coactivator for the AF-2 transactivation domain of steroid, thyroid, retinoid, and vitamin D receptors. Mol Cell Biol 15:2735–2744

Hong WK, Sporn MB (1997) Recent advances in chemoprevention of cancer. Science 278:1073–1077

Horlein AJ, Naar AM, Heinzel T, Torchia J, Gloss B, Kurokawa R, Ryan A, Kamai Y, Soderstrom M, Glass CK, Rosenfeld MG (1995) Hormone-independent repression by the thyroid hormone receptor is mediated by a nuclear receptor co-repressor N-COR. Nature 377:397–404

Horwitz KB, Jackson TA, Bain DL, Richer JK, Takimoto GS, Tung L (1996) Nuclear receptor coactivators and corepressors. J Mol Endocrinol 10:1167–1177

Kamei Y, Xu L, Heinzel T, Torchia J, Kurokawa R, Gloss B Lin S-C, Heyman RA, Rose DW, Glass CK, Rosenfeld MG (1996) A CBP integrator complex mediates transcriptional activation and AP-1 inhibition by nuclear receptors. Cell 85:403–414

Kliewer SA, Lenhard JM, Willson TM, Patel I, Morris DC, Lehmann J (1995) A prostaglandin J2 metabolite binds peroxisome proliferator-activated receptor gamma and promotes adipocyte differentiation. Cell 83:813–819

Kurokawa R, DiRenzo J, Boehm M, Sugarman J, Gloss B, Rosenfeld MG, Heyman RA, Glass CK (1994) Regulation of retinoid signalling by receptor polarity and allosteric control of ligand binding. Nature 371:528–531

Kurokawa R, Soderstrom M, Horlein A, Halachmi S, Brown M, Rosenfeld MG, Glass CK (1995) A co-repressor specifies polarity-specific activities of retinoic acid receptors. Nature 377:451–457

Lala DS, Mukherjee R, Schulman IG, Canan-Koch SS, Dardashti LJ, Nadzan AM, Croston GE, Evans RM, Heyman RA (1996) Activation of specific RXR heterodimers by an antagonist of RXR homodimers. Nature 383:450–453

Lehmann JM, Moore LB, Smith-Oliver TA, Wilkison WO, Willson TM, Kliewer SA (1995) An antidiabetic thiazolidinedione is a high affinity ligand for peroxisome proliferator-activated receptor gamma (PPARγ). J Biol Chem 270:12953–12956

Levin LA, Sturzenbecker LJ, Kazmer S, Bosakowski T, Huselton C, Allenby G, Speck J, Kratzeisen C, Rosenberger M, Lovey A, Grippo JF (1992) 9-Cis retinoic acid sterioisomer binds and activates the nuclear receptor RXRα. Nature 355:359–361

Li C, Schwabe JW R, Banayo E, Evans RM (1997) Coexpression of nuclear receptor partners increases their solubility and biological activities. Proc Natl Acad Sci 94:2278–2283

Lippman SM, Benner SE, Hong WK (1994) Retinoid chemoprevention studies in upper aerodigestive tract and lung carcinogenesis. Cancer Res 54:2025S-2028 s

Lotan R (1996) Retinoids in cancer chemoprevention. FASEB J 10:1031–1039

Mandrup S, Lane MD (1997) Regulating adipogenesis. J Biol Chem 272:5367–5370

Mangelsdorf DJ, Borgmeyer U, Heyman RA, Zhou JY Ong ES Oro AE, Kakizuka A, Evans RM (1992) Characterization of three RXR genes that mediate the action of 9-*cis* retinoic acid. Genes Dev 6:329–344

Mangelsdorf DJ, Evans RM (1995) The RXR heterodimers and orphan receptors. Cell 83:841–850

Mangelsdorf DJ Ong ES, Dyck JA E RM (1990) Nuclear receptor that identifies a novel retinoic acid response pathway. Nature 345:224–229

Mangelsdorf DJ, Thummel C, Beato M, Herrlich P, Schutz G, Umesono K, Kastner P, Mark M, Chambon P, Evans RM (1995) The nuclear receptor superfamily: the second decade. Cell 83:835–839

Mangelsdorf DJ, Umesono K, Kliewer SA, Borgmeyer U Ong ES, Evans RM (1991) A direct repeat in the cellular retinol-binding protein type II gene confers differential regulation by RXR and RAR. Cell 66:555–561

Minucci S, Leid M, Toyama R, Saint-Jeannet J-P, Peterson VJ, Horn V, Ishmael JE, Bhattacharyya N, Dey A, Dawid IB, Ozato K (1997) Retinoid X receptor (RXR) within the RXR-retinoic acid receptor heterodimer binds its ligand and enhances retinoid-dependent gene expression. Mol Cell Biol 17:644–655

Mukherjee R, Davies PJ A, Crombie DL, Bischoff ED, Cesario RM, Jow L, Hamann LG, Boehm MF, Mondon CE, Nadzan AM, Paterniti Jr JR, Heyman RA (1997) Sensitization of diabetic and obese mice to insulin by retinod X receptor agonists. Nature 386:407–410

Onate SA, Tsai SY, Tsai, M-J, O'Malley BW (1995) Sequence and characterization of a coactivator for thr steroid hormone receptor superfamily. Science 270:1354–1347

Oro AE, McKeown M, Evans RM (1990) Relationship between the product of the Drosophila ultraspiracle locus and the vertebrate retinoid X receptor. Nature 347:298–301

Perlmann T, Evans RM (1997) Nuclear receptors in Sicily: All in the famiglia. Cell 90:391–397

Renaud J-P, Rochel N, Ruff M, Vivat V, Chambon P, Gronemeyer H, Moras D (1995) Crystal structure of the RAR-γ ligand-binding domain bound to all-*trans* retinoic acid. Nature 378:681–689

Schulman IG, Chakravarti D, Juguilon H, Romo A, Evans RM (1995) Interactions between the retinoid X receptor and a conserved region of the TATA-binding protein mediate hormone-dependent transactivation. Proc Natl Acad Sci USA 92:8288–8292

Schulman IG, Juguilon H, Evans RM (1996) Activation and repression by nuclear hormone receptors: hormone modulates an equilibrium between active and repressive states. Mol Cell Biol 16:3807–3818

Schulman IG, Li C, Schwabe JW R, Evans RM (1997) The phantom ligand effect: allosteric control of transcription by the retinoid X receptor. Genes Dev 11:299–308

Schwabe JW R (1996) Transcriptional control: How nuclear receptors get turned on. Curr Biol 6:372–374

Suter SL, Nolan JJ, Wallace P, Gumbiner B, Olefsky JM (1992) Metabolic effects of new oral hypoglycemic agent CS-045 in NIDDM subjects. Diabetes Care 15 193–203

Tontonoz P, Hu E, Speigelman BM (1994) Stimulation of adipogenesis in fibroblasts by PPARγ2, a lipid-activated transcription factor. Cell 79:1147–1156

Torchia J, Rose DW, Inostroza J, Kamei Y, Westin S, Glass CK, Rosenfeld MG (1997) The transcriptional co-activator p/CIP binds CBP and mediates nuclear-receptor function. Nature 387:677–684

Umesono K, Murakami KK, Thompson CC, Evans RM (1991) Direct repeats as selective response elements for the thyroid hormone, retinoic acid and vitamin D_3 receptors. Cell 65:1255–1266

Vivat V, Zechel C, Wurtz J-M, Bourguet W, Kagechika H, Umemiya H, Shudo K DM, Gronemeyer H, Chambon H (1997) A mutation mimicking ligand-induced conformational changes yields a constitutive RXR that senses allosteric effects in heterodimers. EMBO J 16:5697–5079

Voegel JJ, Heine MJ S, Zechel C, Chambon P, Gronemeyer H (1996) TIF2, a 160 kDa transcriptional mediator for ligand-dependent activation function AF-2 of nuclear receptors. EMBO J 15:3667–3675

Wagner RL, Apriletti JW, McGrath ME, West BL, Baxter JD, Fletterick RJ (1995) A structural role for hormone in the thyroid hormone receptor. Nature 378:690–697

Willson TM, Cobb JE, Cowan DJ, Wiethe RW, Correa ID, Prakash SR, Beck KD, Moore LB, Kliewer SA, Lehmann JM (1996) The structure-activity relationship between peroxisome-proliferator-activated receptor gamma and the antihyperglycemic activity of thiazolidinediones. J Med Chem 39:665–668

Yao T-P, Segraves WA Oro AE, Mckeown M, Evans RM (1992) Drosophila ultraspiracle modulates ecdysone receptor function via heterodimer formation. Cell 71:63–72

Zamir I, Zhang J, Lazar MA (1997) Stoichiometric and steric principles governing repression by nuclear hormone receptors. Genes Dev 11:835–846

Section C
Differentiation, Proliferation, and Cancer

CHAPTER 8

Growth Control by Retinoids: Regulation of Cell Cycle Progression and Apoptosis

B. HARVAT and A.M. JETTEN

A. Introduction

In multicellular organisms, a balance between differentiation, proliferation and cell death maintains tissue homeostasis in adult tissues and directs normal development during embryonic morphogenesis. Vitamin A is one of the critical factors regulating the molecular events involved in these processes. The requirement of retinoids for normal development and tissue homeostasis has been known since Wolbach and Howe first reported on the defects that occurred in vitamin A-deficient animals in 1926. A large literature has since appeared describing the role of retinoids in promoting proliferation, differentiation or apoptosis in vivo (see also Chaps. 12–15, this volume), in a variety of cell culture systems (see also Chaps. 3, 9, 10, this volume) and more recently through studies using retinoid receptor knock-out strategies (TANEJA et al. 1996; SAPIN et al. 1997). In addition, these investigations have provided insight into the mechanisms by which retinoids induce teratogenic effects and inhibit neoplastic development, and hold promise for new therapeutic applications of retinoids, particularly in cancer.

Significant progress has been made in understanding the molecular mechanisms by which retinoids control proliferation through modulating the action of negative and positive growth factors, including EGF, TGFα, TGFβ, insulin, IL1α, IL6, interferon γ, estrogen and vitamin D_3. Regulation by retinoids can occur at the level of expression of growth factors/cytokines, their respective receptors/binding proteins or downstream genes in the signaling pathway (e.g., KOLLA et al. 1996; MATIKAINEN et al. 1996; BLUTT et al. 1997; SHERBET and LAKSHMI 1997). Details of the mechanisms by which retinoids accomplish these actions have unfolded in a large body of literature that is beyond the scope of this review.

This chapter reviews recent studies concerning the action of various retinoids on the control of cell cycle progression and the regulation of cell death. These processes can be mediated by retinoid receptor-dependent and -independent mechanisms. Little is known about the mechanisms by which retinoids interact with or influence cell cycle regulatory proteins. Although in many cases reports of the effects of retinoids on cell cycle proteins have been descriptive, recent discoveries indicate that retinoids can directly control expression of at least a few critical cell cycle genes. Similarly, the mechanisms by which retinoids regulate apoptosis are not well understood, particularly for retinoid receptor-mediated apoptosis. Recent studies have provided insight, however, into the receptor-independent signals generated by several synthetic retinoids.

B. Mechanisms of Retinoid Action

I. Retinoid Receptor-Dependent Mechanisms

Many of the biological effects and changes in gene expression induced by retinoids are mediated by retinoic acid receptors (RARα, RARβ, and RARγ) and/or retinoid X receptors (RXRα, RARα, and RARγ) (MANGELSDORF et al. 1994; GIGUÈRE 1994; CHAMBON 1996). RARs form heterodimers with RXRs that bind preferentially to retinoic acid response elements (RAREs), while RXRs, in addition to forming heterodimers with many other nuclear receptors, form also homodimers that interact with response elements referred to as RXREs. The retinoid receptor complexes regulate gene expression through interaction with transcription intermediary proteins which function as corepressors, coactivators or integrators (GLASS et al. 1997; SHIBATA et al. 1997). Binding of a retinoid to its receptor typically induces a conformational change that results in the dissociation of a corepressor and/or promotes interaction with coactivators. All-*trans* RA and 4-oxo-retinol bind and activate only RARs, while 9-*cis* RA functions as a ligand for both RARs and RXRs (GIGUÈRE 1994; ACHKAR et al. 1996; CHAMBON 1996). Several synthetic agonists and antagonists for various receptor isotypes have been identified that offer valuable tools for studying nuclear retinoid receptor signaling pathways.

In addition to mediating RARE- or RXRE-dependent transactivation, retinoid receptors can also affect gene expression by inhibiting the action of other transcription factors, including AP-1 (SCHÜLE et al. 1991; SAATCIOGLU et al. 1994). Such interactions can be related to either competition for the same or adjacent DNA binding sites (VAN DE KLUNDERT et al. 1995; CRESTANI et al. 1996), interaction with and sequestration of transcription factors (SCHROEN and BRINCKEROFF 1996) or competition with common transcription intermediary factors (HORLEIN et al. 1995). The antagonism between RA and AP-1 signaling pathways is due at least in part to a block in the JNK signaling cascade that results in an inhibition of c-jun phosphorylation on Ser-63/73 (CAELLES et al. 1997). This phosphorylation is required for AP-1 activity. Synthetic retinoids with selective antiAP-1 activity, such as SR11203 and SR11238, have been identified that bind RARs and exhibit antiAP-1 activity but do not cause RARE/RXRE-dependent transactivation (FANJUL et al. 1994). More details about specific retinoids and the retinoid receptors can be found in Chaps. 5–7, this volume.

II. Retinoid Receptor-Mediated Control of Cell Proliferation

Numerous studies have analyzed the role of specific receptor signaling pathways in growth control (NERVI et al. 1991; DEL SENNO et al. 1993; SCHILTHUIS et al. 1993; LOTAN 1994; AVIS et al. 1995; DAWSON et al. 1995; GUO et al. 1995; HAN et al. 1995; MOGHAL and NEEL 1995; MOTZER et al. 1995; SCHADENDORF et al. 1995; DARWICHE et al. 1996; CHAO et al. 1997; GIANNINI G et al. 1997; DE VOS et al. 1997; WU et al. 1998). These studies have indicated that at least some of the effects of retinoids on growth can be mediated by RAR and/or RXR receptors. Several studies have particularly implicated RARβ in growth-inhibitory effects of retinoids. RARβ is transcriptionally up-regulated during senescence of dermal fibroblasts (LEE X et al. 1995) and mammary epithelial cells (SWISSHELM et al. 1994). RARβ has also been implicated in RA-induced growth inhibition in certain neuroblastoma cell lines (CHEUNG et al. 1996), Caco2 intestinal carcinoma (MCCORMACK et al. 1996) and HeLa cervical carcinoma cells (SI et al. 1996). The DNA-binding domain of RARβ is required for RA-induced growth suppression in HeLa cells (FRANGIONI et al. 1994). Decreased levels of RARβ expression in certain mammary (VAN DER BURGH et al. 1993; SHEIKH et al. 1994; SWISSHELM et al. 1994; SEEWALDT et al. 1995, LIU Y et al. 1996; WIDSCHWENDTER et al. 1997), renal (GERHARZ et al. 1996), lung (NERVI et al. 1991; MOGHAL and NEEL 1995) and head and neck carcinomas (XU et al. 1994), as well as in neuroblastoma (PLUM and CLAGETT-DAME 1995) have been linked to resistance to growth inhibition by RA. However, certain cell lines are refractory despite the presence of RARβ (SUN et al. 1997a). Ectopic expression of RARβ promotes growth arrest in some RARβ-negative carcinoma cell lines (SEEWALDT et al. 1995; LIU Y et al. 1996; LI et al. 1998). In contrast, ectopic expression of RARβ in head and neck squamous carcinoma cell lines induced greater sensitivity to RA-mediated repression of squamous differentiation markers and was correlated with a decrease in tumorigenicity, while exogenous RARγ made them more sensitive to the growth-inhibitory effects of RA (LOTAN 1996). However, a different study found a good correlation between growth-inhibition and induction of RARβ in several RA-treated oral squamous cell carcinoma cell lines (GIANNINI F et al. 1997). In several cases, the loss of RARβ gene expression was related to an inability to activate RARE-dependent transactivation. Defects in this activation can occur at many different levels of transcriptional control (ZHANG et al. 1994; MOGHAL and NEEL 1995). RXR-selective ligand activation of RXR-Nur77 heterodimers and changes in the balance between levels of the nuclear orphan receptors COUP-TF and Nur77 have also been implicated in the control of RARE transactivation, induction of RARβ and RA sensitivity in mammary and lung carcinoma cells (WU Q et al. 1997a,b).

The responsiveness of mammary carcinoma cells to the growth-inhibitory effects of retinoic acid depends on the expression of the estrogen receptor (ER), RARα and RARβ. Most ER-positive cell lines are sensitive to growth-inhibitory effects of RA while ER-negative cells are resistant (RUBIN et al. 1994; SHEIKH et al. 1994; CHEN AC et al. 1997; MANGIAROTTI et al. 1998). Overexpression of RARβ rendered ER-negative MDA-MB-231 cells sensitive to RA-induced G_1 growth arrest (SEEWALDT et al. 1995; LIU Y et al. 1996). Ectopic expression of RARα into mammary carcinoma cells expressing low levels of RARα or estrogen-mediated up-regulation of RARα can also result in increased growth inhibition, presumably by RARα-driven increased expression of RARβ (ROMAN et al. 1993; SHEIKH et al. 1994; VAN DER LEEDE et al. 1995; RISHI et al. 1996). These studies indicate the importance of RARβ in the growth inhibitory effects of retinoids in mammary carcinoma cells and suggest that RARβ functions as a tumor suppressor gene (VAN DER BURG et al. 1993; SHEIKH et al. 1993; SWISSHELM et al. 1994; SEEWALDT et al. 1995; LIU Y et al. 1996).

Expression of retinoid receptors can also be controlled by upstream events in growth factor signaling pathways. For example, in keratinocytes carrying an oncogenic, activated form of Ha-ras, RARα and RARγ levels are markedly decreased as is RARE-mediated transcriptional activation (DARWICHE et al. 1996). Ectopic expression of RARα restores inhibition of cell growth by RA. These results indicate that expression of certain oncogenes in carcinoma cells may suppress retinoid receptor expression and thereby alter the responsiveness to RA.

III. Receptor-Independent Regulation of Cell Proliferation

Some actions of retinoids on growth and apoptosis operate through RAR/RXR-independent mechanisms. RA was recently shown to bind with high affinity to the mannose-6-phosphate (M6P)/insulin-like growth factor-II (IGF-II) receptor, a multifunctional membrane glycoprotein known to activate TGF-β and to function in trafficking of lysosomal enzymes (KANG et al. 1997). Binding of RA to this receptor may be involved in growth suppression and/or induction of apoptosis. The inhibition of cell death in fibroblasts and lymphocytes by the retinol metabolites 14-hydroxy-4,14-retroretinol (14-HRR) and 13,14-dihydro-retinol (DHR) occurs independently of nuclear retinoid receptors but through a novel mechanism that has not yet been identified (for more details see also Chap. 3, this volume) (BUCK et al. 1991; O'CONNELL et al. 1996). 4-Oxo-retinol inhibits the growth of RA-resistant mammary carcinoma cells by what appears to be a novel mechanism (CHEN AC et al. 1997) and several synthetic retinoids, including N-(4-hydroxyphenyl)retinamide (4-HPR) and 6-[3-(1-adamantyl)-4-hydroxyphenyl]-2-naphthalene carboxylic acid (AHPN or also known as CD437) (BERNARD et al. 1992), induce apoptosis mostly by mechanisms that do not involve RARs or RXRs (see below). The arotinoid RO40–8757, which does not bind nuclear retinoid receptors, has also been reported to inhibit proliferation in many cancer cell lines through a yet unknown mechanism (ELIASON et al. 1993; LOUVET et al. 1994). The reader is referred to Fig. 1 for the chemical structures of these retinoids.

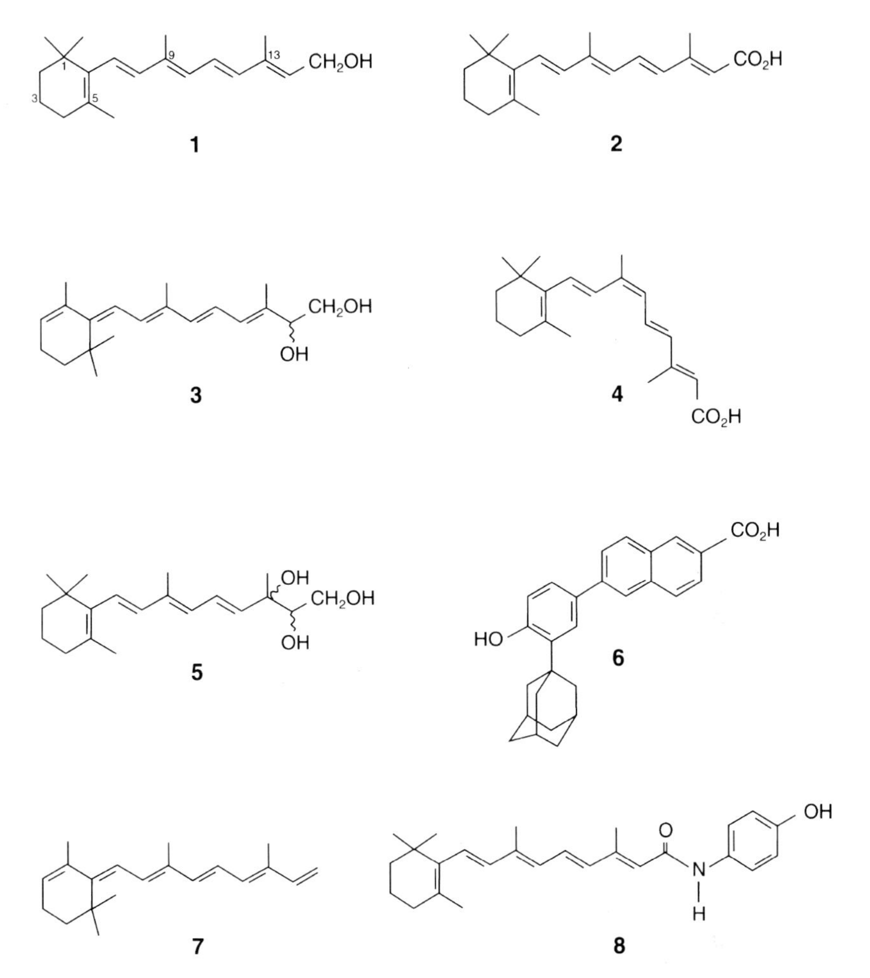

Fig. 1. Chemical structure of several retinoids active in regulating growth and cell death. *1*, Retinol; *2*, all-*trans*-retinoic acid (all-*trans* RA); *3*, 14-hydroxy-4,14-*retro* retinol; *4*, 13-*cis*-retinoic acid (13-*cis* RA); *5*, 13,14-dihydro-retinol; *6*, 6-[3-(1-adamantyl)-4-hydroxyphenyl]-2-naphthalene carboxylic acid (AHPN; also known as CD437); *7*, anhydroretinol; 8. *N*-(4-hydroxyphenyl)retinamide (4-HPR)

Retinoids also form thioester bonds with various proteins and this retinoylation could affect protein activity or function of these proteins (Tournier et al. 1996). However, as yet no evidence has been provided linking any of the changes in proliferation or apoptosis induced by retinoids to retinoylation of specific proteins. Retinoic acid also activates neutral sphyingomyelinase, an enzyme involved in converting sphyingomyelin to ceramide (Riboni et al. 1995). Ceramide has been reported to induce growth arrest and/or apoptosis in a variety of cell types (Obeid and Hannun 1995).

C. Control of Cell Cycle Progression by Retinoids

I. Overview of Cell Cycle Regulation

Regulation of cell cycle progression in mammalian cells by external factors such as retinoids generally occurs in the G_1 phase of the cycle (SHERR and ROBERTS 1995). One critical pathway involved in G_1 arrest controls the activity of the retinoblastoma protein and the E2F family of transcription factors. To assist the reader, a schematic overview of this pathway is provided in Fig. 2. The Rb pathway is regulated by cyclin-dependent kinase (cdk) complexes of cdk2, cdk4 or cdk6 that phosphorylate Rb family members and deactivate their growth inhibitory function (SHERR and ROBERTS 1995). Hypophosphorylated Rb exercises its growth inhibitory activity in part by forming a complex with transcription factors of the E2F family, thereby preventing expression of a number of genes encoding proteins essential for progression through late G_1 and S phase (ADAMS and KAELIN 1995). The activity of these cdks is regulated at the level of their expression, their phosphorylation and the complexes they form with activating cyclins and inhibitory proteins (ckis). Both cyclin and cki levels have been shown to be sensitive to external signals regulating growth (SHERR and ROBERTS 1995). Members of the Ink4 family of ckis, including $p16^{Ink4a}$, $p15^{Ink4b}$, p18 and p19, specifically bind to

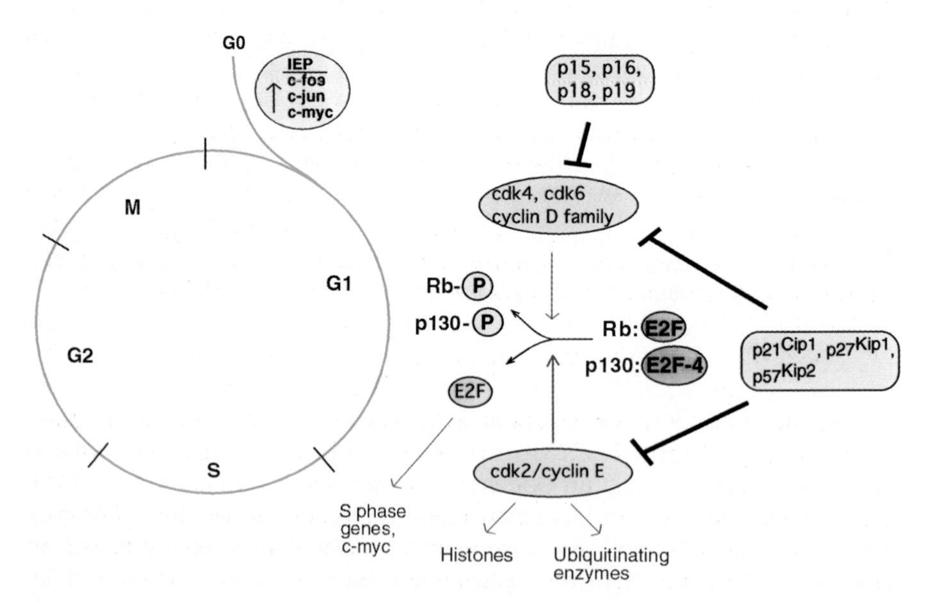

Fig. 2. Schematic view of several biochemical events during the G_1 phase of the cell cycle

and inhibit cdk 4 and cdk6, while members of the Kip family, consisting of $p21^{WAF1/Cip1}$, $p27^{Kip1}$ and $p57^{Kip2}$, have inhibitory activity toward all G_1 cdks (XIONG et al. 1993; SHERR and ROBERTS 1995).

In addition to the Rb pathway, external growth regulatory signals can affect a group of growth-related genes, the immediate early genes (IEG), that are rapidly induced in response to mitogens and include the AP-1 transcription factors c-fos and c-jun as well as c-myc (IMLER and WASYLYK 1989). c-Myc has recently been shown to regulate the transcription of cyclin E and to prevent p27 binding and inactivation of cdk2/cyclin E complexes (AMATI and LAND 1994; VLACH et al. 1996; PEREZ-ROGER et al. 1997). Regulation of c-myc expression is complex, having both an immediate response to mitogens and a late G_1 activation that is regulated by E2F and therefore downstream of Rb (MCMAHON and MONROE 1992; ISHIDA et al. 1994). c-Myc is required for traverse of G_1 into S phase, although its exact function during the transition is not clear. Down-regulation of c-myc has been associated with growth arrest in many different systems (AMATI and LAND 1994).

A number of studies have demonstrated an association of specific cell cycle changes with retinoid-induced growth arrest. Most but not all of these reports show that retinoids induce an accumulation of cells in G_0/G_1, suggesting that retinoids block the progression through G_1 into S phase (DIMARTINO et al. 1990; HASKOVEC et al. 1990; DESBOIS et al. 1991; TONINI et al. 1991; BLOMHOFF et al. 1992; JONK et al. 1992; PREISLER et al. 1992; HUI and YUNG 1993; BURGER et al. 1994; ZHAO et al. 1995; HAGIWARA et al. 1996; WILCKEN et al. 1996; CHAO et al. 1997; NAKA et al. 1997; TOMA et al. 1997; WU S et al. 1997; MANGIAROTTI et al. 1998). In F9 embryonic carcinoma cells, however, RA lengthens the cell cycle in G_1 and G_2 while inducing differentiation (ROSENSTRAUS et al. 1982).

Many studies have failed to carry out a careful examination of whether retinoids directly affect cell cycle regulators and thereby elicit growth arrest or indirectly affect these growth control proteins. A compelling example are the studies in mammary epithelial cells. SEEWALDT et al. (1997) reported that normal human mammary epithelial cells treated with RA arrest in the G_1 phase of the cell cycle and is associated with decreased levels of hyperphosphorylated Rb and down-regulation of cdk4 and cyclins D_1 and E, while p21, p27 or c-myc levels are unaffected. RA treatment of MCF-7 cells likewise produces growth arrest and down-regulation of Rb phosphorylation. However, cyclin D_1, p21, cdk4 and cdk6 levels and activity are reported to either not change or decrease to varying degrees during RA-induced growth arrest of MCF-7 cells (TEIXEIRA and PRATT 1997; ZHOU et al. 1997; ZHU W et al. 1997). Definitive studies that point to specific ways in which retinoids have direct effects on cell cycle regulatory proteins would make interpretation of this data easier and more relevant to the understanding of defects that occur in tumorigenesis. Such studies have begun to appear in the last few years.

II. Cell Cycle Regulatory Targets of Retinoid Action

1. Retinoids Can Affect the Rb Pathway at Several Levels

Retinoids have been shown to affect Rb expression or activity in a number of cell types. The molecular mechanisms by which it achieves these effects are only beginning to be understood. Activation of Rb by promoting its dephosphorylation has been commonly reported in many systems of retinoid-induced growth arrest. Direct evidence for retinoid modulation of Rb phosphorylation was revealed in studies showing up-regulation of the cdk inhibitor $p21^{WAF1/Cip1}$ in HL-60 (JIANG et al. 1994) and U937 cells (LIU M et al. 1996) by RA. In U937 cells, the up-regulation was mediated by RAR/RXR heterodimer binding to an RARE in the p21 promoter. Moreover, studies using specific agonists and antagonists implicated RARα in growth arrest and Rb dephosphorylation in U937 cells (BROOKS et al. 1996), suggesting that it may also mediate the up-regulation of p21 by RA.

YEN and VARVAYANIS (1994) have shown that Rb dephosphorylation in RA-treated HL-60 cells appeared after cells growth arrest and express differentiation markers. They proposed that Rb dephosphorylation is a late event in RA-induced differentiation consistent with a role in maintaining, but not causing, the terminally differentiated state. This interpretation would be consistent with findings in NB4 promyelocytic cells where RARα-mediated differentiation/growth arrest is not restricted to G_1, but Rb phosphorylation is down-regulated even in cells in G_2/M (BROOKS et al. 1997). Similar to HL-60 cells, increases in p21 levels in these cells are preceded by increases in the differentiation marker, CD11b (BOCCHIA et al. 1997). Interestingly, the use of antisense oligonucleotides to block Rb expression rendered U937 cells resistant to differentiation, but not growth arrest, by RA (BERGH et al. 1997). Growth inhibition in the gastric carcinoma cell line TMK-1 is also associated with reduced Rb phosphorylation and cyclin D_1, and an increase in p21 protein (NAKA et al. 1997). Corollary, but not direct evidence, for RA regulation of p27 levels has also been demonstrated in growth inhibition of Epstein-Barr virus infected B lymphocytes (POMPONI et al. 1996). RA treatment of SMS-KCNR neuroblastoma cells leads to G_1 growth arrest, reduction in Rb hyperphosphorylation and decreased cdk activity and associated with increased levels of p27 rather than p21 (MATSUO and THIELE 1998). This action appears to require activation of both RARs and RXRs (GIANNINI G et al. 1997).

In addition to up-regulation of cdk inhibitors, retinoids have been shown to down-regulate G_1 cdk activity by modulating levels of the cyclin partners of the cdks. In HL60 cells, cyclin E levels decrease during RA-induced differentiation (BURGER et al. 1994) while in several mammary carcinoma cell lines both RA and 9-*cis* RA inhibited growth and decreased cyclin D levels (ZHOU et al. 1997; BARDON and RAZANAMAHEFA 1998). The most promising studies concerning mechanisms for these effects come from carcinogen-exposed immortalized human epithelial bronchial cells (BEAS-2B cells), where retinoic acid prevented transformation-associated increases in cyclin D_1 and cyclin E (LANGENFELD et al. 1996, 1997). For cyclin D_1 at least, these affects were associated with increased proteolysis mediated by the ubiquitin degradation pathway and required the PEST instability sequence in the C-terminus of cyclin D_1. Since both cyclin and cki levels are reported to be regulated by ubiquitin-mediated degradation (PAGANO et al. 1995; CLURMAN et al. 1996; WON and REED 1996; DIEHL et al. 1997), definitive studies on the mechanisms by which RA may regulate the proteasome complex or posttranslational modifications of these proteins which target them effectively to the complex could provide significant advances into our understanding of how retinoids prevent progression through the cell cycle.

In some cell types, retinoid-induced growth arrest involves an increase in the absolute levels of Rb protein without necessarily promoting its dephosphorylation. Indirect transcriptional activation and posttranscriptional regulation appear to be mechanisms that are used, because direct transcriptional activation by retinoids has not been described to date. For example, RA and 4-HPR were shown to up-regulate expression of hyperphosphorylated Rb in MCF-7 and T47D breast cancer cells by a mechanism that did not require the transactivating function of RAR, but involved blocking estrogen and progesterone receptor transcriptional activity (KAZMI et al. 1996). Rb mRNA was increased in P19 embryonic carcinoma cells induced to differentiate by RA by a mechanism requiring RARα, although direct activation of the Rb promoter by RARα was not seen (SLACK et al. 1993). The growth arrest in pancreatic and gastrointestinal cancer cell lines by the arotinoid RO40–8757, which does not bind to RARs or RXRs (ELIASON et al. 1993), is unrelated to a specific phase of the cell cycle and causes accumulation of both hyper- and hypophosphorylated forms of Rb (LOUVET et al. 1996). However, a second report on a different panel of pancreatic cancer cell lines showed that RO40–8757-induced G_1 arrest was associated with up-regulation of p21 and p27, and dephosphorylation of Rb (KAWA et al. 1997). Similarly, treatment of non-small-cell lung carcinoma cells with RA initially increased hyperphosphorylated Rb which then converted to the hypophosphorylated form and, in some cell lines, became undetectable (MAXWELL 1994). In this case, the increases in Rb were clearly attributable to a posttranscriptional mechanism as the levels of mRNA did not change significantly.

Decreases in Rb phosphorylation prior to inhibition of kinase activity and accumulation in G_1 have been reported in T-47D mammary carcinoma cells (WILCKEN et al. 1996), suggesting that retinoids may also have an effect on an Rb phosphatase. Evidence for retinoid regulation of a cell cycle-regulated phosphatase has been reported. The phosphatase PP2A becomes temporarily inactive during the G_1 to S phase transition because of methylation of the C-terminal of its catalytic subunit. RA blocks the methylation and therefore keeps PP2A active, presumably preventing progression through the cell cycle (ZHU T et al. 1997).

Taken all together, these findings indicate that accumulation of hypophosphorylated Rb can be a direct downstream target of RA signaling through up-regulation of p21 and/or p27 and perhaps down-regulation of cyclin D_1. Furthermore, although some of these studies indicate that RA is able to promote growth arrest by mechanisms other than activating Rb, most suggest its effects on Rb may be essential for the irreversible growth arrest state associated with RA-induced terminal differentiation. The effect of Rb hypophosphorylation on terminal differentiation/growth arrest has been reinforced by studies which show that a difference in E2F DNA-binding complexes occurs with differentiation (REICHEL 1992; CORBEIL et al. 1995) and affects expression of E2F-regulated genes in a manner specific to RA signaling and not growth arrest in general (GAETANO et al. 1991).

2. Regulation of c-myc Expression

Decreases in c-myc during growth inhibition/differentiation of HL-60 cells by RA have been reported to be mediated at the transcriptional level (BENTLEY and GROUDINE 1986; BARKER and NEWBURGER 1990; FERRARI et al. 1992; YEN and VARVAYANIS 1992; DORE et al. 1993; KIZAKI et al. 1993). A rapid up-regulation of the myc-family member mad, which opposes c-myc by competing for max, its dimeric partner, has also been described in HL-60 and U937 cells (LARSSON et al. 1994). c-Myc and mad expression has been correlated, respectively, with proliferation and growth inhibition/differentiation in a variety of cell types (AYER et al. 1993; HURLIN et al. 1996; ROUSSEL et al. 1996). Recent evidence has indicated that RA can affect c-myc transcription directly. RARs can act synergistically with the CCTC-binding factor (CTCF), a zinc finger transcription factor that mediates transcriptional repression through the c-myc P2 promoter (BURCIN et al. 1997). Mobility shift assays using the E2F response element in the P2 promoter as a probe demonstrated that RA treatment also alters the E2F binding complex to one that includes Rb and represses c-myc expression in HL-60 cells (ISHIDA et al. 1994).

Down-regulation of c-myc or the related proteins N-myc and L-myc has been associated with RA-induced differentiation in a variety of other cell types, including F9 embryonic carcinoma (DEAN et al. 1986; MARTIN et al. 1990), colon carcinoma (TAYLOR et al. 1992) and small cell lung carcinoma cell lines (DOYLE et al. 1989; OU et al. 1996), B16 melanoma (PRASAD et al. 1990), and several neuroblastoma cell lines (THIELE et al. 1985; HAMMERLING et al. 1987; DIMARTINO et al. 1990; SAUNDERS et al. 1995). In the colon carcinoma cell lines the ability of RA to modulate the levels of c-myc is correlated with their sensitivity to the growth inhibitory effects of RA. Reduction in c-myc levels, however, is not a universal mechanism for retinoid action, as c-myc levels have been shown not to change with RA-induced differentiation or growth arrest in other cell types, including TE-671–1A rhabdomyosarcoma (RAMP et al. 1995), glomerular mesangial (SIMONSON 1994), T47D breast carcinoma (WILCKEN et al. 1996) and NTERA-2cl.D1 teratocarcinoma (MILLER et al. 1990) cells.

3. Modulation of AP-1-Mediated Growth Signals

AP-1 consists of a heterodimeric complex of members of the c-fos and c-jun families which are targets of growth factor signaling pathways. In a variety of cell types growth inhibition by retinoids is parallelled by decreased AP-1 activity while no such decrease is observed in RA-resistant cell lines (FANJUL et al. 1994; SIMONSON 1994; VAN DER BURGH et al. 1995; CHEN et al. 1995; FANJUL et al. 1996a; ORIDATE et al. 1996b; MIANO et al. 1996). Moreover, anti-AP-1 selective retinoids have been shown to be growth-inhibitory in several cell types. However, in certain prostate carcinoma and leukemic cell lines, antiAP-selective retinoids have no effect on cell proliferation (DE VOS et al. 1997; KIZAKI et al. 1996). As stated above, the antagonism of AP-1 activity may be due to a block in the JNK signaling cascade which would prevent activation of c-jun-mediated transcription (CAELLES et al. 1997). In support of this hypothesis, the antiproliferative effect of retinoids in mammary carcinoma MCF-7 cells can be overcome by overexpression of c-jun (YANG LM et al. 1997). In glomerular mesangial cells, RA had no effect on AP-1 DNA binding but repressed serum-stimulated transcriptional activation of both c-jun and c-fos (SIMONSON 1994). RA has also been shown to directly down-regulate expression of the c-fos gene in B16 melanoma cells through the serum response element (SRE) of its promoter (BUSAM et al. 1992), although specific involvement of retinoid receptors has not been defined. The down-regulation of c-fos levels may be responsible for the observed down-regulation of c-jun in these cells which coincided with their growth arrest, as c-jun expression is regulated by an AP-1 response element in its promoter (HAN et al. 1992). One potential target for growth inhibitory antiAP-1 effects is the expression of the glutathione-*S*-transferase gene, GSTP1–1, which has been associated with proliferation and malignancy and is repressed in an RAR-dependent manner involving an AP-1 element (XIA et al. 1996).

In some cell systems, particularly those in which retinoids induce both growth arrest and differentiation, retinoids induce AP-1 activity. AP-1 is a target of PKC signal transduction pathways that may be activated by retinoids. In RA-differentiated B16 mouse melanoma cells, RA-mediated increases in PKCα, particularly in the nuclear fraction, are associated with up-regulation of AP-1 activity (GRUBER et al. 1995). Up-regulation of PKCα has also been associated with RA-induced growth inhibition and differentiation in F9 cells (KINDREGAN et al. 1994) and in vascular smooth muscle cells (HALLER et al. 1995; HAYASHI et al. 1995), while PKCε is up-regulated in neuroblastoma differentiation (PONZONI et al. 1993). In F9 cells this has been coordinated with the up-regulation of c-jun expression, which would account for the increased AP-1 activity (KINDREGAN et al. 1994). In contrast, in human lung carcinoma Calu-1 cells grown in serum-free medium, activation of AP-1 by RA was associated with growth stimulation (WAN et al. 1997).

4. Effects on Miscellaneous Cell Cycle Proteins

A number of other proteins that affect progression through the cell cycle have been reported to be affected in RA-induced growth arrest. For example, telomerase activity has been associated with progression through S phase of the cell cycle (Zhu et al. 1996). RA treatment down-regulates telomerase activity in HL-60 and NB4 promyelocytic cells with kinetics that parallel the up-regulation of p21 and Rb hypophosphorylation, but is independent of the Rb growth arrest pathway (Savoysky et al. 1996). Resistance to retinoic acid in certain NB4 subclones has been linked to persistent telomerase activity (Nason-Burchenal et al. 1997).

Expression of the H1 degree histone gene, which has an inverse relationship to cell proliferation presumably through its ability to negatively affect DNA synthesis, has been shown to be directly activated by RA or 9-*cis* RA via a DR-8 response element in its promoter that binds a RAR-RXR heterodimer (Bouterfa et al. 1995). Finally, inhibition of thioredoxin reductase by 13-*cis* RA, which is normally increased during S phase, has been associated with inhibition of DNA synthesis in T-cells transformed by human T lymphotrophic virus I (HTLV-1) (U-Taniguchi et al. 1995).

D. Regulation of Apoptosis

I. Apoptotic Mechanisms

Apoptosis is a genetically regulated process of cell death that plays a role in morphogenesis during embryonic development and in the maintenance of homeostasis in many tissues (Wyllie 1997). In addition, alterations in mechanisms controlling apoptosis can result in either premature cell death or resistance to cell death and have been implicated in many diseases, including cancer and certain neurodegenerative and immune disorders (Thompson 1995; Mountz et al. 1996). Treatments designed to induce or inhibit apoptosis have the potential to change the progression of these diseases.

Apoptosis is characterized by specific morphological changes, including cell shrinkage, loss of cell-cell contacts, surface blebbing and condensation of chromatin in bodies marginal to the nuclear membrane. The plasma membrane and organelles retain their integrity initially, but eventually fragment into small vesicles which can be phagocytosed (Hale et al. 1996; Jehn and Osborne 1997; Wyllie 1997; Rowan and Fisher 1997). In most cell types, apoptosis is accompanied by internucleosomal DNA fragmentation and loss of membrane phospholipid asymmetry, characterized by the translocation of phosphatidylserine from the inner layer to the outer layer of the membrane (Vermes et al. 1995).

One common target of signaling molecules that induce apoptosis is a protease cascade involving the family of caspases that cleave specific Asp-X bonds (Patel et al. 1996; Kumar and Lavin 1996). All known members of this family are synthesized as single-chain proenzymes which require proteolytic processing to produce active, heterodimeric proteases. Some of the end targets of the cascade include poly-ADP-ribose polymerase (PARP), α-fodrin, topoisomerase I and nuclear lamins (EARNSHAW 1995; ZHIVOTOVSKY et al. 1997). Release of cytochrome *c* from the mitochondrial intermembrane space into the cytosol has been closely linked to caspase activation (KLUCK et al. 1997; YANG, LM et al. 1997). However, cytochrome c-independent mechanisms of caspase activation have also been described (CHAUHAN et al. 1997).

The bcl-2 family forms an important group of proteins involved in the regulation of apoptosis. Some of these proteins (Bcl-2, Bcl-X_L) protect cells against apoptosis while others promote apoptosis (Bax, Bad, Bak, Bcl-X_S) (REED 1997). Bcl-2 members compete with each other to form homodimers and selective heterodimers; the balance between the various members of the Bcl-2 family is important in the regulation of apoptosis. Bcl-2 is located predominantly in the outer mitochondrial membrane, the endoplasmic reticulum and the nuclear membrane and prevents apoptosis in part by blocking the release of cytochrome c from mitochondria into the cytosol (KHARBANDA et al. 1997; YANG J et al. 1997). Bcl-2 functions as a proto-oncogene because inappropriate expression can inhibit apoptosis and contribute to the malignant phenotype of cancer cells.

Several other proteins have been associated with apoptosis. p53-mediated apoptosis is a well-described response to DNA damage (LEE and BERNSTEIN 1995; GRASSO and MERCER 1997). The precise mechanisms by which p53 induces apoptosis are not fully understood. Recent studies indicate that p53 induces several genes encoding proteins involved in oxidative stress, suggesting that it may increase reactive oxygen species (POLYAK et al. 1997). Another study (WU GS et al. 1997) has shown that p53 induces the receptor for TRAIL (tumor necrosis factor-related apoptosis-inducing ligand), a cytotoxic factor that leads to activation of the caspase cascade. p53 can also repress bcl-2 expression through a p53-dependent negative response element in the bcl-2 promoter (MIYASHITA et al. 1994).

Expression of transglutaminase II (TGase II) has been correlated with apoptosis in a variety of cells in vivo and in vitro (FESUS et al. 1996; SZONDY et al. 1997), but whether it has a causative function in inducing apoptosis is not clear. The strongest evidence that TGase II has an active role in apoptosis comes from expression of sense or antisense cDNA in SK-N-BE(2) neuroblastoma cells which renders these cells highly susceptible or more resistant, respectively, to spontaneous and RA-induced apoptosis (MELINO et al. 1997). It has been proposed that TGase II may crosslink proteins into a scaffold which temporarily stabilizes the integrity of the dying cell and its apoptotic bodies prior to phagocytosis (DAVIES et al. 1992).

II. Regulation of Apoptosis by Retinoids

The ability of retinoids to control embryonic morphogenesis and homeostasis of adult tissues in many instances involves regulation of apoptosis (SANDERS and WRIDE 1995; BOSMAN et al. 1996; COLE and PRASAD 1997). The potent teratogenic effects of retinoids that produce craniofacial, neural tube and limb defects (discussed in detail in Chaps. 15, 16, this volume; ELMAZAR et al. 1996) stem in part from a selective increase or inhibition of apoptosis at inappropriate times during tissue morphogenesis. This is illustrated by studies showing that RA promotes cell death in the interdigital and marginal regions of the mouse limb that is associated specifically with an increase in cdk-5 and its activator p35 (LEE et al. 1994; KOCHHAR et al. 1993; TAMAGAWA et al. 1995; AHUJA et al. 1997a,b). In hammertoe mutant mice in which this cell death is suppressed, administration of RA results in a rectification of the deformity (ZAKERI and AHUJA 1994; AHUJA et al. 1997a). Prevention of retinoid signaling can cause decreased proliferation and increased apoptosis, as has been demonstrated in retina defects of RARβ2/RARγ2 double knock-out mutant mice (GRONDONA et al. 1996) and testicular degeneration in vitamin A-deficient rats (AKMAL et al. 1998). Conversely, retinoid treatment induces apoptosis and causes abnormal epidermal wound healing in zebrafish (FERRETTI and GERAUDI 1995). Retinoids have also an important role in inhibiting the T cell receptor/Fas-induced apoptotic response to T cell activation (IWATA et al. 1992; YANG et al. 1995), which may explain original observations of an enlarged thymus and increased number of thymocytes observed in mice fed a vitamin A-supplemented diet (SEIFTER et al. 1981).

Despite the known apoptotic effects of retinoids in vivo, details of the mechanisms by which they work and the direct targets of retinoid action in regulating apoptosis are not well understood. Control of apoptosis has been intensively studied in a variety of human carcinoma cell lines in which retinoids either induce or inhibit cell death (ZHANG and JETTEN 1997). This action is dependent on the chemical nature of the retinoid and may involve RAR and/or RXR receptors or be mediated by nuclear receptor-independent mechanisms as has been demonstrated for 4-HPR, retroretinoids and AHPN. In a number of instances, retinoids induce apoptosis only in combination with other signaling factors, such as hydrocortisone (LEE PP et al. 1995), interferon γ (KIKUCHI et al. 1996) or vitamin D_3 (ELSTNER et al. 1996).

1. Receptor-Mediated Induction of Apoptosis

RA has been shown to induce apoptosis in a variety of hematopoietic cell lines and in tumor cells from patients with acute promyelocytic leukemia (ORITANI et al. 1992; TOSI et al. 1994). Some significant advances in understanding the molecular events of RA-induced apoptosis and differentiation have been made using the promyelocytic leukemia cell line HL-60 (COLLINS et al. 1990; MARTIN et al. 1990; NAGY et al. 1995). Treatment of HL-60 cells with RAR-selective retinoids induces granulocyte differentiation but does not induce apoptosis, while treatment with RXR-selective retinoids alone has no effect; however, in combination the two retinoids cause apoptosis and induce TGase II (MARTIN et al. 1990; NAGY et al. 1995; BOEHM et al. 1995). Since HL60 cells express only RARα, these findings indicate that induction of differentiation is dependent on RARα-mediated effects, while induction of cell death requires RARα-dependent induction of differentiation as well as activation of RXR. This interpretation is supported by studies using RA-resistant HL-60R cells which harbor a point mutation in the RARα receptor that confers dominant-negative activity. Expression of wild-type RARα in these cells restores both the differentiative and apoptotic responses to RA (Collins et al. 1990; Mehta et al. 1996). IGF-1 has been reported to inhibit RA-induced apoptotic death in HL-60 cells and to stimulate granulocyte differentiation (Liu et al. 1997). This antagonistic effect of IGF-1 appears to depend on increased phosphatidylinositol-3-kinase activity. A recent study has implicated the protein tyrosine kinases c-Lyn and c-Fgr in the regulation of apoptosis and differentiation in HL-60 cells by RA (Katagiri et al. 1996).

RA-induced apoptosis in HL-60 cells is associated with down-regulation of Bcl-2 and Bcl-X_L at both the mRNA and protein level (Benito et al. 1995) and up-regulation of caspase activity, caspase-1 mRNA and caspase-1 and caspase-3 protein levels (William et al. 1997). Bax expression is not altered. Overexpression of Bcl-2 or Bcl-X_L or use of the caspase inhibitors YVAD-CMK and DEVD-CHO inhibited the induction of apoptosis by RA while RA-induced differentiation was not affected (Park et al. 1994; Naumovski and Cleary 1994; Benito et al. 1996; William et al. 1997). The down-regulation of Bcl-2 in HL-60 cells by RA appears to be mediated by RARα, since RA is unable to down-regulate Bcl-2 in HL-60R cells (Grillier et al. 1997). However, no direct mechanism of transregulation of Bcl-2 or caspase genes by retinoid receptors has been demonstrated to date.

Induction of apoptosis in other leukemic cell lines occur by mechanisms different from those in HL-60 cells. RA induces apoptosis in the promyelocytic leukemia cell line PBL-985 without any requirement for differentiation (Monczak et al. 1997). While activation of both RAR and RXR receptors is required for apoptosis and Bcl-2 is down-regulated, no effect on cell cycle or TGase II, Bcl-X, Ich-1, Bax, Bag or Bak expression was observed after RA treatment. In promyelocytic leukemia NB4 cells that carry a t(15:17) rearrangement resulting in the synthesis of a PML-RARα fusion protein, RA induces differentiation rather than apoptosis (Benedetti et al. 1996). Paradoxically, the RA-induced differentiation is associated with down-regulation of Bcl-2. The PML-RARα fusion protein appears to regulate mechanisms that govern the choice between differentiation, cell survival or apoptosis, because expression of PML-RARα blocks differentiation and promotes cell survival in hematopoietic precursor cells (Bruel et al. 1995; Grignani et al. 1993), but results in growth suppression and cell death in fibroblasts and several hematopoietic cell lines (Ferrucci et al. 1997).

In some mammary carcinoma cell lines, growth inhibition by retinoic acid is associated with induction of apoptosis (van der Burgh et al. 1993; Sheikh et al. 1993; Seewaldt et al. 1995; Liu Y et al. 1996; Chen Y et al. 1997). As discussed earlier in this chapter, the responsiveness of mammary carcinoma cells such as MCF-7 to the growth-inhibitory effects of retinoic acid is dependent on the expression of RARβ. Overexpression of RARβ in MCF-7 cells induced apoptotic cell death (SHAO et al. 1995), while expression of antisense RARβ RNA or treatment with an RARβ-selective antagonist blocked RA-induced apoptosis and growth inhibition in the ER-positive cell line ZR-75–1 (LIU Y et al. 1996; SEEWALDT et al. 1997).

Most lung carcinoma cell lines are rather refractory to the growth-inhibitory effects of RA (VOLLBERG et al. 1992; GERADTS et al. 1993; KALEMKERIAN et al. 1994; SUN et al. 1997a; ADACHI et al. 1998a). However, in some lung carcinoma cell lines retinoids cause growth inhibition and apoptosis, including Calu-6 and H460 cells (ZHANG et al. 1995; LI et al. 1998). RA-induced apoptosis in the rat tracheal cell line SPOC-1 is accompanied by an increase in TGase II expression (ZHANG et al. 1995). The induction of both apoptosis and TGase II occurred with RAR-selective, but not RXR-selective, retinoids and was blocked by an RARα-selective antagonist or expression of a dominant-negative RARα receptor, suggesting a specific role for RARα.

RA-induced apoptosis in a number of embryonal carcinoma cell lines, including F9 and P19, is coupled to induction of differentiation and is retinoid receptor-dependent (ATENCIA et al. 1994; HORN et al. 1996; OKAZAWA et al. 1996). Treatment of P19 cells with RA is accompanied by a reduction in Bcl-2 expression while overexpression of Bcl-2 inhibits RA-induced apoptosis (OKAZAWA et al. 1996). Apoptosis in P19 cells is mediated by RAR-selective but not RXR-selective retinoids, while a combination of the two has a synergistic effect (HORN et al. 1996). Expression of a dominant-negative RXR receptor or addition of an RAR antagonist has an inhibitory effect on RA-induced apoptosis as well as differentiation, suggesting that both processes are mediated by retinoid receptors. Receptors also have a role in RA-induced apoptosis in F9 cells, because RXRα and RARα null F9 cells are resistant to RA-induced apoptosis and do not differentiate into parietal endoderm cells but are still able to differentiate into visceral endoderm cells (CHIBA et al. 1997). F9 cells in which both RXRα and RARγ were knocked-out were also resistant to visceral endoderm differentiation.

Studies of apoptosis and growth inhibition in neuroblastoma cell lines have also revealed a role for retinoid receptors (PIACENTINI et al. 1992; HOWARD et al. 1993). RA causes a G_1 growth arrest in neuroblastoma SK-N-BE(2) cells, induction of apoptosis and increased expression of TGase II. Induction of apoptosis requires the presence of both RARα- and RARγ-selective retinoids, while the expression of TGase II can be induced by either ligand alone (PIACENTINI et al. 1992, 1993; MELINO et al. 1994). The induction of apoptosis in neuroblastoma ND7 cells by RA was shown to be enhanced by agents that activate PKC (MAILHOS et al. 1994).

2. Receptor-Mediated Inhibition of Apoptosis

Retinoids have an important role in preventing apoptosis in lymphocytes and thymocytes, thus regulating lymphocyte function and homeostasis in the thymus (IWATA et al. 1992; YANG et al. 1993). Retinoids specifically inhibit apoptosis mediated by activation of the T-cell receptor (TCR/CD3) and do not inhibit the induction of apoptosis by calcium ionophore or phorbol ester (IWATA et al. 1992). TCR-mediated apoptosis is critical for self-reactive T-cell negative selection in the thymus and is mediated by an upregulation of Fas ligand (FasL) expression, a member of the TNF/NGF family. RA prevents this apoptosis by inhibiting FasL synthesis (YANG et al. 1995). Studies with RAR-selective and RXR-selective retinoids demonstrated that optimal inhibition of FasL expression and apoptosis requires the activation of both the RAR and RXR receptor signaling pathways (BISSONNETTE et al. 1995). 9-*cis* RA was shown to be tenfold more potent than RA in inhibiting apoptosis in T cells (YANG et al. 1993). T cells transfected with RXRβ were more sensitive to 9-*cis* RA rescue from activation-induced death; in contrast, cells transfected with a dominant-negative RXRβ could not be rescued from death with 9-*cis* RA (YANG et al. 1995). In contrast, inhibition of cell activation and apoptosis in B-lymphocytes was shown to be mediated by RARs rather than RXRs (LOMO et al. 1998).

3. Retinoid Receptor Independent Mechanisms

Retroretinols and two synthetic retinoids, 4-HPR (fenretinide) and AHPN (also referred to as CD437) (BERNARD et al. 1992), have been shown to influence apoptosis through RAR/RXR-independent mechanisms (see Fig. 1 above for chemical structures). 14-Hydroxy-4,14-retroretinol and 13,14-dihydro-retinol, metabolites of retinol, prevent cell death in a number of cell systems and function as survival factors (O'CONNELL et al. 1996; CHEN et al. 1997). (The actions of these retro retinoids are also discussed in detail in Chap. 3, this volume.)

The ability of 4-HPR and AHPN to inhibit cell growth and induce apoptosis in a large variety of cell types has generated considerable interest in their potential as chemotherapeutic drugs. A number of studies in animals have demonstrated inhibition of tumor formation by 4-HPR in tissues (VILLA et al. 1993; ZHANG and JETTEN 1997), including prostate (POLLARD et al. 1991; LUCIA et al. 1995), breast (ABOU-ISSA et al. 1993; COHEN et al. 1994), lung (MOON et al. 1992) and ovary (FORMELLI and CLERIS 1993). In addition, 4-HPR has been used in clinical trials as a chemopreventative agent for treatment of breast cancer (VERONESI et al. 1992; COBLEIGH 1994), prostate cancer (KIM et al. 1995; PIENTA et al. 1997) and oral leukoplakia (CHIESA et al. 1992; COSTA et al. 1994). AHPN is very effective in inhibiting growth of human lung tumors in an animal model (LU et al. 1997). Both 4-HPR and AHPN inhibit proliferation and induce apoptosis in many cell lines, including a variety of hematopoietic cell lines (DELIA et al. 1993, 1995; HSU et al. 1997a), lymphoma (TRIZNA et al. 1993; CHAN et al. 1997; PIEDRAFITA and PFAHL 1997; ADACHI et al. 1998b), mammary (PELLEGRINI et al. 1995; SHEIKH et al. 1995; KAZMI et al. 1996; CHEN et al. 1997; SHAO et al. 1995; WIDSCHWENDTER et al. 1997; HSU et al. 1997b), lung (KALEMKERIAN et al. 1995; LU et al. 1997; SUN et al. 1997a,b; ADACHI et al. 1998a), ovarian (SUPINO et al. 1996; CHAO et al. 1997; WU et al. 1998), and cervical carcinoma cell lines (ORIDATE et al. 1997a). 4-HPR is also effective in neuroblastoma (MARIOTTI et al. 1994; DI VINCI et al. 1994; PONZONI et al. 1995), prostate adenocarcinoma (IGAWA et al. 1994; HSIEH et al. 1995), esophageal (MULLER et al. 1997), and head and neck squamous cell carcinoma (ORIDATE et al. 1996a), while AHPN has been shown to inhibit growth of melanoma cells (SCHADENDORF et al. 1996; SPANJAARD et al. 1997).

The O-glucuronide analog of 4-HPR has been found to exhibit a greater chemopreventative activity than 4-HPR (CURLEY et al. 1996) while N-(4-methoxyphenyl)retinamide (4-MPR), which is a major metabolite of 4-HPR, was unable to induce apoptosis in mammary and cervical carcinoma cells (SHEIKH et al. 1995; ORIDATE et al. 1997a). Several AHPN-related compounds, such as CD2325, also induce growth arrest and apoptosis (LU et al. 1997).

Both 4-HPR and AHPN bind RARγ, but not RARα or RXRs, and act as moderate to weak RARγselective ligands (SANI et al. 1995; SHEIKH et al. 1995; FANJUL et al. 1996b; KAZMI et al. 1996; BERNARD et al. 1992; SHAO et al. 1995; SUN et al. 1997a; ADACHI et al. 1998a,b). 4-HPR also weakly binds RARβ. Neither retinoid exhibits any antiAP-1 activity. Although 4-HPR and AHPN are able to cause RARE transactivation through RARγ, a consensus appears to be growing that many of their effects on cell growth and apoptosis are mediated by unique mechanisms that do not involve RAR or RXR receptors (SHEIKH et al. 1995; ORIDATE et al. 1997b). This is corroborated by studies demonstrating that selective retinoids that bind RARγ with higher affinity than AHPN, such as TTAB, do not induce apoptosis and do not antagonize the induction of cell death by AHPN (SHAO et al. 1995; ADACHI et al. 1998a,b). The RARγ-selective antagonist CD2665 also fails to inhibit AHPN-induced apoptosis (SUN et al. 1997b). In addition, 4-HPR and AHPN elicit different effects than RA in a variety of cells. Both retinoids induce apoptosis in many squamous cell, cervical, ovarian and small cell lung carcinoma cell lines, most of which are refractory to RA (KALEMKERIAN et al. 1995; SUPINO et al. 1996; ORIDATE et al. 1996a 1997a). Only small cell lung carcinoma lines are sensitive to 4-HPR (KALEMKERIAN et al. 1995), while AHPN inhibits growth in carcinoma cell lines from all lung tumor types (SUN et al. 1997b; ADACHI et al. 1998a). RA promotes neutrophil differentiation and apoptosis in HL-60 cells, whereas 4-HPR only causes apoptosis (DIPIETRANTONIO et al. 1996). 4-HPR also inhibits clonal growth while AHPN induces apoptosis in an RA-resistant clone of HL-60 cells that does not contain functional RARs (BRIGATI 1995; HSU et al. 1997a). Both synthetic retinoids are able to inhibit certain RA-sensitive, ER-positive as well as RA-refractory, ER-negative mammary carcinoma cell lines (SHEIKH et al. 1995; PELLEGRINI et al. 1995; SHAO et al. 1995; CHEN et al. 1997).

a) Mechanisms of 4-HPR Action in Apoptosis

Two different mechanisms have been recently implicated in 4-HPR-induced apoptosis. 4-HPR increases expression of TGF-β_1 mRNA and secretion of active TGF-β_1 in PC-3 prostate adenocarcinoma cells and an antibody against TGF-β_1 abrogates the induction of apoptosis (ROBERSON et al. 1997). Moreover, TGF-β_1 expression is not increased in 4-HPR-resistant mammary carcinoma BT-20 cells. These studies suggest a role for TGF-β_1 in the induction of apoptosis by 4-HPR in these cells. The up-regulation of TGF-β_1 may also explain certain cell cycle effects seen with 4-HPR treatment of prostate cancer or other cells, including arrest at specific stage(s) of the cell cycle and down regulation of regulatory proteins such as cyclins D and E and c-myc (HSIEH et al. 1995; IGAWA et al. 1994; ROBERSON et al. 1997; DIPIETRANTONIO et al. 1996). The induction of growth arrest by 4-HPR in neuroblastoma cells, however, was not associated with a specific stage of the cell cycle (DI VINCI et al. 1994; PONZONI et al. 1995).

In HL-60 and cervical carcinoma cell lines, 4-HPR increases the production of reactive oxygen species within 1.5h of treatment (DELIA et al. 1997; ORIDATE et al. 1997b). Antioxidants such as N-acetylcysteine, ascorbic acid, alpha-tocopherol and pyrrolidine dithiocarbamate inhibit the induction of apoptosis by 4-HPR and suppress the generation of reactive oxygen species (DELIA et al. 1995, 1997; ORIDATE et al. 1997b). The induction of oxygen radicals is a specific property of 4-HPR, since 4-MPR and several RAR and RXR-selective retinoids were unable to induce apoptosis or generate reactive oxygen species. The inhibition of apoptosis by Bcl-2 could at least in part relate to its role in the antioxidant pathway (HOCKENBERY et al. 1993). However, in several cell lines, 4-HPR treatment does not cause changes in Bcl-2 protein levels and overexpression of Bcl-2 delays but does not block apoptosis (DELIA et al. 1995; ORIDATE et al. 1997a). Reactive oxygen species have been implicated in the induction of apoptosis by a number of other agents, including TGF-β (OHBA et al. 1994). Whether the reported induction of TGF-β in prostatic carcinoma cells by 4-HPR results in the generation of reactive oxygen species and has a role in the induction of apoptosis in these cells has yet to be established.

b) Mechanisms of AHPN Action

The mechanism of action of AHPN is clearly distinct from that of 4-HPR. AHPN-induced apoptosis cannot be blocked by the antioxidant glutathione (SPANJAARD et al. 1997), implying that its effects are unrelated to the generation of reactive oxygen species as is the case for 4-HPR.

Several studies have reported that AHPN activates caspase-3-like activity while having little effect on caspase-1-like activity. AHPN-induced caspase activity has been associated with cleavage of PARP and the transcriptional factor Sp1 (HSU et al. 1997a; ADACHI et al. 1998b; PIEDRAFITA and PFAHL 1997). The latter may provide a mechanism by which AHPN alters transcription of certain genes. However, synthesis of new proteins does not appear to be required for AHPN-induced apoptosis. AHPN-induced apoptosis is not universally associated with down regulation of Bcl-2 or Bcl-X(L) (SHAO et al.

1995; Hsu et al. 1997b; Schadendorf et al. 1996; Adachi et al. 1998a). Overexpression of Bcl-2 or treatment of lymphoma cells with the irreversible caspase inhibitor Z-VAD-FMK block caspase-3-like activity, the increase in annexin V binding and DNA fragmentation, and although they slow down the execution of AHPN-induced apoptosis, they do not prevent cell death and do not interfere with the ability of AHPN to cause growth arrest (Adachi et al. 1998b). These results suggest that AHPN-induced cell death involves both caspase-dependent and -independent mechanisms. Figure 3 presents a schematic view of some of the events that are likely involved in AHPN-induced apoptosis.

The relationship between AHPN-induced growth arrest and cell death is not clearly understood. Treatment of mammary and lung carcinoma cell lines with AHPN causes accumulation of cells in the G_1 phase of the cell cycle and is associated with an increase in the hypophosphorylated form of Rb and decreases in cdk activity and cyclin A levels (Shao et al. 1995; Adachi et al. 1998a). In contrast, treatment of T cell lymphoma Jurkat and Molt-4 cells with AHPN causes accumulation of cells in the S-phase of the cell cycle with no decrease in Rb hyperphosphorylation or cyclin A levels (Adachi et al. 1998b). AHPN induces the expression of a number of growth inhibitory proteins including $p21^{WAF1/CIP1}$, p53, MYD118 and GADD45 (Shao et al. 1995; Liu M et al. 1996; Hsu et al. 1997a; Adachi et al. 1998a). Although AHPN increases p53 expression in some cell lines, the induction of apoptosis by AHPN occurs independently of the increase in p21 or p53 protein (Shao et al. 1995; Adachi et al. 1998a). Interestingly, the AHPN-induced increase in the expression of several mRNAs is regulated at a posttranscriptional level and are due to an increase in mRNA stability. The mechanism of this posttranscriptional control

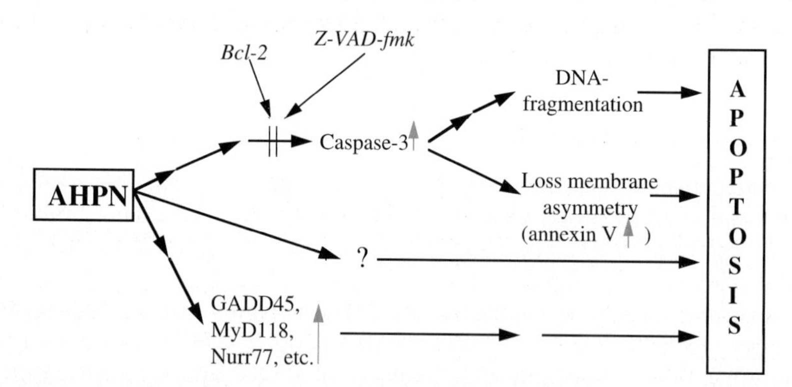

Fig. 3. Pathway of AHPN-induced apoptosis. Induction of caspase-3-like activity plays an important role in the execution of AHPN-induced apoptosis but cells can still undergo cell death in the absence of this activity suggesting that AHPN-induced cell death involves both caspase-dependent and caspase-independent mechanisms. The caspase-independent mechanisms may involve the observed increased expression of GADD45, MyD118, Nur77, or possibly involvement of other cell signaling components

is unknown, but could involve AHPN activation of a specific RNA-binding protein involved in RNA stabilization.

In melanoma cells, AHPN-induced apoptosis was preceded by an increase in AP-1-dependent transactivation that was related to increased expression and appearance of c-fos and c-jun in AP-1/DNA complexes (SCHADENDORF et al. 1996). Whether the increase in AP-1 activity has a causative role in AHPN-mediated apoptosis remains to be established.

E. Summary

Because of their ability to induce terminal differentiation, growth arrest and apoptosis, retinoids have attracted considerable interest for their potential as preventative or therapeutic agents in disease (SAKAI and LANGILLE 1992; FONG et al. 1993; SEGAERT et al. 1997). This function of retinoids may be especially important in the treatment of cancer, where activation of a terminal differentiation program or induction of apoptosis removes cells from the proliferating compartment of the tumor. The synthetic retinoids which mediate apoptosis through apparent nonreceptor mechanisms may prove to be effective in the treatment of cancer without producing the serious side effects associated with RA treatment. The current challenge for biologists has been to determine the specific mechanisms by which retinoids affect these processes. Much still remains to be discovered concerning the direct targets of the various retinoid signaling pathways that regulate cell cycle progression and apoptosis. However, progress is being made and the knowledge gained by these studies is already being used to design more targeted synthetic retinoids that are more effective for treatment of disease and have fewer side effects. Furthermore, understanding the targets of retinoid action may suggest new routes of combination therapy with other cytokines or drugs that promote apoptosis or inhibit cell cycle progression by distinct mechanisms.

References

Abou-Issa H, Curley RW, Panigot MJ, Wilcox KA, Webb TE (1993) In vivo use of N-(4-hydroxyphenyl retinamide)-O-glucuronide as a breast cancer chemopreventative agent. Anticancer Res 13:1431–1436

Achkar CC, Derguini F, Blumberg B, Langston A, Levin AA, Speck J, Evans RM, Bolado J, Nakanishi K, Buck J, Gudas L (1996) 4-Oxo-retinol, a new natural ligand and transactivator of the retinoic acid receptors. Proc Natl Acad Sci USA 93:4879–4884

Adachi H, Preston G, Harvat B, Dawson MI, Jetten AM (1998a) Inhibition of cell proliferation and induction of apoptosis by the retinoid AHPN in human lung carcinoma cellls, Am J Respir Cell Mol Biol 18:323–333

Adachi H, Adams A, Hughes FM, Zhang J, Cidlowski JA, Jetten AM (1998b) Induction of apoptosis by the novel retinoid AHPN in human T cell lymphoma cells involves caspase-dependent lymphoma cells involves caspase-dependent and independent pathways, Cell Death and Diff 5:973–983

Adams PD, Kaelin WGJ (1995) Transcriptional control by E2F, Sem Cancer Biol 6:99–108
Ahuja HS, James W, Zakeri Z (1997a) Rescue of the limb deformity in hammertoe mutant mice by retinoic acid-induced cell death. Dev Dyn 208:466–481
Ahuja HS, Zhu Y, Zakeri Z (1997b) Association of cyclin-dependent kinase 5 and its activator p35 with apoptotic cell death. Dev Genet 21:258–267
Akmal KM, Dufour JM, Vo M, Higginson S, Kim KH (1998) Ligand-dependent regulation of retinoic acid receptor alpha in rat testis: in vivo response to depeletion and repletion of vitamin A, Endocrin 139:1239–1248
Amati B, Land H (1994) Myc-Max-Mad: a transcription factor network controlling cell cycle progression, differentiation and death, Curr Opinion Genetics Dev 4:102–108
Atencia R, Garcia-Sanz M, Unda F, Arechaga J (1994) Apoptosis during RA-induced differentiation of F9 embryonal carcinoma cells. Exp Cell Res 214:663–667
Avis I, Mathias A, Unsworth EJ, Miller MJ, Cuttitta F, Mulshine JL, Jakowlew SB (1995) Analysis of small cell lung cancer cell growth inhibition by 13-*cis*-retinoic acid: importance of bioavailability. Cell Growth Differ 6:485–492
Ayer DE, Kretzner L, Eisenman RN (1993) Mad: a heterodimeric partner for Max that antagonizes Myc transcriptional activity. Cell 72:211–222
Bardon S, Razanamahefa L (1998) Retinoic acid suppresses insulin-induced cell growth and cyclin D1 gene expression in human breast cancer cells, Int J Oncol 12:355–359
Barker KA, Newburger PE (1990) Relationships between the cell cycle and the expression of c-myc and transferrin receptor genes during induced myeloid differentiation. Exp Cell Res 186:1–5
Benedetti L, Grignani F, Scicchitano BM, Jetten AM, Diverio D, Lo Coco F, Avvisati G, Gambacorti-Passerini C, Adamo S, Levin AA, Pelicci PG, Nervi C (1996) Retinoid-induced differentiation of acute promyelocytic leukemia involves PML-RARalpha-mediated increase of type II transglutaminase. Blood 87:1939–1950
Benito A, Grillot D, Nunez G, Fernandez-Luna JL (1995) Regulation and function of Bcl-2 during differentiation-induced cell death in HL-60 promyelocytic cells. Am J Pathol 146:481–490
Benito A, Silva M, Grillot D, Nunez G, Fernandez-Luna JL (1996) Apoptosis induced by erythroid differentiation of human leukemia cell lines is inhibited by Bcl-XL. Blood 87:3837–3843
Bentley DL, Groudine M (1986) A block to elongation is largely responsible for decreased transcription of c-myc in differentiated HL60 cells. Nature 321:702–706
Bergh G, Ehinger M, Olofsson R, Baldertorp B, Johnsson E, Brycke H, Lingren G, Olsson I, Gullberg U (1997) Altered expression of the retinoblastoma tumor-suppressor gene in leukemic cell lines inhibits induction of differentiation but not G1-accumulation. Blood 89:2938–2950
Bernard BA, Bernardon JM, Delecluse C, Martin B, Lenoir MC, Maignan J, Charpentier B, Pilgrim WR, Reichert U, Shroot B (1992) identification of synthetic retinoids with selectivity for human retinoic acid receptor gamma, Biochem Biophys Res Comm 186:977–983
Bissonette RP, Brunner T, Lazarchik SB, Yoo NJ, Boehm MF, Green DR, Heyman RA (1995) 9-*cis* retinoic acid inhibition of activation-induced apoptosis is mediated via regulation of Fas ligand and requires retinoic acid receptor and retinoid X receptor activation. Mol Cell Biol 15:5576–5585
Blomhoff HK, Smeland EB, Erikstein B, Rasmussen AM, Skrede B, Skjonsberg C, Blomhoff R (1992) Vitamin A is a key regulator for cell growth, cytokine production, and differentiation in normal B cells. J Biol Chem 267:23988–23992
Blutt SE, Allegretto EA, Pike JW, Weigel NL (1997) 1,25-dihydroxyvitamin D3 and 9-*cis*-retinoic acid act synergistically to inhibit the growth of LNCaP prostate cells and cause accumulation of cells in G1, Endocrin 138:1491–1497
Bocchia M, Xu Q, Wesley U, Xu Y, Korontsvit T, Loganzo F, Albino AP, Scheinberg DA (1997) Modulation of p53, WAF1/p21 and BCL-2 expression during retinoic acid-induced differentiation of NB4 promyelocytic cells. Leuk Res 21:439–447

Boehm MF, Zhang L, Zhi L, McClurg MR, Berger E, Wagoner M, Mais DE, Suto CM, Davies PJA, Heyman RA, Nadzan AM (1995) Design and synthesis of potent retinoid X receptor selective ligands that induce apoptosis in leukemia cells. J Med Chem 38:3146–3155

Bosman FT, Visser BC, van Oeveren J (1996) Apoptosis: pathophysiology of programmed cell death, Pathol Res Prac 192:676–683

Bouterfa HL, Piedrafita FJ, Doenecke D, Pfahl M (1995) Regulation of H1 (o) gene expression by nuclear receptors through an unusual response element: implications for regulation of cell proliferation. DNA Cell Biol 14:909–919

Brigati C, Ferrari N, Megna M, Roncella S, Cutrona G, Tosetti F, Vidali G (1995) A retinoic acid resistant HL-60 cell clone sensitive to N-(4-hydroxyphenyl) retinamide-mediated clonal growth inhibition. Leuk Lymphoma 17:175–180

Brooks SC III, Kasmer S, Levin AA, Yen A (1996) Myeloid differentiation and retinoblastoma phosphorylation changes in HL-60 cells induced by retinoic acid receptor- and retinoid X receptor-selective retinoic acid analogs. Blood 87:227–237

Brooks SC III, Sturgill R, Choi J, Yen A (1997) An RXR-selective analog attenuates the RAR alpha-selective analog-induced differentiation and non-G1-restricted growth arrest of NB4 cells. Exp Cell Res 234:259–269

Bruel A, Benoit G, De-Nay D, Brown S, Lanotte M (1995) Distinct apoptotic responses in maturation sensitive and resistant t(15;17) acute promyelocytic leukemia NB4 cells. 9-*cis* RA induces apoptosis independent of maturation and Bcl-2 expression, Leukemia. 9:1173–1184

Buck J, Derguini F, Levi E, Nakanishi K, Hammerling U (1991) Intracellular signaling by 14-hydroxy-4,14-retro-retinol. Science 254:1654–1656

Burcin M, Arnold R, Lutz M, Kaiser B, Runge D, Lottspeich F, Filippova GN, Lobanenkov VV, Renkawitz R (1997) Negative protein 1, which is required for function of the chicken lysozyme gene silencer in conjunction with hormone receptors, is identical to the multivalent zinc finger repressor CTCF. Mol Cell Biol 17:1281–1288

Burger C, Wick M, Muller R (1994) Lineage-specific regulation of cell cycle gene expression in differentiating myeloid cells, J Cell Science 107:2047–2054

Busam KJ, Roberts AB, Sporn MB (1992) Inhibition of mitogen-induced c-fos expression in melanoma cells by retinoic acid involves the serum response element. J Biol Chem 267:19971–19977

Caelles C, Gonzalez-Sancho JM, Munoz A (1997) Nuclear hormone receptor antagonism with AP-1 by inhibition of the JNK pathway, Genes Develop 11:3351–3364

Chambon P (1996) A decade of molecular biology of retinoic acid receptors. FASEB J 10:940–954

Chan LN, Zhang S, Shao J, Waikel R, Thompson EA, Chan TS (1997) N-(4-hydroxyphenyl)retinamide induces apoptosis in T lymphoma and T lymphoblastoid leukemia cells. Leuk Lymphoma 25:271–280

Chao W, Hobbs PD, Jong L, Zhang X, Zheng Y, Wu Q, Shroot B, Dawson MI (1997) Effects of receptor class- and subtype-selective retinoids and an apoptosis-inducing retinoid on the adherent growth of the NIH: OVCAR-3 ovarian cancer cell line in culture, Cancer Letters 115:1–7

Chauhan D, Pandey P, Ogata A, Teoh G, Krett N, Halgren R, Rosen S, Kufe D, Kharbanda S, Anderson K (1997) Cytochrome c-dependent and -independent induction of apoptosis in multiple myeloma cells. J Biol Chem 272:29995–29997

Chen JY, Penco S, Ostrowski J, Balaguer P, Pons M, Starrett JE, Reczek P, Chambon P, Gronemeyer H (1995) RAR-specific agonist/antagonists which dissociate transactivation and AP1 transrepression inhibit anchorage-independent cell proliferation, Embo J 14:1187–1197

Chen AC, Guo X, Derguini F, Gudas LJ (1997) Human breast cancer cells and normal mammary epithelial cells: retinol metabolism and growth inhibition by the retinol metabolite 4-oxoretinol, Cancer-Res 57:4642–4651

Chen Y, Derguini F, Buck J (1997) Vitamin A in serum is a survival factor for fibroblasts. Proc Natl Acad Sci USA 94:10205–10208

Cheung B, Hocker JE, Smith SA, Reichert U, Norris MD, Haber M, Stewart BW, Marshall GM (1996) Retinoic acid receptors beta and gamma distinguish retinoid signals for growth inhibition and neuritogenesis in human neuroblastoma cells, Biochem Biophys Res Comm 229:349–354

Chiba H, Clifford J, Metzger D, Chambon P (1997) Specific and redundant functions of retinoid X receptor/retinoic acid receptor heterodimers in differentiation, proliferation, and apoptosis of embryonal carcinoma cells. J Cell Biol 139:735–747

Chiesa F, Tradati N, Marazza M, Rossi N, Boracchi P, Mariani L, Clerici M, Formelli F, Barzan L, Carrassi A, Pastorini A, Camerini T, Giardini R, Zurrida S, Minn FL, Costa A, De Palo G, Veronesi U (1992) Prevention of relapses and new localizations of oral leukoplakias with the synthetic retinoid fenretinide (4-HPR). Preliminary results, Oral Oncol. Eur J Cancer 28B: 97–103

Clurman BE, Sheaff RJ, Thress K, Groudine M, Roberts JM (1996) Turnover of cyclin E by the ubiquitin-proteasome pathway is regulated by cdk2 binding and cyclin phosphorylation. Genes Dev 10:1979–1990

Cobleigh MA (1994) Breast cancer and fenretinide, an analogue of vitamin A. Leukemia 8 Suppl 3: S59–63

Cohen LA, Epstein M, Saa-Pabon V, Meschter C, Zang E (1994) Interactions between 4-HPR and diet in NMU-induced mammary tumorigenesis, Nutr. Cancer 21: 271–283

Cole WC, Prasad KN (1997) Contrasting effects of vitamins as modulators of apoptosis in cancer cells and normal cells: a review, Nutr. Cancer 29:97–103

Collins SJ, Robertson KA, Mueller L (1990) RA-induced granulocytic differentiation of HL-60 myeloid leukemia cells is mediated directly through the RA receptor (RAR-a). Mol Cell Biol 10:2154–2163

Corbeil HB, Whyte P, Branton PE (1995) Characterization of transcription factor E2F complexes during muscle and neuronal differentiation. Oncogene 11:909–920

Costa A, Formelli F, Chiesa F, Decensi A, De Palo G, Veronesi U (1994) Prospects of chemoprevention of human cancers with the synthetic retinoid fenretinide. Cancer Res 54 (Suppl):2032s–2037s

Crestani M, Sadeghpour A, Stroup D, Galli G, Chiang JY (1996) The opposing effects of retinoic acid and phorbol esters converge to a common response element in the promoter of the rat cholesterol 7 alpha-hydroxylase gene (CYP7a). Biochem Biophys Res Comm 225:585–592

Curley RW, Abou-Issa H, Panigot MJ, Clagett-Dame M, Alshafie G (1996) Chemopreventive activities of C-glucuronide/glycoside analogs of retinoid-O-glucuronides against breast cancer development and growth. Anticancer Res 16:757–763

Darwiche N, Scita G, Jones C, Rutberg S, Greenwald E, Tennenbaum T, Collins SJ, DeLuca LM, Yuspa SH (1996) Loss of retinoic acid receptors in mouse skin and skin tumors is associated with activation of the ras(Ha) oncogene and high risk for premalignant progression. Cancer Res 56:4942–4949

Davies PJA, Stein JP, Chiocca EA, Basilion JP, Gentile V, Thomazy V, Fesus L (1992) Retinoid-regulated expression of transglutaminases: links to the biochemistry of programmed cell death. In: (eds) Retinoids in normal development and teratogenesis, Oxford University Press, New York, p 249

Dawson MI, Chao WR, Pine P, Jong L, Hobbs PD, Rudd CK, Quick TC, Niles RM, Zhang XK, Lombardo A, et al (1995) Correlation of retinoid binding affinity to retinoic acid receptor alpha with retinoid inhibition of growth of estrogen receptor-positive MCF-7 mammary carcinoma cells. Cancer Res 55:4446–4451

Dean M, Levine RA, Campisi J (1986) c-myc regulation during retinoic acid-induced differentiation of F9 cells is posttranscriptional and associated with growth arrest. Mol Cell Biol 6:518–524

Delia D, Aiello A, Lombardi L, Pelicci PG, Grignani F, Formelli F, Menard S, Costa A, Veronesi U, Pierotti MA (1993) N-(4-hydroxyphenyl)retinamide induces apoptosis of malignant hemopoietic cell lines including those unresponsive to RA. Cancer Res 53:6036–6041

Delia D, Aiello A, Formelli F, Fontanella E, Costa A, Miyashita T, Reed JC, Pierotti MA (1995) Regulation of apoptosis induced by the retinoid N-(4-hydroxyphenyl) retinamide and effect of deregulated bcl-2. Blood 85:359–367

Delia D, Aiello A, Meroni L, Nicolini M, Reed JC, Pierotti MA (1997) Role of antioxidants and intracellular free radicals in retinamide-induced cell death. Carcinogenesis 18:943–948

del Senno L, Rossi R, Gandini D, Piva R, Franceschetti P, Degli-Uberti EC (1993) Retinoic acid-induced decrease of DNA synthesis and peroxidase mRNA levels in human thyroid cells expressing retinoic acid receptor alpha mRNA. Life Sci 53:1039–1048

Desbois C, Apin B, Guilhot C, Benchaibi M, French M, Ghysdael J, Madjar JJ, Samarut J (1991) v-erbA oncogene abrogates growth inhibition of chicken embryo fibroblasts induced by retinoic acid. Oncogene 6:2129–2135

de Vos S, Dawson MI, Holden S, Le T, Wang A, Cho SK, Chen DL, Koeffler HP (1997) Effects of retinoid X receptor-selective ligands on proliferation of prostate cancer cells. Prostate 32:115–121

Diehl JA, Zindy F, Sherr CJ (1997) Inhibition of cyclin D1 phosphorylation on threonine-286 prevents its rapid degradation via the ubiquitin-proteasome pathway. Genes Dev 11:957–972

DiMartino D, Ponzoni M, Cornaglia-Ferraris P, Tonini GP (1990) Different regulation of mid-size neurofilament and N-myc mRNA expression during neuroblastoma cell differentiation induced by retinoic acid. Cell Mol Neurobiol 10:459–470

Dipietrantonio A, Hsieh TC, Wu JM (1996) Differential effects of retinoic acid (RA) and N-(4-hydroxyphenyl) retinamide (4-HPR) on cell growth, induction of differentiation, and changes in p34cdc2, Bcl-2, and actin expression in the human promyelocytic HL-60 leukemic cells, Biochem Biophys Res Comm 224:837–842

DiVinci A, Geido E, Infusini E, Giaretti W (1994) Neuroblastoma cell apoptosis induced by the synthetic retinoid N-(4-hydroxyphenyl)retinamide. Int J Cancer 59:422–426

Dore BT, Uskokovic MR, Momparler RL (1993) Interaction of retinoic acid and vitamin D3 analogs on HL-60 myeloid leukemic cells. Leuk Res 17:749–757

Doyle LA, Giangiulo D, Hussain A, Park HJ, Yen RW, Borges M (1989) Differentiation of human variant small cell lung cancer cell lines to a classic morphology by retinoic acid. Cancer Res 49:6745–6751

Earnshaw WC (1995) Nuclear changes in apoptosis. Curr Opin Cell Biol 7:337–343

Eliason JF, Kaufmann F, Tanaka T, Tsukaguchi T (1993) Anti-proliferative effects of the arotinoid Ro 40–8757 on human cancer cell lines in vitro. Br J Cancer 67:1293–1298

Elstner E, Linker-Israeli M, Umiel T, Le J, Grillier I, Said J, Shintaku IP, Krajewski S, Reed JC, Binderup L, Koeffler HP (1996) Combination of a potent 20-epi-vitamin D3 analogue (KH 1060) with 9-*cis* retinoic acid irreversibly inhibits clonal growth, decreases bcl-2 expression and induces apoptosis in HL-60 leukemic cells. Cancer Res 56:3570–3576

Elmazar MM, Reichert U, Shroot B, Nau H (1996) Pattern of retinoid-induced teratogenic effects: possible relationship with relative selectivity for nuclear retinoid receptors RARα, RARβ and RARγ, Teratology 53:158–167

Fanjul A, Dawson MI, Hobbs PD, Jong L, Cameron JF, Harlev E, Graupner G, Lu XP, Pfahl M (1994) A new class of retinoids with selective inhibition of AP-1 inhibits proliferation. Nature 372:107–111

Fanjul AN, Bouterfa H, Dawson M, Pfahl M (1996a) Potential role for retinoic acid receptor-gamma in the inhibition of breast cancer cells by selective retinoids and interferons. Cancer Res 56:1571–1577

Fanjul AN, Delia D, Pierotti MA, Rideout D, Qiu J, Pfahl M (1996b) 4-hydroxyphenyl retinamide is a highly selective activator of retinoid receptors. J Biol Chem 271:22441–22446

Ferrari S, Taliafico E, Manfredini R, Grande A, Rossi E, Zucchini P, Torelli G, Torelli U (1992) Abundance of the primary transcript and its processed product of growth-related genes in normal and leukemic cells during proliferation and differentiation. Cancer Res 52:11–16

Ferretti P, Geraudie J (1995) RA-induced cell death in the wound epidermis of regenerating zebrafish fins. Dev Dyn 202:271–283

Ferrucci PF, Grignani F, Pearson M, Fagioli M, Nicoletti I, Pelicci PG (1997) Cell death induction by the acute promyelocytic leukemia-specific PML/RAR alpha fusion protein. Proc Natl Acad Sci USA 94:10901–10906

Fesus L, Madi A, Balajthy Z, Nemes Z, Szondy Z (1996) Transglutaminase induction by various cell death and apoptosis pathways. Experientia 52:942–949

Fong CJ, Sutkowski DM, Braunk EJ, Sherwood ER, Lee C, Kozlowski JM (1993) Effect of retinoic acid on the proliferation and secretory activity of androgen-responsive prostatic carcinoma cells. J Urol 149:1190–1194

Formelli F, Cleris L (1993) Synthetic retinoid fenretinide is effective against human ovarian carcinoma xenograft and potentiates *cis*-platin activity. Cancer Res 53:5374–5376

Frangioni JV, Moghal N, Stuart-Tilley A, Neel BG, Alper SL (1994) The DNA binding domain of retinoic acid receptor beta is required for ligand-dependent suppression of proliferation, J Cell. Science 107:827–838

Gaetano C, Matsumoto K, Thiele CJ (1991) Retinoic acid negatively regulates p34cdc2 expression during human neuroblastoma differentiation. Cell Growth Differ 2:487–493

Geradts J, Chen JY, Russell EK, Yankaskas JR, Nieves L, Minna JD (1993) Human lung cancer cell lines exhibit resistance to retinoic acid treatment, Cell Growth Differ 4:799–809

Gerharz CD, Ramp U, Reinecke P, Ressler U, Dienst M, Marx N, Friebe U, Engers R, Pollow K, Gabbert HE (1996) Effects of TNF-α and retinoic acid on the proliferation of the clear cell and chromophilic types of human renal cell carcinoma in vitro. Anticancer Res 16:1633–1641

Giannini F, Maestro R, Vukosavljevic T, Pomponi F, Boiocchi M (1997) All-*trans*, 13-*cis* and 9-*cis* retinoic acids induce a fully reversible growth inhibition in HNSCC cell lines: implications for in vivo retinoic acid use. Int J Cancer 70:194–200

Giannini G, Dawson MI, Zhang XK, Thiele CJ (1997) Activation of three distinct RXR/RAR heterodimers induces growth arrest and differentiation of neuroblastoma cells. J Biol Chem 272:26693–26701

Giguëre V (1994) RA receptors and cellular retinoid binding proteins: complex interplay in retinoid signaling, Endo Rev 15:61–79

Glass CK, Rose DW, Rosenfeld MG (1997) Nuclear receptor coactivators. Curr Opin Cell Biol 9:222–232

Grignani F, Ferrucci PF, Testa U, Talamo G, Fagioli M, Alcalay M, Mencarelli A, Grignani F, Peschle C, Nicoletti I, Pelicci PG (1993) The acute promyelocytic leukemia-specific PML-RARa Fusion protein inhibits differentiation and promotes survival of meyloid precursor cells. Cell 74:423–431

Grillier I, Umiel T, Elstner E, Collins SJ, Koeffler HP (1997) Alterations of differentiation, clonal proliferation, cell cycle progression and bcl-2 expression in RAR alpha-altered sublines of HL-60. Leukemia 11:393–400

Grasso L, Mercer WE (1997) Pathways of p53-dependent apoptosis. Vitam Horm 53:139–173

Grondona JM, Kastner P, Gansmuller A, Decimo D, Chambon P, Mark M (1996) Retinal dysplasia and degeneration in RARα2/RARγ2 compound mutant mice. Development 122:2173–2188

Gruber JR, Desai S, Blusztajn JK, Niles RM (1995) Retinoic acid specifically increases nuclear PCKa and stimulates AP-1 transcriptional activity in B16 mouse melanoma cells. Exp Cell Res 221:377–384
Guo WX, Gill PS, Antakly T (1995) Inhibition of AIDS-Kaposi's sarcoma cell proliferation following retinoic acid receptor activation. Cancer Res 55:823–829
Hagiwara H, Inoue A, Nakajo S, Nakaya K, Kojima S, Hirose S (1996) Inhibition of proliferation of chondrocytes by specific receptors in response to retinoids, Biochem Biophys Res Comm 222:220–224
Hale AJ, Smith CA, Sutherland LC, Stoneman VE, Longthorne V, Culhane AC, Williams GT (1996) Apoptosis: molecular regulation of cell death. Eur J Biochemistry 236:1–26
Haller H, Lindschau C, Quass P, Distler A, Luft FC (1995) Differentiation of vascular smooth muscle cells and the regulation of protein kinase C-a. Circulation 76:21–29
Hammerling U, Bjelfman C, Pahlman S (1987) Different regulation of N- and c-myc expression during phorbol ester-induced maturation of human SH-SY5Y neuroblastoma cells. Oncogene 2:73–77
Han G, Chang B, Connor MJ, Sidell N (1995) Enhanced potency of 9-*cis* versus all-*trans*-retinoic acid to induce the differentiation of human neuroblastoma cells. Differentiation 59:61–69
Han TH, Lamph WW, Prywes-R (1992) Mapping of epidermal growth factor-, serum-, and phorbol ester-responsive sequence elements in the c-jun promoter. Mol Cell Biol 12:4472–4477
Haskovec C, Lemez P, Neuwirtova R, Wilhelm J, Jarolim P (1990) Differentiation of human myeloid leukemia cell line ML-1 induced by retinoic acid and 1,25-dihydroxyvitamin D3, Neoplasma 37:565–572
Hayashi A, Suzuki T, Tajima S (1995) Modulations of elastin expression and cell proliferation by retinoids in cultured vascular smooth muscle cells, J Biochem Tokyo 117:132–136
Hockenbery DM, Oltvai ZN, Yin XM, Milliman CL, Korsmeyer SJ (1993) Bcl-2 functions in an antioxidant pathway to prevent apoptosis. Cell 75:241–251
Horlein AJ, Naar AM, Heinzel T, Torchia J, Gloss B, Kurokawa R, Ryan A, Kamei Y (1995) Ligand-independent repression by the thyroid hormone receptor mediated by a nuclear receptor co-repressor. Nature 377:397–404
Horn V, Minucci S, Ogryzko VV, Adamson ED, Howard BH, Levin AA, Ozato K (1996) RAR and RXR selective ligands cooperatively induce apoptosis and neuronal differentiation in P19 embryonal carcinoma cells. FASEB J 10:1071–1077
Howard MK, Burke LC, Mailhos C, Pizzey A, Gilbert CS, Lawson WD, Collins MK, Thomas NS, Latchman DS (1993) Cell cycle arrest of proliferating neuronal cells by serum deprivation can result in either apoptosis or differentiation. J Neurochem 60:1783–1791
Hsieh TC, Ng C, Wu JM (1995) The synthetic retinoid N-(4-hydroxyphenyl) retinamide (4-HPR) exerts antiproliferative and apoptosis-inducing effects in the androgen-independent human prostatic JCA-1 cells, Biochem Mol Biol Int 37:499–506
Hsu CA, Rishi AK, Su-Li X, Gerald TM, Dawson MI, Schiffer C, Reichert U, Shroot B, Poirer GC, Fontana JA (1997a) Retinoid induced apoptosis in leukemia cells through a retinoic acid nuclear receptor-independent pathway. Blood 89:4470–4479
Hsu CKA, Rishi AK, Li X, Dawson MI, Reichert U, Shroot B, Fontana JA (1997b) Bcl-XL expression and its downregulation by a novel retinoid in breast carcinoma cells. Exp Cell Res 232:17–24
Hui EK, Yung BY (1993) Cell cycle phase-dependent effect of retinoic acid on the induction of granulocytic differentiation in HL-60 promyelocytic leukemia cells. Evidence for sphinganine potentiation of retinoic acid-induced differentiation. FEBS Lett 318:193–199
Hurlin PJ, Queva C, Koskinen PJ, Steingrimsson E, Ayer DE, Copeland NG, Jenkins NA, Eisenman RN (1996) Mad3 and Mad4: novel Max-interacting transcriptional

repressors that suppress c-myc dependent transformation and are expressed during neural and epidermal differentiation. EMBO J 15:2030

Igawa M, Tanabe T, Chodak GW, Rukstalis DB (1994) N-(4-hydroxyphenyl)retinamide induces cell cycle specific growth inhibition in PC3 cells. Prostate 24:299–305

Imler JL, Wasylyk B (1989) AP1, a composite transcription factor implicated in abnormal growth control, Prog Growth Factor Res 1:69–77

Ishida S, Shudo K, Takada S, Koike K (1994) Transcription from the P2 promoter of human protooncogene myc is suppressed by retinoic acid through an interaction between the E2F element and its binding proteins. Cell Growth Differ 5:287–294

Iwata M, Mukai M, Nakai Y, Iseki R (1992) Retinoic acids inhibit activation-induced apoptosis in T cell hybridomas and thymocytes. J Immunol 149:3302–3308

Jehn BM, Osborne BA (1997) Gene regulation associated with apoptosis. Crit Rev Eukaryot Gene Expr 7:179–193

Jiang H, Lin J, Su ZZ, Collart FR, Huberman E, Fisher PB (1994) Induction of differentiation in human promyelocytic HL-60 leukemia cells activates p21, WAF1/CIP1, expression in the absence of p53. Oncogene 9:3397–3406

Jonk LJ, deJonge ME, Kruyt FA, Mummery CL, van der Saag PT, Kruijer W (1992) Aggregation and cell cycle dependent retinoic acid receptor mRNA expression in P19 embryonal carcinoma cells. Mech Dev 36:165–172

Kalemkerian G P, Jasti RKP, Celano P, Nelkin BD, Mabry M (1994) All-*trans*-retinoic acid alters myc gene expression and inhibits in vitro progression in small cell lung cancer, Cell Growth Differ 5:55–60

Kalemkerian GP, Slusher R, Sakkaraiappan R, Gadgeel S, Mabry M (1995) Growth inhibition and induction of apoptosis by fenretinide in small-cell lung cancer cell lines. J Natl Cancer Inst 87:1674–1680

Kang JX, Li YY, Leaf A (1997) Mannose-6-phosphate/insulin-like growth factor receptor is a receptor for retinoic acid. Proc Natl Acad Sci USA 94:13671–13676

Katagiri K, Yokoyama KK, Yamamoto T, Omura S, Irie S, Katagiri T (1996) Lyn and Fgr protein-tyrosine kinases prevent apoptosis during retinoic acid-induced granulocytic differentiation of HL-60 cells. J Biol Chem 271:11557–11562

Kawa S, Nikaido T, Aiki U, Zhai Y, Kumagaya T, Furihata K, Fujii S and Kiyosawa K (1997) Arotinoid mofarotene (RO40–8757) up-regulates p21 and p27 during growth inhibition of pancreatic cancer cell lines. Int J Cancer 72:906–911

Kazmi SMI, Plante RK, Visconti V, Lau CY (1996) Comparison of N-(4-hydroxyphenyl)retinamide and all-*trans*-retinoic acid in the regulation of retinoid receptor-mediated gene expression in human breast cancer cell lines. Cancer Res 56:1056–1062

Kharbanda S, Pandey P, Schofield L, Israels S, Roncinske R, Yoshida K, Bharti A, Yuan ZM, Saxena S, Weichselbaum R, Nalin C, Kufe D (1997) Role for Bcl-xL as an inhibitor of cytosolic cytochrome C accumulation in DNA damage-induced apoptosis, Proc Natl Acad Sci 94:6939–6942

Kikuchi H, Iizuka R, Sugiyama S, Gon G, Mori H, Arai M, Mizumoto K, Imajoh-Ohmi S (1996) Monocytic differentiation modulates apoptotic response to cytotoxic anti-Fas antibody and tumor necrosis factor alpha in human monoblast U937 cells. Differentiation 60:778–783

Kim JH, Tanabe T, Chodak GW, Rukstalis DB (1995) In vitro anti-invasive effects of N-(4-hydroxyphenyl)-retinamide on human prostatic adenocarcinoma, Anticancer-Res 15:1429–1434

Kindregan HC, Rosenbaum SE, Ohno S, Niles RM (1994) Characterization of conventional protein kinase C (PKC) isotype expression during F9 teratocarcinoma differentiation. Overexpression of PKCa alters the expression of some differentiation-dependent genes. J Biol Chem 269:27756–27761

Kizaki M, Ikeda Y, Tanosaki R, Nakajima H, Morikawa M, Sakashita A, Koeffler HP (1993) Effects of novel retinoic acid compound, 9-*cis*-retinoic acid, on proliferation, differentiation, and expression of retinoic acid receptor-alpha and retinoid X receptor-alpha RNA by HL-60 cells. Blood 82:3592–3599

Kizaki M, Dawson MI, Heyman R, Elster E, Morosetti R, Pakkala S, Chen DL, Ueno H, Chao W, Morikawa M, Ikeda Y, Heber D, Phahl M, Koeffler HP (1996) Effects of novel retinoid X receptor-selective ligands on myeloid leukemia differentiation and proliferation in vitro. Blood 87:1977–1984

Kluck RM, Bossy-Wetzel E, Green DR, Newmeyer DD (1997) The release of cytochrome c from mitochondria: a primary site for Bcl-2 regulation of apoptosis. Science 275:1132–1136

Kochhar DM, Jiang H, Harnish DC, Soprano DR (1993) Evidence that RA-induced apoptosis in the mouse limb bud core mesenchymal cells is gene-mediated. Prog Clin Biol Res 383B: 815–825

Kolla V, Lindner DJ, Ziao W, Borden EC, Kalvakolanu DV (1996) Modulation of interferon (IFN)-inducible gene expression by retinoic acid. Up-regulation of STAT1 protein in IFN-unresponsive cells. J Biol Chem 271:10508–10514

Kumar S, Flavin MF (1996) The ICE family of cysteine proteases as effectors of cell death. Death Diff 3:255–267

Langenfeld J, Lonardo F, Kiyokawa H, Passalaris T, Ahn M, Rusch V, Dmitrovsky E (1996) Inhibited transformation of immortalized human bronchial epithelial cells by retinoic acid is linked to cyclin E down-regulation. Oncogene:1983–1989

Langenfeld J, Kiyokawa H, Sekula D, Boyle J, Dmitrovsky E (1997) Posttranslational regulation of cyclin D1 by retinoic acid: a chemoprevention mechanism. Proc Natl Acad Sci USA 94:12070–12074

Larsson LG, Pettersson M, Oberg F, Nilsson K, Luscher B (1994) Expression of mad, mxi1, max and c-myc during induced differentiation of hematopoietic cells: opposite regulation of mad and c-myc. Oncogene 9:1247–1252

Lee KK, Li FC, Yung WT, Kung JL, Ng JL, Cheah-KS (1994) Influence of digits, ectoderm, and retinoic acid on chondrogenesis by mouse interdigital mesoderm in culture. Dev Dyn 201:297–309

Lee JM, Bernstein A (1995) Apoptosis, cancer and the p53 tumour suppressor gene. Cancer Metastasis Rev 14:149–161

Lee PP, Darcy KM, Shudo K, Ip MM (1995) Interaction of retinoids with steroid and peptide hormones in modulating morphological and functional differentiation of normal rat mammary epithelial cells. Endocrinology 136:1718–1730

Lee X, Si SP, Tsou HC, Peacocke M (1995) Cellular aging and transformation suppression: a role for retinoic acid receptor beta 2. Exp Cell Res 218:296–304

Li Y, Dawson MI, Agadir A, Lee M-O, Jong L, Hobbs PD, Zhang X (1998) Regulation of RARα expression by RAR- and RXR-selective retinoids in human lung cancer cell lines: effect on growth inhibition and apoptosis induction. Int J Cancer 75:88–95

Liu M, Iavarone A, Freedman LP (1996) Transcriptional activation of the human p21WAF1/CIP1 gene by retinoic acid receptor. Correlation with retinoid induction of U937 cell differentiation. J Biol Chem 271:31723–31728

Liu Q, Schacher D, Hurth C, Freund GG, Dantzer R, Kelley KW (1997) Activation of phosphatidylinositol 3'-kinase by insulin-like growth factor-I rescues promyeloid cells from apoptosis and permits their differentiation into granulocytes. J Immunol 159:829–837

Liu Y, Lee MO, Wang HG, Li Y, Hashimoto Y, Klaus M, Reed JC, Zhang X (1996) Retinoic acid receptor beta mediates the growth-inhibitory effect of retinoic acid by promoting apoptosis in human breast cancer cells. Mol Cell Biol 16:1138–1149

Lomo J, Smeland EB, Ulven S, Natarajan V, Blomhoff R, Gandhi U, Dawson MI, Blomhoff HK (1998) RAR-, not RXR, ligands inhibit cell activation and prevent apoptosis in B-lymphocytes. J Cell Physiol 175:68–77

Lotan R (1994) Suppression of squamous cell carcinoma growth and differentiation by retinoids, Cancer Res (Suppl) 54:1987s-1990s

Lotan R (1996) Retinoids and their receptors in modulation of differentiation, development, and prevention of head and neck cancers. Anticancer Res 16:2415–2419

Louvet C, Empereur S, Fagot D, Forgue-Lafitte E, Chastre E, Zimber A, Mester J, Gespach C (1994) The arotinoid Ro 40–8757 has antiproliferative effects in drug-resistant human colon and breast cancer cell lines in vitro. Cancer Lett 85:83–86
Louvet C, Djelloul S, Forgue-Lafitte ME, Mester J, Zimber A, Gespach C (1996) Antiproliferative effects of the arotinoid Ro 40–8757 in human gastrointestinal and pancreatic cancer cell line: combinations with 5-fluorouracil and interferon α. Br J Cancer 74:394–399
Lucia MS, Anzano MA, Slayer MV, Anver MR, Gren DM, Shrader MW, Logsdon DL, Driver CL, Brown CC, Peer CW (1995) Chemopreventive activity of tamoxifen, N-(4-hydroxyphenyl) retinamide, and the vitamin D analogue Ro24–5531 for androgen-promoted carcinomas of the rat seminal vesicle and prostate. Cancer Res 55:5621–5627
Lu XP, Fanjul A, Picard N, Pfahl M, Rungta D, Nared-Hood K, Carter B, Piedrafita J, Tang S, Fabbrizio E, Pfahl M (1997) Novel retinoid-related molecules as apoptosis inducers and effective inhibitors of human lung cancer cells in vivo. Nat Med 3:686–690
Mailhos C, Howard MK, Latchman DS (1994) A common pathway mediates RA and PMA-dependent programmed cell death (apoptosis) of neuronal cells. Brain Res 644:7–12
Mangelsdorf DJ, Umesono K, Evans RM (1994) The retinoid receptors. In: Sporn MB, Roberts AB, Goodman DS (eds) The retinoids. Raven, New York p319–350
Mangiarotti R, Danova M, Alberici R, Pellicciari C (1998) All-*trans* retinoic acid (ATRA)-induced apoptosis is preceded by G1 growth arrest in human MCF-7 breast cancer cells, Brit J. Cancer Res 77:186–191
Mariotti A, Marcora E, Bunone G, Costa A, Veronesi U, Pierotti MA, Della Valle G (1994) N-(4-hydroxyphenyl)retinamide: a potent inducer of apoptosis in human neuroblastoma cells. J Natl Cancer Inst 86:1245–1247
Martin CA, Ziegler LM, Napoli JL (1990) Enhanced expression of c-fos is not obligatory for retinoic acid-induced F9 cell differentiation, Mol Cell Endo 71:27–31
Matikainen S, Ronni T, Hurme M, Pine R, Julkunen I (1996) Retinoic acid activates interferon regulatory factor-1 gene expression in myeloid cells. Blood 88:114–123
Matsuo T, Thiele CJ (1998) p27^{Kip1}: a key mediator of retinoic acid induced growth arrest in the SMS-KCNR human neuroblastoma cell line, Oncogene (in press)
Maxwell SA (1994) Retinoblastoma protein in non-small cell lung carcinoma, cells arrested for growth by retinoic acid. Anticancer Res 14:1535–1540
McCormack SA, Viar MJ, Tague L, Johnson LR (1996) Altered distribution of the nuclear receptor RAR beta accompanies proliferation and differentiation changes caused by retinoic acid in Caco-2 cells, In Vitro Cell. Dev Biol 32:53–61
McMahon SB, Monroe JG (1992) Role of primary response genes in generating cellular responses to growth factors. FASEB J 6:2707–2715
Mehta K, McQueen T, Neamata N, Collins S, Andreeff M (1996) Activation of retinoid receptors RARα and RXRα induces differentiation and apoptosis, respectively in HL-60 cells. Cell Growth Differ 7:179–186
Melino G, Annicchiarico-Petruzzelli M, Piredda L, Candi E, Gentile V, Davies PJ, Piacentini M (1994) Tissue transglutaminase and apoptosis: sense and antisense transfection studies with human neuroblastoma cells. Mol Cell Biol 14:6584–6596
Melino G, Thiele CJ, Knight RA, Piacentini M (1997) Retinoids and the control of growth/death decisions in human neuroblastoma cell lines, J Neurooncol 31:65–83
Miano JM, Topouzis S, Majesky MW, Olson EN (1996) Retinoid receptor expression and all-*trans* retinoic acid-mediated growth inhibition in vascular smooth muscle cells. Circulation 93:1886–1895
Miller WH Jr, Moy D, Li A, Grippo JF, Dmitrovsky E (1990) Retinoic acid induces down-regulation of several growth factors and proto-oncogenes in a human embryonal cancer cell line. Oncogene 5:511–517
Miyashita T, Harigai M, Hanada M, Reed JC (1994) Identification of a p53-dependent negative response element in the bcl-2 gene. Cancer Res 54:3131–3135

Moghal N, Neel BG (1995) Evidence for impaired retinoic acid receptor-thyroid hormone receptor AF-2 cofactor activity in human lung cancer. Mol Cell Biol 15:3945–3959

Monczak Y, Trudel M, Lamph WW, Miller WH (1997) Induction of apoptosis without differentiation by retinoic acid in PBL-985 cells requires the activation of both RAR and RXR. Blood 90:3345–3355

Moon RC, Mehta RG, Detrisac CJ (1992) Retinoids as chemopreventive agents for breast cancer, Cancer Detec Prev 16:73–79

Motzer RJ, Schwartz L, Law TM, Murphy BA, Hoffman AD, Albino AP, Vlamis B, Nanus DM (1995) Interferon a2a and 13-*cis*-retinoic acid in renal cell carcinoma: antitumor activity in a phase II trial and interactions in vitro. J Clin Oncol 13:1950–1957

Mountz JD, Zhou T, Su X, Wu J, Cheng J (1996) The role of programmed cell death as an emerging new concept for the pathogenesis of autoimmune diseases. Clin Immunol Immunopathol 80: S2-S14

Muller A, Nakagawa H, Rustgi AK (1997) Retinoic acid and N-(4-hydroxyphenyl) retinamide suppress growth of esophageal squamous carcinoma cell lines. Cancer Lett 113:95–101

Nagy L, Thomazy VA, Shipley GL, Fesus L, Lamph W, Heyman RA, Chandraratna RA, Davies PJA (1995) Activation of retinoid X receptors induces apoptosis in HL-60 cell lines. Mol Cell Biol 15:3450–3551

Naka K, Yokozaki H, Domen T, Hayashi K, Kuniyasu H, Yasui W, Yasui W, Lotan R, Tahara E (1997) Growth inhibition of cultured human gastric cancer coils by 9-*cis*-retinoic acid with induction of cdk inhibitor Waf1/Cip1/Sdi1/p21 protein. Differentiation 61:313–320

Nason-Burchenal K, Maerz W, Albanell J, Allopenna J, Martin P, Moore MAS, Dmitrovsky E (1997) Common defects of different retinoic acid resistant promyelocytic leukemia cells are persistent telomerase activity and nuclear body disorganization. Differentiation 61:321–331

Naumovski L, Cleary ML (1994) Bcl2 inhibits apoptosis associated with terminal differentiation of HL-60 myeloid leukemia cells. Blood 83:2261–2267

Nervi C, Vollberg TM, George MD, Zelent A, Chambon P, Jetten AM (1991) Expression of nuclear retinoic acid receptors in normal tracheobronchial cells and lung carcinoma cells. Exp Cell Res 195:163–170

Obeid LM, Hannun YA (1995) Ceramide: a stress signal and mediator of growth suppression and apoptosis. J Cell Biochemistry 58:191–198

O'Connell MJ, Chua R, Hoyos B, Buck J, Chen Y, Derguini F, Hammerling U (1996) Retro-retinoids in regulated cell growth and death. J Exp Med 184:549–555

Ohba M, Shibanuma M, Kuroli T, Nose K (1994) Production of hydrogen peroxide by transforming growth factor-β1 and its involvement in induction of erg-1 in mouse osteoblastic cells. J Cell Biol 26:1079–1088

Okazawa H, Shimizu J, Kamei M, Imafuku I, Hamada H, Kanazawa I (1996) Bcl-2 inhibits retinoic acid-induced apoptosis during the neural differentiation of embryonal stem cells. J Cell Biol 132:955–968

Oridate N, Lotan D, Xu XC, Hong WK, Lotan R (1996a) Differential induction of apoptosis by all-*trans* retinoic acid and N-(4-hydroxyphenyl) retinamide in human head and neck squamous cell carcinoma cell lines. Clin Cancer Res 2:855–863

Oridate N, Esumi N, Lotan D, Hong WK, Rochette-Egly C, Chambon P, Lotan R (1996b) Implication of retinoic acid receptor gamma in squamous differentiation and response to retinoic acid in head and neck SqCC/Y1 carcinoma cells. Oncogene 12:2019–2028

Oridate N, Higuchi M, Suzuki S, Shroot B, Hong WK, Lotan R (1997a) Rapid induction of apoptosis in human C33 a cervical carcinoma cells by the synthetic retinoid 6-[3-(1-adamantyl)hydroxyphenyl]-2-naphtalene carboxylic acid (CD437). Int J Cancer 70:484–487

Oridate N, Suzuki S, Higuchi M, Mitchell MF, Hong WK, Lotan R (1997b) Involvement of reactive oxygen species in N-(4-hydroxyphenyl)retinamide-induced apoptosis in cervical carcinoma cells. J Natl Cancer Inst 89:1191–1198

Oritani K, Kaisho T, Nakajima K, Hirano T (1992) RA inhibits interleukin-6-induced macrophage differentiation and apoptosis in a murine hematopoietic cell line, Y6. Blood 80:2298–2305

Ou X, Campau S, Slusher R, Jasti RK, Mabry M, Kalemkerian GP (1996) Mechanism of all-*trans*-retinoic acid-mediated L-myc gene regulation in small cell lung cancer. Oncogene 13:1893–1899

Pagano M, Tam SW, Theodoras AM, Beer-Romero P, Del Sal G, Chau V, Yew PR, Draetta GF, Rolfe M (1995) Role of the ubiquitin-proteasome pathway in regulating abundance of the cyclin-dependent kinase inhibitor p27. Science 269:682–685

Park JR, Robertson K, Hickstein DD, Tsai S, Hockenbery DM, Collins SJ (1994) Dysregulated bcl-2 expression inhibits apoptosis but not differentiation of retinoic acid-induced HL-60 granulocytes. Blood 84:440–445

Patel T, Gores GJ, Kaufmann SH (1996) The role of proteases during apoptosis. FASEB J 10:587–597

Pellegrini R, Mariotti A, Tagliabue E, bressan R, Bunone G, Coradini D, Della Valle G, Formelli F, Cleris L, Radice P (1995) Modulation of markers associated with tumor aggressiveness in human breast cancer cell lines by N-(4-hydroxyphenyl)retinamide, Cell Growth Diff 6:863–869

Perez-Roger I, Solomon DL, Sewing A, Land H (1997) Myc activation of cyclin E/cdk2 kinase involves induction of cyclin E gene transcription and inhibition of p27(Kip1) binding to newly formed complexes. Oncogene 14:2373–2381

Piacentini M, Annicchiarico-Petruzzelli M, Oliverio S, Piredda L, Biedler JL, Melino E (1992) Phenotype-specific "tissue" transglutaminase regulation in human neuroblastoma cells in response to retinoic acid: correlation with cell death by apoptosis. Int J Cancer 52:271–278

Piacentini M, Fesus L, Melino G (1993) Multiple cell cycle access to the apoptotic death programme in human neuroblastoma cells. FEBS Lett 320:150–154

Piedrafita FJ, Pfahl M (1997) Retinoid-induced apoptosis and Sp1 cleavage occur independently of transcription and require caspase activation. Mol Cell Biol 17:6348–6358

Pienta KJ, Esper PS, Zwas F, Krzeminski R, Flaherty LE (1997) Phase II chemoprevention trial of oral fenretinide in patients at risk for adenocarcinoma of the prostate, Am J Clinic Oncol 20:36–39

Plum LA, Clagett-Dame M (1995) 9-*cis*-retinoic acid selectively activates the cellular retinoic acid binding protein-II gene in human neuroblastoma cells. Arch Biochem Biophys 319:457–463

Pollard M, Luckert PH, Sporn MB (1991) Prevention of primary prostate cancer by 4-(hydroxyphenyl)retinamide. Cancer Res 51:3610–3615

Polyak K, Xia Y, Zweier JL, Kinzler KW, Vogelstein B (1997) A model for p53-induced apoptosis. Nature 389:300–305

Pomponi F, Cariati R, Zancai P, DePaoli P, Rizzo S, Tedeschi RM, Pivetta B, DeVita S, Boiocchi M, Dolcetti R (1996) Retinoids irreversibly inhibit in vitro growth of Epstein-Barr virus-immortalized B lymphocytes. Blood 88:3147–3159

Ponzoni M, Lucarelli E, Corrias MV, Cornaglia-Ferraris P (1993) Protein kinase C isoenzymes in human neuroblasts. Involvement of PKC epsilon in cell differentiation. FEBS Lett 322:120–124

Ponzoni M, Bocca P, Chiesa V, Decensi A, Pistoia V, Raffaghello L, Rozzo C, Montaldo PG (1995) Differential effects of N-(4-hydroxyphenyl)retinamide and RA on neuroblastoma cells: apoptosis versus differentiation. Cancer Res 55:853–861

Prasad KN, Cohrs RJ, Sharma OK (1990) Decreased expressions of c-myc and H-ras oncogenes in vitamin E succinate induced morphologically differentiated murine B-16 melanoma cells in culture. Biochem Cell Biol 68:1250–1255

Preisler HD, Raza A, Larson RA (1992) Alteration of the proliferative rate of acute myelogenous leukemia cells in vivo in patients. Blood 80:2600–2603

Ramp U, Gerharz CD, Engers R, Marx N, Gabbert HE (1995) Differentiation induction in the human rhabdomyosarcoma cell line TE-671. A morphological, biochemical and molecular analysis. Anticancer Res 15:181–188

Reed JC (1997) Bcl-2 family proteins and the hormonal control of cell life and death in normalcy and neoplasia. Vitam Horm 53:99–138

Reichel RR (1992) Regulation of E2F/cyclin A-containing complex upon retinoic acid-induced differentiation of teratocarcinoma cells. Gene Expr 2:259–271

Riboni L, Prinetti A, Bassi R, Caminiti A, Tettamanti G (1995) A mediator role of ceramide in the regulation of neuroblastoma Neuro2a cell differentiation. J Biol Chem 270:26868–26875

Rishi AK, Gerald TM, Shao ZM, Li XS, Baumann RG, Dawson MI, Fontana JA (1996) Regulation of the human retinoic acid receptor alpha gene in the estrogen receptor negative human breast carcinoma cell lines SKBR-3 and MDA-MB-435. Cancer Res 56:5246–5252

Roberson KM, Penland SN, Padilla GM, Selvan RS, Kim CS, Fine RL, Roberson CN (1997) Fenretinide: induction of apoptosis and endogenous transforming growth factor beta in PC-3 prostate cancer cells. Cell Growth Differ 8:101–111

Roman SD, Ormandy CJ, Manning DL, Blamey RW, Nicholson RI, Sutherland RL, Clarke CL (1993) Estradiol induction of retinoic acid receptors in human breast cancer cells. Cancer Res 53:5940–5945

Rosenstraus MJ, Sundell CL, Liskay RM (1982) Cell-cycle characteristics of undifferentiated and differentiating embryonal carcinoma cells. Dev Biol 89:5116–5120

Roussel ME, Ashmun RA, Sherr CJ, Eisenman RN, Ayer DE (1996) Inhibition of cell proliferation by the Mad1 transcriptional repressor. Mol Cell Biol 16:2796–2801

Rowan S, Fisher DE (1997) Mechanisms of apoptotic cell death. Leukemia 11:457–465

Rubin M, Fenig E, Rosenauer A, Menendez-Botet, Achkar C, Bentel JM, Yahalom J, Mendelsohn J, Miller WH (1994) 9-Cis retinoic acid inhibits growth of breast cancer cells and down-regulates estrogen receptor RNA and protein. Cancer Res 54:6549–6556

Saatcioglu F, Claret FX, Karin M (1994) Negative transcriptional regulation by nuclear receptors, Sem Cancer Biol 5:347–359

Sakai A, Langille RM (1992) Differential and stage dependent effects of retinoic acid on chondrogenesis and synthesis of extracellular matrix macromolecules in chick craniofacial mesenchyme in vitro. Differentiation 52:19–32

Sanders EJ, Wride MA (1995) Programmed cell death in development. Int Rev Cytol 163:105–173

Sani BP, Shealy YF, Hill DL (1995) N-(4-hydroxyphenyl)retinamide: interactions with retinoid-binding proteins/receptors. Carcinogenesis 16:2531–2534

Sapin V, Dolle P, Hindelang C, Kastner P, Chambon P (1997) Defects of the chorioallantoic placenta in mouse RXRalpha null fetuses. Dev Biol 191:29–41

Saunders DE, Hannigan JH, Zajac CS, Wappler NL (1995) Reversal of alcohol's effects on neurite extension and on neuronal GAP43/B50, N-myc, and c-myc protein levels by retinoic acid. Brain Res Dev Brain Res 86:16–23

Savoysky E, Yoshida K, Ohtomo T, Yamaguchi Y, Akamatsu K, Yamazaki T, Yoshida S, Tsuchiya M (1996) Down-regulation of telomerase activity is an early event in the differentiation of HL60 cells, Biochem Biophys Res Comm 226:329–334

Schadendorf D, Worm M, Jurgovsky K, Dippel E, Reichert U, Czarnetzki BM (1995) Effects of various synthetic retinoids on proliferation and immunphenotype of human melanoma cells in vitro. Recent Results Cancer Res 139:183–193

Schadendorf D, Kern MA, Artuc M, Pahl HL, Rosenbach T, Fichtner I, Nürnberg W, Stüting S, Stebut EV, Worm M, Makki A, Jurgovsky K, Kolde G, Henz BM (1996) Treatment of melanoma cells with the synthetic retinoid CD437 induces apoptosis via activation of AP-1 in vitro, and causes growth inhibition in xenografts in vivo. J Cell Biol 135:1889–1898

Schilthuis JG, Gann AA, Brockes JP (1993) Chimeric retinoic acid/thyroid hormone receptors implicate RARα 1 as mediating growth inhibition by retinoic acid. EMBO J 12:3459–3466

Schroen DJ, Brinckeroff CE (1996) Inhibition of rabbit collagenase (matrix metalloproteinase-1; MMP-1) transcription by retinoid receptors: evidence for binding of RARs/RXRs to the –77 AP-1 site through interactions with c-jun. J Cell Physiol 169:320–332

Schle R, Rangarajan P, Yang N, Kliewer S, Ransone LJ, Bolado J, Verma IM, Evans RM (1991) Retinoic acid is a negative regulator of AP-1-responsive genes. Proc Natl Acad Sci USA 88:6092–6096

Seewaldt VL, Johnson BS, Parker MB, Collins SJ, Swisshelm K (1995) Expression of RA receptor beta mediates RA-induced growth arrest and apoptosis in breast cancer cells. Cell Growth Differ 6:1077–1088

Seewaldt V, Kim JH, Caldwell E, Johnson B, Swisshelm K, Collins S (1997) All-*trans*-retinoic acid mediates G1 arrest but not apoptosis of normal human mammary epithelial cells. Cell Growth Differ 8:631–641

Segaert S, Garmyn M, Degreef H, Bouillon R (1997) Retinoic acid modulates the antiproliferative effect of 1,25-dihydroxyvitamin D3 in cultured human epidermal keratinocytes. J Invest Dermatol 109:46–54

Seifter EG, Rettura G, Levenson SM (1981) Decreased resistance of C3H/HeHa mice to C3HBA tumor transplants: increased resistance due to supplemental vitamin A. J Natl Cancer Inst 67:467–474

Shao Z, Dawson MI, Li XS, Rishi AK, Sheikh MS, Han Q, Ordonez JV, Shroot B, Fontana JA (1995) p53 independent G0/G1 arrest and apoptosis induced by a novel retinoid in human breast cancer cells. Oncogene 11:493–504

Sheikh MS, Shao ZM, Chen JC, Ordonez JV, Fontana JA (1993) Retinoid modulation of c-myc and max gene expression in human breast carcinoma. Anticancer Res 13:1387–1392

Sheikh MS, Shao ZM, Li XS, Dawson M, Jetten AM, Wu S, Conley BA, Garcia M, Rochefort H, Fontana JA (1994) Retinoid-resistant estrogen receptor-negative human breast carcinoma cells transfected with retinoic acid receptor-alpha acquire sensitivity to growth inhibition by retinoids. J Biol Chem 269:21440–21447

Sheikh MS, Shao ZM, Li XS, Ordonez JV, Conley BA, Wu S, Dawson MI, Han QX, Chao WR, Quick T, et-al (1995) N-(4-hydroxyphenyl)retinamide (4-HPR)-mediated biological actions involve retinoid receptor-independent pathways in human breast carcinoma. Carcinogenesis 16:2477–2486

Sherbet GV, Lakshmi MS (1997) Retinoid and growth factor signal transduction. In: Sherbet GV (ed) Retinoids: their physiological function and their therapeutic potential, Adv in Organ Biology. vol 3. JAI, Greenwich, CT p 141

Sherr CJ, Roberts JM (1995) Inhibitors of mammalian G1 cyclin-dependent kinases. Genes Dev 9:1149–1163

Shibata H, Spencer TE, Onata SA, Jenster G, Tsai SY, Tsai MJ, O'Malley BW (1997) Role of co-activators and co-repressors in the mechanism of steroid/thyroid receptor action, Rec Progr. Horm Res 52:141–165

Si SP, Lee X, Tsou HC, Buchsbaum R, Tibaduiza E, Peacocke M (1996) RAR beta 2-mediated growth inhibition in HeLa cells. Exp Cell Res 223:102–111

Simonson MS (1994) Anti-AP-1 activity of all-*trans* retinoic acid in glomerular mesangial cells. Am J Physiol 267: F805-F815

Slack RS, Hamel PA, Bladon TS, Gill RM, McBurney MW (1993) Regulated expression of the retinoblastoma gene in differentiating carcinoma cells. Oncogene 8:1585–1591

Spanjaard RA, Ikeda M, Lee PJ, Charpentier B, Chin WW, Eberlein TJ (1997) Specific activation of retinoic acid receptors (RARs) and retinoic X receptors reveals a unique role for RARgamma in induction of differentiation and apoptosis of S91 melanoma cells. J Biol Chem 272:18990–18999

Sun SY, Yue P, Dawson MI, Shroot B, Michel S, Lamph WW, Heyman RA, Teng M, Chandraratna RA, Shudo K, Hong WK, Lotan R (1997) Differential effects of synthetic nuclear retinoid receptor-selective retinoids on the growth of human non-small cell carcinoma cells. Cancer Res 57:4931–4939

Sun SY, Yue P, Shroot B, Hong WK, Lotan R (1997) Induction of apoptosis in human non-small cell lung carcinoma cells by the novel synthetic retinoid CD-437. J Cell Physiol 173:279–284

Supino R, Crosti M, Clerici M, Warlters A, Cleris L, Zunino F, Formelli F (1996) Induction of apoptosis by fenretinide (4HPR) in human ovarian carcinoma cells and its association with retinoic acid receptor expression. Int J Cancer 65:491–497

Swisshelm K, Ryan K, Lee X, Tsou HC, Peacocke M, Sager R (1994) Down-regulation of retinoic acid receptor α in mammary carcinoma cell lines and its up-regulation in senescing normal mammary epithelial cells. Cell Growth Differ 5:133–141

Szondy Z, Reichert U, Bernardon JM, Michel S, Toth R, Ancian P, Ajzner E, Fesus L (1997) Induction of apoptosis by retinoids and retinoic acid receptor gamma-selective compounds in mouse thymocytes through a novel apoptosis pathway. Mol Pharmacol 51:972–982

Tamagawa M, Morita J, Naruse I (1995) Effects of all-*trans*-retinoic acid on limb development in the genetic polydactyly mouse. J Toxicol Sci 20:383–393

Taneja R, Roy B, Plassat J-L, Zusi CF, Ostrowski J, Reczek PR, Chambon P (1996) Cell-type and promoter-context dependent retinoic acid receptor (RAR) redundancies for RARβ2 and Hoxa-1 activation in F9 and P19 cells can be artefactually generated by gene knock-outs. Proc Natl Acad Sci USA 93:6197–6202

Taylor CW, Kim YS, Childress-Fields KE, Yeoman LC (1992) Sensitivity of nuclear c-myc levels and induction to differentiation-inducing agents in human colon tumor cell lines, Cancer Letters 29:95–105

Teixeira C, Pratt MAC (1997) CKD2 is a target for retinoic acid-mediated growth inhibition in MCF-7 human breast cancer cells. Mol Endocrinol 11:1191–1202

Thiele CJ, Reynolds CP, Israel MA (1985) Decreased expression of N-myc precedes retinoic acid-induced morphological differentiation of human neuroblastoma. Nature 313:404–406

Thompson CB (1995) Apoptosis in the pathogenesis and treatment of disease. Science 267:1456–1462

Toma S, Isnardi L, Raffo P, Dastoli G, DeFrancisci E, Riccardi L, Palumbo R, Bollag W (1997) Effects of all-*trans*-retinoic acid and 13-*cis*-retinoic acid on breast-cancer cell lines: growth inhibition and apoptosis induction. Int J Cancer 70:619–627

Tonini GP, Casalaro A, Cara A, DiMartino D (1991) Inducible expression of calcyclin, a gene with strong homology to S-100 protein, during neuroblastoma cell differentiation and its prevalent expression in Schwann-like cell lines. Cancer Res 51:1733–1737

Tosi P, Visani G, Gibellini D, Zauli G, Ottaviani E, Cenacchi A, Gamberi B, Manfroi S, Marchisio M, Tura S (1994) All-*trans* RA and induction of apoptosis in acute promyelocytic leukemia cells, Leuk Lymphom 14:503–507

Tournier S, Raynaud F, Gerbaud P, Lohmann SM, Anderson WB, Evain-Brion D (1996) Retinoylation of the type II cAMP-binding regulatory subunit of cAMP-dependent protein kinase is increased in psoriatic human fibroblasts. J Cell Physiol 167:196–203

Trizna Z, Benner SE, Shirley L, Furlong C, Hong WK (1993) N-(4-hydroxyphenyl) retinamide is anticlastogenic in human lymphoblastic cell lines. Anticancer Res 13:355–356

Taniguchi Y, Furuke K, Masutani H, Nakamura H, Yodoi J (1995) Cell cycle inhibition of HTLV-1 transformed T cell lines by retinoic acid: the possible therapeutic use of thioredoxin reductase inhibitors. Oncol Res 7:183–189

van der Burg B, van der Leede BM, Kwakkenbos-Isbrucker L, Salverda S, de Laat SW, van der Saag PT (1993) Retinoic acid resistance of estradiol-independent breast cancer cells coincides with diminished retinoic acid receptor function. Mol Cell Endocrinol 91:149–157

van der Burgh B, Slager-Davidov R, van der Leede BM, de Laat SW, van der Saag PT (1995) Differential regulation of AP-1 activity by retinoic acid in hormone-dependent and-independent cancer cells. Mol Cell Endocrinol 112:143–152

Van de Klundert FA, Jansen HJ, Bloemendal H (1995) Negative regulation of a special, double AP-1 consensus element in the vimentin promoter: interference by the retinoic acid receptor. J Cell Physiol 164:85–92

van der Leede BJ, Folkers GE, van den Brink CE, van der Saag PT, van der Burg B (1995) Retinoic acid receptor alpha 1 isoform is induced by estradiol and confers retinoic acid sensitivity in human breast cancer cells. Mol Cell Endocrinol 109:77–86

Vermes I, Haanen C, Steffens-Nakken H, Reutelingsperger C (1995) A novel assay for apoptosis. Flow cytometric detection of phosphatidylserine expression on early apoptotic cells using fluorescein labelled annexin V, J Immunol. Methods 184:39–51

Veronesi U, De Palo G, Costa A, Formelli Marubini E, Del Vecchio M (1992) Chemoprevention of breast cancer with retinoids, J Natl Cancer Inst Monogr 12:93–98

Villa ML, Ferrario E, Trabattoni D, Formelli F, De Palo G, Magni A, Veronesi U, Clerici E (1993) retinoids, breast cancer and NK cells. Br J Cancer 68:845–850

Vlach J, Hennecke S, Alevizopoulos K, Conti D, Amati B (1996) Growth arrest by the cyclin-dependent kinase inhibitor p27Kip1 is abrogated by c-Myc, Embo J 15:6595–6604

Vollberg TM, George MD, Nervi C, Jetten AM (1992) Regulation of Type I and Type II transglutaminase in normal human bronchial epithelial and lung carcinoma cells. Am J Respir Cell Mol Biol 7:10–18

Veronesi U, De palo G, Costa A, Formelli, Marubini E, Del Vecchio M (1992) Chemoprevention of breast cancer with retinoids, J Natl Cancer Inst Monogr 12:93–98

Wan H, Dawson MI, Hong WK, Lotan R (1997) Enhancement of Calu-1 human lung carcinoma cell growth in serum-free medium by retinoids: dependence on AP-1 activation, but not on retinoid response element activation. Oncogene 15:2109–2118

Widschwendter M, Berger J, Daxenbichler G, Muller-Holzner E, Widschwendter A, Mayr A, Marth C, Zeimet AG (1997) Loss of retinoic acid receptor beta expression in breast cancer and morphologically normal adjacent tissue but not in the normal breast tissue distant from the cancer. Cancer Res 57:4158–4161

Wilcken NR, Sarcevic B, Musgrove EA, Sutherland RL (1996) Differential effects of retinoids and antiestrogens on cell cycle progression and cell cycle regulatory genes in human breast cancer cells. Cell Growth Differ 7:65–74

William R, Watson G, Rotstein OD, Parodo J, Bitar R, Hackam D, Marshall JC (1997) Granulocyte differentiation of HL60 cells results in spontaneous apoptosis mediated by increased caspase expression. FEBS Lett 412:603–609

Wolbach SB, Howe PR (1926) Tissue changes following deprivation of fat-soluble vitamin. J Exp Med 42:753–781

Won KA, Reed SI (1996) Activation of cyclin E/CDK2 is coupled to site-specific autophosphorylation and ubiquitin-dependent degradation of cyclin E. EMBO J 15:4182–4193

Wu GS, Burns TF, McDonald ER, Jiang W, Meng R, Krantz ID, Kao G, Gan DD, Zhou JY, Muschel R, Hamilton SR, Spinner NB, Markovitz S, Wu G, el-Deiry WS (1997) Killer/DR5 is a DNA damage-inducible p53-regulated death receptor gene. Nat Genet 17:141–143

Wu S, Donigan A, Platsoucas CD, Jung W, Soprano DR, Soprano KJ (1997) All-*trans*-retinoic acid blocks cell cycle progression of human ovarian adenocarcinoma cells at late G1. Exp Cell Res 232:277–286

Wu SJ, Zhang DM, Donigan A, Dawson MI, Soprano DR, Soprano KJ (1998) Effects of conformationally restricted synthetic retinoids on ovarian tumor cell growth. J Cell Biochemistry 68:378–388

Wu Q, Dawson MI, Zheng Y, Hobbs PD, Agadir A, Jong L, Li Y, Liu R, Lin B, Zhang X (1997) Inhibition of *trans*-retinoic acid-resistant human breast cancer cell growth by retinoid X receptor-selective retinoids. Mol Cell Biol 17:6598–6608

Wu Q, Li Y, Liu R, Agadir A, Lee M, Liu Y, Zhang X (1997) Modulation of retinoic acid sensitivity in lung cancer cells through dynamic balance of orphan receptors nur77 and COUP-TF and their heterodimerization. EMBO J 16:1656–1669

Wyllie AH (1997) Apoptosis: an overview. Br Med Bull 53:451–465

Xia C, Hu J, Ketterer B, Taylor JB (1996) The organization of the human GSTP1-1 gene promoter and its response to retinoic acid and cellular redox status. Biochem J 313:155–161

Xiong Y, Hannon GJ, Zhang H, Casso D, Kobayashi R, Beach D (1993) p21 is a universal inhibitor of cyclin kinases. Nature 366:701–704

Xu XC, Ro JY, Lee JS, Shin DM, Hong WK, Lotan R (1994) Differential expression of nuclear retinoid receptors in normal, premalignant, and malignant head and neck tissues. Cancer Res 54:3580–3587

Yang J, Liu X, Bhalla K, Kim CN, Ibrado AM, Cai J, Peng TI, Jones DP, Wang X (1997) Prevention of apoptosis by Bcl-2: release of cytochrome c from mitochondria blocked. Science 275:1129–1132

Yang LM, Kim HT, Munoz-Medellin D, Reddy P, Brown PH (1997) Induction of retinoid resistance in breast cancer cells by overexpression of cJun. Cancer Res 57:4652–4661

Yang Y, Vacchio MS, Ashwell JD (1993) 9-*cis*-RA inhibits activation-driven T-cell apoptosis: implications for retinoid X receptor involvement in thymocyte development, Proc Natl Acad Sci 90:6170–6174

Yang Y, Minucci S, Ozato K, Heyman RA, Ashwell JD (1995) Efficient inhibition of activation-induced Fas ligand up-regulation and T cell apoptosis by retinoids requires occupancy of both retinoid X receptors and retinoic acid receptors. J Biol Chem 270:18672–18677

Yen A, Varvayanis S (1992) RB tumor suppressor gene expression responds to DNA synthesis inhibitors, In Vitro Cell. Dev Biol 28 a: 669–672

Yen A, Varvayanis S (1994) Late dephosphorylation of the RB protein in G2 during the process of induced cell differentiation. Exp Cell Res 214:250–257

Zakeri ZF, Ahuja HS (1994) Apoptotic cell death in the limb and its relationship to pattern formation. Biochem Cell Biol 72:603–613

Zhang L, Jetten AM (1997) Retinoids and apoptosis. In: Sherbet GV, Bittar EE (eds) Advances in Organ Biology, Retinoids: their physiological function and therapeutic potential, vol 3. JAI, Greenwich, Conn, p 161

Zhang LX, Mills KJ, Dawson MI, Collins SJ, Jetten AM (1995) Evidence for the involvement of RA receptor RARα-dependent signaling pathway in the induction of tissue transglutaminase and apoptosis by retinoids. J Biol Chem 270:6022–6029

Zhang XK, Liu Y, Lee MO, Pfahl M (1994) A specific defect in the retinoic acid response associated with human lung cancer cell lines. Cancer Res 54:5663–5669

Zhao Z, Zhang ZP, Soprani DR, Soprano KJ (1995) Effect of 9-*cis*-retinoic acid on growth and RXR expression in human breast cancer cells. Exp Cell Res 219:555–561

Zhivotovsky B, Gahm A, Orrenius S (1997) Two different proteases are involved in the proteolysis of lamin during apoptosis, Biochem Biophys Res Comm 233:96–101

Zhou Q, Stetler-Stevenson M, Steeg PS (1997) Inhibition of cyclin D expression in human breast carcinoma cells by retinoids in vitro. Oncogene 15:107–115

Zhu T, Matsuzawa S, Mizuno Y, Kamibayashi C, Mumby MC, Andjelkovic N, Hemmings BA, Onoe K, Kikuchi K (1997) The interconversion of protein phosphatase 2 a between PP2A1 and PP2A0 during retinoic acid-induced granulocytic differentiation and a modification on the catalytic subunit in S phase of HL-60 cells. Arch Biochem Biophys 339:210–217

Zhu W, Jones CS, Kiss A, Matsukuma K, Amin S, De Luca LM (1997) Retinoic acid inhibition of cell cycle progression in MCF-7 human breast cancer cells. Exp Cell Res 234:293–299

Zhu X, Kumar R, Mandal M, Sharma N, Sharma HW, Dhingra U, Sokoloski JA, Hsiao R, Narayanan R (1996) Cell cycle-dependent modulation of telomerase activity in tumor cells, Proc Natl Acad Sci 93:6091–6095

CHAPTER 9

Retinoids and Differentiation of Normal and Malignant Hematopoietic Cells

A. AGADIR and C. CHOMIENNE

A. Introduction

Retinoids, both natural vitamin A derivatives and synthetic analogs, have been shown to be essential in vertebrate development, and in the regulation of cellular proliferation and differentiation in adult life. They have been found to inhibit the growth of and induce differentiation of a number of leukemic cells, as well as to inhibit the proliferation of both normal and malignant cells. Retinoids mediate their effects through their interaction with two nuclear receptor classes: the retinoic acid receptors (RARs) and the retinoid X receptors (RXRs), which function as ligand-inducible transcription factors. Data are now accumulating which place retinoic acid, its receptors and intermediary proteins in the transcriptional complex of myeloid master genes which control the proliferation and differentiation of specific myeloid lineages. To assist the reader in understanding the actions of retinoids in regulating myeloid cell proliferation and differentiation, a brief review of retinoid metabolism and actions that is directly relevant for understanding the information presented in this chapter is provided below. The reader is also referred to earlier chapters of this volume for more details regarding retinoid metabolism and actions.

Ultimately, the major source of retinoic acid (RA) is from its precursor retinol, which is stored as retinyl ester in the liver. Retinol is released from the liver into the bloodstream, bound to plasma retinol-binding protein (RBP) (BLANER 1989). Within the cell, retinol and retinal bound to cellular retinol-binding proteins (CRBP I and CRBP II) serve as substrates for dehydrogenases which catalyze the formation of retinoic acid (POSCH et al. 1991, 1992; GIGUERE et al. 1994; DONOVAN et al. 1995; DUESTER 1996). Similarly, RA bound to cellular RA-binding proteins (CRABP I and CRABP II) can serve as a substrate for enzymes of the cytochrome P450 family that can convert RA to other more polar metabolites (FIORELLA and NAPOLI 1991; NAPOLI 1996). The cellular retinoid-binding proteins (CRBP I and II, CRABP I and II) belong to a superfamily of proteins that comprises the fatty acid-binding proteins and the CRBPs and CRABPs. Each of these proteins binds one molecule of a hydrophobic ligand and is thought to be actively involved in the metabolism of the hydrophobic ligands (CLARKE and ARMSTRONG 1989; GIGUERE 1994). RA metabolism and signaling pathway are summarized in Fig. 1.

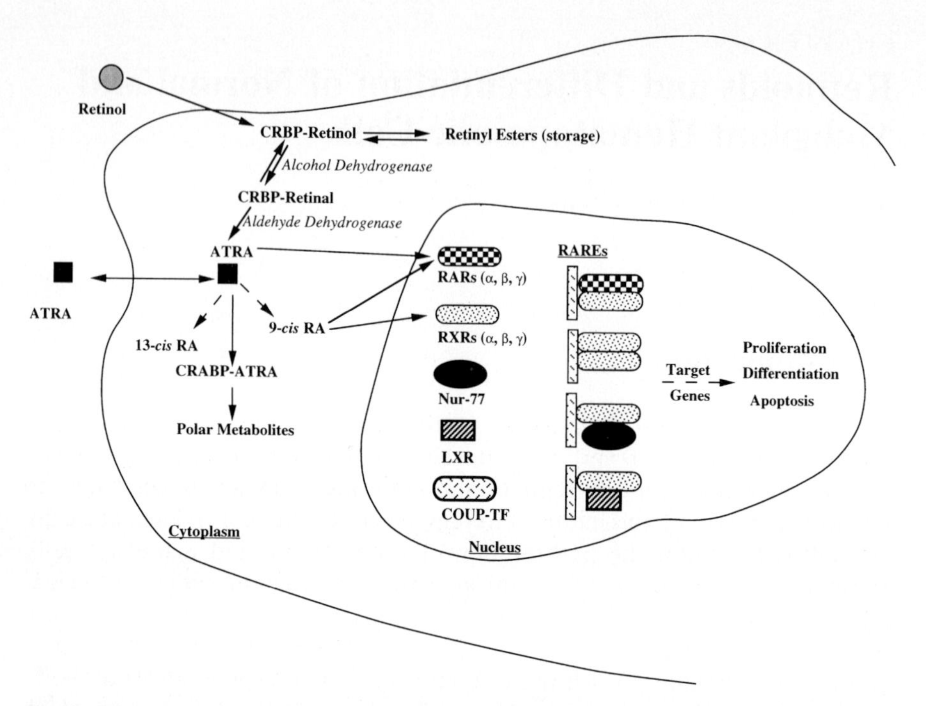

Fig. 1. Diagrammatic summary of retinoid metabolism and action within the cell. *ATRA*, All-*trans*-retinoic acid; *CRBP-Retinol*, retinol bound to CRBP-I; *CRBP-Retinal*, retinal bound to CRBP-I; *9-*cis *RA*, 9-*cis*-retinoic acid; *13-*cis *RA*, 13-*cis*-retinoic acid; RARs, retinoic acid receptors; *RXRs*, retinoid X receptors; *RAREs*, retinoic acid response elements

Retinoids modulate the specific expression of their target genes upon binding to two distinct families of nuclear retinoid receptors, the retinoic acid receptors (RARs) and the retinoid X receptors (RXRs) (CHAMBON 1994, 1996). All-*trans* RA (ATRA) is a high-affinity ligand only for the RARs, whereas 9-*cis* RA a stereoisomer of ATRA, is a high-affinity ligand for both RARs and RXRs (HEYMAN et al. 1992; LEVIN et al. 1992; ALLEGRETTO et al. 1993). Each receptor family includes three distinct members α, β, and γ, and each member has several isoforms that result from differential promoter usage and alternative splicing of the receptor transcripts. Each RAR and RXR subtype is expressed in specific patterns among tissues and is thought to have a specific profile of gene-regulating activity. A schematic representation of the structure of the human retinoid receptors RARs and RXRs families is shown in Fig. 2. The RARs and RXRs are hormone-dependent transcription factors that belong to the nuclear receptor superfamily which includes the receptors for steroid hormones, mineralocorticoids, vitamin D, and thyroid hormones (ZHANG and PFAHL 1993; CHAMBON 1994; MANGELSDORF and EVANS 1995). These receptors consist of six domains (A–F), the A/B region is responsible for ligand independent transactivation, the C domain containing two zinc

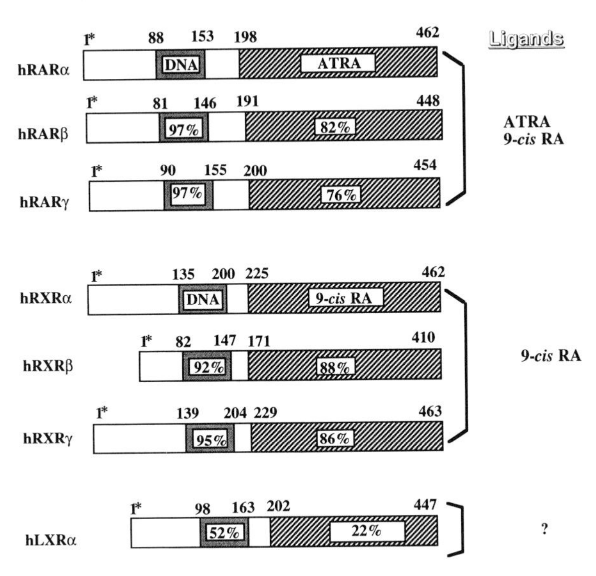

Fig. 2. Schematic representations for the human RARs and RXRs. *Numbers above*, amino acid residue numbers. The DNA binding domain is indicated for hRARα and hRXRα; *percentages* (values given in hRARβ or hRARγ and hRXRβ or hRXRγ), proportion homology of the domain with that of either hRARα or hRXRα. *ATRA*, the ligand binding domain of the receptor in hRARα; *percentages* (values given in this domain for hRARβ and hRARγ), proportion homology of the domain with that of hRARα. *9-*cis *RA*, the ligand binding domain of hRXRα; *percentages* (values given in this domain for hRXRβ and hRXRγ), proportion homology of the domain with that of RXRa. *Below*, schematic representation of hLXRα, another ligand-dependent transcription factor that is closely related to the RARs and RXRs. percentages (values given within the DNA binding domain and the ligand binding domain of the hLXRα), proportion homology of the domain with hRARα

fingers is responsible for DNA binding, and the E domain for ligand binding and ligand-dependent transactivation (Chambon 1994; Mangelsdorf and Evans 1995).

To control the expression of RA-responsive genes, RARs interact with RXRs forming heterodimers that bind to specific DNA sequences designated as retinoic acid response elements (RAREs) that generally consist of 6-base pair repeated motifs that are either a direct or invert repeats of the sequence (5′AGGTCA) separated by two (DR2) or five (DR5) base pairs (Chambon 1994; Mangelsdorf and Evans 1995). In addition, RXR can heterodimerize not only with RAR but also with other members of the nuclear receptor family including thyroid hormone receptors, vitamin D receptors, and other orphan receptors whose ligands have not been yet identified such as NGFI-B/Nur77 and OR-1 (Zhang and Pfahl 1993; Chambon 1994; Mangelsdorf and Evans 1995; Leblanc and Stunnenberg 1995; Wu et al. 1996). Activation of RXR has been shown to be required for the function of the RXR-containing heterodimers, such as RXR/Nur (Forman et al. 1995; Perlmann and Jansson

1995) and RXR/LXR (liver X receptor) (WILLY et al. 1995). Diverse biological responses can be generated due to the multiplicity of receptor combinations resulting from these heterodimerizations. Gene transcriptional activation by RXR/RAR heterodimers is mainly activated by RAR-selective ligands, while transactivation by RXR/Nur77 is mediated by RXR specific ligands (FORMAN et al. 1995; PERLMANN and JANSSON 1995).

Recent studies have led to the identification and cloning of genes encoding coactivator and corepressor molecules which mediate the effect of ligand dependent transcription factors. None appear as yet specific for one nuclear receptor: SCR-1 (steroid receptor coactivator 1) otherwise called TIF2 (transcriptional intermediary factor 2) and CBP/p300 (cAMP response element binding protein), both act as coactivator whereas N-Cor and SMRT function as corepressors (GLASS et al. 1997). Recently, as described for other nuclear receptors, it has been clearly demonstrated that the phosphorylation of RARs is important for their transcriptional role (ROCHETTE-EGLY et al. 1995).

B. Retinoids and Hematopoiesis

The role of retinoids in embryogenesis also suggests a capital role for retinoids in normal cell differentiation and proliferation. It is now clear that retinoids play an important role in normal blood cell differentiation. It is also clear that some leukemias arise through chromosomal damage which effects the gene for RARα. Thus, impaired retinoid signaling is associated with some forms of leukemia.

I. Effect of Retinoic Acid on Cellular Proliferation, Differentiation, and Apoptosis of Hematopoietic Leukemic Cells

1. Effect on Myeloid Leukemic Cell Differentiation

Early studies revealed that cultured leukemic cells could be induced to differentiate after treatment with various agents (SACHS 1978). It was subsequently shown that in vitro treatment of several human myeloid leukemic cell lines (HL-60, THP-1, U-937), murine teratocarcinoma F9 cells and various prostate, mammary and neuroblastoma cell lines with RA results in inhibition of cellular proliferation and induction of cellular differentiation (LOTAN et al. 1980; BREITMAN et al. 1980, 1981; KOEFFLER 1986; TOBLER et al. 1986; CHOMIENNE et al. 1986a,b; COLLINS et al. 1990). The effects of retinoids on cell proliferation and differentiation are both structure and dose dependent (CHOMIENNE et al. 1986a,b; TOBLER et al. 1986). Thus, ATRA and 13-*cis* RA are equally effective in inducing differentiation of HL-60 and U-937 myeloid leukemic cell lines (BREITMAN et al. 1981; CHOMIENNE et al. 1986a,b), whereas other retinoids such as the retinyl ester etretinate are less effective

(Chomienne et al. 1986a,b). 9-*Cis* RA, the ligand of the RXR receptor, has been shown to be superior for inducing HL-60 cell differentiation to RARα-specific retinoids (Calabresse et al. 1995) and liposomal preparations of ATRA (Chomienne et al. 1997).

Interestingly, in leukemia cell lines, RA may induce the lineage of hematopoietic differentiation. Thus, RA induces the monoblastic cell line U-937 along the monocytic cell line (Chomienne et al. 1986b) whereas a erythroid leukemic cell line is induced towards red blood cells (Yamashita et al. 1997). To date this observation has not been observed to hold in normal myeloid progenitor cells, and thus should be regarded as a particularity of leukemia cell lines.

2. Effect on Acute Promyelocytic Leukemic Cells

RA can induce specifically the entire leukemic cell population of acute promyelocytic leukemic (APL) t(15;17) cells to differentiate into mature granulocytes (Breitman et al. 1981; Chomienne et al. 1989, 1990). However, RA treatment is ineffective in inducing other AML subtypes. In APL cells, 13-*cis* RA is only effective in inducing differentiation at higher concentrations, i.e., $10^{-6}\,M$ while ATRA is effective at $10^{-7}\,M$. This provides a possible explanation for the limited in vivo efficacy of 13-*cis* RA observed in the treatment of APL (Fontana et al. 1986; Chomienne et al. 1989, 1990; Runde et al. 1992). Interestingly, APL cells do not follow an identical structure-dose response as observed for HL-60 cells. For example, although RARα analogs are the most potent inducers of differentiation in HL-60 cells, 9-*cis* RA and ATRA are both superior to 13-*cis* RA. Combinational studies using various receptor specific analogs have shown, using the NB4 cell line, that RARα ligands can induce two separate events within these cells; one allowing RXR agonists to act and another inducing transcriptional activity of RARα (Chen et al. 1996).

Induction of differentiation of myeloid leukemic cell lines by retinoids is enhanced by the presence of certain cytokines (Dubois et al. 1994a). In fact, a strong correlation between the expression of certain cytokines and differentiation induction with ATRA was observed in APL patients. Patients who expressed tumor necrosis factor α (TNFα), interleukin (IL 6, IL-8, and IL-1 (this is the case of most APL patients) responded to ATRA therapy, whereas patients who have no TNFα, no IL-3, no granulocyte colony stimulating factor (G-CSF) and no granulocyte-macrophage colony stimulating factor (GM-CSF) respond poorly to ATRA (Dubois et al. 1994a). Moreover, results from our group have shown that some of these parameters are modulated during ATRA-induced APL differentiation, i.e., cathepsin G expression and IL-8 secretion are decreased and GM-CSF and adhesion molecule (CD11b, CD15, CD45RO) expression are increased (de Gentille et al. 1994; Dubois et al. 1994b; Seale et al. 1996).

Cells that harbor chromosomal translocations resulting in rearrangements of the RARα gene but involving chromosomes other than chromosome 15, such as t(11;17) and t(5;17), do not differentiate in the presence of ATRA (LICHT et al. 1995; GUIDEZ et al. 1994; REDNER et al. 1996)

3. Effect on Induced Myeloid Leukemic Cell Apoptosis

Although neutrophils are known to be eliminated from the peripheral blood by apoptosis and HL-60 cells treated with ATRA gives rise to apoptosis in up to 40% fo the cells (MARTIN et al. 1990), ATRA fails to induce apoptosis in vitro in APL cells (TOSI et al. 1994; CALABRESSE et al. 1995). These differentiated leukemic cells can no longer produce leukemic clones when grown only in soft agar (CHOMIENNE et al. 1990) and a significant decrease in bcl-2 protein in these cells strongly suggests that differentiation has initiated some control over cell death even though no morphological or biochemical criteria for apoptosis are noted (CHOMIENNE et al. 1992). However, a topoisomerase II inhibitor, etoposide, does induce apoptosis in APL cells. This implies that other properties or features of APL cells, such as the PML-RAR α (GRIGNANI et al. 1993) or growth factors (DUBOIS et al. 1994a) may prevent effective induction of apoptosis in vitro by ATRA (NAGY et al. 1998).

II. Retinoic Acid as a Novel Myeloid Differentiation Factor

Several studies attempted to establish a linkage between retinoid receptors and ATRA-induced granulocytic differentiation in APL. Results from our group and others showed that ATRA enhances granulocytic differentiation of total and CD34+ normal bone marrow progenitor cells (DOUER and KOEFFLER 1982; GRATAS et al. 1993; MICLEA and CHOMIENNE 1994) (see Fig. 4a). This effect is similar to that induced by G-CSF, in as much as stem cells are committed towards the granulocytic lineage, with a concomitant decrease in erythroid and monocytic colonies. Alone, RA has no proliferation or differentiating effect but acts in synergy with SCF and IL-3. As has been observed for other cell models, retinoid-dependent induction shows a structure-dose dependence. Normal myeloid progenitors have a similar response to various retinoids that is similar to the HL-60 cell line. However, clonogenic cells are more sensitive to high concentrations of RA. Thus, effective induction enhancing concentrations of RA are between 10^{-7} and $10^{-8}\,M$. G-CSF and RA have antagonistic effects in normal CD34^{+} cells (RUSTEN et al. 1996) but once differentiation by RA is triggered, treatment with a combination of G-CSF and RA allows achievement of maximal differentiated and functional cells. However, the more recent literature clearly indicates that retinoids are inducers of granulocytic differentiation. This differentiation response observed for human cells appears to hold up in mouse models but for the chicken model, some differences are observed in as much as retinoids also may trigger erythroid differentiation (GADRILLON et al. 1989).

III. Retinoic Acid Control of Myeloproliferative Growth

Induction of differentiation by retinoids is linked to little inhibition of cell growth. HL-60 cells passed in culture for 6 days accumulate in the G1 phase of the cell cycle (AGADIR et al. 1995b). In fresh APL blasts, inhibition of cell growth is practically nonexistent and for normal myeloid progenitors, colony growth implies a proliferative step. Thus, although retinoids may initially promote cell proliferation before differentiation, retinoids clearly enhance cell survival when differentiation is observed. However, apart from being a differentiation inducer, retinoids may also have a specific and distinct inhibitory effect on cell growth without inducing apparent differentiation. This has been observed in solid tumor cell lines but also has been noted in very immature myeloid cell lines such as KG-1 (DOUER and KOEFFLER 1982). This observation could be related to the anti-AP-1 effect of retinoids (DESBOIS et al. 1991; NAGPAL et al. 1995).

The inhibition of erythroid and monocytic clonogenic growth obtained upon retinoid treatment of normal myeloid progenitors has also been observed in myeloproliferative disorders, leading to the proliferation of either the erythroid (NIKOLOVA et al. 1998; NOTARIO et al. 1996), myeloid (chronic myeloid leukemia) or monocytic (JCML and CMML) lineages (BARUCHEL et al. 1993; CAMBIER et al. 1996; CORTES et al. 1997).

C. Retinoic Acid Signaling Pathways in Hematopoietic Cells

I. Differential Expression of Nuclear Retinoic Acid Receptors and Retinoid-Binding Proteins

Specific efficacy of retinoid action in a given tissue is clearly related to the presence of specific retinoic acid isomers at required concentrations and to the presence of specific retinoic acid receptors. In situ hybridization studies and Northern and Western analyses have clearly indicated that both RARα and RXRα receptors are preferentially expressed in hematopoietic cells (CHOMIENNE et al. 1995). Use of receptor specific retinoid agonists and antagonists confirms their importance. The fact that RARα is strongly expressed both in CD34^{+} cells and in differentiated polymorphonuclear cells suggests that along with serving as an inducer of differentiation, RA may also participate in the normal functions of leukocytes. Inhibition of HL-60 induced differentiation with a dominant RARα negative receptor (COLLINS et al. 1990) and of lymphomyeloid murine cells (TSAI et al. 1992) confirms the predominant role of RARα in RA induced differentiation in hematopoietic cells. Cellular retinoid-binding proteins, such as CRABP I and CRABP II are also expressed in myeloid cells (CORNIC et al. 1992). This suggests a possible role for these binding proteins in retinoid signaling within hematopoietic cells.

II. Alteration of RARα in Hematopoietic Malignancies

1. Molecular Characteristics of APL

APL is designated as AML3 in the FAB cytological classification (BENNETT et al. 1976, 1980). This disease is characterized by a predominance in the bone marrow of abnormal promyelocytes with Auer rods in their cytoplasm. A balanced chromosomal translocation t(15;17) is found in the leukemic cells of 95% of the patients diagnosed with APL (LARSON et al. 1984). The t(15;17) translocation involves the RARα gene on chromosome 17 and the PML gene on chromosome 15 (BORROW et al. 1990; DE THÉ et al. 1990; ALCALAY et al. 1991; PANDOLFI et al. 1991) and results in the formation of PML-RARα and the reciprocal RARα-PML fusion genes which can encode several chimeric proteins depending on the breakpoint position within the PML gene (PANDOLFI et al. 1992; KASTNER et al. 1992) (see Fig. 3). The breakpoint on chromosome 17 is always located in the second intron of the RARα gene, while on chromosome 15 three breakpoints named Bcr (breakpoint cluster region) have been identified (Bcr_1, intron 6; Bcr_2, exon 6; Bcr_3, intron 3) in the PML gene.

Several studies have been carried out to determine the role of PML-RARα fusion protein in the pathogenesis and/or RA sensitivity of APL. Transient transfection studies have shown that PML-RARα may antagonize the function of the wild-type RARα protein (DE THÉ et al. 1991; KAKIZUKA et al. 1991; PANDOLFI et al. 1991; KASTNER et al. 1992; PEREZ et al. 1993). When PML-RARα was transiently transfected in HL-60 cells, PML-RARα inhibits at low

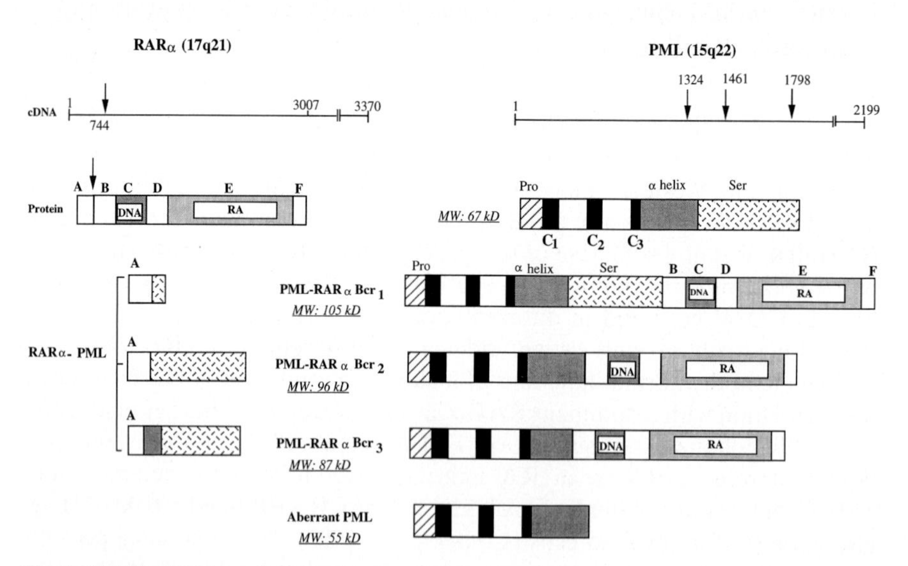

Fig. 3. RARα-PML and PML-RARα fusion proteins that can be formed from the t(15;17) chromosomal translocation and the molecular characteristics of these proteins

RA concentrations the granulocytic differentiation pathway, whereas treatment with high RA concentrations bring about relief of this inhibition (ROUSSELOT et al. 1994). As with wild-type RARα, PML-RARα can form heterodimers with RXRα and thus interfere with the normal functions of the wild-type receptors and this may disturb other regulatory pathways requiring RXRα (PEREZ et al. 1993). Based on these data, it appears that PML-RARα is responsible for both the inhibition of wild-type functions and the response to RA therapy (MILLER et al. 1992; GRIGNANI et al. 1993; DYCK et al. 1994; KOKEN et al. 1994; ROUSSELOT et al. 1994; WEIS et al. 1994).

The fact that these chimeric proteins are sufficient to produce a leukemia has now been confirmed by the generation of transgenic PML-RARα mice bearing a similar leukemia to that observed in humans and demonstrating a similar sensitivity to RA (BROWN et al. 1997). Figure 4 provides a schematic summary of the molecular events giving rise to APL and of how ATRA or 9-*cis* RA may act to bring about remission in APL patients.

2. Alterations of RARα in Other Malignancies

In a small subset of APL cases (1%–2%), two other chromosomal translocations t(11;17)(q23-q21), and t(5;17) have been described. The t(11;17) translocation fuses the RARα locus with the PLZF (promyelocytic leukemia zinc finger) gene on chromosome 11 (q23) (CHEN Z et al. 1994; CHEN SJ et al. 1993; ZELENT 1994). The t(5;17) translocation involves the RARα locus and a known gene on chromosome 5 termed NPM (nucleophosmin) (REDNER et al. 1996). The NPM gene, encodes a nuclear phosphoprotein possibly involved in ribosome processing or assembly, and has also been implicated in the t(2;5) translocation (MORRIS et al. 1994; RABBITTS 1994). Although the partner genes of RARα in both t(15;17) and t(11;17) encode putative transcription factors, the two genes are of different origins: PLZF belongs to the Kruppel family (CHEN SJ et al. 1993; CHEN et al. 1994), whereas PML is a member of a family encoding zinc finger proteins, including other transcription factors (Rpt-1, Rfp) (KAKIZUKA et al. 1991; DE THÉ et al. 1991; GODDARD et al. 1991; KASTNER et al. 1992). Furthermore, PML is ubiquitously expressed (GODDARD et al. 1991), whereas PLZF expression is restricted to the hematopoietic system (CHEN Z et al. 1994). Functional studies have shown that PLZF-RARα fusion protein acts as does PML-RARα by antagonizing wild-type RARα receptors function and sequestering RXRα from RAR/RXR heterodimers (CHEN Z et al. 1994). The observation from these three chromosomal translocations observed in APL patients that the RARα gene is the common target suggests that the disturbance or the loss of this retinoid signaling pathway plays a critical role in the pathogenesis of APL. However, screening for RARα rearrangements in lymphoid and myeloid disorders has proven that alterations of the RARα gene are not frequent (PADUA et al. 1997).

In mouse knock-out models lacking one of the RAR forms, no abnormal or striking hematopoietic defect has been observed. As has been the case for

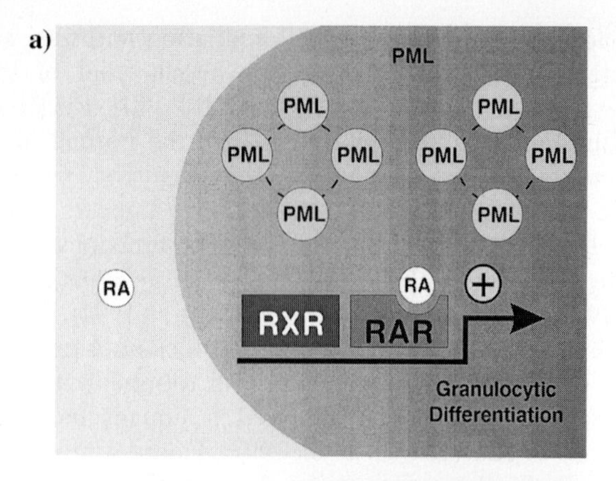

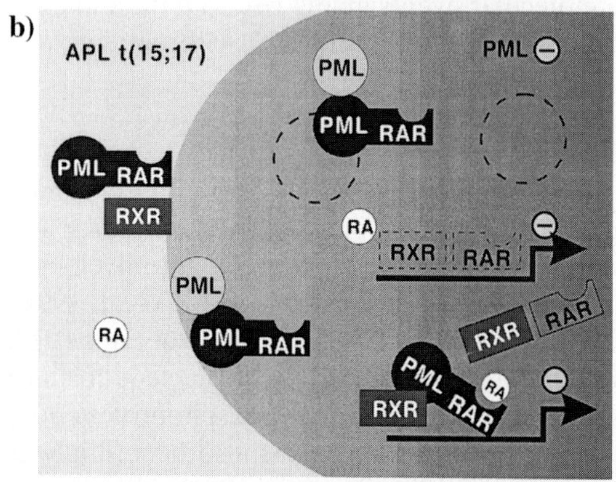

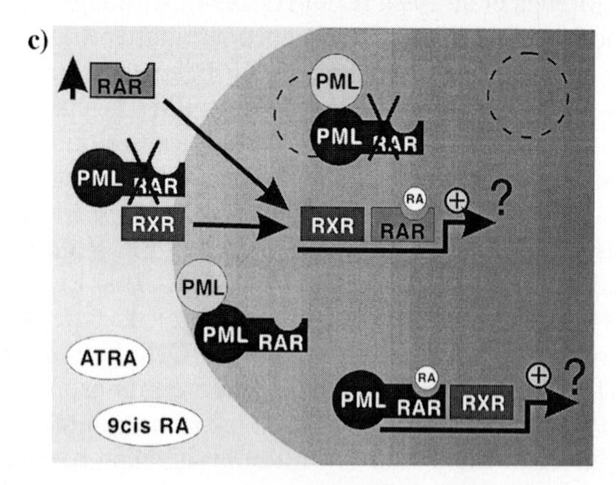

Fig. 4a–c. Hypothetical scheme for explaining the actions of ATRA and 9-*cis* RA in bringing about remission in APL resulting from the t(15;17) chromosomal translocation. **a** The actions RXR, RAR, and RA in normal differentiating hematopoietic cells. **b** Normal RXR, RAR, RA actions are blocked by the presence of PML-RAR fusion protein with the leukemia cells. **c** Possible basis for explaining how treatment with phamacological doses of ATRA or 9-*cis* RA can bring about remission in APL patients

other knock out models of transcriptionally active myeloid proteins, this may reflect that the absence of the RAR is not a crucial parameter for development of APL and underlines the redundancy of the retinoic acid receptors and of other hematopoietic factors (KASTNER et al. 1995).

D. Retinoic Acid as a Therapeutic Agent in Hematopoietic Malignancies

I. Acute Promyelocytic Leukemia

APL represents approximately 10%–15% of all acute nonlymphoid leukemias. Based on the observation that malignant cells could be reversed to a normal phenotype (SACHS 1978), ATRA differentiation therapy was initiated at the Shanghais Institute of Hematology (HUANG et al. 1987a,b, 1988; WANG et al. 1990) and in France (CHOMIENNE et al. 1989; CASTAIGNE et al. 1990; DEGOS et al. 1990) in APL patients experiencing a first relapse. Complete remission rates of about 90% were obtained when patients were treated with a daily dose of 45 mg/m^2 of ATRA. Subsequently, several clinical trials in other countries were successfully carried out (WARRELL et al. 1991; OHNO et al. 1993) and ATRA is now a standard treatment in APL cases.

It was demonstrated that this response in APL patients was not due to cytotoxicity but to in vivo differentiation of APL blasts into neutrophils, leading to a progressive elimination of the leukemia cells and their replacement by normal polyclonal hematopoietic cells (HUANG et al. 1988; CASTAIGNE et al. 1990; WARRELL et al. 1993). Rapid improvement in coagulopathy was also seen in patients receiving ATRA instead of initial worsening observed with conventional chemotherapy. Furthermore, studies have shown that differentiation of leukemia cells by retinoids, could be enhanced by the addition of cytokines in vitro are correlating well *with* in vivo results obtained from APL patients.

Unfortunately, ATRA treatment has two major adverse secondary effects. The first occurs in newly diagnosed APL patients, where a rapid increase in white blood cell (WBC) number is seen in approximately one third of the patients. This is accompanied by the clinical signs of retinoic acid syndrome (see Sect. D.1.2) which has proven to be fatal to some patients (CASTAIGNE et al. 1990; FENAUX et al. 1992; FRANKEL et al. 1992). Intensive chemotherapy brings about a reduction in leukocyte cell count and allows most patients to enter complete remission (FENAUX et al. 1992). Recently, our group tried to reduce the incidence of hyperleukocytosis and of the ATRA syndrome in APL patients, by treating patients with low-dose ATRA (25 mg/m^2 per day or even 15 mg/m^2 per day). These doses proved effective in differentiating APL blasts and the complete remission rates obtained were identical to those obtained with the dose of 45 mg/m^2 per day. However, the incidence of hyperleukocytosis remained similar to that seen at 45 mg/m^2 per day (CASTAIGNE et al. 1993). Experience has shown that combination of ATRA and chemotherapy helps improve complete remission rates and more importantly reduces the incidence of relapse (FENAUX et al. 1992, 1995; FRANKEL et al. 1994). This suggests that ATRA and chemotherapy have an additive or synergistic effect on the elimination of leukemia cells.

The second adverse effect of ATRA treatment occurs in patients who have achieved complete remission after treatment with either ATRA alone or ATRA and low-dose chemotherapy for maintenance and who subsequently develop resistance to ATRA and relapse. In fact, within a few months after achievement of complete remission almost all patients relapsed (CASTAIGNE et al. 1990; DEGOS et al. 1990; OHASHI et al. 1992; WANG et al. 1990). These findings prompted clinicians to administer treatment combining ATRA and intensive chemotherapy in APL.

1. Retinoid Resistance in APL

In the ongoing APL93 trial (including 414 patients) no primary resistance to ATRA was noted. This underlines the necessity for a precise diagnosis of APL confirmed by cytogenetic and/or molecular data (FENAUX et al. 1997), since progress has been made in the treatment of APL patients with ATRA through improved management of both retinoic acid acquired resistance and the ATRA syndrome. APL patients routinely relapse after ATRA treatment alone. Secondary resistance to ATRA was observed in almost all patients experiencing early relapse after ATRA therapy (within 3 months to 1 year) (DELVA et al. 1993; WARRELL et al. 1994). This resistance could be related to a feedback mechanism that reduces ATRA concentrations in blood and tissues. Data have been reported that suggest that alterations in ATRA catabolism may account for the inability of leukemia cells to sustain adequate RA levels. During the first induction therapy with ATRA, the plasma ATRA concentration achieved in APL patients is within those used in vitro to induce differentiation (LEFEBVRE et al. 1991; MUINDI et al. 1992a,b). However, continuous administration of ATRA leads to a progressive decline in ATRA plasma concentrations (MUINDI et al. 1992a,b). This may be related to a specific induction of enzymes aimed at protecting the cell from high retinoid concentrations. We have observed in both normal and malignant myeloid cells an increase in CRABP II protein levels (DELVA et al. 1993; CORNIC et al. 1992, 1994). CRABP II levels return to undetectable levels 3 months after ATRA withdrawal, when patients may again become sensitive to ATRA in vivo (DELVA et al. 1993; CORNIC et al. 1994). This observation suggested that CRABP II may play a role in the development of ATRA resistance.

Alternatively, the secondary resistance to ATRA may develop in APL patients via genetic mechanisms including the selection of ATRA insensitive clones during continuous treatment with the drug (DUPREZ et al. 1992; DERMIME et al. 1993). This resistance could also be due to an altered function of RARα or PML-RARα as shown by in vitro studies of HL-60 and NB4 leukemia cell lines (COLLINS et al. 1990; ROBERSON et al. 1992; LI et al. 1994).

To date, altered RARα or PML-RARα function has only been observed in leukemia cell lines in which resistance was induced in vitro. However, recently mutations in the ligand binding domain of RARα gene has been described for APL patients after ATRA treatment. Since these mutations were not noted at diagnosis, it is possible that they arose after ATRA treatment and that this alters RARα function and/or activity (Li et al. 1997).

Several therapeutic strategies aimed at overcoming resistance to ATRA have been studied. The addition of inhibitors of cytochrome P450 enzymes (ketoconazole, liarozole) (Miller et al. 1993) or interferon to ATRA treatment regimens to help restore effective ATRA plasma concentrations (Rigas et al. 1993; Warrell et al. 1993; Miller et al. 1995) have been described in the literature. The most effective so far, for overcoming resistance to ATRA has been the cessation of retinoic acid therapy as soon as complete remission is obtained (i.e., 1 month). In the APL93 trial, RA was not reintroduced during the chemotherapy consolidation (approximately 3 months). Interestingly, when retinoic acid was reintroduced in the maintenance therapy in a low and alternative schedule (15 days every 3 months) a beneficial effect on event free survival was noted. This suggests that if induction of ATRA resistance is well managed, continuous retinoic acid therapy of APL may be effective (Fenaux et al. 1997).

Another possibility for eliminating ATRA resistance and improving APL management, is the use of selective synthetic retinoids that act as RARα agonists. One such retinoid, the aromatic retinoid AM80 which was shown effective in APL resistant patients (Tukeshita et al. 1996). 9-*Cis* RA is as effective as ATRA for inducing complete remission in vivo and may have the advantage of not inducing decreased 9-*cis* RA plasma levels during treatment (Miller et al. 1995). However, longer follow-up of patients treated with 9-*cis* RA is needed to confirm the absence of acquired RA resistance (Miller et al. 1995). Since some cytokines such as G-CSF can increase the differentiating effect of ATRA on APL cells in vivo, the combined use of cytokines and RA in therapy may offer another possible regimen for avoiding the acquisition of RA resistance (Vaichus et al. 1993; Nakajima et al. 1994).

2. Retinoic Acid Syndrome and Other Side Effects of RA Therapy

An increase in the WBC count is sometimes associated with clinical use of ATRA and termed ATRA or retinoic acid syndrome (Fenaux et al. 1992; Frankel et al. 1992; Ohno et al. 1993; Warrell et al. 1993; Wiley and Firkin 1995; Degos et al. 1995). A more precise definition of the clinical signs grouped under the name of the ATRA syndrome was given by Frankel et al. (1992). This syndrome occurs in about 30% of APL patients treated with ATRA and in the first trials one third of the patients died of complications from ATRA syndrome. The syndrome combines fever, respiratory distress, weight gain, lower extremity edema, pleural or pericardial effusions, hypotension, and sometimes renal failure (Frankel et al. 1992). These signs are preceded by an increase in the WBC counts in the majority of cases, but some patients develop symptoms with normal WBC counts (Degos et al. 1995). Studies have shown that cytokines may have a stimulatory effect on the proliferation of these cells, i.e., induction of IL-1β and secretion of G-CSF may contribute to hyperleukocytosis in vivo (Dubois et al. 1994a,b). Two approaches are used to prevent ATRA syndrome. One approach consists of adding chemotherapy at the beginning of ATRA treatment for patients presenting with high WBC counts or when the WBC number increases during ATRA therapy (Fenaux et al. 1992, 1993, 1994; Ohno et al. 1994). The second approach consists of adding corticosteroids (dexamethasone) as soon as the first symptoms appear (Frankel et al. 1992, 1994, Warrell et al. 1994; Wiley and Firkin 1995).

Other side effects can be observed during ATRA therapy, these include dryness of lips and mucosa, increases in serum transglutaminases and triglycerides and bone pain. These effects do not require treatment discontinuation (Degos et al. 1995). Intracranial hypertension associated with headaches have been reported. This effect is moderate in adult but can be fatal in children (Mahmoud et al. 1993). Fever, is also another side effect of ATRA therapy and can be reversed after ATRA discontinuation or upon addition of dexamethasone (Fenaux et al. 1992, 1993). Finally, another hyperinflammatory reaction during ATRA therapy, the Sweets syndrome, has been reported (Shirono et al. 1995; Christ et al. 1996). This syndrome consists of infiltrates of the skin and internal organs by neutrophylic granulocytes, fever, and painful erythematous cutaneous eruptions, prominent musculosketal involvement, and sterile pulmonary infiltration and intercurrent proteinuria (Christ et al. 1996). This syndrome responds within 48 h to glucocorticoids (Sweet 1964; Cohen et al. 1988; von den Dreisch 1994).

3. Mechanisms of retinoic acid induced differentiation in APL patients

In APL cells, ATRA treatment rapidly increases the expression of the RARα gene (Chomienne et al. 1991; Rousselot et al. 1994; Agadir et al. 1995b) and a lessening of PML-RARα mRNA and protein is also noted. This may be followed by the release of RXRα which was sequestered by PML-RAR in the cytoplasm. This allows for enhanced formation of RXRα-RARα heterodimers and for degradation of PML-RARα, bringing about a release of PML. PML then can relocalize to its nuclear bodies (Daniel et al. 1993; Dyck et al. 1994; Koken et al. 1994; Weis et al. 1994). Any mechanism describing RA actions must also take into account the other players in retinoic acid signaling such as nuclear receptor protein ratios and the coordination of transcriptional coactivators and corepressors (Fig. 4c).

II. Retinoids as Therapeutic Alternative to Myeloproliferative Disorders

The treatment of myeloproliferative disorders has so far relied mainly on the inhibition of the abnormal proliferation by cytostatic agents, but these treatments are disappointing. In JCML preliminary results have shown that ATRA can reduce WBC counts and organomegaly in approximately one half of patients (CASTELBERRY et al. 1994). We have tested the efficacy of all-*trans* RA in the adult type (CMML) as this disorder has many features in common with JCML (CAMBIER et al. 1997). Half of the patients showed a significant response but ATRA syndrome was observed in patients with high WBC needing the addition of Hydroyurea (CAMBIER et al. 1996). In chronic myeloid leukemia, RA treatment has also been tried either alone or in association with Interferon α (CORTES et al. 1997). In all studies carried out to date, no significant response or complete remission has been noted. Whether this lack of a response is due to a cellular unresponsiveness to retinoic acid response or to a lack of retinoic acid bioavailability remains to be established.

E. Summary and Perspectives

The discovery that RA is an inducer of leukemic granulocytic differentiation has led to a wide range of novel research leading to new differentiation therapies and to the identification of novel signaling pathway in hematopoietic tissues and novel genes in leukemogenesis. The recent finding that a partner of the retinoic acid nuclear signaling pathway is also implicated in the ETO protein linked to the AML2t(8;21) leukemia further links retinoic acid to other myeloid-specific proteins. Unraveling the actions of the numerous proteins involved in modulating the actions of retinoic acid as an inducer of granulocytic differentiation will not only enlighten our view of granulopoiesis but will also help define the molecular basis for other diseases and provide further targeted therapies.

References

Agadir A, Cornic M, Lefebvre P, Gourmel B, Jerome M, Degos L, Fenaux P, Chomienne C (1995a) All-*trans* retinoic acid pharmacokinetics and bioavailability in acute promyelocytic leukemia: Intracellular concentrations and biological response relationship. J Clin Oncol 13:2517–2523

Agadir A, Cornic M, Jerome M, Menot ML, Cambier N, Gaub MP, Gourmel B, Lefebvre P, Degos L, Chomienne C (1995b) Characterization of nuclear retinoic acid binding activity in sensitive leukemic cell lines: cell specific uptake of ATRA and RARα protein modulation Biochem Biophys Res Commun 213:112–122

Alcalay M, Zangrilli D, Pandolfi PP, Longo L, Mencarelli A, Giacomucci A, Rocchi M, Biondi A, Rambaldi A, Lo Coco F (1991) Translocation breakpoint of acute promyelocytic leukemia. Proc Natl Acad Sci USA 88:1977–1981

Allegretto EA, McClurg MR, Lazarchik SB, Clemm DL, Kerner SA, Elgort MG, Boehm MF, White SK, Pike JW, Heyman RA (1993) Transactivation properties of retinoic acid and retinoid X receptors in mammalian cells and yeast. Correlation with hormone binding and effects of metabolism. J Biol Chem 268:26625–26633

Baruchel A, Menot ML, Balitrand N, Leblanc T, Heberlin C, Donadieu J, Bordigoni P, Rohrlich P, Vilmer E, Degos L, Najean Y, Schaison G, Chomienne C (1993) All-trans retinoic acid inhibits CFU-M (monocytic) colony formation and self renewal in juvenile chronic myelogenous leukemia. Blood 82 [Suppl 11]:197a

Bennett JM, Catowsky D, Daniel MT, Flandrin G, Galton D, Gralnick M, Sultan C (1976) Proposals for the classification of the acute leukemias. Br J Haematol 33:451–458
Bennett JM, Catowsky D, Danier MT, Flandrin G, Galton D, Gralnick M, Sultan C (1980) A variant form of hypergranular promyelocytic leukemia (M3). Ann Intern Med 92:261
Blaner WS (1989) Retinol-binding protein: the serum transporter protein for vitamin A. Endocr Rev 10:308–316
Borrow J, Goddard AD, Sheer D, Soloman E (1990) Molecular analysis of acute promyelocytic leukemia breakpoint cluster region on chromosome 17. Science 249:1577–1580
Breitman TR, Collins SJ, Keene BR (1981) Terminal differentiation of human promyelocytic leukemic cells in primary culture in response to retinoic acid. Blood 57:1000–1004
Breitman TR, Selonick SE, Collins SJ (1980) Induction of differentiation of the human promyelocytic leukemia cell line (HL-60) by retinoic acid. Proc Natl Acad Sci USA 77:2936–2940
Brown D, Kogan S, Lagasse E, Weissman I, Alcalay M, Pelicci PG, Atwater S, Bishop M (1997) A PML-RARα transgene initiates murine acute promyelocytic leukemia. Proc Natl Acad Sci USA 94:255–2556
Calabresse C, Barbey S, Venturini L, Balitrand N, Degos L, Fenaux P, Chomienne C (1995) In vitro treatment with retinoids or the topoisomerase inhibitor, VP16, evidences different functional apoptotic pathways in acute promyelocytic leukemic cells. Leukemia 9:2049–2057
Cambier N, Wattel E, Menot M.L. Guerci A, Chomienne C, Fenaux P (1996) All-trans retinoic acid in adult chronic myelomonocytic leukemia: results of a pilot study. Leukemia 10:1164–1167
Cambier N, Baruchel A, Schlageter MH, Menot ML, Wattel E, Fenaux P, Chomienne C (1997) Chronic myelomonocytic leukemia: from biology to therapy. Hematol Cell Ther 39: 41–48
Castaigne S, Lefebvre P, Chomienne C, Suc E, Rigal-Huguet F, Gardin C, Delmer A, Archimbaud E, Tilly H, Janvier M (1993) Effectiveness and pharmacokinetics of low-dose all-*trans* retinoic acid (25 mg/m^2) in acute promyelocytic leukemia. Blood 82:3241–3249
Castaigne S, Chomienne C, Daniel MT, Berger R, Fenaux P, Degos L (1990) All-*trans* retinoic acid as a differentiating therapy for acute promyelocytic leukemias. I. Clinical results. Blood 76:1704–1709
Castelberry RP, Emmanuel PD, Zuckerman KS, et al (1994) A pilot study of isoretinoid in the treatment of juvenile chronic myelogenous leukemia. N Engl J Med 331:1680–1684
Chambon P (1994) The retinoid signaling pathway: molecular and genetic analyses, Semin Cell Biol 5:115–125
Chambon P (1996) A decade of molecular biology of retinoic acid receptors. FASEB J 10:940–954
Chang KS, Stass SA, Chu DT, Deaven L, Trujillo JM, Freireich EJ (1992) Characterization of a fusion cDNA (RARA/myl) transcribed from the t(15;17) translocation breakpoint in acute promyelocytic leukemia. Mol Cell Biol 12:800–810
Chen SJ, Zelent A, Tong JH, Yu HQ, Wang ZY, Derre J, Berger R, Waxman S, Chen Z (1993) Rearrangements of the retinoic acid receptor alpha and promyelocytic leukemia zinc finger genes resulting from t(11;17) (q23;q21) in a patient with acute promyelocytic leukemia. J Clin Invest 91:2260–2267
Chen JY, Clifford J, Zusi C, Starrett J, Torlatoni D, Ostrowski J, Reczek PR, Chambon P, Gronemeyer H (1996) Two distinct actions of retinoid-receptors ligands. Nature 382:819–822
Chen Z, Guidez F, Rousselot P, Agadir A, Chen SJ, Wang ZY, Degos L, Waxman S, Zelent A, Chomienne C (1994) PLZF-RAR fusion protein generated from the translocation t(11;17)(q23;q21) displays altered transactivation properties

against the wild-type retinoic acid receptors, Proc Natl Acad Sci USA 91:1178–1182

Chomienne C, Guidez F, Miclea JM, Bastie JN, Delva L, Rousselot P, Ballerini P (1995) Récepteurs nucléaires et ontogénèse du tissu hématopoïétique. CR Soc Biol 189:493–501

Chomienne C, Balitrand N, Abita JP (1986a) Inefficacy of the synthetic aromatic retinoid etretinate and of its free acid on the in vitro differentiation of leukemic cells. Leuk Res 10:1079–1081

Chomienne C, Balitrand N, Ballerini P, Castaigne S, de Thé H, Degos L (1991) All-*trans* retinoic acid modulates the retinoic acid receptor alpha in promyelocytic cells. J Clin Invest 88:2150–2154

Chomienne C, Balitrand N, Cost H, Degos L, Abita JP (1986b) Structure-activity relationships of the aromatic retinoids on the differentiation of the human histiocytic lymphoma cell line U-937. Leuk Res 10:1301–1305

Chomienne C, Ballerini P, Balitrand N, Amar M, Bernard JF, Bolvin P, Daniel MT, Berger R, Castaigne S, Degos L (1989) Retinoic acid therapy for promyelocytic leukemia. Lancet 147:746–747

Chomienne C, Ballerini P, Balitrand N, Daniel MT, Fenaux P, Castaigne S, Degos L (1990) All trans retinoic acid in promyelocytic leukemias. II. In vitro studies structure function relationship. Blood 76:1710–1717

Chomienne C, Barbey S, Balitrand N, Degos L, Sachs L (1992) Regulation of Bcl-2 and cell death by all-*trans* retinoic acid in acute promyelocytic leukemia cells. Proc Am Assoc Cancer Res 33:41

Chomienne C, Balitrand N, Guidez F, Barbet C, Clabresse C, Guidez F, Delva L, Bastie JN (1997) Retinoic acid receptor alpha analogs and ATRA liposomal preparations induce maximal differentiation of acute promyelocytic primary leukemic cells. Anticancer Res 17:3940

Christ E, Linka A, Jacky E, Speich R, Marineck B, Schaffner A (1996) Sweet s syndrome involving the musculoskeletal system during treatment of promyelocytic leukemia with all-*trans* retinoic acid. Leukemia 10:731–734

Clarke SD, Armstrong MK (1989) Cellular lipid binding proteins: expression, function, and nutritional regulation. FASEB J 3:2480–2487

Cohen PR, Talpaz M, Kurzrock R (1988) Malignancy-associated Sweet's syndrome: review of the world literature. J Clin Oncol 6:1887–1897

Collins SJ, Robertson KA, Mueller L (1990) Retinoic acid-induced granulocytic differentiation of HL-60 myeloid leukemia cells is mediated directly through the retinoic acid receptor (RARα). Mol Cell Biol 10:2154–2163

Cornic M, Delva L, Castaigne S, Balitrand N, Degos L, Chomienne C (1994) In vitro all-*trans* retinoic acid (ATRA) sensitivity and cellular retinoic acid binding protein (CRABP) levels in relapse leukemic cells after remission induction by ATRA in acute promyelocytic leukemia. Leukemia 8:914–917

Cornic M, Delva L, Guidez F, Balitrand N, Degos L, Chomienne C (1992) Induction of retinoic acid-binding protein in normal and malignant human myeloid cells by retinoic acid in acute promyelocytic leukemia patients. Cancer Res 52:3329–3334

Cortes JK, O'Brien S, Beran M, Estey E, Keating M, Talpaz M (1997) A pilot study of all-trans retinoic acid in patients with Philadelphia chromosome -positive chronic myelogenous leukemia. Leukemia 11:929–932

Daniel MT, Koken M, Romagne O, Barbey S, Bazarbachi A, Stadler M, Guillemin MC, Degos L, Chomienne C, de Thé H (1993) PML protein expression in haematopoietic and APL cells. Blood 82:1858–1867

de Gentile A, Toubert ME, Dubois C, Krawice I, Schlageter MH, Balitrand N, Castaigne S, Degos L, Rain JD, Najean Y, Chomienne C (1994) Induction of high-affinity GM-CSF receptors during all-*trans* retinoic acid treatment of acute promyelocytic leukemia. Leukemia 8:1758–1762

de Thé H, Chomienne C, Lanotte M, Degos L, Dejean A (1990) The t(15;17) translocation in acute promyelocytic leukemia fuses the retinoic acid receptor a gene to a novel transcribed locus. Nature 347:558–561

de Thé H, Lavau C, Marchio A, Chomienne C, Degos L, Dejean A (1991) The PML-RARa fusion mRNA generated by the t(15;17) translocation in acute promyelocytic leukemia encodes a functionally altered RAR. Cell 66:675–684
Degos L, Chomienne C, Daniel MT, Berger R, Dombret H, Fenaux P, Castaigne S (1990) Treatment of first relapse in acute promyelocytic leukemia with all-*trans* retinoic acid. Lancet 336:1440–1441
Degos L, Dombret H, Chomienne C, Daniel MT, Miclea JM, Chastang C, Castaigne S, Fenaux P (1995) All-*trans* retinoic acid as a differentiating agent in the treatment of acute promyelocytic leukemia. Blood 85:2643–2653
Delva L, Cornic M, Balitrand N, Guidez F, Miclea JM, Delmer A, Teillet F, Fenaux P, Castaigne S, Degos L, Chomienne C (1993) Resistance to all-*trans* retinoic acid (ATRA) therapy in relapsing acute promyelocytic leukemia: a study of in vitro ATRA sensitivity and cellular retinoic acid binding protein levels in leukemic cells. Blood 82:2175–2181
Dermime S, Grignani F, Clerici M, Nervi C, Sozzi G, Talamo GP, Marchesi E, Formelli F, Parmiani G, Pellici PG, Gambaccorti-Passerinic C (1993) Occurrence of resistance to retinoic acid in acute promyelocytic leukemia cell line NB4 is associated with altered expression of the PML/RAR alpha protein. Blood 82:1573–1577
Desbois C, Aubert D, Legrand C, Pain B, Samarut J (1991) A novel mechanism of action for v-ErbA: abrogation of the inactivation of transcription factor AP-1 by retinoic acid and thyroid hormone receptors. Cell 67:731–740
Donovan M, Olofsson B, Gustafson A-L, Dencker L, Eriksson U (1995) The cellular retinoic acid binding proteins. J Steroid Biochem Mol Biol 53:459–465
Douer D, Koeffler P (1982) Retinoic acid enhances colony-stimulating factor-induced clonal growth of normal human myeloid progenitor cells in vitro. Exp Cell Res 138:193–198
Dubois C, Schlageter MH, De Gentile A, Guidez F, Balitrand N, Toubert ME, Krawice I, Fenaux P, Castaigne S, Najean Y, Degos L, Chomienne C (1994a) Hematopoietic growth factor expression and ATRA sensitivity in acute promyelocytic leukemia. Blood 83:3264–3270
Dubois C, Schlageter MH, De Gentile A, Guidez F, Balitrand N, Toubert ME, Krawice I, Fenaux P, Castaigne S, Najean Y, Degos L, Chomienne C (1994b) Modulation of Il-6 and Il-1β and G-CSF secretion by all-*trans* retinoic acid in acute promyelocytic leukemia. Leukemia 8:1750–1757
Duester G (1996) Involvement of alcohol dehydrogenase, short-chain dehydrogenase/reductase, aldehyde dehydrogenase, and cytochrome P450 in the control of retinoid signaling by activation of retinoic acid synthesis. Biochemistry 35:12221–12227
Duprez E, Ruchaud S, Houge G, Martin-Thouvenin V, Valensi F, Kastner P, Berger R, Lanotte M (1992) A retinoid acid resistant t(15;17) acute promyelocytic leukemia cell line: isolation, morphological, immunological and molecular features. Leukemia 6:1281–1287
Dyck JA, Maul GG, Miller WH Jr, Chen JD, Kakizuka A, Evans RM (1994) A novel macromolecular structure is a target of the promyelocyte-retinoic acid receptor oncoprotein. Cell 76:333–43
Fenaux P, Castigne S, Dombret H, Archimbaud E, Duarte M, Morel P, Lamy T, Tilly H, Guerci A, Maloisel F, Bordessoule D, Sadoun A, Tiberghien P, Fegueux N, Daniel MT, Chomienne C, Degos L (1992) All-*trans* retinoic acid followed by intensive chemotherapy gives a high complete remission rate and may prolong remissions in newly diagnosed acute promyelocytic leukemia: A pilot study on 26 cases. Blood 80:2176–2181
Fenaux P, Chastang C, Chomienne C, Castaigne S, Sanz M, Link H, Lowenberg B, Fey M, Eric A-B, Degos L, The European APL Group (1995) Treatment of newly diagnosed acute promyelocytic leukemia (APL) by all-*trans*retinoic acid (ATRA) combined with chemotherapy: the European experience. Leukemia Lymphoma 16:431–437

Fenaux P, Le Deley MC, Castaigne S, Archimbaud E, Chomienne C, Link H, Guerci A, Duarte M, Daniel MT, Bowen D, Huebner G, Bauters F, Feguex N, Fey M, Sanz M, Lowenberg B, Maloisel F, Auzanneau G, Sadou A, Gardin C, Bastion Y, Ganser A, Jacky E, Dombret H, Chastang C, Degos L and the European APL 91 Group (1993) Effect of all-*trans* retinoic acid in newly diagnosed acute promyelocytic leukemia. Results of multicenter randomised trials. Blood 82:3241–3249

Fenaux P, Chastang C, sanz M, Thomas X, Dombret H, Link H, Guerci A, Fegueux N, San Miguel J, Rayon C, Zittoun R, Gardin C, Maloisel F, Fey M, Travade P, Reiffers J, Stamatoulas A, Stoppa AM, Caillot D, Lefrere F, Hayat M, Castaigne S, Chomienne C, Degos L, and the European APL Groups (1997) ATRA followed by chemotherapy (CT) vs ATRA plus CT and the role of maintenance therapy in newly diagnosed acute promyelocytic leukemia (APL). First interim results of APL 93 trials. Blood 90:122a

Fenaux P, Wattel E, Archimbaud E, Sanz M, Hecquet B, Guerci A, Link H, Feguex N, Fey M, Castaigne S, Degos L (1994) Prolonged follow up confirms that all-*trans* retinoic acid (ATRA) followed by chemotherapy reduces the risk of relapse in newly diagnosed acute promyelocytic leukemia (APL). Blood 84:666–667

Fiorella PD, Napoli JL (1991) Expression of cellular retinoic acid binding protein (CRABP) in *Escherichia coli*. Characterization and evidence that holo-CRABP is a substrate in retinoic acid metabolism. J Biol Chem 266:16572–16579

Fontana JA, Rogers JS, Durham JP (1986) The role of 13-*cis* retinoic acid in the remission induction of a patient with acute promyelocytic leukemia. Cancer 57:209–217

Forman BM, Umesono K, Chen J, Evans RM (1995) Unique response pathways are established by allosteric interactions among receptors. Cell 81:541–550

Francis PA, Rigas JR, Muindi J, Kris MG, Young CW, Warrel RP (1992) Modulation of all-*trans* retinoic acid (RA) pharmacokinetics by a cytochrome p450 enzyme system inhibitor. Proc Am Soc Clin Oncol 11:108

Frankel SR, Eardley A, Lauwers G, Weiss M, Warrell R (1992) The "retinoic acid syndrome" in acute promyelocytic leukemia. Ann Int Med 117:292–296

Frankel SR, Eardley A, Heller G, Berman E, Miller WH Jr, Dmitrovsky E, Warrell RP Jr (1994) All-trans retinoic acid for acute promyelocytic leukemia. Results of the New York Study. Ann Intern Med 120:278–286

Gandrillon O, Jurdic P, Pain B, Desbois C, Madjar JJ, Moscovici MG, Moscovici C, Samarut J (1989) Expression of the v-erbA product, an altered nuclear hormone receptor, is sufficient to transform erythrocytic cells in vitro. Cell 58:115–121

Giguere V (1994) Retinoic acid receptors and cellular retinoic binding proteins: Complete interplay in retinoic signaling. Endocrine Rev 15:61–79

Glass CK, Rose WR, Rosenfeld MG (1997) Nuclear receptor coactivators. Current Opin Cell Biol 9:222–232

Goddard AD, Borrow J, Freemont P, Soloman E (1991) Characterization of a zinc finger gene disrupted by the t(15;17) in acute promyelocytic leukemia. Science 254:1371–1374

Gratas C, Menot ML, Dresch C, Chomienne C (1993) Retinoids support granulocytic but not erythroid differentiation of myeloid progenitors in normal bone marrow cells. Leukemia 7:1156–1162

Grignani F, Fagioli M, Alcalcay M, Long L, Pandolfi P, Donti E, Biondi A, Lo Coco F, Pellici P (1994) Review: Acute promyelocytic leukemia: from genetics to treatment. Blood 83:10–25

Grignani F, Ferrucci PF, Testa U, Talamo G, Fagioli M, Alcalay M, Mencarelli A, Pescehele C, Nicoletti L, Pellici PG (1993) The acute promyelocytic leukemia-specific PML/RAR alpha fusion protein inhibits differentiation and promotes survival of myeloid precursor cells. Cell 74:423–431

Groopman J, Ellman L (1979) Acute promyelocytic leukemia. Am J Hematol 7:395–408

Guidez F, Huang W, Tong JH, Dubois C, Balitrand N, Waxman S, Michaux JL, Martiat P, Degos L, Chen Z, Chomienne C (1994) Poor response to all-*trans* retinoic acid therapy in a t(11;17) PLZF/RARα patient. Leukemia 8:312–317

Heyman RA, Mangelsdorf DJ, Dyck JA, Stein RB, Eichele G, Evans RM, Thaller C (1992) 9-*cis* retinoic acid is a high affinity ligand for the retinoid X receptor. Cell 68:397–406
Huang ME, Ye YC, Chen SR (1987b) All-*trans* retinoic acid with or without low dose cytosine arabinoside in acute promyelocytic leukemia. Chin Med J 100:949–953
Huang ME, Ye YC, Wang ZY (1987a) Treatment of 4 APL patients with ATRA, Chin. J Intern Med 26:330–332
Huang ME, Ye YI, Chen SR, Chai JR, Lu JX, Zhoa L, Gu LJ, Wang ZY (1988) Use of all-*trans* retinoic acid in the treatment of acute promyelocytic leukemia. Blood 72:567–571
Kakizuka A, Miller WH Jr, Umesono K, Warrell RP Jr, Frankel SR, Murty WVS, Dmitrovsky E, Evans RM (1991) Chromosome translocation t(15;17) in human acute promyelocytic leukemia fuses RARα with a novel putative transcription factor, PML. Cell 66:663–674
Kakizuka et al. (1992) ••
Kastner P, Mark M, Chambon P (1995) Nonsteroid nuclear receptors: What are genetic studies telling us about their role in real life?. Cell 83:859–869
Kastner P, Perez A, Lutz Y, Rochette-Egly C, Gaub MP Durand B, Lanotte M, Berger R, Chambon P (1992) Structure, localisation and transcriptional properties of two classes of retinoic acid receptor a fusion proteins in acute promyelocytic leukemia (APL): Structural similarities with a new family of oncoproteins. EMBO J 11:629–642
Kizaki M, Ikeda Y, Tanosaki R, Nakajima H, Morikawa M, Sakashita A, Koeffler P (1993) Effects of novel retinoic acid compound, 9-cis retinoic acid on proliferation, differentiation and expression of retinoic acid receptor-a and retinoid X receptor-alpha RNA by HL-60 cells. Blood 82: 3592–3599
Koeffler HP (1986) Human acute myeloid leukemia lines: models of leukemogenesis Sem Hematol 23:223–236
Koken MHM, Puvion-Dutilleul F, Guillemin MC, Viron A, Linares-Cruz G, Stuurman N, de Jong L, Szostecki C, Calvo F, Chomienne C, Degos L, Puvion E, de Thé H (1994) The t(15;17) translocation alters a nuclear body in a retinoic acid reversible fashion. EMBO J 13:1073–1083
Lanotte M, Martin-Thouvenin V, Najman S, Ballerini P, Valensi F, Berger R (1991) NB4, a maturation inducible cell line with t(15;17) marker isolated from a human acute promyelocytic leukemia (M3). Blood 77:1080–1086
Larson RA, Kondo K, Vardiman JW, Butler ARE, Golomb HM, Rowley JD (1984) Evidence for a 15;17 translocation in every patient with acute promyelocytic leukemia. Am J Med 76:827–841
Leblanc BP, Stunnenberg HG (1995) 9-*cis* retinoic acid signaling: changing partners causes some excitement. Genes Dev 9:1811–1816
Lefebvre P, Thomas G, Gourmel B, Agadir A, Castaigne S, Dreux C, Degos L, Chomienne C (1991) Pharmaconkinetics of oral all-trans retinoic acid in patients with acute promyelocytic leukemia. Leukemia 12: 1054–1058
Levin AA, Sturzenbecker LJ, Kazmer S, Bosakowski T, Huselton C, Allenby G, Speck J, Kratzeisen C, Rosenberger M, Lovey A, Grippo JF (1992) 9-*cis* retinoic acid stereoisomer binds and activates the nuclear receptor RXRα. Nature 355:359–361
Li YP, Said F, Gallagher RE (1994) Retinoic acid-resistant HL-60 cells exclusively contain mutant retinoic acid receptor-alpha. Blood 83:3298–3302
Li YP, Andersen J, Zelent A, Sreeenivas R, Paletta E, Tallman MS, Wiernik PH, Gallagher RE (1997) RARα1/RARα2-PML mRNA expression in acute promyelocytic leukemia cells: a molecular and laboratory-clinical correlative study. Blood 90: 306–312
Licht JD, Chen A, Goy A, Wu Y, Scott AA, DeBlasio A, Miller WH, Zelenetz AD, Willman CL, Head DR, Chomienne C, Chen Z, Chen SJ, Zelent A, MacIntrye E, Veil A, Cortes J, Kantarjian H, Waxman S (1995) Clinical and molecular charac-

terization of acute promyelocytic leukemia associated with translocation (11;17). Blood 85:1083–1094
Lotan R, Kramer RH, Neumann G, Lotan D, Nicolson GL (1980) Retinoic acid-induced modifications in the growth and cell surface components of a human carcinoma (HeLa) cell line. Exp Cell Res 130:40–414
Mahmoud HH, Hurwitz CA, Roberts WM, Santana VM, Ribeiro RC, Krance RA (1993) Tretinoin toxicity in children with acute promyelocytic leukaemia. Lancet 342:1394–1395
Mangelsdorf DJ, Evans RM (1995) The RXR heterodimers and orphan receptors. Cell 83:841–850
Martin SJ, Bradley JG, Cotter TG (1990) HL-60 cells induced to differentiate towards neutrophils subsequently die via apoptosis. Clin Exp Immunol 79:448–453
Miclea JM, Chomienne C (1994) Effect of all-*trans* retinoic acid on $CD34^+$ human myeloid cells. Leukemia 8:214
Miller VA, Rigas JR, Kris MG, Muindi JRF, Young CW, Warrel RP (1993) Use of liarozole, a novel inhibitor of cytochrome p450 oxidases, to modulate catabolism of all-*trans* retinoic acid, Proc Am Soc Clin Oncol 12:159
Miller WH Jr, Kakizuka A, Frankel SR, Warrell RP Jr, Deblsio A, Levine K, Evans RM, Dmitrovsky E (1992) Reverse transcription polymerase chain reaction for the rearranged retinoic acid receptor a clarifies diagnosis and detects minimal residual disease in acute promyelocytic leukemia by reverse transcription polymerase chain reaction assay for the PML/RARa fusion mRNA. Blood 82:1689–1694
Miller WH, Jakubowski A, Tong WA, Miller VA, Rigas JR, Bendetti F, Gill GM, Truglia JA, Ulm E, Shirley M, Warrrell RP (1995) 9-cis retinoic acid induces complete remission, but does not reverse clinically acquired retinoid resistance in acute promyelocytic leukemia. Blood 85: 3021–3027
Morris SW, Kirstein MN, Valentine MB, Dittmer KG, Shapiro DN, Saltman DL, Look AT (1994) Fusion of a kinase gene, ALK, to a nucleolar protein gene, NPM, in non-Hodgkin's lymphoma. Science 263:1281–1284
Muindi J, Frankel SR, Huselton C, De Grazia F, Garland WA, Young CW, Warrell Jr RP (1992a) Clinical pharmacology of oral all-*trans* retinoic acid in patients with acute promyelocytic leukemia. Cancer Res 52:2138–2142
Muindi J, Frankel SR, Miller Jr WH, Jakubowski A, Scheinberg DA, Young C, Dmitrovsky E, Warrell Jr RP (1992b) Continuous treatment with all-*trans* retinoic acid causes a progressive reduction in plasma drug concentrations: implications for relapse and retinoid resistance in patients with acute promyelocytic leukemia. Blood 79:299–303
Nagpal S, Athanikar J, Chandraratna RA (1995) Separation of transactivation and AP1 antagonism functions of retinoic acid receptor alpha. J Biol Chem 270:923–927
Nagy L, Thomazy VA, Heyman RA, Davies P (1998) Retinoid-induced apoptosis in normal neoplastic tissues. Cell Death Diff 5:11–19
Nakajima K, Hatake K, Miyata T, Muroi K, Komatsu N, Miura Y (1994) Acute promyelocytic leukemia, tretinoin, and granulocyte conony-stimulating factor. Lancet 343:173–174
Napoli JL (1996) Retinoic acid biosynthesis and metabolism. FASEB J 10:993–1001
Nikolova G, Menot ML, Descryver F, Dussault A, Schlageter MH, Rain JD, Chomienne C (1998) L'acide rétinoïque inhibe la pousse spontanée des progéniteurs érythroïdes et myéloïdes du sang péripherique des sujets polyglobuliques, Hematologie 132
Notario A, Rolandi ML, Mazzuchelli I (1996) Peripheral blood and bone marrow changes after treatment with ATRA and G-CSF in AML, APL and blast crisis following Vasquez's disease. Haematologica 10(7):81, (3):261–264
Ohashi H, Ichikawa A, Takagi H, Hotta T, Naoe T, Ohno R, Saito H (1992) Remission induction of acute promyelocytic leukemia by all-*trans*-retinoic acid: Molecular

evidence of restoration of normal hematopoiesis after differentiation and subsequent extinction of leukemic clone. Leukemia 6:859–862

Ohno R, Tanimoto M, Murakami H, Kanamaru A, Asou N, Kobayshi T, Kuriyama K, Ohmoto E, Sakamaki H, Tsubaki K, Hiraoka A, Yamada O, Oh H, Furusawa S, Matsuda S (1994) Treatment of newly diagnosed acute promyelocytic leukemia (APL) with all-*trans* retinoic acid (ATRA). Proc Am Soc Clin Oncol 13:303

Ohno R, Yoshida H, Fukutani H (1993) Multi-institutional study of all-*trans* retinoic acid as a differentiation therapy of refractory acute promyelocytic leukemia. Leukemia 7:1722–1727

Padua RA, Parrado A, Jordan D, Dupas S, Bastard C, Smith M, White D, McKenna S, Wittaker JA, Bentley P, Burnett A, Chomienne C (1997) Identification of retinoic acid receptor alpha (RARα) gene rearrangments in myeloid, lymphoid leukaemias and lymphomas. Anticancer Res 17:3951

Pandolfi PP, Alcalay M, Fagioli M, Zangrilli D, Mencarelli A, Diverio D, Biondi A, LoCoco F, Rambladi A, Grignani F, Rochette-Egly C, Gaub MP, Chambon P, Pelicci PG (1992) Genomic variability and alternative splicing generate multiple PML/RARα transcripts that encode aberrant PML protein and PML/RARα isoforms in acute promyelocytic leukemia. EMBO J 11:1397–1407

Pandolfi PP, Gargani F, Alcay M, Mencarelli A, Biondi A, LoCoco F, Pelicci PG, (1991) Structure and origin of the acute promyelocytic leukemia myl/RARα cDNA and characterization of its retinoid binding and transactivation properties. Oncogene 6:1285–1292

Perez A, Kastner P, Sethi S, Lutz Y, Reibel C, Chambon P (1993) PML-RAR homodimers: Distinct DNA binding properties and heteromeric interactions with RXR. EMBO J 12:3171–3182

Perlmann T, Jansson L (1995) A novel pathway for vitamin A signaling mediated by RXR heterodimerization with NGFI-B and NURRI. Genes Dev 9:769–782

Posch KC, Boerman MHEM, Burns RD, Napoli JL (1991) Holocellular retinol binding protein as a substrate for microsomal retinal synthesis. Biochemistry 30:6224–6230

Posch KC, Burns RD, Napoli JL (1992) Biosynthesis of all-*trans* retinoic acid from retinal. Recognition of retinal bound to cellular retinol binding protein (type I) as substrate by a purified cytosolic dehydrogenase. J Biol Chem 267:19676–19682

Rabbitts TH (1994) Chromosome translocations in human cancer. Nature 372:143–149

Redner RL, Rush EA, Faas S, Rudert WA, Corey SJ (1996) The t(5;17) variant of acute promyelocytic leukemia expresses a nucleophosmin-retinoic acid receptor fusion. Blood 87:882–886

Redner RL, Corey SJ, Rush EA (1997) Differentiation of t(5;17) variant acute promyelocytic leukemic blasts by all-*trans* retinoic acid. Leukemia 11:1014–1016

Rigas JR, Francis PA, Muindi JRF, Kris MG, Huselton C, DeGrazia F, Orazem JP, Young CW, Warrell RP (1993) Constitutive variability in the pharmacokinetics of the natural retinoid, all-*trans* retinoic acid and its modulation by ketoconazole. J Natl Cancer Inst 85:3021–3027

Roberson KA, Emami B, Collins S (1992) Retinoic acid-resistance HL-60 cells harbor a point mutation in the retinoic acid receptor ligand-binding domain that confers dominant negative activity. Blood 80:1885–1889

Rochette-Egly C, Oulad-Abdelghani M, Staub A, Pfister V, Scheuer I, Chambon P, Gaub MP (1995) Phosphorylation of the retinoic acid receptor-a by protein kinase A. Mol Endocrinol 9:860–871

Rousselot P, Hardas B, Patel A, Guidez F, Gaken J, Castaigne S, Dejean A, de Thé H, Degos L, Farzaneh F, Chomienne C (1994) The PML/RARA gene product of the t(15;17) translocation inhibits retinoic acid induced differentiation and mediated transactivation in human myeloid cells. Oncogene 18:545–551

Runde V, Aul C, Sudhoff T, Heyll A, Schneider W (1992) Retinoic acid in the treatment of acute promyelocytic leukemia: inefficacy of the 13-*cis* isomer and induc-

tion of complete remission by the all-*trans* isomer complicated by thromboembolic events. Ann Haematol 64:270–272
Rusten LS, Dybedal I, Blomhoff HK, Blomhoff R, Smeland EB, Jacobsen SEW (1996) The RAR-RXR as well as the RXR-RXR pathway is involved in signaling growth inhibition of human CD34$^+$ erythroid progenitor cells (1996). Blood 87: 1728–1736
Sachs L (1978) Control of normal differentiation and the phenotypic reversion of malignancy in myeloid leukemia cells. Nature 274:535–539
Seale J, Delva L, Renesto P, Balitrand N, Dombret H, Scrobohaci ML, Degos L, Paul P, Chomienne C (1996) All-*trans* retinoic acid rapidly decreases cathepsin G synthesis and mRNA expression in acute promyelocytic leukemia. Leukemia 10:95–101
Shirono K, Kiyofuji C, Tsuda H (1995) Sweet's syndrome in a patient with acute promyelocytic leukemia during treatment with all-*trans* retinoic acid. Int J Hematol 62:183–187
Sweet RD (1964) An acute febrile neutrophilic dermatosis. Br J Dermatol 74:349–356
Teboul M, Enmark E, Li Q, Wikstrom AC, Pelto-Huikko M, Gustafsson JA (1995) OR-1, a member of the nuclear receptor superfamily that interacts with the 9-*cis* retinoic acid receptor. Proc Natl Acad Sci USA 92:2006–2100
Tobler A, Dawson MI, Koeffler HP (1986) Retinoids: structure-function relationship in normal and leukemic hematopoiesis in vitro. J Clin Invest 78:303–309
Tosi P, Visani G, Gibellini D, Zauli G, Ottaviani E, Cenacchi A, Gamberi B, Manfroi S, Marchisio M, Tura S (1994) All-trans retinoic acid and induction of apoptosis in acute promyelocytic leukemia cells. Leuk Lymphoma 14:503-507
Tsai MJ, Bartelmez S, Heyman R, Damm K, Evans R, Collins SJ (1992) A mutated retinoic acid receptor alpha exhibiting dominant negative activity alters the lineage development of a multipotent hematopoietic cell line. Gene Dev 6:2258–2269
Tukeshita A, Shibata Y, Shinjo K, Yanagi M, Tobita T, Ohnishi K, Miyawaki S, Shudo K, Ohno R (1996) Successful treatment of relapse of acute promyelocytic leukemia with a new synthetic retinoid, Am80. Ann Int Med 124:893–896
Vaickus L, Villalona-Calero MA, Caliguiri T (1993) Acute proganulocytic leukemia (APL): possible in vivo differentiation by granulocyte colony-stimulating factor (G-CSF). Leukemia 7:1680–1681
von den Driesch P (1994) Sweet's syndrome (acute febrile neutrophilic dermatosis). J Am Acad Dermatol 31:535–556
Wang ZY, Sun GL, Lu JX, Gu LJ, Huang ME, Chen SR (1990) Treatment of acute promyelocytic leukemia with all-*trans* retinoic acid in China. Nouv Rev Fr Hematol 32:34–36
Warrell Jr RP, Frankel SR, Miller WH, Scheinberg DA, Itri LM, Hittelman WN, Vyas R, Andreff A, Tafudi A, Jakubowski A, Gabrilove J, Gordon M, Dmitrovsky E (1991) Differentiation therapy of acute promyelocytic leukemia with tretinoin (all-*trans* retinoic acid). N Engl J Med 324:1385–1393
Warrell RP Jr, de Thé H, Wang ZY, Degos L (1993) Acute promyelocytic leukemia. N Engl J Med 329:177–189
Warrell RP, Maslak P, Eardley A, Heller G, Miller WH, Frankel SR (1994) Treatment of acute promyelocytic leukemia with all-*trans* retinoic acid: an update of the New York experience. Leukemia 8:926–933
Weis K, Rambaud S, Lavau C, Jansen J, Carvalho T, Carmo-Fonseca M, Lamond A, Dejean A (1994) Retinoic acid regulates aberrant nuclear localization of PML-RARa in acute promyelocytic leukemia cells. Cell 76:345–356
Wiley JS, Firkin FC (1995) Reduction of pulmonary toxicity by prednisolone prophylaxis during all-*trans* retinoic acid treatment of acute promyelocytic leukemia. Study Group Leukemia 9:774–778
Willy PJ, Mangelsdorf DJ (1997) Unique requirements for retinoid-dependent transcriptional activation by the orphan receptor LXR. Genes Dev 11:289–298

Willy PJ, Umesono K, Ong ES, Evans RM, Heyman RA, Mangelsdorf DJ (1995) LXR, a nuclear receptor that defines a distinct retinoid response pathway. Genes Dev 9:1033-1045
Wu Q, Li Y, Liu R, Agadir A, Lee M-O, Liu Y, Zhang X-K (1997) Modulation of retinoic acid sensitivity in lung cancer cells through dynamic balance of orphan receptors nur77 and COUP-TF and their heterodimerization. EMBO J 16:1656–1669
Yashamita T, Wakao H, Futaki M, Nakahata T, Miyajima A, Asano S (1997) Retinoids induce sustained activation of the JAK 2/STAT 5 and erythroid differentiation in a murine erythroleukemia cell line. Abst 1379
Yu V, Naar AM, Rosenfeld MG (1992) Transcriptional regulation by the nuclear receptor superfamily. Current Opin Biotechnol 3:597–602
Zelent A (1994) Translocation of the RARα locus to the PML or PLZF gene in acute promyelocytic leukemia. Br J Haematol 86:451–460
Zhang XK, Hoffmann B, Tran PBV, Graupner G, Pfahl M (1992) Retinoid X receptor is an auxiliary protein for thyroid hormone and retinoic acid receptors. Nature 355:441–446
Zhang X-K, Pfahl M (1993) Regulation of retinoid and thyroid hormone action through homodimeric and heterodimeric receptors. Trends Endocrinol Metab 4:156–162

CHAPTER 10

The Retinoids: Cancer Therapy and Prevention Mechanisms

K. NASON-BURCHENAL and E. DMITROVSKY

I. Summary

The retinoids, synthetic and natural derivatives of vitamin A, are ligands activating nuclear transcription factors that are members of the steroid receptor superfamily. Ligand-retinoid receptor interactions lead to activation or repression of target genes which in turn signal retinoid biological actions. There is a convergence of basic scientific and clinical findings in the retinoid field. As a fuller understanding of the retinoid signaling pathway has evolved, beneficial clinical effects were found using the retinoids in cancer therapy and prevention. Underscoring this convergence is the tight link that often exists between observed clinical retinoid responses and activation or repression of specific retinoid receptors. This chapter discusses this convergence through an analysis of retinoid-based cancer therapy and prevention mechanisms. Attention focuses on how an understanding of the retinoid signaling pathway provides insight into mechanisms responsible for retinoid clinical responses.

II. Retinoid Clinical Activities

Vitamin A is essential for vision and for fertility. It plays an important role in development and in regulation of the growth and maturation states of normal and neoplastic cells. Vitamin A also exerts major effects on immune function. In this chapter, attention is placed on the retinoid role in cancer therapy and prevention. While the teratogenic effects of the retinoids are recognized (GUDAS et al. 1994; ROTHMAN et al. 1995) and limit their use in women of child-bearing years, discussion of these biological effects are beyond the scope of this chapter. The therapeutic role of the retinoids in nonneoplastic diseases is also outside the scope of this chapter.

The retinoids have reported clinical activity in cancer therapy and prevention. Clinical activity is reported in several settings. These include treatment of the premalignant lesions oral leukoplakia (HONG et al. 1986; LIPPMAN et al. 1993), actinic keratosis (LIPPMAN et al. 1987), cervical dysplasia (MEYSKENS et al. 1994), and xeroderma pigmentosum (KRAEMER et al. 1988). Reduction in second hepatocellular (MUTO et al. 1996) or aerodigestive tract cancers (HONG et al. 1990; PASTORINO et al. 1993) follows retinoid treatments. Some established malignancies respond to retinoid-based therapy and include:

acute promyelocytic leukemia (APL) (Huang et al. 1988; Castaigne et al. 1990; Warrell et al. 1991), juvenile chronic myelogenous leukemia (Castleberry et al. 1994), and mycosis fungoides (Warrell et al. 1983). In combination with interferon-α2a, 13-*cis*-retinoic acid (13-*cis*-RA) is reported as active in treating squamous cell cancers of the skin (Lippman et al. 1992a) and cervix (Lippman et al. 1992b) or disseminated kidney cancer (Motzer et al. 1995). Nutritional intervention trials in Linxian, China with vitamin and mineral combinations have also affected cancer incidence and disease specific mortality (Blot et al. 1993). The following is a summary of the retinoid clinical activity in cancer therapy and prevention:

Treatment of premalignant lesions
- Oral leukoplakia
- Cervical dysplasia
- Xeroderma pigmentosum
- Actinic keratosis

Chemoprevention
- Second lung cancers
- Second head and neck cancers
- Second hepatocellular carcinomas

Treatment of overt malignancies
- Acute promyelocytic leukemia
- Juvenile chronic myelogenous leukemia
- Mycosis fungoides

Combination therapy
- Squamous cell skin cancer
- Squamous cell cervical cancer
- Kidney cancer

The observed reduction in second malignancies of the aerodigestive tract or liver following retinoid treatments raised the prospect of retinoid-based prevention of cancers in high-risk individuals. To begin to test this possibility, patients who were at risk for aerodigestive tract cancers due to tobacco use were treated with carotenoid-based prevention regimens. Unfortunately, clinical cancer prevention activity was not observed in this setting with the tested carotenoid (Alpha Tocopheol, Beta Carotene Cancer Prevention Study Group 1994; Omenn et al. 1996; Hennekens et al. 1996). The combination of beta carotene and vitamin A was also found not to reduce lung cancer incidence (Omenn et al. 1996). Given the high cost and long duration of such clinical prevention trials, alternative approaches need to be developed to assess which of the retinoids exhibit cancer prevention activity. This chapter summarizes the present knowledge about the retinoid signaling pathway to understand how retinoids act in cancer therapy and prevention. Key elements of the retinoid signaling pathway are identified, as reviewed in Fig. 1. The way in

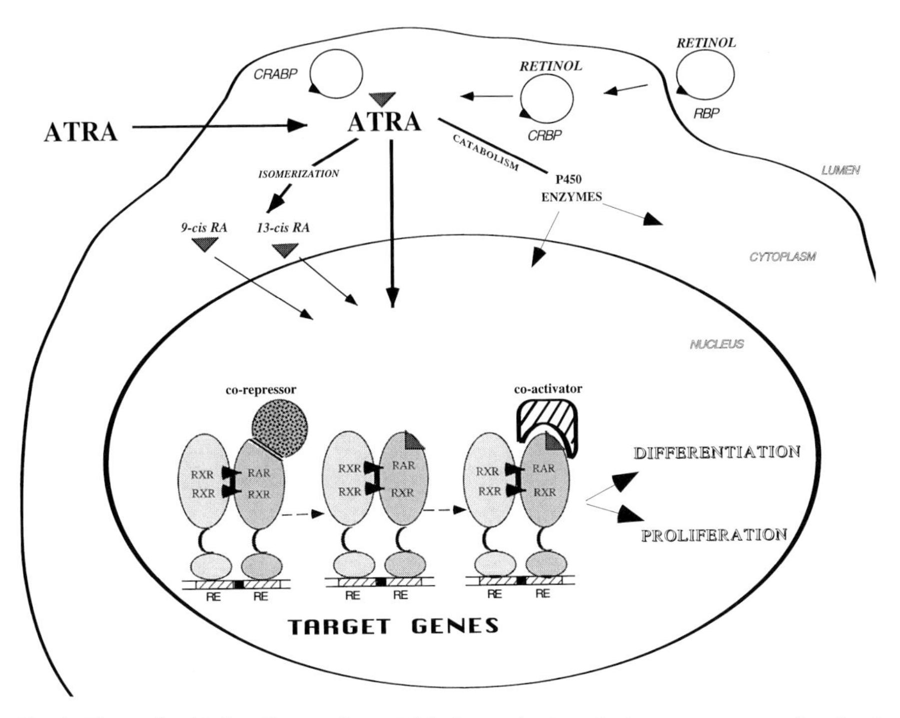

Fig. 1. The retinoid signaling pathway. This figure depicts the key components involved in all-*trans*-retinoic acid (*ATRA*) signaling. These components include the retinol-binding protein (*RBP*) present in the lumen, the cytosolic retinoid binding proteins (*CRABP* and *CRBP*), catabolic and modifying enzymes present in the cytoplasm, and the nuclear retinoid receptors (*RARs* and *RXRs*) which interact with defined DNA response elements (*RE*). Also depicted are important protein-protein interactions found between retinoid receptors and coactivators or corepressors that contribute to the ligand-dependent activation of target genes which directly transmit retinoid growth and differentiation signals. *Closed triangles*, ligands. Two stereoisomers of ATRA are depicted, including 9-*cis*-retinoic acid (9-*cis*-RA) and 13-*cis*-retinoic acid (13-*cis*-RA). (copyright by the American Society of Clinical Oncology Educational Book)

which these components play a critical role in clinical retinoid responses is discussed.

III. Mechanisms of Retinoid Action

Important components of the retinoid signaling pathway include cytoplasmic and nuclear retinoid receptors. Cytosolic and nuclear receptors are isolated and sequenced. These are structurally and functionally distinct receptors. Cytosolic retinoid binding proteins are identified (Stoner and Gudas 1989; Giguère et al. 1990). These cytosolic receptors can bind retinoids but not DNA. Their precise functions in the retinoid signaling pathway are being defined. These receptors may serve roles as intracellular retinoid storage sites.

The nuclear retinoid receptors are members of the steroid receptor superfamily, including vitamin D receptor (MCDONNEL et al. 1987), estrogen receptor (GREEN and KUMAR 1987), and thyroid hormone receptors (WEINBERGER et al. 1986), among others. These are hormone dependent transcription factors. The retinoid nuclear receptors bind specific retinoids through ligand binding domains and interact with DNA through DNA binding domains. These DNA-receptor and ligand-receptor interactions transcriptionally activate or repress target genes that mediate retinoid growth and differentiation effects. There are two families of nuclear retinoid receptors: retinoic acid receptors (RARs) and retinoid X receptors (RXRs). Three RARs exist with multiple isoforms: RARα (PETKOVICH et al. 1987; GIGUÈRE et al. 1987), RARβ (BRAND et al. 1988), and RARγ (KRUST et al. 1989). RARα is located on the long arm of chromosome 17 (MATTEI et al. 1988) at the site of the t(15;17) acute promyelocytic leukemia breakpoint. The chromosomal locations of the other RARs are known (MATTEI et al. 1991).

Three RXRs exist: RXRα (MANGELSDORF et al. 1990), RXRβ (HAMADA et al. 1989; MANGELSDORF et al. 1992), and RXRγ (MANGELSDORF et al. 1992; LEID et al. 1992). These retinoid receptors were known as orphan receptors until a physiological ligand was found: 9-*cis*-retinoic acid (9-*cis*-RA) (HEYMAN et al. 1992; LEVIN et al. 1992). The RXRs heterodimerize with other steroid receptor superfamily members, as reported (HAMADA et al. 1989; MANGELSDORF et al. 1992; LEID et al. 1992; HEYMAN et al. 1992; LEVIN et al. 1992; MANGELSDORF et al. 1991; KLIEWER et al. 1992; ZHANG et al. 1992a). Heterodimerization plays a role in cooperation between receptor-dependent signaling pathways. Different ligands are reported to favor homodimer versus heterodimer formation. For instance, 9-*cis*-RA favors formation of RXR-RXR homodimers versus RAR-RXR heterodimers (ZHANG et al. 1992b). The biological effects of these distinct dimerization patterns may depend on the cell context examined. Understanding the function of specific retinoid receptors in signaling the biological effects of the retinoids is assisted by development of agonists and antagonists for specific retinoid receptors which have been studied during induced tumor cell differentiation (LEHMANN et al. 1991, 1992; CHIBA et al. 1997). It is now recognized protein-protein interactions with coregulators have major effects on signaling retinoid receptor actions through associations with stimulatory coactivators or inhibitory corepressors that interact with the basal transcriptional machinery (CHEN and EVANS 1995; CHAKRAVARTI et al. 1996; KUROKAWA et al. 1995; OÑATE et al. 1995; GLASS et al. 1997, and references therein).

While synthetic retinoids are engineered to activate or inhibit transcription of specific retinoid receptors (LEHMANN et al. 1991, 1992), others exist that antagonize transcription of AP-1 (FANJUL et al. 1994). Some retinoids signal effects through receptor independent mechanisms. One example is N-(4-hydroxyphenyl)retinamide (4HPR). Although able to transcriptionally activate RARγ (Fanjul et al. 1996), 4HPR is viewed as predominantly signaling effects through receptor independent mechanisms (Delia et al. 1993, 1997; Sheikh et al. 1995; Oridate et al. 1996, 1997). 4HPR preferentially signals apoptosis even in retinoid-resistant cells (Delia et al. 1993). That 4HPR triggers apoptosis through novel mechanisms is supported by the finding its biological effects are signaled through induction of reactive oxygen species in some cell contexts (Delia et al. 1997; Oridate et al. 1997). Another retinoid, designated AHPN, also appears to signal through retinoid receptor independent mechanisms (Shao et al. 1995).

A tight link exists between expression of specific retinoid receptors and preclinical or clinical retinoid responses, as summarized below. Preclinical or clinical findings are displayed that indicate a link between retinoid biological responses and expression of specific retinoid receptors. Examples of a link between expression of specific retinoid receptors and beneficial clinical responses are found in oral leukoplakia where basal RARβ expression is repressed and 13-*cis*-retinoic acid induces RARβ expression in responding lesions. In acute promyelocytic leukemia, expression in leukemic cells of the t(15;17) rearranged receptor, PML/RARα, identifies those patients responsive to all-*trans*-retinoic acid therapy (ATRA) (in some cases an association between expression of specific receptors and clinical or preclinical responses has not yet been reported, indicated below by "–"):

- Oral leukoplakia: RARβ
- Squamous cell lung cancer: RARβ
- Transformed bronchial epithelium: RARα + RARβ
- Acute promyelocytic leukemia: PML/RARα
- Myeloid leukemia (HL-60): RARα
- Kidney cancer: RARβ
- Breast cancer: RARα + RARβ
- Male germ cell cancer: RARγ
- Cervical cancer: –
- Squamous cell skin cancer: –
- Mycosis fungoides: –
- Juvenile chronic myelogenous leukemia: –
- Aerodigestive tract chemoprevention: –
- Hepatocellular carcinoma chemoprevention: –

One example is illustrated in the premalignant lesion oral leukoplakia. In this example, an association between retinoid-induced RARβ mRNA expression and favorable clinical response to 13-*cis*-RA treatment is reported (Lotan et al. 1995). Another example is acute promyelocytic leukemia where expression of the fusion product of the t(15;17) rearrangement, PML/RARα, predicts clinical response to (ATRA)-treatment (Miller et al. 1992). Other examples of a link between preclinical retinoid responses and expression of specific retinoid receptors are presented above. These include results of transfection experiments revealing RARα mediates retinoid maturation response in myeloid leukemia (COLLINS et al. 1990), RARβ triggers retinoid growth suppressive response in squamous cell lung carcinoma cells (HOULE et al. 1993) and RARγ mediates retinoid growth and differentiation responses in human embryonal carcinomas (MOASSER et al. 1994, 1995b; SPINELLA et al. 1998). In transformed human bronchial epithelial cells, RARα and RARβ are mediators of retinoid-dependent growth suppression (AHN et al. 1995a). These findings and others obtained from transfection experiments provide a rationale for use of retinoid-receptor selective agonists to treat defined neoplastic diseases.

IV. Retinoids and Cancer Chemoprevention

A major therapeutic potential role for the retinoids is found in cancer prevention. The concept of cancer chemoprevention, first coined by SPORN et al. (1976), formulated the therapeutic rationale for treatments at early stages of carcinogenesis, before overt malignancies develop. Chemoprevention strategies can be directed towards the initiation, promotion, or progression stages of carcinogenesis. Several carcinogenesis stages might be targeted, including invasive or metastatic stages. If treatment strategies are successful in reversing or preventing progression of early neoplastic lesions, the major clinical consequences of malignancies would be reduced.

Based on preclinical findings (WOLBACH and HOWE 1925), epidemiological evidence (reviewed in HONG et al. 1994), data derived from experimental animal models (MOON et al. 1994), and successful clinical trials treating premalignant lesions or preventing second malignancies (HONG et al. 1986; LIPPMAN et al. 1993; LIPPMAN et al. 1987; KRAEMER et al. 1988; MUTO et al. 1996; HONG et al. 1990; PASTORINO et al. 1993), there is a strong rationale for use of retinoids in cancer chemoprevention. As reported by WOLBACH and HOWE in 1925, vitamin A is essential for epithelial cell homeostasis. Studies with vitamin A deficient rodents indicate squamous metaplasia develops. These metaplastic lesions are reminiscent of those detected in active smokers. Notably, when this vitamin A deficiency is repleted, the squamous metaplasia is reversed. This finding, when coupled with the epidemiological evidence of an inverse relationship between site-specific cancer incidences and serum vitamin A or β-carotene levels (reviewed in HONG et al. 1994), provides a strong rationale to conduct retinoid-based cancer prevention clinical trials. That β-carotene appears inactive in prevention of aerodigestive tract tumors in studied patients (ALPHA TOCOPHEROL, BETA CAROTENE CANCER PREVENTION STUDY GROUP 1994; OMENN et al. 1996; HENNEKENS et al. 1996) underscores the need to have an earlier assessment of chemoprevention response than is currently available when clinical outcome is the only end point used to assess the potential therapeutic impact. If clinical outcome is the sole outcome used to evaluate response, progress in retinoid-based cancer prevention will not likely be rapid.

Reliable "intermediate markers" are needed as surrogate endpoints for chemoprevention trials. Ideally, these would reflect early changes in affected epithelium useful to monitor effective clinical chemoprevention responses. Intermediate markers may identify individuals at high-risk for malignant transformation of precursor lesions. Alternatively, absence of such epithelial changes would identify those lesions least likely to develop into overt malignancies. In 1953 SLAUGHTER and colleagues proposed the existence of carcinogen-exposed epithelial "fields" of altered epithelial cells preceding development of invasive malignancies. The field cancerization hypothesis and the concept of multistep carcinogenesis are established and well defined especially in colon carcinogenesis (FEARON and VOGELSTEIN 1990). While these concepts have guided thinking in clinical cancer prevention, it is perhaps surprising that it remains to be seen which genetic changes are rate-limiting for maintenance or progression of premalignant lesions. These rate-limiting changes may differ for each carcinogenic program. In support of this view, frequent and distinct genetic changes are reported in diverse epithelial sites including the colon (FEARON and VOGELSTEIN 1990), head and neck (CALIFANO et al. 1996), lung (HUNG et al. 1995; RUSCH et al. 1995), and others. Which pathways are essential for development or maintenance of the transformed phenotype need to be determined since these are attractive therapeutic targets for cancer prevention.

Some changes associated with epithelial cell transformation are altered by effective retinoid chemoprevention treatments. Other changes predict clinical retinoid resistance. One example of an association between expression of a specific retinoid receptor and beneficial retinoid clinical response is found in the repressed expression of RARβ present in premalignant aerodigestive tract lesions, including oral leukoplakia (LOTAN et al. 1995). After 13-*cis*-RA treatment of patients with oral leukoplakia, RARβ mRNA expression is induced in clinically responsive lesions (LOTAN et al. 1995). In contrast, aberrant p53 protein expression in oral leukoplakia lesions predicts clinical retinoid resistance (LIPPMAN et al. 1995). While mechanisms responsible for retinoid resistance triggered by aberrant p53 expression are not yet identified, the finding of clinical retinoid resistance occurring in oral leukoplakia lesions overexpressing p53 illustrates how genetic or immunohistochemical analyses of premalignant epithelial lesions can reveal those lesions likely to progress to an irreversibly transformed phenotype or those responsive to retinoid-based chemoprevention (LIPPMAN et al. 1995). As microarray technologies advance, perhaps those expressed mRNA species indicative of the transformed phenotype will be determined in patient-derived specimens.

One useful strategy to reveal chemoprevention mechanisms is to exploit clinical responses to guide basic scientific studies. While retinoids are reported to reduce second malignancies of the head and neck (HONG et al. 1990), lung (PASTORINO et al. 1993), or liver (MUTO et al. 1996), carotenoid-based cancer chemoprevention of aerodigestive tract tumors in high risk individuals has not been SUCCESSFUL (ALPHA TOCOPHEROL, BETA CAROTENE CANCER PREVENTION

STUDY GROUP 1994; OMENN et al. 1996; HENNEKENS et al. 1996). Perhaps differences in activated or repressed target genes triggered by retinoid but not carotenoid treatments account for these observed differences in clinical responses. Analysis of mechanisms responsible for these differences should identify pathways linked to effective cancer chemoprevention. Another approach taken in pursuit of retinoid-based cancer therapy or prevention mechanisms is to study relevant in vitro models.

V. In Vitro Models for Retinoid Activity in Cancer Therapy and Prevention

To understand how retinoids act in cancer therapy and prevention, in vitro models reflective of clinical retinoid activities are available for study. Hematopoietic and nonhematopoietic models exist and are useful to reveal retinoid-dependent growth and differentiation signals. Several in vitro models are often used to evaluate retinoid growth and differentiation signals in human tumors. These include: the multipotent human embryonal carcinoma cell line NTERA-2 clone D1 (abbreviated NT2/D1), the APL cell line NB4, and the immortalized human bronchial epithelial cell line BEAS-2B, among others. The distinct features of these models are discussed below.

A. The Multipotent NT2/D1 Human Embryonal Carcinoma Line

The NT2/D1 human embryonal carcinoma cell line is multipotent, differentiating into a neuronal phenotype and other lineages after ATRA-treatment (ANDREWS 1984; ANDREWS et al. 1984). ATRA-treatment represses NT2/D1 tumorigenicity (DMITROVSKY et al. 1990a). Hexamethylene bisacetamide (HMBA) induces a nonneuronal phenotype (MILLER et al. 1994) with repressed tumorigenic potential. The undifferentiated NT2/D1 cells express the stage specific embryonal antigen, SSEA-3 (ANDREWS et al. 1990; FENDERSON et al. 1987). Following ATRA or HMBA treatments, SSEA-3 marker expression declines (MILLER et al. 1994). The neuronal antigen recognized by the A2B5 monoclonal antibody is induced by ATRA-treatment (MILLER et al. 1994; FENDERSON et al. 1987) but not by HMBA-treatment (MILLER et al. 1994). This in vitro NT2/D1 differentiation model is clinically relevant since maturation is common in clinical germ cell tumors. This is reflected in the frequent appearance of mature teratoma in some clinical germ cell tumors (HONG et al. 1977).

When compared to F9 murine teratocarcinomas, the human NT2/D1 line has distinct features. These include: (a) requirement for high ATRA doses to efficiently signal terminal differentiation (KURIE et al. 1993b); (b) a multipotent response to ATRA (ANDREWS 1984; ANDREWS et al. 1984); (c) distinct phenotypes induced following combined treatments with dibutyryl cAMP and ATRA or combined treatments with phorbol ester and ATRA (KURIE et al.

1993a,c, 1994); (d) unique immunophenotypic maturation markers (Fenderson et al. 1987); and (e) a marker chromosomal abnormality in NT2/D1 and other human germ cell tumors, an isochromosome of 12p (Dmitrovsky et al. 1990b; Bosl et al. 1989; Murty et al. 1990).

Findings obtained using the NT2/D1 embryonal carcinoma cell line are more reflective of the differentiation potential of human germ cell tumors than studies performed using available murine teratocarcinoma models. To determine which retinoid receptor is responsible for terminal differentiation of these human embryonal carcinoma cells, retinoid-resistant NT2/D1 cells were derived via mutagenization and selection in increasing ATRA doses. Notably, all ATRA-resistant clones derived and a spontaneously retinoid resistant embryonal carcinoma cell line, N2102Ep, aberrantly expressed RARγ (Moasser et al. 1996). Over-expression of wild-type RARγ in a retinoid resistant NT2/D1 clone, designated NT2/D1-R1 and having altered RARγ expression, reversed retinoid resistance but did not restore the complete retinoid differentiation response (Moasser et al. 1994, 1995b). Over-expression of RARγ in parental NT2/D1 cells leads to terminal growth suppression and induction of a mesenchymal phenotype, not a neuronal phenotype (Moasser et al. 1995b).

In search for other retinoid receptors which might trigger terminal differentiation in these human embryonal carcinoma cells, each of the RARs and the RXRs were individually transfected into parental NT2/D1 cells. Results reveal of the RARs only RARγ induces terminal NT2/D1 morphological maturation and growth suppression (Moasser et al. 1995b; Spinella et al. 1998). Of the RXRs, only RXRβ triggers in these cells morphological maturation and terminal growth suppression (Spinella et al. 1998). The phenotypes induced by ATRA-treatment versus RARγ or RXRβ transfection are distinct. To determine whether cooperation between RAR and RXR pathways are required to elicit the full retinoid response, RARγ and RXRβ were cotransfected into NT2/D1 cells. This yielded a neuronal phenotype reminiscent of that following RA-treatment (Spinella et al. 1998). Thus, these findings indicate cooperation between RAR and RXR pathways is required to trigger the full retinoid response in human embryonal carcinoma cells. Perhaps an analogous cooperation is active in other cell contexts.

Whether retinoid-treatment of chemotherapy-resistant clinical germ cell cancers signals tumor cell differentiation or antitumor responses was studied using receptor nonselective retinoids including 13-*cis*-RA and ATRA (Gold et al. 1984; Moasser et al. 1995a). While extensive teratoma formation was identified in some chemotherapy-resistant germ cell cancer patients treated with 13-*cis*-RA and stable disease was noted in others, objective antitumor responses were not reported following treatment with 13-*cis*-RA or ATRA (Gold et al. 1984; Moasser et al. 1995a). Whether retinoid receptor selective agonists that activate RARγ or RXRβ trigger major antitumor responses through germ cell tumor differentiation is the subject of future clinical work.

B. Acute Promyelocytic Leukemia In Vitro Models

ATRA-treatment of APL cases that contain the t(15;17) rearrangement induces leukemic cell maturation and transient complete remissions (HUANG et al. 1988; CASTAIGNE et al. 1990; WARRELL et al. 1991). Paradoxically, these clinical ATRA responses are associated with expression of the t(15;17) fusion product, PML/RARα (MILLER et al. 1992). To understand how ATRA-treatment triggers differentiation in APL, the NB4 APL cell line (LANOTTE et al. 1991) has proven useful to study. This is an APL line containing the t(15;17) rearrangement and expressing PML/RARα. Variant translocations also exist and involve RARα rearrangements (CHEN et al. 1993; REDNER et al. 1996; WELLS et al. 1997). In NB4 APL cells and in clinical APL cases, the PML/RARα fusion transcript is expressed in leukemic cells with the known or putative transcription factors: PML and RARα. These are the products of the nonrearranged alleles from chromosomes 15 and 17, respectively. Findings indicate PML/RARα is a ligand-dependent transcription factor with dominant-negative functions exerted on PML, RARα, both PML and RARα, or down-stream pathways (DE THÉ et al. 1991; KAKIZUKA et al. 1991; KASTNER et al. 1992). Transfection experiments conducted in the leukemic cell lines, U937, K562, or HL-60, which lack PML/RARα expression, indicate inhibitory effects of PML/RARα (ROUSSELOT et al. 1994; GRIGNANI et al. 1993, 1995) on these leukemic cells.

In addition to inhibitory effects on leukemic cell maturation by engineered PML/RARα expression, inhibitory effects on the PML pathway are reported. Within t(15;17) APL cells, PML is associated with nuclear bodies, designated as PML oncogenic domains (PODs). ATRA-treatment reverses this abnormal POD distribution pattern (DYCK et al. 1994; KOKEN et al. 1994; WEISS et al. 1994). Antagonism of RARα-dependent transcription also occurs when PML/RARα is cotransfected in transient transfection assays (DE THÉ et al. 1991; KAKIZUKA et al. 1991; KASTNER et al. 1992). That PML/RARα triggers leukemogenesis is shown through transgenic lines evaluating PML/RARα expression in myeloid cells. While all reported lines develop impairment of myelopoiesis, in some lines leukemia occurs after a long latency in a subset of mice (EARLY et al. 1996; GRISOLANO et al. 1997; BROWN et al. 1997; HE et al. 1997). That a long latency is required for the development of overt leukemia suggests a second hit is needed to trigger leukemogenesis. The nature of this additional oncogenic event is not yet known.

The NB4 line is useful to evaluate mechanisms of retinoic acid response in an APL cell line that reflects key aspects of the biology of this leukemia and of clinical retinoid response (LANOTTE et al. 1991). As discussed, this line contains the t(15;17) balanced rearrangement and expresses PML/RARα (LANOTTE et al. 1991; KAKIZUKA et al. 1991). Before ATRA-treatment, NB4 cells have an immature myeloid phenotype and are highly proliferative and tumorigenic. After ATRA-treatment, a mature myeloid phenotype is induced with expression of the epitope recognized by the myelomonocytic immunophenotypic marker, MO-1. These differentiated cells have acquired ability to reduce nitrotetrazolium blue, a marker of mature myeloid function. These ATRA-differentiated NB4 cells have a more limited growth potential than undifferentiated NB4 cells. These differentiated cells have negligible clonal growth potential or tumorigenic capacity (LANOTTE et al. 1991; NASON-BURCHENAL et al. 1997; K. N.-B., unpublished observations). Thus, this line reflects key aspects of the ATRA differentiation response observed in APL cells of ATRA-treated t(15;17) patients.

To comprehensively evaluate the PML/RARα role in ATRA-signaling in the NB4 APL cell line, the growth and differentiation effects of transcription factors involved in retinoid signaling of differentiation have been evaluated. Using gain of function experiments, the products of the nonrearranged alleles, RARα and PML were each over-expressed in NB4 cells (AHN et al. 1995b; MU et al. 1994). When RARα or PML are individually over-expressed in NB4 cells, growth suppression is signaled. For RARα over-expression, maturation response is enhanced (AHN et al. 1995b). These findings indicate reducing PML/RARα levels relative to the nonrearranged alleles, RARα or PML, antagonizes the inhibitory effects of PML/RARα. Building on this view is the finding that engineered over-expression of RXRα in NB4 cells causes marked growth suppression of transfectants (NASON-BURCHENAL et al. 1997). Since RXR is a known heterodimerization partner of PML/RARα, these findings suggest sequestration of PML/RARα antagonizes the leukemogenic effects of PML/RARα. These gain of function studies, when coupled with the findings obtained using transgenic mice (EARLY et al. 1996; GRISOLANO et al. 1997; BROWN et al. 1997; HE et al. 1997), indicate the inhibitory effect of PML/RARα is overcome by ATRA-treatment.

To complement these studies, loss of function experiments were conducted using molecular genetic strategies to target PML/RARα expression. Two approaches taken include: homologous recombination at the PML/RARα locus in NB4 cells and hammerhead ribozymes engineered to inactivate PML/RARα mRNA in APL cells (PACE et al. 1994; NASON-BURCHENAL et al. 1998a,b). PML/RARα targeting experiments indicate persistent PML/RARα expression is required for maintenance of basal promyelocytic leukemic cell growth. This is shown to be the case for retinoid sensitive NB4 cells and retinoid resistant NB4 cells that harbor a mutation within the ligand binding domain of the RARα portion of PML/RARα (NASON-BURCHENAL et al. 1998a,b). Targeting of PML/RARα yields a lethal phenotype associated with prominent induction of apoptosis (NASON-BURCHENAL et al. 1998a,b). Taken together, these loss of function experiments (NASON-BURCHENAL et al. 1998a,b) and gain of function experiments using engineered PML/RARα expression in myeloid leukemic cells (ROUSSELOT et al. 1994; GRIGNANI et al. 1993, 1995), indicate PML/RARα acts as an antiapoptotic transcription factor. ATRA-treatment overcomes these antiapoptotic effects. While a full explanation for the clinical retinoid response in t(15;17) APL cases does not yet exist, the existence of cellular and transgenic models and useful molecular genetic tools such as hammerhead ribozymes to target PML/RARα expression should provide additional mechanistic insight into APL biology. Application of these reagents could lead to a resolution of the apparent paradoxic response to ATRA-treatment in APL cases. In addition to the PML/RARα role in leukemogenesis, PML/RARα cofactors and the histone deacetylase complex appear to play important roles in retinoid response (LIN et al. 1998; GRIGNANI et al. 1998). The existence of PML "knock-out" mice provides an opportunity to learn how PML or its target genes contribute to leukemogenesis (WANG et al. 1998).

Another approach to understand important mechanisms responsible for ATRA-maturation response in APL is to derive ATRA-resistant APL cells and to explore acquired defects accounting for retinoid resistance. Several groups reported ATRA-resistant NB4 APL lines (NASON-BURCHENAL et al. 1997; DUPREZ et al. 1992; DERMINE et al. 1993; SHAO et al. 1997; RUCHAUD et al. 1994). Common defects of different retinoic acid resistant promyelocytic leukemia cells include persistent telomerase activity and nuclear body disorganization (NASON-BURCHENAL et al. 1997). In some ATRA-resistant APL lines, mutations within the ligand binding domain of the RARα portion of PML/RARα are found (NASON-BURCHENAL et al. 1998a; SHAO et al. 1997). This suggests a dominant-negative PML/RARα contributes to acquired retinoid resistance. Alternatively, a role for second messenger pathways are implicated in overcoming retinoid resistance (RUCHAUD et al. 1994). That persistent expression of an altered PML/RARα is required for maintenance of leukemic cell growth is shown through targeting of the PML/RARα transcript derived from an NB4 retinoid resistant line that contains a ligand binding domain mutation in the RARα portion of PML/RARα (NASON-BURCHENAL et al. 1998a).

During the ATRA-treatment of t(15;17) APL patients, the majority of treated patients achieve clinical remissions. Yet, retinoid resistance occurs following single-agent ATRA therapy. Several hypotheses are offered to account for this retinoid resistance including induction of retinoic acid-binding proteins (CORNIC et al. 1992) and pharmacological mechanisms of resistance (MUINDI et al. 1992). These clinical mechanisms of APL ATRA-resistance need not be mutually exclusive. Perhaps those mechanisms shown as active in the in vitro setting of ATRA-resistant NB4 cells will be active in vivo in APL patients resistant to ATRA-based therapy. Whether mutations are frequently found in the ligand binding domain of the RARα portion of PML/RARα of APL cells derived from patients exhibiting retinoid resistance needs to be learnt. Future work will determine whether ATRA-resistant cell lines are useful to pursue clinically relevant mechanisms to overcome retinoid resistance in APL patients.

While ATRA-treatment induces remissions in t(15;17) APL cases, the PLZF-RARα variant translocation cases appear retinoid resistant (CHEN et al. 1993). The clinical retinoid responses of APL variant translocation cases need to be studied in greater depth. For t(15;17) APL cases, ATRA-treatment combined with chemotherapy induces long-term disease-free survival superior to that achieved with combination chemotherapy alone (DEGOS et al. 1995; FENAUX et al. 1993; FRANKEL et al. 1994; TALLMAN et al. 1997). The optimal scheduling and doses used for combining ATRA with chemotherapy regimens need to be determined.

C. An In Vitro Lung Cancer Prevention Model

Given the long duration and large size of clinical cancer chemoprevention trials using retinoids or other prevention agents, there is a need for in vitro models predictive of clinical cancer prevention activities. While reduction in second malignancies is reported for aerodigestive tract and hepatocellular cancers (MUTO et al. 1996; HONG et al. 1990; PASTORINO et al. 1993), the optimal retinoids active in each of these clinical settings have not yet been identified. This view is supported by clinical trials showing β-carotene treatment is ineffective in preventing lung cancers from arising in high risk individuals (ALPHA TOCOPHEROL, BETA CAROTENE CANCER PREVENTION STUDY GROUP 1994; OMENN et al. 1996; HENNEKENS et al. 1996).

To determine which retinoids are most active in lung cancer prevention, an in vitro model (REDDEL et al. 1988) was adapted to study how carcinogens mediate malignant transformation of immortalized human bronchial epithelial cells. This model uses T antigen-immortalized human bronchial epithelial cells and exposure of these cells to tobacco-derived carcinogens to assess whether retinoids antagonize transformation and to reveal pathways responsible for effective in vitro prevention (LANGENFELD et al. 1996). This model has revealed that ATRA inhibits malignant transformation of these epithelial cells by carcinogens (LANGENFELD et al. 1996). This in vitro cancer prevention activity is linked to a delay in the G_1-S cell cycle transition signaled through a decline in expression of cyclin E (LANGENFELD et al. 1996) and cyclin D (LANGENFELD et al. 1997) proteins. This ATRA-mediated decline in cyclin expression is signaled through posttranslational mechanisms activating cyclin proteolysis via a proteasome-dependent degradation pathway (LANGENFELD et al. 1997).

This model appears useful to identify transformation pathways activated by carcinogen exposure. It also reveals which pathways are inhibited by effective retinoid-based prevention in the in vitro setting. Whether these transformation pathways are antagonized during effective clinical chemoprevention is not yet known. How other retinoid dependent mechanisms are engaged during clinical cancer chemoprevention is being explored. Retinoid receptor-selective agonists and antagonists exist that may have reduced pharmacological toxicities as compared to receptor nonselective retinoids. As retinoid receptor-dependent pathways are identified as active in signaling cancer prevention, these agents will become available for testing in appropriate clinical trials. Perhaps the retinoids will ultimately prove most useful as pharmacological tools to identify anticarcinogenic pathways that are more optimally targeted by other therapeutic agents. Of course, nonretinoid prevention agents exist with potential clinical cancer chemoprevention activities. These agents affect estrogen-dependent signaling, inducible cyclooxygenase, or other pathways relevant to cancer prevention, as reviewed (HONG and SPORN 1997). Cooperation between the retinoids and other chemoprevention agents may augment clinical cancer prevention activities.

VI. Summary and Future Directions

The retinoids, synthetic and natural derivatives of vitamin A, are pharmacological agents active in cancer therapy and prevention. There is a convergence of basic and clinical findings in the retinoid field. Clinical progress in this field has occurred along with a fuller understanding of which nuclear retinoid receptors or coregulators transmit retinoid growth and differentiation signals. That some clinical responses are observed as linked to expression or induction of specific retinoid receptors indicate how receptor-selective pathways are responsible for these clinical actions. These findings, when coupled with available preclinical data, provide a strong rationale for use of retinoid receptor-selective agonists as therapeutic agents for treatment of neoplastic diseases.

One way to optimize the use of receptor selective or nonselective retinoids is to explore biological effects in in vitro models that reflect activity observed in clinical cancer therapy or prevention. The retinoids may eventually prove most useful to highlight key growth, differentiation, or anticarcinogenic pathways that are more favorably targeted by other agents. Whether receptor selective retinoids reduce pharmacological toxicities, as compared with receptor-nonselective retinoids, will be learned. How clinical efficacy of these retinoids are affected by their increased selectivity is not yet known. Identification of target genes that transmit retinoid biological effects is needed. These target genes might be distinct for different neoplastic settings. The pharmacological regulation of target genes could alter the aberrant growth or differentiation of neoplastic cells.

How the retinoids are clinically active in combination therapy with other ligands of the steroid-receptor superfamily, cytokines, growth factor agonists or antagonists, radiation therapy, chemotherapeutic agents, or other agents needs to be further studied. In vitro experiments indicate cooperation is already observed between second messenger pathways and the retinoid signaling pathway (KURIE et al. 1993a,c, 1994; RUCHAUD et al. 1994). Clinical cooperation between the retinoid and interferon signaling pathways is established in the treatment of squamous cell cancers of the skin, cervix, and in disseminated kidney cancer (LIPPMAN et al. 1992a,b; MOTZER et al. 1995). How best to use retinoids as single agents or as part of combination therapy regimens is under study. The clinical challenge is to understand how the retinoids act and when these agents should be used in the clinic. Pursuit of basic mechanisms responsible for retinoid biological effects will optimize their use in cancer therapy and prevention.

Acknowledgements. This work was supported partly by NIH R01-CA54494 (E.D.), NIH R01-CA62275 (E.D.), NIH P01-CA29502 (K.N.-B.), and ACS grant RPG-90-019-08-DDC (E.D.).

References

Ahn M-J, Langenfeld J, Moasser MM, Rusch V, Dmitrovsky E (1995a) Growth suppression of transformed human bronchial epithelial cells by all-*trans*-retinoic acid occurs through specific retinoid receptors. Oncogene 11:2357–2364

Ahn M-J, Nason-Burchenal K, Moasser MM, Dmitrovsky E (1995b) Growth suppression of acute promyelocytic leukemia cells having increased expression of the non-rearranged alleles: RARα or PML. Oncogene 10:2307–2314

Alpha Tocopherol, Beta Carotene Cancer Prevention Study Group (1994) The effect of vitamin E and beta carotene on the incidence of lung cancer and other cancers in male smokers. N Engl J Med 330:1029–1035

Andrews PW (1984) Retinoic acid induces neuronal differentiation of a cloned human embryonal carcinoma cell line in vitro. Devel Biol 103:285–293

Andrews PW, Damjanov I, Simon D, Banting GS, Carlin C, Dracopoli NC, Fogh J (1984) Pluripotent embryonal carcinoma clones derived from the human teratocarcinoma cell line tera-2: differentiation in vivo and in vitro. Laboratory Investigation 50:147–162

Andrews PW, Nudelman E, Hakomori S-I, Fenderson BA (1990) Different patterns of glycolipid antigens are expressed following differentiation of TERA-2 human embryonal carcinoma cells induced by retinoic acid, hexamethylene bisacetamide (HMBA), or bromodeoxyuridine (BUdR). Differentiation 43:131–138

Blot WJ, Li J-Y, Taylor PR, Guo W, Dawsey S, Wang G-Q, Yang CS, Zheng S-F, Gail M, Li G-Y, Yu Y, Liu B-Q, Tangrea J, Sun Y-H, Liu F, Fraumeni JF Jr, Zhang Y-H, Li B (1993) Nutrition intervention trials in Linxian, China: supplementation with specific vitamin/mineral combinations, cancer incidence, and disease – specific mortality in the general population. J Natl Can Inst 85:1483–1491

Bosl GJ, Dmitrovsky E, Reuter VE, Samaniego F, Rodriguez E, Geller NL, Chaganti, RSK (1989) Isochromosome of chromosome 12p: clinically useful marker from male germ cell tumors. J Natl Canc Inst 81:1874–1878

Brand N, Petkovitch M, Krust A, Chambon P, de Thé H, Marchio A, Tiollais P, Dejean A (1988) Identification of a second human retinoic acid receptor. Nature 332:850–853

Brown D, Kogan S, Lagasse E, Weissman I, Alcalay M, Pelicci PG, Atwater S, Bishop JM (1997) A PMLRARα transgene initiates murine acute promyelocytic leukemia. Proc Natl Acad Sci (USA) 94:2551–2556

Califano J, van der Riet P, Westra W, Nawroz H, Clayman G, Piantadosi S, Corio R, Lee D, Greenberg B, Koch W, Sidransky D (1996) Genetic progression model for head and neck cancer: Implications for field cancerization. Cancer Res 56:2488–2492

Castaigne S, Chomienne C, Daniel MT, Ballerini P, Berger R, Fenaux P, Degos L (1990) All-*trans* retinoic acid as a differentiation therapy for acute promyelocytic leukemia. I. Clinical results. Blood 76:1704–1709

Castleberry RP, Emanuel PD, Zuckerman KS, Cohn S, Strauss L, Byrd RL, Homans A, Chaffee S, Nitschke R, Gualtieri RJ (1994) A pilot study of isotretinoin in the treatment of juvenile chronic myelogenous leukemia. N Eng J Med 331:1680–1684

Chakravarti D, LaMorte VJ, Nelson MC, Nakajima T, Schulman IG, Juguilon H, Montminy M, Evans RM (1996) Role of CBP/P300 in nuclear receptor signaling. Nature 383:99–103

Chen Z, Brand NJ, Chen A, Chen SJ, Tong JH, Wang ZY, Waxman S, Zelent A (1993) Fusion between a novel Krüppel-like zinc finger gene and the retinoic acid receptor-α locus due to a variant t(11;17) translocation associated with acute promyelocytic leukemia. EMBO J 12:1161–1167

Chen JD, Evans RM (1995) A transcriptional co-repressor that interacts with nuclear hormone receptors. Nature 377:454–457
Chiba H, Clifford J, Metzger D, Chambon P (1997) Distinct retinoid X receptor-retinoic acid receptor heterodimers are differentially involved in the control of expression of retinoid target genes in F9 embryonal carcinoma cells. Molecular and Cellular Biology 17:3013–3020
Collins SJ, Robertson KA, Mueller L (1990)Retinoic acid-induced granulocytic differentiation of HL-60 myeloid leukemia cells is mediated directly through the retinoic acid receptor (RAR-α). Molecular and Cellular Biology 10:2154–2163
Cornic M, Delva L, Guidez F, Balitrand N, Degos L, Chomienne C (1992) Induction of retinoic acid-binding protein in normal and malignant human myeloid cells by retinoic acid in acute promyelocytic leukemia patients. Cancer Res 52:3329–3334
Degos L, Dombret H, Daniel M-T, Miclea J-M, Chastang C, Castaigne S, Fenaux P (1995) All-*trans*-retinoic acid as a differentiation agent in the treatment of acute promyelocytic leukemia. Blood 85:2643–2653
Delia D, Aiello A, Lombardi L, Pelicci PG, Grignani F, Grignani, F, Formelli F, Menard S, Costa A, Veronesi U, Pierotti MA (1993) N-(4-Hydroxyphenyl)retinamide induces apoptosis of malignant hemopoietic cell lines including those unresponsive to retinoic acid. Cancer Res 53:6036–6041
Delia D, Aiello A, Meroni L, Nicolini M, Reed JC, Pierotti MA (1997) Role of antioxidants and intracellular free radicals in retinamide-induced cell death. Carcinogenesis 18:943–948
Dermime S, Grignani F, Clerici M, Nervi C, Sozzi G, Talamo GP, Marchesi E, Formelli F, Parmiani G, Pelicci PG, Gambacorti-Passerini C (1993) Occurrence of resistance to retinoic acid in the acute promyelocytic cell line NB4 is associated with altered expression of the pml/RARα protein. Blood 82:1573–1577
de Thé H, Lavau C, Marchio A, Chomienne C, Degos L, Dejean A (1991) The PML-RARα fusion mRNA generated by the t(15;17) translocation in acute promyelocytic leukemia encodes a functionally altered RAR Cell 66:675–684
Dmitrovsky E, Moy D, Miller WH Jr, Li A, Masui H (1990a) Retinoic acid causes a decline in TGF-α expression, cloning efficiency, tumorigenicity in a human embryonal cancer cell line. Oncogene Res 5:233–239
Dmitrovsky E, Murty VVVS, Moy D, Miller WH Jr, Nanus D, Albino AP, Samaniego F, Bosl G, Chaganti RSK (1990b) Isochromosome 12p in non-seminoma cell lines: karyological amplification of c-ki-ras$_2$ without point-mutational activation. Oncogene 5:543–548
Duprez E, Ruchaud S, Houge G, Martin-Thouvenin V, Valensi F, Kastner P, Berger R, Lanotte M (1992) A retinoic acid "resistant" t(15;17) acute promyelocytic leukemia cell line: isolation, morphological, immunological, molecular features. Leukemia 6:1281–1287
Dyck JA, Maul GG, Miller WH Jr, Chen JD, Kakizuka A, Evans RM (1994) A novel macromolecular structure is a target of the promyelocyte-retinoic acid receptor oncoprotein. Cell 76:333–343
Early E, Moore MAS, Kakizuka A, Nason-Burchenal K, Martin P, Evans RM, Dmitrovsky E (1996) Transgenic expression of PML/RARα impairs myelopoiesis. Proc Natl Acad Sci (USA) 93:7900–7904
Fanjul AN, Delia D, Pierotti MA, Rideout D, Qiu J, Pfahl M (1996) 4-Hydroxyphenyl retinamide is a highly selective activator of retinoid receptors. J Biol Chem 271:22441–22446
Fanjul A, Dawson MI, Hobbs PD, Jong L, Cameron JF, Harlev E, Graupner G, Lu X-P, Pfahl M (1994) A new class of retinoids with selective inhibition of AP-1 inhibits proliferation. Nature 372:107–111
Fearon ER and Vogelstein B (1990) A genetic model for colorectal tumorigenesis. Cell 61:759–767
Fenaux P, LeDeley MC, Castaigne S et al. (1993) Effect of all-*trans*-retinoic acid in newly diagnosed acute promyelocytic leukemia. Results of the multicenter randomized trial. Blood 82:3241–3249

Fenderson BA, Andrews PW, Nudelman E, Clausen H, Hakomori S-I (1987) Glycolipid core structure switching from globo- to lacto- and ganglio-series during retinoic acid-induced differentiation of TERA-2 derived human embryonal carcinoma cells. Devel Biol 122:21–34
Frankel SR, Eardley A, Heller G, Berman E, Miller WH Jr, Dmitrovsky E, Warrell RP Jr (1994) All-*trans*-retinoic acid for acute promyelocytic leukemia: Results of the New York study. Ann Int Med 120:278–286
Giguère D, Lyn S, Yip P, Siu C-H, Amin S (1990) Molecular cloning of cDNA encoding a second cellular retinoic acid-binding protein. Proceedings of the National Academy of Sciences (USA) 87:6233–6237
Giguère V, Ong ES, Segui P, Evans RM (1987) Identification of a receptor for the morphogen retinoic acid. Nature 330:624–629
Glass CK, Rose DW, Rosenfeld MG (1997) Nuclear receptor coactivators. Current Opinions in Cell Biology 9:222–232
Gold EJ, Bosl GJ, Itri LM (1984) Phase II trial of 13-*cis*-retinoic acid in patients with advanced non-seminomatous germ cell tumors. Cancer Treat Rep 68:1287–1288
Green S, Kumar WP (1987) Human estrogen receptor cDNA: Sequence, expression, and homology to c-erb-A. Nature 325:75–78
Grignani F, De Matteis S, Nervi C, Tomassoni L, Gelmetti V, Cioce M, Fanelli M, Ruthardt M, Ferrara FF, Zamir I, Seiser C, Grignani F, Lazar MA, Minucci S, Pelicci PG (1998) Fusion proteins of the retinoic acid receptor-α recruit histone deacetylase in promyelocytic leukemia. Nature 391:815–818
Grignani F, Ferrucci PF, Testa U, Talamo G, Fagioli M, Alcalay M, Mencarelli A, Grignani F, Peschle C, Nicoletti I, Pelicci PG (1993) The acute promyelocytic leukemia-specific PML-RARα fusion protein inhibits differentiation and promotes survival of myeloid precursor cells. Cell 74:423–431
Grignani F, Testa U, Fagioli M, Barberi T, Masciulli R, Mariani G, Peschle C, Pelicci PG (1995) Promyelocytic leukemia-specific PML-retinoic acid receptor fusion protein interferes with erythroid differentiation of human erythroleukemia K562 cells. Cancer Res 55:440–443
Grisolano JL, Wesselschmidt RL, Pelicci PG, Ley TJ (1997) Altered myeloid development and acute leukemia in transgenic mice expressing PML-RARα under control of cathepsin G regulatory sequences. Blood 89:376–387
Gudas LJ, Sporn MB, Roberts AB (1994) Cellular biology and biochemistry of the retinoids. In Sporn MB, Roberts AB, Goodman DS (eds). The Retinoids: Biology, Chemistry, and Medicine (second edition). New York, NY. Raven Press, Ltd. p. 443
Hamada K, Gleason SL, Levi B-Z, Hirschfield S, Appella E, Ozato K (1989) H-2RIIBP, a member of the nuclear hormone receptor family that binds to both the regulatory element of major histocompatibility class I genes and the estrogen response element. Proceedings National Academy of Sciences (USA) 86:8289–8293
He L-Z, Tribioli C, Rivi R, Peruzzi D, Pelicci PG, Soares V, Cattoretti G, Pandolfi PP (1997) Acute leukemia with promyelocytic features in PML/RARα transgenic mice. Proc Natl Acad Sci (USA) 94:5302–5307
Hennekens CH, Buring JE, Manson JE, Stampfer M, Rosner B, Cook NR, Belenger C, La Motte F, Gaziano JM, Ridker PM, Willett W, Peto R (1996) Lack of effect of long-term supplementation with beta carotene on the incidence of malignant neoplasms and cardiovascular disease. N Eng J Med 334:1145–1149
Heyman RA, Mangelsdorf DJ, Dyck JA, Stein RB, Eichele G, Evans RM, Thaller C (1992) 9-*cis* retinoic acid is a high affinity ligand for the retinoid X receptor. Cell 68:397–406
Hong WK, Endicott J, Itri LM, Doos W, Batsakis JG, Bell R, Fofonoff S, Byers R, Atkinson EN, Vaughan C, Toth BB, Kramer A, Dimery IW, Skipper P, Strong S (1986) 13-*cis*-retinoic acid in the treatment of oral leukoplakia. N Eng. J Med 315:1501–1505
Hong WK and Itri LM. Retinoids and human cancer (1994) In Sporn MB, Roberts AB, Goodman DS (Eds). The Retinoids (Second Edition). New York, NY, Raven Press Ltd, p. 597

Hong WK, Lippman SM, Itri LM, Karp DD, Lee JS, Byers RM, Schantz SP, Kramer AM, Lotan R, Peters LJ, Dimery IW, Brown BW, Goepfert H (1990) Prevention of second primary tumors with isotretinoin in squamous-cell carcinoma of the head and neck. N Eng J Med 323:795–801

Hong WK, Sporn MB (1997) Recent advances in chemoprevention of cancer. Science 278:1073–1077

Hong WK, Wittes RE, Hajdu ST, Cvitkovic E, Whitmore WF, Golbey RB (1977) The evolution of mature teratoma from malignant testicular tumors. Cancer 40:2987–2992

Houle B, Rochette-Egly C, Bradley WEC (1993) Tumor-suppressive effect of the retinoic acid receptor β in human epidermoid lung cancer cells. Proceedings of the National Academy of Sciences (USA) 90:985–989

Huang M-E, YE Y-C, Chen S-R, Chai J-R, Lu J-X, Zhoa L, Gu L-J, Wang Z-Y (1988) Use of all-*trans* retinoic acid in the treatment of acute promyelocytic leukemia. Blood 72:567–572

Hung J, Kishimoto Y, Sugio K, Virmani A, McIntire DD, Minna JD, Gazdar AF (1995) Allele-specific chromosome 3p deletions occur at an early stage in the pathogenesis of lung carcinoma. JAMA 273:558–563

Kakizuka A, Miller WH Jr, Umesono K, Warrell RJ Jr, Frankel SR, Murty VVVS, Dmitrovsky E, Evans RM (1991) Chromosomal translocation t(15;17) in human acute promyelocytic leukemia fuses RARα with a novel putative transcription factor, PML. Cell 66:663–674

Kastner P, Perez A, Lutz Y, Rochette-Egly C, Gaub M-P, Durand B, Lanotte M, Berger R, Chambon P (1992) Structure, localization, and transcriptional properties of two classes of retinoic acid receptor α fusion proteins in acute promyelocytic leukemia (APL): structural similarities with a new family of oncoproteins. EMBO J 11:629–642

Kliewer SA, Umesono K, Mangelsdorf DJ, Evans RM (1992) Retinoid X receptor interacts with nuclear receptors in retinoic acid, thyroid hormone and vitamin D_3 signaling. Nature 355:446–449

Koken MHM, Puvion-Dutilleul F, Guillemin MC, Viron A, Linares-Cruz G, Stuurman M, de Jong L, Szostecki C, Calvo F, Chomienne C, Degos L, Puvion E, de Thé H (1994) The t(15;17) translocation alters a nuclear body in a retinoic acid-reversible fashion. EMBO J 13:1073–1083

Kraemer KH, DiGiovanna JJ, Moshell AN, Tarone RE, Peck G (1988) Prevention of skin cancer in xeroderma pigmentosum with the use of oral isotretinoin. N Eng J Med 318:1633–1637

Krust A, Kastner P, Petkovich M, Zelent A, Chambon P (1989) A third human retinoic acid receptor, hRAR-γ. Proceedings National Academy of Sciences (USA) 86:5310–5314

Kurie JM, Allopenna J, Dmitrovsky E. Retinoic acid stimulates protein kinase A-associated G proteins during human teratocarcinoma differentiation (1994) Biochim et Biophys Acta 1222:88–94

Kurie JM, Brown P, Salk E, Scheinberg D, Birrer M, Deutsch P, Dmitrovsky E (1993a) Cooperation between retinoic acid and phorbol esters enhances human teratocarcinoma differentiation. Differentiation 54:115–122

Kurie JM, Buck J, Eppinger TM, Moy D, Dmitrovsky E (1993b) 9-*cis* and all-*trans* retinoic acid induce a similar phenotype in human teratocarcinoma cells. Differentiation 54:123–129

Kurie JM, Younes A, Miller WH Jr, Burchert M, Chiu C-F, Kolesnick R, Dmitrovsky E (1993c) Retinoic acid stimulates the protein kinase C pathway before activation of its β-nuclear receptor during human teratocarcinoma differentiation. Biochim et Biophys Acta 1179:203–207

Kurokawa R, Söderström M, Horlein A, Halachmi S, Brown M, Rosenfeld MG, Glass CK (1995) Polarity-specific activities of retinoic acid receptors determined by a co-repressor. Nature 377:451–454

Langenfeld J, Kiyokawa H, Sekula D, Boyle J, Dmitrovsky E (1997) Posttranslational regulation of cyclin D1 by retinoic acid: a chemoprevention mechanism. Proc Natl Acad Sci (USA) 94:12070–12074
Langenfeld J, Lonardo F, Kiyokawa H, Passalaris T, Ahn M-J, Rusch V, Dmitrovsky E (1996) Inhibited transformation of immortalized human bronchial epithelial cells by retinoic acid is linked to cyclin E down-regulation. Oncogene 13:1983–1990
Lanotte M, Martin-Thouvenin V, Najman S, Balerini P, Valensi F, Berger R (1991) NB4, a maturation inducible cell line with t(15;17) marker isolated from a human acute promyelocytic leukemia (M3). Blood 77:1080–1086
Lehmann JM, Dawson MI, Hobbs PD, Husmann M, Pfahl M (1991) Identification of retinoids with nuclear receptor subtype-selective activities. Cancer Res 51:4804–4809
Lehmann JM, Jong L, Fanjul A, Cameron JF, Lu XP, Haefner P, Dawson MI, Pfahl M (1992) Retinoids selective for retinoid X receptor response pathways. Science 258:1944–1946
Leid M, Kastner P, Lyons R, Nakshatri H, Saunders M, Zacharewski T, Chen J-Y, Staub A, Garnier J-M, Mader S, Chambon P (1992) Purification, cloning and RXR identity of the HeLa cell factor with which RAR or TR heterodimerizes to bind target sequences efficiently. Cell 68:377–395
Levin AA, Sturzenbecker LJ, Kazmer S, Bosakowski T, Huselton C, Allenby G, Speck J, Kratzeisen CL, Rosenberger M, Lovey A, Grippo JF (1992) 9-*cis* retinoic acid stereoisomer binds and activates the nuclear receptor RXRα. Nature 355:359–361
Lin RJ, Nagy L, Inoue S, Shao W, Miller WH Jr, Evans RM (1998) Role of the histone deacetylase complex in acute promyelocytic leukemia. Nature 391:811–814
Lippman SM, Batsakis JG, Toth BB, Weber RS, Lee JJ, Martin JW, Hays GL, Goepfert H, Hong WK (1993) Comparison of low-dose isotretinoin with beta carotene to prevent oral carcinogenesis. N Eng J Med 328:15–20
Lippman SM, Kavanagh JJ, Paredes-Espinoza M, Delgadillo-Madrueno F, Paredes-Casillas P, Hong WK, Holdener E, Krakoff IH (1992b) 13-*cis*-retinoic acid plus interferon α-2a: Highly active systemic therapy for squamous cell carcinoma of the cervix. J Natl Can Inst 84:241–245
Lippman SM, Kessler JF, Meyskens FL Jr (1987) Retinoids as preventive and therapeutic anticancer agents. Cancer Treat Rep 71:391–405; 493–515
Lippman SM, Parkinson DR, Itri LM, Weber RS, Schantz SP, Ota DM, Schusterman MA, Krakoff IH, Gutterman JU, Hong WK (1992a) 13-*cis*-retinoic acid and interferon α-2a: effective combination therapy for advanced squamous cell carcinoma of the skin. J Natl Can Inst 84:235–241
Lippman SM, Shin DM, Lee JJ, Batsakis JG, Lotan R, Tainsky MA, Hittelman WN, Hong WK (1995) p53 and retinoid chemoprevention of oral carcinogenesis. Cancer Res 55:16–19
Lotan R, Xu X-C, Lippman SM, Ro JY, Lee JS, Lee JJ, Hong WK (1995) Suppression of retinoic acid receptor-β in premalignant oral lesions and its up-regulation by isotretinoin. N Eng J Med 332:1405–1410
Mangelsdorf DJ, Borgmeyer U, Heyman RA, Zhou JY, Ong ES, Oro AE, Kakizuka A, Evans RM (1992) Characterization of three RXR genes that mediate the action of 9-*cis*-retinoic acid. Genes and Development 6:329–344
Mangelsdorf DJ, Ong ES, Dyck JA, Evans RM (1990) Nuclear receptor that identifies a novel retinoic acid response pathway. Nature 345:224–228
Mangelsdorf DJ, Umesono K, Kliewer SA, Borgemeyer U, Ong ES, Evans RM (1991) A direct repeat in the cellular retinol-binding protein type II gene confers differential regulation of RXR and RAR Cell 66:555–561
Mattei MG, Petkovich M, Mattei J-F, Brand N, Chambon P (1988) Mapping of the human retinoic acid receptor to the q21 band of chromosome 17. Human Genetics 80:186–188
Mattei M-G, Rivière M, Krust A, Ingvarsson S, Vennström B, Islam, MQ, Levan G, Kautner P, Zelent A, Chambon P, Szpirer J, Szpirer C (1991) Chromosomal assign-

ment of the retinoic acid receptor (RAR) genes in the human, mouse, rat genomes. Genomics 10:1061–1069

McDonnel DP, Mangelsdorf DJ, Pike JW, Haussler MR, O'Malley BW (1987) Molecular cloning of the complementary DNA encoding the avian receptor for vitamin D. Science 235:1214–1217

Meyskens FL Jr, Surwit E, Moon TE, Childers JM, Davis JR, Dorr RT, Johnson CS, Alberts DS (1994) Enhancement of regression of cervical intraepithelial neoplasia II (moderate dysplasia) with topically applied all-*trans*-retinoic acid: a randomized trial. J Natl Canc Inst 86:539–543

Miller WH Jr, Kakizuka A, Frankel SR, Warrell RP Jr, DeBlasio A, Levine K, Evans RM, Dmitrovsky E (1992) Reverse transcription polymerase chain reaction for the rearranged retinoic acid receptor α clarifies diagnosis and detects minimal residual disease in acute promyelocytic leukemia. Proceedings of the National Academy of Sciences (USA) 89:2694–2698

Miller WH Jr, Maerz WJ, Kurie J, Moy D, Baselga J, Lucas DA, Grippo JF, Masui H, Dmitrovsky E (1994) All-*trans*-retinoic acid and hexamethylene bisacetamide (HMBA) regulate TGF-α and Hst-1/kFGF expression in differentiation sensitive but not in resistant human teratocarcinomas. Differentiation 55:145–152

Moasser MM, DeBlasio A, Dmitrovsky E (1994) Response and resistance to retinoic acid are mediated through the retinoic acid nuclear receptor γ in human teratocarcinomas. Oncogene 9:833–840

Moasser MM, Khoo K-S, Maerz WJ, Zelenetz A, Dmitrovsky E (1996) Derivation and characterization of retinoid-resistant human embryonal carcinoma cells. Differentiation 60:251–257

Moasser MM, Motzer RJ, Khoo K-S, Lyn P, Murphy BA, Bosl GJ, Dmitrovsky E (1995a) All-*trans*-retinoic acid for treating germ cell tumors in vitro activity and results of a phase II trial. Cancer 76:680–686

Moasser MM, Reuter VE, Dmitrovsky E (1995b) Overexpression of the retinoic acid receptor γ directly induces terminal differentiation of human embryonal carcinoma cells. Oncogene 10:1537–1543

Moon RC, Mehta RG, Rao KVN (1994) Retinoids and cancer in experimental animals. In Sporn MB, Roberts AB, Goodman DS (Eds). The Retinoids (Second Edition). New York, NY, Raven Press Ltd, p. 573

Motzer RJ, Schwartz L, Law TM, Murphy BA, Hoffman AD, Albino AP, Vlamis V, Nanus DM (1995) Interferon alfa-2a and 13-*cis*-retinoic acid in renal cell carcinoma: antitumor activity in a phase II trial and interactions in vitro J Clin Onc 13:1950–1957

Mu Z-M, Chin K-V, Liu J-H, Lozano G, Chang K-S (1994) PML, a growth suppressor disrupted in acute promyelocytic leukemia. Mol Cell Biol 14:6858–6857

Muindi J, Frankel SR, Miller WH Jr, Jakubowski A, Scheinberg DA, Young CW, Dmitrovsky E, Warrell RJ Jr (1992) Continuous treatment with all-*trans*-retinoic acid causes a progressive reduction in plasma drug concentrations: implication for relapse in retinoid "resistance" in patients with acute promyelocytic leukemia. Blood 79:299–303

Murty VVVS, Dmitrovsky E, Bosl GJ, Chaganti RSK (1990) Nonrandom chromosome abnormalities in testicular and ovarian germ cell tumor cell lines. Cancer Genet Cytogenet 50:67–73

Muto Y, Moriwaki H, Ninomiya M, Adachi S, Saito A, Takasaki KT, Tanaka T, Tsurumi K, Okuno M, Tomita E, Nakamura T, Kojima T for the Hepatoma Prevention Study Group (1996) Prevention of second primary tumors by an acyclic retinoid, polyprenoic acid, in patients with hepatocellular carcinoma. N Eng J Med 334:1561–1567

Nason-Burchenal K, Allopenna J, Bègue A, Stéhelin D, Dmitrovsky E, Martin P (1998a) Targeting of PML/RARα is lethal to retinoic acid resistant promyelocytic leukemia cells. Blood 92:1758–1767

Nason-Burchenal K, Maerz W, Albanell J, Allopenna J, Martin P, Moore MAS, Dmitrovsky E (1997) Common defects of different retinoic acid resistant promyelocytic leukemia cells are persistent telomerase activity and nuclear body disorganization. Differentiation 61:321–331

Nason-Burchenal K, Takle G, Pace U, Flynn S, Allopenna J, Martin P, George ST, Goldberg AR, Dmitrovsky E (1998b) Targeting the PML/RARα translocation product triggers apoptosis in promyelocytic leukemia cells. Oncogene 17:1759–1768

Omenn GS, Goodman GE, Thornquist MD, Balmes J, Cullen MR, Glass A, Keogh JP, Meyskens FL Jr, Valanis B, Williams JH Jr, Barnhart S, Hammar S (1996) Effects of a combination of beta carotene and vitamin A on lung cancer and cardiovascular disease. N Eng J Med 334:1150–1155

Oñate SA, Tsai SY, Tsai M-J, O'Malley BW (1995) Sequence and characterization of a coactivator for the steroid hormone receptor superfamily. Science 270:1354–1356

Oridate N, Lotan D, Xu X-C, Hong WK, Lotan R (1996) Differential induction of apoptosis by all-*trans*-retinoic acid and N-(4-hydroxyphenyl)retinamide in human head and neck squamous cell carcinoma cell lines. Clinical Cancer Res 2:855–863

Oridate N, Suzuki S, Higuchi M, Mitchell MF, Hong WK, Lotan R (1997) Involvement of reactive oxygen species in N-(4-hydroxyphenyl)retinamide-induced apoptosis and cervical carcinoma cells. J Natl Canc Inst 89:1191–1198

Pace U, Bockman JM, MacKay BJ, Miller WH Jr, Dmitrovsky E, Goldberg AR (1994) A ribozyme which discriminates in vitro between PML/RARα, the t(15;17)-associated fusion RNA of acute promyelocytic leukemia, and PML and RARα, the transcripts from the nonrearranged alleles. Cancer Res 54:6365–6369

Pastorino U, Infante M, Maioli M, Chiesa G, Buyse M, Firket P, Rosmentz N, Clerici M, Soresi E, Valente M, Belloni PA, Ravasi G (1993) Adjuvant treatment of stage I lung cancer with high-dose vitamin. A J Clin Oncol 11:1216–1222

Petkovitch M, Brand NJ, Krust A, Chambon P (1987) A human retinoic acid receptor which belongs to the family of nuclear receptors. Nature 330:444–450

Reddel RR, Ke Y, Gerwin BI, McMenamin MG, Lechner JF, Su RT, Brash DE, Park J-B, Rhim JS, Harris CC (1988) Transformation of human bronchial epithelial cells by infection by SV40 or adenovirus-12 SV40 hybrid virus, or transfection via strontium phosphate coprecipitation with the plasmid containing SV40 early region genes. Cancer Res 48:1904–1909

Redner RL, Rush EA, Faas S, Rudert WA, Corey SJ (1996) The t(5;7) variant of acute promyelocytic leukemia expresses a nucleophosmin-retinoic acid receptor fusion. Blood 87:882–886

Rothman JK, Moore LL, Singer MR, Nguyen U-S, Mannino S, Milunsky (1995) Teratogenicity of high vitamin A intake. N Eng J Med 333:1369–1373

Rousselot P, Hardas B, Patel A, Guidez F, Gäken J, Castaigne S, Dejean A, de Thé H, Degos L, Farzaneh F, Chomienne C (1994) The PML-RARα gene product of the t(15;17) translocation inhibits retinoic acid-induced granulocytic differentiation and mediated transactivation in human myeloid cells. Oncogene 9:545–551

Ruchaud S, Duprez E, Gendron MC, Houge G, Genieser HG, Jastorff B, Doskeland SO, Lanotte M (1994) Two distinctly regulated events, priming and triggering, during retinoid-induced maturation and resistance of NB4 promyelocytic leukemia cell line. Proc Natl Acad Sci (USA) 91:8428–8432

Rusch V, Klimstra D, Linkov I, Dmitrovsky E (1995) Aberrant expression of p53 or the epidermal growth factor receptor is frequent in early bronchial neoplasia, and coexpression precedes squamous cell carcinoma development. Cancer Res 55:1365–1372

Shao W, Benedetti L, Lamph WW, Nervi C, Miller WH Jr (1997) A retinoid-resistant acute promyelocytic leukemia subclone expresses a dominant-negative PML-RARα mutation. Blood 89:4282–4289

Shao Z-M, Dawson MI, Li XS, Rishi AK, Sheikh MS, Han Q-X, Ordonez JV, Shroot B, Fontana JA (1995) p53 independent G_0/G_1 arrest and apoptosis induced by a novel retinoid in human breast cancer cells. Oncogene 11:493–504

Sheikh MS, Shao Z-M, Li X-S, Ordonez JV, Conley BA, Wu S, Dawson MI, Han Q-X, Chao W-R, Quick T, Niles RM, Fontana JA (1995) N-(4-hydroxyphenyl)retinamide (4-HPR)-mediated biological actions involve retinoid receptor-independent pathways in human breast carcinoma Carcinogenesis 16:2477–2486

Slaughter DP, Southwick HW, Smejkal W (1953) "Field cancerization" in oral stratified squamous epithelium. Clinical implications of multicentric origin. Cancer 6:963–968

Spinella MJ, Kitareewan S, Mellado B, Sekula D, Khoo K-S, Dmitrovsky E (1998) Specific retinoid receptors cooperate to signal growth suppression and maturation of human embryonal carcinoma cells. Oncogene 8:3471–3480

Sporn MB, Dunlop NM, Newton DL, Smith JM (1976) Prevention of chemical carcinogenesis by vitamin A and its synthetic analogs (retinoids). Fed Proc 35:1332–1338

Stoner CM and Gudas LJ (1989) Mouse cellular retinoic acid binding protein: Cloning, complementary DNA sequence, and messenger RNA expression during the retinoic acid-induced differentiation of F9 wild type and RA-3–10 mutant teratocarcinoma cells. Cancer Res 49:1497–1504

Tallman MS, Andersen JW, Schiffer CA, Appelbaum FR, Feusner JH, Ogden A, Shepherd L, Willman C, Bloomfield CD, Rowe JM, Wiernik PH (1997) All-*trans*-retinoic acid in acute promyelocytic leukemia. N Engl J Med 337:1021–1028

Wang ZG, Delva L, Gaboli M, Rivi R, Giorgio M, Cordon-Cardo C, Grosveld F, Pandolfi PP (1998) Role of PML in cell growth and the retinoic acid pathway. Science 279:1547–1551

Warrell RP Jr, Coonley CJ, Kempin SJ, Myskowski P, Safai B, Itri LM (1983) Isotretinoin in cutaneous T-cell lymphoma. Lancet 2:629

Warrell RP Jr, Frankel SR, Miller WH Jr, Scheinberg DA, Itri LM, Hittelman WN, Vyas R, Andreeff M, Tafuri A, Jakubowski A, Gabrilove J, Gordon MS, Dmitrovsky E (1991) Differentiation therapy of acute promyelocytic leukemia with tretinoin (all-*trans*-retinoic acid). N Eng J Med 324:1385–1393

Weinberger C, Thompson CC, Ong ES, Lebo R, Gruol DJ, Evans RM (1986) The c-erb-A gene encodes a thyroid hormone receptor. Nature 324:641–646

Weis K, Rambaud S, Lavau C, Jansen J, Carvalho T, Carmo-Fonseca M, Lamond A, Dejean A (1994) Retinoic acid regulates aberrant nuclear localization of PML-RARα in acute promyelocytic leukemia cells. Cell 76:345–356

Wells RA, Catzavelos C, Kamel-Reid S (1997) Fusion of retinoic acid receptor α to NuMA, the nuclear mitotic apparatus protein, by a variant translocation in acute promyelocytic leukemia. Nature Genetics 17:109–113

Wolbach SB and Howe PR (1925) Tissue changes following deprivation of fat-soluble vitamin A. J Exp Med 42:753–777

Zhang X-K, Hoffman B, Tran PB-V, Graupner G, Phahl M (1992a) Retinoid X receptor is an auxiliary protein for thyroid hormone and retinoic acid receptors. Nature 355:441–446

Zhang X-K, Lehmann J, Hoffmann B, Dawson MI, Cameron J, Graupner G, Hermann T, Tran P, Pfahl M (1992b) Homodimer formation of retinoid X receptor induced by 9-*cis* retinoic acid. Nature 358:587–591

CHAPTER 11

Aberrant Expression and Function of Retinoid Receptors in Cancer

X.-C. Xu and R. Lotan

A. Introduction

Retinoids are a group of structural and functional analogs of vitamin A. They regulate a number of fundamental physiological processes including vision, reproduction, metabolism, differentiation, bone development, and pattern formation during embryogenesis (De Luca 1991; Gudas et al. 1994; Lotan 1995a). Certain retinoids are capable of modulating cell growth, differentiation, and apoptosis and suppressing carcinogenesis in a variety of tissue types (e.g., lung, skin, mammary gland, prostate, bladder) in animal models (Moon et al. 1994; Lotan 1996) and in clinical trials with patients with premalignant or malignant lesions of the oral cavity, cervix, bronchial epithelium, skin, and other organs (Kraemer et al. 1992; Hong and Itri 1994; Lotan 1996; Hong and Sporn 1997; Moon et al. 1997). Some retinoids exhibit antitumor activity against fully malignant cells in vitro as reflected by suppression of proliferation and induction of differentiation or apoptosis (Amos and Lotan 1991; Lotan et al. 1991; Gudas et al. 1994; Lotan 1995a). Consequently, retinoids are considered as potential therapeutic agents as well (Smith et al. 1992). The ability of all-*trans*-retinoic acid (ATRA) to induce differentiation of acute promyelocytic leukemia cells into granulocytes is the basis for its therapeutic activity in patients and ATRA is currently used for therapy of this type of cancer (Degos 1997; Soignet et al. 1997; Tallman et al. 1997; see Chap. 11, Sect. III).

The effects of retinoids are thought to be mediated mainly by nuclear retinoid receptors, which are members of the steroid hormone receptor superfamily. As with other members of this family, the retinoid receptors are ligand-activated, DNA-binding *trans*-acting, transcription-modulating proteins. The retinoid receptors include two types: retinoic acid receptors (RARs), which bind to all-*trans*-RA and 9-*cis* RA with similar affinities, and retinoid X receptors (RXRs), which bind 9-*cis* RA (Mangelsdorf et al. 1994; Chambon 1996). In humans, RARα mRNA is expressed in most of the tissues. In contrast, RARβ expression is more limited; it is prevalent in neural tissues and undetectable in skin. RARγ is expressed predominantly in the skin (Mangelsdorf et al. 1994; Chambon 1996). RXRs, as a whole, are widely expressed in adult tissues. RXRβ is found in almost all tissues without a striking pattern, while RXRα is abundant in the liver, kidney, spleen, and skin. In contrast, RXRγ

appears to be rather restricted to muscle and brain (Mangelsdorf et al. 1992; 1994). The RARα, RARβ, and RARγ genes have been localized to chromosomes 17q21, 3p24, and 12q13, respectively. The RXRα, RXRβ, and RXRγ genes have been mapped to chromosome 9q34.3, 6p21.3 and 1q22–23, respectively (Chambon 1996). Each type of receptor includes three subtypes: α, β, and γ, which possess distinct amino- and carboxy-terminal domains. RARs can form heterodimers with RXRs; The heterodimers can bind to specific DNA sequence-RA response elements (RAREs), characterized by direct repeats of (A/G)GGTCA separated by five nucleotides (DR5), with RXR bound in the 5′ and RAR in the 3′ position (Mangelsdorf et al. 1994; Chambon 1996). In the absence of all-*trans*-RA, corepressor is associated with the complex, thus mediating transcription repression. After addition of all-*trans*-RA, the CoR is dissociated and a coactivator is recruited. The complex is then able to activate transcription (Perlmann and Vennstroem 1995; Hoerlein et al. 1995; Kurokawa et al. 1995; Chen and Evans 1995). RXRs can also form homodimers that bind to DR1 RXREs as well as heterodimers with several other members of the steroid receptor family including vitamin D receptor, thyroid hormone receptor, peroxisome proliferator activator receptor (Mangelsdorf et al. 1994; Chambon 1996).

Because each receptor subtype exhibits distinct pattern of expression during embryonal development and different distributions in adult tissues, each is thought to regulate the expression of a distinct set of genes as well as a common set of genes (Mangelsdorf et al. 1994; Chambon 1996). Further details on nuclear retinoid receptors and their mechanisms of actions are reviewed in Chap. 7, Sect. I.

Epidemiological studies have indicated that a relationship exists between vitamin A status and cancer development (Hong and Itri 1994). This association suggests that retinoid-dependent signaling pathways play a role in suppression of carcinogenesis. The majority of the population in the developed countries is vitamin A sufficient, therefore, it is unlikely that vitamin A deficiency plays an important role in cancer development. However, because nuclear retinoid receptors are the proximate mediators of many effects of retinoids on gene expression, it is plausible to assume that changes in their expression and function may cause aberrations in the response of cells to retinoids and thereby alter the regulation of cell growth, differentiation, and the expression of the transformed phenotype. Further, altered expression or function of nuclear retinoid receptors could abrogate retinoid signal transduction pathways and enhance cancer development. Therefore, investigations of the expression patterns of retinoid receptors in normal, premalignant, and malignant tissues may provide important insight into the roles of these receptors in cancer development and the response of these tissues to retinoid treatment.

In this chapter, we summarized the reports on the expression of receptors in cultured untransformed and tumor cell lines and in embryos. Few reports described the expression of nuclear retinoid receptors in specimens of premalignant and malignant tissues in vivo.

B. Aberrant nuclear retinoid receptor expression and function in cultured cancer cell lines

I. Expression of Nuclear Retinoid Receptors in Lung Cancer Cell Lines

Normal airway epithelium expresses all three RARs in the following pattern: RARα > RARβ > RARγ. ATRA increased RARβ expression in normal epithelium, whereas RARα and RARγ remained unaffected by treatment. Lung cancer cell lines usually express mRNAs of three RXRs; however, they expressed varying level of RARα and RAR$\gamma$and RARβ expression is usually suppressed (Gebert et al. 1991; Geradts et al. 1993; Houle et al. 1991; Nervi et al. 1991; Sun et al. 1997; Zhang et al. 1996). The majority of lung cancer cell lines exhibit resistance to ATRA-induced growth inhibition and also exhibit a lack of RARβ inducibility by ATRA. Expression of an exogenous RARβ in NSCLC cell lines resulted in increased response of the cells to the growth inhibitory effects of ATRA (our unpublished data). Furthermore, lung carcinoma cells expressing a transfected RARβ exhibited decreased tumorigenicity in nude mice (Houle et al. 1993) and transgenic mice expressing anti-sense RAR-β_2 developed lung cancer (Berard et al. 1996). These findings indicate that RARβ may play an important role in lung carcinogenesis and response to ATRA.

II. Expression of Nuclear Retinoid Receptors in Head and Neck Squamous Cell Carcinoma Cell Lines

Abnormal expression of retinoic acid receptors in human oral and epidermal squamous cell carcinoma cell lines was reported by Hu et al. (1991). RARα mRNA was detected in all of the normal cell strains and in the head and neck squamous cell carcinoma (HNSCC) cell lines grown in the presence or absence of ATRA, whereas RARγ mRNA was expressed a lower level in 6 of 9 SCC cell lines than in their normal cell counterparts. In contrast, normal epithelial cells in the oral cavity expressed RARβ however, expression of this receptor was lost in 7 of 9 SCC cell lines derived from the oral cavity. Abnormal low expression of RARβ was suggested to contribute to neoplastic progression in stratified squamous epithelia.

Zou et al. (1994) examined the effects of retinoids on the growth and differentiation of four HNSCC cell lines. ATRA was found to inhibit proliferation of 3 of 4 cell lines and suppress squamous differentiation markers keratin CK1, transglutaminase type I (TGase I), and involucrin mRNAs and proteins to varying degrees in 3 of 4 cell lines. All cell lines expressed mRNAs for RARα, RARβ, RARγ, and RXRα. However, RARβ mRNA levels were very low in the 1483 and SqCC/Y1 cells that expressed the highest levels of squamous differentiation markers. ATRA treatment increased the mRNA levels of the three RARs in most of the cell lines but had no effect on the RXR mRNAs. The results showed that in HNSCC cells there was a decrease in RARβ and that there were no simple relationships between the expression of nuclear retinoid receptors and response to RA-induced growth inhibition (Zou et al. 1994). Further studies have demonstrated that the removal of serum from the growth medium of HNSCC lines 1483 and SqCC/Y1 resulted in a decrease in RARβ mRNA level and concurrent increases in the expression of keratin K1 and TGase I. ATRA increased RARβ and decreased K1 and TGase I mRNA levels in serum-free medium. Transcriptional activation of reporter genes by means of retinoid response elements (RARE and RXRE) indicated that the RXR-RAR pathway predominates over the RXR homodimer pathway in the 1483 cells. Among several synthetic retinoids with preference for binding to specific nuclear retinoid receptors, those that induced RARβ also suppressed K1. The inverse association between RARβ expression and K1 and TGase I levels implicates this receptor in suppression of keratinization in oral epithelial cells (Zou et al. 1999). Indeed, expression of exogenous RARβ in HNSCC SqCC/Y1 cells enhanced sensitivity to the suppressive effects of ATRA on squamous differentiation; TGase I level was suppressed after a 3-day treatment with 0.1 μM ATRA in four of five transfected clones, whereas it was not suppressed in vector only transfected cells even by 1 μM ATRA. None of the RARβ transfected clones showed an increased growth inhibition by ATRA. These findings suggest that in SqCC/Y1 cells, RARβ mediate suppression of squamous differentiation by ATRA without altering its growth inhibitory effects (Wan et al. 1999). A recent analysis of receptor expression in a panel of primary cultures of human oral biopsies from nonkeratinizing sites revealed that the loss of RARβ expression among cells derived from dysplastic lesions correlated strongly with the transition from senescence to immortality during oral cancer progression (McGregor et al. 1997). Thus, loss of RARβ expression may be associated with abnormal keratinization in nonkeratinizing oral epithelial cells as well as with oral carcinogenesis.

III. Expression of Nuclear Retinoid Receptors in Esophageal Cancer Cell Lines

The expression of retinoid receptors in seven human esophageal cancer (HEC) cell lines was analyzed by northern blotting and by reverse transcriptase polymerase chain reaction (RT-PCR) (Xu et al. 1999a). All seven HEC cell lines were found to express RARα and RARγ when grown in monolayer cultures in medium supplemented with 10% fetal bovine serum, but RARβ

was expressed only in five of the seven lines. Expression of RARα and RARγ showed minor changes in these cells after addition of ATRA, however, RARβ expression was up-regulated by ATRA in five of seven cell lines. The two cell lines TE-1 and TE-8 that were negative for RARβ before ATRA treatment failed to upregulate this receptor after treatment. The five cell lines in which RARβ was upregulated by ATRA also showed 25%–70% growth inhibition. In contrast, the two cell lines without RAR-β up-regulation failed to show any growth inhibition by ATRA. Furthermore, the five cell lines, which expressed RARβ failed to form colonies in soft agar, whereas the two cell lines, which expressed no RARβ did form colonies. Treatment with ATRA (1 μM) failed to inhibit colony formation in TE1 and TE8 cells in soft agar. These results suggest that the loss of RARβ expression in some esophageal cancer cells may be associated with resistance to ATRA.

IV. Expression of Nuclear Retinoid Receptors in Breast Cancer Cell Lines

Aberrant expression of nuclear retinoid receptors in breast cancer was reported by Roman et al. (1992). RARα was detected by northern blotting in all 11 cell lines and the level of its expression was greater in estrogen receptor (ER) positive cell lines than in ER negative ones. RARβ was expressed in 7 of 11 cell lines preferentially in ER negative lines, while RARγ was expressed in all 11 cell lines at similar levels in both ER positive and ER negative cell lines. Treatment with 1 μM ATRA failed to modulate RARα and RARγ expression, whereas RARβ mRNA increased from barely detectable to low levels in T-47D line (Roman et al. 1992). swisshelm et al. (1994) demonstrated that most breast cancer cell lines lost RARβ expression, whereas RARα, RARγ, and RXRβ were expressed variably in both normal and cancer cells. Interestingly, RARβ expression increased in senescing normal mammary epithelial cells during culture and ATRA was able to induce RARβ expression in these normal cells but not in several breast cancer cells. In some breast cancer cell lines, in which ATRA was able to up-regulate RARβ, this effect correlated with the growth-inhibitory effects of retinoids (Seewaldt et al. 1995). Several groups demonstrated by stable transfection that RARβ can mediated ATRA-induced growth inhibition in breast cancer cells (Li et al. 1995; Seewaldt et al. 1995, 1997; Liu et al. 1996). RARβ-transfected clones showed either G1 arrest or apoptosis when treated with ATRA. Other studies, however, showed RARα mediation of ATRAs effects in ER positive breast cancer cells (Sheikh et al. 1993; Fitzgerald et al. 1997). In some ER negative cell lines RARα expression was undetectable but could be induced by estradiol after transfection of estrogen receptor (Sheikh et al. 1993). These results suggest that loss of retinoic acid receptor function may be associated with the status of ER and could be a critical event in breast carcinogenesis.

V. Expression of Nuclear Retinoid Receptors in Normal Skin and Skin Cancer Cell Lines

Darwiche et al. (1996) found that both RARα and RARγ proteins were decreased in mouse skin keratinocytes overexpressing an oncogenic v-*ras*(Ha) gene, and RARα was suppressed in a benign keratinocyte cell line carrying a mutated c-*ras*(Ha) gene. Transfection of an RARα expression vector into benign tumor cells inhibited DNA synthesis in response to ATRA. Keratinocytes transduced with the v-*ras*(Ha) oncogene v-*ras*(Ha)-keratinocytes exhibited a suppressed ability of ATRA to transactivate a reporter gene driven by DR5 RARE. Both RARα and RARγ were down-regulated in cultured keratinocytes by 12-*O*-tetradecanoylphorbol-13-acetate, implicating protein kinase C in the regulation of retinoid receptors. Indeed, blocking protein kinase C function in v-*ras*(Ha)-keratinocytes with bryostatin restored RARα protein expression. These findings suggest that modulation of RARs could contribute to the neoplastic phenotype in mouse skin carcinogenesis.

Expression of nuclear retinoid receptors has been analyzed in normal human skin keratinocytes. The predominant receptors are RARγ and RXRα. Small amounts of RARα were found but no RARβ was detected by northern blotting and immunoprecipitation (Elder et al. 1991; Redfern and Todd 1992; Vollberg et al. 1992; Fisher et al. 1994). RARα and RARγ mRNAs were detected in similar abundance in undifferentiated keratinocytes, however, differentiation in vitro resulted in lowering of RARα mRNA relative to RAR γ mRNA. Treatment of keratinocytes with ATRA did not induce expression of RARβ mRNA. Similarly, three squamous cell carcinoma cell lines derived from human skin expressed RARα and RAR γ transcripts, but not RARβ transcripts. Transfection of keratinocytes with reporter plasmids indicated that the endogenously expressed RARs could activate transcription through the RARβ response element in a concentration-dependent manner.

RARs have been implicated in the regulation of epidermal keratinocyte differentiation (Fisher et al. 1996) and a truncated RARγ was found to act as a dominant negative suppressor of differentiation of keratinocytes (Aneskievich and Fuchs 1991).

VI. Expression of Nuclear Retinoid Receptors in Cervical Cancer Cell Lines

RARβ is expressed in normal cells of the uterine cervix and its expression was neither detectable nor inducible by ATRA in cervical cancer cell lines. The constitutively expressed RARα and RARγ were not modulated by ATRA (Lancillotti et al. 1995; Geisen et al. 1997). In HeLa cells, RARα, RARγ, and RXRα were expressed but RARβ was not detected (Bartsch et al. 1992). An analysis of 9 cervical carcinoma cell lines revealed that they expressed mRNAs for RARα, RARγ, and RXRα, however, RARβ mRNA was detected only in ME180, MS751, C4I, and CaSki. Treatment with retinoic acid increased RARα

level in C4I, HT-3, and CaSki; RAR-β in HT-3; and RARγ in HT-3 and C4I cells. There appeared to be no correlation between the responses of the cells to these agents and the presence or type of nuclear retinoid receptor, human papilloma virus, or mutant p53 in the cells (LOTAN et al. 1996). The HPV-immortalized cell line, ECE16-1, expressed RARα, RARγ, and RXRα in retinoid-free medium, but no RARβ was expressed in this cell line. ATRA treatment increased RARβ expression in ECE16-1 cells, whereas the other receptors were not regulated by four different retinoids (AGARWAL et al. 1996b). In another study, however, RARβ expression was not inducible or only slightly increased by ATRA treatment in CaSki, MS751, and ME180 cells (GEISEN et al. 1997). RAR-selective retinoids were able to suppress the proliferation of cervical cancer cells in a concentration dependent manner, while RXR-selective retinoids failed to show any growth inhibitory effects (AGARWAL et al. 1996; LOTAN et al. 1995b). However, combination of RAR and RXR-selective retinoids showed more than additive effects on growth of cervical cancer cells (LOTAN et al. 1995b).

VII. Expression of Nuclear Retinoic Acid Receptors in Embryonal Carcinoma Cell Lines

ATRA can induce differentiation of several pluripotential embryonal carcinoma cells including the P19 line, which differentiates into neuronal or muscle cells. A mutant clone of P19 (RAC65) which was selected to be resistant to ATRA was found to express a mutant RARα. As a result of a rearrangement the mutant RARα gene encodes a truncated protein that has lost the last 70 amino acids in the RA-binding carboxy terminal domain. Although the mutant receptor acted as a dominant repressor of transcription from a transiently transfected RARE-reporter construct; the transfection of this mutant RARα into wild-type cells failed to confer ATRA resistance, suggesting that P19-RAC65 cells may have additional mutation(s) affecting RA signaling (PRATT et al. 1990).

Truncated versions of the three RARs created deliberately were compared for their potency as inhibitors of ATRA-mediated gene transcription and of P19 cell differentiation. Interestingly, only truncated RARα was effective dominant negative repressor of transcription, whereas truncated RARβ and RARγ were not. To be effective in this assay, the truncated RARα had to have an intact DNA-binding domain and to heterodimerize with RXRs or RARs. The heterodimers prevent binding of normal receptors to RARE (COSTA and MCBURNEY 1996).

VIII. Expression of Nuclear Retinoid Receptors in Myeloid Leukemia

ATRA induces granulocytic differentiation of HL-60 early myeloid leukemia cells. However, most human leukemia cell lines fail to undergo differentiation after exposure to ATRA. To determine whether this resistance is related to altered retinoid receptor structure or function, ROBERTSON et al. (1991) compared receptor expression in sensitive HL-60 cells and resistant K562 cells and found that the K562 cells possessed much lower level of RARα (80 per cell) than the ATRA-sensitive HL-60 cells (550 per cell). Retroviral-mediated transduction of RARα cDNA into K-562 increased the number of RARα to 2000 per cell and rendered the cells ATRA-sensitive in that these cells exhibited diminished cell proliferation (accumulation of cells in G_0/G_1) and a decreased c-myc after ATRA treatment. Although a slight increase was observed in hemoglobin production the cells did not undergo significant erythroid, megakaryocytic, or myeloid differentiation. These results indicate that decreased level of RARα may lead to ATRA resistance and that restoration of RARα may result in sensitivity to growth inhibitory but not necessarily to differentiation-inducing effects of ATRA (ROBERTSON et al. 1991).

A mutant subclone of HL-60 (designated HL-60R) that exhibits relative resistance to ATRA was found to express a mutant RARα with markedly reduced affinity for ATRA. The mutant has a point mutation (C $\rightarrow$ T) in codon 411 of RARα. This mutation generates a termination codon resulting in the truncation of 52 amino acids at the COOH terminal domain of RARα. This mutated RARα exerts a dominant negative activity on *trans*-activating function of the normal RARα (ROBERTSON et al. 1992). Various retinoids, vitamin D3 analogs, hexamethylene-bis-acetamide, or dimethylsulfoxide, which induce differentiation of HL-60 wild-type cells failed to induce significant differentiation of HL-60R. These results indicate that in the HL-60 cells differentiation is directly mediated by RARα and mutation in this receptor can abrogate differentiation not only by retinoids but also by other agents (GRILLIER et al. 1997).

Interestingly, NB4 cells derived from a patient with acute promyelocytic leukemia exhibit a characteristic defective RARα resulting from the t(15;17) chromosomal translocation, which fuses a gene called PML with RARα (see Sect. C.I).

C. Aberrant Nuclear Retinoid Receptor Expression in Specimens from Normal, Premalignant, and Malignant Tissues

I. Expression of Aberrant RARα in Acute Promyelocytic Leukemia

Acute promyelocytic leukemia (APL) is a distinct type of acute myeloid leukemia that is characterized by blast cell morphology (M3 type in the FAB nomenclature) and t(15;17) translocation that fuses the promyelocytic leukemia (PML) gene on chromosome 15 to the retinoic acid receptor alpha (RARα) gene on chromosome 17 (DE THE et al. 1990). This results in the formation of PML-RARα and RARα-PML chimeric genes. The PML-RARα is transcribed to produce a chimeric protein comprised of a part of the PML protein in the amino terminus and a part of RARα in the carboxy terminus. Paradoxically, despite the RARα abnormality, ATRA is able to induce differentiation in APL cells, both in vitro and in patients, and this is followed by apoptosis, whereas other types of leukemia with intact retinoid receptors fail to respond to ATRA. The formation of PML-RARα is thought to be the cause for leukemogenesis because the chimeric protein acts as a dominant negative oncogene it antagonizes retinoid signaling and disrupts PML function. These effects block differentiation and promote survival (FERUCCI et al. 1997; SCHULMAN and EVANS 1997). PML, which has been localized to nuclear matrix associated bodies, contains a RING, B-box, and coiled-coil domains that are thought to mediate protein-protein interactions and may be critical for the formation and stabilization of the multiprotein complex that includes at least five additional proteins and is called POD that build the nuclear bodies. In APL, the formation of PML-RARα and aberrant PML proteins leads to destabilization of the protein complexes and disruption of nuclear bodies. PML-RARα can bind RXR and limit its availability to interact with intact RARs, vitamin D receptor, thyroid hormone receptor, and other heterodimerization partners and the disregulated interactions might lead to the block in differentiation at the promyelocyte stage characteristic of APL. Treatment with ATRA leads to degradation of PML-RARα and reassembly of the nuclear bodies (GRIMWADE and SOLOMON 1997).

Other cytogenetic variants of APL were found to have distinct translocations that also disrupt RARα. These include leukemias associated with t(5;17) and t(11;17) translocations, which fuse the NPM and PLZF genes to RARα, respectively. Unlike APL with t(15,17), these leukemias are resistant to ATRA. Although PLZF-RARα inhibits differentiation of hematopoietic precursor cell lines in response to various differentiation inducing agents such as PML-RARα, the PLZF-RARα protein does not increase sensitivity of hematopoietic precursor cells to ATRA and does not restores sensitivity in ATRA-resistant hematopoietic cells. Further, PLZF-RARα has either no effect or an inhibitory activity on retinoid signaling, whereas PML-RARα enhances ATRA signaling (RUTHARDT et al. 1997). The mechanism underlying the difference between the two chimeric proteins has been unraveled recently when it was found that whereas both PML-RARα and PLZF-RARα fusion proteins can act as transcriptional repressors and are able to interact with nuclear receptor transcriptional corepressors, such as SMRT or NCoR, the complex of the corepressors with PLZF-RARα, but not with PML-RARα are insensitive to ATRA and do not dissociate. Thus, unlike PML-RARα, PLZF-RARα cannot become transcription activator in the presence of ATRA (HE et al. 1998).

II. Expression of Nuclear Retinoid Receptors in Adjacent Tissues and Head and Neck Squamous Cell Carcinoma

Most tissue specimens available for retrospective studies are formalin-fixed, paraffin-embedded surgical specimens of solid tumors. The analysis of retinoid receptors in such samples required an adaptation of in situ hybridization to localize mRNA because the antibodies available were not monospecific as they cross reacted with several protein bands when analyzed by western blotting. A nonradioactive in situ hybridization method with digoxigenin-labeled antisense riboprobes of RARα, β, and γ and RXRα, β, and γ was used to analyze receptor expression in surgical specimens (Xu et al. 1994a). An analysis of histological sections of specimens from normal volunteers and HNSCC patients revealed that specimens from normal volunteers expressed all of these receptors. Similar levels of RARγ and RXRα and RXRβ mRNAs were detected in most of the adjacent normal, hyperplastic, dysplastic, and malignant tissues. RARα mRNA was detected in 94% of adjacent normal tissues and hyperplastic tissues and in 87% of dysplasia and 77% of HNSCCs. In contrast, RARβ mRNA was detected in about 70% of adjacent normal and hyperplastic lesions, and its expression decreased further to 56% of dysplastic lesions and to 35% of HNSCCs. The difference in RARβ level in carcinoma and adjacent normal tissues was significant (Xu et al. 1994b). These results indicated that decreased expression of RARβ occurs in dysplastic lesions and may be associated with HNSCC development.

A retrospective study of retinoid receptor expression in tissue specimens taken by punch biopsy from oral premalignant lesions (oral leukoplakia) from patients without cancer before ($n = 52$) and after ($n = 39$) a 3-month treatment revealed that RARα, RARγ, RXRα, RXRβ, and RXRγ were all expressed in these lesions at similar levels. In contrast, RAR-β mRNA was detected in only 21 of 52 oral premalignant lesions. The suppression of RARβ expression in the premalignant lesions was significant. Thirty-five of the 39 specimens available after treatment expressed RARβ. Eighteen of 22 cases (82%) in which RARβ level increased responded clinically to 13-*cis*-RA. These results showed that a selective loss of RARβ mRNA expression occurs at early stages of oral carcinogenesis and demonstrated that RARβ can be up-regulated by isotretinoin in humans. Furthermore, the association between the increase in RAR-β expression and clinical response suggests that RARβ have a role in mediating retinoid response (Lotan et al. 1995c).

III. Expression of Nuclear Retinoic Acid Receptors in Hamster Cheek-Pouch Mucosa During 7,12-Dimethylbenz[a]anthracene-Induced Carcinogenesis

To determine how early during oral carcinogenesis the expression of RARβ begins to decrease, we used hamster cheek-pouch carcinogenesis as model. The cheek-pouch carcinogenesis of the Syrian golden hamster involved using 0.5% DMBA three times a week for 16 weeks (Gimenez-Conti and Slaga 1993). Hyperplastic lesions developed in the cheek-pouch mucosa after two week-DMBA treatment, dysplastic and malignant lesions after four to eight week-treatment. For the detection of RARs expression, the animals were killed in 2, 4, 8, 16 weeks and formalin-fixed, paraffin-sections were prepared from the cheek-pouch mucosa. RAR-γ expression showed no remarkable changes during the 16-week treatments, whereas RARα expression decreased in part of the adjacent normal and malignant lesions only in later stages. In contrast, the expression of RARβ was decreased already after a 4-eek treatment with DMBA and completely lost in the tumors (Xu et al. 1998).

IV. Expression of Nuclear Retinoid Receptors in Adjacent Bronchial Epithelium and Non-small Cell Lung Cancer

Northern blot analysis of 9 tissue specimens from primary non-small cell lung cancer patients showed low RARβ expression in 30% of samples (Gebert et al. 1991). In surgical specimens from normal lung tissue and non-small cell lung cancers, all six receptors of RARs and RXRs were expressed in at least 89% of normal control bronchial specimens from 17 patients without a primary lung cancer and in distant normal bronchial specimens from NSCLC patients (Xu et al. 1997a). RARα, RXRα, and RXRγ were expressed in more than 95% of the 79 NSCLC specimens. In contrast, RARβ, RARγ, and RXRγ expression was detected in only 42%, 72% and 76% of NSCLC, respectively. These data, together with previously cell lines studies described above (B.I), suggest that the loss of RARβ expression may contribute to lung carcinogenesis.

Recently, we analyzed by in situ hybridization RARβ mRNA expression in bronchial biopsies from heavy smokers at baseline and after 6 months of treatment with 13-*cis*-RA or placebo to determine whether the expression of these receptors is modulated by chemopreventive intervention (Xu et al. 1999b). RARβ expression, which we detected previously in 90% of bronchial specimens from nonsmokers, was lost at baseline in at least one of six biopsies of smokers in 30 of 35 subjects in the 13-*cis*-RA group and in 24 of 33 subjects in the placebo group, or 85.7% and 72.7%, respectively. After 6 months of 13-*cis*-RA treatment, the number of RARβ positive cases increased from 5/35 to 13/35 (2.6-fold) such that the percentage of cases with an aberrant RARβ expression decreased to 62.9% (22/35), which is a significant difference from baseline expression (p = 0.01). In the placebo group there was no significant difference in RARβ expression between baseline and 6 months. RARβ expression was not related to smoking status or reversal of squamous metaplasia. These results indicate that RARβ is an independent marker of response to 13-*cis*-RA and may serve as an intermediate biomarker in chemoprevention trials of upper aerodigestive tract cancers (Xu et al. 1999b).

V. Expression of Nuclear Retinoid Receptors in Adjacent Normal, Premalignant, and Malignant Breast Tissues

The expression of nuclear retinoid receptors in breast tissue specimens was analyzed by using a nonradioactive in situ hybridization method

(Widschwendter et al. 1997; Xu et al. 1997b). Widschwendter et al. demonstrated that RARα was expressed in all specimens from 14 patients, whereas RARγ was expressed in all normal breast tissues but in only 11 of 14 tumors. RARβ was found in all cases of distant normal breast tissues, but it was completely lost in 13 of 14 tumor lesions and the morphologically adjacent normal tissues. Somewhat different results were obtained by Xu et al. (1997b), who found that the adjacent normal tissues expressed all the nuclear retinoid receptors (RARα, RARβ, RARγ, and RXRα) in nearly all cases. The expression of RARα, RARγ, and RXRα in ductal carcinoma in situ (DCIS) was also similar to normal tissue; however, the expression of RARβ was significant lower in DCIS than in normal, especially in the poorly differentiated DCIS. The expression of RARγ and RXRα in invasive cancer was similar to DCIS and adjacent normal tissues, whereas the expression of RARα and RARβ decreased to 83.6% and 51.6%, respectively in invasive carcinomas. The decreased expression of RARα in invasive carcinomas compared to normal as well as DCIS was marginally statistically significant, whereas the decrease in RARβ expression compared to normal as well as DCIS was significant. These results suggest that loss of RARβ expression may be involved in mammary carcinogenesis but at later stages than in aerodigestive tract carcinogenesis.

Lawrence et al. (1998) used a radioactive in situ hybridization, which enables one to count grains after autoradiography, and found that the relative expression levels of RXR in 58 biopsy specimens was higher than in the normal ductal/lobular units in each section in 66% of noncomedo DCIS and 88% of comedo DCIS lesions, both of which are associated increase in breast cancer risk. In contrast, only 8% of lesions that confer little or no increase in breast cancer risk overexpressed RXR mRNA. Overexpression of RARα, but not other RARs, was detected in only 36% of cases. Immunohistochemistry performed on 52 DCIS specimens demonstrated that the predominant overexpressed form of RXR was RXRα. It appears, therefore, that RXRα overexpression may be involved in breast oncogenesis

Immunostaining of specimens from 33 breast cancers revealed nuclear staining in the tumor cells as well as in the adjacent normal cells. RARα protein was expressed at significantly higher levels in tumors with greater proliferative activity as determined by immunostaining of adjacent sections with Ki-67 antibody. It is not clear whether the higher RARα expression is responsible for increased proliferation or is the consequence of the proliferation but the data suggest that RARα expression may be altered during tumor progression. A previous study relying on northern blotting (Roman et al. 1992) suggested a positive correlation between RARα mRNA levels and estrogen receptor status of breast tumors; however, no such relationship was detected at the protein level (van der Leede et al. 1996). This issue remains to be resolved because a recent study by Han et al. (1997), which also used immunohistochemistry, did find that among ten ER positive and nine ER negative breast cancer specimens, RARα protein levels were higher in the ER positive samples. The question was raised as to whether the decrease in RARα protein levels and loss of RA-mediated growth inhibition in ER-negative tumor plays a role in the increased metastatic potential of ER-negative tumors.

VI. Expression of Nuclear Retinoid Receptors in Normal Skin and Skin Cancers

Downregulation of RARs was suggested to be an essential component of the mechanism of mouse skin tumor carcinogenesis and promotion (Kumar et al. 1994; Darwiche et al. 1995). Darwiche et al. (1995; 1996) found that the mRNA levels of RARα and RARγ are decreased in benign mouse epidermal tumors relative to normal skin and are almost absent in carcinomas. An analysis of receptors at the protein level by western blotting demonstrated that RARα was slightly reduced in papillomas promoted with 12-*O*-tetradecanoylphorbol-13-acetate (low risk) and markedly reduced or absent in papillomas promoted by mezerein (high risk). However, mezerein also caused substantial reduction in RARα in nonmalignant skin. RARγ was not detected in tumors from either protocol and was greatly reduced in skin treated by either promoter.

Several studies showed a decreased or lost expression of RARγ in some skin cancers or premalignant lesions (Elder et al. 1991; Finzi et al. 1992). In situ hybridization experiments with normal sun-exposed epidermis and adjacent squamous cell carcinoma showed positive hybridization with antisense riboprobes for RARα and RARγ, but not RARβ in normal epidermis (Xu et al. 1996). Expression of RARα and RARγ was not uniform throughout the epidermal layers. RARα and RARγ receptor expression was much higher in the spinous and granular layers in comparison to the basal layer. This pattern of expression was present in all normal skin samples examined. RARβ was not found in any of the samples examined. Sections hybridized with antisense riboprobes for RXRα, RXRβ, and RXRγ showed expression of all three RXRs in normal epidermis. As with RARα and RARγ, the RXRs were expressed at higher levels in the spinous and granular layers than in the basal layer of the epidermis.

Consecutive sections of skin SCC were analyzed for the expression of nuclear retinoid receptors. In comparison to adjacent normal epidermis, SCCs exhibited marked decreases in the expression of RARα, RARγ, RXRα, RXRβ, and RXRγ. Again, RARβ was not detected in any SCC cells. The results showed that RARα and RARγ expression was decreased in 16 (94%) cases of the SCCs; RXRα, RXRβ, and RXRγ expression was decreased in 15 (88%), 12 (70%), and 14 (82%) cases of the SCCs, respectively. No increased expression of these receptors was found in any SCCs examined.

Low level of nuclear retinoid receptor expression was also seen in basal cell carcinoma (BCC). All BCCs examined were found to express the nuclear retinoid receptors at much lower levels than normal epidermis (RAR-β was not detected in any BCC). However, because these tumors are histologically similar to the nonkeratinizing basal cells of the epidermis, in situ staining was compared to the epidermal basal layer. The majority of the BCCs expressed RARα and RARγ and RXRα, RXRβ, and RXRγ at levels comparable to the cells of the basal epidermis.

VII. Expression of the Nuclear Retinoid Receptors in Normal Uterine Cervix and Cervical Cancer

The expression of nuclear retinoid receptors in normal cervical tissues of experimental animals was first reported by DARWICHE et al. (1994). In mouse cervix, both ectocervical and endocervical epithelia have been reported to express RARs and RXRs at varying level. RARα, RARγ, RXRα, and RXRβ were associated with the cervical stratified squamous subjunctional epithelium. The columnar epithelium expressed high levels of RARα, RARβ, RXRα, and RXRβ mRNAs, whereas RXRγ was undetectable in cervical epithelium by in situ hybridization (DARWICHE et al. 1994). RARβ, RXRα, and RXRβ mRNAs were down-modulated in vitamin A deficiency mice.

In surgical specimens from cervical cancer patients, RARs were expressed in all adjacent normal cervical epithelia, but their expression was reduced in the lesions, including cervical squamous cell carcinoma and cervical intraepithelial neoplasia I, II, and III (XU et al. 1999c).

D. Mechanisms of Altered Receptor Expression and Function In Vitro and In Vivo

Most abnormalities in retinoid receptors have been found in RARα and RARβ. The defects in RARα were mostly the result of point mutations and rearrangements. The most well studied are the translocation t(15;17) that leads to the fusion of PML with RARα and t(11;17) that results in fusion of PLZF and RARα (DE THE 1996; see Sect. C.I). Point mutations in RARα were reported for mutants of HL-60 cells (ROBERTSON et al. 1992) and P19 cells (COSTA and MCBURNEY 1996) selected for resistance to ATRA. It is not clear whether such mutations occur in spontaneous tumors or in APL patients who develop resistance while undergoing treatment with retinoids.

The mechanisms responsible for the decreased expression of RARβ are less well understood although this change occurs in a variety of premalignant and malignant lesions. Rearrangements in the RARβ gene were found in 3 of 33 lung cancer cell lines (GEBERT et al. 1991) and alterations in the coding region of RARβ were detected in 6 of 28 lung cancer cell lines (GERADTS et al. 1993). These in vitro findings could not be confirmed in 9 primary lung cancer samples (GEBERT et al. 1991) and in 15 epidermoid lung tumors (HOULE et al. 1991). Southern blot analysis of DNA from HNSCC cell lines that failed to express RARβ mRNA did not reveal any gross rearrangements or deletions (HU et al. 1991). No mutations were found in the promoter region of the RARβ gene in HeLa cells (BARTSCH et al. 1992). Lastly, the upregulation of

RARβ by 13-*cis* RA treatment in vivo indicates that the RARβ gene was not lost (Lotan et al. 1995c). Taken together, these results indicate that it is unlikely that changes in gene structure (mutations, gene rearrangement) are the reason for the decreased RARβ expression in either tumors or premalignant lesions and other mechanisms should be considered.

Another possible mechanism to be considered is suppression of RARβ gene expression. Downregulation of the RARβ gene expression could be caused by overexpression of RARγ1 in epithelial cell lines (Miquel et al. 1993), or by the expression of a dominant negative RAR (Kruyt et al. 1991; Damm et al. 1993) or RXR (Blanco et al. 1996), which could antagonize the transactivation of RARβ RARE. Yet another mechanism could be the loss of cofactors such as the E1A-like factor, which may be required for the activation of RARβ gene transcription by interacting with other factors that bind to the cyclic AMP response elements found in the RARβ promoter (Kruyt et al. 1992, 1993). Methylation of CpG islands in the promoter has been demonstrated by Cote and Momparler (1997) in DLD-1 colon carcinoma cells and treatment with the hypomethylating agent 5-aza-2′-deoxycytidine increased RARβ expression. Cote et al. (1998) have identified the specific 5-methylcytosine positions in the region of –46 to +251 bp from the transcription start site of RARβ2. Six of 14 primary human colon adenocarcinoma tumor samples showed hypermethylation of this promoter region of RARβ.

Because transcription of RARβ, as with any other target gene for retinoid receptors, depends on corepressors and coactivators, it is plausible that in some cells the suppression of expression of this receptor may be due to aberrant expression of cofactors. Indeed, studies with lung cancer cell lines indicated that novel RAR thyroid hormone receptor AF-2 specific cofactors, which are necessary for high levels of transcription of RARβ may be frequently inactivated in human lung cancer (Moghal and Neel 1995). RARβ expression may depend on the relative expression of the orphan receptors COUP-TF, which sensitizes RA responsiveness of RAREs by repressing their basal transactivation activity, and nur77, which enhances ligand-independent transactivation of RAREs and desensitizes their RA responsiveness. In human lung cancer cell lines, COUP-TF is highly expressed in RA-sensitive cell lines while nur77 expression is associated with RA resistance. Stable expression of COUP-TF in nur77-positive, RA-resistant lung cancer cells enhances the inducibility of RARβ gene expression and growth inhibition by RA. These observations indicate that expression levels and interactions between nur77 and COUP-TF are important determinants of RARβ expression and induction by ATRA in human lung cancer cells (Wu et al. 1997).

Finally, RARβ expression may depend on the cellular level of retinoids because this receptor has a classical DR5 RARE in its promoter and is inducible by retinoids. This conclusion is supported by the finding that RARβ expression is selectively reduced in several organs during vitamin A deficiency and is enhanced by retinoic acid, as demonstrated in studies with rats (Verma et al. 1992; Kato et al. 1992). Although it is unlikely that persons in the United

States are vitamin A deficient due to severe malnutrition or other causes, one cannot exclude the possibility that some premalignant tissues are vitamin A deficient because of reduced uptake of vitamin A from serum or due to abnormally elevated rates of catabolism of intracellular retinoids. Indeed, we have demonstrated that the binding of anti-retinoic acid antibodies to histological sections of premalignant lesions was much lower than to normal oral mucosa, suggesting that the lesions had lower levels of retinoids (Xu et al. 1995).

Regardless of the exact mechanism, the aberrations in the level and function of RARβ may simulate vitamin A deficiency in affected tissues and enhance carcinogenesis (Gebert et al. 1991; Houle et al. 1991). Because most of the cells and tissues that express little or no RARβ still express RARα, RARγ, and at least one of the RXRs, which could mediate the transactivation of retinoid-responsive genes, it is not clear why the suppression of RARβ level could lead to or enhance transformation. It is possible that RARβ regulates specific genes that are important for suppression of carcinogenesis. This conclusion is supported by the finding discussed earlier that transfection of RARβ decreases the tumorigenicity of a human lung cancer cell line (Houle et al. 1993) and the more recent demonstration that transgenic mice expressing antisense RARβ2 develop lung cancers spontaneously (Berard et al. 1996).

Acknowledgements. Studies from the authors' laboratories summarized in this chapter were supported in part by United States Public Health Service grants PO1 CA 52051, U19 CA68437, and R29 CA74835 from the National Cancer Institute and P50 DE11906 from the National Institutes of Dental Research and Craniofacial Diseases.

References

Agarwal C, Chandraratna RA, Teng M, Nagpal S, Rorke EA, Eckert RL (1996) Differential regulation of human ectocervical epithelial cell line proliferation and differentiation by retinoid X receptor- and retinoic acid receptor-specific retinoids. Cell Growth Differ 7:521–530

Amos B, Lotan, R (1991) Retinoid sensitive cells and cell lines. In Packer L (ed) Methods in enzymology: retinoids, vol 190. Academic, Orlando, p 217

Aneskievich BJ, Fuchs E (1991) Terminal differentiation in keratinocytes involves positive as well as negative regulation by retinoic acid receptors and retinoid X receptors at retinoid response elements. Mol Cell Biol 12:4862–4871

Bartsch D, Boye B, Baust C, zur Hausen H, Schwarz E (1992) Retinoic acid-mediated repression of human papillomavirus 18 transcription and different ligand regulation of the retinoic acid receptor beta gene in non-tumorigenic and tumorigenic HeLa hybrid cells. EMBO J 11:2283–2291

Berard J, Laboune F, Mukuna M, Masse S, Kothary R, Bradley WE (1996) Lung tumors in mice expressing an antisense RARbeta2 transgene. FASEB J 10:1091–1097

Blanco JC, Dey A, Leid M, Minucci S, Park BK. Jurutka PW, Haussler MR, Ozato K (1996) Inhibition of ligand induced promoter occupancy in vivo by a dominant negative RXR. Genes Cells 1:209–221

Chambon P (1996) A decade of molecular biology of retinoic acid receptors. FASEB J 10:940–954

Chen JD, Evans RM (1995) A transcriptional co-repressor that interacts with nuclear hormone receptors. Nature 377:454–457

Costa SL, McBurney MW (1996) Dominant negative mutant of retinoic acid receptor alpha inhibits retinoic acid-induced P19 cell differentiation by binding to DNA. Exp Cell Res 225:35–43
Cote S, Momparler RL (1997) Activation of the retinoic acid receptor beta gene by 5-aza-2′-deoxycytidine in human DLD-1 colon carcinoma cells. Anticancer Drugs 8:56–61
Cote S, Sinnett D, Momparler RL (1998) Demethylation by 5-aza-2′-deoxycytidine of specific 5-methylcytosine sites in the promoter region of the retinoic acid receptor beta gene in human colon carcinoma cells. Anticancer Drugs 9:743–750
Damm K, Heyman RA, Umesono K, Evans RM (1993) Functional inhibition of retinoic acid response by dominant negative retinoic acid receptor mutants. Proc Natl Acad Sci USA 90:2989–2993
Darwiche N, Celli G, De Luca LM (1994) Specificity of retinoid receptor gene expression in mouse cervical epithelia. Endocrinology 134:2018–2025
Darwiche N, Celli G, Tennenbaum T, Glick AB, Yuspa SH, De Luca LM (1995) Mouse skin tumor progression results in differential expression of retinoic acid and retinoid X receptors. Cancer Res 55:2774–2782
Darwiche N, Scita G, Jones C, Rutberg S, Greenwald E, Tennenbaum T, Collins SJ, De Luca LM, Yuspa SH (1996) Loss of retinoic acid receptors in mouse skin and skin tumors is associated with activation of the *ras*(Ha) oncogene and high risk for premalignant progression.. Cancer Res 56:4942–4949
Degos L (1997) Differentiation therapy in acute promyelocytic leukemia: European experience. J Cell Physiol 173:285–287
De Luca LM (1991) Retinoids and their receptors in differentiation, embryogenesis, and neoplasia. FASEB J 5:2924–2933
de The H, Chomienne C, Lanotte M, Degos L, Dejean A (1990) The t(15;17) translocation of acute promyelocytic leukaemia fuses the retinoic acid receptor alpha gene to a novel transcribed locus. Nature 347:558–561
de The H (1996) Altered retinoic acid receptors. FASEB J 10:955–960
Elder JT, Fisher GJ, Zhang QY, Eisen D, Krust A, Kastner P, Chambon P, Voorhees JJ (1991) Retinoic acid receptor gene expression in human skin. J Invest Dermatol 96:425–433
Ferrucci PF, Grignani F, Pearson M, Fagioli M, Nicoletti I, Pelicci PG (1997) Cell death induction by the acute promyelocytic leukemia-specific PML/RARalpha fusion protein. Proc Natl Acad Sci U S A 94:10901–10906
Finzi E, Blake MJ, Celano P, Skouge J, Diwan R (1992) Cellular localization of retinoic acid receptor-gamma expression in normal and neoplastic skin. Am J Pathol 140:1463–1471
Fisher GJ, Talwar HS, Xiao JH, Datta SC, Reddy AP, Gaub MP, Rochette-Egly C, Chambon P, Voorhees JJ (1994) Immunological identification and functional quantitation of retinoic acid and retinoid X receptor proteins in human skin. J Biol Chem 269:20629–20635
Fisher GJ, Voorhees JJ (1996) Molecular mechanisms of retinoid actions in skin. FASEB J 10:1002–1013
Fitzgerald P, Teng M, Chandraratna RA, Heyman RA, Allegretto EA (1997) Retinoic acid receptor alpha expression correlates with retinoid- induced growth inhibition of human breast cancer cells regardless of estrogen receptor status. Cancer Res 57:2642–2650
Gebert JF, Moghal N, Frangioni JV, Sugarbaker DJ, Neel BG (1991) High frequency of retinoic acid receptor beta abnormalities in human lung cancer. Oncogene 6:1859–1868 (erratum in 7:821)
Geisen C, Denk C, Gremm B, Baust, C, Karger A, Bollag W, Schwarz E (1997) High-level expression of the retinoic acid receptor beta gene in normal cells of the uterine cervix is regulated by the retinoic acid receptor alpha and is abnormally down-regulated in cervical carcinoma cells. Cancer Res 57:1460–1467

Geradts J, Chen JY, Russell EK, Yankaskas JR, Nieves L, Minna JD (1993) Human lung cancer cell lines exhibit resistance to retinoic acid treatment. Cell Growth Differ 4:799–809

Gimenez-Conti IB, Slaga TJ (1993) The hamster cheek pouch carcinogenesis model. J Cell Biochem 17F[Suppl]:83–90

Grillier I, Umiel T, Elstner E, Collins SJ, Koeffler HP (1997) Alterations of differentiation, clonal proliferation, cell cycle progression and bcl-2 expression in RAR alpha-altered sublines of HL-60. Leukemia 11:393–400

Grimwade D, Solomon E (1997) Characterisation of the PML/RAR alpha rearrangement associated with t(15;17) acute promyelocytic leukaemia. Curr Top Microbiol Immunol 220:81–112

Gudas LJ, Sporn MB, Roberts AB (1994) Cellular biology and biochemistry of the retinoids. In Sporn MB, Roberts AB, Goodman DS (eds) The Retinoids. Raven, New York, p 443

Han QX, Allegretto EA, Shao ZM, Kute TE, Ordonez J, Aisner SC, Rishi AK, Fontana JA (1997) Elevated expression of retinoic acid receptor-alpha (RAR alpha) in estrogen-receptor-positive breast carcinomas as detected by immunohistochemistry. Diagn Mol Pathol 6:42–48

He LZ, Guidez F, Tribioli C, Peruzzi D, Ruthardt M, Zelent A, Pandolfi PP (1998) Distinct interactions of PML-RARalpha and PLZF-RARalpha with co-repressors determine differential responses to RA in APL. Nat Genet 18:126–135

Hong WK, Itri LM (1994) Retinoids and human cancer. In Sporn MB, Roberts AB, Goodman DS (eds) The Retinoids. Raven, New York, p 597

Hong WK, Sporn MB (1997) Recent advances in chemoprevention of cancer. Science 278:1073–1077

Hoerlein AJ, Naar AM, Heinzel T, Torchia J, Gloss B, Kurokawa R, Ryan A, Kamei Y, Soderstrom M, Glass CK, et al. (1995) Ligand-independent repression by the thyroid hormone receptor mediated by a nuclear receptor co-repressor. Nature 377:397–404

Houle B, Leduc F, Bradley WE (1991) Implication of RARB in epidermoid (Squamous) lung cancer. Genes Chromosomes Cancer 3:358–366

Houle B, Rochette-Egly C, Bradley WE (1993) Tumor-suppressive effect of the retinoic acid receptor beta in human epidermoid lung cancer cells. Proc Natl Acad Sci USA 90:985–989

Hu L, Crowe DL, Rheinwald JG, Chambon P, Gudas LJ (1991) Abnormal expression of retinoic acid receptors and keratin 19 by human oral and epidermal squamous cell carcinoma cell lines. Cancer Res 51:3972–3981

Kato S, Mano H, Kumazawa T, Yoshizawa Y, Kojima R, Masushige S (1992) Effect of retinoid status on alpha, beta and gamma retinoic acid receptor mRNA levels in various rat tissues. Biochem J 286:755–760

Kraemer KH, DiGiovanna JJ, Peck GL (1992) Chemoprevention of skin cancer in xeroderma pigmentosum. J Dermatol 19:715–718

Kruyt FAE, van den Brink CE, Defize LHK, Donath MJ, Kastner P, Kruijer W, Chambon P, van der Saag PT (1991) Transcriptional regulation of retinoic acid receptor β in retinoic acid-sensitive and -resistant P19 embryocarcinoma cells. Mech Devel 33:171–178

Kruyt FAE, Folkers GE, van den Brink CE, van der Saag PT (1992) A cyclic AMP response element is involved in retinoic acid-dependent RARβ2 promoter activation. Nuc Acid Res 20:6393–6399

Kruyt FAE, Folkers GE, Wlhout AJM, van der Leede BJ M, van der Saag PT (1993) E1 a functions as a coactivator of retinoic acid-dependent retinoic acid receptor-β2 promoter activation. Mol Endocrinol 7:604–615

Kumar R, Shoemaker AR, Verma AK (1994) Retinoic acid nuclear receptors and tumor promotion: decreased expression of retinoic acid nuclear receptors by the tumor promoter 12- O-tetradecanoylphorbol-13-acetate. Carcinogenesis 15:701–705

Kurokawa R, Soderstrom M, Horlein A, Halachmi S, Brown M, Rosenfeld MG, Glass CK (1995) Polarity-specific activities of retinoic acid receptors determined by a co-repressor. Nature 377:451–454

Lancillotti F, Giandomenico V, Affabris E, Fiorucci G, Romeo G, Rossi GB (1995) Interferon alpha-2b and retinoic acid combined treatment affects proliferation and gene expression of human cervical carcinoma cells. Cancer Res 55:3158–3164

Lawrence JA, Merino MJ, Simpson JF, Manrow RE, Page DL, Steeg PS (1998) A high-risk lesion for invasive breast cancer, ductal carcinoma in situ, exhibits frequent overexpression of retinoid X receptor. Cancer Epidemiol Biomarkers Prev 7:29–35

Li XS, Shao ZM, Sheikh MS, Eiseman JL, Sentz D, Jetten AM, Chen JC, Dawson MI, Aisner S, Rishi AK, et al. (1995) Retinoic acid nuclear receptor beta inhibits breast carcinoma anchorage independent growth. J Cell Physiol 165:449–458

Liu Y, Lee MO, Wang HG, Li, Y. Hashimoto Y, Klaus M, Reed JC, Zhang X (1996) Retinoic acid receptor beta mediates the growth-inhibitory effect of retinoic acid by promoting apoptosis in human breast cancer cells. Mol Cell Biol 16:1138–1149

Lotan R (1995a) Cellular Biology of the retinoids. In Degos L, Parkinson DR (eds), Retinoids in Oncology. Springer, Berlin Heidelberg New York, p 27

Lotan R, Dawson MI, Zou CC, Jong L, Lotan D, Zou CP (1995b) Enhanced efficacy of combinations of retinoic acid- and retinoid X receptor-selective retinoids and alpha-interferon in inhibition of cervical carcinoma cell proliferation. Cancer Res 55:232–236

Lotan R, Xu XC, Lippman SM, Ro JY, Lee JS, Lee JJ, Hong WK (1995c) Suppression of retinoic acid receptor-beta in premalignant oral lesions and its up-regulation by isotretinoin. N Engl. J Med 332:1405–1410

Lotan R (1996) Retinoids in cancer chemoprevention. FASEB J 10:1031–1039

Mangelsdorf DJ, Borgmeyer U, Heyman RA, Zhou JY, Ong ES, Oro AE, Kakizuka A, Evans RM, (1992) Characterization of three RXR genes that mediate the action of 9-cis retinoic acid. Genes Dev 6:329–344

Mangelsdorf DJ, Umesono K, Evans RM (1994) The retinoid receptors. In Sporn MB, Roberts AB, Goodman DS (eds) The retinoids. Raven, New York, p 319

McGregor F, Wagner E, Felix D, Soutar D, Parkinson K, Harrison PR (1997) Inappropriate retinoic acid receptor-beta expression in oral dysplasias: correlation with acquisition of the immortal phenotype. Cancer Res 57:3886–3889

Miquel C, Clusel C, Semat A, Gerst C, Darmon M (1993) Retinoic acid receptor isoform RARγ1: an antagonist of the transactivation of the RARβ RARE in epithelial cell lines and normal human keratinocytes. Mol Biol Rep 17:35–45

Moghal N, Neel BG (1995) Evidence for impaired retinoic acid receptor-thyroid hormone receptor AF-2 cofactor activity in human lung cancer. Mol Cell Biol 15:3945–3959

Moon RC, Mehta RG, Rao KVN (1994) Retinoids and cancer in experimental animals. In Sporn MB, Roberts AB, Goodman DS (eds) The retinoids. Raven, New York, p537

Moon TE, Levine N, Cartmel B, Bangert JL, Rodney S. Dong Q, Peng YM, Alberts DS (1997) Effect of retinol in preventing squamous cell skin cancer in moderate- risk subjects: a randomized, double-blind, controlled trial. Southwest Skin Cancer Prevention Study Group. Cancer Epidemiol Biomarkers Prev 6:949–956

Nervi C, Vollberg TM, George MD, Zelent A, Chambon P, Jetten AM (1991) Expression of nuclear retinoic acid receptors in normal tracheobronchial cells and in lung carcinoma cells. Exp Cell Res 195:163–170

Perlmann T, Vennstrom B (1995) The sound of silence. Nature 377:387–388

Pratt MA, Kralova J, McBurney MW (1990) A dominant negative mutation of the alpha retinoic acid receptor gene in a retinoic acid-nonresponsive embryonal carcinoma cell. Mol Cell Biol 10:6445–6453

Robertson KA, Mueller L, Collins SJ (1991) Retinoic acid receptors in myeloid leukemia: characterization of receptors in retinoic acid-resistant K-562 cells. Blood 77:340–347

Robertson KA, Emami B, Collins SJ (1992) Retinoic acid-resistant HL-60R cells harbor a point mutation in the retinoic acid receptor ligand-binding domain that confers dominant negative activity. Blood 80:1885–1889

Redfern CP, Todd C (1992) Retinoic acid receptor expression in human skin keratinocytes and dermal fibroblasts in vitro. J Cell. Science 102:113–121

Roman SD, Clarke CL, Hall RE, Alexander IE, Sutherland RL (1992) Expression and regulation of retinoic acid receptors in human breast cancer cells. Cancer Res 52:2236–2242

Ruthardt M, Testa U, Nervi C, Ferrucci PF, Grignani F, Puccetti E, Grignani F, Peschle C, Pelicci PG (1997) Opposite effects of the acute promyelocytic leukemia PML-retinoic acid receptor alpha (RAR alpha) and PLZF-RAR alpha fusion proteins on retinoic acid signalling. Mol Cell Biol 17:4859–4869

Schulman IG, Evans RM (1997) Retinoid receptors in development and disease. Leukemia 11 Suppl 3:376–377

Seewaldt VL, Johnson BS, Parker MB, Collins SJ, Swisshelm K (1995) Expression of retinoic acid receptor beta mediates retinoic acid- induced growth arrest and apoptosis in breast cancer cells. Cell Growth Differ 6:1077–1088

Seewaldt VL, Kim JH, Caldwell LE, Johnson BS, Swisshelm K, Collins SJ (1997) All-trans-retinoic acid mediates G1 arrest but not apoptosis of normal human mammary epithelial cells. Cell Growth Differ 8:631–641

Sheikh MS, Shao ZM, Chen JC, Hussain A, Jetten AM, Fontana JA (1993) Estrogen receptor-negative breast cancer cells transfected with the estrogen receptor exhibit increased RAR alpha gene expression and sensitivity to growth inhibition by retinoic acid. J Cell Biochem 53:394–404

Smith MA, Parkinson DR, Cheson BD, Friedman MA (1992) Retinoids in cancer therapy. J Clin Onncol 10:839–864

Soignet S, Fleischauer A, Polyak T, Heller G, Warrell RP Jr (1997) All-trans retinoic acid significantly increases 5-year survival in patients with acute promyelocytic leukemia: long-term follow-up of the New York study. Cancer Chemother Pharmacol 40 [Suppl]:S25–S29

Sun SY, Yue P, Dawson MI, Shroot B, Michel S, Lamph WW, Heyman RA, Teng M, Chandraratna RA, Shudo K, Hong WK, Lotan R (1997) Differential effects of synthetic nuclear retinoid receptor-selective retinoids on the growth of human non-small cell lung carcinoma cells. Cancer Res 57:4931–4939

Swisshelm K, Ryan K, Lee X, Tsou HC, Peacocke M, Sager R (1994) Down-regulation of retinoic acid receptor beta in mammary carcinoma cell lines and its up-regulation in senescing normal mammary epithelial cells. Cell Growth Differ 5:133–141

Tallman MS, Andersen JW, Schiffer CA, Appelbaum FR, Feusner JH, Ogden A, Shepherd L, Willman C, Bloomfield CD, Rowe JM, Wiernik PH (1997) All-trans-retinoic acid in acute promyelocytic leukemia. N Engl J Med 337:1021–1028

van der Leede BM, Geertzema J, Vroom TM, Decimo D, Lutz Y, van der Saag PT, van der Burg B (1996) Immunohistochemical analysis of retinoic acid receptor-alpha in human breast tumors: retinoic acid receptor-alpha expression correlates with proliferative activity. Am J Pathol 148:1905–1914

Verma AK, Shoemaker A, Simsiman R, Denning M, Zachman RD (1992) Expression of retinoic acid nuclear receptors and tissue transglutaminase is altered in various tissues of rats fed a vitamin A-deficient diet. J Nutr 122:2144–2152

Vollberg TM, Sr. Nervi C, George MD, Fujimoto W, Krust A, Jetten AM (1992) Retinoic acid receptors as regulators of human epidermal keratinocyte differentiation. Mol Endocrinol 6:667–676

Wan H, Oridate N, Lotan D, Hong WK, Lotan R (1999) Overexpression of retinoic acid receptor β in head and neck squamous cell carcinoma cells increases their sensitivity to suppression of squamous differentiation by retinoids. Cancer Res (in press)

Widschwendter M, Berger J, Daxenbichler G, Muller-Holzner E, Widschwendter A, Mayr A, Marth C, Zeimet AG (1997) Loss of retinoic acid receptor beta expression in breast cancer and morphologically normal adjacent tissue but not in the normal breast tissue distant from the cancer. Cancer Res 57:4158–4161

Wu Q, Li Y, Liu R, Agadir A, Lee MO, Liu Y, Zhang X (1997) Modulation of retinoic acid sensitivity in lung cancer cells through dynamic balance of orphan receptors nur77 and COUP-TF and their heterodimerization. EMBO J 16:1656–1669

Xu XC, Clifford JL, Hong WK, Lotan R (1994a) Detection of nuclear retinoic acid receptor mRNA in histological tissue sections using nonradioactive in situ hybridization histochemistry. Diagnostic Molecular Pathology 3:122–131

Xu XC, Ro JY, Lee JS, Shin DM, Hong WK, Lotan R (1994b) Differential expression of nuclear retinoid receptors in normal, premalignant, and malignant head and neck tissues. Cancer Res 54:3580–3587

Xu XC, Zile MH, Lippman SM, Lee JS, Lee JJ, Hong WK, Lotan R (1995) Anti-retinoic acid (RA) antibody binding to human premalignant oral lesions, which occurs less frequently than binding to normal tissue, increases after 13-cis-RA treatment in vivo and is related to RA receptor beta expression. Cancer Res 55:5507–5511

Xu XC, Wong WYL, Goldberg LH, Baer S, Wolf JE, Lotan R (1996) Suppression of nuclear retinoid receptor expression in squamous cell carcinomas of the skin. Proc Am Assoc Cancer Res 37:1588

Xu XC, Sozzi G, Lee JS, Lee JJ, Pastorino U, Pilotti S, Kurie JM, Hong WK, Lotan R (1997a) Suppression of retinoic acid receptor beta in non-small-cell lung cancer in vivo: implications for lung cancer development [see comments]. J Natl Cancer Inst 89:624–629

Xu XC, Sneige N, Liu X, Nandagiri R, Lee JJ, Lukmanji F. Hortobagyi G, Lippman SM, Dhingra K, Lotan R (1997b) Progressive decrease in nuclear retinoic acid receptor beta messenger RNA level during breast carcinogenesis. Cancer Res 57:4992–4996

Xu XC, Liu X-M, Gimenez-Conti IB, Conti CJ, Lotan R (1998) Loss of RAR-beta expression during DMBA-induced hamster cheek-pouch carcinogenesis. Proc Amer Assoc Cancer Res 39:784

Xu XC, Liu X, Tahara E, Lippman SM, Lotan R (1999a) Expression and up-regulation of retinoic acid receptor-beta is associated with retinoid sensitivity and colony formation in esophageal cancer cell lines. Cancer Res 59:2477–2483

Xu XC, Lee JS, Morice RC, Liu XM, Lee JJ, Lippman SM, Hong WK, Lotan R (1999b) Decreased expression of nuclear retinoid receptor beta in bronchial epithelium of heavy smokers and its up-regulation during chemoprevention trial with 13-*cis*-retinoic acid. J Natl Cancer Inst (in press)

Xu XC, Mitchell MF, Silva E, Jetten A, Lotan R (1999c) Decreased expression of retinoic acid receptors, transforming growth factor-beta, involucrin, and cornifin in cervical intraepithelial neoplasia. Clin Cancer Res 59:2477–2483

Zhang XK, Liu Y, Lee MO (1996) Retinoid receptors in human lung cancer and breast cancer. Mutat Res 350:267–277

Zou CP, Clifford JL, Xu XC, Sacks PG, Chambon P, Hong WK, Lotan, R (1994) Modulation by retinoic acid (RA) of squamous cell differentiation, cellular RA-binding proteins, and nuclear RA receptors in human head and neck squamous cell carcinoma cell lines. Cancer Res 54:5479–5487

Zou CP, Hong WK, Lotan R (1999) Expression of retinoic acid receptor β is associated with inhibition of keratinization in human head and neck squamous carcinoma cells. Differentiation 64:123–132

Section D
Development and Teratogenesis

CHAPTER 12

Genetic and Molecular Approaches to Understanding the Role of Retinoids in Mammalian Spermatogenesis

A.I. PACKER and D.J. WOLGEMUTH

A. Introduction

I. Historical Perspective of the Role of Vitamin A in Spermatogenesis

It has been known for many years that vitamin A is required by the testis for spermatogenesis to occur and by the epididymis for the maturation of spermatozoa (WOLBACH and HOWE 1925; HOWELL et al. 1963; reviewed in ESKILD and HANSSON 1994). Animals maintained on a vitamin A deficient diet which has been supplemented with retinoic acid are relieved of virtually all the symptoms of vitamin A deficiency, except that they are blind and sterile (e.g., HOWELL et al. 1963). Although genetic studies would indicate that retinoic acid, not retinol, is the active retinoid functioning in spermatogenesis (KASTNER et al. 1995), dietary retinol is clearly required. This requirement for dietary retinol, not retinoic acid, for the maintenance of spermatogenesis is likely due to the existence of the blood-testis barrier preventing retinol from being available to the adluminal part of the tubule. This barrier is perhaps more accurately referred to as the Sertoli cell barrier, since it is the specialized junctions in the Sertoli cells that sequester the meiotic and postmeiotic compartment from the circulation.

Gene knockout strategies have begun to elucidate the key players in this regulation of spermatogenesis by the retinoids. In particular, null mutations of $RAR\alpha$ and $RXR\beta$ have resulted in male sterility, with specific defects in spermatogenesis (LUFKIN et al. 1993; KASTNER et al. 1996).

II. Focus of This Review

The effort to catalogue the sites of expression of the major retinoid receptors and binding proteins in the testis is well underway, and should be completed in the near future as specific antisera that are useful for immunohistochemical studies become available. In addition to this descriptive work (see Sects. B, C, Table 1), the major advances in our understanding of the role of retinoids in spermatogenesis in the past 5 years, as indicated above, have come through the genetic manipulation of key proteins involved in retinoid function. This review focuses on the importance of gene targeting, transgenesis, and other molecular genetic approaches that promise to add to our understanding over

Table 1. Summary of the expression of retinoid receptors and cellular binding proteins in the testis and male reproductive tract

Gene	Testis cells/tissue	RA responsive?
CRBP-I	*Sertoli cells*, low in peritubular cells, germ cells – spermatogonia, primary spermatocytes, Leydig cells, epididymis, vas deferens	Low in vitamin A deficiency; induced by vitamin A in vitamin A deficient rats
CRABP-I	Leydig cells (?), germ cells – gonocytes and spermatogonia, epididymis, vas deferens, seminal vesicles	(–)
CRABP-II	Fetal and early postnatal, *Sertoli* and *Leydig cells*	
RARα	Sertoli, Leydig, peritubular cells, germ cells – spermatogonia, highest in *early, round spermatids*, lower in *pachytene*, epididymis	RARα2 induced by vitamin A in vitamin A deficient rats
RARα1	Germ cells and Sertoli cells	
RARα2	Sertoli cells	
RARβ	Sertoli cells	Transient induction within 24 h of RA injection
RARγ	Testis, type A spermatogonia, pachytene	Induced by vitamin A and RA in vitamin A deficient rats
RXRα	Early, round spermatids (?)	
RXRα2 RXRα3	Testis specific	
RXRβ	*Sertoli cells*	
RXRγ	Testis (very low levels)	

Italics, cells in which expression of the protein has been detected. Most of the data were obtained from expression studies in the adult rat. Reference sources: AKMAL et al. 1996, 1997; BROCARD et al. 1996; DE ROOIJ et al. 1994; ESKILD et al. 1991; ESKILD and HANSSON 1994; GAEMERS et al. 1996, 1997; KIM and GRISWOLD 1990; HUANG et al. 1994; KASTNER et al. 1996; KATO et al. 1985; RAJAN et al. 1990, 1991; VAN PELT et al. 1992; ZHAI et al. 1997; ZHENG et al. 1996.

the next years (see Sects. E, F.II, F.III). Further, we emphasize the importance of identifying the roles of retinoids in both the germ cell and somatic cell lineages in the testis (see particularly Sect. FI). Analyses of the vitamin A deficient testis (see Sect. D) suggest that retinoids are required in each lineage and may play a role in establishing the complex pattern of intercellular signaling between and among the various cell types in the seminiferous tubule.

B. Serum and Cellular Retinoid Binding Proteins

I. Sites of Synthesis and Action

1. Retinol Binding Protein and Transthyretin

As described earlier, the failure of spermatogenesis in vitamin A deficient rodents indicates that the testis is a major site of retinoid action. Of particular interest with regard to the potential mechanisms of action of retinoids during male germ cell development is the observation that retinol supplementation can rescue normal spermatogenesis (HUANG and HEMBREE 1979; UNNI et al. 1983), but all *trans*-retinoic acid, one of the active metabolites of vitamin A, cannot (HOWELL et al. 1963; HUANG and HEMBREE 1979). Although there has been one report of partial rescue by retinoic acid, this involved direct intratesticular injection in high replicate doses (VAN PELT and DE ROOIJ 1991), a decidedly nonphysiological paradigm. The architecture of the seminiferous tubule – particularly the Sertoli cell blood-testis barrier – prevents diffusion of metabolites into the adluminal compartment, and it has been confirmed that there is little uptake of radiolabeled retinoic acid by the rat testis (BLANER and OLSON 1994). Since less than 1% of the retinoic acid in the rat testis is derived from the plasma pool (KURLANDSKY et al. 1995), testicular retinoic acid must be derived from retinol that is taken up from the serum.

As is the case at other sites of retinol transport, storage, and metabolism, it has been assumed that testicular retinol is found in a complex with a variety of serum and cellular retinoid binding proteins, rather than as a free molecule. The two major serum binding proteins, retinol binding protein (RBP) and transthyretin (TTR), have indeed been shown to participate in the transport of serum retinol to Sertoli cells. This appears to occur via the peritubular myoid cells which surround the seminiferous tubules. RBP protein has been localized to the interstitial space (MCGUIRE et al. 1981; KATO et al. 1985), and in vitro studies suggest that retinol is delivered to the peritubular cells in a complex with both RBP and TTR, where it is then resecreted bound to RBP synthesized by the peritubular cells themselves (DAVIS and ONG 1995). Several in vitro experiments have demonstrated uptake of tritiated retinol from retinol-RBP or retinol-RBP-TTR complexes by cultured Sertoli cells (BISHOP and GRISWOLD 1987; SHINGLETON et al. 1989), suggesting a model whereby serum retinol is chaperoned by RBP and TTR first to the peritubular cells, and finally to the Sertoli cells (reviewed in ESKILD and HANSSON 1994; KIM and AKMAL 1996) where it can be esterified and stored or converted to retinoic acid.

2. Cellular Retinol and Retinoic Acid Binding Proteins

Of the two cellular retinol binding proteins (CRBP I and II), only CRBP I has been found in the testis, with expression restricted to the Sertoli and peritubular cells (KATO et al. 1985; PORTER et al. 1985; BLANER et al. 1987). CRBP I protein in Sertoli cells is cytoplasmic and varies in abundance and subcellular localization throughout the seminferous tubule cycle, with high levels in the basal area at stages III-V and low levels, mostly in the apical processes, at stages VII-VIII (KATO et al. 1985; RAJAN et al. 1990). Recently, CRBP I mRNA has also been observed, at low levels, in spermatogonia and primary spermatocytes (ZHAI et al. 1997). As for the cellular retinoic acid binding proteins (CRABP I and II), CRABP I is found in gonocytes and, later, in the cytoplasm of spermatogonia (Rajan et al. 1991; Zheng et al. 1996). CRABP II has an entirely distinct pattern of expression in the rat testis, with CRABP II protein detected as early as the day 19 fetus, and relatively high levels of CRABP II mRNA found at postnatal day 4, decreasing at later times until postnatal day 20 (Zheng et al. 1996). Immunostaining of fetal rat testes revealed expression in fetal Leydig and Sertoli cells (Zheng et al. 1996). CRABP II is not expressed in the mature testis (Bailey and Sui 1990).

II. Potential Functions During Spermatogenesis

A role for RBP and TTR in transporting retinol to the peritubular and Sertoli cells was outlined above. The patterns of expression of the cellular retinoid binding proteins are more clearly suggestive in regard to their roles in retinoid metabolism and function in spermatogenesis. For example, Sertoli cells preincubated with retinol were found to take up more retinol and to convert more of it to retinyl esters (Bishop and Griswold 1987). The degree of retinol uptake corresponds to the level of CRBP I in the cells (Shingleton et al. 1989), which is itself regulated by levels of retinoic acid due to the presence of a retinoic acid response element in its promoter (Husmann et al. 1992). CRBP I appears to regulate the uptake of retinol-RBP and to balance the esterification and storage of retinol (mostly in the form of retinyl palmitate) with the oxidation of retinol to retinal and retinoic acid (Napoli et al. 1991). The significance of its relative abundance at stages III-V and its polarized distribution is unclear, but presumably corresponds either to an increased requirement for Sertoli cell retinoic acid at these stages or an increased requirement for retinol transport to germ cells. Of potential interest is the movement of pachytene spermatocytes through the blood-testis barrier at stages III-IV (Eskild and Hansson 1994), raising the possibility that the remodeling of the seminiferous epithelium that occurs might in some way be retinoid-dependent (see discussion of *Stra6* in Sect. F(III)). Although disruption of the Sertoli cell tight junctions has been reported in vitamin A deficient rats (Huang et al. 1988), other investigators have found ultrastructurally normal tight junctions in electron microscopic studies of the testes from such animals (Ismail and Morales 1992).

The presence of CRABP I exclusively in the cytoplasm of gonocytes and spermatogonia suggests that it is being actively retained in the cytoplasm, since its small size would ordinarily allow diffusion into the nucleus (Zheng et al. 1996). CRABP I might then have a role in sequestering retinoic acid in the cytoplasm of these mitotically dividing germ cells, thereby preventing ligand-dependent activation of the nuclear retinoic acid receptors, as has been suggested for CRABP I in other cells and tissues (Fiorella and Napoli 1991; Boylan and Gudas 1992; Fogh et al. 1993). This role for CRABP I is consistent with the observation that, with the exception of the division of A_1 into A_2 spermatogonia (see Sect. D), the spermatogonial population is relatively unaffected by vitamin A deficiency (DE ROOIJ et al. 1994).

The pattern of expression of CRABP II in fetal and prepubertal Sertoli cells overlaps almost exactly with the developmental timing of Sertoli cell proliferation in the rat testis (ORTH 1982), which is maximal at day 20 of gestation and decreases until postnatal day 21. These and other observations on the correlation of CRABP II expression and sites of retinoic acid synthesis led ZHENG et al. (1996) to propose that CRABP II is involved in the retinoic acid-dependent autocrine or paracrine regulation of Sertoli cell proliferation. In addition, they proposed that Leydig cell CRABP II may be involved in the secretion of retinoic acid in the fetal and neonatal rat testis.

C. Receptors

I. Retinoic Acid and Retinoid X Receptors and Their Ligands

The two major active isoforms of retinoic acid – all-*trans* and 9-*cis* retinoic acid – exert their biological effects through a family of nuclear receptor/transcription factors. There are two classes of retinoid receptors-the retinoic acid receptors (RARs) and the retinoid X receptors (RXRs) (reviewed in LEID et al. 1992; CHAMBON 1995). Each class of receptors is comprised of three major subtypes (α, β, and γ) whose distribution and function is partly overlapping and partly unique (reviewed in KASTNER et al. 1995). The functional distinction between the two classes stems to some degree from their different relative affinities for the two retinoid ligands. RARs are activated by either all-*trans* or 9-*cis* retinoic acid while RXRs are exclusively activated by 9-*cis* retinoic acid.

II. Regulation of Transcription by Ligand-Activated Retinoid Receptors

The RARs and RXRs activate or repress transcription of target genes by binding to conserved elements termed retinoic acid response elements (RAREs) (reviewed in CHAMBON 1996). The receptors appear to bind to the RARE as RAR/RXR heterodimers or as RXR homodimers, and given the number of isoforms for each subtype ($\alpha1$, $\alpha2$, $\beta1$–$\beta4$, $\gamma1$, $\gamma2$) there are up to 48 potential heterodimers that could be formed. In addition, the RXRs also function as heterodimeric partners for other nuclear receptors, including the thyroid hormone receptor and the vitamin D receptor (reviewed in MANGELSDORF and EVANS 1995; PFAHL and CHYTIL 1996). The RAREs consist of two direct repeats [PuG(G/T)TCA] usually separated by a 1, 2, or 5 nucleotide spacer. RAREs have been found in promoter or enhancer elements adjacent to a large number of genes, including those encoding growth factors and their receptors, hormones, protein kinases, components of the extracellular matrix, enzymes involved in intermediary metabolism, proto-oncoproteins, and HOX proteins (GUDAS et al. 1994). The broad scope of retinoid-dependent gene activation helps to explain a large body of classical data on the wide array of effects observed in animal models and in humans upon vitamin A deficiency or excess. As such, an essential first step toward elucidating the function(s) of retinoids in a particular cell type or tissue is to determine the pattern of expression of the receptors.

III. RAR and RXR Expression in the Testis

Most attempts at localizing receptor expression in the testis have utilized in situ hybridization and Northern blotting approaches to examine expression at the mRNA level. By Northern analysis, KIM and GRISWOLD (1990) found a 1.8 kb and a 4.7 kb RARα transcript in pachytene spermatocytes and a 2.7-kb RARα transcript in Sertoli cells, while ESKILD (ESKILD et al. 1991) found 4 kb and 7 kb RARα transcripts in round spermatids. KIM and AKMAL (1996) reported RARα expression in Sertoli cells by in situ hybridization, both in normal prepubertal animals and in vitamin A deficient animals before and after retinol replenishment. They also observed expression in round spermatids, particularly at stages VII-VIII of the seminiferous epithelium. Finally, DE ROOIJ and colleagues (1994) reported RARα expression in Sertoli cells and in type A spermatogonia following retinol replenishment. Although immunohistochemical studies of testicular RARα expression will probably resolve the issue (see below), it seems likely that RARα is expressed in Sertoli cells and, to some extent, in alternatively spliced forms in both pre- and postmeiotic germ cells. It is worth noting that RARα mRNA levels appear to be higher in Sertoli cells at postnatal day 10 than at day 20 (KIM and AKMAL 1996), suggesting a similar profile of expression as that observed for CRABP II. Whether this pattern implies a role for retinoids in Sertoli cell proliferation (ZHENG et al. 1996; see above) or in Sertoli cell differentiation (KIM and AKMAL 1996) will require further functional analysis. RARβ mRNA expression appears to be restricted to a 2.9 kb transcript in Sertoli cells (VAN PELT et al. 1992) and RARγ to type A spermatogonia (DE ROOIJ et al. 1994) and pachytene spermatocytes (HUANG et al. 1994). RARβ expression is also significantly influenced by retinoid status (GAEMERS et al. 1996), most likely due to an RARE in its 5' flanking region (DE THE et al. 1990). All three RXRβ subtypes have been found in the testis (KASTNER et al. 1996; GAEMERS et al. 1997), with RXRα in round spermatids and RXRβ restricted to the Sertoli cells (KASTNER et al. 1996). Two novel, testis-specific RXRα isoforms have also been identified in the mouse that are upregulated at puberty (BROCARD et al. 1996). Finally, RXRγ mRNA has been detected at very low levels in the normal mouse testis by ribonuclease protection assay (GAEMERS et al. 1997).

The availability of receptor subtype-specific antisera has begun to allow the identification of retinoid receptor proteins in the testis. A recent immunohistochemical study demonstrated that RARα protein is present in Sertoli cells, germ cells in meiotic prophase, and in elongating spermatids in the adult rat testis (AKMAL et al. 1997). This pattern is consistent with that observed for the RARα mRNA, and the authors conclude that RARα function may be important in Sertoli cells, in meiotic prophase, and in the transition from round to elongating spermatids (AKMAL et al. 1997).

D. The Vitamin A Deficient Testis

As indicated in the Introduction, prior to the cloning, characterization, and genetic manipulation of the retinoid receptors, the potential role of retinoic acid in the testis was assessed by examining spermatogenesis in animals deprived of dietary vitamin A. The failure of spermatogenesis and the resulting sterility have been studied in great detail (WOLBACH and HOWE 1925; MASON 1933; THOMPSON et al. 1964; DE ROOIJ et al. 1989; GRISWOLD et al. 1989; reviewed in KIM and WANG 1993) and more recent analyses have begun to address the cellular defects underlying the degeneration of the seminiferous epithelium (WANG and KIM 1993a; DE ROOIJ et al. 1994).

There appear to be at least three major defective transitions in spermatogenesis in the vitamin A deficient testis. The first is a failure of the production of A_2 spermatogonia from A_1 spermatogonia (ISMAIL et al. 1990; VAN PELT and DE ROOIJ 1990; DE ROOIJ et al. 1994). Thus, to a large extent, the depletion of the seminiferous tubules is due to an inhibition of maturation of the differentiating spermatogonia. The A_2, A_3, A_4, and B spermatogonia are unaffected, but some degeneration of In spermatogonia is observed (DE ROOIJ et al. 1994). In addition, vitamin A deficiency appears to affect both the onset and the progression of meiotic prophase, with a delay in the transition of spermatocytes from preleptotene to leptotene and the appearance of abnormal synaptonemal complexes in zygotene and pachytene (DE ROOIJ et al. 1994). Spermatid degeneration has also been reported (SOBHON et al. 1979; MITRANOND et al. 1979) along with a disruption of Sertoli cell-spermatid associations (HUANG et al. 1988), and a delay in spermiation (HUANG and MARSHALL 1983; MORALES and GRISWOLD 1991). Repopulation of the testis in response to retinol supplementation occurs through the synchronous production of A_2 spermatogonia in all of the seminiferous tubules (DE ROOIJ et al. 1994).

Many questions remain to be answered with regard to the vitamin A deficient testis. For example, the major events in spermatogenesis have been likened to "transcriptional checkpoints" (SASSONE-CORSI 1997), given the undoubted requirement for rapid changes in gene expression to facilitate the remarkable morphological transitions that accompany mitosis, meiosis, and spermiogenesis. Can the arrest observed at the A_1/A_2 transition and the partial arrest and degeneration observed in meiotic prophase be attributed to the loss of specific retinoic acid-dependent transcriptional checkpoints? Or does the depletion of testicular retinoic acid have a more global effect that indirectly brings about the loss of particular germ cells? To what extent does the vitamin A deficient testis accurately reflect the roles of retinoic acid in spermatogenesis? It is important to remember that postnatal vitamin A deficiency might be masking the possible importance of retinoic acid in the very early development and maturation of the seminiferous epithelium. Given the expression of RARα and CRABP II in very early Sertoli cells (see Sect. C), retinoid signaling may indeed have a role in establishing the Sertoli cell population. Clearly, careful comparisons with mice with genetic deficiencies in one or more components of the retinoid signaling pathway are essential (see Sect. E), as are experiments to extend the vitamin A deficient regime to earlier stages in the development of the male germ line.

E. Genetic Analysis of Retinoid Function

I. Overview of Mouse Mutants

Over the last decade, the cloning of many of the components of the retinoid signaling pathway has allowed for systematic functional tests of each through the technique of gene targeting. Each of the receptor subtypes has been deleted in mice, as have many of the retinoid binding proteins, and the resulting phenotypes have been analyzed and compared with the developmental and postnatal defects observed in vitamin A deficiency. Deletion of the *RARs* and *RXRs* has yielded several major findings (LUFKIN et al. 1993; LOHNES et al. 1993, 1994; MENDELSOHN et al. 1994; KASTNER et al. 1994, 1996; reviewed in KASTNER et al. 1995; see Table 2). First, mice carrying deletions of particular receptors have significant developmental and postnatal abnormalities, confirming the importance of the receptors in mediating retinoic acid signaling in vivo. These include, but are not limited to, decreased viability, growth deficiency, male sterility, and homeotic transformations of cervical vertebrae in *RAR*α mutant mice (LUFKIN et al. 1993; LOHNES et al. 1994), and in *RAR*γ mutant mice (LOHNES et al. 1993), cardiac and ocular abnormalities in *RXR*α mutant mice (KASTNER et al. 1994; SUCOV et al. 1994; DYSON et al. 1995), and male sterility in *RXR*β mutant mice (KASTNER et al. 1996).

Despite these significant phenotypes, it is now clear that many of the defects observed in vitamin A deficiency – particularly those in embryonic development – are not found in single receptor knockout mice. Mice lacking two receptor genes, however, do exhibit a wide range of congenital malformations. This is most apparent in *RAR*α/β*2* double-knockout embryos, which exhibit respiratory tract, cardiac, ureter, and genital tract abnormalities, among others (LOHNES et al. 1994; MENDELSOHN et al. 1994). These results suggest the presence of a certain degree of functional redundancy, in a laboratory setting, between different retinoid receptors, as well as the possibility of compensatory upregulation of receptor expression in the absence of a particular *RAR* or *RXR*. Finally, certain developmental defects were observed in particular

Table 2. Summary of retinoic acid receptor phenotypes versus vitamin A deficiency (VAD) phenotypes (adapted from KASTNER et al. 1995)

RAR KO Phenotypes	Mutant(s)	Fetal VAD	Postnatal VAD	Not observed in VAD
Respiratory tract defects	RARα/β2	*		
Cardiac defects	RARα/β2; RARα/γ; RXRα	*		
Diaphragmatic hernia	RARα/β2	*		
Ureter abnormalities	RARα/β2; RARα/γ	*		
Genital tract defects	RARα/β2; RARα/γ	*		
Ocular abnormalities	RARα/β2; RARα/γ; RXRα	*		
Male sterility	RARα; RXRβ		*	
Growth deficiency	RARα; RARγ		*	
Poor viability	RARα; RARγ		*	
Loss of photoreceptors	–		*	
Exencephaly	RARα/γ			*
Skeletal abnormalities	RARα/β2; RARα/γ; RARβ2/γ			*
Lens agenesis	RARα/γ			*
Retinal dysplasia	RARβ2/γ2			*
Glandular defects	RARγ; RARα/γ; RARβ2/γ			*
Webbed digits	RARα; RARγ			*
Kidney agenesis	RARα/β2			*
Anal canal agenesis	RARα/β2			*

double mutant combinations that were not reported in fetal vitamin A deficiency, including exencephaly and glandular defects and lens agenesis in *RARα/γ* mutants (LOHNES et al. 1994), and kidney agenesis in *RARα/β2* mutants (MENDELSOHN et al. 1994). As suggested by KASTNER and colleagues (1995), this reveals the difficulty in achieving a state of vitamin A deficiency that is compatible with fetal viability.

II. Mutations in Retinoid Receptors and Binding Proteins with no Testicular Phenotype

As indicated above, mutations either in *RARα* or in *RXRβ* result in male sterility (detailed below). However, mice carrying mutations in *RARβ, RXRα,* and *RXRγ* are fertile, and no abnormalities in spermatogenesis have been reported. Since none of these receptor subtypes is the only receptor expressed in a particular cell type in the seminiferous tubule, it may be that RARα and RXRβ are the primary signaling receptors in the germ cells and Sertoli cells, respectively. In addition, since RARβ expression in the Sertoli cells is apparently unable to compensate for the absence of RXRβ, this suggests that 9-*cis* retinoic acid is an essential retinoic acid isomer in Sertoli cells. This is supported by the recent finding of a putative 9-*cis* retinol dehyrdrogenase that is transcribed at a relatively high level in the human testis (MERTZ et al. 1997).

The retinoid binding proteins also appear to be dispensable for development and adult life. Deletions of CRABP I and II, either singly, or in combination, have no apparent effect on development or adult life, including fertility (GORRY et al. 1994; LAMPRON et al. 1995). The same holds true for TTR-deficient mice, although they do have markedly lower levels of serum retinol (WOLF 1995). The targeted deletions of the RBP and CRBP genes have not yet been reported, but the analysis of retinol metabolism in these mice, under both normal and stressed conditions, should shed light on their physiological roles.

III. Mutations in Retinoid Receptors that Result in Male Sterility

1. RARs

Mutation of the *RARα* gene in mice was shown to result in neonatal lethality (Table 2), and in those mice that survived to adulthood, in male sterility (LUFKIN et al. 1993). Although there has been no detailed histological analysis of the etiology of the sterility phenotype, the degeneration of the seminiferous epithelium in *RARα* null mutant mice of ~4–6 months of age was reported as being similar to that seen in mice maintained on a vitamin A deficient diet. Since these animals will have been essentially deficient in *RARα* function from conception, it is not known when the defects in spermatogenesis are first manifested. An example of the defects that can be seen as early as 8 weeks after birth are shown in Fig. 1.

Mutations in the *RARγ* gene also resulted in male sterility, but for very different reasons. In these mice, there is a squamous metaplasia of the glandular epithelia of the prostate and seminal vesicles (LOHNES et al. 1993), an abnormality which was noted to also be symptomatic of vitamin A deficiency. In the double *RARα/γ* mutant, the genital ducts are severely abnormal. The vas deferens and seminal vesicles are agenic or never form.

A transgenic approach in which a dominant negative *RARα* receptor transgene was expressed under the direction of the mouse mammary tumor virus (MMTV) promoter also resulted in male sterility (COSTA et al. 1997). The MMTV promoter is known to be active only in adult animals, in a subset of tissues which includes the epididymis and other tissues of the male reproductive tract. In the male transgenics ($MMTV/RAR\alpha^{DN}$) there was aberrant differentiation of the epithelial cells in the epididymis, suggesting that the infertility resulted from improper maturation and transport of sperm in the epididymis (COSTA et al. 1997). As it has been shown that *RARα* is normally expressed in a region-specific pattern in the epididymis (AKMAL et al. 1996), it will be important to determine if the sterility in *RARα* null mice involves

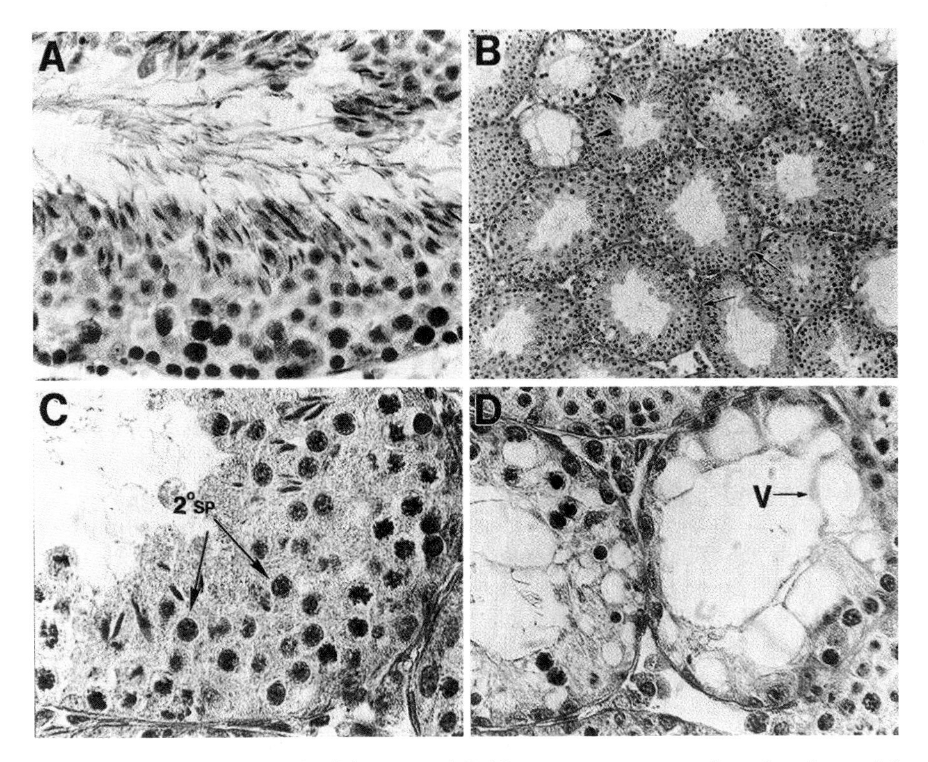

Fig. 1A–D. Morphology of wild-type and *RARα* mutant testes at 8 weeks of age. All sections are stained with hematoxylin and eosin. **A** High magnification view of a section from a wild-type mouse testis showing the full range of spermatogenic cell types. **B** Low magnification view of a section from an *RARα* mutant testis. Relatively normal seminiferous tubules (*arrows*) can be seen as well as degenerating tubules (*arrowheads*). **C** High magnification view of a section from an *RARα* mutant testis showing meiotic germ cells up to and including secondary spermatocytes (2°sp). **D** High magnification view of a section from an *RARα* mutant test showing a highly vacuolated (*v*) tubule with very few germ cells remaining

epididymal abnormalities in addition to the testicular defects (LUFKIN et al. 1993; Fig. 1).

2. RXRs

Recent studies have revealed that mutation of *RXRβ* also results in male sterility, due to oligo-astheno-teratozoospermia (KASTNER et al. 1996), while the females appear to be fully fertile. In these mice, there was a failure of spermatid release within the seminiferous epithelium and very few spermatozoa were found in the epididymis. Those spermatozoa that did form exhibited abnormalities in the acrosome and tail. The Sertoli cells accumulated lipids, histochemically characterized as unsaturated triglycerides. As the animals aged, the testes progressively degenerated, eventually being composed of acellular lipid-filled tubules. In situ hybridization analysis revealed that in the adult testis, *RXRβ* is expressed exclusively in Sertoli cells (KASTNER et al. 1996). The pattern of expression and pathology of the mutation led the authors to speculate that the primary site of function for the RXRβ protein during spermatogenesis is in the Sertoli cells. They also speculate as to the possible role of the known RXR heterodimerizing protein PPAR in the resulting phenotype.

F. Where Do We Go From Here?

I. Assessment of Lineage Specificity of Receptor Function in the Testis

As discussed above, *RAR*α is broadly expressed throughout embryogenesis and in adult tissues, including the testis, wherein existing data suggest its expression in both the germinal and somatic lineages. Furthermore, the results of targeted mutagenesis in the mouse have revealed that while one of the major sites of *RAR*α function is clearly the testis, it is not clear whether both the somatic and germinal lineages require its function. To address this question, it would be necessary to ablate *RAR*α function specifically in one lineage or the other. Two approaches, which take advantage of recent developments in quite different technologies might afford the opportunity to accomplish this. One possibility would be to take advantage of the spermatogonial transplantation techniques developed by Brinster and colleagues (CLOUTHIER et al. 1996). In this approach, the recipient testis is first made deficient of spermatogonial stem cells and then is reconstituted with stem cells from a donor testis. The efficacy of this technique has been assessed by using, as the source of donor cells, transgenic strains of mice that carry readily identifiable markers, such as lacZ. Theoretically, this approach could be applied to dissecting lineage-specificity of function of genes expressed in the germ cell lineage, disruption of which does not affect the spermatogonial stages. Donor, null-mutant stem cells would be transferred to wild type, recipient testes which are devoid of germ cells. Rescue of spermatogenesis would indicate that the gene in question, such as *RAR*α, was not absolutely required for spermatogenesis to proceed.

A second strategy would be to achieve cell-specific ablation using the 'Cre recombinase/*loxP* recognition sequence' system for generating conditional mutations. This approach is made feasible by the availability of spermatogenic stage-specific promoters that are well characterized with regard to their specificity and activity in transgenic mice (WOLGEMUTH and WATRIN 1991; WOLGEMUTH et al. 1995). The strategy would be to generate transgenic lines carrying germ line-specific regulatory sequences driving *Cre* recombinase These lines would then be mated with mice in which portions of the endogenous *RAR*α gene have been replaced with *RAR*α into which *loxP* recognition sites have been inserted. These mice would be produced by targeted homologous recombination in embryonic stem cells and the subsequent generation of chimeric mice. These chimeras would in turn be used to generate mice homozygous for *loxP* sequences on both alleles of *RAR*α. The feasibility of such an approach is enhanced by the existence of exquisitely stage- and cell-specific promoters and the fact that this cell lineage is one in which mutations are not likely to lead to lethality in the animals.

Ideally, one could propose to use this approach to accomplish the reciprocal experiment, that of generating mice in which *RARα* function is disrupted specifically only in Sertoli cells. One possible strategy would be to mate the *RARα /loxP* mice with transgenic mice carrying a Sertoli-cell specific promoter driving Cre recombinase. Although not totally Sertoli-cell specific (as there is expression in fetal liver), one possible candidate is the gene for androgen binding protein (ABP). Reventos et al. (1993) have reported the expression of the rat *ABP* gene in transgenic mice and have reported that its expression in the adult mouse is limited to the testis (among the tissues examined). The exact sequences responsible for this tissue specific expression have not been further delineated, but may nonetheless serve as a useful reporter for these studies.

As an alternative approach, one could consider expressing a dominant negative form of RARα in specific spermatogenic stages in transgenic mice and ask if this inhibits the progression of spermatogenesis. This could be accomplished by directing *RARα* cDNAs with C-terminal deletions by use of the spermatogenic stage-specific promoters. *RARα* function in cultured cells can be suppressed by introducing a dominant negative mutant construct into cells (Tsai et al. 1992). A mutated RARα was generated by deleting 59 amino acids from the C-terminal region. When this truncated RARα ($RAR\alpha^{dn}$) was overexpressed by 6-fold as compared to the wild-type RARα, it suppressed normal RARα function, which can be monitored by the inability to activate transcription of a reporter gene (Tsai et al. 1992).

II. Identification of In Vivo Cofactors

There is a growing body of evidence to suggest that RARs and RXRs do not interact directly with the transcription machinery, but are complexed with transcriptional intermediary factors (TIFs) that are required for receptor-dependent transactivation or transrepression. A number of these intermediary proteins have been identified in yeast and in mammalian cells that bind to the vitamin D and estrogen receptors, as well as to the retinoid receptors (Voegel et al. 1996; le Douarin et al. 1996a; reviewed in le Douarin et al. 1996b; Chambon 1996). Some of these proteins appear to act as bridges between the nuclear receptors and the general transcription machinery, while others appear to be involved in chromatin remodeling rather than in activation of transcription per se. For example, in the latter category, TIF1α interacts in vitro with both RARα and RXRα, but experiments in which nuclear receptors and TIF1α are coexpressed have failed to demonstrate transactivation of reporter genes (le Douarin et al. 1996b). TIF1α has been shown to interact with HP1, a protein belonging to the chromatin organization modifier superfamily, suggesting a role in forming inactive, condensed heterochromatin, or active, open euchromatin structures (le Douarin et al. 1996a). Another intermediary factor, mSUG1, interacts with TBP and TAFII30, both components of the basal transcriptional machinery, suggesting that it is directly involved in the transactivation function of nuclear receptors. Interestingly, mSUG1 binds to RARα, but not to RXRα (LE DOUARIN et al. 1996b), raising the possibility that there is as much complexity in receptor-TIF interactions as there is in other aspects of retinoid metabolism and function. Indeed, the identification of other cofactors such as TIF2 and SRC-1 (VOEGEL et al. 1996), each with unique affinities for the retinoid receptors, for "downstream" transcription factors, and for chromatin, suggests that the role of retinoids in any given tissue will be influenced to a great extent by the presence or absence of these intermediary proteins. The analysis of TIF expression and function in vivo promises to be an important area of investigation, not least in spermatogenesis, where chromatin remodeling is an essential aspect of postmeiotic germ cell maturation.

III. Potential Targets of Retinoid Action During Spermatogenesis

Another largely unexplored area in the study of the retinoid-dependent development of the male germline is the identification of downstream targets of the retinoid receptors. Although retinoic acid response elements have been identified in regions flanking a number of genes (GUDAS et al. 1994), little progress has been made in identifying target genes whose altered testicular expression in states of dietary vitamin A deficiency or receptor mutation might be responsible for the observed degeneration of the seminiferous epithelium and, ultimately, sterility. The candidate target genes include (a) genes whose expression overlaps to some degree with the receptors and whose deletion or misexpression affects spermatogenesis, (b) genes expressed in the testis for which there is evidence that their testicular expression is affected by retinoids, and (c) genes expressed in the testis for which there is evidence that their expression or activity is regulated by retinoids in other tissues or in cultured cell lines.

1. Mutations Affecting Spermatogenesis

The bone morphogenetic proteins (BMPs) are members of the transforming growth factor-β superfamily of secreted polypeptide growth factors. They are widely expressed during embryogenesis and detailed analyses of their expression patterns suggest that they are involved in patterning and organogenesis throughout development (reviewed in HOGAN 1996). Recently, male mice with a targeted mutation in *Bmp8b* have been shown to be sterile (ZHAO et al. 1996). Histological analysis of the mutant testes indicates that there are two distinct defects in spermatogenesis: a reduction in spermatogonial proliferation and subsequent delay in differentiation, and increased apoptosis of pachytene spermatocytes. Although increased germ cell apoptosis is not observed in the *RARα* or *RXRβ* mutants (SASSONE-CORSI 1997), the effect on spermatogonia may be similar to that seen in vitamin A deficient and *RARα*-deficient animals. Interestingly, while *Bmp8b* expression is found in all germ cells at low levels in prepubertal animals, expression becomes restricted to stage VI-VIII round spermatids in the adult (ZHAO et al. 1996), overlapping with RARα mRNA expression (ESKILD et al. 1991).

Another potential target of retinoic acid in the testis is desert hedgehog (dhh). Dhh, a secreted factor homologous to sonic hedgehog (shh) and indian hedgehog (ihh), is expressed in Sertoli cells beginning at E11.5 and continuing in the adult (BITGOOD et al. 1996). Male mice lacking *dhh* are sterile, the seminiferous tubules exhibiting a gross germ cell deficiency that is first observed in the embryonic and early postnatal testis, consistent with a reduction in germ cell proliferation (BITGOOD et al. 1996). *Dhh* deficiency is also associated with pachytene spermatocyte cell death or a block in spermatid differentiation, depending on the genetic background. Each aspect of this phenotype is reminiscent of the dysregulation of spermatogenesis observed in the vitamin A deficient testis. Although it is not known if retinoid receptors are expressed and functional in the early embryonic pre-Sertoli cells, further expression studies and the examination of spermatogenesis in mice that are vitamin A deficient throughout embryogenesis may answer this question. That *dhh* might be retinoid-responsive is suggested by the retinoic acid-dependent upregulation of the *shh* gene in the chick limb bud (RIDDLE et al. 1993) and the *ihh* gene in F9 cells (BECKER et al. 1997). It is also of note that several factors that mediate shh signaling or are regulated by shh in the chick limb bud are also expressed in the testis. These include the downstream effectors of shh, Gli1 and Gli3, expressed in spermatogonia (PERSENGIEV et al. 1997), and potential targets, including basic fibroblast growth factor in Leydig cells, round and elongated spermatids, and Sertoli cells (HAN et al. 1993), and the aforementioned Bmp8b, expressed in germ cells (ZHAO et al. 1996). Expression of patched, a hedgehog receptor, has been reported in fetal Leydig cells (BITGOOD et al. 1996), but its expression in postnatal germ cells has not been established. If conserved in the testis, the elucidation of retinoid-dependent, hedgehog-mediated signaling pathways could yield valuable insights into retinoid function during spermatogenesis.

2. Genes Whose Testicular Expression Is Affected by Retinoids

Several investigators have examined gene expression in testes following withdrawal and replenishment of retinol, in an attempt to identify retinoic acid-responsive genes during the repopulation of the seminiferous tubules. Among those genes whose expression is affected are the epidermal growth factor (EGF) and insulin-like growth factor-I (IGF-I) genes (BARTLETT et al. 1990). In particular, testicular EGF levels were elevated at stages IX-II following retinol replenishment, roughly coincident with the timing of spermatogonial division. In addition, genes related to cell cycle progression are also induced by retinol replenishment of vitamin A deficient rat testes. The threonine/tyrosine phosphatase *cdc25a*, whose activity is required for entry into S phase (Jinno et al. 1994), is expressed in both pre– and postmeiotic rat germ cells and is upregulated in response to retinol (Mizoguchi and Kim 1997). The upregulation in rat Sertoli cells of immediate early genes such as *c-jun* (a component of the transcription factor AP-1) and *c-myb* as early as one hour after retinol supplementation suggests that vitamin A-dependent transcription in Sertoli cells is one of the earliest steps in the reestablishment of spermatogenesis (Page et al. 1996). Male germ cell-associated kinase (*mak*) transcripts are downregulated by retinol in spermatogonia and upregulated in pachytene spermatocytes and round spermatids, suggesting compartment-specific roles for this retinoid-responsive gene (Wang and Kim 1993b). Finally, transferrin mRNA levels are increased in the rat testis by retinol (Hugly and Griswold 1987), but decreased by retinoic acid (Huggenvik et al. 1984), and α–inhibin protein is increased by retinol (Zhuang et al. 1997). Although the upregulation of many of these genes is likely to be an indirect effect of renewed proliferation following retinol replenishment, a few – particularly those that are rapidly induced (*c-jun* and *c-myb*) – may in fact be direct targets of RARs or RXRs.

Perhaps the most intriguing examples of genes whose testicular expression is affected by retinoids are *Stra6* and *Stra8*, genes originally identified in a subtractive hybridization screen for retinoic acid-inducible genes in P19 embryonal carcinoma cells (Bouillet et al. 1995, 1997). *Stra6* appears to encode a novel integral membrane protein whose expression in the embryo and adult includes localization to blood-organ barriers in several tissues, including the brain, eye, and testis (Bouillet et al. 1997). In the testis, *Stra6* in found in a spermatogenic cycle-dependent pattern in the basal plasma membranes of Sertoli cells, particularly in stage VI-VII tubules. Interestingly, in *RAR*α mutant testes, *Stra6* is expressed in almost all of the tubules, suggesting that RARα is required for the selective expression of *Stra6* at stage VI-VII (Bouillet et al. 1997). *Stra8*, encoding a novel cytoplasmic phosphoprotein, is restricted to pre-meiotic germ cells at stages VI-VII (Oulad-Abdelghani et al. 1996), and it too is expressed in almost all of the tubules in *RAR*α mutant testes (Bouillet et al. 1997). Conditional targeted deletion of these genes in the testis should reveal their roles as retinoid-responsive targets in the seminiferous epithelium. In addition, the use of subtractive hybridization and differential display strategies should also continue to be useful in identifying such targets, particularly as it is applied to the *RAR*α and *RXR*β mutant testes.

3. Other Retinoid-Responsive Genes

There are a few genes that are expressed in the testis for which there is evidence that their expression or activity in cultured cells in vitro is affected by retinoic acid. For example, cyclin D1 protein degradation is stimulated by retinoic acid in immortalized human bronchial epithelial cells (LANGENFELD et al. 1997). Retinoic acid also inhibits the expression of cyclin D3 and the kinase activities of Cdk2 and Cdk4 in human breast carcinomal cells in vitro (ZHOU et al. 1997). Although it is difficult to extrapolate from results in cell lines, each of these genes is expressed in testicular cells that are likely to be influenced by retinoids: *cyclin D1* in Sertoli cells (RAVNIK et al. 1995), *cyclin D3* in round spermatids (RAVNIK et al. 1995), *Cdk2* in spermatocytes (RHEE and WOLGEMUTH 1995) and *Cdk4* in peripheral germ cells (RHEE and WOLGEMUTH 1995). While the roles of these cell cycle-related genes in the testis are unknown, their expression and activity in testicular cells from mutant and vitamin A deficient mice is of considerable interest.

Acknowledgements. We thank our colleagues, in particular Drs. William S. Blaner, Max Gottesman, Cathy Mendelsohn, David Talmage, and Monica Peacocke for stimulating discussions. This work was supported in part by grants from the NIH (DK54057, DK52444) and from the USDA (97–35200–4291).

References

Akmal KM, Dufour JM, Kim KH (1996) Region-specific localization of retinoic acid receptor-α expression in the rat epididymis. Biol Reprod 54:1111–1119
Akmal KM, Dufour JM, Kim KH (1997) Retinoic acid receptor α-gene expression in the rat testis: potential role during the prophase of meiosis and in the transition from round to elongating spermatids. Biol Reprod 56:549–556
Bailey JS, Sui C (1990) Unique tissue distribution of two distinct cellular retinoic acid binding proteins in neonatal and adult rat. Biochim Biophys Acta 1033:267–272
Bartlett JM, Spiteri-Grech J, Nieschlag E (1990) Regulation of insulin-like growth factor I and stage-specific levels of epidermal growth factor in stage synchronized rat testes. Biol Reprod 127:747–758
Becker S, Wang ZJ, Massey H, Arauz A, Labosky P, Hammerschmidt M, St-Jacques B, Bumcrot D, McMahon A, Grabel L (1997) A role for Indian hedgehog in extraembryonic endoderm differentiation in F9 cells and the early mouse embryo. Dev Biol 187:298–310
Bishop PD, Griswold MD (1987) Uptake and metabolism of retinol in cultured Sertoli cells: evidence for a kinetic model. Biochemistry 26:7511–7518
Bitgood MJ, Shen L, McMahon AP (1996) Sertoli cell signaling by Desert hedgehog regulates the male germline. Curr Biol 6:298–304
Blaner WS, Galdieri M, Goodman DS (1987) Distribution and levels of cellular retinol- and retinoic acid-binding protein in various types of rat testis cells. Biol Reprod 36:130–137
Blaner WS, Olson JA (1994) Retinol and retinoic acid metabolism. In: Sporn MB, Roberts AB, Goodman DS (eds) The retinoids, 2nd edn. Raven, New York, p 244
Bouillet P, Oulad-Abdelghani M, Vicaire S, Garnier JM, Schuhbaur B, Dolle P, Chambon P (1995) Efficient cloning of cDNAs of retinoic acid-responsive genes in P19 embryonal carcinoma cells and characterization of a novel mouse gene, STRA1 (mouse LERK-2). Dev Biol 170:420–433
Bouillet P, Sapin V, Chazaud C, Messaddeq N, Decimo D, Dolle P, Chambon P (1997) Developmental expression pattern of Stra6, a retinoic acid-responsive gene encoding a new type of membrane protein. Mech Dev 63:173–186
Boylan JF, Gudas LJ (1992) The level of CRABP I expression influence the amounts and types of all *trans*-retinoic acid metabolites in F9 teratocarcinoma stem cells. J Biol Chem 267:21486–21491

Brocard J, Kastner P, Chambon P (1996) Two novel RXR alpha isoforms from mouse testis. Biochem Biophys Res Commun 229:211–218

Chambon P (1995) The molecular and genetic dissection of the retinoid signaling pathway. Rec Prog. Horm Res 50:317–332

Chambon P (1996) A decade of molecular biology of retinoic acid receptors. FASEB J 10:940–954

Clouthier DE, Avarbock MR, Maika SD, Hammer RE, Brinster RL (1996) Rat spermtogenesis in mouse testis. Nature 381:418–421

Costa SL, Boekelheide K, Vanderhyden BC, Seth R, McBurney MW (1997) Male infertility caused by epididymal dysfunction in transgenic mice expressing a dominant negative mutation of retinoic acid receptor alpha. Biol Reprod 56:985–990

Davis JT, Ong DE (1995) Retinol processing by the peritubular cell from rat testis. Biol Reprod 52:356–364

De Rooij DG, Van Dissel-Emiliani FMF, Van Pelt AMM (1989) Regulation of spermatogonial proliferation. In: Ewing LL, Robaire B (eds) Regulation of testicular function: signaling molecules and cell-cell communication. Ann NY Acad Sci 564:140–153

DeRooij DG, van Pelt AMM, Van de Kant HJG, van der Saag PT, Peters AHFM, Heyting C, de Boer P (1994) Role of retinoids in spermatogonial proliferation and differentiation and the meiotic prophase. In: Bartke A (ed) Function of somatic cells in the testis. Springer, Berlin Heidelberg New York, p 345

De The H, del Mar Vivanco-Ruiz M, Tiollais P, Stunnenberg H, Dejean A (1990) Identification of a retinoic acid responsive element in the retinoic acid receptor β-gene. Nature 343:177–180

Dyson E, Sucov HM, Kubalak SW, Schmid-Schonbein GW, DeLano FA, Evans RM, Ross J Jr, Chien KR (1995) Atrial-like phenotype is associated with embryonic ventricular failure in retinoid X receptor α–/– mice. Proc Natl Acad Sci USA 92:7386–7390

Eskild W, Ree AH, Levy FO, Jahnsen T, Hansson V (1991) Cellular localization of messenger RNAs for retinoic acid receptor-α, cellular retinol-binding protein, and cellular retinoic acid-binding protein in rat testis-evidence for germ cell-specific mRNAs. Biol Reprod 44:53–61

Eskild W, Hansson V (1994) Vitamin A functions in the reproductive organs. In: Blomhoff R (ed) Vitamin A in health and science. Dekker, New York, p 531

Fiorella PD, Napoli JL (1991) Expression of cellular retinoic acid binding protein (CRABP) in Escherichia coli. Characterization and evidence that holo-CRABP is a substrate in retinoic acid metabolism. J Biol Chem 266:16572–16579

Fogh K, Voorhees JJ, Astrom A (1993) Expression, purification, and binding properties of human cellular retinoic acid binding protein type I and type II. Arch Biochem Biophys 300:751–755

Gaemers IC, Van Pelt AMM, Van der Saag PT, De Rooij DG (1996) All-*trans*-4-oxo-retinoic acid: a potent inducer of in vivo proliferation of growth-arrested A spermatogonia in the vitamin A deficient mouse testis. Endocrinology 137:479–485

Gaemers IC, Van Pelt AMM, Van der Saag PT, Hoogerbrugge JW, Themmen APT, De Rooij DG (1997) Effect of retinoid status on the messenger ribonucleic acid expression of nuclear retinoid receptors α, β, and γ, and retinoid X receptors α, β, and γ in the mouse testis. Endocrinology 138:1544–1551

Gorry P, Lufkin T, Dierich A, Rochette-Egly C, Decimo D, Dolle P, Mark M, Durand B, Chambon P (1994) The cellular retinoic acid binding protein I is dispensible. Proc Natl Acad Sci USA 91:9032–9036

Griswold MD, Bishop PD, Kim KH, Ping R, Siiteri JE, Morales C (1989) Function of vitamin A in normal and synchronized seminiferous tubules. In: Ewing LL, Robaire B (eds) Regulation of testicular function: signaling molecules and cell-cell communication. Ann NY Acad Sci 564:154–172

Gudas LJ, Sporn MB, Roberts AB (1994) Cellular biology and biochemistry of the retinoids. In: Sporn MB, Roberts AB, Goodman DS (eds) The retinoids in biology, chemistry, and medicine. Raven, New York, p 443

Han IS, Sylvester SR, Kim KH, Schelling ME, Venkateswaran S, Blanckaert VD, McGuinness MP, Griswold MD (1993) Basic fibroblast growth factor is a testicular germ cell product which may regulate Sertoli cell function. Mol Endocrinol 7:889–897

Hogan BLM (1996) Bone morphogenetic proteins: multifunctional regulators of vertebrate development. Genes Dev 10:1580–1594

Howell JM, Thompson JN, Pitt GAJ (1963) Histology of the lesions produced in the reproductive tract of animals fed a diet deficient in vitamin A alcohol but containing vitamin A acid, I. The male rat. J Reprod Fertil 5:159–167

Huang HFS, Hembree WC (1979) Spermatogenic response to vitamin A in vitamin A deficient rats. Biol Reprod 21:891–904

Huang HFS, Marshall GR (1983) Failure of spematid release under various vitamin A states – an indication of delayed spermiation. Biol Reprod 28:1163–1172

Huang HFS, Yang CS, Meyenhofer M, Gould S, Boccabella AV (1988) Disruption of sustentacular (Sertoli) cell tight junctions and regression of spermatogenesis in vitamin A deficient rats. Acta Anat (Basel) 133:10–15

Huang HFS, Li MT, Pogach LM, Qian L (1994) Messenger ribonucleic acid of rat testicular retinoic acid receptors: developmental pattern, cellular distribution, and testosterone effect. Biol Reprod 51:541–550

Huggenvik J, Sylvester SR, Griswold MD (1984) Control of transferrin mRNA synthesis in Sertoli cells. Ann NY Acad Sci 438:1–7

Hugly S, Griswold MD (1987) Regulation of levels of specific Sertoli cell mRNAs by vitamin A. Dev Biol 121:316–324

Husmann M, Hoffmann B, Stump DG, Chytil F, Pfahl M (1992) A retinoic acid responsive element from the rat CRBP I promoter is activated by an RAR/RXR heterodimer. Biochem Biophys Res Commun 187:1558–1564

Ismail N, Morales C, Clermont Y (1990) Role of spermatogonia in the stage-synchronization of the seminiferous epithelium in vitamin A deficient rats. Am. J Anat 188:57–63

Ismail N, Morales C (1992) Effects of vitamin A deficiency on the inter-Sertoli cell tight junctions and on the germ cell population. Microsc Res Techn 20:43–49

Jinno S, Suto K, Nagata A, Igarashi M, Kanaoka Y, Nojima H, Okayama H (1994) Cdc25 a is a novel phosphatase functioning early in the cell cycle. EMBO J 13:1549–1556

Kastner P, Grondona J, Mark M, Gansmuller A, LeMeur M, Decimo D, Vonesch JL, Dolle P, Chambon P (1994) Genetic analysis of RXR-α-developmental function: convergence of RXR and RAR signaling pathways in heart and eye morphogenesis. Cell 78:987–1003

Kastner P, Mark M, Chambon P (1995) Nonsteroid nuclear receptors: what are genetic studies telling us about their role in real life. Cell 83:859–869

Kastner P, Mark M, Leid M, Gansmuller A, Chin W, Grondona JM, Decimo D, Krezel W, Dierich A, Chambon P (1996) Abnormal spermatogenesis in RXR-β-mutant mice. Genes Dev 10:80–92

Kato M, Sung WK, Kato K, Goodman DS (1985) Immunohistochemical studies on the localization of cellular retinol-binding protein in rat testis and epididymis. Biol Reprod 32:173–189

Kim KH, Griswold MD (1990) The regulation of retinoic acid receptor messenger RNA levels during spermatogenesis. Mol Endocrinol 4:1679–1688

Kim KH, Wang ZQ (1993) Action of vitamin A on testis: role of the Sertoli cell. In: Griswold MD, Russel L (eds) The Sertoli cell. Cache River, Clearwater, p 514

Kim KH, Akmal KM (1996) Role of vitamin A in male germ-cell development. In: Cellular and molecular regulation of testicular cells. Springer, Berlin Heidelberg New York, p 83

Kurlandsky SB, Gamble MV, Ramakrishnan R, Blaner WS (1995) Plasma delivery of retinoic acid to tissues in the rat. J Biol Chem 270:17850–17857

Lampron C, Rochette-Egly C, Gorry P, Dolle P, Mark M, Lufkin T, LeMeur M, Chambon P (1995) Mice deficient in cellular retinoic acid binding protein II (CRABP II) or in both CRABP I and CRABP II are essentially normal. Development 121:539–548

Langenfeld J, Kiyokawa H, Sekula D, Boyle J, Dmitrovsky E (1997) Posttranslational regulation of cyclin D1 by retinoic acid: a chemoprevention mechanism. Proc Natl Acad Sci USA 94:12070–12074

Le Douarin B, Nielsen AL, Garnier J-M, Ichinose H, Jeanmougin F, Losson R, Chambon P (1996a) A possible involvement of TIF1α and TIF1β in the epigenetic control of transcription by nuclear receptors. EMBO J 15:6701–6715

Le Douarin B, Vom Baur E, Zechel C, Heery D, Heine M, Vivat V, Gronemeyer H, Losson R, Chambon P (1996b) Ligand-dependent interaction of nuclear receptors with transcriptional intermediary factors (mediators). Phil Trans R Soc Lond B 351:569–578

Leid M, Kastner P, Chambon P (1992) Multiplicity generates diversity in the retinoic acid signaling pathways. Trends Biochem Sci 176:427–433

Lohnes D, Kastner P, Dierich A, Mark M, LeMeur, Chambon P (1993) Function of retinoic acid receptor γ (RAR-γ) in the mouse. Cell 73:643–658

Lohnes D, Mark M, Mendelsohn C, Dolle P, Dierich A, Gorry P, Gansmuller A, Chambon P (1994) Function of the retinoic acid receptors (RARs) during development. I. Craniofacial and skeletal abnormalities in RAR double mutants. Development 120:2723–2748

Lufkin T, Lohnes D, Mark M, Dierich A, Gorry P, Gaub M-P, LeMeur M, Chambon P (1993) High postnatal lethality and testis degeneration in retinoic acid receptor α (RAR-α) mutant mice. Proc Natl Acad Sci USA 90:7225–7229

Mangelsdorf DJ, Evans RM (1995) The RXR heterodimers and orphan receptors. Cell 83:841–850

Mason KE (1933) Differences in testes injury and repair after vitamin A deficiency, vitamin E deficiency and inanition. Am. J Anat 52:153–239

McGuire BW, Orgebin-Crist M-C, Chytil F (1981) Autoradiographic localization of serum retinol-binding protein in rat testis. Endocrinology 108:658–667

Mendelsohn C, Lohnes D, Decimo D, Lufkin T, LeMeur M, Chambon P, Mark M (1994) Function of the retinoic acid receptors (RARs) during development. II. Multiple abnormalities at various stages of organogenesis in RAR double mutants. Development 120:2749–2771

Mertz JR, Shang E, Piantedosi R, Wei S, Wolgemuth DJ, Blaner WS (1997) Identification and characterization of a stereospecific human enzyme that catalyzes 9-*cis* retinol oxidation. J Biol Chem 272:11744–11749

Mitranond V, Sobhon P, Tosukhowong P, Chindaduangrat W (1979) Cytological changes in the testes of vitamin A deficient rats. I. Quantitation of germinal cells in the seminiferous tubules. Acta Anat 103:159–168

Mizoguchi S, Kim KH (1997) Expression of cdc25 phosphatases in the germ cells of the rat. Biol Reprod 56:1474–1481

Morales CR, Griswold MD (1991) Variations in the level of transferrin and SGP-2 mRNAs in Sertoli cells of vitamin A deficient rats. Cell Tissue Res 263:125–130

Napoli JL, Posch KP, Fiorella PD, Boerman MH (1991) Physiological occurrence, biosynthesis and metabolism of retinoic acid: evidence for roles of cellular retinol-binding protein (CRBP) and cellular retinoic acid-binding protein (CRABP) in the pathway of retinoic acid homeostasis. Biomed Pharmacother 45:131–143

Orth JM (1982) Proliferation of Sertoli cells in fetal and postnatal rats. Anat Rec 203:485–492

Oulad-Abdelghani M, Bouillet P, Decimo D, Gansmuller A, Heyberger S, Dolle P, Bronner S, Lutz Y, Chambon P (1996) Characterization of a premeiotic germ cell-specific cytoplasmic protein encoded by Stra8, a novel retinoic acid-responsive gene. J Cell Biol 135:469–477

Page KC, Heitzman DA, Chernin MI (1996) Stimulation of c-jun and c-myb in rat Sertoli cells following exposure to retinoids. Biochem Biophys Res Comm 222:595–600

Persengiev SP, Kondova II, Millette CF, Kilpatrick DL (1997) Gli family members are differentially expressed during the mitotic phase of spermatogenesis. Oncogene 14:2259–2264

Pfahl M, Chytil F (1996) Regulation of metabolism by retinoic acid and its nuclear receptors. Annu Rev Nutr 16:257–283

Porter SB, Ong DE, Chytil F, Orgebin-Crist M-C (1985) Localization of cellular retinol-binding protein and cellular retinoic acid binding-protein in rat testis and epididymis. J Androl 6:197–212

Rajan N, Sung WK, Goodman DS (1990) Localization of cellular retinol-binding protein mRNA in rat testis and epididymis and its stage-dependent expression during the cycle of the seminiferous epithelium. Biol Reprod 43:835–842

Rajan N, Kidd GL, Talmage DA, Blaner WS, Suhara A, Goodman DS (1991) Cellular retinoic acid binding protein messenger RNA levels in rat tissues and localization in rat testis. J Lipid Res 32:1195–1204

Ravnik SE, Rhee K, Wolgemuth DJ (1995) Distinct patterns of expression of the D-type cyclins during testicular development in the mouse. Dev Genet 16:171–178

Reventos J, Sullivan PM, Joseph DR, Gordon JW (1993) Tissue-specific expression of the rat androgen-binding protein/sex hormone-binding globulin gene in transgenic mice. Mol Cell Endocrinol 96:69–73

Rhee K, Wolgemuth DJ (1995) The distinct patterns of expression of several Cdk family genes during mouse germ cell development suggest specific functions during both cell division and differentiation. Dev Dynam 204:406–420

Riddle RD, Johnson RL, Laufer E, Tabin C (1993) Sonic hedgehog mediates the polarizing activity of the ZPA. Cell 75:1401–1416

Sassone-Corsi P (1997) Transcriptional checkpoints determining the fate of male germ cells. Cell 88:163–166

Shingleton JL, Skinner MK, Ong DE (1989) Retinol esterification in Sertoli cells by lecithin-retinol acyltransferase. Biochemistry 28:9647–9653

Sobhon P, Mitranond V, Tosukhowong P, Chindaduangrat W (1979) Cytological changes in the testes of vitamin A deficient rats. II. Ultrastructural study of the seminiferous tubules. Acta Anat 103:169–183

Sucov HM, Dyson E, Gumeringer CL, Price J, Chien KR, Evans RM (1994) RXR-α-mutant mice establish a genetic basis for vitamin A signaling in heart morphogenesis. Genes Dev 8:1007–1018

Thompson JN, Howell JM, Pitt GAJ (1964) Vitamin A and reproduction in rats. Proc R Soc Lond Biol 159:510–535

Tsai S, Bartelmez S, Hyeman R, Damm K, Evans R, Collins SJ (1992) A mutated retinoic acid receptor α exhibiting dominant-negative activity alters the lineage development of a multipotent hematopoietic cell line. Genes Dev 6:2258–2269

Unni E Rao, MRS, Ganguly J (1983) Histological and ultrastructural studies on the effect of vitamin A depletion and subsequent repletion with vitamin A on germ cells and Sertoli cells in rat testis. Indian. J Exp Biol 21:180–192

Van Pelt AMM, De Rooij DG (1990) Synchronization of the seminiferous epithelium after vitamin A replacement in vitamin A deficient mice. Biol Reprod 43:363–367

Van Pelt AMM, De Rooij DG (1991) Retinoic acid is able to reinitiate spermatogenesis in vitamin A deficient rats and high replicate doses support the full development of spermatogenic cells. Endocrinology 128:697–704

Van Pelt AMM, Van den Brink CE, De Rooij DG, Van der Saag PT (1992) Effects of retinoids on retinoic acid receptor mRNA levels in the vitamin A deficient rat testis. Endocrinology 131:344–350

Voegel JJ, Heine MJS, Zechel C, Chambon P, Gronemeyer H (1996) TIF2, a 160 kDa transcriptional mediator for the ligand-dependent activation function AF-2 of nuclear receptors. EMBO J 15:3667–3675

Wang Z, Kim KH (1993a) Vitamin A deficient testis germ cells are arrested at the end of S phase of the cell cycle: a molecular study of the origin of synchronous spermatogenesis in regenerated seminiferous tubules. Biol Reprod 48:1157–1165
Wang ZQ, Kim KH (1993b) Retinol differentially regulates male germ cell-associated kinase (mak) messenger ribonucleic acid expression during spermatogenesis. Biol Reprod 49:951–964
Wolbach BS, Howe PR (1925) Tissue changes following deprivation of fat-soluble A vitamin. J Exp Med 42:753–777
Wolf G (1995) Retinol transport and metabolism in transthyretin knockout mice. Nutr Rev 53:98–99
Wolgemuth DJ, Watrin F (1991) List of cloned mouse genes with unique expression patterns during spermatogenesis. Mam. Genome 1:283–288
Wolgemuth DJ, Rhee K, Wu S, Ravnik SE (1995) Genetic regulation of the mammalian spermatogenic cell cycle. Reprod Fertil Dev 7:669–683
Zhai Y, Higgins D, Napoli JL (1997) Coexpression of the mRNAs encoding retinol dehydrogenase isozymes and cellular retinol-binding protein. J Cell Physiol 173:36–43
Zhao G-Q, Deng K, Labosky PA, Liaw L, Hogan BLM (1996) The gene encoding bone morphogenetic protein 8B is required for the initiation and maintenance of spermatogenesis in the mouse. Genes Dev 10:1657–1669
Zheng WL, Bucco RM, Schmitt MC, Wardlaw SA, Ong DE (1996) Localization of cellular retinoic acid-binding protein (CRABP) II and CRABP in developing rat testis. Endocrinology 137:5028–5035
Zhou Q, Stetler-Stevenson M, Steeg PS (1997) Inhibition of cyclin D expression in human breast carcinoma cells by retinoids in vitro. Oncogene 15:107–115
Zhuang YH, Blauer M, Ylikomi T, Tuohimaa P (1997) Spermatogenesis in the vitamin A deficient rat: possible interplay between retinoic acid receptors, androgen receptor and inhibin alpha-subunit. J Steroid Biochem Mol Biol 60:67–76

CHAPTER 13

The Role of Retinoids in Vertebrate Limb Morphogenesis: Integration of Retinoid- and Cytokine-Mediated Signal Transduction

H.-C. Lu, C. Thaller, and G. Eichele

A. Introduction

Vitamins, important accessory molecules required for a large number of physiological processes, were discovered at the turn of this century. It has long been recognized that in the absence of vitamin A, the growth of animals is adversely affected (McCollum and Davis 1913; Osborne and Mendel 1913). In the 1940s and 1950s, Warkany and colleagues studied the development of vitamin A deficient rats (e.g., Wilson et al. 1953). These investigations resulted in the identification of a broad spectrum of developmental defects in offspring of vitamin A deficient rats such as heart, lung, thymus and digestive tract dysmorphogenesis. Studies in the 1920s by Wolbach and Howe (1925) provided the first link between vitamin A status and tissue histology. These investigators showed that a deficiency in vitamin A causes squamous metaplasia and keratinization of most columnar epithelia of the body. Subsequently, Fell and Mellanby (1953) showed that high concentrations of vitamin A suppressed keratinization of epithelia and caused this tissue to differentiate into a mucus-secreting, ciliated epithelium.

Although the requirement for vitamin A had been documented for almost 80 years, the mechanism of action of this lipophilic molecule in growth and differentiation has remained enigmatic for a long time. In 1987 two important advances were reported. First, it was shown that all-*trans*-retinoic acid was present in the embryo in nanomolar concentrations (Thaller and Eichele 1987) and second, that the biological effects of all-*trans*-retinoic acid are mediated by specific nuclear receptors that act as ligand-dependent transcription factors (Giguère et al. 1987; Petkovich et al. 1987). All subsequent research pertaining to the role of vitamin A compounds in embryonic development is built on these two findings. The literature dealing with the mechanism of action, the physiology and pathophysiology of retinoids is vast and has been reviewed repeatedly over the past few years (Hofmann and Eichele 1994; Mangelsdorf et al. 1994; Sporn et al. 1994; Beato et al. 1995; Conlon 1995; Kastner et al. 1995; Leblanc and Stunnenberg 1995; Mangelsdorf and Evans 1995; Means and Gudas 1995; Chambon 1996; Fisher and Voorhees 1996; Minucci and Ozato 1996; Eichele 1997; Pazin and Kadonaga 1997; Rogers 1997). The present review focuses on the role of retinoids in development with emphasis on the developing vertebrate limb. In particular, we wish to illustrate the important point that no signaling molecule, including retinoids, acts by itself in a biological process. Cells in embryos and in adult tissues respond to multiple signals which they integrate in an appropriate manner.

This chapter is organized as follows. First, we provide a general overview of retinoids in vertebrate development with emphasis on recent advances. Additional information on some of these issues are found in other chapters of this volume. Many morphogenetic processes are affected by retinoids and the developing vertebrate limb has been a useful experimental system to study the relationship between retinoid signaling and other signaling pathways. Thus, we discuss in more detail how retinoids are thought to act in vertebrate limb development and how they integrate with other signaling molecules, such as Sonic Hedgehog protein and Hox proteins. The chapter concludes with a review of recent advances in understanding the role of retinoids in limb regeneration.

B. An Overview of Retinoid Signal Transduction in the Vertebrate Embryo

Three components constitute retinoid signaling: (a) synthesis, transport, and degradation of the biologically active retinoic acid at the appropriate time and location; (b) the retinoid receptors RARs and RXRs and their coactivators and corepressors; (c) transactivation of retinoid-regulated target genes. At any of three these stepping stones there is an opportunity for other signaling molecules to modulate the retinoid signaling pathway.

Several biologically active retinoids (e.g., all-*trans*-retinoic acid, 9-*cis*-retinoic acid, 4-oxo-retinoic acid, 3,4-didehydro-retinoic acid) have been found in vertebrate embryos (Thaller and Eichele 1987; Durston et al. 1989; Chen et al. 1992, 1994; Hogan et al. 1992; Creech Kraft et al. 1994; Scott et al. 1994; Horton and Maden 1995; Costaridis et al. 1996; Stratford et al. 1996). Maternally provided precursors in the form of retinol and retinyl esters are locally converted to the biologically active carboxylated retinoids by what is believed specific dehydrogenases (see Sect. B.II). However, pharmacological studies have shown that exogenously provided retinoic acid can also enter the embryo via the maternal circulation (e.g., Hummler et al. 1994). The mechanism of uptake of these precursors by embryonic cells is currently unclear but because retinoids are hydrophobic molecules, they can probably pass through the plasma membrane. Additionally, there could be specific cell surface receptors that bind retinoid-charged carrier proteins (Dew and Ong 1997). Once retinoids are inside cells, they bind to specific cellular retinoid binding proteins (Ong et al. 1993; Mangelsdorf et al. 1994). These proteins are expressed in embryos in often intriguing patterns (Maden et al. 1988, 1989; Dollé et al. 1989; Ruberte et al. 1991, 1992). Enzymes capable of converting precursors to retinoic acid are found inside cells (Duester 1996; Napoli 1996), and cellular retinoid binding proteins have been shown to fine-tune the reactions catalyzed by these enzymes (Ong et al. 1993; Napoli 1996). However, cellular binding proteins do not directly mediate retinoid signaling. This task is carried out by two families of retinoid receptors (LEID et al. 1992), the retinoic acid receptors (RARα, RARβ, RARγ and their isoforms) and the retinoid X receptors (RXRα, RXRβ, RXRγ and their isoforms (reviewed in Chap. 5; MANGELSDORF et al. 1994; MANGELSDORF and EVANS 1995).

These receptors are ligand dependent transcription factors that belong to the nuclear receptor superfamily. Accordingly, RARs and RXRs modulate gene expression by binding to specific DNA sequences, the retinoic acid responsive elements (RAREs), which are located in the control regions of target genes. Embryos lacking RARs or RXRs as a result of a gene knock-out usually display embryonic defects (see Sect. B.III). Recent advances in retinoid receptor research have made it clear that these receptors act by a complex mechanism involving ligand-induced conformational changes, heterodimer formation and interaction with coactivators and corepressors, whose role lies in modifying the chromatin structure by acetylation and de-acetylation of histones (reviewed in PAZIN and KADONAGA 1997). It is worth noting that other transcription factors act by very similar mechanisms. The result of the ligand-mediated transactivation is the turning on or off of genes that encode effector proteins. Knowing the identity of these retinoid target genes is the basis for understanding how retinoids exert their action in embryogenesis. Many such genes have now been identified and there are some very conclusive findings regarding the role of retinoids in the regulation of *Hox* genes (see Sects. B.IV, C.III).

I. Retinoid Requirement in Development

Retinoids have striking effects on the development of vertebrate embryos. An excess or deficiency of retinoic acid or of its biosynthetic precursors, leads to abnormal development (HOFMANN and EICHELE 1994; CONLON 1995; MADEN et al. 1996 and references therein). A broad spectrum of malformations is induced by exogenous retinoic acid, such as cleft palate, cardiac malformations, and thymus hypoplasia or aplasia, and defects in the central nervous system (reviewed in MEANS and GUDAS 1995). When rat or quail embryos develop under vitamin A deficient condition, pronounced abnormalities are found in the development of the cardiovascular tissue, the eye, the cranial neural crest cells (and their derivatives), the genitourinary tract, the diaphragm, and the lung (MADEN et al. 1996; see Chap. 15, this volume). The deficiency data in conjunction with the retinoid receptor mouse mutants (see Sect. B.III) make a compelling argument for the requirement of retinoids in embryogenesis.

II. Retinoic Acid Synthesizing Enzymes and Degrading Enzymes in Embryos

Retinoic acid is produced from its precursors retinol and retinal by a two-step dehydrogenation. In the first reaction a retinol dehydrogenase (RoDH) converts retinol to retinal. This reaction is reversible and represents the rate-limiting step in retinoic acid biosynthesis. Subsequently, retinal is irreversibly converted to retinoic acid by a retinal dehydrogenase (RalDH). While these reactions were demonstrated in adult tissues in the 1950s (Dowling and Wald 1960) and in embryos (Thaller and Eichele 1990), only recently have some of the responsible enzymes been characterized at a molecular level. The complicated picture that emerges is that there are several RoDH isozymes and RalDH isozymes that can use natural retinoids as substrates (see Duester 1996, Napoli 1996 for reviews). An additional twist in retinoid metabolism is that cellular retinoid binding proteins, which are widely expressed in embryos, modulate retinoid metabolism (see Napoli 1996 for a review).

Several groups, most notably those of Duester, Napoli and Dräger, have isolated RoDHs and RalDHs (Zgombic-Knight et al. 1995; Chai and Napoli 1996 and references therein; Bhat et al. 1995; Wang et al. 1996; Zhao et al. 1996; Penzes et al. 1997 and references therein). The expression patterns of some of these enzymes during embryogenesis have been described. An *RoDH* designated as *class I alcohol dehydrogenase* (*class I ADH*) is expressed in embryonic kidney at E 9.5 and *class III ADH* transcripts are found throughout the embryo between E 6.5 and 9.5 (Ang et al. 1996). In contrast, *class IV ADH* shows localized expression in many embryonic tissues (Ang et al. 1996; Ang and Duester 1997). The gene is expressed in the mesoderm of the primitive streak as early as E7.5. Interestingly, the streak region also expresses two retinal dehydrogenases, *class I ALDH* (Ang and Duester 1997), and *RALDH-2* (Niederreither et al. 1997). This suggests that this region is a source of retinoic acid in the early embryo and such regionally produced retinoic acid may regulate the expression of retinoic acid-responsive *Hox* genes in the early embryo (Hogan et al. 1992) (see Sect. B.IV).

In older embryos, *class IV ADH*, *class I ALDH*, and *RALDH-2* are expressed in complex patterns. *Class IV ADH* is found in craniofacial primordia, the dorsal neural fold, in neural crest and its derivatives, in somites, in paraxial mesoderm and in the proximal portion of the forelimb bud (Ang et al. 1996). *Class I ALDH* mRNA is found in craniofacial mesenchyme, in somites and in paraxial mesoderm (Ang and Duester 1997). The expression of *RALDH-2* has been well characterized (Niederrheither et al. 1997). Sites of expression include undifferentiated somites, the optic vesicle, the lateral walls of the coelom, interdigital tissue of the differentiating limb, the tooth buds, the inner ear and the pituitary gland.

Niederreither et al. (1999) showed that mouse embryos lacking the *RALDH-2* gene have major developmental abnormalities and die mid gestation. Posterior mesodermal structures are severely affected, limb buds never form, and defects in heart and craniofacial structures are observed. These data demonstrate a role of retinoic acid in the morphogenesis in the early stages of embryogenesis.

Retinoic acid degradation is thought to initiate with the formation of polar intermediates such as 4-OH-retinoic acid and 4-oxo-retinoic acid. Whether these molecules are intermediates of catabolism or have specific physiological functions is an open question (PIJNAPPEL et al. 1993). WHITE et al. (1996, 1997) and RAY et al. (1997) have identified a P450 enzyme, termed $P450^{RAI}$. $P450^{RAI}$ is a member of a novel cytochrome P450 family referred to as CYP26. $P450^{RAI}$ enzyme produces 4-hydroxy-retinoic acid, 4-oxo-retinoic acid and 18-OH-retinoic acid from retinoic acid (WHITE et al. 1997) and the expression of *P450*^{RAI} is inducible by retinoic acid (RAY et al. 1997; FUJII et al. 1997; WHITE et al. 1997). FUJII et al. (1997) have cloned a P450 enzyme which they believe to be different from $P450^{RAI}$, chiefly because their enzyme produces 5,8-epoxy-retinoic acid. However, sequence comparison of the mouse proteins described by RAY et al. and FUJII et al. shows that the proteins are identical. Further research will be required to resolve the issue of which metabolites are generated by $P450^{RAI}$.

The expression of *P450*^{RAI} has been described in some detail in the mouse embryo (FUJII et al. 1997). Expression initiates around E6. At E7, extraembryonic and embryonic endoderm and embryonic mesoderm express this gene. By E7.25 transcripts are detected in the primitive streak region, i.e., in the region that corresponds to the trunk. By E7.5, expression in the primitive streak region has greatly diminished, but *P450*^{RAI} mRNA is now found in the anterior part of the embryo in all three germ layers. By E8.5, the gene is highly expressed in the tailbud mesoderm, and the caudal neural plate, but transcripts are also found at the opposite end of the embryo: the hindbrain, the foregut, and the first pharyngeal arch. By E 9.5 to 10.5 transcripts are found in the caudal most part of the embryo and in neural crest cells that contribute to the cranial ganglia. These cells are known to be sensitive to excess retinoic acid and one role of $P450^{RAI}$ may be to keep retinoic acid levels low in these cells. Transcripts of *P450*^{RAI} are also found in the limb bud as well as in the interdigital space. Of note, treatment of embryos with retinoic acid resulted in an abolishment of the caudal expression domain of *P450*^{RAI} and at the same time, in an up-regulation of expression of this gene in the cranial region. Thus in the embryo, this gene is differentially regulated by retinoic acid. It will be interesting to compare the expression patterns of the retinoic acid producing enzymes (*RoDH* and *RalDH*) with that of *P450*^{RAI}. The present data suggest that in the early embryo (E7.5 and before) the two types of enzymes are coexpressed in the primitive streak region. This is somewhat of a paradox, because it would mean that locally produced retinoic acid is at once degraded. A future task will be to determine whether additional retinoid metabolism enzymes exist and where and when these are expressed precisely.

III. RARs and RXRs Mediate Retinoic Acid Signal Transduction In Vivo and Are Required for Development

Null mutations of *RAR*α, *RAR*β, or *RAR*γ have been generated in the mouse (reviewed by KASTNER et al. 1995). Mice lacking *RAR*α or *RAR*γ exhibit some of the defects displayed by the fetuses of retinoid deficient rats, including poor viability, growth deficiency, and male sterility. These mutants also display congenital malformations, which are confined to a small subset of the tissues expressing retinoid receptors. Pairwise combination of *RAR* null mutants have been generated by cross-mating and these test for possible functional redundancy within the *RAR* family. In contrast to *RAR* single mutants, most of the double mutants exhibit a dramatically reduced viability. Furthermore, almost all of the embryonic retinoid deficiency syndromes appear in the different combination of *RAR* double mutants. These findings demonstrate that retinoic acid is the retinoid derivative relevant during embryogenesis and that the function of retinoids are mediated by RARs.

RXR mutant mice also exhibit morphological defects. *RXRα* null mutants display a thin ventricular wall and die from cardiac failure at around embryonic day 10.5 (Kastner et al. 1994; Sucov et al. 1994; Dyson et al. 1995). This phenotype is similar to the "spongy myocardium" dysmorphogenesis in retinoid deficient fetuses. This result suggests that RXRα plays a role in retinoic acid signal transduction. $RXR\alpha^{-/-}$ mice also display ocular abnormalities, notably malformations of the anterior segment of the eye and shortening of the ventral retina (Kastner et al. 1994). The latter defect is an abnormality also present in retinoid deficient embryos. However, none of the *RAR* single or double mutants have this defect, suggesting that ventral retina development requires retinoid ligand bound to RXRα, and that signaling in this case does not require RARs. Interestingly, mutation of the *Drosophila RXR* homolog, *ultraspiracle* (*usp*), results in a shorter ventral retina (Oro et al. 1992), which may reflect a functional conservation of *RXR*/*usp* signaling in eye development.

Having mice lacking the various *RARs* and *RXRs* makes it possible to demonstrate that RAR/RXR heterodimers are required to transduce the retinoid signal, as in vitro studies had suggested (reviewed in Kastner et al. 1995). Examination of $RXR\alpha^{-/-}$ mice and of various *RXRα*/*RAR* double mutants, has revealed a convergence between *RXR*- and *RAR*-dependent signaling pathways. The severity of several ocular abnormalities present in RXRα null mutants increases in a graded manner with the successive removal of the two alleles of either *RARβ* or *RARγ* (Kastner et al. 1994). Furthermore, the inactivation of only one *RXRα* allele from a *RARγ* null genetic background can cause eye defects identical to those observed in *RXRα* null mutants. In addition, new abnormalities such as persistent *truncus arteriosus*, not seen in *RXRα* null mutants, are generated in *RXRα*/*RAR* double mutants. Since all of these abnormalities are present in retinoid deficient fetuses, the synergistic effects of RXRα and RAR, in conjunction with the fact that RXR-RAR heterodimers bind much more tightly to RAREs than either receptor homodimer alone, argue that RAR/RXR heterodimers act as functional units transducing the retinoic acid signal in vivo. In summary, retinoid receptor mutation studies and the analysis of retinoid deficient embryos make it very clear that retinoids have a critical role in vertebrate morphogenesis.

Several studies have shown that RXR forms heterodimers not only with RARs but with several other nuclear receptors including thyroid hormone receptor, the peroxisome proliferator activated receptors (PPAR) and LXR (reviewed in LEBLANC and STUNNENBERG 1995; MANGELSDORF and EVANS 1995). MUKHERJEE et al. (1997) showed that RXR:PPARγ and RXR:LXR heterodimers have dual-ligand responsiveness. When submaximal levels of a RXR-specific agonist and a PPARγ agonist are present, there is a synergistic transcriptional response (MUKHERJEE et al. 1997). These results demonstrate that the heterodimer RXR:PPARγ responds to ligands that bind to either receptor subunit, the maximal activation requires both ligands, and the response with both ligands is synergistic. Thus retinoids can work together with other signaling molecules through the heterodimerization of RXR with various nuclear receptors. Indeed, in *ob/ob* and *db/db* mice, two genetic models of non-insulin-dependent diabetes mellitus (NIDDM), RXR agonists function as insulin sensitizers, markedly decreasing serum glucose, triglycerides and insulin. The combination of RXR and PPARγ ligands has additive and synergistic effects in stimulating transcription and reducing the degree of hyperglycemia and hypertiriglyceridemia in animal models of NIDDM (MUKHERJEE et al. 1997). Thus it is possible to use RXR-specific agonists to treat certain endocrine disorders by altering transcriptional activity of different RXR-containing receptor heterodimers.

IV. Retinoid Target Genes

As predicted from the diverse biological effects of retinoids, consensus binding sites for RARs and RXRs have been found in many genes which participate in diverse biological processes (Table 1) (GUDAS et al. 1994; MANGELSDORF et al. 1994). Of particular interest is the finding that such enhancer elements are present in several *Hox* genes, *Hoxa-1*, *Hoxb-1*, *Hoxd-4*, and *Hoxd-11*, which are therefore direct targets of retinoid signaling. The body plan of animals is specified, in part, through the action of the conserved *Hox/HOM-C* genes, a family of transcription factors containing a homeodomain (reviewed in McGINNIS and KRUMLAUF 1992; KRUMLAUF 1994). These genes have been found in all animals. When mutated, they are often associated with homeotic transformations, in which one body part is replaced by another homologous body part (e.g., a leg by a wing). These homeotic phenotypes, consistently associated with *Hox/HOM-C* genes in different organisms, strongly suggest that *Hox* genes are involved in the patterning process. *Hoxa-1* contains two RAREs in the 3′ untranslated region (LANGSTON and GUDAS 1992), *Hoxb-1* contains three RAREs, one in the 5′ noncoding region and two in the 3′ noncoding region (MARSHALL et al. 1994; STUDER et al. 1994; OGURA and EVANS 1995a,b; LANGSTON et al. 1997). These RAREs confer retinoic acid inducibility to these *Hox* genes which can be demonstrated by treatment with exogenous retinoic acid. During neurulation, the anterior expression boundary of *Hoxb-1* in the neural tube is located at rhombomere 4 (e.g., SUNDIN et al. 1990).

Table 1. Retinoids target genes

Gene acronym	Name of the gene	Type of RARE	Reference
mRARβ2	Mouse retinoic acid receptor β_2	DR5	Sucov et al. (1990)
hRARβ2	Human retinoic acid receptor β_2	DR5	de Thé et al. (1990)
mRARα2	Mouse retinoic acid receptor α_2	DR5	Leroy et al. (1991)
hRARα2	Human retinoic acid receptor α_2	DR5	Leroy et al. (1991)
hRARγ2	Human retinoic acid receptor γ_2	DR5	Lehmann et al. (1992)
mCP-H	Mouse complement factor H	DR5	Muñoz-Cánoves et al. (1990)
hADH3	Human class I alcohol dehydrogenase 3	DR5	Duster et al. (1991)
hCMV-IEP	Human cytomegalovirus immediate early promoter	DR5	Ghazal et al. (1992)
hMGP	Human matrix Gla protein	DR5	Cancela and Price (1992)
mHoxa-1	Mouse Hoxa1	DR5	Langston and Gudas (1992)
hHoxA1	Human Hoxa1	DR5	Langston et al. (1997)
cHoxa-1	Chick Hoxa1	DR5	Langston et al. (1997)
mHoxb-1	Mouse Hoxb1	DR2, DR5	Marshall et al. (1994), Studer et al. (1994), Ogura and Evans (1995a,b), Langston et al. (1997)
mHoxd-4	Mouse Hoxd4	DR5	Pöpperl and Featherstone (1993)
zShh	Zebrafish sonic hedgehog	DR5	Chang et al. (1997)
mCRBPI	Mouse cellular retinol binding protein type I	DR2	Smith et al. (1991)
rCRBPI	Rat cellular retinol binding protein type I	DR2	Husmann et al. (1992)
hApoAI	Human apolipoprotein AI	DR1, DR2	Rottman et al. (1991); Williams et al. (1992)
mCRABPII	Mouse cellular retinoic acid binding protein type II	DR1, DR2	Durand et al. (1992)
bGH	Bovine growth hormone	Palindrome	de Verneuil and Metzger (1990)
hOST	Human osteocalcin	Palindrome	Schüle et al. (1990)
xVitA2	*Xenopus* vitellogenin A2	Palindrome	de Verneuil and Metzger (1990); Forman et al. (1992)
rGH	Rat growth hormone	Complex	Umesono et al. (1988); Williams et al. (1992)
mLamB1	Mouse laminin B1	Complex	Vasios et al. (1989); Williams et al. (1992)
hOXY	Human oxytocin	Complex	Richard and Zingg (1991); Lipkin et al. (1992)
hMCAD	Human medium chain acyl-coenzyme A dehydrogenase	Complex	Raisher et al. (1992)
mTGase	Mouse tissue transglutaminase	Complex	Nagy et al. (1996)
cOVAL	Chick ovoalbumin	DR1	Kliewer et al. (1992)
rAOX	Rat acyl-CoA oxidase	DR1	Kliewer et al. (1992)
mMHCI	Mouse major histocompatibility complex class I	DR1	Nagata et al. (1992)
rPEPCK	Rat phospho*enol*pyruvate carboxykinase	DR1	Lucas et al. (1991)
hHBV	Human hepatitis B virus	DR1	Huan and Siddiqui (1992)
mHoxd-11	Mouse Hoxd11	Complex	Gérard et al. (1996)
cHoxd-11	Chick Hoxd11	Complex	Gérard et al. (1996)

Rhombomeres are distinct segments found in the hindbrain of vertebrate embryos. Treatment with retinoic acid shifts the *Hoxb-1* boundary to more anterior position. This anterior shifting seems to be dependent on the presence of a RAR and a RXR ligand (Lu et al. 1997a). In other words, unless both types of ligands are provided to the embryo, the anterior shift of the *Hoxb-1* expression boundary does not occur. This could mean that both receptors need to have a ligand bound for *Hoxb-1* activation. A similar synergism between an RAR and a RXR ligand and thus between RAR and RXR has been reported for *RARβ*, another retinoic acid target gene. Studies in cultured cells also exhibit ligand synergism (Roy et al. 1995).

The transactivation of *Hox* genes by exogenous retinoic acid and the expression of reporter genes regulated by RARE containing *Hox* control regions suggest that retinoids have an essential role in the expression of these particular *Hox* genes. This is strongly supported by studies in which one of the RAREs of *Hoxa-1* was mutated using gene targeting methods. Dupé et al. (1997) have replaced an upstream RARE of the *Hoxa-1* gene with a neutral element not capable of binding retinoid receptors. The in vivo function of the RARE in *Hoxa-1* was then addressed by examining the consequence of this mutation. It was found that the establishment of the characteristic anterior expression boundary of *Hoxa-1* at the level of rhombomere 4 in the hindbrain was markedly delayed in mutant mice. Rhombomeres are the site of neural crest cell migration which in the hindbrain region contributes to the formation of cranial ganglia and cranial nerves (Noden 1988; Lumsden and Krumlauf 1996). The loss of the RARE of *Hoxa-1* resulted in phenotypic effects; in a significant fraction of homozygous embryos, the proximal portion of the IXth cranial nerve was absent, a defect also seen in mice in which the *Hoxa-1* gene had been totally deleted (Carpenter et al. 1993; Mark et al. 1993). Taken together, these findings demonstrate a direct role of endogenous retinoic acid in the regulation of *Hoxa-1* expression and, thus, in the specification of tissues whose development is controlled by this gene. This reinforces the general view that retinoids are important for establishing the body plan, in part, through the regulation of *Hox* genes.

Another gene, *Sonic hedgehog* (*Shh*), which is of great interest to developmental biology (Hammerschmidt et al. 1997), is partly regulated by retinoic acid. Akimenko and Ekker (1995) found that systemic application of retinoic acid to zebrafish embryos resulted in the ectopic expression of the *Shh* gene in the developing pectoral fin buds and in floor plate cells. Normally, *Shh* is expressed at the posterior margin of the fin bud, but ectopic retinoic acid induces a second patch of *Shh* expression at the anterior margin of the bud. Analyses of the *Shh* regulatory region have identified a functional RARE in the *Shh* promoter (Chang et al. 1997). In addition, this promoter also contains two binding sites for the winged-helix DNA-binding protein HNF3β, a factor required for proper notochord development (Ang and Rossant 1994; Weinstein et al. 1994). This is a good illustration of the notion that retinoid signal transduction is coupled with other signaling pathways.

Interestingly, there are several genes (Table 1) involved in retinoid signal transduction that are regulated by retinoids directly (autoregulation). Their targets include the cellular retinoid-binding proteins (for example, CRBPI, CRABPII) and enzymes involved in vitamin A transport and metabolism (for example, human class I alcohol dehydrogenase 3 and human apolipoprotein AI, $P450^{RA1}$), as well as the retinoic acid receptors, RARs. The ability to autoregulate the synthesis of the signaling molecule and its receptors provides a sensitive mechanism to control the transcription of retinoid target genes. Retinoids also modulate the lipid and carbohydrate metabolism through their target genes (Table 1), such as human medium chain acyl-coenzyme A dehydrogenase and rat phospho*enol*pyruvate carboxykinase.

C. Retinoids in Vertebrate Limb Development

The developing vertebrate limb provides a useful model system for the study of pattern formation (reviewed in Tickle and Eichele 1994; Tabin 1995; Johnson and Tabin 1997). Most research on limb development has been carried out in the chick embryo, which is readily accessible to experimental and transgenic manipulations. The regions in the embryo that give rise to the forelimbs are examples of so-called embryonic fields that develop in a largely autonomous fashion. When tissue encompassing the chick wing field (Fig. 1, dark gray area in day 2 embryo) is transplanted into a neutral environment such as the coelomic cavity, an ectopic wing develops (Hamburger 1938). Thus the wing field contains all the signaling regions and tissues required for growth, differentiation, and patterning of a wing. Signaling regions in the limb primordium were first identified through classical grafting and extirpation experiments (reviewed in Tickle and Eichele 1994; Johnson and Tabin 1997), and more recently several signaling molecules expressed in these regions have been discovered. This has enormously advanced our understanding of the mechanism by which those signaling regions are established, how they act and how the various molecules orchestrate specific aspects of vertebrate limb patterning and growth. It is important to emphasize that the signaling mechanisms that underlie limb development also operate in other morphogenetic processes such as in the neural tube (Roelink et al. 1994), the somites (Johnson et al. 1994), and the craniofacial primordia (Chiang et al. 1996; Helms et al. 1997).

Several lines of evidence suggested that retinoids play a role during limb development. (a) Ectopic application of retinoic acid to the anterior margin of a chick limb bud induces the formation of a duplicated wing in which digits are arranged in a mirror-symmetrical manner across the anteroposterior axis (Tickle et al. 1982). (b) Systemic administration of retinoic acid to pregnant mice can induce the formation of embryos with extra limbs (Rutelege et al. 1994; Niederreither et al. 1996). (c) Endogenous retinoids have been found

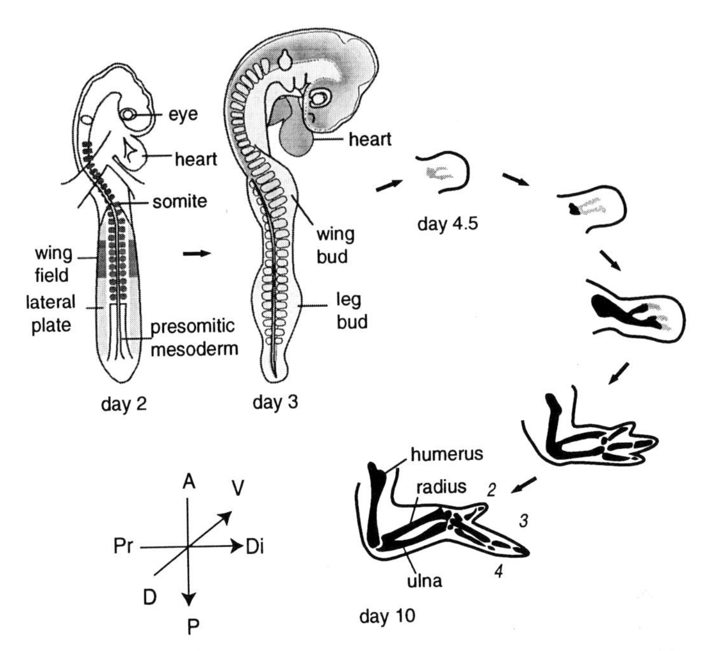

Fig. 1. Chick limb development. The wing fields (*dark gray*) are located in the lateral plate mesoderm (*light gray*) of a day 2 chick embryo. By day 3 limb buds are clearly seen, and at day 4.5, chondrocytes in limbs begin to condense (*gray region*) and eventually these condensates will form the skeletal elements (*black*). *2, 3, 4,* Digits. The orientation of anteroposterior (*A-P*) axis, dorsoventral (*D-V*) axis, and proximodistal (*Pr-Di*) axis is indicated

in developing limb buds (THALLER and EICHELE 1987, 1990) and retinoid receptors are expressed in these tissues (SMITH and EICHELE 1991; HAYAMIZU and BRYANT 1994; SELEIRO et al. 1995). (d) *RARα/RARγ*, *RARα/RXRα*, or *RARβ/RXRα* mutants display defects in limb bones and digits (LOHNES et al. 1994; KASTNER et al. 1997). (e) Embryos lacking the *RALDH-2* gene do not develop limb buds, and *Fgf8* and *Fgf10*, markers that delineate the nacent limb bud are not expressed (NIEDERREITHER et al. 1999). (f) Blocking retinoid signal transduction by application of retinoid receptor antagonists during early limb development impedes limb patterning and results in truncated limbs that can be rescued by exogenous retinoic acid (HELMS et al. 1996; LU et al. 1997b). Similarly limb outgrowth is blocked when retinoic acid synthesis inhibitors are applied to the limb primordium (STRATFORD et al. 1996). Because of the accessibility for manipulation and the detailed knowledge of the signaling molecules involved in limb development (see Sects. C.II, C.III), the developing chick limb provides a unique model system to study retinoic acid action and the interactions between the retinoid signaling pathway and other signaling cascades.

I. Embryology of the Vertebrate Limb

Limbs develop from small buds protruding from the body wall that consist of mesenchymal cells encased in an ectodermal hull (reviewed in Tickle and Eichele 1994; Tabin 1995; Johnson and Tabin 1997). Figure 1 illustrates the process of limb formation in the chick wing. By day 3, wing and leg buds are clearly discernable. The initial formation of limb buds is not caused by an increase in cell proliferation at the sites where buds develop, but by a selective decrease in proliferation in the interlimb flank tissue (Searls and Janners 1971). Once limb buds have formed, they rapidly grow at the tip and while growth continues, mesenchymal cells at the base of the bud start to terminally differentiate into various tissues that constitute a limb. All vertebrate limbs exhibit typical axial organization. The proximodistal axis extends between the

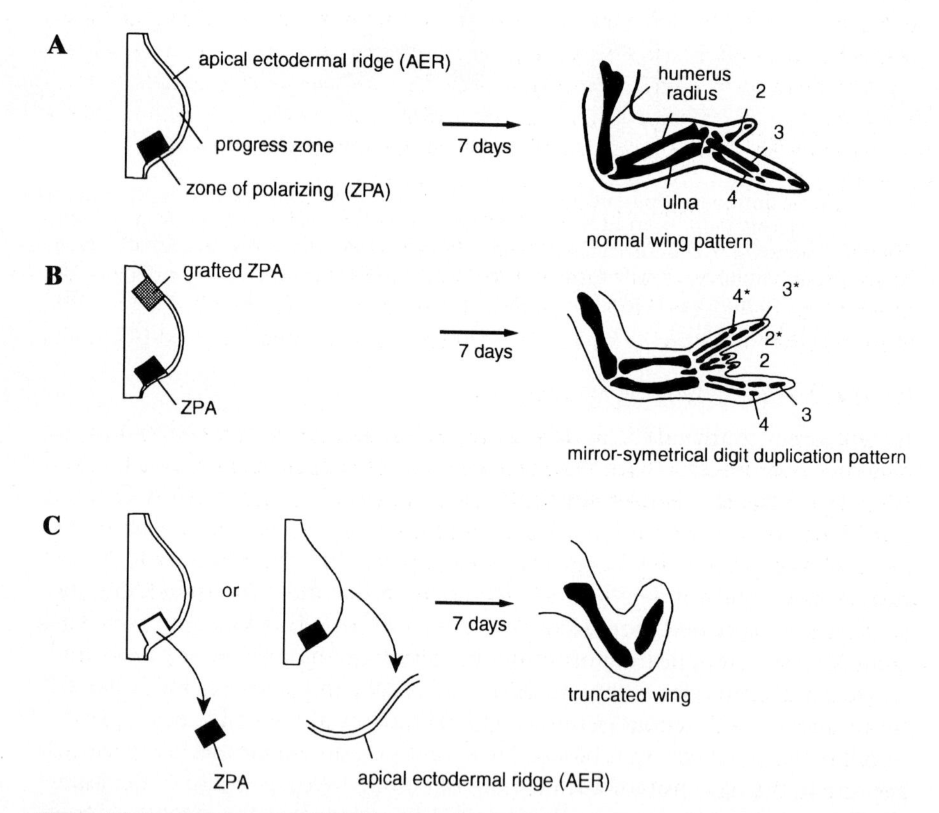

Fig. 2. The apical ectodermal ridge (*AER*), the zone of polarizing activity (*ZPA*) and the progress zone are three regions important for limb development. *A*, Location of the ZPA, the AER, and the progress zone in the developing wing bud. Normal wing skeletal pattern (*right*) that forms from the bud (*left*). *B*, A ZPA grafted into the anterior margin of a host wing bud will lead to a mirror-symmetrical digit duplication pattern. *Asterisks*, the induced digits. *C*, The removal of the ZPA or the AER will lead to a truncated wing

shoulder and the tip of the fingers. The anteroposterior axis spans between the little finger and the thumb. And the dorsoventral axis extends between the back of the hand and the palm.

Experimental manipulations of chick limbs have led to the identification of two major signaling regions that mediate patterning and growth of developing limb buds (Fig. 2A). The apical ectodermal ridge (AER) is a specialized epithelial structure which is located at the margin of the limb bud. The primary function of the AER is to provide signals that through interaction with the underlying mesenchyme promote limb outgrowth (SAUNDERS 1948; SUMMERBELL 1974; ROWE and FALLON 1982). When the AER is removed, the limb stops growing (Fig. 2C). The second signaling region consists of a small group of posterior mesenchymal cells, known as the zone of polarizing activity (ZPA, SAUNDERS and GASSELING 1968). When a ZPA is grafted to the anterior margin of a host wing bud, the graft instructs the host tissue to form an ectopic set of digits which are arranged in a mirror-symmetrical manner with respect to the endogenous digits (Fig. 2B). When the ZPA is removed, the limb fails to develop a distinct pattern along the anteroposterior axis (EICHELE 1989; PAGAN et al. 1996). The third region of interest in the limb bud is called the progress zone which resides at the distal tip of the bud (SUMMERBELL et al. 1973). Cells in the progress zone receive growth signals from the AER and the ZPA (Fig. 2A). Because of the continuous proliferation at the tip, the limb grows out. As the progress zone 'moves' distally, the cells are 'left behind' proximally, stop dividing and begin to terminally differentiate into cartilage and other tissues. The ZPA, the AER and the progress zone are all required for limb development; removing any of them results in a truncated limb (Fig. 2C).

II. The Zone of Polarizing Activity

As noted above, the ZPA acts as an organizer for the anteroposterior axial pattern of the developing limb (reviewed in TICKLE and EICHELE 1994; TABIN 1995; JOHNSON and TABIN 1997). We discuss the ZPA in more detail since there is evidence that its formation is dependent on retinoids. The basic idea is that the ZPA is a source of a morphogen (WOLPERT 1969). This hypothetical morphogen would be a diffusible substance released by the ZPA, thereby setting up a concentration gradient between the posterior and anterior limb bud margin. In such a gradient cells of the limb bud are exposed to different morphogen concentrations. Cells closest to the ZPA experience high morphogen concentration and would become a digit 4 while cells some distance from the ZPA (at the low end of the gradient), would see less morphogen and form a digit 3 and 2, respectively. A key aspect of the morphogen model is that morphogens act in a concentration dependent fashion and that they act over a long distance. Evidence for concentration dependence is seen when an increasing number of ZPA cells are grafted to the anterior wing bud margin, digit duplications occur that reflect the number of grafted ZPA cells. About 50 cells induce a digit 2, 80 cells induce a digit 3, while 200 cells induce a digit 4 (TICKLE 1981). Evidence for long-range effects of the ZPA was provided by HONIG

(1981). He intercalated nonZPA tissue between the ZPA graft and the host limb bud and found that the ZPA signal is transmitted through a tissue block of 10–20 cells diameter width.

Historically, retinoic acid was thought to be the ZPA morphogen for the following reasons. Ectopic application of retinoic acid to the anterior margin of a stage 20 limb bud leads to the same type of digit duplication pattern as a ZPA graft (Tickle et al. 1982) and the extent of digit duplication is dose-dependent (Tickle et al. 1985). The concentration of retinoic acid required for inducing a full set of extra digits is in the same concentration range as the endogenous concentration of retinoic acid in the ZPA region (Thaller and Eichele 1990; Thaller et al. 1993). Retinoic acid also diffuses over several hundred micrometers (Thaller and Eichele 1987) and retinoic acid is enriched in the ZPA (Thaller and Eichele 1987). RARs and RXRs, the mediators of retinoids action, are expressed in the limb buds (Smith and Eichele 1991; Hayamizu and Bryant 1994; Seleiro et al. 1995). A particularly striking demonstration of the morphogenetic effect of exogenously applied retinoic acid is seen when a retinoic acid-releasing bead is implanted into a wing bud whose ZPA has been surgically removed (Eichele 1989; Tamura et al. 1990) (Fig. 2C). Normally, such truncated wing buds give rise to a wing with only a humerus and a radius, but when a retinoic acid-releasing bead is implanted at the anterior margin, a complete set of wing structures develops. The polarity of the digit pattern in such "rescued" wings is reversed. Instead of the normal sequence of digits 234 (Fig. 1), a 432 pattern is generated with digit 4 at the anterior wing margin. Thus retinoic acid alone, in the absence of an endogenous ZPA, can provide patterning information (Saunders and Gasseling 1968) to the anterior tissue of limb bud. Furthermore, the expression of genes which are involved in limb development are induced by a ZPA graft and also by ectopic application of retinoic acid. Examples of genes induced by retinoic acid or a ZPA graft in a host limb bud are *bmp-2*, *Hoxd-11*, *-12*, *-13*, and *fgf-4* (reviewed in Tickle and Eichele 1994; Tabin 1995; Johnson and Tabin 1997).

However, it is now believed that retinoic acid does not directly act as a ZPA morphogen but induces an ectopic ZPA. Initial clues for this came from studies by Noji et al. (1991) and Wanek et al. (1991), but most compelling is the finding that ectopic application of retinoic acid induces the expression of *Shh* (Riddle et al. 1993; Helms et al. 1994). This gene was isolated in screens for vertebrate homologs of the *Drosophila* segment polarity gene *hedgehog* (Echelard et al. 1993; Krauss et al. 1993; Riddle et al. 1993; Roelink et al. 1994). *Shh* is highly expressed in the ZPA as well as other signaling centers. SHH protein is sufficient to induce digit pattern duplication when applied at the anterior margin of a limb bud (Riddle et al. 1993; Lopez-Martinez et al. 1995), and loss-of-function mutation of the *Shh* gene results in truncated limbs (Chiang et al. 1996). The extent of extra digit pattern duplication induced by ectopic SHH depends on the number of *Shh*-expressing cells or on the amount of SHH protein applied. Thus there is a clear dose-dependence between pattern and the concentration of SHH. In both chick and mouse limb buds, reduced levels of *Shh* expression are correlated with absence of posterior digits (PARR and MCMAHON 1995; YANG and NISWANDER 1995). These types of experiments led to the idea that SHH might be the ZPA morphogen.

YANG et al. (1997) have examined the diffusion of SHH in the chick limb bud. When cells expressing a fusion protein consisting of the active moiety of SHH (SHH-N) and an integral membrane protein (CD4) are implanted into the anterior margin of chick wing bud, mirror-image digit pattern duplications are obtained in a dose-dependent fashion (YANG et al. 1997). Since the fusion protein is firmly attached to the membrane it cannot diffuse. Based on this, Yang and colleagues conclude that the diffusion of SHH protein is not necessary to exert the dose-dependent and long range effects. If SHH is not a direct long-range signal, how may the dose-dependent properties of SHH in limb patterning be explained? YANG et al. propose that other signaling molecules are activated in limb bud cells in response to SHH, or that cooperation between SHH and other signaling molecules leads to dose-dependent digit respecification and long-range effects on cells. The bone morphogenetic proteins such as BMP-2 and/or BMP-7 are plausible candidates for this function.

The *bmp-2* gene, which encodes a secreted molecule and is a member of the TGFβ family, is also expressed in the ZPA region. The expression domain of *bmp-2* is slightly larger than that of *Shh* and *bmp-2* can be induced by ectopic retinoic acid as well as by exogenous SHH (FRANCIS et al. 1994). *bmp-2* is the homolog of *Drosophila decapentaplegic* (*dpp*) and *dpp* appears to act as a morphogen for patterning of the *Drosophila* wing disc (LECUIT et al. 1996; NELLEN et al. 1996). When *bmp-2* is ectopically expressed at the anterior margin of the chick wing bud, it induces an additional digit 2 (DUPREZ et al. 1996). This modest response lends some support to the idea that BMP-2 is a morphogen, but more research is needed to investigate this possibility. A putative sequence of events might be that retinoic acid first induces a ZPA which expresses *Shh* which then activates the *bmp-2* gene.

III. Retinoids and the Generation of the ZPA

The previous section summarized evidence for the ability of locally applied retinoic acid to induce an ectopic ZPA in the chick wing bud. If applied retinoic acid can induce a ZPA, will endogenous retinoic acid play a role in the establishment of the native ZPA during early limb development? Tissue with polarizing activity is found in the flank of the embryo, well before limb buds appear and long before *Shh* is expressed (HORNBRUCH and WOLPERT 1991). Such flank tissue, just as a ZPA, causes digit pattern duplications when grafted to a wing bud. HELMS et al. (1996) have found that flank tissue with polarizing activity is very efficient in converting retinal to retinoic acid. Conversion is especially high in cells that constitute the wing field (see day 2 embryo in Fig. 1). The wing field tissue is approximately 10 times more effective in converting retinal to retinoic acid than wing bud tissue. These data reveal a correlation between retinoic acid synthesis and the establishment of a ZPA and suggests a role of retinoic acid in the formation of this signaling tissue. This notion was further tested by blocking retinoid signaling through application of retinoid receptor antagonists. Small beads loaded with an RAR and an RXR antagonist were placed at the boundaries of the wing field prior to the formation of wing buds. Such treatment frequently abolishes wing development (Helms et al. 1996). In most instances, wings are obtained that lack a hand plate and one or both forearm skeletal elements. It is important to note that despite antagonist treatments, wing buds still form. Certain molecular markers that reflect outgrowth, e.g., fibroblast growth factor 8 (*fgf-8*), are expressed in treated wing buds. However, none of the genes that characterize the anteroposterior patterning process or serve as markers for the ZPA are expressed in antagonist-treated wing buds. Thus, such buds are devoid of expression of *Shh*, and *bmp-2* (Helms et al. 1996; Lu et al. 1997b). The absence of expression of the ZPA signal, *Shh,* is a strong indication that without retinoic acid signaling, a ZPA cannot form. In fact, when the posterior mesenchyme (site of the ZPA) of antagonist-treated buds are grafted to a normal wing bud, digit pattern duplications usually fail to form, whereas in control transplants such duplications invariably develop (Lu et al. 1997b).

Stratford et al. (1996) have treated the wing-forming region and nascent wing buds with disulfiram, an inhibitor of aldehydedehydrogenases that convert retinal to retinoic acid (Vallari and Pietruszko 1982). In the majority of cases no wings developed and the *Shh* gene failed to be expressed. At least initially, limb bud outgrowth is not greatly affected since the *fgf-8* gene is expressed in disulfiram-exposed buds. When somewhat older chick wing buds that already express *Shh* are treated with disulfiram, this gene and also *fgf-4* (encodes another fibroblast growth factor) in the AER, are down-regulated. Interestingly, such inhibitor treated buds form relatively normal wings, raising the possibility that *Shh* is no longer required at this developmental stage. It should also be pointed out that retinoid receptor antagonists are not capable of down regulating *Shh* expression in older limb buds (Lu, Thaller and Eichele, unpublished data). Similarly, Tanaka et al. (1996) show that citral, another inhibitor of retinoic acid synthesis (Connor and Smith 1987; Connor 1988), has little effect on the maintenance of *Shh* expression, but their experimental approach does not address the question of whether retinoic acid has a role in the induction of this gene. Taken together, two different strategies of inhibiting retinoic acid signaling provide evidence that retinoic acid is required for proper limb patterning. The requirement for retinoic acid is not for outgrowth per se since *fgf-8*, a marker for outgrowth, is expressed in inhibitor treated buds. However, expression of the anteroposterior patterning gene *Shh* is inhibited by both types of blocking agents. As discussed below, in the absence of *Shh,* the reciprocal signaling between the ZPA and the AER fails to occur and eventually this results in a termination of growth of the bud, and a truncated limb forms. Thus the absence of retinoic acid signaling has ultimately two consequences. First, a lack of specification of digit identity as a result of the absence of the ZPA. Second, a failure of continued limb bud growth, because the ZPA-AER reciprocal signaling cannot occur.

As noted above, *Shh* from zebrafish contains a RARE in its promoter region indicating that the expression of this gene is retinoic acid dependent (CHANG et al. 1997). Does retinoic acid thus directly induce *Shh* expression in the nascent limb buds? Studies with applied retinoic acid show that the *Shh* gene is turned on by retinoic acid within approximately 24h, which is slow for a direct response. This raises the possibility that *Shh* is not a direct target of retinoic acid in the chick limb and that there are intermediary factors that transduce the retinoic acid signal. Alternatively, to activate *Shh*, retinoid receptors may require coactivators that are themselves retinoic acid-induced. Potential candidate coactivators may be the *Hox*-genes, in particular *Hoxb-6* and *Hoxb-8*. Both these genes are expressed in the lateral plate mesoderm, and the domain of *Hoxb-8* expression and of tissue with ZPA activity are coextensive (LU et al. 1997b). Retinoid receptor antagonists released into the wing field down regulate *Hoxb-8* expression. Moreover, ectopically applied retinoic acid can induce the expression of *Hoxb-8* very quickly, even in the absence of protein synthesis (LU et al. 1997b). Finally, studies in transgenic mice show that ectopic expression of *Hoxb-8* throughout the forelimb region creates an ectopic ZPA. The development of such an ectopic ZPA is suggested by the appearance of a patch of *Shh* expression at the anterior bud margin, and furthermore, by the formation of digit pattern duplications similar to those seen in chick wing buds that received a ZPA graft (CHARITÉ et al. 1994). Whether *Hoxb-8* or a related *Hox* gene is the endogenous factor that provides a link between retinoic acid and *Shh* expression in the ZPA is an open question. *Hoxb-6* is also retinoic acid inducible suggesting that this gene may also be part of this signaling pathway. It also remains to be demonstrated whether these *Hox* genes indeed act as coactivators of retinoid receptors, as speculated. It should be emphasized that none of the retinoid receptor mutants either alone or in combination have led to limb phenotypes that would be expected from a reduction or loss of the ZPA. This may be due to the fact that there is redundancy and for such effects to occur embryos would have to be devoid of all *RARs* and *RXRs*. Although a definite genetic proof is still lacking, the various experiments discussed in this section support the hypothesis that retinoids are required for the establishment of a ZPA.

To summarize the findings discussed in this section: around day 2 (Hamburg-Hamilton stage 14), retinoic acid is synthesized in the lateral plate mesoderm of the presumptive wing region. At this stage there is not yet a limb bud. The resulting increase in local retinoic acid concentration in this region triggers an RAR/RXR mediated signal transduction cascade that results in transactivation of retinoic acid target genes such as *Hoxb-6* and *Hoxb-8*. Expression of *Hox* genes initiates directly or indirectly the expression of *Shh* in the posterior mesenchyme of the limb bud. The expression of *bmp-2*, a gene downstream of *Shh*, is also induced. This establishes the ZPA and a graded signal emanating from the ZPA, possibly BMPs (Yang et al. 1997), assigns distinct identities to the cells of the limb bud mesenchyme. This process of cell fate specification provides the basis for the subsequent formation of distinct skeletal elements that characterize the vertebrate limb.

A pattern can only be established if the limb bud grows out. Thus, simultaneous with the above specification processes, a signaling cascade regulating limb bud outgrowth is initiated. Fibroblast growth factors (FGFs) are candidates for mediating cell proliferation in this system, because ectopic application of FGF-1, -2, -4, -8, -10 protein to the interlimb flank induces the formation of an ectopic limb (Cohn et al. 1995; Vogel et al. 1996; Crossley et al. 1996a; Ohuchi et al. 1997). Around day 2, when retinoic acid induces *Hoxb-8*, *fgf-10* begins to be expressed in the mesenchyme of the nascent limb bud and maintains this tissue in a state of high proliferation (Ohuchi et al. 1997). *fgf-10* then turns on the expression of *fgf-8* in the nascent AER (Crossley and Martin 1995; Crossley et al. 1996a). After the AER is formed, *fgf-4* begins to be expressed in the posterior portion of the AER. FGF-2,-4, and –8 in the AER and FGF-2 in the progress zone, are thought to maintain cell proliferation in the progress zone and at the same time maintain the expression of *Shh* in the ZPA. Conversely, *Shh* maintains the expression of FGFs in the AER. Thus there is a cross signaling between the AER and the ZPA mediated by SHH from the ZPA signaling to the AER and FGF-4 from the AER signaling to the ZPA (Laufer et al. 1994; Niswander et al. 1994).

IV. Retinoids in Limb Differentiation

Retinoids also influence the later stages of limb development. Embryonic vitamin A deficiency in rats results in a reduction or deletion of long bones and in fused digits (Warkany and Nelson 1940). Truncations and deletions of the long bones (the humerus, radius, and ulna in the forelimb) and digit deletions and fusions, occur after retinoic acid treatment of mammalian embryos (Kochhar 1973, 1985). Double mutations in *RARα/γ* genes result in the loss of radius, the first digit, the central carpal bone, and the prepollex (Lohnes et al. 1994). These defects may reflect a requirement for retinoic acid to generate the proper amount of limb mesenchyme, since a deficit in this mesenchyme leads to a preferential loss of anterior skeletal elements (Alberch and Gale 1983). Such a deficit is also suggested by the generalized size reduction in the carpals in *RARα/γ* double mutants. The phalanges of both the forelimbs and hind limbs of 18.5 dpc *RARα/γ* double mutants were consistently malformed, appearing bulbous with poorly defined boundaries. The perichondrial cells, surrounding the emerging blastemata, are believed to be important for the directional growth of the blastemata and in determining their final shape (Shubin and Alberch 1986 and refs therein). Since *RARα* and *RARγ* are both expressed in the perichondrial region of blastemal condensations (Dollé et al. 1989), the loss of these receptors could conceivably alter the functional integrity of the phalangeal perichondrium, thus leading to an altered morphology.

Interdigital webbing was observed with an incomplete penetrance in *RARα* and *RARγ* mutants. A mild interdigital webbing was also observed at a low frequency in $RXR\alpha^{+/-}$ mutants (KASTNER et al. 1997). The frequency of webbing increased markedly upon inactivation of one *RARα* or *RARγ* allele within the $RXR\alpha^{+/-}$ background (KASTNER et al. 1997). In addition, a significant proportion of these double heterozygotes displayed an extensive webbing which was never observed in *RAR* or *RXR* single mutants. These defects suggest that retinoid receptors play a role in cell death of the interdigital regions.

D. Retinoids in Limb Regeneration

The striking effects of retinoids on pattern formation in limbs was first discovered not in the chick wing bud but in regenerating limbs of tadpoles of toads (NIAZI and SAXENA 1978). It has long been known that limbs and tails of amphibians can regenerate following amputation. In mammals this ability is largely lost, although the distal tip of fingers can regenerate (REGINELLI et al. 1995 and references therein).

The molecular mechanisms underlying regeneration are still elusive, but progress has been made in understanding this process from the analysis of limb regeneration in newt (see BROCKES 1994, 1997 for a review). Newts are capable of regenerating amputated limbs, even as adults. As a first reaction to amputation, epithelial cells from the wound circumference migrate to seal the wound and form a transient wound epidermis which is essential for regeneration. The cells of the mesenchyme underlying the wound epidermis dedifferentiate and form a blastema in which cells proliferate. The first critical feature of regeneration is the ability of differentiated cells to reenter the cell cycle from a postmitotic state. This reentry into the cell cycle may be the result of hyperphosphorylation of retinoblastoma (Rb) protein, which controls the G_1-S check point. The factors that alter the phosphorylation state of Rb in newt blastema are presently unknown. A second critical aspect of regeneration is that the blastema gives rise only to missing structures. Thus, amputation at wrist level leads to the regeneration of a hand while amputation at shoulder level results in the generation of a whole arm. Thus blastema cells know their position (they have "positional identity") along the proximodistal limb axis and regenerate only structures that are distal to the site of amputation. There is a remarkable conceptual similarity between limb development in the embryo and regeneration. First, both processes require growth and second, cells in the blastema and in the limb bud must create an ordered pattern of tissues that constitute an anatomically correct limb. Therefore, it is likely that regeneration involves, at least in part, the reactivation of developmental programs.

When distal limb blastemas of newt or axolotl limbs are exposed to retinoids (retinyl palmitate, retinol or retinoic acid) for 2 days, the blastemal cells are respecified to a more proximal identity (Maden 1982; Thoms and Stocum 1984; reviewed in Hofmann and Eichele 1994; Brokes 1997). For example, if the limb is amputated at the level of the wrist, retinoid treatment can evoke the formation of an entire limb and not merely the hand, as would normally be the case. This results in duplications of the skeletal pattern along the proximodistal axis. In the example mentioned above, the limb would consist of an upper arm, a forearm, a second upper arm, a second forearm and a hand. The extent of duplication is dependent on the amount of retinoid present at the time of blastema formation (Maden 1982). High doses result in more extensive proximodistal duplications than low doses. Retinoids also have effects on the anteroposterior and dorsoventral axes of regenerating limbs (e.g., Ludolph et al. 1990; Monkemeyer et al. 1992). A series of classical experiments suggest that the positional identity of blastemal cells is encoded at the cell surface and ultimately retinoids may alter this surface.

How do applied retinoids exert their effects on regeneration and do they play a role in normal regeneration? There is as yet no conclusive evidence that retinoids are required for normal regeneration but there are several promising leads. First, endogenous retinoids are present in limb blastema (Scadding and Maden 1994). Second, several retinoid receptors are expressed in the blastema (Ragsdale et al. 1989, 1992; Hill et al. 1993). Third, a retinoic acid reporter gene introduced into the blastema is differentially activated along the proximodistal axis (Brockes 1992; reviewed in Brockes 1994). Specifically, a retinoid reporter gene is activated approximately four fold higher when placed in proximal versus distal blastema cells. This suggests that proximal cells are characterized by higher retinoid activity than the distal ones. This agrees with the above finding that increasing the dose of applied retinoid leads to progressively more proximal regenerates. Fourth, the wound epidermis overlying the blastema releases retinol, 9-*cis*-retinoic acid and didehydroretinoids (Viviano et al. 1995). Retinoids in and secreted by the wound epidermis are thought to have two roles (Viviano et al. 1995). First, these retinoids may maintain the epithelium. It has long been known that retinoids are critical for the maintenance of all epithelia. Second, secreted retinoids may affect the blastemal mesenchyme and be involved in the dedifferentiation process, the specification of positional identity of blastemal cells or both.

Five different RARs have been characterized in newt. To investigate their individual contributions to the proximalizing effect of retinoic acid, a chimeric RAR was generated whose ligand-binding domain was substituted by the ligand binding domain of the thyroid hormone (T_3) receptor (Pecorino et al. 1996). Plasmids expressing the chimeric receptor together with an alkaline phosphatase marker gene were transfected into mesenchymal cells derived from distal blastemas by particle bombardment. Transfected cells were subsequently grafted to a stump of a proximally amputated leg and then exposed to T_3 or to dimethylsulfoxide (control). The use of a T_3-activated chimeric receptor allows a selective activation of only the chimeric transgene without activating endogenous retinoid receptors present in the blastema. If the transfected chimeric RAR does not confer the proximalizing effect of the ligand, the cells carrying the transgene are found throughout the regenerating limb. However, one would predict that transgenic cells are found proximally if they contained the chimeric RAR which confers the proximalizing effect of retinoic acid. It was found that the RARδ2 chimeric receptor (newt homologue of the mammalian RARγ2) was necessary and sufficient to confer the proximalizing activity of retinoic acid. All other chimeric constructs, including the closely related RARδ1 isoform, could not mediate the proximalizing effect of retinoic acid. Of note, in these experiments fewer than 1% of cells were transfected with the RARδ2 chimera. Thus a transfected cell is surrounded by many normal cells, yet the transfected cell can change its positional identity upon T_3 treatment of the blastema. This suggests that retinoic acid acts directly on positional identity. Recall, in the developing limb, retinoic acid appears to first induce an ectopic ZPA which in turn specifies cell position (see Sect. C.III). In summary, exogenous retinoids have profound effects on limb regeneration and evidence is accruing suggesting a role of endogenous retinoids in this process. As to the nature of target genes, little is known, but it is encouraging that proximalization is mediated by only one receptor, RARδ2, which should facilitate the search for target genes.

Another startling effect of retinoids on amphibian regeneration is that following tail amputation, retinoids can cause multiple limbs to develop instead of a tail (Mohanty-Hejmadi et al. 1992; Maden 1993). There can be up to nine supernumerary limbs, with the length of treatment, the dose and the stage determining the number of additional limbs (Mohanty-Hejmadi et al. 1992; Maden 1993). The development of limbs instead of a tail is an example of a homeotic transformation, an event in which the identity of one body part is transformed into that of another body part. How this occurs at a molecular level and whether these mechanisms relate to those described for the newt limb regeneration, remain to be determined.

References

Akimenko M-A, Ekker M (1995) Anterior duplication of the Sonic hedgehog expression pattern in the pectoral fin buds of zebrafish treated with retinoic acid. Dev Biol 170:243–247

Alberch P, Gale EA (1983) Size-dependence during the development of the amphibian foot: colchincine induced digital loss or reduction. J Embryol Exp Morphol 76:177–197

Ang HL, Duester G (1997) Initiation of retinoid signaling in primitive streak mouse embryos: spatiotemporal expression patterns of receptors and metabolic enzymes for ligand synthesis. Dev Dyn 208:536–543

Ang HL, Hayamizu TF, Zgombic-Knight M, Duester G (1996) Retinoic acid synthesis in mouse embryos during gastrulation and craniofacial development linked to class IV alcohol dehydrogenase gene expression. J Biol Chem 271:9526–9534

Ang SL, Rossant J (1994) HNF-3β is essential for node and notochord formation in mouse development. Cell 78:561–574

Beato M, Herrich P, Schütz G (1995) Steroid hormone receptors: many actors in search of a plot. Cell 83:851–857
Bhat PV, Labrecque J, Boutin J-M, Lacroix A, Yoshida A (1995) Cloning of a cDNA encoding rat aldehyde dehydrogenase with high activity for retinal oxidation. Gene 166:303–306
Brockes JP (1992) Introduction of a retinoic reporter gene into the urodele limb blastema. Proc Natl Acad Sci USA 89:11386–11390
Brockes JP (1994) New approaches to amphibian limb regeneration. Trends Genet 10:169–173
Brockes JP (1997) Amphibian limb regeneration: rebuilding a complex structure. Science 276:81–87
Cancela ML, Price PA (1992) Retinoic acid induces matrix Gla protein gene expression in human cells. Endocrinology 130:102–108
Carpenter EM, Goddard JM, Chisaka O, Manley NR, Capecchi MR (1993) Loss of HoxA-1 (Hox-1.6) function results in the reorganization of the murine hindbrain. Development 118:1063–1075
Chai X, Napoli J (1996) Cloning of a rat cDNA encoding retinol dehydrogenase isozyme type III. Gene 169:219–222
Chambon P (1996) A decade of molecular biology of retinoic acid receptors. FASEB J 10:940–954
Chang B-E, Blader P, Fischer N, Ingham PW, Strähle U (1997) Axial (HNF3β) and retinoic acid receptors are regulators of the zebrafish sonic hedgehog promoter. EMBO 16:3955–3964
Charité J, Graaff WD, Shen S, Deschamps J (1994) Ectopic expression of Hoxb-8 causes duplication of the ZPA in the forelimb and homeotic transformation of axial structures. Cell 78:589–601
Chen Y, Huang L, Solursh M (1994) A concentration gradient of retinoids in the early xenopus laevis embryo. Dev Biol 161:70–76
Chen Y-P, Huang L, Russo AF, Solursh M (1992) Retinoic acid is enriched in Hensen's node and is developmentally regulated in the early chicken embryo. Proc Natl Acad Sci USA 89:10056–10059
Chiang C, Litingtung Y, Lee E, Young KE, Corden JL, Westphal H, Beachy PA (1996) Cyclopia and defective axial patterning in mice lacking sonic hedgehog gene function. Nature 383:407–413
Cohn MJ, Izpisúa-Belmonte JC, Abud H, Heath JK, Tickle C (1995) Fibroblast growth factors induce additional limb development from the flank of chick embryos. Cell 80:739–746
Conlon RA (1995) Retinoic acid and pattern formation in vertebrates. Trends Genet 11:314–319
Connor MJ (1988) Oxidation of retinol to retinoic acid as a requirement for biological activity in mouse epidermis. Cancer Res 48:7038–7040
Connor MJ, Smith MH (1987) Terminal-group oxidation of retinol by mouse epidermis. Inhibition in vitro and in vivo. Biochem J 244:489–492
Costaridis P, Horton C, Zeitlinger J, Holder N, Maden M (1996) Endogenous retinoids in the zebrafish embryo and adult. Dev Dyn 205:41–51
Creech Kraft J, Schuh T, Juchau M, Kimelman D (1994) The retinoid X receptor ligand, 9-cisretinoic acid, is a potential regulator of early Xenopus development. Proc Natl Acad Sci USA 91:3067–3071
Crossley PH, Martin GR (1995) The mouse Fgf8 gene encodes a family of polypeptides and is expressed in regions that direct outgrowth and patterning in the developing embryo. Development 121:439–451
Crossley PH, Minowanda G, MacArthur CA, Martin GR (1996a) Roles for FGF8 in the induction, initiation, and maintenance of chick limb development. Cell 84:127–136
de Thé H, Mar Vivanco-Ruiz MD, Tiollais P, Stunnenberg H, DeJean A (1990) Identification of a retinoic acid response element in the retinoic acid receptor β gene. Nature 343:177–180

De Verneuil, Metzger (1990)
Dew SE, Ong DE (1997) Absorption of retinol from the retinol: retinol-binding complex by small intestinal gut sheets from the rat. Arch Biochem Biophys 338:233–236
Dollé P, Ruberte E, Kastner P, Petkovich M, Stoner CM, Gudas LJ, Chambon P (1989) Differential expression of genes encoding α, β and γ retinoic acid receptors and CRABP in the developing limbs of the mouse. Nature 342:702–705
Dowling JE, Wald G (1960) The biological function of vitamin A acid. Proc Natl Acad Sci USA 46:587–608
Duester G (1996) Involvement of alcohol dehydrogenase, short-chain dehydrogenase/reductase, aldehyde dehydrogenase, and cytochrome P450 in the control of retinoid signaling by activation of retinoic acid synthesis. Biochemistry 35:12221–12227
Dupé V, Davenne M, Brocard J, Dollé P, Mark M, Dierich A, Chambon P, Rijli FM (1997) In vivo functional analysis of the Hoxa-1 3′ retinoic acid response element (3′ RARE). Development 124:399–410
Duprez DM, Kostakopoulou K, Francis-West PH, Tickle C, Brickell PM (1996) Activation of Fgf-4 and HoxD gene expression by BMP-2 expression cells in the developing chick limb. Development 122:1821–1828
Durand B, Saunders M, Leroy P, Leid M, Chambon P (1992) All-*trans* and 9-*cis* retinoic acid induction of CRABPII transcription is mediated by RAR-RXR heterodimers bund to DR1 and DR2 repeated motifs. Cell 71:73–85
Durston AJ, Timmermans JPM, Hage WJ, Hendriks HFJ, de Vries NJ, Heideveld M, Nieuwkoop PD (1989) Retinoic acid causes an anteroposterior transformation in the developing central nervous system. Nature 340:140–144
Duster G, Shean ML, McBride MS, Stewart MJ (1991) Retinoic acid responsive element in the human alcohol dehydrogenase gene ADH3: implications for regulation of retinoic acid synthesis. Mol Cell Biol 11:1638–1646
Dyson E, Sucov HM, Kubalak SW, Schmid-Schonbein GW, DeLano FA, Evans RM, Ross J Jr, Chien KR (1995) Atrial like phenotype is associated with embryo in ventricular failure in retinoid X receptor α –/– mice. Proc Natl Acad Sci USA 92:7386–7390
Echelard Y, Epstein DJ, St-Jacques B, Shen L, Mohler J, McMahon JA, McMahon AP (1993) Sonic hedgehog, a member of a family of putative signaling molecules, is implicated in the regulation of CNS polarity. Cell 75:1417–1430
Eichele G (1989) Retinoic acid induces a pattern of digits in anterior half wing buds that lack the zone of polarizing activity. Development 107:863–868
Eichele G (1997) Retinoids: from hindbrain patterning to Parkinson disease. Trends Genet 13:943–945
Fell HB, Mellanby E (1953) Metaplasia produced in cultures of chick ectoderm by high vitamin A. J Physiol 119:470–488
Fisher GJ, Voorhees JJ (1996) Molecular mechanism of retinoid actions in skin. FASEB J 10:1002–1013
Forman et al. (1992)
Francis PH, Richardson MK, Brickell PM, Tickle C (1994) Bone morphogenetic proteins and a signaling pathway that controls patterning in the developing chick limb. Development 120:209–218
Fujii H, Sato T, Kaneko S, Gotoh O, Fulii-Kuriyama Y, Osawa K, Kato S, Hamada H (1997) Metabolic inactivation of retinoic acid by a novel P450 differentially expressed in developing mouse embryo. EMBO 16:4163–4173
Gérard M, Chen J-Y, Gronemeyer H, Chambon P, Duboule D, Zákány J (1996) In vivo targeted mutagenesis of a regulatory element required for positioning the Hoxd-11 and Hoxd-10 expression boundaries. Genes Dev 10:2326–2334
Ghazal P, DeMattei C, Giulietti E, Kliewer SA, Umesono K, Evans RM (1992) Retinoic acid receptors initiate induction of the cytomegalovirus enhancer in embyonal cells. Proc Natl Acad Sci USA 89:7630–7634

Giguère V, Ong ES, Segui P, Evans RM (1987) Identification of a receptor for the morphogen retinoic acid. Nature 330:624–629
Gudas LJ, Sporn MB, Roberts AB (1994) Cellular biology and biochemistry of the retinoids. In: Sporn MB, Roberts AB, Goodman DS (eds) The retinoids. Raven, New York, pp 443–520
Hamburger V (1938) Morphogenetic and axial self-differentiation of transplanted limb primordia of 2-day chick embryos. J Exp Zool 77:379–400
Hammerschmidt M, Brook A, McMahon AP (1997) The world according to hedgehog. Trends Genet 13:14–21
Hayamizu TF, Bryant SV (1994) Reciprocal changes in Hox D13 and RARβ2 expression in response to retinoic acid in chick limb buds. Dev Biol 166:123–132
Helms J, Kim CH, Thaller C, Eichele G (1996) Retinoic acid signaling is required during early limb development. Development 122:1385–1394
Helms J, Thaller C, Eichele G (1994) Relationship between retinoic acid and sonic hedgehog, two polarizing signals in the chick wing bud. Development 120:3267–3274
Helms JA, Kim CH, Hu D, Minkoff R, Thaller C, Eichele G (1997) Sonic hedgehog participates in craniofacial morphogenesis and is down-regulated by teratogenic doses of retinoic acid. Dev Biol 187:25–35
Hill DS, Ragsdale CWJ, Brockes JP (1993) Isoform-specific immunological detection of newt retinoic acid receptor δ1 in normal and regenerating limbs. Development 117:937–945
Hofmann C, Eichele G (1994) Retinoids in development. In: Sporn MB, Roberts AB, Goodman DS (eds) The retinoids. Raven, New York, pp 387–441
Hogan BLM, Thaller C, Eichele G (1992) Evidence that Hensen's node is a site of retinoic acid synthesis. Nature 359:237–241
Honig LS (1981) Positional signal transmission in the developing chick limb. Nature 291:72–73
Hornbruch A, Wolpert L (1991) The spatial and temporal distribution of polarizing activity in the flank of the pre-limb-bud stages in the chick embryos. Development 111:725–731
Horton C, Maden M (1995) Endogenous distribution of retinoids during normal development and teratogenesis in the mouse embryo. Dev Dyn 202:312–323
Huan M-E, Siddiqui A (1992) Retinoid X receptor RXR-α binds to and *trans*-activates the hepatitis B virus enhancer. Proc Natl Acad Sci USA 89:9059–9063
Hummler H, Hendrickx AG, Nau H (1994) Maternal pharmacokinetics, metabolism, and embryo exposure following a teratogenic dosing regimen with 13-*cis*-retinoic acid (isotretinoin) in the cynomolgus monkey. Teratology 50:184–193
Husmann M, Hoffmann B, Stump DG, Chytil F, Pfahl M (1992) A retinoic acid response element from the rat CRBPI promoter is activated by an RAR/RXR heterodimer. Biochem Biophys Res Commun 187:1558–1564
Johnson RL, Tabin CJ (1997) Molecular models for vertebrate limb development. Cell 90:979–990
Johnson et al (1994)
Kastner P, Grondona JM, Mark M, Gansmuller A, LeMeur M, Decimo D, Vonesch J-L, Dollé P, Chambon P (1994) Genetic analysis of RXRα developmental function: convergence of RXR and RAR signaling pathways in heart and eye morphogenesis. Cell 78:987–1003
Kastner P, Mark M, Chambon P (1995) Nonsteroid nuclear receptors: what are genetic studies telling us about their role in real life? Cell 83:859–869
Kastner P, Mark M, Ghyselinck N, Krezel W, Dupé V, Grondona JM, Chambon P (1997) Genetic evidence that the retinoid signal is transduced by heterodimeric RXR/RAR functional units during mouse development. Development 124:313–326

Kliewer SA, Umesono K, Heyman RA, Mangelsdorf DJ, Dyck JA, Evans RM (1992) Retinoid X receptor-COUP-TF interactions modulate retinoic acid signaling. Proc Natl Acad Sci USA 89:1448–1452
Kochhar DM (1973) Limb development in mouse embryos. Teratology 7:289–299
Kochhar DM (1985) Skeletal morphogenesis: comparative effects of a mutant gene and a teratogen. Prog Clin Biol Res 171:267–281
Krauss S, Concordet JP, Ingham PW (1993) A functionally conserved homolog of the Drosophila segment polarity gene hh is expressed in tissues with polarizing activity in zebrafish embryos. Cell 75:1431–1444
Krumlauf R (1994) Hox genes in vertebrate development. Cell 78:191–201
Langston AW, Gudas LJ (1992) Identification of a retinoic acid responsive enhancer 3′ of the murine homeobox gene Hox-1.6. Mech Dev 39:217–227
Langston AW, Thompson JR, Gudas LJ (1997) Retinoic acid-responsive enhancers located 3′ of the HoxA and HoxB homeobox gene clusters. J Biol Chem 272:2167–2175
Laufer E, Nelson CE, Johnson RL, Morgan BA, Tabin C (1994) Sonic hedgehog and Fgf-4 act through a signaling cascade and feedback loop to integrate growth and patterning of the developing limb bud. Cell 79:993–1003
Leblanc BP, Stunnenberg HG (1995) 9-*cis* retinoic acid signaling: changing partners causes some excitement. Genes Dev 9:1811–1816
Lecuit T, Brook WJ, Ng M, Calleja M, Sun H, Cohen SM (1996) Two distinct mechanisms for long-range patterning by Decapentaplegic in the Drosophila wing. Nature 381:387–392
Lehmann JM, Zhang X-K, Pfahl M (1992) RAR gamma2 expression is regulated through a retinoic acid response element embedded in Sp1 sites. Mol Cell Biol 12:2976–2985
Leid M, Kastner P, Chambon P (1992) Multiplicity generates diversity in the retinoic acid signaling pathways. Trends Biochem Sci 176:427–433
Leroy P, Nakshatri H, Chambon P (1991) Mouse retinoic acid receptor $\alpha 2$ isoform is transcribed from a promoter that contains a retinoic acid response element. Proc Natl Acad Sci USA 88:10138–10142
Lipkin SM, Nelson CA, Glass CK, Rosenfeld MG (1992) A negative retinoic acid response element in the rat oxytocin promoter restricts transcriptional stimulation by heterologous transactivation domains. Proc Natl Acad Sci USA 89:1209–1213
Lohnes D, Mark M, Mendelsohn C, Dollé P, Dierich A, Gorry P, Gansmuller A, Chambon P (1994) Function of the retinoic acid receptors (RARs) during development (I) Craniofacial and skeletal abnormalities in RAR double mutants. Development 120:2723–2748
Lopez-Martinez A, Chang DT, Chiang C, Porter JA, Ros MA, Simandl BK, Beachy PA, Fallon JF (1995) Limb patterning activity and restricted posterior localization of the amino-terminal product of Sonic hedgehog cleavage. Curr Biol 5:791–796
Lu et al (1997a)
Lu H-C, Revelli J-P, Goering L, Thaller C, Eichele G (1997b) Retinoids signaling is required for the establishment of a ZPA and for the expression of Hoxb-8, a mediator of ZPA formation. Development 124:1643–1651
Lucas PC, Forman BM, Samuels HH, Granner DK (1991) Specificity of a retinoic acid response element in the phosphoenolpyruvate carboxykinase gene promoter: consequences of both retinoic acid and thyroid hormone receptor binding. Mol Cell Biol 11:5164–5170
Ludolph DC, Cameron JA, Stocum DL (1990) The effect of retinoic acid on positional memory in the dorsoventral axis of regenerating axolotl limbs. Dev Biol 140:41–52
Lumsden A, Krumlauf R (1996) Patterning the vertebrate neuraxis. Science 274:1109–1115
Maden M (1982) Vitamin A and pattern formation in the regenerating limb. Nature 295:672–675

Maden M (1993) The homeotic transformation of tails into limbs in Rana temporaria by retinoids. Dev Biol 159:379–391

Maden M, Gale E, Kostetskii I, Zile M (1996) Vitamin A-deficient quail embryos have half a hindbrain and other neural defects. Curr Biol 6:417–426

Maden M, Ong DE, Summerbell D, Chytil F (1988) Spatial distribution of cellular protein binding to retinoic acid in the chick limb bud. Nature 335:733–735

Maden M, Ong DE, Summerbell D, Chytil F (1989) The role of retinoid-binding proteins in the generation of pattern in the developing limb, the regenerating limb and the nervous system. Development 107 [Suppl]:109–119

Mangelsdorf DJ, Evans RM (1995) The RXR heterodimers and orphan receptors. Cell 83:841–850

Mangelsdorf DJ, Umesono K, Evans RM (1994) The retinoid receptors. In: Sporn MB, Roberts AB, Goodman DS (eds) The retinoids. Ravern, New York, pp 319–349

Mark M, Lufkin T, Vonesch JL, Ruberte E, Olivo JC, Gorry P, Lumsden A, Chambon P (1993) Two rhombomeres are altered in Hoxa-1 mutant mice. Development 119:319–338

Marshall H, Studer M, Pöpperl H, Aparicio S, Kuroiwa A, Brenner S, Krumlauf R (1994) A conserved retinoic acid response element required for early expression of the homebox gene Hoxb-1. Nature 370:567–571

McCollum EV, Davis M (1913) The necessity of certain lipids in the diet during growth. J BIol Chem 15:167–175

McGinnis W, Krumlauf R (1992) Homeobox genes and axial patterning. Cell 68:283–302

Means, AL, Gudas LJ (1995) The roles of retinoids in vertebrate development. Annu Rev Biochemistry 64:201–233

Minucci S, Ozato K (1996) Retinoid receptors in transcriptional regulation. Curr Opin Gen Dev 6:567–574

Mohanty-Hejmadi P, Dutta SK, Mahapatra P (1992) Limbs generated at site of tail amputation in marbled balloon frog after vitamin A treatment. Nature 355:352–353

Monkemeyer J, Ludolph DC, Cameron J-A, Stocum DL (1992) Retinoic acid-induced change in anteroposterior positional identity in regenerating axolotl limbs is dose-dependent. Dev Dyn 193:286–294

Mukherjee R, et al (1997) Sensitization of diabetic and obese mice to insulin by retinoids X receptor agonists. Nature 386:407–410

Muñoz-Cánoves P, Vik DP, Tack BF (1990) Mapping of a retinoic acid-responsive element in the promoter region of the complement factor H gene. J Biol Chem 265:20065–20068

Nagata T, Segars JH, Levi B-Z, Ozato K (1992) Retinoic acid-dependent transactivation of major histocompatibility complex class I promoters by the nuclear hormone receptor H-2RIIBP in undifferentiated embryonal carcinoma cells. Proc Natl Acad Sci USA 89:937–941

Nagy L, et al (1996) Identification and characterization of a versatile retinoid response element (retinoic acid receptor response element-retinoid X receptor response element) in the mouse tissue transglutaminase gene promoter. J Biol Chem 271:4355–4365

Napoli JL (1996) Retinoic acid biosynthesis and metabolism. FASEB 10:993–1001

Nellen D, Burke R, Struhl G, Basler K (1996) Direct and long-range action of a DPP morphogen gradient. Cell 85:357–368

Niazi IA, Saxena S (1978) Abnormal hind limb regeneration in tadpoles of the toad. Bufo andersoni, exposed to excess vitamin A. Folia Biol (Krakow) 26:3–8

Niederreither K, McCaffery P, Dräger UC, Chambon P, Dollé P (1997) Restricted expression and retinoic acid-induced downregulation of the retinal dehydrogenase type 2 (RALDH-2) gene during mouse development. Mech Dev 62:67–78

Niederreither K, Subbarayan V, Dollé P, Chambon P (1999) Embryonic retinoic acid synthesis is essential for early mouse post-implantation development. Nat Genet 21:444–448

Niederreither K, Ward SJ, Dollé P, Chambon P (1996) Morphological and molecular characterization of retinoic acid-induced limb duplication in mice. Dev Biol 176:185–198
Niswander L, Jeffrey S, Martin GR, Tickle C (1994) A positive feedback loop coordinates growth and patterning in the vertebrate limb. Nature 371:609–612
Noden DM (1988) Interactions and fates of avian craniofacial mesenchyme. Development 103 [Suppl]:121–140
Noji et al. (1991)
Ogura T, Evans RM (1995a) A retinoic acid-triggered cascade of HOXB1 gene activation. Proc Natl Acad Sci USA 92:387–391
Ogura T, Evans RM (1995b) Evidence for two distinct retinoic acid response pathways for HOXB1 gene regulation. Proc Natl Acad Sci USA 92:392–396
Ohuchi H, Nakagawa T, Yamamoto A, Agara A, Ohata T, Ishimaru Y, Yoshioka H, Kuwana T, Nohno T, Yamasaki M, Itoh N, Noji S (1997) The mesenchymal factor, FGF10, initiates and maintains the outgrowth of the chick limb bud through interaction with FGF8, an apical ectodermal factor. Development 124:2235–2244
Ong DE, Newcomer ME, Chytil F (1993) Cellular retinoid-binding proteins. In: Sporn MB, Roberts AB, Goodman DS (eds) The retinoids. Raven, New York
Oro AE, McKeown M, Evans RM (1992) The Drosophila retinoid X receptor homolog ultraspiracle functions in both female reproduction and eye morphogenesis. Development 115:449–462
Osborne TB, Mendel LB (1913) The relation of growth to the chemical constituents of the diet. J Biol Chem 15:311–326
Pagan SM, Ros MA, Tabin C, Fallon JF (1996) Surgical removal of limb bud Sonic hedgehog results in posterior skeletal defects. Dev Biol 180:35–40
Parr B, McMahon A (1995) Dorsalizing signal Wnt-7a required for normal polarity of D-V and A-P axes of mouse limb. Nature 374:350–353
Pazin MJ, Kadonaga JT (1997) What's up and down with histone deacetylation and transcription. Cell 89:325–328
Pecorino LT, Entwistle A, Brockes JP (1996) Activation of a single retinoic acid receptor isoform mediates proximodistal respecification. Curr Biol 6:563–569
Penzes P, Wang X, Sperkova Z, Napoli JL (1997) Cloning of a rat cDNA encoding retinal dehydrogenase isozyme type I and its expression in E. coli. Gene 191:167–172
Petkovich M, Brand NJ, Krust A, Chambon P (1987) A human retinoic acid receptor which belongs to the family of nuclear receptors. Nature 330:444–450
Pijnappel WW, Hendriks HF, Folkers GE, van den Brink CE, Dekker EJ, Edelenbosch C, van der Saag PT, Durston AJ (1993) The retinoid ligand 4-oxo-retinoic acid is a highly active modulator of positional specification. Nature 366:340–344
Pöpperl H, Featherstone MS (1993) Identification of a retinoic acid response element upstream of the murine Hox-4.2 gene. Mol Cell BIol 13:257–265
Ragsdale CW, Jr, Gates PB, Hill DS, Brockes JP (1992) Delta retinoic acid receptor isoform d1 is distinguished by its exceptional N-terminal sequence and abundance in the limb regeneration blastema. Mech Dev 40:99–112
Ragsdale CWJ, Petkovich M, Gates PB, Chambon P, Brockes JP (1989) Identification of a novel retinoic acid receptor in regenerative tissues of the newt. Nature 341:654
Raisher BD, Guilick T, Zhang Z, Strauss AW, Moore DD, Kelly DP (1992) Identification of a novel retinoid-responsive element in the promoter region of the medium chain acyl-coenzyme A dehydrogenase gene. J Biol Chem 267:20264–20269
Ray WJ, Bain G, Yao M, Gottlieb DI (1997) CYP26, a novel mammalian cytochrome P450, is induced by retinoic acid and defines a new family. J Biol Chem 272:18702–18708
Reginelli AD, Wang Y-Q, Sassoon D, Muneoka K (1995) Digit tip regeneration correlates with regions of Msx1 (Hox 7) expression in fetal and newborn mice. Development 121:1065–1076

Richard S, Zingg HH (1991) Identification of a retinoic acid response element in the human oxytocin promoter. J Biol Chem 266:21428–21433
Riddle RD, Johnson RL, Laufer E, Tabin C (1993) Sonic hedgehog mediates the polarizing activity of the ZPA. Cell 75:1401–1416
Roelink H, Augsburger A, Heemskerk J, Korzh V, Norlin S, Ruiz i Altaba A, Tanabe Y, Placzek M, Edlund T, Jessell TM, Dodd J (1994) Floor plate and motor neuron induction by vhh-1, a vertebrate homolog of hedgehog expressed by the notochord. Cell 76:761–775
Rogers MB (1997) LIfe-and-death decisions influenced by retinoids. Curr Top Dev Biol 35:1–46
Rottman JN, Widom RL, Nadal-Ginard B, Mahdavi V, Karathanasis SK (1991) A retinoic acid-response element in the apolipoprotein AI gene distinguishes between two different retinoic acid response pathways. Mol Cell Biol 11:3814–3820
Rowe DA, Fallon JF (1982) The proximodistal determination of skeletal parts in the developing chick leg. J Embryol Exp Morphol 68:1–7
Roy B, Taneja R, Chambon P (1995) Synergistic activation of retinoic acid (RA)-responsive genes and induction of embryonal carcinorma cell differentiation by an RA receptorα (RARα)-, RARβ-, or RARγ-selective ligand in combination with a retinoid X receptor-specific ligand. Mol Cell Biol 15:6481–6487
Ruberte E, Dollé P, Chambon P, Morriss-Kay G (1991) Retinoic acid receptors and cellular retinoid binding proteins II. Their differential pattern of transcription during early morphogenesis in mouse embryos. Development 111:45–60
Ruberte E, Friederich V, Morriss-Kay G, Chambon P (1992) Differential distribution patterns of CRABP I and CRABP II transcripts during mouse embryogenesis. Development 115:973–978
Rutelege JC, Shourbaji AG, Hughes LA, Polifka JE, Cruz YP, Bishop JB, Generoso WM (1994) Limb and lower-body duplications induced by retinoic acid in mice. Proc Natl Acad Sci USA
Saunders JW (1948) The proximo-distal sequence of origin of the parts of the chick wing and the role of the ectoderm. J Exp Zool 108:363–403
Saunders JWJ, Gasseling MT (1968) Ectodermal-mesenchymal interactions in the origin of limb symmetry. In: Fleischmajer R, Billingham RE (eds) Epithelial-mesenchymal interactions. Williams and Wilkins, Baltimore, pp 78–97
Scadding S, Maden M (1994) Retinoic acid gradients during limb regeneration. Dev Biol 162:608–617
Schüle R, Umesono K, Mangelsdorf DJ, Bolado J, Pike JW, Evans RM (1990) Jun-Fos and receptors for vitamins A and D recognize a common response element in the human osteocalcin gene. Cell 61:497–504
Scott WJJ, Walter R, Tzimas G OSJ, Nau H, Collins MD (1994) Endogenous status of retinoids and their cytosolic binding proteins in limb buds of chick vs mouse embryos. Dev Biol 165:397–409
Searls RL, Janners MY (1971) The initiation of limb bud outgrowth in the embryonic chick. Dev Biol 24:198–213
Seleiro EAP, Rowe A, Brickell PM (1995) The chicken retinoid-X-receptor-α gene and its expression in the developing limb. Roux's Arch. Dev Biol 204:244–249
Shubin NH, Alberch P (1986) A morphogenetic approach to the origin and basic organization of the tetrapod limb. Evol Biol 20:319–387
Smith SM, Eichele G (1991) Temporal and regional differences in the expression of distinct retinoic acid receptor-β transcripts in the chick embryo. Development 111:245–252
Smith WC, Nakshatri H, Leroy P, Rees J, Chambon P (1991) A retinoic acid response element is present in the mouse cellular retinol binding protein I (mCRBPI) promoter. EMBO J 10:2223–2230
Sporn MB, Roberts AB, Goodman DS (1994) Book the retinoids. Raven, New York
Stratford T, Horton C, Maden M (1996) Retinoic acid is required for the formation of outgrowth in the chick limb bud. Curr Biol 6:1124–1133

Studer M, Pöpperl H, Marshall H, Kuroiwa A, Krumlauf R (1994) Role of a conserved retinoic acid response element in rhombomere restriction of Hoxb-1. Science 265:1728–1731

Sucov HM, Dyson E, Gumeringer CL, Price J, Chien KR, Evans RM (1994) RXRα mutant mice establish a genetic basis for vitamin A signaling in heart morphogenesis. Genes Dev 8:1007–1018

Sucov HM, Murakami KK, Evans RM (1990) Characterization of an autoregulated response element in the mouse retinoic acid receptor type β gene. Proc Natl Acad Sci USA 87:5392–5396

Summerbell D (1974) A quantitative analysis of the effect of excision of the AER from the chick limb bud. J Embryol Exp Morphol 32:651–660

Summerbell D, Lewis J, Wolpert L (1973) Positional information in chick limb morphogenesis. Nature 224:492–496

Sundin OH, Busse HG, Rogers MB, Gudas LJ, Eichele G (1990) Region-specific expression in early chick and mouse embryos of Ghox-lab and Hox 1.6 (Hoxa-1), vertebrate homeobox-containing genes related to Drosophila labial. Development 108:47–58

Tabin C (1995) The initiation of the limb bud: growth factors, Hox genes, and retinoids. Cell 80:671–674

Tamura K, Kagechika H, Hashimoto Y, Shudo K, Oshugi K, Ide K (1990) Synthetic retinoids, retinobenzoic acids, Am80, Am580 and Ch55 regulate morphogenesis in the chick limb bud. Cell Differ Dev 32:17–26

Tanaka M, Tamura K, Ide H (1996) Citral, an inhibitor of retinoic acid synthesis, modifies chick limb development. Dev Biol 175:239–247

Thaller C, Eichele G (1987) Identification and spatial distribution of retinoids in the developing chick limb bud. Nature 327:625–628

Thaller C, Eichele G (1990) Isolation of 3,4-didehydroretinoic acid, a novel morphognetic signal in the chick wing bud. Nature 345:815–819

Thaller C, Hoffmann C, Eichele G (1993) 9-*cis*-retinoic acid, a potent inducer of digit pattern duplications in the chick wing bud. Development 118:957–965

Thoms SD, Stocum DL (1984) Retinoic acid-induced pattern duplication in regenerating urodele limb. Dev Biol 103:319–328

Tickle C (1981) The number of polarizing region cells required to specify additional digits in the developing chick wing. Nature 289:295–298

Tickle C, Alberts B, Wolpert L, Lee J (1982) Local application of retinoic acid to the limb bud mimics the action of the polarizing region. Nature 296:564–566

Tickle C, Eichele G (1994) Vertebrate limb development. Annu Rev Cell Biol 10:121–152

Tickle C, Lee J, Eichele G (1985) A quantitative analysis of the effect of all-*trans*-retinoic acid on the pattern of chick wing development. Dev Biol 109:82–95

Umesono K, Giguère V, Glass CK, Rosenfeld MG, Evans RM (1988) Retinoic acid and thyroid hormone induce gene expression through a common responsive element. Nature 336:262–265

Vallari RC, Pietruszko R (1982) Human aldehyde dehydrogenase: mechanism of inhibition by disulphiram. Science 216:637–639

Vasios GW, Gold JD, Petkovich M, Chambon P, Gudas LJ (1989) A retinoic acid-responsive element is present in the 5′ flanking region of the laminin B1 gene. Proc Natl Acad Sci USA 86:9099–9103

Viviano CM, Horton CE, Maden M, Brockes JP (1995) Synthesis and release of 9-*cis* retinoic acid by the urodele wound epidermis. Development 121:3753–3762

Vogel A, Rodriguez C, Izpisúa-Belmonte J-C (1996) Involvement of FGF-8 in initiation, outgrowth and patterning of the vertebrate limb. Development 122:1737–1750

Wanek et al. (1991)

Wang X, Penzes P, Napoli JL (1996) Cloning of a cDNA encoding an aldehyde dehydrogenase and its expression in Escherichia coli. J Biol Chem 271:16288–16293

Warkany J, Nelson RC (1940) Appearance of skeletal abnormalities in the offspring of rats reared on a deficient diet. Science 92:383–384

Weinstein DC, Ruiz i Altaba A, Chen WS, Hoodless P, Prezioso VR, Jessell TM, Darnell JEJ (1994) The winged-helix transcription factor HNF3β is required for notochord development in the mouse embryo. Cell 575–588:

White JA, Beckett-Jones B, Guo Y-D, Dilworth FJ, Bonasoro J, Jones G, Petkovich M (1997) cDNA cloning of human retinoic acid-metabolizing enzyme (hP450RAI) identifies a novel family of cytochromes P450 (CYP26). J Biol Chem 272:18538–18541

White JA, Guo Y-D, Baetz K, Beckett-Jones B, Bonasoro J, Hsu KE, Dilworth FJ, Jones G, Petkovich M (1996) Identification of the retinoic acid-inducible all-*trans*-retinoic acid 4-hydroxylase. J Biol Chem 271:29922–29927

Williams GR, Harney JW, Moore DD, Larsen PR, Brent GA (1992) Differential capacity of wild type promoter elements for binding and *trans*-activation by retinoic acid and thyroid hormone receptors. Mol Endocrinol 6:1527–1537

Wilson JG, Roth CB, Warkany J (1953) An analysis of the syndrome of malformations induced by maternal vitamin A deficiency. Effects of restoration of vitamin A at various times during gestation. Am. J Anat 92:189–217

Wolbach SB, Howe PR (1925) Tissue changes following deprivation of fat-soluble A vitamin. J Exp Med 42:753–777

Wolpert L (1969) Positional information and the spatial pattern of cellular differentiation. J Theoret Biol 25:1–47

Yang Y, et al. (1997) Relationship between dose, distance and time in Sonic Hedgehog-mediated regulation of anteroposterior polarity in the chick limb. Development 124:4393–4404

Yang Y, Niswander L (1995) Interaction between the signaling molecules WNT7a and SHH during vertebrate limb development: dorsal signals regulate anteroposterior patterning. Cell 80:939–947

Zgombic-Knight M, Ang HL, Foglio MH, Duester G (1995) Cloning of the mouse class IV alcohol dehydrogenase (retinol dehydrogenase) cDNA and tissue-specific expression patterns of the murine ADH gene family. J Biol Chem 270:10868–10877

Zhao D, McCaffery P, Ivins KJ, Neve RL, Hogan P, Chin WW, Dräger UC (1996) Molecular identification of a major retinoic-acid-synthesizing enzyme, a retinal-specific dehydrogenase. Eur J Biochemistry 240:15–22

CHAPTER 14

Retinoids in Neural Development

M. MADEN

The evidence for a crucial role for vitamin A and its derivatives, the retinoids, in the development and maintenance of the nervous system is now beyond doubt, and this chapter presents the data to support this contention. The evidence comes from several types of experiment: the induction of neural differentiation in embryonal carcinoma cells; the effects of excess retinoids; the effects of a deficiency of retinoids; the distribution of retinoid-binding proteins and retinoid receptors and the results of interference with their signalling functions; the detection of endogenous retinoids; and the distribution of retinoid synthesising enzymes. Each of these topics are reviewed below.

Indeed, one of the most striking observations concerning the induction of differentiation by RA in embryonal carcinoma cells is that the type of differentiated cell induced depends on the concentration of RA applied. At low doses cardiac muscle is induced, at intermediate doses skeletal muscle is induced, and at high doses neurons and astroglia are induced (EDWARDS and MCBURNEY 1983). This immediately raises the possibility that in the embryo different concentrations of endogenous RA might be responsible for the induction of the different parts of the embryo such as the central nervous system, or the mesoderm, or the heart. As is discussed below, both in the central nervous system (CNS) and the peripheral nervous system (PNS), development does indeed depend upon the presence of RA and that its levels must be very tightly controlled. Exactly where and how the RA is generated we do not yet know, but its presence is an absolute requirement.

A. The Induction of Neuronal Differentiation in Culture

I. Embryonal Carcinoma Cells and Neuroblastoma Cells

Prior to the observation described above concerning the concentration dependence of neuronal differentiation in P19 cells (EDWARDS and MCBURNEY 1983), it had been shown that RA (or RA plus cAMP) induced neurons and glia in several other lines of embryonal carcinoma cells which were derived from the mouse (JONES-VILLENEUVE et al. 1982; KUFF and FEWELL 1980; LIESI et al. 1983; MCBURNEY et al. 1982) and it was subsequently shown to be the case in carcinoma cell lines derived from the human (ANDREWS 1984; THOMPSON et al. 1984). The same is true in neuroblastoma cells (SIDELL 1982). Although one might expect cells with a neural origin to differentiate into neurons, it could obviously be important from a therapeutic standpoint that cell division in malignant neural cells can be inhibited and the cells returned to the fully differentiated state by RA (PAHLMAN et al. 1984; SHEA et al. 1985; SIDELL 1982; SIDELL et al. 1983).

The use of cell cultures has been valuable for providing the beginnings of an understanding of the mechanism of action of RA during the induction of neurons and glia. It also provides a valuable resource for those studies on neural development and differentiation which cannot be performed on the whole animal because of its inaccessibility to experimental intervention. After RA treatment and the induction of neuronal differentiation, markers such as neurofilament proteins are, of course, induced along with the morphological change into neurons, but a wide range of nuclear, cytoplasmic, cell surface and extracellular proteins are also induced.

Concerning transcription factors, for example, ATBF1 (IDO et al. 1994) and AP-2 (PHILLIP et al. 1994) are strongly induced, the mammalian homologue of the *Drosophila* neural determination gene *achete-scute*, MASH1, is induced to high levels whereas MASH2 is repressed (JOHNSON et al. 1992), Pax-3 is induced (GOULDING et al. 1991) and the homeobox genes *Hox2.3*, *Hox1.1*, *Hox2.1* and *En-1* are induced (DESCHAMPS et al. 1987; DESCHAMPS et al. 1987). Similarly, the levels of the *src* gene are increased (LYNCH et al. 1986) whereas those of the N-*myc* gene and the c-*myb* gene are decreased (AMATRUDA et al. 1985; THIELE et al. 1985, 1988). The major cytoplasmic changes seem to involve cytoskeletal proteins. Thus the microtubule-associated proteins MAP2 and tau are induced (PLEASURE et al. 1992; FISCHER et al. 1985), as is β-tubulin (BAIN et al. 1995), GAP-43 (FINLEY et al. 1996; PLEASURE et al. 1992), along with the ubiquitous finding of the induction of neurofilament proteins. Interestingly, in cells transfected with antisense MAP2 RNA neuronal differentiation can no longer be induced by RA (DINSMORE and SOLOMON 1991), thus attesting to the importance of these cytoskeletal proteins in RA-induced differentiation. Enzymes such as those of those involved in the pathway of catecholamine synthesis are induced (SHARMA and NOTTER 1988) (but see below) and also proteins associated with synapses such as synaptophysin and synapsin (FINLEY et al. 1996; PLEASURE et al. 1992). Concerning the cell surface, cell surface glycolipids change (ANDREWS et al. 1986), there is a shift towards the synthesis of more complex gangliosides (LEVINE and FLYNN 1986), Thy-1 and N-CAM are up-regulated (THOMPSON et al. 1984), the low and high-affinity cell surface receptors for nerve growth factor (NGF) both increase (HASKELL et al. 1987) as does NGF mRNA itself (WION et al. 1987).

Recent experiments demonstrate the importance of cell-cell signalling via secreted factors during RA induction of neuronal differentiation. Activin A is a secreted signalling molecule and it inhibits the RA induction in P19 cells

(VAN DEN EIJNDEN-VAN RAAIJ et al. 1991). Most excitingly, several of the *Wnt* genes are induced in P19 cells and the *Wnts* are also secreted modulators of cell-cell interactions. When one *Wnt* in particular, *Wnt*-1, was overexpressed then several other *Wnt* genes were induced without RA and the loss of a particular cell surface antigen, SSEA-1, was also observed (SMOLICH and PAPKOFF 1994). Despite the fact that all the cells in the RA-treated culture dish would receive a RA signal, these observations mean that further intercellular signalling takes place via *Wnt*-1, leading to MAP2 induction (see above) and neural differentiation. This type of analysis will gradually lead to the unfolding of the molecular pathway which begins with RA and ends with the appearance of the fully differentiated neuron.

Changes in the extracellular matrix are also bound to be important for neurite outgrowth. Thus, two low molecular weight, heparin-binding proteins have been described which are induced by RA and which stimulate neurite outgrowth. One is called RI-HB, for retinoic acid-induced heparin binding protein (RAULAIS et al. 1991) and the other is called MK, for midkine (MURAMATSU and MURAMATSU 1991; MURAMATSU 1993; NURCOME et al. 1992). MK is certainly found on the surfaces of neural cells and thus could fulfil a natural function, but it is also widely expressed in other regions of the embryo and is thus not unique to neurite producing cells (KADOMATSU et al. 1990; MURAMATSU et al. 1993). The same is true for RIHB which is found in the very early embryo throughout the blastodisc, in all of the germ layers following gastrulation, and following neurulation its expression *decreases* in the neural tube before it does in other tissues (COCKSHUT et al. 1994). However, RA also induces an increase in the α^1/β^1 integrin receptor, but does not change the α^3/β^1 integrin receptor (ROSSINO et al. 1991). Since the α^1/β^1 integrin receptor is the major laminin receptor and these RA-treated cells subsequently demonstrate an increased neurite response to laminin (in contrast to NGF, insulin or phorbol acetate treated cells) these results suggest a mechanism of action via extracellular matrix molecules. Indeed, laminin itself is up-regulated by RA in EC cells (WANG et al. 1985).

After this host of molecular changes, the cells become true neurons and undergo electrophysiological maturation. Glutamate receptor channels are induced, synaptic proteins such as synaptophysin and synapsin appear and synapses can be seen in the electron microscope, the neurotransmitters for both inhibitory and excitatory synaptic transmission are produced and the neurons generate action potentials and establish electrical connectivity (BAIN et al. 1995; FINLEY et al. 1996; MACPHERSON et al. 1997; STAINES et al. 1994; TURETSKY et al. 1993; YOUNKIN et al. 1993). It is interesting that, at least in one case using NTera2 cells, that the differentiated neurons show the immunocytochemical characteristics of CNS neurons rather than PNS neurons (PLEASURE et al. 1992). Further characterisation of these neurons with regard to neurotransmitter phenotype has produced differing results. As mentioned above, some report the induction of a catecholamine phenotype in P19 cells (SHARMA and NOTTER 1988) while others have found that a cholinergic phenotype is induced in P19 cells (MCBURNEY et al. 1988), PC12 cells (MATSUOKA et al. 1989) and human neuroblastoma cells (CASPER and DAVIES 1989; SIDELL et al. 1984). A more detailed study of these characteristics in P19 cells has revealed that the induced neuronal cells are highly diverse with several neurotransmitter phenotypes induced in the same population (GABA, neuropeptide Y, serotonin and others) (STAINES et al. 1994). Thus we may conclude that there is no one single neurotransmitter phenotype which is induced by RA.

II. Dissociated or Explanted Neuronal Cells

In light of the above effects on undifferentiated cells it might be expected that RA would induce either more or longer neurites in neuronal cells in culture and indeed, both effects are seen. For example, chick embryo dorsal root ganglia (DRG) explanted into normal medium exhibited no neurite extension, but when treated with a relatively high concentration of RA (10^{-5} *M*), neurites were produced (HASKELL et al. 1987). Using mouse neonatal DRG, the same concentration of RA stimulated not more, but longer neurites (QIUNN and DE BONI 1991). In dissociated embryonic rat spinal cord cultures, RA promoted the survival of the neurons and increased neuritic density (WUARIN et al. 1990). In particular, RA increased the survival of cholinergic and met-enkephalinergic neurons and had no effect on GABAergic neurons whereas the effect on neurite outgrowth was seen on most neurons regardless of their neurotransmitter phenotype (WUARIN and SIDELL 1991). In fact RA not only can increase neurite number and length but can also change the differentiated phenotype with regard to the neurotransmitter expressed. Thus in cultured newborn rat sympathetic neurons, RA increased choline acetyltransferase activity and the level of acetylcholine while reducing the activity of two catecholamine synthetic enzymes normally associated with sympathetic neurons (BERRARD et al. 1993).

It would seem that cultured CNS neurons are more sensitive to RA than the above PNS neurons from dorsal root ganglia since a typical concentration of 10^{-8} *M* or 10^{-9} *M* RA is used for CNS effects whereas 10^{-5} *M* produced maximal effect in the studies using DRG referred to above. For example, in dissociated embryonic chick neural tube, we have observed a maximal stimulation of neurite number at 10^{-9} *M* RA (Maden and Jones, unpublished data). In these studies effects on both neurite length *and* neurite number were seen at the same time, with different concentrations of RA producing maximal effect: 10^{-9} *M* for neurite number and 10^{-7} *M* for neurite length. Similarly, in cultures of embryonic mouse cerebra, 10^{-8} *M* maximally stimulated axons and dendrites (VED and PIERINGER 1993). Interestingly, a greater stimulation was seen when RA and thyroid hormone were added together which fits very well with the known interactions of the retinoic acid receptors and the thyroid hormone receptors (see Chap. 5). Again, low concentrations of RA were optimal at inducing neurite outgrowth in explants of larval axolotl spinal cord (HUNTER et al. 1991), foetal mouse spinal cord and human spinal cord (QUINN and DE BONI 1991). In studies using embryonic rat cerebellar explants both all-*trans*-RA and 9-*cis*-RA stimulated neurite outgrowth at low concentrations (YAMAMOTO et al. 1996).

Studies on the mechanism of action of RA on neural tissue have been less extensive but have revealed similar effects on the cellular machinery to those described above for embryonal carcinoma and neuroblastoma cells. In embryonic chick sympathetic neurons, for example, RA induces the high-affinity NGF receptor which then permit the neurons to become NGF responsive (RODRIGUEZ-TEBAR and ROHRER 1991). These cells express RABβ, RARγ and RXRγ mRNAs and, in particular, increasing levels of RABβ are correlated with the development of NGF responsiveness (PLUM and CLAGETT-DAME 1996). This effect of RA seems not to occur in cultured trigeminal sensory neurons of the mouse embryo, however (PAUL and DAVIES 1995). The survival of embryonic chick ciliary neurons is enhanced because RA interacts with a heparin-binding growth factor (HENDRY and BELFORD 1991), a similar molecule to the MK and RI-HB proteins described above.

Surprisingly, all the above studies have been performed using all-*trans*-RA as the ligand. Only recently have any other retinoic acids been used. Didehydroretinoic acid is the most prevalent retinoic acid in the chick embryo (SCOTT et al. 1994; THALLER and EICHELE 1990; MADEN et al. 1998) and is equally effective as tRA in supporting the survival and axon outgrowth of embryonic chick sympathetic neurons (REPA et al. 1996).

III. Neural Crest Cells

The neural crest is a group of migratory cells which arises from the dorsal surface of the neural tube and migrates out into the body of the embryo along well defined pathways. The crest cells give rise to an amazing array of differentiated structures in the body including the skull and connective tissue of the head, the autonomic and sensory neurons of the peripheral nervous system, melanophores, Schwann cells and glandular tissues (LE DOUARIN 1982). In tissue culture, neural crest cells differentiate into a wide variety of cell types which reflects their derivatives in the embryo (BAROFFIO et al. 1991; BAROFFIO et al. 1988). However, 1–10 nM retinoic acid stimulates the differentiation of neurons in such cultures by selectively promoting neuronal precursor proliferation and survival (HENION and WESTON 1994). Furthermore, RA stimulates one particular type of neuron, the adrenergic neuron, to differentiate (DUPIN and LE DOUARIN 1995; ROCKWOOD and MAXWELL 1996a) and this effect seems specific to the RARα pathway because the RARα-selective compound AM580 also induces this effect whereas 9-*cis*-RA does not (ROCKWOOD and MAXWELL 1996a). However, this does not square with the observation that thyroid hormone antagonises the action of RA on adrenergic differentiation (Rockwood and Maxwell 1996b) as such an effect implies a pathway of action via the RXR/TH heterodimer. A further curious comparison is that these observations of the stimulation of adrenergic differentiation by RA in neural crest cells contrast with the opposite picture, that of the stimulation of cholinergic differentiation which emerges from the effect of RA on neuroblastoma cells (see above). Since neuroblastoma cells are derived from the neural crest we might have expected them to behave similarly after RA treatment.

B. The Effects of Excess Retinoids on the CNS

The teratogenicity of retinoids on the CNS and its derivatives was first recognised in 1953 when Cohlan reported a high level of exencephaly and a low level of eye defects in rat embryos after treating the mothers with vitamin A. Since that time the many reports on the teratogenicity of retinoids on the CNS have confirmed and extended these findings which can be summarised as microphthalmia, encephalocoel, exencephaly, spina bifida, microcephaly. In addition, there are also characteristic defects of the head and neck which can be attributed to effects on the neural crest and which are relevant to our discussion here – hypoplastic maxilla or mandible, microtia, thyroid and thymus malformations (Alles and Silik 1990; Langman and Welch 1967; Shenfelt 1972; Sulik et al. 1995; Yasuda et al. 1986). Sadly, the same spectrum of CNS abnormalities are seen in humans after inadvertent administration of 13-*cis*-RA to the embryo (Die-Smulders et al. 1995; Lammer and Armstrong 1992; Lammer et al. 1985; Rosa et al. 1986). Two of the abnormalities that have been seen in humans, facial palsies and cerebellar defects are particularly relevant to the effects seen in the experiments on vertebrate embryos summarised below.

It is important to bear in mind that the CNS does not merely consist of a neural tube from which motoneurons grow, but the neural crest migrates out from its dorsal aspect to form many structures of the embryo, as described above. The sensory peripheral nervous system (and in the head this consists of the cranial nerves and ganglia) are derived from the neural crest and are particularly relevant to the discussion below. Clearly, a localised defect in the formation of the CNS has an effect on the neural crest from that region which results in aberrant cranial nerve and ganglia formation. This is indeed the case, as is described below.

Classical teratological studies have moved on from descriptions of defects to experimental studies examining the patterns of gene activity and their alteration soon after the teratogenic insult. These have provided surprising revelations and it seems that there are three effects on patterning in the CNS. It further seems that these three different phenotypes are caused either by differences in concentration (*Xenopus*) or differences in stage of treatment (rat and mouse) or treatment with different RA isoforms (zebrafish).

I. Effect 1: Posteriorisation

The first effect is generally referred to as posteriorisation of the embryo and is reflected in the fact that the front end of the embryo (forebrain) is missing, the domains of expression of anterior genes are extinguished and the domains of expression of posterior genes are expanded. This was first observed by DURSTON et al. (1989) using the *Xenopus* embryo and treating them with RA between the late blastula and early neurula stages. They noted that the embryos failed to develop anterior neural structures such as the forebrain, midbrain and eyes (Fig. 1, C). Identical effects are seen with axolotl embryos (MADEN et al. 1992a). Since the total volume of CNS tissue, estimated from sections of treated embryos was the same as in controls, Durston et al. concluded that the missing anterior neural structures had been respecified to form posterior neural structures and when patterns of gene activity were analysed this conclusion was reinforced because anterior-specific genes were repressed

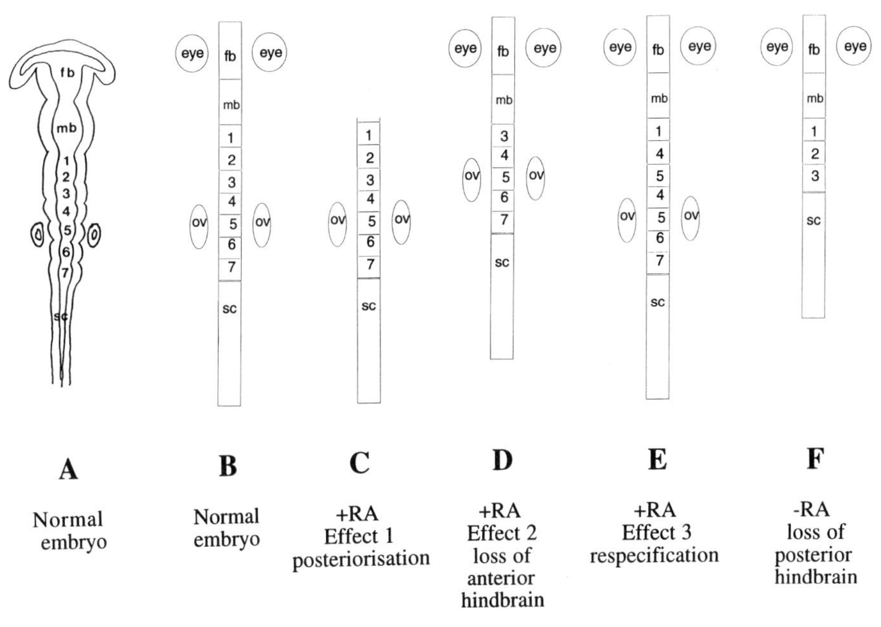

Fig. 1. Summary of the effects of excess and deficiency of RA on the embryonic CNS. *A*, Camera lucida drawing of the chick embryo CNS soon after neurulation to showing the subdivisions into forebrain (*fb*), midbrain (*mb*), hindbrain rhombomeres (*1–7*) and spinal cord (*sc*). Adjacent to rhombomere 5 are the otic vesicles which will form the ears. *B*, Cartoon representation of the embryonic CNS drawn in *A*, also showing the eyes which develop from the forebrain. *ov*, Otic vesicles. *C*, Effect 1: excess RA causes loss of the anterior structures namely, eyes, forebrain and midbrain. *D*, Effect 2: excess RA causes loss of rhombomeres 1 and 2 and a resulting anterior shift of the otic vesicles and rest of the hindbrain. *E*, Effect 3: excess RA causes a respecification of rhombomeres 2 and 3 into 4 and 5, thus repeating those structures. *F*, A deficiency of RA in the quail embryo causes the loss of the posterior rhombomeres 4–7

after RA treatment and posterior genes were up-regulated (CHO and DE ROBERTIS 1990; DEKKER et al. 1992; LEROY and DE ROBERTIS 1992; LOPEZ and CARRASCO 1992; LOPEZ et al. 1995; SIVE et al. 1990). Further experiments revealed that RA could mimic normal development by generating the correct spatial pattern of expression of two genes, XlHbox 6 and XIF 6 in isolated explants of *Xenopus* animal caps (SHARPE 1991). This has lead to the suggestion that there is an endogenous gradient of RA within the neural plate with a high point at the posterior end (see Sect. F.VI on interfering with RAR signalling).

The *Xenopus* embryo has been valuable in asking whether this posteriorising effect is one which acts directly on the CNS during its specification or earlier on the mesoderm and thus only secondarily on the CNS. Several experiments have been conducted using *Xenopus* to specifically answer this question. Isolated parts of the embryo have been treated with RA and then recombined or embryos have been treated before or after the induction of the nervous system and the general conclusion from these experiments is that RA acts *both* directly on the CNS and indirectly via the underlying mesoderm (RUIZ I ALTABA and JESSELL 1991; RUIZ I ALTABA and JESSELL 1991; SIVE and CHENG 1991; SIVE et al. 1990). This explains why the same general effects on anterior neural structures are seen irrespective of whether the embryos are treated with RA before or after neural induction.

The "anterior amputation" phenotype has been seen in all vertebrates embryos studied. It occurs in zebrafish after treatment with 9-*cis*-RA rather than all-*trans*-RA (ZHANG et al. 1996), and in rat and mouse embryos (AVANTAGGIATO et al. 1996; CUNNINGHAM et al. 1994; SIMEONE et al. 1995). In rat embryos, where the effect is seen after treatment at the late streak stage, there is a progressive loss of anterior structures with increasing length of exposure to RA (and hence an increased total dose), exactly as seen in *Xenopus* embryos. In mouse embryos this effect also peaks after treatment at mid- to late-streak stage and was referred to as atelencephalic microcephaly – absence of anterior sense organs, reduction in the anterior brain volume accompanied by an increase in the hindbrain mass and poor differentiation of the neuroepithelium (SIMEONE et al. 1995). The loss of *Emx1*, *Emx2* and *Dlx1* gene expression domains (normally expressed in the forebrain) and the anteriorisation of the *Wnt-1*, *En-1*, *En-2*, *Otx2* and *Pax-2* (normally expressed more posteriorly in the midbrain and hindbrain) confirmed the morphological observations (AVANTAGGIATO et al. 1996). *Otx2* is considered to be an important gene in determination of the forebrain as a targeted mutation of the gene in mouse leads to the loss of anterior neural structures, a phenotype which is remarkable similar to the RA phenotype of effect 1 (ANG et al. 1996). Furthermore, the down-regulation and shift of the expression border of *Otx2* is also seen in the chick embryo (BALLY-CUIF et al. 1995) and the *Xenopus* embryo (PANNESE et al. 1995).

The general conclusion from these studies is that RA induces a stage-specific repatterning of the anterior CNS by altering the endogenous morphogenetic signals and genes used to regionalise the CNS (Fig. 1, A).

II. Effect 2: Loss of Anterior Hindbrain

The external phenotype of embryos showing effect 2 has long been recognised: an abnormally rostral position of the otocyst and a shortened preotic hindbrain (MORRISS 1972). It results from the loss of a section of CNS tissue in the posterior midbrain/anterior hindbrain region (Fig. 1, D) and is a universal teratological finding not only in rat and mouse embryos (CUNNINGHAM et al. 1994; LEE et al. 1995; LEONARD et al. 1995; MORRISS 1972; MORRISS-KAY et al. 1991; SIMEONE et al. 1995), but also in chick (LOPEZ et al. 1995; SUNDIN and EICHELE 1992), *Xenopus* (LOPEZ et al. 1995; PAPALOPULU et al. 1991) and zebrafish (HOLDER and HILL 1991; ZHANG et al. 1996). In *Xenopus*, effect 2 is seen at a lower dose of RA than that needed for effect 1; in zebrafish it is seen when using all-*trans*-RA rather than 9-*cis*-RA; and in rat and mouse embryos it is stage dependent – treatment at the foregut pocket stage for effect 2 rather than the late streak stage for effect 1 (CUNNINGHAM et al. 1994; MORRISS-KAY et al. 1991).

The results with zebrafish most clearly illustrate this phenotype (HOLDER and HILL 1991). Figure 2A shows a normal 24-h zebrafish embryo stained with an antibody to the engrailed protein from *Drosophila*. The immunoreactive cells in the CNS form a stripe of tissue which includes the posterior midbrain plus the anterior hindbrain. Also stained are some single cells in the somites. After RA treatment (Fig. 2B) the embryo is virtually identical – the single cells in the somites are still present – except for the midbrain/hindbrain stripe of engrailed +ve cells which has completely disappeared. As a result of this presumed tissue loss, more posterior CNS regions and the otic vesicle are shifted anteriorly (Fig. 1, D).

However, in mammals it is not so clear that there is a complete loss of tissue as is implied in the zebrafish result above. The normal hindbrain of all vertebrates is composed of a series of bulges, or rhombomeres (Fig. 1, A) and there are seven of them in all. Many genes, in addition to *engrailed*, have been used as markers to try to understand what happens to the anterior hindbrain after RA treatment, notably *Hox* genes and others such as *Krox-20* which are expressed in a segment specific fashion thus allowing the molecular identification of individual rhombomeres. After RA treatment of embryos the typical effect on a variety of *Hox* genes is to induce an anterior spread of their expression patterns into the midbrain and forebrain regions where they are not normally expressed and then they retract posteriorly leaving behind an aberrant expression pattern (CONLON and ROSSANT 1992). There is a clear disruption of rhombomeric segmentation (PAPALOPULU et al. 1991; LOPEZ et al. 1995) and a loss of one or two rhombomeres (HOLDER and HILL

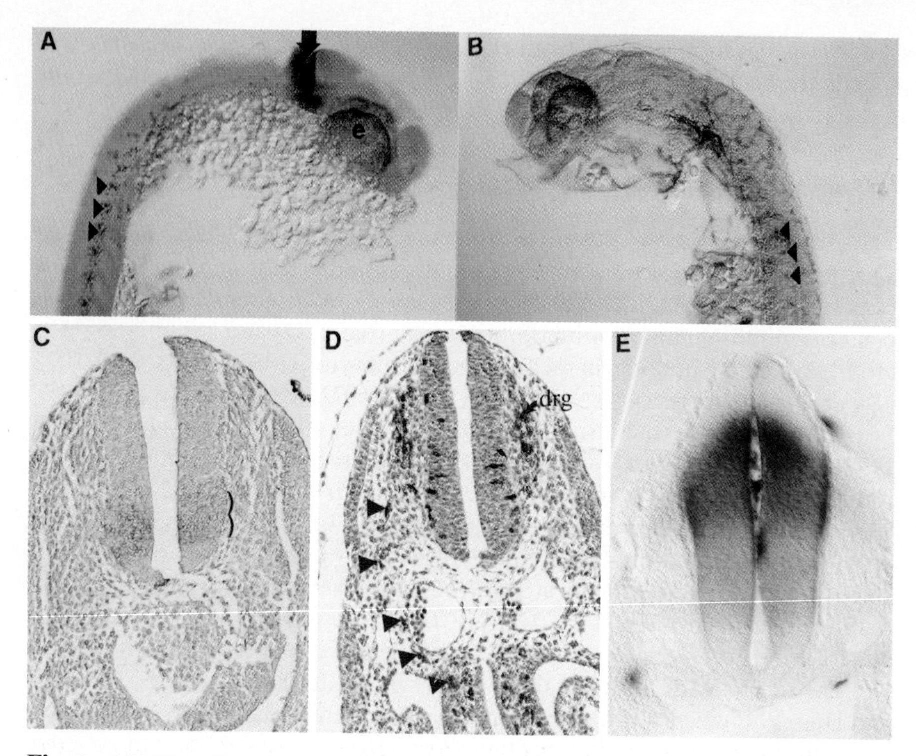

Fig. 2. A,B The effects of excess RA on the zebrafish embryo hindbrain (effect 2 in Fig. 1). **A** Normal 24-h zebrafish stained with an antibody to the *Drosophila* engrailed protein. A section of the anterior hindbrain and posterior midbrain is stained (*arrow*) as well as a few somite cells in each segment down the body (*arrowheads*). *e*, eyes. **B** A 24-h embryo treated at gastrulation with $1.5 \times 10^{-7} M$ RA. Most of the embryo is normal: the eyes are present, the engrailed positive cells are present in the somites (*arrowheads*), but the segment of engrailed positive anterior hindbrain is completely missing. **C–E** comparative expression patterns of the three retinoid binding proteins in the day 10.5 embryonic mouse spinal cord. **C**, CRBP protein is present in the future motoneurons of the cord (*bracket*). **D** CRABP I protein is present in future commissural interneurons of the cord as well as in the neural crest derived dorsal root ganglia (*drg*) and in the migrating neural crest cells (*arrowheads*) which are travelling ventrally to form the enteric ganglia around the developing gut. **E** CRABP II mRNA is present in the alar part of the cord but excluding the most dorsal cells and in a curious diagonal shape

1991; LEONARD et al. 1995; MORRISS-KAY et al. 1991; PAPALOPULU et al. 1991; WOOD et al. 1994) accompanying the anterior shift of the otocyst and some have suggested that the hindbrain anterior to the otocyst takes on the gene expression characteristics of a single large rhombomere 4 (CONLON and ROSSANT 1992; WOOD et al. 1994). However, when the later anatomy of the head is studied including the neural crest derivatives, the picture is slightly more complicated.

Thus, the two other characteristic features of this effect 2 (anterior shift of the otic vesicle) are the fusion of the trigeminal ganglion (derived from rhombomere 2 neural crest) and facial ganglion (derived from rhombomere 4 neural crest) and the fusion of the first and second branchial arches (Cunningham et al. 1994; Kessel 1993; Lee et al. 1995; Leonard et al. 1995; Pratt et al. 1987; Seegmiller et al. 1991). The fusion of ganglia and arches could be brought about by the loss of rhombomere 3. This would result the two streams of neural crest which normally arise as separate entities, one from rhombomere 2 and one from rhombomere 4, coming together to form a fused ganglion and migrating ventrally en masse, thus causing the arches to fuse. In support of this interpretation, the anterior stripe of expression of the *Krox-20* gene which is normally expressed in rhombomere 3 is absent in RA treated embryos (Morriss-Kay et al. 1991; Papalopulu et al. 1991; Wood et al. 1994).

In addition, there is no doubt that in such RA treated embryos the migration of neural crest cells is either inhibited or mis-directed. It was first shown in culture that retinoids inhibit the migration of crest cells (Smith-Thomas et al. 1987; Thorogood et al. 1982) and this provides an explanation for the observed teratological phenomena associated with the neural crest derivatives – hyoplastic maxilla or mandible, microtia, thyroid or thymus malformations. Furthermore, it has also been possible to directly observe an inhibition or alteration of migration in vivo. In mouse embryos treated with 13-*cis*-RA (the active component of Accutane) the neural crest either does not leave the neuroepithelium or remains near to it and the eventual abnormalities seen in these mouse embryos were strikingly similar to those reported in humans including ear and thymus defects (Pratt et al. 1987; Webster et al. 1986). A similar failure of crest to migrate occurs in the chick embryo (Moro Balbas et al. 1993). In rat embryos treated with retinyl palmitate, preotic neural crest cells have been observed to migrate more slowly and their pathway was shifted more anteriorly (Morriss and Thorogood 1978). Lee et al. (1995) using rat embryos directly observed the mis-migration of DiI labelled anterior hindbrain crest into the second branchial arch (as well as populating the first arch as normal) and a mis-migration of rhombomere 4 neural crest anteriorly to contribute to the trigeminal ganglion (as well as populating the facial ganglion as normal). In the chick embryo a similar mis-migration of DiI-labelled crest in two directions from rhombomere 4 has been seen. Mis-routing in a rostral direction resulted in an aberrant contribution to the trigeminal ganglion and caused a fusion of the trigeminal and facial ganglia and mis-routing in a caudal direction resulted in completely aberrant facial nerve tracts (Gale et al. 1996). The latter result seems particularly relevant to the observations of facial nerve palsies in Accutane exposed children.

Thus it may be concluded that effect 2 is characterised by an abnormal segmentation of the anterior hindbrain (at least a loss of rhombomere 3, but possibly more tissue than that), fusion of the trigeminal and facial ganglia, fusion of the first and second branchial arches, mis-migration of the neural crest.

III. Effect 3: Transformation of Anterior Hindbrain

The third phenotype is a fascinating combination of both effects 1 and 2 in that the anterior hindbrain is affected (effect 2) and it is posteriorised (effect 1). The revelation of this third phenotype specifically came about due to the variety of molecular and neuroanatomical markers which are available for the identification of individual rhombomeres. In mouse embryos, transgenic animals were generated which contained *LacZ* reporter genes to reveal the endogenous expression of various *Hox* genes (MARSHALL et al. 1992). *Hoxb*-1, for example, is a gene which in the day 9.5 mouse embryo has become restricted to one rhombomere, rhombomere 4 and to the cranial nerve which derives from rhombomere 4, the facial nerve. After treatment of late streak stage embryos with relatively low doses of RA the expression of *Hoxb*-1 at first shifts anteriorly (as is typical for *Hox* genes – (CONLON and ROSSANT 1992)), but then becomes progressively restricted to two stripes in the hindbrain rather than the normal one. The second stripe is rostral to the normal rhombomere 4 stripe and is in rhombomere 2. The normal cranial nerve which emerges from rhombomere 2, the trigeminal nerve, is also changed in character and comes to resemble the normal rhombomere 4 nerve, the facial. We are led to the conclusion that rhombomere 2 has been transformed into a rhombomere 4 and, in addition, the authors concluded that the intervening rhombomere 3 comes to resemble a rhombomere 5 (MARSHALL et al. 1992). Thus the normal sequence of rhombomeres of 1, 2, 3, 4, 5, 6, 7 has been transformed to 1, 4, 5, 4, 5, 6, 7 (Fig. 1, E). Others have concluded on the basis that the gene expression in rhombomere 3 does not change, that the sequence of rhombomeres is 1, 4, 3, 4, 5, 6, 7 (WOOD et al. 1994). However, further neuroanatomical data have confirmed that a change does occur in rhombomere 3. Axons of motoneurons in rhombomere 3 normally project anteriorly and exit with the trigeminal nerve from rhombomere 2. After RA treatment these axons now project posteriorly and join the facial nerve to exit in rhombomere 4 (KESSEL 1993). This clearly indicates a change in motoneuron phenotype in rhombomere 3 and would have a significant effect on the motor innervation of the face, a further observation of relevance to the appearance of facial palsies in Accutane exposed children.

The difference between effect 2 and effect 3 in the mouse is the stage of development that the embryos have reached when treatment is performed (LEE et al. 1995; SIMEONE et al. 1995; WOOD et al. 1994). However, in the zebrafish embryo the difference is due to dose as effect 3 is seen with a slightly lower dose of RA than that used to generate effect 2 (HILL et al. 1995). Remarkably, this happens only by changing the treatment dose from 1.5×10^{-7} M to 1×10^{-7} M. In lower vertebrates there is a unique neuron present in rhombomere 4 called the Mauthner neuron which has aided these studies. Thus lower dose treatment of embryos during gastrulation results in the appearance of ectopic Mauthner neurons in rhombomere 2, suggesting at least a partial transformation of phenotype. In contrast to the mouse embryo the neurons in rhombomere 3 seem unaffected and it has therefore been concluded that the normal sequence of rhombomeres has been transformed into 1, 4, 3, 4, 5, 6, 7 (HILL et al. 1995). This duplication of Mauthner neurons is also seen in *Xenopus* embryos (MANNS and FRITZSCH 1992).

That this dramatic transformation of rhombomere phenotype is caused by the RA induction of aberrant *Hox* gene expression has recently been elegantly demonstrated by showing that the overexpression of a single *Hox* gene mimics the RA phenotype of these embryos. Thus there is another gene, *Hoxa*-1, which behaves identically to *Hoxb*-1, in that two stripes of gene expression appear (in rhombomeres 2 and 4) rather than the normal one (in rhombomere 4) after RA treatment (ALEXANDRE et al. 1996; CONLON and ROSSANT 1992). *Hoxa*-1 is up-regulated within 1 h of RA treatment (ALEXANDRE et al. 1996) and both *Hoxa*-1 and *Hoxb*-1 have retinoic acid response elements (RAREs) in their upstream elements (LANGSTON and GUDAS 1992; MARSHALL 1994; STUDER 1994). It was shown that ectopic expression of *Hoxa*-1 either by injecting the mRNA into zebrafish eggs (ALEXANDRE et al. 1996) or by creating transgenic mouse embryos (ZHANG et al. 1994) results in the same patterns of ectopic *Hoxb*-1 and *Hoxa*-2 expression and the same down-regulation of the anterior *Krox-20* stripe in rhombomere 3 that is caused by RA. Furthermore, the effect on the anatomy of the zebrafish embryo is exactly the same after ectopic *Hoxa*-1 expression or RA treatment – duplicated Mauthner neurons; abnormal development of the jaw apparatus (Meckel's cartilage); fusion of the two streams of neural crest from rhombomere 2 and rhombomere 4 into one stream; fusion of the trigeminal and facial ganglia (ALEXANDRE et al. 1996).

C. The Effects of RA on Associated CNS Structures

There are three other systems or cell types associated with the CNS which are affected by RA during their development. The first is the differentiation of the retina. The effects of excess RA (Sect. B) and a deficiency of RA (Sect. D) on the developing eye as a whole has been referred to several times, generating such abnormalities as microphthalmia or anophthalmia. In addition, however, the differentiation of specific cell types within the retina is affected by RA. RA promotes the development of photoreceptor cells from rat embryos and 9-*cis*-RA is more effective than all-*trans*-RA (KELLEY et al. 1994). The same phenomenon is seen in chick embryo photoreceptors although, in this case, retinol is more effective than RA in stimulating photoreceptor development (STENKAMP et al. 1993).

The second system is the stimulation of regeneration of mammalian auditory hair cells by RA. Corti's organ contains the hair cells and in experiments using neonatal rat pups the hair cells were first killed by the administration of ototoxic drugs. After 7 days of RA treatment (10^{-8} *M*), 78% of the cells had been regenerated whereas without RA, no cells regenerated (LEFEBVRE et al. 1993). In an undamaged organ of Corti, RA increased the number of cells that developed as hair cells resulting in large regions of supernumerary hair cells (KELLEY et al. 1993) and the organ itself contains RA. This raises the exciting possibility of the regeneration of hair cells in the human. The induction of differentiation of the secretory epithelium and the early sensory epithelium by RA was also seen in cultured otic vesicles of the chick along with the suppression of *c-fos* induced proliferation (REPRESA et al. 1990).

A third system concerns the differentiation of oligodendrocytes, which are cells that myelinate CNS axons. In contrast to the hair cells described above, RA inhibits the differentiation of oligodendrocytes from rat embryonic spinal cord in vitro (NOLL and MILLER 1994). Since the migration of oligodendrocytes away from their place of origin in the developing spinal cord into the white matter is a precisely timed event during CNS development, these authors proposed that one of the functions of the high levels of RA in the spinal cord (see below) is to inhibit the differentiation of oligodendrocytes so that migration takes place. A rather different effect has been reported by BARRES et al. (1994) on oligodendrocytes from the rat optic nerve. In this case, RA slowed the cell-cycle time and thus promoted their differentiation, and there may therefore be a difference in mechanism between the spinal cord, where RA is present in abundance (see below) and the brain where RA is probably present at far lower levels.

D. The Effects of a Deficiency of Retinoids on the CNS

If retinoids are an important component of the developing CNS rather than being simple teratogenic agents, then depleting the embryo of vitamin A should result in abnormal CNS development. Furthermore, if the effects of excess retinoids described in Sects. A and B are meaningful, we might expect an absence of retinoids to result in (a) defective neurite outgrowth; and/or (b) defective anteroposterior organisation (perhaps loss of posterior structures?); and/or (c) defective hindbrain. The following discussion shows that these predictions are quite well borne out.

I. Deprivation of Retinoids

Vitamin deficiencies, particularly of vitamin A, were the first dietary means of producing congenital malformations of the embryo. Initial experiments, performed as long ago as the 1930s on pigs (HALE 1933) resulted in a litter of pigs being born without eyes. Such malformations have, over the subsequent decades, been repeatedly seen in similar experiments using rats, rabbits, cattle and sheep, complemented with observations on humans. The abnormalities in the mammalian CNS include hydrocephalus, spina bifida, anopthalmia, microphtalmia, retina defects and in the head and neck region, ear and teeth abnormalities, cleft palate and hare lip (KALTER and WARKANY 1959; WARKANY 1945; WILSON et al. 1953). More detailed studies on the precise anatomical defects in these mammalian studies were not, at that time, performed and now that there is a wide spectrum of molecular and neuroanatomical markers available, such as those referred to above, we should be able to reap a rich harvest from this type of experiment. For example, DICKMAN et al. (1997) have generated retinoid deficient embryos by a dietary method which allows one to produce a sudden, acute retinoid deficiency during a selected gestational window. Using rat embryos, a deficiency created from embryonic day 11.5 to day 13.5 resulted in an underdeveloped hindbrain, absent cranial flexure, microphthalmia, foreshortening of the snout and branchial arches. In the eye, the lens showed apoptosis and the retina failed to invaginate to form the optic nerve. There was considerable apoptosis in the frontonasal mass and maxilla, but not the mandible or hyoid. The cranial nerves were defective, again due to extensive apoptosis and the vagus nerve was absent. The neural tube was reduced in size and cellularity and the brain had reductions in differentiating neuronal populations due to reduced proliferation.

This striking appearance of apoptosis in neural crest derived structures accords well with the studies described above (see Sects. A, B) and was previously seen in deficiency studies using the quail embryo. It transpires that it is easier to produce vitamin A deficient embryos in chickens and quails than in mammals and the original studies of such embryos observed a failure of cardiovascular development (THOMPSON et al. 1969; HEINE et al. 1985; DERSCH and ZILE 1993). Examining the CNS in these quail embryos, we have observed three defects (MADEN et al. 1996).

The first defect is that the posterior part of the hindbrain is completely missing. Tissue sections through these deficient embryos revealed that instead of the normal seven rhombomeres usually present (Fig. 1, A,B) there were only three (Fig. 1, F). By using the expression patterns of the genes *Krox-20*, *Fgf-3*, *Hoxa-2* and *Hoxb-1* it was revealed that the rhombomeres present were 1, 2 and 3. Thus in the absence of retinoids, rhombomeres 4, 5, 6 and 7 fail to develop. Our most recent studies (MADEN et al. 1997) have shown that the loss of this tissue comes about very early in development, as the neural tube forms. Soon after gastrulation, at the 5 somite stage, we observed a band of cells in the mesoderm in the region of the first somite undergoing programmed cell death. This was followed 2h later by a band of cell death in the neuroepithelium of the prospective hindbrain. Thus the absence of the posterior hindbrain is caused by a highly localised region of cell death occurring as the hindbrain becomes specified at the 7-somite stage. This is the time when the first segmentation event takes place in the neuroepithelium and we presume that this loss of tissue in the A- embryo is caused by an aberrant pattern of expression of the genes which are responsible for the regionalisation of the anteroposterior axis of the embryo.

The second defect in the CNS of the deficient quail embryos is that as differentiation of neurons in the neural tube begins, axons do not appear within the surrounding mesoderm as one expects and the axon trajectories within the neural tube itself are unorganised and chaotic. The normal neurofilament antibody staining of quail embryos should show a typical pattern of localisation within the mantle layer of the neural tube and in the ventral motor and dorsal sensory axon tracts in the periphery. In the deficient embryo there is a delay of several stages in the onset of neurofilament protein expression and the pattern within the neural tube never approaches that of normal. No dorsal sensory axons or ventral motor axons expressing neurofilament proteins can be seen in the periphery and within the neural tube itself there are far fewer expressing neurons than normal. Furthermore, some axons within the neural tube seem to display a chaotic trajectory, for example, by crossing the dorsal midline, a phenomenon which never happens in the normal embryo. When cultured in minimal medium the neural tubes from deficient embryos do not extend neurites, but when RA is added to the medium then neurites appear (MADEN et al. 1998) confirming that this is a specific effect of retinoid deficiency.

The third defect seen in these embryos is widespread apoptosis in the neural crest, as is seen in rat embryos. This neural crest cell death begins at about stage 14 and affects all neural crest cells throughout the embryo whether they have finished migrating or are still migrating. Our studies with the TUNEL technique have revealed streams of apoptotic cells which coincide with the known pathways of migration of neural crest cells (MADEN et al. 1996).

A novel technique recently described for generating retinoid deficiency by a non-dietary means is to administer antisense oligodeoxynucleotides for retinol binding protein (RBP) mRNA by injecting them into the yolk sac cavity of cultured mouse embryos (BAVIK et al. 1996). Such treatment at the 3–5 somite stage results in accumulation of blood islands on the yolk sac due to lack of vitelline vessels, an identical result to that seen in deficient chick and quail embryos (DERSCH and ZILE 1993; THOMPSON et al. 1969). Exencephaly was produced in the embryos and the eyes were defective – the lens placode failed to appear and the optic vesicle failed to invaginate. The expression of three genes was investigated. TGF-β and *sonic hedgehog* were both down-regulated in oligo-treated embryos whereas there was no effect on *Wnt-3a*.

II. Inhibition of RA Synthesis

An additional method of depriving embryos of RA is to inhibit its synthesis with drugs. Citral, for example, is a competitive inhibitor of aldehyde dehydrogenases and inhibits RA synthesis in the mouse (CONNOR and SMIT 1987; MCCAFFERY et al. 1992) and zebrafish (MARSH-ARMSTRONG et al. 1994) embryo. The result of treating zebrafish embryos with citral for 2 h at neurula stages is that the ventral half of the eye fails to develop (MARSH-ARMSTRONG et al. 1994). This observation complements the finding that RA treatment of zebrafish embryos at the early somite stage results in bifurcated eyes which have two retinas due to the overgrowth of the ventral part (HYATT et al. 1992). Treatment of *Xenopus* embryos with citral also produces eye defects (microphthalmia) along with laterally expanded heads, changes in pigmentation, heart defects and blood pools (SCHUH et al. 1993). In chick embryos citral has been used to generate microphthalmia (ABRAMOVICI et al. 1978) thereby emphasising the role of retinoids in eye development. Treatment of mouse embryos with citral at a specific period of development inhibits the olfactory system, another part of the embryo which depends on RA for its development (ANCHAN et al. 1997).

Another drug, disulphiram, inhibits aldehyde dehydrogenases (MARSH-ARMSTRONG et al. 1995; MCCAFFERY et al. 1992; VALLARI and PIETRUSZKO 1982) and can therefore be used in the same way as citral to deplete the embryo of RA. Treatment of zebrafish embryos with this compound produced embryos with short tails and wavy notochords, and the effect on the CNS was to shorten the spinal cord (MARSH-ARMSTRONG et al. 1995).

E. Endogenous RA in the CNS

The most direct method for determining the presence or absence of retinoids in the CNS is to use high performance liquid chromatography (HPLC). Individual retinoids can be identified according to their elution time and co-elution with known standards. This method requires a considerable amount of starting material and presents no problems for measuring adult CNS tissue, but for individual components of the embryo this presents a serious drawback. To overcome this, a novel technique has been developed involving a reporter construct which is either transfected into cells onto which pieces of embryo can then be placed or used to generate transgenic embryos. These experiments are now described.

I. HPLC Data

Few measurements of endogenous retinoids in the CNS have been made. In lower vertebrates, the larval axolotl spinal cord contains all-*trans*-RA, all-*trans*-retinol and 13-*cis*-RA (HUNER et al. 1991). In higher vertebrates, the mouse embryonic day 10.5 and day 13 CNS shows an interesting differential distribution. The forebrain, midbrain and hindbrain contain extremely low levels of all-*trans*-RA, but the neural tube which will develop into the spinal cord contains the highest level in the whole embryo (HORTON and MADEN 1995). Therefore, somewhere in the posterior part of the hindbrain or at the hindbrain/spinal cord junction there is a sudden step from very low levels to very high levels of RA. This has been confirmed by two other methods (see below) and thus seems to be a consistent finding suggesting that there is an important boundary of pattern formation here in the embryo with regard to RA. This HPLC study also showed that a dose of RA which is required to produce teratogenic effects on the embryo (see Sect. B) raises the endogenous levels of RA in the CNS between 200x (spinal cord) and 1300x (forebrain and midbrain), a massive increase over normal physiological levels.

The developing eye in the mouse embryo has also been the subject of HPLC studies. Whole retinas from varying aged embryos and adult mice were used to detect all-*trans*-RA and 13-*cis*-RA (McCaffery et al. 1993). Surprisingly high levels of 13-*cis*-RA was measured, about 25%–50% of the levels of all-*trans*-RA. The total retinoic acid concentration of embryonic retinas was 500 nM and this value decreased four-fold in the adult. Cytosolic fractions of dorsal and ventral retinas were also examined for retinoic acid synthesis from retinal and it was found that the dorsal part gave a tenfold higher peak than the ventral part (McCaffery et al. 1992). However, the converse was found using the F9 reporter cells described below: ventral halves gave higher responses than dorsal halves. This contradiction was interpreted as a dorsoventral difference in the ability of the enzymes to metabolise at high retinal levels and that the ventral pathway became substrate inhibited in the HPLC studies. This study highlights the caution that must be imposed upon the interpretation of data from only one type of analysis – we need more than one of these assays to be able to come to a realistic conclusion.

No HPLC measurements have been made earlier in embryonic life than day 10.5 (see above), and to answer the question of when RA generation begins in the CNS other methodologies have therefore been used.

II. Transgenic Embryos

Using a reporter construct containing a retinoic acid response element (RARE) linked to a minimal promoter and the β-galactosidase gene, transgenic zebrafish and mouse embryos have been created to indicate when and where RA is made in the embryo. In the zebrafish, even though the construct was injected into the one- and two-cell embryo, the first localisation was not observed until the surprisingly late stage of 18–21 somites (Marsh-Armstrong et al. 1995). Perhaps this reflects the limited sensitivity of this particular transgene construct in the zebrafish embryo. The RA-induced β-galactosidase expression was, however, concentrated in the anterior trunk region in tissues which included the spinal cord and floor plate. Even though the time course of appearance of this activity was later than might be expected the pattern corresponded well with the mice results described below.

Four groups have generated transgenic mice with the RARβ promoter construct (Mendelsohn et al. 1991; Reynolds et al. 1991; Rossant et al. 1991; Shen et al. 1992) and in the mouse embryo, activation of the reporter construct began earlier than the zebrafish embryo, at the neural plate stage, but again, it was only in the posterior half of the embryo. When the neural tube had completed closure there was a sharp anterior boundary of expression in the hindbrain, consistent with the pattern seen in endogenous RA measurements, and expression ceased at about the level of the posterior neuropore (although it did not extend so far posteriorly in the studies of Shen et al. and Mendelsohn et al.). In the head region only the optic eminences and the olfactory mesenchyme showed any transgene activity and, later, the eye itself (both lens and retina) was positive. This confirms that the forebrain, midbrain and most of the hindbrain develops in the absence of RA with a strong presence of RA in the spinal cord. Within the cross-section of the spinal cord the data generally showed that the transgene activity was initially present throughout the dorsoventral axis, but gradually became more concentrated in the dorsal part or the alar plate. However, the ventral motoneurons were positive, but there seems to be some disagreement between these studies as to whether dorsal root ganglia are regions of positive transgene activity. Some found this to be true (Mendelsohn et al. 1991), some suggest only the surrounding meningeal layer is positive (Reynolds et al. 1991) and some find them to be entirely negative (Rossant et al. 1991).

All these studies agree, however, that the activity of the transgene within the spinal cord which is expressed throughout the anteroposterior extent of the cord at day 10.5 becomes restricted to two distinct anterior and posterior domains at day 12.5 of development (Colbert et al. 1993). These domains correspond to the brachial and lumbar enlargements where the limb buds will develop, that is C1–T2 and T13 – S2 with a negative region in between (Colbert et al. 1995) and are "hot spots" of retinoic acid synthesis (see below) (McCaffery and Drager 1994b).

The association of high RA levels with the cervical and lumbar enlargements is a fascinating one as these are the areas where the limbs will grow out and there are more motoneurons here than elsewhere in the cord. It suggests that RA may contribute to the enhanced neuronal survival found here or may be responsible for the regulation of differentiation of particular classes of motoneuronal types. If it is also true that retinoids are important for the maintenance of these classes of neurons or perhaps motoneurons as a whole, it may further be the case that loss of retinoid function either through lack of synthesis or lack of receptor expression could be responsible for the onset of neurodegenerative diseases such as motoneuron disease.

III. Reporter Cells

A similar type of assay has been used for individual pieces of the embryo which can be placed on lawns of reporter cells or the tissue homogenised and the homogenate added to the medium. The reporter cells concerned are F9 cells (Wagner et al. 1992) or L-C_2A_5 cells (Colbert et al. 1993) which have been transfected with the RAREβ-galactosidase construct and the cells respond to the release of RA from the embryonic tissue by turning blue after histochemical staining for *LacZ* exactly as whole embryos described above do. Using this technique in the zebrafish embryo a similar picture emerged to the transgenic studies (MARSH-ARMSTRONG et al. 1995). There was a gradient of RA along the trunk (whole trunk sections were used rather than the CNS itself because of the small size) with the highest levels in the anterior regions (at the level of somites 7–9) and the hindbrain region showed only very low levels. Our studies of the chick embryo (MADEN et al. 1998) have produced very similar results to those of the zebrafish and the transgenic mice described above. After neural tube formation no activation of F9 cells occurs when forebrain, midbrain or hindbrain neuroepithelium is placed onto the cells, but high activation is found from the spinal part of the neural tube. No difference could be detected along the length of the spinal cord at these early stages, but it appears later, as described below for the mouse embryo. The absence of very low levels of detectable retinoids in the forebrain has also been observed by WAGNER et al. (1992) in rat embryos and LAMANTIA et al. (1993) in mouse embryos. A more detailed study of the developing olfactory system has revealed that the retinoid generating tissue involved in this case is the mesenchyme which lies between the ectodermal olfactory placode and the olfactory bulb of the forebrain. When citral is applied to the embryo, olfactory development is inhibited (ANCHAN et al. 1997), confirming a role for endogenous RA here.

In the mouse embryo, a similar picture emerged when whole embryos were placed onto these F9 reporter cells (ANG et al. 1996). RA could first be detected at late streak stages (day 7.5) arising from the embryonic tissue, not the extraembryonic tissue. One day later RA was detected adjacent to the trunk region with an anterior border in the hindbrain/spinal cord region and a posterior border towards the posterior end but not reaching it. The highest region of activity was located in the prospective forelimb bud region. When the head was dissected into its component parts no activity could be detected in the anterior hindbrain region or the posterior midbrain–anterior forebrain region, but there was retinoid activity in the craniofacial region where the optic eminences arise. A comparison of several day 11 mouse embryonic tissues has shown that the cerebral vesicles and mesencephalon have very low levels of activity whereas the spinal cord has very high levels (MCCAFFERY et al. 1992). The eye was also studied in detail and the ventral retina was shown to have a higher level of activity than the dorsal retina (see above).

The uniform levels of retinoids detected in the early spinal cord changes later during development such that there is a concentration of retinoid activity in the brachial and lumbar eminences of the CNS as assayed by the F9 reporter cells, exactly as found in the transgenic mice (COLBERT et al. 1993; MCCAFFERY and DRAGER 1994b). These "hot spots" of retinoic acid synthesis continue up to postnatal day 13 and are generated by local concentrations of a retinal dehydrogenase (COLBERT et al. 1993; MCCAFFERY and DRAGER 1994b).

In the cross-section of the spinal cord WAGNER et al. (1992) showed that the ventral floor plate activated the F9 reporter cells, but so too did the dorsal spinal cord in rat embryos. This accords with experiments on the synthesis of RA by the chick embryo spinal cord where it was shown that the rate of synthesis by the ventral floor plate was not very different from that of the dorsal spinal cord (WAGNER et al. 1990). However, studies in the mouse embryo have shown that at levels intermediate between the brachial and lumbar eminences there was an equivalent synthesis of RA in dorsal and ventral halves, but at the "hot spots" themselves there was a vast excess of RA generated in the ventral halves (MCCAFFERY and DRAGER 1994b). This is not the result that was obtained in the transgenic animals nor in the synthesis experiments in the chick embryo, in fact quite the opposite, and to reconcile the data these authors have suggested that the levels of RA are so high in the ventral half of the cord at the eminences that the reporter is turned off.

IV. RA Synthesising Enzymes

A crucial piece of evidence to support all of the above data would be to ask where are the enzymes that synthesise RA located in the CNS? Unfortunately we do not yet know enough about the full panoply of enzymes involved, but some exciting beginnings have been made.

Several enzymes have been described in the developing eye. Initially, an asymmetrically distributed protein, concentrated in the dorsal retina, was found in protein studies and this proved to be aldehyde dehydrogenase, AHD-2 (MCCAFFERY et al. 1991). The same enzyme is also exclusively located in the dorsal part of the developing eye of the chick embryo (GODBOUT et al. 1996). Surprisingly, on the F9 reporter cells the ventral retina of the mouse embryo generated more RA than the dorsal retina suggesting the presence of an additional enzyme in the ventral part. When enzyme inhibition studies were performed, certain inhibitors acted selectively on the dorsal or ventral retina enzymes at particular concentrations (MCCAFFERY et al. 1992) and it is now clear that there are several different aldehyde dehydrogenases in the retina. The dorsal enzyme is AHD-2, a basic enzyme, and there are three ventral enzymes V1, V2 and V3 which are acidic enzymes and whose relative activity varies with developmental age (MCCAFFERY et al. 1993). Similarly, the contribution to total RA synthesis of the retina from the dorsal and ventral halves shifts from a 5%–10% contribution from the dorsal half at the beginning of eye development to 100% post-natally. The aldehyde dehydrogenase which is responsible for generating the "hot spots" of RA synthesis in the embryonic mouse spinal cord is an acidic one (MCCAFFERY and DRAGER 1994b) and equates to the ventral V2 enzyme present in the retina. On the other hand, the aldehyde dehydrogenase which is responsible for the anteroposterior distribution of RA in the zebrafish trunk is equivalent to the enzyme in the dorsal retina, AHD-2 (MARSH-ARMSTRONG et al. 1995).

The expression of one of the ventral V2 enzyme, known as RALDH-2, has been studied in mouse development where it has been cloned (NIEDERREITHER et al. 1997). Surprisingly, in view of the distribution of endogenous RA in the developing CNS, there is no expression of RALDH-2 in the neurectoderm or CNS associated structures up to day 11.5 of development apart from the optic vesicle and otic vesicle. Then the two expression domains in the spinal cord at the brachial and lumbar eminences appeared (the "hot spots" of RA synthesis) and within these regions the motoneurons were labelled. In the facial and branchial arch regions the trigeminal ganglion and nerve seemed to be expressing RALDH-2 and later still (day 14.5–15.5) expression was also detected in the periocular mesenchyme. However, in the zymography assay this V2 enzyme can first be detected at neural plate stages on late day 7 of mouse embryogenesis (DRAGER and MCCAFFERY 1995) and its absence in the anterior part of the neural plate accords precisely with the distribution of endogenous RA measured above. So the fact that the mRNA cannot be detected until day 11.5 suggests this may not be the relevant enzyme.

Much later during development and extending post-natally, this enzyme is also expressed at high levels in the choroid plexus of the fourth ventricle in the developing rat brain (YAMAMOTO et al. 1996). It is suggested that the RA generated by this enzyme is responsible for the growth and differentiation of the cerebellum.

A final and fascinating location for the AHD-2 enzyme was found in the adult mouse brain, in the axons of the dopaminergic neurons of the striatum and substantia nigra (MCCAFFERY and DRAGER 1994a). This suggests not only an involvement of RA in the maintenance of axonal functioning in this system, but that a retinoic acid generating enzyme can be transported down an axon to generate RA and hence act at a considerable distance from the cell body. A further twist to this tale is that adults who ingest disulphiram (Antabuse) as a deterrent for alcoholics show brain distrubances such as Parkinsonism and catatonia (FISHER 1989; KRAUSS et al. 1991; LAPLANE et al. 1992).

The enzyme which precedes AHD-2 in the synthesis of retinoic acid from retinol is alcohol dehydrogenase and there are 5 classes of this enzyme, ADH, in humans. The mouse only has three of these classes, namely I, III and IV (ZGOMBIC-KNOGHT et al. 1995), and of these three only class IV is correlated with the above measurements of endogenous RA in the early embryo (ANG et al. 1996). This group also investigated the expression patterns of an aldehyde dehydrogenase enzyme ALDH-1 and found it to be present in very similar domains to ADH-IV thus suggesting that these two act in concert to locally synthesise RA from retinol via retinal (ANG and DUESTER 1997). However, there were some differences between the distribution of these enzymes and that of endogenous RA, notably in the cranial neuroepithelium which contains both of these enzymes, but has no detectable RA.

V. Which Cells in the Nervous System Synthesise RA?

The above detection of AHD-2 in the axons of dopaminergic neurons clearly suggests that neurons can make RA. However, an interesting alternative which has been previously hypothesised is that it is the glial cells which make the RA and the axons take it up for use by the neurons in the same way that some growth factors operate. This idea came from experiments on dissociated cultures of embryonic rat spinal cords (WUARIN et al. 1990). These authors found that in such cultures, both retinol and RA promotes the survival and differentiation of neurons. However, in pure neuronal cultures retinol has no effect, only RA is effective, suggesting that retinol is converted to RA by the glial cells. Using cortical neurons or astrocytes they then showed that when retinol was added to these cultures RA could only be detected in the medium from astrocytes, not in the medium from neurons. Similarly, in the cells of the retina, the glial cells known as Muller cells can convert retinol to RA and, again, it is suggested that the adjacent amacrine neurons which contain CRABP take up and use this locally synthesised RA (EDWARDS et al. 1992).

Another cell type found to contain relatively high levels of RALDH-2 is the ependyma derived choroid plexus (YAMAMOTO et al. 1996) and the neural crest-derived meninges (MCCAFFERY and DRAGER 1995). Perhaps, then, representatives of all the cell types found in the CNS can synthesise RA depending on their location and developmental age.

F. Binding Proteins and Receptors in the Developing CNS

The retinol binding protein CRBP I and the retinoic acid binding proteins CRABP I and CRABP II are cytoplasmic proteins involved in the uptake, sequestering and metabolism of their respective ligands, retinol and RA (see Chap. 4). They are present in the CNS with intriguing distributions which suggest specific functions for retinol and RA in the development of specific neuronal populations (see comparative expressions in Fig. 2C–E).

I. CRBP I

In the amphibian and chick embryo, CRBP I protein has been localised to a specialised region of the developing neural tube, the ventral floor plate, using an antibody raised against rat CRBP I (HUNTER et al. 1991; MADEN and HOLDER 1991; MADEN et al. 1989a). The floor plate is an important organising region consisting of radial glial cells which is responsible for establishing the dorsoventral pattern of neuronal specification and neurite outgrowth within the cord. In the developing mammalian CNS, however, the same antibody recognises the early differentiating motoneurons as they are specified (Maden et al. 1990 – Fig. 2C) and then the ventral roots of the spinal cord, suggesting a role for retinoids in motoneuron specification. The mRNA for CRBP I is expressed in this location in the mammalian embryo, as well as in the ependyma of the later spinal cord and the roof plate (Perez-Castro et al. 1989; Ruberte et al. 1993). Earlier in neural development, however, at the early somite stage of the mouse embryo CRBP I mRNA is more widely expressed in the neural tube from posterior levels up to the hindbrain, but expression is very low in the forebrain (Ruberte et al. 1991). Thus in mammals a widespread expression of CRBP I in the neural tube early on, becomes gradually restricted during development to motoneurons.

In the developing brain CRBP I is found in the lamina terminalis (Maden et al. 1990), the corpus striatum, hypothalamus, olfactory tubercle, mesencepahlic nucleus of the trigeminal, paraventricular nucleus of the thalamus, the Purkinje cell layer of the cerebellum, most motor nuclei and in the choroid plexus (Ruberte et al. 1993). CRBP I is also expressed in the developing eye both throughout the neural retina and the pigmented retina layers (Dolle et al. 1990; Perez-Castro et al. 1989) and is abundant in the inner ear and the anterior pituitary (Dolle et al. 1990).

In the face and branchial arches CRBP I is again widespread, being present in all the branchial arch mesenchyme and showing a somewhat complementary distribution to that of CRABP I in the facial derivatives (Dolle et al. 1990) – CRABP I is most abundant in the rostral part of the frontonasal mesenchyme and in the distal part of the mandibular arch mesenchyme whereas CRBP I is most abundant in the central regions of the facial mesenchyme and in the caudal half of the mandibular arch. A complementarity between CRBP I and CRABP I was also noted by Gustafson et al. (1993) in immunocytochemical studies, but with a rather different distribution of CRBP I to that observed in the above in situ hybridisation studies. Gustafson reported that CRBP I protein was present in the ectoderm of the frontonasal region and branchial arches and CRABP I protein was present in the underlying mesenchyme, suggesting a role for retinol/retinoic acid in ectodermal/mesenchymal interactions. Such a specific distribution of CRBP I in the ectoderm has not been observed in any other studies either with antibodies or in situ probes, and therefore the significance of this report remains unclear.

II. CRABP I

CRABP I has a more restricted distribution than CRBP I. CRABP I is present in a population of interneurons in the developing chick and mammalian neural tube known as the commissural neurons (Dencker et al. 1990; Maden and Holder 1991; Maden et al. 1989b, 1990; Momoi et al. 1989; Perez-Castro et al. 1989; Ruberte et al. 1992, 1993; Shiga et al. 1995; Vaessen et al. 1990) which are located in the mantle layer of the intermediate region of the neural tube and send their axons ventrally towards the ventral floor plate (Fig. 2D). The trajectory of the axons of these commissural neurons is brought about by the release of a chemoattractant from the ventral floor plate (PLACZEK et al. 1990). In the chick the presence of CRBP in the floor plate suggested to us that retinol might be taken up by the cells of the floor plate, metabolised to retinoic acid and released where it could form a dorsoventral gradient to act as a chemoattractant for the CRABP positive commissural axons (MADEN and HOLDER 1991, 1992). However, this chemoattractant has now been identified as a molecule called netrin (KENNEDY et al. 1994) and, in addition, there is no evidence that the ventral part of the neural tube generates RA at a higher rate than the dorsal part (see Sect. E.III). Indeed, the opposite seems to be the case. Nevertheless, we have shown in vitro that neurites from cultured chick neural tube neurons respond to a gradient of RA and turn and grow up the gradient (Maden and Jones, unpublished data).

In the developing brain CRABP I is not expressed in the forebrain, but is expressed in some neurons of the midbrain roof which will become the mesencephalic trigeminal nucleus (LEONARD et al. 1995; MADEN et al. 1990, 1991, 1992; DENCKER et al. 1990; PEREZ-CASTRO et al. 1989; RUBERTE et al. 1992; REBERTE et al. 1993; VAESSEN et al. 1989). In the developing hindbrain of mammalian embryos CRABP I shows an interesting rhombomere specific expression pattern. By day 9.5 of mouse development CRABP I protein and mRNA is not expressed in rhombomere 1, expressed lightly in rhombomere 2, not in rhombomere 3, then heavily expressed in rhombomeres 4, 5, 6 and 7 (LEONARD et al. 1995; MADEN et al. 1992b). Earlier in development, at day 8 (late headfold/early somite stages) when the CNS is just forming, there is a discrete band of CRABP I labelling which is just posterior to the preotic sulcus, adjacent to the early stripe of *Krox-20* expression in presumptive rhombomere 3. This band of CRABP I expression marks the presumptive posterior rhombomeres of the hindbrain from rhombomere 4 posteriorly (LEONARD et al. 1995; MADEN et al. 1992b). When these embryos are treated with a teratogenic dose of RA then the band of CRABP I expression is anteriorised and spreads strongly into the presumptive midbrain and forebrain region just as *Hox* genes do (see Sect. B.II). Presumably therefore RA via CRABP I plays some role in the specification of this posterior hindbrain region as this is the region which is missing in quail embryos deprived of RA altogether (see Sect. D.I).

This differential distribution of CRABP I in the hindbrain rhombomeres has proved of value in identifying individual rhombomeres in mutant mice, such as *kreisler* (MCKAY et al. 1994) and *Hoxa*-1 null mice (MARK et al. 1993). In the chick embryo this rhombomere specific expression throughout the thickness of the neuroepithelium is not found. Instead there is expression of CRABP I in individual reticulospinal neurons as they differentiate (MADEN et al. 1991). However, since these neurons differentiate in a rhombomere specific fashion, a rhombomeric sequence of CRABP I-positive neurons is seen during differentiation of the hindbrain – first rhombomere 4, then rhombomere 6 and posteriorly, then rhombomere 2 and anteriorly.

Because the neural crest also expresses CRABP I, the streams of neural crest which migrate from the various rhombomeres in the embryonic mouse hindbrain reflect the expression levels of CRABP I in the rhombomeres from which they originate. Thus the stream of crest from rhombomere 2 lightly expresses CRABP I and gives rise to the lightly expressing mesenchyme of the mandibular arch which it generates. The heavily expressing remaining streams give rise to heavily expressing crest and mesenchyme of the remaining branchial arches. Thus there is a differential expression of CRABP I protein within the branchial arches of the mouse embryo (Maden et al. 1992b) and the mesenchyme of the frontonasal mass and the maxillary process also express CRABP I strongly. The cranial nerves and ganglia which derive from the neural crest in the hindbrain also express CRABP I (Ruberte et al. 1991). In the chick embryo there is also a difference in relative levels of CRABP I between the branchial arches, but it is different from the mouse embryo. Whereas in the mouse embryo it is the mandibular arch (arch 1) which lightly expresses CRABP I, in the chick embryo, the mandibular arch is the strongest and it is the hyoid arch (arch 2) which lightly expresses it (Maden et al. 1991; Vaessen et al. 1990).

The neural crest migrating from the spinal part of the neural tube also expresses CRABP I and it thus serves as a very useful marker for studies of the migration routes and the differentiated products of this cell type (Fig. 2D) (Dencker et al. 1990; Leonard et al. 1995; Maden et al. 1989b, 1990, 1992b). These neural crest derivatives which all express CRABP I strongly are dorsal root ganglia, Schwann cells, sympathetic ganglia, enteric ganglia and, of course, the mesenchyme of the branchial arches as described above.

In other CNS derivatives such as the eye, whereas CRBP I was expressed throughout the retina and lens, CRABP I is only expressed in the centre of the neural retina and spreads out as differentiation of ganglion cells proceeds (McCaffery et al. 1993). The otic vesicle expresses CRABP I (Maden et al. 1991; Vaessen et al. 1990), perhaps in the presumptive organ of Corti (Dolle et al. 1990) which is known to express CRABP at later stages (Kelley et al. 1993).

In *Xenopus* a CRABP has been cloned which has properties of both CRABP I and CRABP II and whose size is twice that of its murine counterparts (Ho et al. 1994). During neural development stages this CRABP is present in an anterior domain *and* a posterior domain. The anterior domain consists of expression in the developing telencephalon, then a non-expressing diencephalon region, followed by a strongly expressing mesencephalon and rhombencephalon region. The posterior domain is in the tail bud and the cranial neural crest also express this xCRABP. In contrast, a different CRABP of normal size (15 kDa) has also been cloned in *Xenopus* by Dekker et al. (1994). It has a surprisingly similar expression pattern – an anterior (light expression in the forebrain and midbrain, strong expression in the hindbrain) and a posterior (tail bud) domain, as well as the cranial neural crest. RA treatment of embryos enhanced the anterior domain and repressed the posterior domain. Most interestingly, overexpression of this CRABP produced the typical RA induced defects – anterior abnormalities such as reduced or fused eyes, strongly reduced forebrain and midbrain, loss of hindbrain segmentation; and posterior defects such as kinked or shortened tails. Furthermore the expression of *Hoxb*-4 and *Hoxb*-8 was enhanced to the same degree as RA treatment, suggesting a role for this xCRABP in the control of AP patterning in the *Xenopus* CNS.

III. CRABP II

In the early mouse embryo CNS CRABP II transcripts are more widespread than those of CRABP I. As the neural tube forms at early somite stages when there is a small band of CRABP I expression in the presumptive hindbrain from rhombomere 4–6, CRABP II expression begins slightly further posteriorly and is not tightly localised to a band as its expression extends over the length of the embryo and gradually diminishes with no clear posterior boundary (LEONARD 1995; LYN and GIGUERE 1994; RUBERTE et al. 1992). Thereafter CRABP II also extends rostrally into the midbrain and forebrain and a rhombomere specific expression pattern develops as in the case of CRABP I. Rhombomeres 2, 4 and 6 are the most intensely labelled, rhombomere 5 the least intensely labelled and the remaining rhombomeres labelled to moderate levels. The midbrain expression virtually disappears whereas the telencephalic vesicles label intensely. The cranial neural crest is also CRABP II positive including the frontonasal mesenchyme and the crest which fills the branchial arches.

The expression of CRBAP II in the spinal part of the neural tube is also not as precisely localised as that of CRABP I as it shows expression throughout the neural tube early on in mouse CNS development and then, according to Ruberte et al. (RUBERTE et al. 1993), becomes somewhat localised to the presumptive motoneurons of the ventral cord as well as being present in newly formed neuroblasts in the dorsal cord. However, in the study of Leonard (LEONARD 1995), there was no expression in the motoneurons and instead there was a clear and strong localisation of CRABP II mRNA to the alar half of the spinal cord in a curious diagonal shape excluding the most dorsal part of the neural tube (Fig. 2E). CRABP II is also expressed by the trunk neural crest cells during their migration and differentiation and is thus found in the dorsal root ganglia. In the later developing brain CRABP II is expressed in the lateral walls of the diencephalon and the optic stalk, posterior mesencephalon, pontine nuclei, thalamic nuclei, geniculate body, the motor nuclei of the hindbrain and the choroid plexus (RUBERTE et al. 1992; RUBERTE et al. 1993).

Thus it can be seen that the binding proteins display both overlapping and unique (see particularly Fig. 2C–E) distributions during CNS development which may provide neuronal cells with differing concentrations of RA or specific retinoids or may allow them to respond to different concentrations of RA or specific retinoids.

IV. RARs and RXRs

In early CNS development of the mouse embryo, RARα is expressed throughout the neural tube and anteriorly up to the rhombomere 3/4 border, RARβ is expressed in a similar domain up to the rhombomere 6/7 border and RARγ is expressed in the open neural tube and not the closed neural tube (RUBERTE et al. 1991). In the chick embryo RARβ is similarly expressed, but the anterior border is one rhombomere more anterior, at the 5/6 boundary (SMITH 1994; SMITH and EICHELE 1991). Thus it seems that the early forebrain, midbrain and anterior hindbrain develops without any RARs which fits with the observation there is no endogenous RA in these regions at these stages (see Sect. C). However, it seems that there is a weak and ubiquitous expression of RXRα and RXRβ in mouse embryos (DOLLE et al. 1994), and therefore these two receptors may be expressed in the developing anterior CNS as part of this ubiquity. The fact that there is a boundary between RARβ expression in the closed neural tube and RARγ in the open neural tube suggests a role for these receptors and RA itself in the process of neurulation and it is therefore significant that the expression levels of these receptors is down-regulated in a mouse mutant, *curly tail*, which displays neural tube defects (CHEN et al. 1995).

Later in development (from day 11.5) RXRγ is co-expressed along with RARβ in two parts of the forebrain, the dorsal hypothalamus and the corpus striatum (DOLLE et al. 1994) and RARα is also expressed in the corpus striatum (RUBERTE et al. 1993). RXRγ is expressed in the organ of Corti and the spiral ganglion in the developing ear and in the neural retina of the eye (DOLLE et al. 1994). In the developing spinal cord RARβ becomes localised to a subset of the developing motoneurons (RUBERTE et al. 1993) along with the co-expression of RXRγ (DOLLE et al. 1994) and in the chick embryo RARβ is also expressed in the motoneurons (MUTO et al. 1991). As the cervical and lumbar enlargements develop (the "hot spots" of RA synthesis, see Sect. E), RARβ expression intensifies in these regions and is down-regulated in the intervening segments (COLBERT et al. 1995)

Concerning the neural crest and its derivatives, RARα is expressed in the migrating neural crest at a higher level than the other RARs and it is subsequently expressed in the anterior branchial arches of the mouse embryo (RUBERTE et al. 1991) and weakly and uniformly over the facial region (OSUMI-YAMASHITA et al. 1990). RARβ is expressed in the neural crest derived mesenchyme around the developing eyes and in the frontonasal mesenchyme and absent from the maxillary process, present in the caudal half of the mandibular arch and absent from the rest of the arches (RUBERTE et al. 1990). RARγ is expressed throughout the frontonasal and branchial arch mesenchyme which give rise to facial cartilages and bones (OSUMI-YAMASHITA et al. 1990; RUBERTE et al. 1990). In the chick embryo it is RXRγ which is a specific marker for neural crest and its subsequent differentiated products such as the sensory peripheral nervous system (ROWE et al. 1991). RARβ, as with the mouse embryo, is expressed by a sub-set of chick neural crest cells and in the anterior facial primordia particularly in the mesenchyme around the eyes. The border between expressing and non-expressing cells is between the maxillary and mandibular primordia (ROWE et al. 1992; SMITH 1994).

In the *Xenopus* embryo there is an interesting reciprocal distribution between two isoforms of RARγ with RARγ1 being expressed in the anterior end of the embryo including the brain and neural crest in the branchial arches whereas the RARγ2 isoform is expressed at the tail end of the embryo including the floor plate of the posterior spinal cord (PFEFFER and DE ROBERTIS 1994). The levels of most of the other receptors which have been cloned in *Xenopus* drop sharply before gastrulation and therefore would not be expressed in the CNS (BLUMBERG et al. 1992).

In zebrafish embryos, RARα is strongly expressed in the neuroepithelium of the hindbrain with an anterior border corresponding to the region of rhombomere 6/7. RARγ is expressed up to the midbrain/hindbrain border in the ventral part of the neural tube with the highest transcript level being in the posterior hindbrain (JOORE et al. 1994). This receptor is also expressed in the head mesenchyme which is presumably of neural crest origin. Interestingly, in the zebrafish there are not three RXRs as might be expected, but five: α, β, γ and two novel ones, δ and ε (JONES et al. 1995). The two novel RXRs do not bind 9-*cis*-RA, although we do not yet know anything about their expression patterns.

V. Knockouts of Binding Proteins and Receptors

In the light of the binding protein and receptor distribution described above we might suggest some clear functions for these molecules, for example, CRABP I in the control of commissural neurite outgrowth, RARβ in motoneuron development, RARγ to function in neural tube closure, RXRγ in the eye, ear and individual forebrain regions etc. Amazingly, when they were knocked out by homologous recombination in mouse embryos the resulting embryos were, with one exception, perfectly normal with regard to the nervous system (for review see KASTNER et al. 1995). Thus CRABP I null mutants, CRABP II null mutants and CRABP I/CRABP II double null mutants are not only normal in their CNS, but in the rest of the body too except for an extra post-axial digit (FAWCETT et al. 1995; GORRY et al. 1994; LAMPRON et al. 1995). Similarly for RARα, RARγ and RXRβ null mutants (for review see KASTNER et al. 1995). Only the RARβ and RXRα knockouts have any defects at all related to the CNS and these are primarily ocular defects. One report reveals that one RARβ null mouse embryo had fused proximal portions of the glossopharyngeal and vagus nerves (cranial nerves IX and X) (LUO et al.

1995), but this defect did not appear in another study (GHYSELINCK et al. 1997). However, both reports showed ocular defects in the form of an abnormal retrolenticular mass of pigmented tissue adherent to the lens, a folded retina and cataracts. The RXRα null mutants also had ocular defects in which the ventral retina was shortened and the anterior segment of the eye was malformed (abnormal cornea, unfused eyelids, abnormal lens, abnormal conjunctiva) (KASTNER et al. 1994). Widespread eye defects including those of the anterior eye as just described, malformation of the mesenchymal structures such as the sclera, shortened ventral retina, retinal dysplasia, microphthalmia and post-natal retinal degeneration are all seen when double RAR null mutants are created such as RARα/γ or RARβ/γ (GHYSELINCK et al. 1997; GRONDONA et al. 1996; LONHES et al. 1994). Further evidence for a role for RA and RAR/RXRs in lens development comes from the identification of a RARE which binds RAR/RXR heterodimers in the lens specific gF-crystallin promoter (TINI et al. 1993). Interestingly, the ventral retina thus seems to be a particularly RA sensitive region – it can be missing in certain knockout combinations, as described above RA excess duplicates the ventral retina and RA deficiency prevents its development.

The only other CNS defect seen in RARα/γ double mutants was a persistent opening of the hindbrain (LOHNES et al. 1994). This resulted in the abnormal folding of the forebrain ventricles and the degeneration of the hindbrain neuroepithelium, both as secondary consequences of the loss of ventricular hydrostatic pressure. Interestingly the absence of the motor nucleus of the abducens nerve, derived from rhombomeres 5 and 6 was also noted, but it was not clear whether this was a primary or secondary defect.

Neural crest abnormalities are also seen double null mutants: periocular mesenchymal abnormalities in RARβ/γ mutants (GHYSELINCK et al. 1997; GRONDONA et al. 1996); abnormalities of the craniofacial and branchial arch skeleton in all combination mutants (GHYSELINCK et al. 1997; LOHNES et al. 1994); abnormalities of the thyroid, parathyroid and thymus glands (derived from the hindbrain neural crest) in RARα/γ and RARα/β mutants (MENDELSOHN et al. 1994); and some heart defects (aorticopulmonary septation and aortic arch abnormalities) due to defective cardiac neural crest in RARα/γ and RARα/β mutants (MENDELSOHN et al. 1994). These defects thus confirm the role of the RARs in neural crest migration and/or survival.

VI. Disruption of Function

So we are left with the conclusion from the knockouts that neither the CRABPs, RARs nor the RXRs are involved in CNS development except the eye, which is a rather baffling conclusion in the light of all the above data on excess RA and RA deficiency. However, clear patterning functions do emerge from experiments where, rather than getting rid of the receptors, they are overexpressed as functioning receptors or overexpressed as dominant negatives. Two such experiments have been performed, both in *Xenopus* embryos, and they have revealed very precise effects on CNS patterning in the anteroposterior axis and on primary neurogenesis.

In the first experiment, Blumberg et al. (1997) injected either a dominant negative or a constitutively active RARα1 receptor into the *Xenopus* embryo to assess the effects of increasing or decreasing retinoid signalling on AP patterning. AP patterning was assayed by the expression patterns of various marker genes. The dominant negative receptor (inhibiting signalling) led to the anteriorisation of the embryos – expansion of forebrain and anterior midbrain, shorter tails, reduced overall length and an expansion of forebrain marker gene domains. The genes used as markers were *Otx2*, which expanded its domain, *engrailed-2* and *Krox-20*, which both underwent a posterior shift, and *Xlim-1,* which being a posterior marker underwent a down-regulation. Thus the embryo had been anteriorised. Conversely, the constitutively active receptor phenocopied RA treatment and led to a decrease in the *Otx2* domain (reduced forebrain size) and a decrease and anterior shift in the *engrailed-2* and *Krox-20* domains. These experiments conform precisely to the model whereby RA is a posteriorising signal for the CNS during development.

The second experiment (Sharpe and Goldstone 1997) relates to neuronal differentiation rather than patterning. Injection of a dominant negative RARα construct prevented the development of primary sensory neurons in the dorsal spinal cord and the embryos could no longer respond to a sensory stimulus. This result was also seen in the experiments of Blumberg et al. (1997). Conversely, injection of the normal RARα along with RXRβ (the active heterodimer) resulted in the formation of ectopic primary neurons lying adjacent to the neural tube. These experiments thus confirm a role for RA both in AP patterning and in the formation of particular groups of primary neurons.

VII. Receptor Selective Agonists

A final technique for investigating the role of individual receptors in various developmental processes including CNS development is to use receptor selective agonists. Elmazar et al. (1996), for example, treated mouse embryos at two different stages with a RARα ligand (CD 336), a RARβ ligand (CD 2019) or a RARγ ligand (CD437). However, no abnormalities of the CNS were reported apart from exencephaly and facial defects such as micrognathia or ear defects which appeared with the RARα ligand.

In *Xenopus* and zebrafish, by contrast, receptor selective agonists do produce characteristic effects on the CNS. In *Xenopus*, ligands selective for both the RARs and RXRs cause the typical posteriorisation defects of the CNS (Fig. 1, C) and the induction of characteristic genes such as *Hoxa*-1 and *Xlim*-1 (Minucci et al. 1996). These authors further found that only the RAR selective ligands caused abnormalities in zebrafish embryos as embryos were unaffected by treatment with two RXR selective ligands, SR 11237 or Ro25-

6603. However, Zhang et al. (1996) reported that the RXR selective ligand SR11217 does cause the posteriorisation CNS defect characterised as a loss of forebrain and midbrain structures, and thus there is clearly some contradiction in these results.

Instead of activating selective receptors with agonists the opposite strategy can also be used – inactivate selective receptors with antagonists to prevent RA signalling. This has been performed in *Xenopus* and chick embryos using Ro41–5253, an antagonist of the RARα receptor. Whereas RA treatment caused an anterior expansion of the expression of *Hoxb*-7 and typical anterior CNS defects, treatment with Ro41–5253 caused a reduction and a posterior shift of *Hoxb*-7 expression, but anterior defects were still seen – loss of eyes, disorganisation of branchial arches and loss of definition of rhombomeres. Similarly, both RA and Ro41–5253 treatment caused a drastic reduction in cranial neural crest cells. So although the gene expression data behaved as expected, the teratology did not seem to follow the same rules.

G. Conclusions

I have reviewed the data which reveal the crucial role that retinoids play in the development of the CNS in vertebrate embryos. This evidence includes the induction of neural differentiation of neurite outgrowth in pluripotent cells and cultured neural cells; the effects of excess and deficiency of RA on the developing CNS; the detection of endogenous RA and RA synthesising enzymes; the presence of retinoid binding proteins and receptors in the CNS and the effect of interfering with their signalling properties. From this data precise roles for retinoids can be proposed both in the AP patterning of the early neural plate and later in neurite outgrowth. Interestingly, recent data has pointed to a role for retinoids in the maintenance of neurons in the adult CNS and this will surely have an impact in future considerations of neurodegenerative diseases such as Parkinson's disease or motoenuron disease.

References

Abramovici A, Kam J, Liban E, Barishak RY (1978) Incipient histopathological lesions in citral-induced microphthalmos in chick embryos. Dev Neurosci 1:177–185

Alexandre D, Clarke JDW, Oxtoby E, Yan Y-L, Jowett T, Holder N (1996) Ectopic expression of Hoxa-1 in the zebrafish alters the fate of the mandibular arch neural crest and phenocopies a retinoic acid-induced phenotype. Development 122:735–746

Alles AJ, Sulik KK (1990) Retinoic acid-induced spina bifida: evidence for a pathogenic mechanism. Development 108:73–81

Amatruda TT, Sidell N, Ranyard J, Koeffler HP (1985) Retinoic acid treatment of human neuroblastoma cells is associated with decreased N-myc expression. Biochem Biophys Res Comm 126:1189–1195

Anchan RM, Drake DP, Haines CF, Gerwe EA, LaMantia A-S (1997) Disruption of local retinoid-mediated gene expression accompanies abnormal development in the mammalian olfactory pathway. J Comp Neurol 379:171–184

Andrews PA (1984) Retinoic acid induces neuronal differentiation of a cloned human embryonal carcinoma cell line in vitro. Dev Biol 103:28–293

Andrews PW, Gonczol E, Plotkin SA, Dignazio M, Oosterhuis JW (1986) Differentiation of TERA-2 human embryonal carcinoma cells into neurons and HCMV permissive cells. Induction by agents other than retinoic acid. Differentiation 31:119–126

Ang HL, Deltour L, Hayamizu TF, Zgombic-Knight M, Duester G (1996) Retinoic acid synthesis in mouse embryos during gastrulation and craniofacial development linked to class IV alcohol dehydrogenase gene expression. J Biol Chem 271:9526–9534

Ang HL, Duester G (1997) Initiation of retinoid signalling in primitive streak mouse embryos: spatiotemporal expression patterns of receptors and metabolic enzymes for ligand synthesis. Dev Dynam 208:536–543

Ang S-L, Jin O, Rhinn M, Daigle N, Stevenson L, Rossant J (1996) A targeted mouse OTX2 mutation leads to severe defects in gastrulation and formation of axial mesoderm and to deletion of rostral brain. Development 122:243–252

Avantaggiato V, Acampora D, Tuorto F, Simeone A (1996) Retinoic acid induces stage-specific repatterning of the rostral central nervous system. Dev Biol 175:347–357

Bain G, Kitchens D, Yao M, Huettner JE, Gottlieb DI (1995) Embryonic stem cells express neuronal properties in vitro. Dev Biol 168:342–357

Bally-Cuif L, Gulisano M, Broccoli V, Boncinelli E (1995) c-otx2 is expressed in two different phases of gastrulation and is sensitive to retinoic acid treatment in chick embryos. Mech of Dev 49:49–63

Baroffio A, Dupin E, Le Douarin NM (1991) Common precursors for neural crest and mesectodermal derivatives in the cephalic neural crest. Development 112:301–305

Baroffio A, Dupin E, LeDouarin NM (1988) Clone-forming ability and differentiation potential of migratory neural crest cells. Proc Natl Acad Sci USA 85:5325–5329

Barres BA, Lazar MA, Raff MC (1994) A novel role for thyroid hormone, glucocroticoids and retinoic acid in timing oligodendrocyte development. Development 120:1097–1108

Bavik C, Ward SJ, Chambon P (1996) Developmental abnormalities in cultured mouse embryos deprived of retinoic acid by inhibition of yolk-sac retinol binding protein synthesis. Proc Natl Acad Sci USA 93:3110–3114

Berrard S, Faucon Biguet N, Houhou L, Lamouroux A, Mallet J (1993) Retinoic acid induces cholinergic differentiation of cultured newborn rat sympathetic neurons. J Neurosci Res 35:382–389

Blumberg B, Bolado J, Moreno TA, Kintner C, Evans RM, Papalopulu N (1997) An essential role for retinoid signalling in anteroposterior neural patterning. Development 124:373–379

Blumberg B, Mangelsdorf DJ, Dyck JA, Bittner DA, Evans RM, De Robertis EM (1992) Multiple retinoid-responsive receptors in a single cell: families of retinoid "X" receptors and retinoic acid receptors in the Xenopus egg. Proc Natl Acad Sci USA 89:2321–2325

Casper D, Davies P (1989) Stimulation of choline acetyltransferase activity by retinoic acid and sodium butyrate in a cultured human neuroblastoma. Brain Res 478:74–84

Chen W-H, Morriss-Kay GM, Copp AJ (1995) Genesis and prevention of spinal neural tube defects in the curly tail mutant mouse: involvement of retinoic acid and its nuclear receptors RAR-β and RAR-γ. Development 121:681–691

Cho KWY, De Robertis EM (1990) Differential activation of Xenopus homeobox genes by mesoderm-inducing growth factors and retinoic acid. Genes Dev 4:1910–1916

Cockshut AM, Jonet L, Jeanny J-C, Vigny M, Raulais D (1994) Retinoic acid induced heparin-binding protein expression and localization during gastrulation, neurulation, and organogenesis. Dev Dynam 200:198–211
Cohlan SQ (1953) Excessive intake of vitamin A as a cause of congenital abnormalities in the rat. Science 117:535–536
Colbert MC, Linney E, LaMantia AS (1993) Local sources of retinoic acid coincide with retinoid-mediated transgene activity during embryonic development. Proc Natl Acad Sci USA 90:6572–6576
Colbert MC, Rubin WW, Linney E, LaMantia E-S (1995) Retinoid signalling and the generation of regional and cellular diversity in the embryonic mouse spinal cord. Dev Dynam 204:1–12
Conlon RA, Rossant J (1992) Exogenous retinoic acid rapidly induces anterior ectopic expression of murine Hox-2 genes in vivo. Development 116:357–368
Connor MJ, Smit MH (1987) Terminal-group oxidation of retinol by mouse epidermis. Biochem J 244:489–492
Cunningham ML, MacAuley A, Mirkes PE (1994) From gastrulation to neurulation: transition in retinoic acid sensitivity identifies distinct stages of neural patterning in the rat. Dev Dynam 200:227–241
Dekker E-J, Pannese M, Houtzager E, Boncinelli E, Durston A (1992) Colinearity in the Xenopus laevis Hox-2 complex. Mech of Dev 40:3–12
Dekker E-J, Vaessen M-J, van den Berg C, Timmermans A, Godsave S, Holling T, Niewkoop P, van Kessel AG, Durston AD (1994) Overexpression of a cellular retinoic acid binding protein (xCRABP) causes anteroposterior defects in developing Xenopus embryos. Development 120:973–985
Dencker L, Annerwall E, Busch C, Eriksson U (1990) Localization of specific retinoid-binding sites and expression of cellular retinoic acid-binding protein (CRABP) in the early mouse embryo. Development 110:343–352
Dersch H, Zile MH (1993) Induction of normal cardiovascular development in the vitamin A-deprived quail embryo by natural retinoids. Dev Biol 160:424–433
Deschamps J, de Laaf R, Joosen L, Meijlink F, Destree O (1987) Abundant expression of homeobox genes in mouse embryonal carcinoma cells correlates with chemically induced differentiation. Proc Natl Acad Sci USA 84:1304–1308
Deschamps J, de Laaf R, Verrijzer P, de Gouw M, Destree O, Meijlink F (1987) The mouse Hox2.3 homeobox-containing gene: regulation in differentiating pluripotent stem cells and expression pattern in embryos. Differentiation 35:21–30
Dickman ED, Thaller C, Smith SM (1997) Temporally-regulated retinoic acid depletion produces specific neural crest, ocular and nervous system defects. Development 124:3111–3121
Die-Smulders CEM, Sturkenboom MCJM, Veraart J, van Katwijk C, Sastrowijoto P, van der Linden E (1995) Severe limb defects and craniofacial anomalies in a fetus conceived during acitretin therapy. Teratology 52:215–219
Dinsmore JH, Solomon F (1991) Inhibition of MAP2 expression affects both morphological and cell division phenotypes of neuronal differentiation. Cell 64:817–826
Dolle P, Fraulob V, Kastner P, Chambon P (1994) Developmental expression of murine retinoid X receptor (RXR) genes. Mech of Dev 45:91–104
Dolle P, Ruberte E, Leroy P, Morriss-Kay G, Chambon P (1990) Retinoic acid receptors and cellular retinoid binding proteins. I. A systematic study of their differential pattern of transcription during mouse organogenesis. Development 110:1133–1151
Drager UC, McCaffery P (1995) Retinoic acid synthesis in the developing spinal cord. In: H. Weiner, et al (eds) Enzymology and Molecular Biology of Carbonyl Metabolism. Plenum, New York, p 185
Dupin E, LeDouarin NM (1995) Retinoic acid promotes the differentiation of adrenergic cells and melanocytes in quail neural crest cultures. Dev Biol 168:529–548

Durston AJ, Timmermans JPM, Hage WJ, Hendricks HFJ, de Vries NJ, Heideveld M, Nieuwkoop PD (1989) Retinoic acid causes an anteroposterior transformation in the developing nervous system. Nature 340:140–144

Edwards MKS, McBurney MW (1983) The concentration of retinoic acid determines the differentiated cell types formed by a teratocarcinoma cell line. Dev Biol 98:187–191

Edwards RB, Adler AJ, Dev S, Claycomb RC (1992) Synthesis of retinoic acid from retinol by cultured rabbit Muller cells. Exp Eye Res 54:481–490

Elmazar MMA, Reichert U, Shroot B, Nau H (1996) Pattern of retinoid-induced teratogenic effects: possible relationship with relative selectivity for nuclear retinoid receptors RARα, RARβ, and RARγ. Teratology 53:158–167

Fawcett D, Pasceri P, Fraser R, Colbert M, Rossant J, Giguere V (1995) Postaxial polydactyly in forelimbs of CRABP-II mutant mice. Development 121:671–679

Finley MFA, Kulkarni N, Huettner JE (1996) Synapse formation and establishment of neuronal polarity by P19 embryonic carcinoma cells. J Neurosci 16:1056–1065

Fischer I, Shea TB, Sapirstein VS, Kosik KS (1985) Expression and distribution of microtubule-associated protein 2 (MAP2) in neuroblastoma and primary neuronal cells. Devl. Brain Res 25:99–109

Fisher CM (1989) "Catatonia" due to disulphiram toxicity. Arch Neurol 46:798–804

Gale E, Prince V, Lumsden A, Clarke J, Holder N, Maden M (1996) Late effects of retinoic acid on neural crest and aspects of rhombomere identity. Development 122:783–793

Ghyselinck NB, Dupe V, Dierich A, Massaddeq N, Garnier J-M, Rochette-Egly C, Chambon P, Mark M (1997) Role of the retinoic acid receptor beta (RARβ) during mouse development. Int J Dev Biol 41:425–447

Godbout R, Packer M, Poppema S, Dabbath L (1996) Localization of cytosolic aldehyde dehydrogenase in the developing chick retina: in situ hydridisation and immunohistochemical analyses. Dev Dynam 205:319–331

Gorry P, Lufkin T, Dierich A, Rochette-Egly C, Decimo D, Dolle P, Mark M, Durand B, Chambon P (1994) The cellular retinoic acid binding protein I is dispensable. Proc Natl Acad Sci USA 91:9032–9036

Goulding MD, Chalepakis G, Deutsch U, Erselius JR, Gruss P (1991) Pax-3, a novel murine DNA binding protein expressed during early neurogenesis. EMBO J 10:1135–1147

Grondona JM, Kastner P, Gansmuller A, Decimo D, Chambon P, Mark M (1996) Retinal dysplasia and degeneration in RARβ2/RARγ2 compound mutant mice. Development 122:2173–2188

Gustafson A-L, Dencker L, Eriksson U (1993) Non-overlapping expression of CRBP I and CRABP I during pattern formation of limbs and craniofacial structures in the early mouse embryo. Development 117:451–460

Hale F (1933) Pigs born without eye balls. J Hered 24:105–106

Haskell BE, Stach RW, Werrbach-Perez K, Perez-Polo JR (1987) Effect of retinoic acid in nerve growth factor receptors. Cell Tissue Res 247:67–73

Heine UI, Roberts AB, Munoz EF, Roche NS, Sporn MB (1985) Effects of retinoid deficiency on the development of the heart and vascular system of the quail embryo. Virchow's Arch B Cell Pathol 50:135–152

Hendry IA, Belford DA (1991) Retinoic acid potentiates the neurotrophic but not the mitogenic action of the class 1 heparin-binding growth factor (HBGF-1). Brain Res 542:29–34

Henion PD, Weston JA (1994) Retinoic acid selectively promotes the survival and proliferation of neurogenic precursors in cultured neural crest populations. Dev Biol 161:243–250

Hill J, Clarke JDW, Vargesson N, Jowett T, Holder N (1995) Exogenous retinoic acid causes specific alterations in the development of the midbrain and hindbrain of the zebrafish embryo including positional respecification of the Mauthner neuron. Mech of Dev 50:3–16

Ho L, Mercola M, Gudas LJ (1994) Xenopus laevis cellular retinoic acid-binding protein: temporal and spatial expression pattern during early embryogenesis. Mech of Dev 47:53–64

Holder N, Hill J (1991) Retinoic acid modifies development of the midbrain-hindbrain border and affects cranial ganglion formation in zebrafish embryos. Development 113:1159–1170

Horton C, Maden M (1995) Endogenous distribution of retinoids during normal development and teratogenesis in the mouse embryo. Dev Dynam 202:312–323

Hunter K, Maden M, Summerbell D, Eriksson U, Holder N (1991) Retinoic acid stimulates neurite outgrowth in the amphibian spinal cord. Proc Natl Acad Sci USA 80:5525–5529

Hyatt GA, Schmitt EA, Marsh-Armstrong NR, Dowling JE (1992) Retinoic acid-induced duplication of the zebrafish retina. Proc Natl Acad Sci USA 89:8293–8297

Ido A, Miura Y, Tamaoki T (1994) Activation of ATBF1, a multiple-homeodomain zinc-finger gene, during neuronal differentiation of murine embryonal carcinoma cells. Dev Biol 163:184–187

Johnson JE, Zimmerman K, Saito T, Anderson DJ (1992) Induction and repression of mammalian achaete-scute homologue (MASH) gene expression during neuronal differentiation of P19 embryonal carcinoma cells. Development 114:75–87

Jones BB, Ohno CK, Allenby G, Boffa MB, Levin AA, Grippo JF, Petkovich M (1995) New retinoid X receptor subtypes in zebra fish (Danio rerio) differentially modulate transcription and do not bind 9-*cis* retinoic acid. Mol Cell Biol 15:5226–5234

Jones-Villeneuve EMV, McBurney MW, Rogers KA, Kalnins VI (1982) Retinoic acid induces embryonal carcinoma cells to differentiate into neurons and glial cells. J Cell Biol 94:253–262

Joore J, van der Lans GBLJ, Lanser PH, Vervaart JMA, Zivkovic D, Speksnijder JE, Kruijer W (1994) Effects of retinoic acid on the expression of retinoic acid receptors during zebrafish embryogenesis. Mech of Dev 46:137–150

Kadomatsu K, Huang R-P, Suganuma T, Murata F, Muramatsu T (1990) A retinoic acid responsive gene MK found in the teratocarcinoma system is expressed spatially and temporally controlled manner during mouse embryogenesis. J Cell Biol 110:607–616

Kalter H, Warkany J (1959) Experimental production of congenital malformations in mammals by metabolic procedure. Physiol Rev 39:69–115

Kastner P, Grondona JM, Mark M, Gansmuller A, LeMeur M, Decimo D, Vonesch J-L, Dolle P, Chambon P (1994) Genetic analysis of RXRα developmental function: convergence of RXR and RAR signalling pathways in heart and eye morphogenesis. Cell 78:987–1003

Kastner P, Mark M, Chambon P (1995) Nonsteroid nuclear receptors: what are genetic studies telling us about their role in real life? Cell 83:859–869

Kelley MW, Turner JK, Reh TA (1994) Retinoic acid promotes differentiation of photoreceptors in vitro. Development 120:2091–2102

Kelley MW, Xu X-M, Wagner MA, Warchol ME, Corwin JT (1993) The developing organ of Corti contains retinoic acid and forms supernumerary hair cells in response to exogenous retinoic acid in culture. Development 119:1041–1053

Kennedy TE, Serafini T, de la Torre J, Tessier-Lavigne M (1994) Netrins are diffusable chemotropic factors for commissural axons in the embryonic spinal cord. Cell 78:425–435

Kessel M (1993) Reversal of axonal pathways from rhombomere 3 correlates with extra Hox expression domains. Neuron 10:379–393

Krauss JK, Mohandjer M, Wakhloo AK, Mundinger F (1991) Dystonia and akinesia due to pallidoputaminal lesions after disulphiram treatment. Movement Disorders 6:166–170

Kuff EI, Fewell JW (1980) Induction of neural-like andacetylcholinesterase activity in cultures of F9 teratocarcinoma cells treated with retinoic acid and dibutyryl cyclic adenosine monophosphate. Dev Biol 77:103–115

LaMantia AS, Colbert MC, Linney E (1993) Retinoic acid induction and regional differentiation prefigure olfactory pathway formation in the mammalian forebrain. Neuron 10:1035–1048

Lammer EJ, Armstrong DL (1992) Malformations of hindbrain structures among humans exposed to isotretinoin (13-*cis*-retinoic acid) during early embryogenesis. In: Morriss-Kay G (ed) Retinoids in normal and development and teratogenesis, Oxford University Press, Oxford, p 281

Lammer EJ, Chen DT, Hoar RM, Agnish AD, Benke PJ, Braun JT, Curry CJ, Fernhoff PM, Grix AW, Lott IT, Richard JM, Sun SC (1985) Retinoic acid embryopathy. New Eng. J Med 313:837–841

Lampron C, Rochette-Egly C, Gorry P, Dolle P, Mark M, Lufkin T, LeMeur M, Chambon P (1995) Mice deficient in cellular retinoic acid binding protein II (CRABP II) or in both CRABP I and CRABPII are essentially normal. Development 121:539–548

Langman J, Welch GW (1967) Effect of vitamin A on development of the central nervous system. J Comp Neurol 128:1–16

Langston AW, Gudas LJ (1992) Identification of a retinoic acid responsive enhancer 3′ of the murine homeobox gene Hox-1.6. Mech of Dev 38:217–228

Laplane D, Attal N, Sauron B, de Billy A, Dubois B (1992) Lesions of basal ganglia due to disulphiram neurotoxicity. J Neurol Neurosurg Psychiatry 55:925–929

Le Douarin NM (1982) The NEURAL Crest. Cambridge University Press, Cambridge

Lee YM, Osumi-Yamashita N, Ninomiya Y, Moon CK, Eriksson U, Eto K (1995) Retinoic acid stage-dependently alters the migration pattern and identity of hindbrain neural crest cells. Development 121:825–837

Lefebvre PP, Malgrange B, Staecker H, Moonen G, van de Water TR (1993) Retinoic acid stimulates regeneration of mammalian auditory hair cells. Science 260:692–695

Leonard L, Horton C, Maden M, Pizzey JA (1995) Anteriorization of CRABP-I expression by retinoic acid in the developing mouse central nervous system and its relationship to teratogenesis. Dev Biol 168:514–528

Leonard LA (1995) Cellular retinoic acid binding proteins in normal vertebrate development and retinoic acid-induced teratogenesis. PhD Thesis, University of London

Leroy P, De Robertis EM (1992) Effects of lithium chloride and retinoic acid on the expression of genes from the Xenopus Hox 2 complex. Dev Dynam. 194:21–32

Levine JM, Flynn P (1986) Cell surface changes accompanying the neural differentiation of an embryonal carcinoma cell line. J Neurosci 6:374–3384

Liesi P, Rechardt L, Wartiovaara J (1983) Nerve growth factor induced adrenergic neuronal differentiation in F9 teratocarcinoma cells. Nature 306:265–267

Lohnes D, Mark M, Mendelsohn C, Dolle P, Dierich A, Gorry P, Gansmuller A, Chambon P (1994) Function of the retinoic acid receptors (RARs) during development (I) Craniofacial and skeletal abnormalities in RAR double mutants. Development 120:2723–2748

Lopez SL, Carrasco AE (1992) Retinoic acid induces changes in the localization of homeobox proteins in the antero-posterior axis of Xenopus laevis embryos. Mech of Dev 36:153–164

Lopez SL, Dono R, Zeller R, Carrasco AE (1995) Differential effects of retinoic acid and a retinoid antagonist on the spatial distribution of the homeoprotein Hoxb-7 in vertebrate embryos. Dev Dynam 204:457–471

Luo J, Pasceri P, Conlon RA, Rossant J, Giguere V (1995) Mice lacking all isoforms of retinoic acid receptor β develop normally and are susceptible to the teratogenic effects of retinoic acid. Mech of Dev 53:61–71

Lyn S, Giguere V (1994) Localization of CRABP I and CRABP II mRNA in the early mouse embryo by whole-mount in situ hybridisation: implications for teratogenesis and neural development. Dev Dynam 199:280–291

Lynch SA, Brugge JS, Levine JM (1986) Induction of altered c-src product during neuronal differentiation of embryonal carcinoma cells. Science 234:873–876

MacPherson PA, Jones S, Pawson PA, Marshall KC, McBurney MW (1997) P19 cells differentiate into glutaminergic and glutamate-responsive neurons in vitro. Neurosci 80:487–499
Maden M, Holder N (1991) The involvement of retinoic acid in the development of the vertebrate central nervous system. Development (Suppl) 2:87–94
Maden M, Holder N (1992) Retinoic acid and development of the central nervous system. BioEssays 14:431–438
Maden M, Ong DE, Summerbell D, Chytil F (1989a) The role of retinoid-binding proteins in the generation of pattern in the developing limb, the regenerating limb and the nervous system. Development (Suppl):109–119
Maden M, Ong DE, Summerbell D, Chytil F, Hirst EA (1989b) Cellular retinoic acid-binding protein and the role of retinoic acid in the development of the chick embryo. Dev Biol 135:124–132
Maden M, Ong DE, Chytil F (1990) Retinoid-binding protein distribution in the developing mammalian nervous system. Development 109:75–80
Maden M, Hunt P, Eriksson U, Kuriowa A, Krumlauf R, Summerbell D (1991) Retinoic acid-binding protein, rhombomeres and the neural crest. Development 111:35–44
Maden M, Gale E, Horton C, Smith JC (1992a) Retinoid-binding proteins in the developing vertebrate nervous system. In: Morriss-Kay G (ed) Retinoids in Normal Development and Teratogenesis. Oxford University Press Oxford p 119
Maden M, Horton C, Graham A, Leonard L, Pizzey J, Siegenthaler G, Lumsden A, Eriksson U (1992b) Domains of cellular retinoic acid-binding protein I (CRABP I) expression in the hindbrain and neural crest of the mouse embryo. Mech of Dev 37:13–23
Maden M, Gale E, Kostetskii I, Zile M (1996) Vitamin A-deficient quail embryos have half a hindbrain and other neural defects. Current Biol 6:417–426
Maden M, Graham A, Gale E, Rollinson C, Zile M (1997) Positional apoptosis during vertebrate CNS development in the absence of endogenous retinoids. Development 124:2799–2805
Maden M, Gale E, Zile M (1998) The role of vitamin A in the development of the central nervous system. J Nutr 128:471S–475S
Manns M, Fritzsch B (1992) Retinoic acid affects the organization of reticulospinal neurons in developing Xenopus. Neurosci Let 139:253–256
Mark M, Lufkin T, Vonesch J-L, Ruberte E, Olivo J-C, Dolle P, Gorry P, Lumsden A, Chambon P (1993) Two rhombomeres are altered in Hoxa-1 mutant mice. Development 119:319–338
Marsh-Armstrong N, McCaffery P, Gilbert W, Dowling JE, Drager UC (1994) Retinoic acid is necessary for development of the ventral retina in zebrafish. Proc Natl Acad Sci USA 91:7286–7290
Marsh-Armstrong N, McCaffery P, Hyatt G, Alonso L, Dowling JE, Gilbert W, Drager UC (1995) Retinoic acid in the anteroposterior patterning of the zebrafish trunk. Roux's Arch. Dev Biol 205:103–113
Marshall H, Studer M, Popperl H, Aparicio S, Kuriowa A, Brenner S, Krumlauf R (1994) A conserved retinoic acid response element required for early expression of the homeobox gene Hoxb-1. Nature 370:567–571
Marshall H, Nonchev S, Sham MH, Muchamore I, Lumsden A, Krumlauf R (1992) Retinoic acid alters hindbrain Hox code and induces transformation of rhombomeres 2/3 into a 4/5 identity. Nature 360:737–741
Matsuoka I, Mizuno N, Kurihara K (1989) Cholinergic differentiation of clonal rat pheochromocytoma cells (PC12) induced by retinoic acid: increase of choline acteyltransferase activity and decrease of tyrosine hydroxylase. Brain Res 502:53–60
McBurney MW, Jones-Villeneuve EMV, Edwards MKS, Anderson PJ (1982) Control of muscle and neuronal differentiation in a cultured embryonal carcinoma cell line. Nature 299:165–167
McBurney MW, Reuhl KR, Ally AI, Nasipuri S, Bell JC, Craig J (1988) Differentiation and maturation of embryonal carcinoma-derived neurons in cell culture. J Neurosci 8:1063–1073

McCaffery P, Drager UC (1994a) High levels of a retinoic acid-generating dehydrogenase in the meso-telencephalic dopamine system. Proc Natl Acad Sci USA 91:7772–7776

McCaffery P, Drager UC (1994b) Hot spots of retinoic acid synthesis in the developing spinal cord. Proc Natl Acad Sci USA 91:7194–7197

McCaffery P, Drager UC (1995) Retinoic acid synthesizing enzymes in the embryonic and adult vertebrate. In: Weiner H, et al (eds) Enzymology and molecular biology of carbonyl metabolism. Plenum New York, p173

McCaffery P, Lee M-O, Wagner MA, Sladek NE, Drager U (1992) Asymmetrical retinoic acid synthesis in the dorsoventral axis of the retina. Development 115:371–382

McCaffery P, Posch KC, Napoli JL, Gudas L, Drager UC (1993) Changing patterns of the retinoic acid system in the developing retina. Dev Biol 158:390–399

McCaffery P, Tempst P, Lara G, Drager UC (1991) Aldehyde dehydrogenase is a positional marker in the retina. Development 112:693–702

McKay IJ, Muchamore I, Krumlauf R, Maden M, Lumsden A, Lewis J (1994) The deaf Kreisler mouse: a hindbrain segmentation mutant. Development 120:2199–2211

Mendelsohn C, Lohnes D, Decimo D, Lufkin T, LeMeur M, Chambon P, Mark M (1994) Function of the retinoic acid receptors (RARs) during development (II) Multiple abnormalities at various stages of organogenesis in RAR double mutants. Development 120:2749–2771

Mendelsohn C, Ruberte E, LeMeur M, Morriss-Kay G, Chambon P (1991) Developmental analysis of the retinoic acid-inducible RAR-$\beta 2$ promoter in transgenic animals. Development 113:723–734

Minucci S, Saint-Jeannet J-P, Toyama R, Scita G, DeLuca LM, Taira M, Levin AA, Ozato K, Dawid IB (1996) Retinoid X receptor-selective ligands produce malformations in Xenopus embryos. Proc Natl Acad Sci USA 93:1803–1807

Momoi MY, Hayasaka M, Hanaoka K, Momoi T (1989) The expression of cellular retinoic acid binding protein CRABP in the neural tube and notochord of mouse embryo. Proc Japan Acad 65:9–12

Moro Balbas JA, Gato A, Alonso Revuelta MI, Pastor JF, Represa JJ, Barbosa E (1993) Retinoic acid induces changes in the rhombencephalic neural crest cells migration and extracellular matrix composition in chick embryos. Teratology 48:197–206

Morriss GM (1972) Morphogenesis of the malformations induced in rat embryos by maternal hypervitaminosis A. J Anat 113:241–250

Morriss GM, Thorogood PV (1978) An approach to cranial neural crest migration and differentiation in mammalian embryos. In: Johnson M (ed) Development in Mammals. p 363

Morriss-Kay GM, Murphy P, Hill RE, Davidson DR (1991) Effects of retinoic acid excess on expression of Hox-2.9 and Krox-20 and on morphological segmentation in the hindbrain of mouse embryos. EMBO J 10:2985–2995

Muramatsu H, Muramatsu T (1991) Purification of recombinant midkine and examination of its biological activities: functional comparison of new heparin binding factors. Biochem Biophys Res Comm 177:652–658

Muramatsu H, Shirahama H, Yonezawa S, Maruta H, Muramatsu T (1993) Midkine, a retinoic acid-inducible growth/differentiation factor: immunochemical evidence for the function and distribution. Dev Biol 159:392–402

Muramatsu T (1993) Midkine (MK), the product of a retinoic acid responsive gene, and pleitrophin constitute a new protein family regulating growth and differentiation. Int J Dev Biol 37:183–188

Muto K, Noji S, Nohno T, Koyama E, Myokai F, Nishijima K, Saito T, Taniguchi S (1991) Involvement of retinoic acid and its receptor β in differentiation of motoneurons in chick spinal cord. Neurosci Lett 129:39–42

Niederreither K, McCaffery P, Drager UC, Chambon P, Dolle P (1997) Restricted expression and retinoic acid-induced downregulation of the retinal dehydrogenase type 2 (RALDH-2) gene during mouse development. Mech of Dev 62:67–78

Noll E, Miller RH (1994) Regulation of oligodendrocyte differentiation: a role for retinoic acid in the spinal cord. Development 120:649–660

Nurcome V, Fraser N, Herlaar E, Heath JK (1992) MK: a pluripotential embryonic stem-cell-derived neuroregulatory factor. Development 116:1175–1183

Osumi-Yamashita N, Noji S, Nohno T, Koyama E, Doi H, Eto K, Taniguchi S (1990) Expression of retinoic acid receptor genes in neural crest-derived cells during mouse facial development. FEBS Lett 264:71–74

Pahlman S, Ruusala AI, Abrahamson L, Mattsson MEK, Esscher T (1984) Retinoic acid-induced differentiation of cultured human neuroblastoma cells: a comparison with phorbol ester-induced differentiation. Cell. Differentiation 14:135–144

Pannese M, Polo C, Andreazzoli M, Vignali, R, Kablar B, Barsacchi G, Boncinelli E (1995) The Xenopus homologue of Otx2 is a maternal homeobox gene that demarcates and specifies anterior body regions. Development 121:707–720

Papalopulu N, Clarke JDW, Bradley L, Wilkinson D, Krumlauf R, Holder N (1991) Retinoic acid causes abnormal development and segmental patterning of the anterior hindbrain in Xenopus embryos. Development 113:1145–1158

Paul G, Davies AM (1995) Trigeminal sensory neurons require extrinsic signals to switch neurotrophin dependence during the early stages of target field innervation. Dev Biol 171:590–605

Perez-Castro AV, Toth-Rogler LE, Wei LN, Nguyen-Huu MC (1989) Spatial and temporal pattern of expression of the cellular retinoic acid-binding protein and the cellular retinol-binding protein during mouse embryogenesis. Proc Natl Acad Sci USA 86:8813–8817

Pfeffer PL, De Robertis EM (1994) Regional specificity of RARγisoforms in Xenopus development. Mech of Dev 45:147–153

Phillip J, Mitchell PJ, Malipiero U, Fontana A (1994) Cell type-specific regulation of expression of transcription factor AP-2 in neuroectodermal cells. Dev Biol 165:602–614

Placzek M, Tessier-Lavigne M, Jessell T, Dodd J (1990) Orientation of commissural axons in vitro in response to a floor plate-derived chemoattractant. Development 110:19–30

Pleasure SJ, Page C, Lee VM-Y (1992) Pure, postmitotic, polarized human neurons derived from NTera2 cells provide a system for expressing exogenous proteins in terminally differentiated neurons. J Neurosci 12:1802–1815

Plum LA, Clagett-Dame M (1996) All-trans retinoic acid stimulates and maintains neurite outgrowth in nerve growth factor-supported developing chick embryonic sympathetic neurons. Dev Dynam 205:52–63

Pratt RM, Goulding EH, Abbott BD (1987) Retinoic acid inhibits migration of cranial neural crest cells in the cultured mouse embryo. J Craniofac Gen. Dev Biol 7:205–217

Quinn SDP, De Boni U (1991) Enhanced neuronal regeneration by retinoic acid of murine dorsal root ganglia and of fetal murine and human spinal cord in vitro. In Vitro Cell. Dev Biol 27 a:55–62

Raulais D, Lagente-Chevallier O, Guettet C, Duprez D, Courtois Y, Vigny M (1991) A new heparin binding protein regulate by retinoic acid from chick embryo. Biochem Biophys Res Comm 174:708–715

Repa JJ, Plum LA, Tadikonda PK, Clagett-Dame M (1996) All-*trans* 3,4-didehydroretinoic acid equals all-*trans* retinoic acid in support of chick neuronal development. FASEB J 10:1078–1084

Represa J, Sanchez A, Miner C, Lewis J, Giraldez F (1990) Retinoic acid modulation of the early development of the inner ear is associated with the control of c-fos expression. Development 110:1081–1090

Reynolds K, Mezey E, Zimmer A (1991) Activity of the β-retinoic acid receptor promoter in transgenic mice. Mech of Dev 36:15–29

Rockwood JM, Maxwell GD (1996a) An analysis of the effects of retinoic acid and other retinoids on the development of adrenergic cells from the avian neural crest. Exp Cell Res 223:250–258

Rockwood JM, Maxwell GD (1996b) Thyroid hormone decreases the number of adrenergic cells that develop in neural crest cultures and can inhibit the stimulatory action of retinoic acid. Dev. Brain Res 96:184–191
Rodriguez-Tebar A, Rohrer H (1991) Retinoic acid induces NGF-dependent survival response and high-affinity NGF receptors in immature chick sympathetic neurons. Development 112:813–820
Rosa FW, Wilk AL, Kelsey FO (1986) Teratogen update: vitamin A congeners. Teratology 33:355–364
Rossant J, Zirngibl R, Cado D, Shago M, Giguere V (1991) Expression of a retinoic acid response element-hsplacZ transgene defines specific domains of transcriptional activity during mouse embryogenesis. Genes Dev 5:1333–1344
Rossino P, Defillippi P, Silengo L, Tarone G (1991) Up-regulation of the integrin a1/b1 in human neuroblastoma cells differentiated by retinoic acid: correlation with increased neurite outgrowth responses to laminin. Cell Reg 2:1021–1033
Rowe A, Eager NSC, Brickell PM (1991) A member of the RXR nuclear receptor family is expressed in neural-crest-derived cells of the developing chick peripheral nervous system. Development 111:771–778
Rowe A, Richman JM, Brickell PM (1992) Development of the spatial pattern of retinoic acid receptor-β transcripts in embryonic chick facial primordia. Development 114:805–813
Ruberte E, Dolle P, Chambon P, Morriss-Kay G (1991) Retinoic acid receptors and cellular retinoid binding proteins. II Their differential pattern of transcription during early morphogenesis in mouse embryos. Development 111:45–60
Ruberte E, Dolle P, Krust A, Zelent A, Morriss-Kay G, Chambon P (1990) Specific spatial and temporal distribution of retinoic acid receptor gamma transcripts during mouse embryogenesis. Development 108:213–222
Ruberte E, Friederich V, Morriss-Kay G, Chambon P (1992) Differential distribution patterns of CRABP I and CRABP II transcripts during mouse embryogenesis. Development 115:973–987
Ruberte E, Friederich V, Chambon P, Morriss-Kay G (1993) Retinoic acid receptors and cellular retinoid binding proteins. III Their differential transcript distribution during mouse nervous system development. Development 118:267–282
Ruiz i Altaba A, Jessell T (1991) Retinoic acid modifies mesodermal patterning in early Xenopus embryos. Genes Dev 5:175–187
Ruiz i Altaba A, Jessell TM (1991) Retinoic acid modifies the pattern of cell differentiation in the central nervous system of neurula stage Xenopus embryos. Development 112:945–958
Schuh TJ, Hall BL, Kraft JC, Privalsky ML, Kimelman D (1993) v-erbA and citral reduce the teratogenic effects of all-*trans*-retinoic acid and retinol, respectively, in Xenopus embryogenesis. Development 119:785–798
Scott WJ, Walter R, Tzimas G, Sass JO, Nau H, Collins MD (1994) Endogenous status of retinoids and their cytosolic binding proteins and limb buds of chick vs mouse embryos. Dev Biol 165:397–409
Seegmiller RE, Harris C, Luchtel DL, Juchau MR (1991) Morphological differences elicited by two weak acids, retinoic and valproic, in rat embryos grown in vitro. Teratology 43:133–150
Sharma S, Notter MFD (1988) Characterisation of a neurotransmitter phenotype during neuronal differentiation of embryonal carcinoma cells. Dev Biol 125:246–254
Sharpe CR (1991) Retinoic acid can mimic endogenous signals involved in transformation of the Xenopus nervous system. Neuron 7:239–247
Sharpe CR, Goldstone K (1997) Retinoid receptors promote primary neurogenesis in Xenopus. Development 124:515–523
Shea TB, Fischer I, Sapirstein VS (1985) Effect of retinoic acid on growth and morphological differentiation of mouse NB2a neuroblastoma cells in culture. Dev. Brain Res 21:307–314

Shen S, van den Brink CE, Kruijer W, van der Saag PT (1992) Embryonic stem cells stably transfected with mRARβ 2-lacZ exhibit specific expression in chimeric embryos. Int J Dev Biol 36:465–476

Shenfelt RE (1972) Morphogenesis of malformations in hamsters caused by retinoic acid: relation to dose and stage at treatment. Teratology 5:103–118

Shiga T, Guar VP, Yamaguchi K, Oppenheim RW (1995) The development of interneurons in the chick embryo spinal cord following in vivo treatment with retinoic acid. J Comp Neurol 360:463–474

Sidell N (1982) Retinoic acid-induced growth inhibition and morphologic differentiation of human neuroblastoma cells in vitro. J Natl Cancer Inst 68:589–596

Sidell N, Altman A, Haussler MR, Seeger RC (1983) Effects of retinoic acid (RA) on the growth and phenotypic expression of several human neuroblastoma cell lines. Exp Cell Res 148:21–30

Sidell N, Lucas CA, Kreutzberg GW (1984) Regulation of acetylcholinesterase activity by retinoic acid in a human neuroblastoma cell line. Exp Cell Res 155:305–309

Simeone A, Avantaggiato V, Moroni MC, Mavilio F, Arra C, Cotelli F, Nigro V, Acampora D (1995) Retinoic acid induces stage-specific antero-posterior transformation of rostral central nervous system. Mech of Dev 51:83–98

Sive HL, Cheng PF (1991) Retinoic acid perturbs the expression of Xhox.lab genes and alters mesodermal determination in Xenopus laevis. Genes Dev 5:1321–1332

Sive HL, Draper BW, Harland RM, Weintrub H (1990) Identification of a retinoic acid-sensitive period during primary axis formation in Xenopus laevis. Genes Dev 4:932–942

Smith SM (1994) Retinoic acid receptor isoform β2 is an early marker for alimentary tract and central nervous system positional specification in the chicken. Dev Dynam 200:14–25

Smith SM, Eichele G (1991) Temporal and regional differences in the expression pattern of distinct retinoic acid receptor-β transcripts in the chick embryo. Development 111:245–252

Smith-Thomas L, Lott I, Bronner-Fraser M (1987) Effects of isotretinoin on the behavior of neural crest cells in vitro. Dev Biol 123:276–281

Smolich BD, Papkoff J (1994) Regulated expression of Wnt family members during neuroectodermal differentiation of P19 embryonal carcinoma cells: over expression of Wnt-1 perturbs normal differentiation-specific properties. Dev Biol 166:300–310

Staines WA, Morassutti DJ, Reuhl KR, Ally AI, McBurney MW (1994) Neurons derived from P19 embryonal carcinoma cells have varied morphologies and neurotransmitters. Neurosci 58:735–751

Stenkamp DL, Gregory JK, Adler R (1993) Retinoid effects in purified cultures of chick embryo retina neurons and photoreceptors. Invest Ophthalmol Vis Sci 34:2425–2436

Studer M, Popperl H, Marshall H, Kuriowa A, Krumlauf R (1994) Role of a conserved retinoic acid response element in rhombomere restriction of Hoxb-1. Science 265:1728–1732

Sulik KK, Dehart DB, Rogers JM, Shernoff N (1995) Teratogenicity of low doses of all-*trans* retinoic acid in presomite mouse embryos. Teratology 51:398–403

Sundin O, Eichele G (1992) An early marker of axial pattern in the chick embryo and its respecification by retinoic acid. Development 114:841–852

Thaller C, Eichele G (1990) Isolation of 3,4-didehydroretinoic acid, a novel morphogenetic signal in the chick wing bud. Nature 345:815–819

Thiele CJ, Cohen PS, Israel MA (1988) Regulation of c-myb expression in human neuroblastoma cells during retinoic acid-induced differentiation. Mol Cell Biol 8:1677–1683

Thiele CJ, Reynolds CP, Israel MA (1985) Decreased expression of N-myc precedes retinoic acid-induced morphological differentiation of human neuroblastoma. Nature 313:404–406

Thompson JN (1969) Vitamin A in development of the embryo. Am J Clin Nutr 22:1063–1069
Thompson JN, Howell JM, Pitt GAJ, McLaughlin CI (1969) The biological activity of retinoic acid in the domestic fowl and the effects of vitamin A deficiency on the chick embryo. Br J Nutr 23:471–490
Thompson S, Stern PL, Webb M, Walsh FS, Engstrom W, Evans EP, Shi W-K, Hopkins B, Graham CF (1984) Cloned human teratoma cells differentiate into neuron-like cells and other cell types in retinoic acid. J Cell Sci 72:37–64
Thorogood P, Smith L, Nicol A, McGinty R, Garrod D (1982) Effects of vitamin A on the behaviour of migratory neural crest cells in vitro. J Cell Sci 57:331–350
Tini M, Otulakowski G, Breitman ML, Tsui L-C, Giguere V (1993) An everted repeat mediates retinoic acid induction of the γF-crystallin gene: evidence of a direct role for retinoids in lens development. Genes Dev 7:295–307
Turetsky DM, Huettner JE, Gottlieb DI, Goldberg MP, Choi DW (1993) Glutamate receptor-mediated surrents and toxicity in embryonal carcinoma cells. J Neurobiol 24:1157–1169
Vaessen M-J, Kootwijk E, Mummery C, Hilkens J, Bootsma D, van Kessel AD (1989) Preferential expression of cellular retinoic acid binding protein in a subpopulation of neural cells in the developing mouse embryo. Differentiation 40:99–105
Vaessen M-J, Meijers JHC, Bootsma D, van Kessel AD (1990) The cellular retinoic acid-binding protein is expressed in tissues associated with retinoic acid-induced malformations. Development 110:371–378
Vallari RC, Pietruszko R (1982) Human aldehyde dehydrogenase: mechanism of inhibition by disulphiram. Science 216:637–639
van den Eijnden-van Raaij AJM, van Achterberg TAE, van der Kruijssen CMM, Piersma AH, Huylebroeck D, de Laat SW, Mummery CL (1991) Differentiation of aggregated murine P19 embryonal carcinoma cells is induced by a novel visceral endoderm-specific FGF-like factor and inhibited by activin A. Mech of Dev 33:157–166
Ved HS, Pieringer RA (1993) Regulation of neuronal differentiation by retinoic acid alone and in cooperation with thyroid hormone or hydrocortisone. Dev Neurosci 15:49–53
Wagner M, Han B, Jessell TM (1992) Regional differences in retinoid release from embryonic neural tissue detected by an in vitro reporter assay. Development 116:55–66
Wagner M, Thaller C, Jessell TM, Eichele G (1990) Polarizing activity and retinoid synthesis in the floor plate of the neural tube. Nature 345:819–822
Wang S-Y, LaRosa GJ, Gudas LJ (1985) Molecular cloning of gene sequences transcriptionally regulated by retinoic acid and dibutyryl cyclic AMP in cultured mouse teratocarcinoma cells. Dev Biol 107:75–86
Warkany J (1945) Manifestations of prenatal nutritional deficiency. Vitamins Hormones 3:73–103
Webster WS, Johnston MC, Lammer EJ, Sulik KK (1986) Isotretinoin embryopathy and the cranial neural crest: an in vivo and in vitro study. J Craniofac Gen. Dev Biol 6:211–222
Wilson JG, Roth CB, Warkany J (1953) An analysis of the syndrome of malformations induced by maternal vitamin A deficiency. Effects of restoration of vitamin A at various times during gestation. Am. J Anat 92:189–217
Wion D, Houlgatte R, Barbot N, Barrand P, Dicou E, Brachet P (1987) Retinoic acid increases the expression of NGF gene in mouse L cells. Biochem Biophys Res Comm 149:510–514
Wood H, Pall G, Morriss-Kay G (1994) Exposure to retinoic acid before or after the onset of somitogenesis reveals separate effects on rhombomeric segmentation and 3′ HoxB gene expression domains. Development 120:2279–2285
Wuarin L, Sidell N (1991) Differential susceptibilities of spinal cord neurons to retinoic acid-induced survival and differentiation. Dev Biol 144:429–435

Wuarin L, Sidell N, De Vellis J (1990) Retinoids increase perinatal spinal cord neuronal survival and astroglial differentiation. Int J Devel Neurosci 8:317–326
Yamamoto M, McCaffery P, Drager UC (1996) Influence of the choroid plexus on cerebellar development: analysis of retinoic acid synthesis. Dev. Brain Res 93:182–190
Yasuda Y, Okamoto M, Konishi H, Matsuo T Kihara T, Tanimura T (1986) Developmental anomalies induced by all-*trans* retinoic acid in fetal mice: I. Macroscopic findings. Teratology 34:37–49
Younkin DP, Tang C-M, Hardy M, Reddy UR, Shi Q-Y, Pleasure SJ, Lee VM, Pleasure D (1993) Inducible expression of neuronal glutamate receptor channels in the NT2 human cell line. Proc Natl Acad Sci USA 90:2174–2178
Zgombic-Knight M, Ang HL, Foglio MH, Duester G (1995) Cloning of the mouse class IV alcohol dehydrogenase (retinol dehydrogenase) cDNA and tissue-specific expression patterns of the murine ADH gene family. J Biol Chem 270:10868–19877
Zhang M, Kim H-J, Marshall H, Gendron-Maguire M, Lucas DA, Baron A, Gudas LJ, Gridley T, Krumlauf R, Grippo JF (1994) Ectopic Hoxa-1 induces rhombomere transformation in mouse hindbrain. Development 120:2431–2442
Zhang Z, Balmer JE, Lovlie A, Fromm SH, Blomhoff R (1996) Specific teratogenic effects of different retinoic acid isomers and analogs in the developing anterior central nervous system of zebrafish. Dev Dynam 206:73–86

CHAPTER 15

Avian Embryo as Model for Retinoid Function in Early Development

M.H. ZILE

A. Introduction

The importance of the micronutrient vitamin A throughout the life cycle has been well established in numerous studies (MOORE 1957; WOLF 1984). It has more recently become quite clear that the requirement for vitamin A begins with embryonic life, and that retinoids, the vitamin A active forms that exert vitamin A function via their nuclear receptor, are critical signaling molecules during development, their absolute essentialness being best illustrated in the avian embryo model: vitamin A deficiency is embryolethal (DERSCH and ZILE 1993; THOMPSON 1969).

It is the early years of vitamin A research dealing with establishing the requirements of this micronutrient across the species and carefully recording the deficiency symptoms and teratologies, to which we owe much of our understanding of the function of this vitamin in embryogenesis and fetal development. Not many of us at the present time are involved in this absolutely essential type of work, nowadays often referred to as "descriptive studies." However, it is the critical observations of such studies that we consult, refer to and quote as supporting evidence for our present day results obtained using molecular biology and gene knockout approaches to elucidate the physiological function of vitamin A in development.

It is clear from a broad body of data that vitamin A is required for many different aspects of life (BLOMHOFF 1994; GUDAS et al. 1994; HOFMANN and EICHELE 1994; MOORE 1957; WOLF 1984). However, the molecular mechanism(s) of action of this pleiotropic micronutrient are still under intense investigation. In the last two decades much has been learned about vitamin A function from cell and molecular biology research and from studies on development. The concept of vitamin A as a powerful differentiating agent evolved from the early work of the vitamin A deficient rat (WOLBACH 1954) and was later substantiated by the classical tissue culture studies of FELL and MELLANBY (1953). This led to extensive research with cells, organ culture and animal models (BROCKES 1989; GUDAS et al. 1994). The elucidation of the role of vitamin A in growth and differentiation received an enormous stimulus from the cancer field in the 1970ies and 1980ies, which continues to be a discipline contributing significant new information on the molecular roles of retinoids (GUDAS et al. 1994; HONG and ITRI 1994; MOON et al. 1994; ROBERTS and SPORN 1984).

The regulation of somatic functions by vitamin A is now generally ascribed to retinoic acid, the biologically most active form of vitamin A, a recognized regulator of cell division and differentiation in tissues of ectodermal, endodermal and mesodermal origin (GUDAS et al. 1994; ROBERTS and SPORN 1984; WOLF 1984). However, it was the discovery of the nuclear receptors for retinoic acid (RARs) and the retinoid-X receptors (RXRs), that provided a breakthrough in the understanding how vitamin A active molecules can exert an effect at the gene level (MANGELSDORF et al. 1995; PFAHL and CHYTIL 1996). Retinoic acid and other bioactive retinoids are capable of producing alterations in the expression of a diverse group of genes (GUDAS et al. 1994). These findings have opened new avenues for the elucidation of the many and sometimes conflicting biological effects of vitamin A. Retinoic acid is now generally recognized as an important signaling molecule which, as a ligand to its nuclear receptors, the RARs, alters gene expression at the level of transcription (GUDAS et al. 1994; MANGELSDORF et al. 1995; PFAHL and CHYTIL 1996; ROBERTS and SPORN 1984). Another endogenous vitamin A active molecule is 9-*cis*-retinoic acid which can activate both the RARs and RXRs (KASTNER et al. 1994; LEBLANC and STUNNENBERG 1995). It is important to keep in mind that the function of vitamin A in visual cycle is via a nongenomic mechanism involving retinal as a prosthetic group on visual proteins (WALD 1968). The reviews on animal developmental models also show that much remains to be learned about the mechanism of action of vitamin A in reproduction.

This chapter presents the available knowledge of vertebrate models for the study of the physiological function of vitamin A in development, with an emphasis on gene regulation and on the avian model systems for studies of early embryonic development.

B. Teratogenic Effects of Vitamin A Deficiency and Excess

It was recognized as early as the 1930s that maternal insufficiency of vitamin A during pregnancy results in incomplete pregnancies, fetal death and severe congenital malformations (HALE 1937; MASON 1935). In subsequent studies WILSON and coworkers (1953) identified a spectrum of congenital abnormalities that result from lack of vitamin A during gestation (BATES 1983; see also THOMPSON 1969); these abnormalities have been observed in different animal species (MORRISS 1972; reviewed in MADEN 1994). The major target tissues of vitamin A deficiency include the heart, the ocular tissues, and the circulatory, urogenital and respiratory systems. Including vitamin A in the diet of the vitamin A deficient female during specific times of pregnancy prevented the occurrence of these abnormalities, suggesting that vitamin A is required at various distinct stages of development, and clearly establishing the important role of vitamin A in normal embryonic development. At this time critical information is lacking as to how maternally derived vitamin A is processed by the embryo to generate physiological amounts of the bioactive retinoids in a developmental stage- and tissue-specific fashion required for normal embryonic and fetal development.

In the human the teratogenic effects of vitamin A deficiency during gestation are often masked by general malnutrition and it is not possible to link an abnormality to a single nutrient or micronutrient (GERSTER 1997). Nevertheless, a pattern has emerged that suggests that children born of mothers with xerophthalmia, a symptom of severe vitamin A deficiency, frequently have ocular abnormalities, and that premature neonates with low vitamin A stores are at high risk of respiratory abnormalities (SHENAI et al. 1995). It has also been suggested that vitamin A supplementation contributes to a the reduction in neural tube defects (WALLINGFORTH and UNDERWOOD 1986).

Embryonic malformations also result from the presence of excessive amounts of vitamin A during development as was established using pharmacological levels. COHLAN (1954) was the first to describe congenital malformations in rats caused by the administration of excess vitamin A during pregnancy. Numerous studies and clinical observations of the teratological effects of excess retinoids have followed. Major target tissues include the heart, skull, skeleton, limbs, central nervous system, brain, eyes and craniofacial structures (BROCKES 1989; COHLAN 1954; HELMS et al. 1997; KOCHHAR et al. 1993; KOCHHAR and CHRISTIAN 1997; MOORE 1957; MORRISS and STEELE 1972; NAU et al. 1994; ROSA 1993; SHENEFELT 1972). The typical pattern of malformations caused by treatment with excess vitamin A in its general features resembles that caused by vitamin A deficiency (GERSTER 1997; MADEN 1994) and includes the central nervous system, eye and ear abnormalities, craniofacial, limb and heart defects; also affected is the urogenital system. In animal studies of fertility and reproductive performance the main adverse effects of excess vitamin A include an increased number of stillbirths and decreased neonatal survival (KAMM et al. 1984; KISTLER and HUMMLER 1985; TEELMANN 1989); however, the pathology has not been examined. LAMMER and coworkers (1985; LAMMER 1988) analyzed 154 human pregnancies in which the mothers had been exposed to retinoids during pregnancy, and found that 46% of those taken to delivery resulted in newborn with some abnormality; about half of those with major malformations did not survive. As many as 1000 cases of congenital deformities may be linked to an exposure of the mother to retinoids (ELLIS and VOORHEES 1987). As a result of the overwhelming evidence that prenatal exposure to excess vitamin A and retinoids can result in congenital abnormalities (BOYD 1989; KOCHHAR et al. 1993; LAMMER et al. 1985; LAMMER 1988; ROSA 1993; SHALITA 1988; STRAUSS et al. 1988), treatment with retinoids is not permitted during pregnancy or when there is a possibility of pregnancy (CHAN et al. 1996).

The overlap of the teratology symptoms of vitamin A deficiency and excess clearly indicates common targets and a critical role for vitamin A in the development of many organs. At this time it is not clear what is the link between the teratological actions of vitamin A and the role of this vitamin in regulating normal embryonic and fetal development.

C. Mammalian Models

Recent advances in molecular developmental biology have provided new tools that have been successfully applied to examine the mechanisms of both normal and abnormal development. Studies with exogenously applied retinoids have been conventionally used to manipulate and perturb normal embryonic development (HOFFMANN and EICHELE 1994; SHENEFELT 1972) and have provided valuable information on the effects of retinoids at various stages of development. In fact, it is well known that almost every organ or tissue system can be affected by retinoic acid if the embryo is treated with it at a critical time in development (OSMOND et al. 1991; SHENEFELT 1972). While these observations do not provide a mechanistic explanation for the role of vitamin A in development, when viewed together with similar abnormalities that are reported for the vitamin A deficient embryo and fetus, they clearly point to common normal developmental pathways that are severely disrupted by an imbalance in vitamin A status. It is important to recall that population studies estimate that approximately 3% of all children born in the United States have a major malformation at birth (HOFFMAN 1995; SHEPARD 1986) and that 70% of these are of unknown etiology (WILSON 1977). Growth retardation and prenatal or early postnatal death also occur at a high ratio in our population (SELEVAN 1981; WILCOX et al. 1985), but information is not available as to their possible causes. In humans, exposure to certain agents at almost any stage during the preorganogenesis period induces fetal malformations, many of them resembling those with unknown etiologies (KIMMEL et al. 1993; RUTLEDGE and GENEROSO 1989). An examination of these data in respect to congenital abnormalities caused by either a lack or an excess of vitamin A, may provide clues to whether some of these malformations are linked to a disruption in critical vitamin A dependent developmental events. More and more clinical data confirm that it is important to look at the nutritional and dietary contributions to birth defects. Poor diet or harmful dietary components, including environmental pollutants (ZILE 1992) and ethanol (DUESTER 1994; GRUMMER et al. 1993; SULIK 1984; TWAL and ZILE 1997) may be ingested during pregnancy, a time when vitamin A nutrition is very critical because maternal tissues are continuously being depleted of vitamin A so as to insure an adequate supply of it to the embryo and fetus (WALLINGFORD and UNDERWOOD 1986). While nutritionists are well aware of the marginal vitamin A status in developing countries (UNDERWOOD 1994), little is known of populations at risk in the industrialized countries. The study by DUITSMAN et al. (1995) is significant as it identifies a population of pregnant women in the United States at high risk of vitamin A inadequacy. As the discussions in this chapter show, the critical vitamin A dependent developmental events are initiated very early in embryogenesis (during gastrulation and neurulation) and coincide with the first 2–3 weeks of human pregnancy. They may be unknowingly severely compromised if maternal vitamin A intake is marginal or if there is interference with vitamin A function.

A recent approach to answering questions about the functions of vitamin A in development has been the use of transgenic mice with changes in retinoid receptor gene structure (BOYLAN et al. 1995; CHAMBON 1993; GIGUERE et al. 1996; KASTNER et al. 1994; SUCOV et al. 1994). Many of the abnormalities in these mutant mice resemble those observed in the fetuses from the vitamin A deficient animals reported earlier. The molecular and genetic dissection of the retinoid signaling pathway has provided valuable insights into the pleiotropic effects of vitamin A and has demonstrated the critical role of retinoid receptors in vertebrate ontogenesis. However, the interpretation of the data has been hampered by receptor redundancy (LUO et al. 1995). Thus, the receptor knockouts alone cannot provide definitive answers to retinoid function. These studies, however, underscore the critical importance of the retinoid ligands: evolution has provided a redundancy of receptors to serve as back-up systems for the essential functions of vitamin A.

Another important approach for the examination of the molecular mechanisms of retinoid action in developmental regulation is the use of in vivo vitamin A deficient animals, in which gene expression can be studied by regulating the presence of the receptor ligands, the retinoids. With rodent models it is possible to target vitamin A deficiency to distinct gestational windows and address its role in fetal development (SMITH et al. 1998; THOMPSON 1969; WELLIK and DELUCA 1995, 1996; WELLIK et al. 1997; WHITE and CLAGETT-DAME 1996). Rat embryos made deficient in retinoids during gestational days 11.5–13.5 exhibit specific cardiac, limb, ocular and central nervous system abnormalities, some of which are partially similar to those reported in retinoid receptor knockout mice, while others are novel (SMITH et al. 1998). Studies on the role of retinoic acid in mammalian reproduction have revealed that marginal vitamin A status during development severely impairs fetal lung development (WELLIK et al. 1997). However, for studies of the function of vitamin A in development, the rodent models are limited to addressing later gestational stages. Additionally, there is uncertainty of the exact vitamin A status of the embryo, since the embryo depends on the maternal supply for all nutrients, and the mother must have vitamin A for her own survival. At this time it is also not clear if the molecular function of vitamin A during the later stages of gestation is via retinoic acid as the active ligand for retinoid receptors, or if retinol itself or another active metabolite of retinol is involved by an unidentified mechanism.

D. Avian Embryo "Retinoid Ligand Knockout" Model

Studies with exogenous retinoids inducing developmental changes and expression of retinoid receptors and binding proteins have been very important in identifying retinoid-sensitive developmental events (DICKMAN and SMITH 1996; GUDAS et al. 1994; HOFMANN and EICHELE 1994; MADEN 1994; MADEN and HOLDER 1992; OSMOND et al. 1991; PFAHL and CHYTIL 1996; SHENEFELT 1972; SMITH et al. 1997; WIENS et al. 1992). However, these approaches may not reflect the physiological function(s) of vitamin A in early vertebrate development. In order to examine the molecular mechanisms of vitamin A action in normal biological processes it is important to complement the above non-physiological studies with models in which gene expression and morphological changes can be studied by regulating the presence of vitamin A compounds within the physiological response range.

An ideal system for studies of early embryonic development is the vitamin A deficient avian embryo, which is completely dependent on vitamin A for its early development and without vitamin A, dies at day 3 of embryonic life (DERSCH and ZILE 1993; THOMPSON 1969). In the vitamin A deficient avian embryo the earliest gross developmental changes are evident at the time of formation of the cardiovascular system and are manifested in an abnormal, closed but beating heart, an absence of large extra-embryonal blood vessels, an abnormal central nervous system and an early embryonic death. Administration of vitamin A active compounds during the early stages "rescues" the embryo and results in normal development (Fig. 1) (DERSCH and ZILE 1993; HEINE et al. 1985; THOMPSON 1969). Recent studies provide convincing evidence that this model offers a unique opportunity to elucidate the physiological functions of vitamin A in early embryonic development (MADEN et al. 1998; ZILE 1998). The model is based on the important discovery by THOMPSON (1969) that vitamin A is required for early embryonic development. Birds are maintained on a semisynthetic diet containing methyl retinoate (HEINE et al. 1985; THOMPSON 1969) or retinoic acid (DERSCH and ZILE 1993) as their only source for vitamin A activity. These forms of vitamin A are not transferred from the hen to the egg nor to the embryo as demonstrated by analyses of the eggs and embryos obtained from quail hens fed the above diet (DONG and ZILE 1995). Thus, an avian embryo "retinoid ligand knockout" model has been developed, providing an ideal system to study the physiological function of retinoids during early development in the absence of maternal environment.

The data obtained using this model attest to its unique potential for elucidating the physiological function of vitamin A in early embryonic development (MADEN et al. 1998; ZILE 1998). Furthermore, the avian embryo is an excellent experimental system for the elucidation of all vertebrate early development. A major advantage is its accessibility for experimental manipulations, even at the earliest stages; other advantages include the fact that the early avian embryo (first 3 days) develops within a 2-dimensional plane, and that its size is relatively large (DIETERLEN-LIEVRE 1997). Studies with the avian embryo have made major contributions to the understanding of body organization, axis, cell migration, cell-cell interactions, patterning, morphogenesis and organogenesis. Widely used models are chimeras resulting from exchanges of

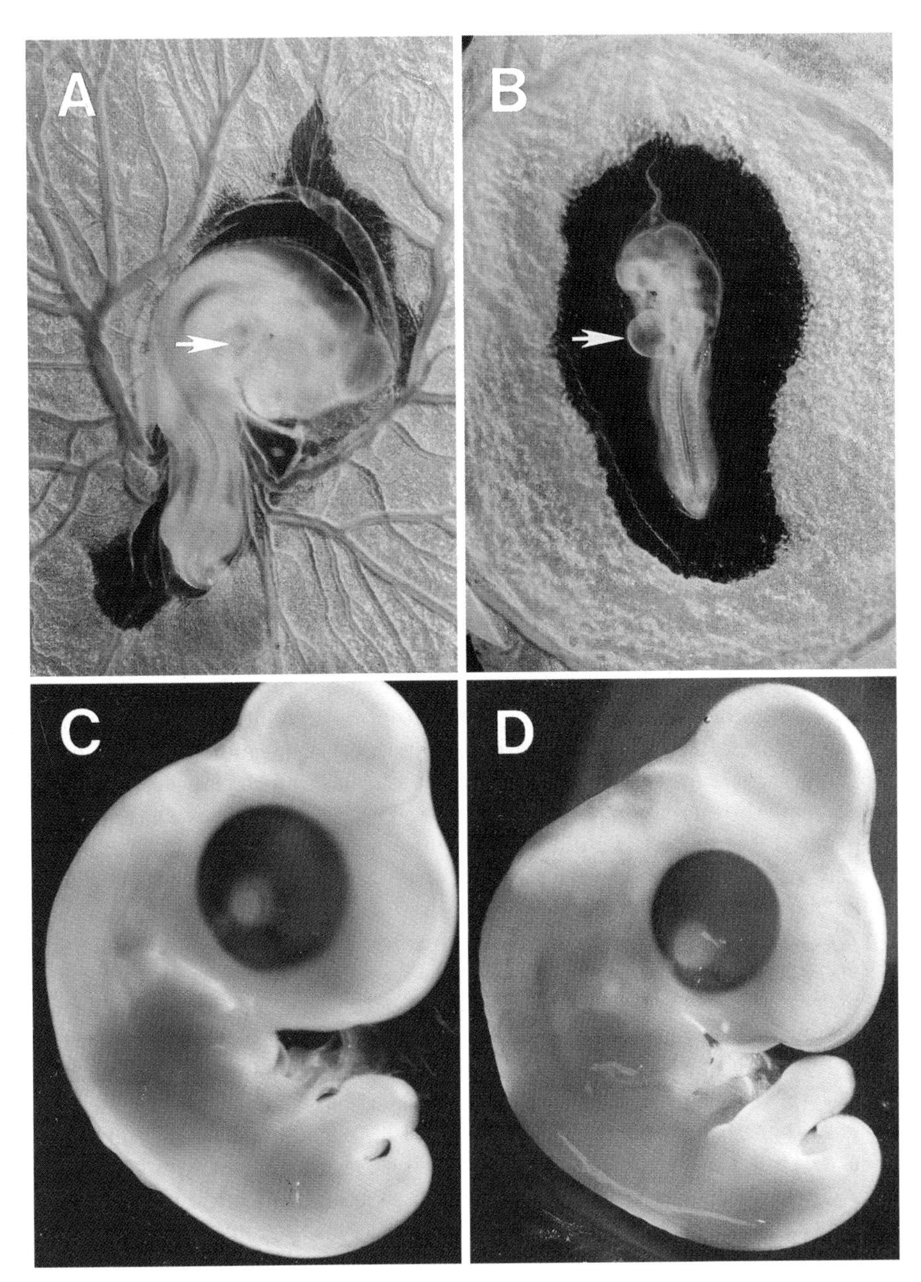

Fig. 1. Normal (**A**) and vitamin A deficient (**B**) quail embryo at 3 days of *in ovo* development. Note the abnormal "ballooning" heart as *situs inversus* in the vitamin A deficient embryo. **C** Normal quail embryo at 5 days of *in ovo* development. **D** Vitamin A deficient quail embryo injected with 1 μg retinol at 20 h of development then allowed to develop to 5 days. The "rescued" embryo appears normal in all developmental aspects. *Arrows*, heart. All views are dorsal

rudiments between chick and quail embryos, a technique pioneered by LEDOUARIN (1969; LEDOUARIN et al. 1995). Other techniques include the ablation of specific embryonic regions, implantation of microbeads with test substances (BRONNER-FRASER 1996). Availability of simple cell labeling techniques has added to the success of the avian model (BRONNER-FRASER 1996; DIETERLEN-LIEVRE 1997; FISHER and SCHOENWOLF 1983; WEDDEN et al. 1989). Reporter cell techniques are now applied to examine embryonic cell migration. It is foreseeable that avian somatic gene transfer using retroviral vectors will be an alternative to the gene knockouts in mice. Clearly, the vitamin A deficient avian embryo provides a unique opportunity to examine the retinoid-regulated developmental pathways and thus to significantly contribute to the field of vertebrate development.

E. Cardiogenesis

As early as the 1930s it was recognized that maternal insufficiency of vitamin A during pregnancy results in fetal death or abnormalities in the offspring that include abnormal heart development (MASON 1935; WILSON and WARKANY 1949; WILSON et al. 1953). THOMPSON (1969) made the important discovery that vitamin A is required for early avian embryonic cardiovascular development. In subsequent studies HEINE et al. (1985) determined that vitamin A is required for establishing cardiovascular circulation. These observations were confirmed and extended further by work in our laboratory (DERSCH and ZILE 1993), using the quail embryo retinoid ligand knockout model. We have further established that there is a critical time within the first 22–28h in quail embryogenesis when the vitamin A active form is required to initiate events that lead to normal development (DERSCH and ZILE 1993).

The early development of the vertebrate heart has been summarized by HAMBURGER and HAMILTON (1951), with excellent recent reviews of the expanding field (FISHMAN and CHIEN 1997; ICARDO 1996; KIRBY 1993; LITTLE and RONGISH 1995). Our studies with the quail embryo ligand knockout model suggest that vitamin A is critically involved in the formation of the primitive heart, likely via the retinoid receptors. The early avian embryo expresses all the genes for retinoid receptors (ZILE et al. 1997). RARβ and its isoform have been identified in the chick embryo (SMITH 1994; SMITH and EICHELE 1991). RARβ2 is expressed in the heart forming region and is downregulated in the absence of vitamin A (KOSTETSKII et al. 1996, 1998). Recently it was demonstrated that other retinoid receptors are also expressed in the heart forming regions (KOSTETSKII et al. 1998). It is logical to think that retinoid receptors participate in the transcriptional regulation of cardiogenesis.

Since there are no master regulators of the heart, the interaction among many factors in the cardiac gene program is very important (DOEVENDANS and VAN BILSEN 1996; EVANS 1997; FISHMAN and CHIEN 1997; LAVERRIERE et al. 1994). Among the few transcription factors expressed predominantly in the heart are the GATA-4/5/6, a subfamily of three transcription factors recently described in the developing heart and gut of the chicken (DOEVENDANS and VANBILSEN 1996; EVANS 1997; LAVERRIERE et al. 1994). GATA-4 expression has been reported to be retinoic acid responsive (ARCECI et al. 1993), and GATA-4 induction by retinoids was found to be abolished in RARγ knockouts (BOYLAN et al. 1995), suggesting that this and likely other GATA genes may be transcriptionally regulated by retinoids. Studies with the quail embryo model show that GATA-4 is severely downregulated in vitamin A deficient quail embryos (ZILE 1998), thus the embryonic retinoid ligand knockout model provides a promising investigative approach for the identification of signaling molecules that function upstream of GATA-4/5/6. Furthermore, the abnormal morphology of the heart in the vitamin A deficient quail embryo may be explained by an effect on downstream cardiac-specific genes that are regulated by GATA-4. Among other genes considered to be essential for heart development are the homeobox class of transcription factor genes (BODMER 1995; CHAN-THOMAS et al. 1993; EVANS et al. 1995; HARVEY 1996; KERN et al. 1995; LINTS et al. 1993), thought to be the key regulators of mammalian body plan formation. In the vitamin A deficient quail embryo, the expression of one of the homeobox genes, Msx-1 was significantly altered at various sites, including the heart forming region (CHEN et al. 1995). This may be explained by the presence of the retinoic acid responsive element, RARE in the Msx-1 gene. Another homeobox gene, the Nkx 2-5, a homolog of the *Drosophila* homeobox gene *tinman* (EVANS et al. 1995; HARVEY 1996; LINTS et al. 1993; SCHULTHEISS et al. 1995), is of great interest since its homozygous mutant mice are reported to have early heart tube defects (BODMER 1995; KERN et al. 1995) that resemble vitamin A deficiency in the early quail embryo (DERSCH and ZILE 1993; KOSTETSKII et al. 1996; THOMPSON 1969; TWAL et al. 1995). The chicken homolog of Nkx 2-5 has been recently cloned (SCHULTHEISS et al. 1995), and we have shown that the expression of this cardiac muscle specific gene is an excellent marker for early vertebrate heart formation (ZILE 1998).

F. Left-Right Asymmetry

In the vertebrate embryo the first localized and permanent expression of asymmetry is the handedness of the heart. Relatively little is known about the positional and molecular cues that lead to the acquisition of morphological characteristics that define the position of the heart. While there have been several hypotheses as to the genes involved in cardiac L-R axis specification, this important problem in biology has not yet been resolved. It is therefore of great interest that in the majority (75%) of the retinoid ligand knockout quail embryos the heart is on the wrong side, i.e., in *situs inversus* position (DERSCH and ZILE 1993; TWAL et al. 1995; ZILE 1998). Our work has shown that as late as the neurulation stage 8, but not later, the anticipated *situs inversus* can be "rescued" by administration of vitamin A active compounds (DERSCH and ZILE

1993; ZILE et al. 1998). This is a challenging observation as it can not be explained by the recently proposed molecular pathway determining cardiac L-R asymmetry. This pathway involves three genes, cAct-RIIa, Shh and cNR-1, transiently expressed asymmetrically along the axis in the gastrula stage embryo; their ectopic expression causes cardiac L-R randomization (LEVIN et al. 1995, 1997). The interpretation of this work initially was that heart sidedness is the reflection of the asymmetric expression of certain genes during gastrulation. However, this idea is not compatible with the evidence obtained in our laboratory. Using the quail embryo retinoid ligand knockout model, we have shown that the asymmetric expression of Shh in the quail Hensen's node is not altered by vitamin A deficiency (CHEN et al. 1996) and that the *situs inversus* observed in the vitamin A deficient quail embryo can be prevented by the administration of vitamin A active compounds downstream of the Shh and cNR-1 signaling pathway (DERSCH and ZILE 1993; KOSTETSKII et al. 1996; TWAL et al. 1995; ZILE 1998; ZILE et al. 1998), before the appearance of the first morphological asymmetry. We are continuing to explore the role of retinoids in regulating heart asymmetry with the recently identified *snail*-related zinc finger gene cSnR (ISAAC et al. 1997), which has been shown to control cardiac L-R asymmetry downstream of Shh and-cNR-1. The expression of this gene may be linked to the vitamin A regulated cardiac asymmetry (ZILE et al. 1998). Asymmetrically localized proteins sensitive to retinoid status have also been identified (SMITH et al. 1997; TSUDA et al. 1996; WUNSCH et al. 1994).

Clearly, the cardiac gene program involves numerous factors. The sequential expression of asymmetric signaling molecules during gastrulation is only a part of the complex gene circuitry of cardiogenesis, that is followed by yet another set of regulatory events during neurulation. The retinoid-deficient avian embryo is an ideal experimental system to obtain important information regarding the developmental regulation of vertebrate asymmetries. It is interesting to note, that cardiac abnormalities also result when embryos are treated with excess retinoids (HOFMANN and EICHELE 1994; SHENEFELT 1972; WIENS et al. 1992). Duplicate hearts, *cardia bifida*, abnormal looping and *situs inversus* can be produced by implanting excess retinoic acid containing microbeads in specific precardiac regions of chick embryos (OSMOND et al. 1991; SMITH et al. 1997). While such studies do not provide answers to the normal physiological mechanism of action of vitamin A in cardiogenesis, they confirm the potent morphogenetic effects of retinoic acid and point to the embryonal sites sensitive to retinoids, sites also likely to be sensitive to the lack of this signaling molecule, as shown by our work with the vitamin A deficient quail embryo model.

G. Vasculogenesis

The vitamin A deficient avian embryo lacks a vascular link between the embryo to its extraembryonal tissue where the blood forms (DERSCH and ZILE

1993; HEINE et al. 1985; THOMPSON 1969; TWAL et al. 1995; ZILE 1998). Initially, the embryonal and extraembryonal vascular systems appear to develop normally in the vitamin A deficient embryo, however, the omphalomesenteric veins that connect the heart to its extra embryonal circulation do not form. Using our retinoid ligand knockout embryo model we have observed that the vascular network is not maintained in the vitamin A deficient embryo (ZILE 1998). Work is in progress to determine the involvement of vitamin A in vasculogenesis.

H. Central Nervous System

Retinoic acid is known to play an important role in the development of the CNS (EICHELE 1997; HOFMANN and EICHELE 1994; MADEN 1994; MADEN and HOLDER 1992; MADEN et al. 1996, 1997 1998). It induces the formation of neurites in neuroblastoma cells and embryonal carcinoma cells, and induces neurites in explanted or dissociated dorsal root ganglia and spinal cord. It is present endogenously in the developing CNS, and when applied in excess to embryos, deletes or respecifies parts of hindbrain, as well as alters neural crest cell migration. The likely targets of retinoic acid action in the CNS include pattern specification, neurite outgrowth and the neural crest (MADEN et al. 1996, 1997 1998). The effects of exogenously administered retinoic acid on CNS morphology and gene expression have been extensively investigated (GALE et al. 1996; HENION and WESTON 1994; HOFMANN and EICHELE 1994; MADEN and HOLDER 1992; THOROGOOD et al. 1982). Several homeobox genes that are involved in CNS specification are known to have retinoic acid responsive elements (MARSHALL et al. 1994; MORRISS-KAY et al. 1991; POPPERL and FEATHERSTONE 1993; STUDER et al. 1994) and are obvious candidates for retinoid regulated specification of the embryonic CNS. The retinoid ligand knockout embryo has been a useful model system to examine the anatomy and the expression domains of a variety of genes in respect to the function of vitamin A in neural development, and it has been verified that the CNS requires vitamin A for normal development (MADEN et al. 1996, 1997, 1998). In the embryos lacking retinoids, the posterior hindbrain is completely missing, the neural tube fails to extend neurites into periphery, and neural crest cells do not survive. Furthermore, the expression patterns of Hoxb-1, Hoxb-4, Krox-20 and FGF-3 are altered, suggesting that segmentation in the myelencephalon is disrupted in the absence of vitamin A (MADEN et al. 1996). Cell death studies of the neuroepithelium suggest that retinoids and Msx-2 may be involved in apoptosis (MADEN et al. 1997). A more detailed discussion on these studies can be found in Chap. 14 (this volume). GOODMAN (1994, 1995, 1996) presents the hypothesis that dysregulation of vitamin A homeostasis during early development leads to anomalies linked to risk factors for schizophrenia. The many brain, craniofacial and digit anomalies that are seen in schizophrenics are suggestive of vitamin A deficiency and excess congenital anomalies (Goodman 1996). It may be possible to apply the embryonic quail retinoid ligand knockout model to address early developmental aspects dealing with this syndrome.

I. Ethanol-Induced Retinoid Depletion

Maternal alcohol abuse results in fetal alcohol syndrome (FAS) in the offspring which includes mental retardation, craniofacial and heart deformities, and low birth weight (Veghely et al. 1978). The exact etiology of FAS is not understood. Recent studies have suggested that FAS might be explained by ethanol causing deficiency of the vitamin A active form, retinoic acid in the embryo. It has been proposed that ethanol is a competitive inhibitor of retinol in the biogeneration of retinoic acid from retinol via alcohol dehydrogenase (ADH), causing a deficiency of retinoic acid in target tissues that require this vitamin A active form for normal embryonic development (Deltour et al. 1996; Duester 1991, 1994 1998). Indeed, it has been demonstrated in the mouse model that ethanol treatment results in a decrease in endogenous retinoic acid (Deltour et al. 1996). Another possibility is that ethanol induces P450 enzymes and retinol- and retinal dehydrogenases which enhance the breakdown of retinol as well as retinoic acid thus contributing to vitamin A depletion (reviewed in Twal and Zile 1997). The abnormalities resulting from vitamin A deficiency in embryos and fetuses are similar to many of the major defects observed in FAS (Jones et al. 1973; Mason 1935; Ruckman 1990; Veghely et al. 1978).

The avian embryo model is well suited to examine the effects of ethanol on early development and to assess if there is a link of FAS to an interference by ethanol in vitamin A metabolism. A quail embryo culture system was developed to test the hypothesis that an exposure to ethanol produces vitamin A deficiency syndrome in the embryo. The retinoid ligand knockout quail embryo provided a useful control system to verify the vitamin A deficient phenotype. When normal embryos were exposed to ethanol, cardiovascular and neural tube abnormalities developed, resembling those observed in the vitamin A deficient embryos grown under the same culture conditions (Twal and Zile 1997). Furthermore, the ethanol-induced abnormalities could be prevented by including retinoic acid in the ethanol-containing medium. While these studies did not provide direct evidence of an interference of retinol metabolism by ethanol, supporting evidence was obtained by adding citral, a competitive inhibitor of aldehyde dehydrogenase, the enzyme responsible for conversion of retinal to retinoic acid (Drager et al. 1998; Raner et al. 1996); citral produced abnormalities similar to those of the vitamin A deficient phenotype. Citral has also been used to induce retinoic acid depletion in the chick limb bud (Tanaka et al. 1996). Retinol accumulation and retinoid receptor depletion was reported in fetuses of rats consuming ethanol (Grummer et al. 1993).

It is clear from recent studies (DERSCH and ZILE 1993; MADEN et al. 1996, 1997, 1998; TWAL and ZILE 1997; ZILE 1998), that retinoic acid, the active form of vitamin A, is required during critical early stages in cardiac cell differentiation, as well as for neural crest cell proper migration. All these processes are affected by chronic exposure of the embryo to ethanol. Thus, the quail embryo model provides a valuable in culture and *in ovo* system to further clarify the link of FAS-induced congenital abnormalities to disturbances in endogenous vitamin A homeostasis during embryogenesis.

J. Retinoid Metabolism

It is becoming more and more clear that the function of vitamin A in early development is inevitably linked to vitamin A metabolism. In all known organisms the availability of free retinoic acid, the most potent metabolite of vitamin A, is very limited and likely under strict metabolic controls. This metabolic regulation is very critical in embryonic development, where exogenous retinoic acid is known to have a teratogenic effect on almost every developing tissue or organ system (AVANTAGGIATO et al. 1996; HOFMANN and EICHELE 1994; OSMOND et al. 1991; SHENEFELT 1972; WOOD et al. 1994; ZHANG et al. 1996). During normal embryonic development only small by physiologically effective amounts of retinoic acids are present in the embryo, as demonstrated by direct analysis of the early quail embryos (DONG and ZILE 1995), yet the yolk of the egg contains vitamin A stores sufficient for the development of the embryo through hatching. Analysis of retinoid metabolites by HPLC in neurulation stage quail embryos revealed the presence of the bioactive RAR ligands all-*trans*-retinoic acid and didehydroretinoic acid, as well as the upstream metabolites retinal, retinols and retinyl esters, attesting to the enzymatic capability of the early embryo to fully process vitamin A (DONG and ZILE 1995). Retinoic acid and 4-oxoretinoic acid have been identified in the *Xenopus* embryo (DURSTON et al. 1989; PIJNAPPEL et al. 1993); retinoic acid and 3,4-didehydroretinoic acid have been identified in the chick limb bud (HOFMANN and EICHELE 1994; THALLER and EICHELE 1987, 1990). Endogenous distribution of retinoids during normal development and teratogenesis has been examined in the mouse (HORTON and MADEN 1995) and zebrafish (COSTARIDIS et al. 1996) embryos. However, direct analyses of endogenous embryonal retinoids is difficult. More commonly used are indirect methods that measure retinoid activity rather than the retinoids themselves. CHEN et al. (1992, 1996) used a reporter-gene construct to measure retinoic acid activity in Hensen's node and in quail embryos; HOGAN et al. (1992) demonstrated the ability of the murine node to generate retinoic acid from retinol; localization of retinoic acid in Hensen's node and in the heart-forming regions of quail embryos was demonstrated with an antiretinoic acid monoclonal antibody (TWAL et al. 1995). It appears that in the developing vertebrate embryo retinoic acid is distributed in patterns that are tightly regulated in a spatiotemporal as well as in a concentration specific framework, with retinoic acid found at high levels at some sites while undetectable at others (CHEN et al. 1992; CREECH-KRAFT et al. 1994; DURSTON et al. 1989; MCCAFFERY and DRAGER 1994; YAMAMOTO et al. 1996). It has been proposed that these patterns are generated by the developmentally regulated spatiotemporal expression of retinoic acid generating enzyme systems which convert retinol to retinal and finally to retinoic acid (ANG and DUESTER 1997; ANG et al. 1996; DRAGER et al. 1998; DUESTER 1998; MCCAFFERY and DRAGER 1994). Using a reporter cell bioassay, retinoic acid biosynthesis in the mouse embryo has been detected beginning with the primitive streak formation (ANG et al. 1996), its spatiotemporal localization matching that of retinol- and retinal oxidizing enzymes, alcohol dehydrogenase-IV and aldehyde dehydrogenase, respectively (DUESTER 1998). A correlation of endogenous retinoic acid levels with that of aldehyde dehydrogenase has been reported in studies of mouse and zebrafish embryos in tissues including the trunk and spinal cord (MARSH-ARMSTRONG et al. 1995; MCCAFFERY and DRAGER 1994), the cerebellar area (YAMAMOTO et al. 1996), and the eye (DRAGER et al. 1998; MCCAFFERY and DRAGER 1994). The generation of retinoic acid by these enzyme systems is hypothesized to result in specific retinoic acid diffusion gradients into the surrounding tissues, a concept that originated in studies with tissues such as the zone of polarizing activity in the chick limb bud (HOFMANN and EICHELE 1994; SCADDING and MADEN 1994) and the chick organizer tissue, the Hensen's node (HOGAN et al. 1992). However, the retinoic acid diffusion gradient hypothesis still awaits verification. What is indisputable is that the biogeneration of the vitamin A active form(s) is the initial and critical developmental event in the initiation of the retinoid regulated signaling pathways, inseparably linking the function of vitamin A to its metabolism. The avian embryo retinoid ligand knockout model is a valuable tool to examine retinoid homeostasis during early development.

K. Future Directions

The embryonic quail retinoid ligand knockout model is uniquely suited for the elucidation of the physiological function of vitamin A in vertebrate early embryonic development. It is clear from available evidence that disruption or dysregulation of vitamin A function during preorganogenesis developmental stages can lead to congenital abnormalities in the newborn, to developmental defects incompatible with life or to an early loss of the embryo or fetus. While the molecular and genetic dissection of the retinoid signaling pathway has provided valuable insights into the pleiotropic functions of vitamin A and have demonstrated the critical role of retinoid receptors in vertebrate ontogenesis, the heterodimerization of these receptors with nonretinoid receptors can result in a large number of receptor combinations, severely complicating the interpretation of the results of such studies. The vitamin A deficient quail embryo model, in contrast, can provide much clearer insights into developmental events dependent solely on retinoids and, in conjunction with other model systems, can contribute to a multifaceted, integrated approach for advancing the field. The nutritionally generated retinoid ligand knockout avian embryo offers a simple and uniquely valuable model system to study retinoid function at the very beginnings of life without an interference of maternal environment, and with all the experimental advantages inherent to an avian model system. It is clear from the studies described in this chapter on the vitamin A deficient quail embryo, that a majority of critical early developmental events cannot proceed normally in the absence of retinoids, but that a repletion with active retinoids rescues the embryo. This opens up many avenues for exploring retinoid signaling pathways in early development using this model system and applying the already well-established avian embryo manipulation techniques. It is expected that studies with the quail embryo retinoid ligand knockout model will provide important new knowledge about the function of retinoids in early avian development and thus also greatly contribute to the basic knowledge of vertebrate embryonic development. Most importantly, this unique embryo model may provide clues to the origin of some yet unidentified etiologies of congenital abnormalities.

References

Ang HL, Deltour L, Hayamizu TF, Zgombic-Knight M, Duester G (1996) Retinoic acid synthesis in mouse embryos during gastrulation and craniofacial development linked to class IV alcohol dehydrogenase gene expression. J Biol Chem 271:9526–9534

Ang HL, Duester G (1997) Initiation of retinoid signaling in primitive streak mouse embryos: spatiotemporal expression patterns of receptors and metabolic enzymes for ligand synthesis. Dev Dyn 208:536–543

Arceci RJ, King AA, Simon MC, Orkin SH, Wilson DB (1993) Mouse GATA-4: a retinoic acid-inducible GATA-binding transcription factor expressed in endodermally derived tissues and heart. Mol Cell Biol 13:2235–2246

Avantaggiato V, Acampora D, Tuorto F, Simeone A (1996) Retinoic acid induces stage-specific repatterning of the rostral central nervous system. Dev Biol 175:347–357

Bates CJ (1983) Vitamin A in pregnancy and lactation. Proc Nutr Soc 42:65–79

Bodmer R (1995) Heart development in Drosophila and its relationship to vertebrates. Trends Cardiovasc Med 5:21–28

Blomhoff R (1994) Overview of vitamin A metabolism and function. In: Blomhoff R (ed) Vitamin A in health and disease. Dekker, New York, pp 1–35

Boyd AS (1989) An overview of retinoids. Am J Med 86:568–574

Boylan JF, Luftkin T, Achkar CC, Taneja R, Chambon P, Gudas LJ (1995) Targeted disruption of retinoic acid receptor alpha (RAR alpha) and RAR gamma results in receptor-specific alterations in retinoic acid-mediated differentiation and retinoic acid metabolism. Mol Cell Biol 15:843–851

Brockes JP (1989) Retinoids, homeobox genes, and limb morphogenesis. Neuron 2:1285–1294

Bronner-Fraser M (1996) Methods in avian embryology, vol 51. In: Bronner-Fraser M (ed) Methods in cell biology. Academic, New York

Chambon P (1993) The molecular and genetic dissection of the retinoid signaling pathway. Gene 135:223–228

Chan A, Hanna M, Abbott M, Keane RJ (1996) Oral retinoids and pregnancy. Med J Aust 165:164–167

Chan-Thomas PS, Thompson RP, Robert B, Yacoub MH, Barton PJR (1993) Expression of homeobox genes Msx-1 (Hox-7) and Msx-2(Hox-8) during cardiac development in the chick. Dev Dyn 197:203–216

Chen YP, Huang L, Russo AF, Solursh M (1992) Retinoic acid is enriched in Hensen's node and is developmentally regulated in the early chicken embryo. Proc Natl Acad Sci USA 89:10056–10059

Chen Y, Kostetskii I, Zile MH, Solursh M (1995) A comparative study of Msx-1 expression in early normal and vitamin A-deficient avian embryos. J Exp Zool 272: 299–310

Chen Y, Dong D, Kostetskii I, Zile MH (1996) Hensen's node from vitamin A-deficient quail embryo induces chick limb bud duplication and retains its normal asymmetric expression of sonic hedgehog (Shh). Dev Biol 173:256–264

Cohlan SQ (1954) Congenital abnormalities in the rat produced by excessive intake of vitamin A during pregnancy. Pediatrics 13:556–567

Costaridis P, Horton C, Zeitlinger J, Holder N, Maden M (1996) Endogenous retinoids in the zebrafish embryo and adult. Dev Dyn 205:41–51

Creech Kraft J, Schuh T, Juchau M, Kimelman D (1994) The retinoid X receptor ligand, 9-*cis*-retinoic acid, is a potential regulator of early Xenopus development. Proc Natl Acad Sci USA 91:3067–3071

Deltour L, Ang HL, Duester G (1996) Ethanol inhibition of retinoic acid synthesis as a potential mechanism for fetal alcohol syndrome. FASEB J 10:1050–1057

Dersch H, Zile MH (1993) Induction of normal cardiovascular development in the vitamin A-deprived quail embryo by natural retinoids. Dev Biol 160:424–433

Dickman E, Smith SM (1996) Selective regulation of cardiomyocyte gene expression and cardiac morphogenesis by retinoic acid. Dev Dyn 206:39–48

Dieterlen-Lievre F (1997) Symposium: current advances in avian embryology and incubation. Avian models in developmental biology. Poultry Sci 76:78–82

Doevedans PA, Vanbilsen M (1996) Transcription factors and the cardiac gene programme. Int J Biochem Cell Biol 28:387–403

Dong D, Zile MH (1995) Endogenous retinoids in the early avian embryo. Biochem Biophys Res Commun 217:1026–1031

Drager UC, Wagner E, McCaffery P (1998) Aldehyde dehydrogenase in the generation of retinoic acid in the developing vertebrate: a central role of the eye. J Nutr 128:463S–466S

Duester G (1991) A hypothetical mechanism for fetal alcohol syndrome involving ethanol inhibition of retinoic acid synthesis at the alcohol dehydrogenase step. Alcohol Clin Exp Res 15:568–572

Duester G (1994) Are ethanol-induced birth defects caused by functional retinoic acid deficiency? In: Blomhoff R (ed) Vitamin A in health and disease. Dekker, New York, p 343

Duester G (1998) Alcohol dehydrogenase as a critical mediator of retinoic acid synthesis from vitamin A in the mouse embryo. J Nutr 128:459S–462S

Duitsman PK, Cook LR, Tanumihardjo SA, Olson JA (1995) Vitamin A inadequacy in socioeconomically disadvantaged pregnant Iowan women as assessed by the modified relative dose response (MRDR) test. Nutr Res 15:1263–1276

Durston AJ, Timmermans JPM, Hage WJ, Hendriks HFJ, De Vries NJ, Heideveld M, Nieuwkoop PD (1989) Retinoic acid causes an anteroposterior transformation in the developing central nervous system. Nature 340:140–144

Eichele G (1997) Retinoids: from hindbrain patterning to Parkinson disease. Trends Gen 13:343–345

Ellis CN, Voorhees JJ (1987) Etretinate therapy. J Am Acad Dermatol 16:267–291

Evans SM, Yan W, Murillo MP, Ponce J, Papalopulu N (1995) tinman, a Drosophila homeobox gene required for heart and visceral mesoderm specification, may be represented by a family of genes in vertebrates: XNkx-2.3, a second vertebrate homologue of tinman. Development 121:3889–3899

Evans T (1997) Regulation of cardiac gene expression by GATA-4/5/6. Trends Cardiovasc Med 7:75–83

Fell HB, Mellanby E (1953) Metaplasia produced in cultures of chick ectoderm by high vitamin A. J Physiol (Lond) 119:470–488
Fisher M, Schoenwolf GC (1983) The use of early chick embryos in experimental embryology and teratology: improvements in standard procedures. Teratology 27:65–72
Fishman MC, Chien KR (1997) Fashioning the vertebrate heart: earliest embryonic decisions. Development 124:2099–2117
Gale E, Prince V, Lumsden A, Clarke J, Holder N, Maden M (1996) Late effects of retinoic acid on neural crest and aspects of rhombomere identify. Development 122:783–794
Gerster H (1997) Vitamin A – functions, dietary requirements and safety in humans. Int J Vit Nutr Res 67:71–90
Giguere V, Fawcett D, Luo J, Evans RM, Sucov HH (1996) Genetic analysis of the retinoid signal. Ann NY Acad Sci 785:12–22
Goodman AB (1995) Chromosomal locations and modes of action of genes of the retinoid (Vitamin A) system support their involvement in the etiology of schizophrenia. Am J Med Genet (Neuropsychiatr Genet) 60:335–348
Goodman AB (1994) Retinoid dysregulation as a cause of schizophrenia. Am J Psychiatry 151:452–453
Goodman AB (1996) Congenital anomalies in relatives of schizophrenic probands may indicate a retinoid pathology. Schizophrenia Res 19:163–170
Grummer MA, Langhough RE, Zachman RD (1993) Maternal ethanol ingestion effects on fetal rat brain. Vitamin A as a model for fetal alcohol syndrome. Alcohol Clin Exp Res 17:592–597
Gudas LJ, Sporn MB, Roberts AB (1994) Cellular biology and biochemistry of the retinoids. In: Sporn MB, Roberts AB, Goodman DS (eds) The retinoids: biology, chemistry, and medicine, 2nd edn. Raven, New York, pp 443–520
Hale F (1937) Relation of maternal vitamin A deficiency to microphthalmia in pigs. Texas State. J Med 33:228–232
Hamburger V, Hamilton HL (1951) A series of normal stages in the development of the chick embryo. J Morphol 88:49–92
Harvey RP (1996) NK-2 homeobox genes and heart development. Dev Biol 178: 203–216
Heine UI, Roberts AB, Munoz EF, Roche NS, Sporn MB (1985) Effects of retinoid deficiency on the development of the heart and vascular system of the quail embryo. Virchows Arch [Cell Pathol] 50:135–152
Helms JA, Kim CH, Hu D, Minkoff R, Thaller C, Eichele G (1997) Sonic hedgehog participates in craniofacial morphogenesis and is down-regulated by teratogenic doses of retinoic acid. Dev Biol 187:25–35
Henion PD, Weston JA (1994) Retinoic acid selectively promotes the survival and proliferation of neurogenic precursors in cultured neural crest cell populations. Dev Biol 161:243–250
Hofmann C, Eichele G (1994) Retinoids in development. In: Sporn MB, Roberts AB, Goodman DS (eds) The retinoids: biology, chemistry, and medicine, 2nd edn. Raven, New York, pp 387–441
Hoffman JIE (1995) Incidence of congenital heart disease: I. Postnatal incidence. Pediatr Cardiol 16:103–113
Hogan BLM, Thaller C, Eichele G (1992) Evidence that Hensen's node is a site of retinoic acid synthesis. Nature 359:237–241
Hong WK, Itri LM (1994) Retinoids and human cancer. In: Sporn MB, Roberts AB, Goodman DS (eds) The retinoids: biology, chemistry and medicine, 2nd edn. Raven, New York, pp 597–630
Horton C, Maden M (1995) Endogenous distribution of retinoids during normal development and teratogenesis in the mouse embryo. Dev Dyn 202:312–323
Icardo JM (1996) Developmental biology of the heart. J Exp Zool 275:144–161
Isaac A, Sargent MG, Cooke J (1997) Control of vertebrate left-right asymmetry by a Snail-related zinc finger gene. Science 275:1301–1304

Jones KL, Smith DW, Ulleland CN, Streissguth AP (1973) Pattern of malformation in offspring of chronic alcoholic mothers. Lancet 7815:1267–1271
Kamm JJ, Ashenfelter KO, Ehmann CW (1984) Preclinical and clinical toxicology of selected retinoids. In: Sporn MB, Roberts AB, Goodman DS (eds) The retinoids. Academic, New York, pp 287–326
Kastner P, Grondona JM, Mark M, Gansmuller A, Lemeur M, Decimo D, Vonesch J-L, Dolle P, Chambon P (1994) Genetic analysis of RXRα developmental function: convergence of RXR and RAR signaling pathways in heart and eye morphogenesis. Cell 78:987–1003
Kern MJ, Argao EA, Potter SS (1995) Homeobox genes and heart development. Trends Cardiovasc Med 5:47–54
Kimmel CA, Generoso WM, Thomas RD, Bakshi KS (1993) Contemporary issues in toxicology. A new frontier in understanding the mechanisms of developmental abnormalities. Toxicol Appl Pharmacol 119:159–165
Kirby ML (1993) Cellular and molecular contributions of the cardiac neural crest to cardiovascular development. Trends Cardiovasc Med 3:18–23
Kistler A, Hummler H (1985) Teratogenesis and reproductive safety evaluation of the retinoid etretin (Ro10–1670) Arch Toxicol 58:50–56
Kochhar DM, Jiang H, Soprano DR, Harnish DC (1993) Early embryonic cell response in retinoid-induced teratogenesis. In: Livrea MA, Packer L (eds) Retinoids. Progress in research and clinical applications. Dekker, New York, pp 383–396
Kochhar DM, Christian MS (1997) Tretinoin: a review of the nonclinical developmental toxicology experience. J Am Acad Dermatol 36:S47–S59
Kostetskii I, Linask KL, Zile MH (1996) Vitamin A deficiency and the expression of retinoic acid receptors during cardiogenesis in early quail embryo. Roux's Arch. Dev Biol 205:260–271
Kostetskii I, Yuan SY, Kostetskaia E, Linask KK, Blanchet S, Seleiro E, Michaille JJ, Brickell P, Zile M (1998) Initial retinoid requirement for early avian development coincides with retinoid receptor coexpression in the precardiac fields and induction of normal cardiovascular development. Dev Dyn 213:188–198
Lammer EJ (1988) Developmental toxicity of synthetic retinoids in humans. In: Liss AR (ed) Transplacental effects of fetal health, pp 193–202
Lammer EJ, Chen DT, Hoar RM, Agnish ND, Benke PJ, Braun JT, Curry CJ, Fernhoff PM, Grix AW, Lott IT, Richard JM, Sun SC (1985) Retinoic acid embryopathy. N Engl J Med 313:837–841
Laverriere AC, McNeil C, Mueller C, Poelmann RE, Burch JBE, Evans T (1994) GATA-4/5/6, a subfamily of three transcription factors transcribed in developing heart and gut. J Biol Chem 269:23177–23184
Leblanc BP, Stunnenberg HG (1995) 9-*cis* retinoic acid signaling: changing partners causes some excitement. Genes Dev 9:1811–1816
LeDouarin NM (1969) Particularités du noyau interphasique chez la caille japonaise. Utilisation de ces particularités comme "marquage biologique" dans les recherches. Bull Biol Fr Belg 103:435–452
LeDouarin NM, Dieterlen-Lièvre F, Teillet MAT (1995) Quail-chick transplantations. In: Moses PB (ed) Methods in cell biology, vol 51. Academic, New York, pp 23–59
Levin M, Johnson RL, Stern CD, Kuehn M, Tabin C (1995) A molecular pathway determining left-right asymmetry in chick embryogenesis. Cell 82:803–814
Levin M, Pagan S, Roberts DJ, Cooke J, Kuehn MR, Tabin CJ (1997) Left/right patterning signals and the independent regulation of different aspects of situs in the chick embryo. Dev Biol 189:57–67
Lints TJ, Parsons LM, Hartley L, Lyons I, Harvey RP (1993) Nkx-2.5: a novel murine homeobox gene expressed in early heart progenitor cells and their myogenic descendants. Development 119:419–431
Little CD, Rongish BJ (1995) The extracellular matrix during heart development. Experientia 51:873–882

Luo J, Pasceri P, Conlon RA, Rosant J, Giguere V (1995) Mice lacking all isoforms of retinoic acid receptor β develop normally and are susceptible to the teratogenic effects of retinoic acid. Mech Dev 53:61–71

Maden M, Gale E, Kostetskii I, Zile MH (1996) Vitamin A-deficient quail embryos have half a hindbrain and other neural defects. Curr Biol 6:417–426

Maden M, Gale E, Zile MH (1998) The role of vitamin A in the development of the central nervous system. J Nutr 128:471S–475S

Maden M, Graham A, Gale E, Rollinson C, Zile MH (1997) Positional apoptosis during vertebrate CNS development in the absence of endogenous retinoids. Development 124:2799–2805

Maden M (1994) Role of retinoids in embryonic development. In: Blomhoff R (ed) Vitamin A in health and disease. Dekker, New York, pp 289–322

Maden M, Holder N (1992) Retinoic acid and development of the central nervous system. Bioessays 14:431–438

Mangelsdorf DJ, Thummel C, Beato M, Herrlich P, Schutz G, Umesono K, Blumberg B, Kastner P, Mark M, Chambon P, Evans RM (1995) The nuclear receptor superfamily: the second decade. Cell 83:835–839

Marsh-Armstrong N, McCaffery P, Hyatt G, Alonso L, Dowling JE, Gilbert W, Drager UC (1995) Retinoic acid in the anteroposterior patterning of the zebrafish trunk. Roux's Arch. Dev Biol 205:103–113

Marshall H, Studer M, Popperl H, Aparicio S, Kuriowa A, Brenner S, Krumlauf R (1994) A conserved retinoic acid response element required for early expression of the homeobox gene Hoxb-1. Nature 370:567–571

Mason KE (1935) Foetal death, prolonged gestation, and difficult parturition in the rat as a result of vitamin A deficiency. Am. J Anat 57:303–349

McCaffery P, Drager UC (1994) Hotspots of retinoic acid synthesis in the developing spinal cord. Proc Natl Acad Sci USA 91:7194–7197

Moon RC, Mehta RG, Rao KVN (1994) Retinoids and cancer in experimental animals. In: Sporn MB, Roberts AB, Goodman DS (eds) The retinoids: biology, chemistry, and medicine, 2nd edn. Raven, New York, pp 573–595

Moore T (1957) Vitamin A. Elsevier, New York

Morriss GM (1972) Morphogenesis of the malformations induced in rat embryos by maternal hypervitaminosis A. J Anat 113:241–250

Morriss GM, Steele CE (1972) Comparison of the effects of retinal and retinoic acid on postimplantation rat embryos in vitro. Teratology 15:109–119

Morriss-Kay GM, Murphy P, Hill RE, Davidson DR (1991) Effects of retinoic acid excess on expression of Hox 2.9 and Krox-20 and of morphological segmentation in the hindbrain of mouse embryos. EMBO J 10:2985–2995

Nau H, Chahoud I, Dencker L, Lammer EJ, Scott WJ (1994) Teratogenicity of vitamin A and retinoids. In: Blomhoff R (ed) Vitamin A in health and disease. Dekker, New York, pp 615–664

Osmond MK, Butler AJ, Voon FCT, Bellairs R (1991) The effects of retinoic acid on heart formation in the early chick embryo. Development 113:1405–1417

Pfahl M, Chytil F (1996) Regulation of metabolism by retinoic acid and its nuclear receptors. Annu Rev Nutr 16:257–283

Pijnappel WWM, Hendriks HFJ, Folkers GE, Van Den Brink CE, Dekker EJ, Edelenbosch C, Van Der Saag PT, Durston AJ (1993) The retinoid ligand 4-oxo-retinoic acid is a highly active modulator of positional specification. Nature 366:340–344

Popperl H, Featherstone MS (1993) Identification of a retinoic acid response element upstream or the murine Hox-4.2 gene. Mol Cell Biol 13:257–265

Raner GM, Vax ADN, Coon MJ (1996) Metabolism of all-*trans*, 9-*cis*, and 13-*cis* isomers of retinal by purified isozymes of microsomal cytochrome P450 and mechanism-based inhibition of retinoid oxidation by citral. Mol Pharmacol 49:515–522

Roberts AB, Sporn MB (1984) Cellular biology and biochemistry of retinoids. In: Sporn MB, Roberts AB, Goodman, DS (eds) The retinoids, vol 2. Academic, Orlando, pp 209–286
Rosa FW (1993) Retinoid embryopathy in humans. In: Koren G (ed) Retinoids in clinical practice: the risk-benefit ratio. Dekker, New York, pp 77–109
Ruckman RN (1990) Cardiovascular defects associated with alcohol, retinoic acid, and other agents. Ann NY Acad Sci 588:281–288
Rutledge JC, Generoso WM (1989) Fetal pathology produced by ethylene oxide treatment of the murine zygote. Teratology 39:563–572
Schultheiss TM, Xydas S, Lassar AB (1995) Induction of avian cardiac myogenesis by anterior endoderm. Development 121:4203–4214
Scadding S, Maden M (1994) Retinoic acid gradients during limb regeneration. Dev Biol 162:608–617
Selevan SG (1981) Design considerations in pregnancy outcome studies of occupational populations. Scand J Work Environ Health 7:76–82
Shalita AR (1988) Lipid and teratogenic effects of retinoids. J Am Acad Dermatol 19:197–198
Shenai JP, Rush MG, Parker RA, Chytil F (1995) Sequential evaluation of plasma retinol-binding protein response to vitamin A administration in very-low-birth weight neonates. Biochem Mol Med 54:67–74
Shenefelt RE (1972) Morphogenesis of malformations in hamsters caused by retinoic acid: relation to dose and stage at treatment. Teratology 5:103–118
Shepard TH (1986) Human teratogenicity. Adv Pediatr 33:225–268
Smith SM (1994) Retinoic acid receptor isoform β_2 is an early marker for alimentary tract and central nervous system positional specification in the chicken. Dev Dyn 200:14–15
Smith SM, Dickman ED, Power SC, Lancman J (1998) Retinoids and their receptors in vertebrate embryogenesis. J Nutr 128:467S–470S
Smith SM, Dickman ED, Thompson RP, Sinning AR, Wunsch AM, Markwald RR (1997) Retinoic acid directs cardiac laterality and the expression of early markers of precardiac asymmetry. Dev Biol 182:162–171
Smith SM, Eichele G (1991) Temporal and regional differences in the expression patterns of distinct retinoic acid receptor-β-transcripts in the chick embryo. Development 111:245–252
Strauss JS, Cunningham WJ, Leyden JJ, Pochi PE, Shalita AR (1988) Isotretinoin and teratogenicity. J Am Acad Dermatol 19:353–354
Studer M, Popperl H, Marshall H, Kuriowa A, Krumlauf R (1994) Role of conserved retinoic acid response element in rhombomere restriction of Hoxb-1. Science 265:1728–1732
Sucov HM, Dyson E, Gumeringer CL, Price J, Chien KR, Evans RM (1994) RXRα mutant mice establish a genetic basis for vitamin A signaling in heart morphogenesis. Genes Dev 8:1007–1018
Sulik KK (1984) Critical periods for alcohol teratogenesis in mice, with special reference to the gastrulation stage of embryogenesis. Mechanisms of alcohol damage in utero. Pitman, London, pp 124–141 (Ciba Foundation symposium 105)
Tanaka M, Tamura K, Ide H (1996) Citral, an inhibitor of retinoic acid synthesis, modifies chick limb development. Dev Biol 175:239–247
Teelmann K (1989) Retinoids: toxicology and teratogenicity to date. Pharmacol Ther 40:29–43
Thaller C, Eichele G (1987) The identification and spatial distribution of retinoids in the developing chick limb bud. Nature 327:625–628
Thaller C, Eichele G (1990) Isolation of 3,4-didehydroretinoic acid, a novel morphogenetic signal in the chick wing bud. Nature 345:815–819
Thompson JN (1969) The role of vitamin A in reproduction. In: DeLuca HF, Suttie JW (eds) The Fat Soluble Vitamins. The University of Wisconsin, Madison, pp 267–281

Thorogood P, Smith L, Nicol A, McGinty R, Garrod D (1982) Effects of vitamin A on the behaviour of migratory neural crest cells in vitro. J Cell Sci 57:331–350

Tsuda T, Philp N, Zile MH, Linask K (1996) Left-right asymmetric localization of flectin in the extracellular matrix during heart looping. Dev Biol 173:39–50

Twal W, Roze L, Zile MH (1995) Anti-retinoic acid monoclonal antibody localizes all-*trans*-retinoic acid in target cells and blocks normal development in early quail embryo. Dev Biol 168:225–234

Twal WO, Zile MH (1997) Retinoic acid reverses ethanol-induced cardiovascular abnormalities in quail embryos. Alcohol Clin Exp Res 21:1137–1143

Underwood BA (1994) Vitamin A in human nutrition: public health considerations. In: Sporn MB, Roberts AB, Goodman DS (eds) The retinoids: biology, chemistry and medicine, 2nd edn. Raven, New York, pp 211–227

Veghely PV, Osztovics M, Kardos G, Leisztner L, Szaszovszky E, Igali S, Imrei J (1978) The fetal alcohol syndrome: symptoms and pathogenesis. Acta Paediat Acad Sci Hung 19:171–189

Wald G (1968) The molecular basis of visual excitation. Nature 219:800–807

Wallingford JC, Underwood BA (1986) Vitamin A deficiency in pregnancy, lactation and the nursing child. In: Bauernfeind JC (ed) Vitamin A deficiency and its control Academic, Orlando, Fl, pp 101–152

Wedden SE, Pang K, Eichele G (1989) Expression pattern of homeobox-containing genes during chick embryogenesis. Development 105:639–650

Wellik DM, DeLuca HF (1995) Retinol in addition to retinoic acid is required for successful gestation in vitamin A-deficient rats. Biol Reprod 53:1392–1397

Wellik DM, DeLuca HF (1996) Metabolites of all-*trans*-retinol in day 10 conceptuses of vitamin A-deficient rats. Arch Biochem Biophys 330:355–362

Wellik DM, Norback DH, DeLuca HF (1997) Retinol is specifically required during midgestation for neonatal survival. Am J Physiol 272 (Endocrinol Metab 35: E25–E29

White JC, Clagett-Dame M (1996) Abnormal development of the optic cup, olfactory pit, cranial ganglia and anterior cardinal vein in vitamin A-deficient rat embryos. Soc Neurosci Abstr 22:990 part 2

Wiens DJ, Mann TK, Fedderson DE, Rathmell WK, Franck BH (1992) Early heart development in the chick embryo: effects of isotretinoin on cell proliferation, α-actin synthesis, and development of contractions. Differentiation 51:105–112

Wilcox AJ, Weinberg CR, Wehmann RE, Armstrong EG, Canfield RE, Nisula BC (1985) Measuring early pregnancy loss: Laboratory and field methods. Fertil Steril 44:366–374

Wilson JG (1977) Embryotoxicity of drugs in man. In: Wilson JG, Fraser FC (eds) Handbook of teratology. Plenum, New York, pp 309–355

Wilson JG, Roth CB, Warkany J (1953) An analysis of the syndrome of malformations induced by maternal vitamin A deficiency. Effects of restoration of vitamin A at various times during gestation. Am. J Anat 92:189–217

Wilson JG, Warkany J (1949) Aortic arch and cardiac anomalies in the offspring of vitamin A deficient rats. Am. J Anat 85:113–155

Wolbach SB (1954) Effects of vitamin A deficiency and hypervitaminosis A in animals. In: Sebrell WH, Harris RS (eds) The vitamins, vol I. Academic, New York, pp 106–136

Wolf G (1984) Multiple functions of vitamin A. Physiol Rev 64:873–938

Wood H, Gurman P, Morriss-Kay G (1994) Exposure to retinoic acid before or after the onset of somitogenesis reveals separate effects on rhombomeric segmentation and 3′ HoxB gene expression domains. Development 120:2279–2285

Wunsch AM, Little CD, Markvald RR (1994) Cardiac endothelial heterogeneity defines valvular development as demonstrated by the diverse expression of JB3, an antigen of the endocardial cushion tissue. Dev Biol 165:585–601

Yamamoto M, McCaffery P, Drager UC (1996) Influence of the choroid plexus on cerebellar development: analysis of retinoic acid synthesis. Dev. Brain Res 93:182–190

Zhang Z, Balmer JE, Lovlie A, Fromm SH, Blomhoff R (1996) Specific teratogenic effects of different retinoic acid isomers and analogs in the developing anterior central nervous system of zebrafish. Dev Dyn 206:73–86

Zile MH (1992) Vitamin A homeostasis endangered by environmental pollutants. Proc Soc Exp Biol Med 201:141–153

Zile MH (1998) Vitamin A and embryonic development: an overview. J Nutr 128:455S-458S

Zile MH, Kostetskii I, Yuan S (1998) Vitamin A specifies heart asymmetry during avian cardiogenesis. FASEB J 12:A319

Zile MH, Yuan S, Kostetskii I (1997) Expression patterns of retinoid receptors in early quail embryos are altered in vitamin A deficiency. FASEB J 11:A412

CHAPTER 16

Retinoid Receptors, Their Ligands, and Teratogenesis: Synergy and Specificity of Effects

H. NAU and M.M.A. ELMAZAR

A. Introduction

Vitamin A and metabolites (retinoids) control numerous processes which are critical for reproduction and development such as differentiation, proliferation, apoptosis and morphogenesis (DE LUCA et al. 1995). Evidence has accumulated since the discovery of the retinoid receptors a decade ago (PETKOVIC et al. 1987; GIGUERE et al. 1987) that these nuclear receptors may be involved in many of these processes. Retinoid receptors are ligand-activated transcription factors which control the expression of a number of target genes involved in development such as growth factors, growth factor receptors, cell adhesion molecules, intercellular matrix molecules, other transcription factors such as *hox* genes, some hormones and cytokines, as well as other receptors of the hormone receptor superfamily; furthermore, retinoid pathways themselves are influenced via the control of expression of retinoid binding proteins, metabolizing enzymes and autoregulation of retinoid receptors (KASTNER et al. 1995; CHAMBON 1996; MORRIS-KAY and SOKOLOVA 1996; VAN DER SAAG 1996; NAGPAL and CHANDRARATHNA 1996; SMITH et al. 1998).

Many, if not most of these pleiotropic effects are mediated by multiple signaling pathways which are generated by the interaction of retinoid receptors with their respective retinoid ligands. There are two families of nuclear retinoid receptors (Fig. 1). The RAR-family, consisting of three members RARα, RARβ and RARγ, interacts with all-*trans*-retinoic acid; On the other hand, 9-*cis*-retinoic acid can interact with both the RAR-family and the second group of retinoid receptors, the RXR-family which also consists of three members, RXRα, RXRβ and RXRγ (LEVIN et al. 1992; HEYMAN et al. 1992; ALLENBY et al. 1993). Thus, the multiplicity of receptors and their ligands, as well as their highly specific spatial-temporal distribution opens the possibilities of numerous receptor-ligand combinations, especially if receptor dimers are also considered. In vitro studies have indicated that heterodimers between members of the RAR and RXR families play important roles in differentiation (DURAND et al. 1992; ZHANG and PFAHL 1993).

Because of the apparent importance of retinoids and their nuclear receptors for development (MADEN 1994; NAU et al. 1994), it is reasonable to hypothesize that a disturbance of the complex retinoid signaling pathways may be involved in teratogenesis. Indeed, both vitamin A deficiency (WILSON et al.

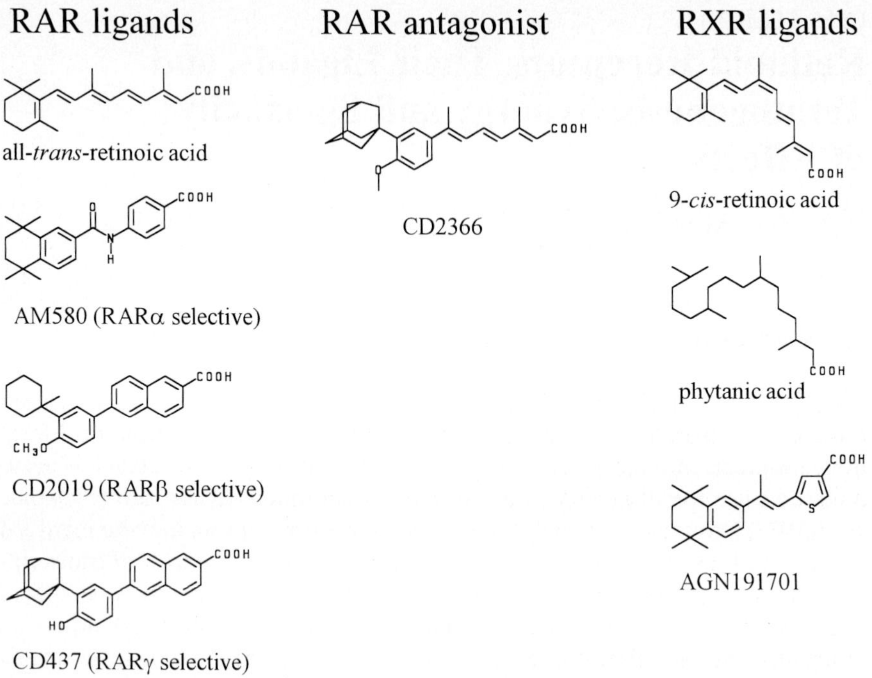

RAR ligands

all-*trans*-retinoic acid

AM580 (RARα selective)

CD2019 (RARβ selective)

CD437 (RARγ selective)

RAR antagonist

CD2366

RXR ligands

9-*cis*-retinoic acid

phytanic acid

AGN191701

Fig. 1. Structures of retinoids which interact with the receptor families RARs or RXRs and which are discussed in this review

1953) as well as excess of vitamin A or retinoids (LAMMER et al. 1985; ROSA et al. 1986; NAU 1993; COBERLY et al. 1996) result in major developmental effects. The pattern of defects observed following hypo- and hypervitaminosis A are quite distinct, although in some cases overlapping. The hallmark effects of vitamin A deficiency are defects of the eye, the heart and the reproductive system, while CNS defects (exencephaly, spina bifida, retarded mental development), craniofacial defects (ear defects), thymus and heart malformations as well as limb defects (latter in experimental animals) are among the most prominent defects observed following exposure to vitamin A/retinoid excess. Experimental and clinical studies have shown, however, that numerous structures in the developing organism can be affected by hypo- and hypervitaminosis A/retinoid excess, depending on the substance administered, the time period of administration or deficiency, and the degree of exposure or deficiency.

This review evaluates the evidence for participation of the retinoid receptor signaling pathways in teratogenesis.

B. Ligands of the Retinoid Receptors

I. Endogenous Ligands of the Retinoid Receptors

There are two major sources of vitamin A: food of animal origin provides predominantly retinyl esters which – following a complex series of absorption, metabolism and distribution processes – are stored mostly in the liver (OLSON 1987; BIESALSKI 1997). Food of plant origin provides carotenoids; one of these, β-carotene, can be centrally cleaved to retinal (for detailed description of pathways of absorption, distribution and storage see Chap. 4, this book). Oxidation of retinol to retinal, and further oxidation of retinal leads to retinoic acid; the all-*trans*-isomer of retinoic acid is the most important ligand of the RARs in mammalian species (Fig. 1).

Hydrolysis of the retinyl esters in food of animal origin (e.g., liver) has produced all-*trans*-retinol as the major isomer, although a low proportion of other isomers (9-*cis*-, 13-*cis*-, 9,13-di-*cis*-retinol) were also found in liver (BRINKMANN et al. 1995). In plasma, all-*trans*-retinol is by far the most predominant retinol isomer, and other isomers are very low, if at all present (T. ARNHOLD, unpublished results from this laboratory). The reason for the predominance of all-*trans*-retinol in blood is probably the fact that this isomer is bound with high affinity to its carrier protein, retinol binding protein (RBP); other isomers, which are not efficiently bound, are more rapidly metabolized or excreted. It is therefore surprising that all-*trans*-retinoic acid is not the predominant retinoic acid isomer in plasma of some mammalian species (rabbit, monkey, human) (NAU et al. 1994); especially in the human, 13-*cis*- and 13-*cis*-4-oxo-retinoic acids are prominent plasma metabolites (NAU 1990; 1994; ECKHOFF and NAU 1990; ECKHOFF et al. 1991), presumably because of their extensive binding to plasma proteins. If these *cis*-metabolites are produced via isomerization at the retinol, retinal- or retinoic acid level is not known, and probably several pathways are possible here (see below).

All-*trans*-didehydro-retinoic acid is also a ligand for the RARs (SANI et al. 1997). Its importance as an endogenous receptor ligand is questionable, however, because it does not occur in significant amounts in most mammalian species. Low levels have been observed in the *Xenopus* embryo (CREECH KRAFT et al. 1994). This retinoid may play an important role in chick embryo development, since its concentrations in the limb bud of this species were fivefold higher than those of all-*trans*-retinoic acid (SCOTT et al. 1994).

9-*Cis*-retinoic acid, the ligand for the RARs as well as the RXRs, was reported to occur in mouse liver (4 ng/g) and mouse kidney (30 ng/g) in considerable concentrations (HEYMAN et al. 1992). Further reports on tissue levels of this isomer are very sparse or controversial. In our laboratory, 9-*cis*-retinoic acid was identified in human plasma following consumption of liver (ARNHOLD et al. 1996), or, at lower concentrations, after vitamin A supplementation. After liver consumption, levels reached 2.7 ± 1.1 ng/ml plasma, and then decreased within a few hours to levels at or below the analytical detection limit of

0.2 ng/ml plasma. At present, the precise concentrations of 9-*cis*-retinoic acid in mammalian blood or tissue, including the embryo, are not known and are likely to be very low. It is therefore difficult to evaluate the role of this isomer in the retinoid signaling pathways, in particular since 9-*cis*-retinoic acid may induce RXR-RXR-homodimers (ZHANG et al. 1992), and RXR may also function as "silent partners" without ligand activation.

All-*trans*-4-oxo-retinol and all-*trans*-4-oxo-retinal have recently been proposed as ligands and transactivators for RARs in *Xenopus* and in some in vitro system (F9 cell differentiation assay) (BLUMBERG et al. 1996; ACHKAR et al. 1996). The relevance of these retinoids in the mammalian retinoid pathways is not known. All-*trans*-retinol was also suggested as RAR ligand and transactivator (REPA et al. 1993); This finding was not further explored. Furthermore, 9,13-di-*cis*-retinoic acid was shown to activate RARα in vitro (IMAI et al. 1997); the relevance of these findings to the in vivo situation is not known.

Phytanic acid was identified as another ligand and transactivator of the RXRs (LEMOTTE et al. 1996) (Fig. 1). This compound is an oxidation product of phytol and phytol esters are occurring in a variety of food (e.g., as chlorophyll). Phytanic acid is present in human blood in considerable concentrations (micromolar amounts). Extremely high concentrations (almost millimolar levels) can be found in some disease states such as Refsum disease or Zellweger syndrome, where phytanic acid α-oxidation is deficient (see WATKINS et al. 1996). We presently investigate if interaction of phytanic acid with retinoid pathways may pose a teratogenic risk.

The importance of retinoid ligands for development was exemplified by "ligand knock-out" approaches. Vitamin A deficiency resulted in a drastic decrease in all retinoids in the embryo, and major effects on reproduction and development were induced (see above). Application of various retinoids during specific developmental periods revealed the efficiency of "rescue" of embryonic structures by a particular retinoid (DICKMANN et al. 1997; WELLIK et al. 1997; ZILE 1998). It was shown in the vitamin A deficient rat that all-*trans*-retinoic acid can substitute for vitamin A in all instances except for reproduction (spermatogenesis) and vision: The embryos of vitamin A deficient dams which were "rescued" by all-*trans*-retinoic acid developed normally, but the offspring were blind and sterile. Retinoic acid cannot be reduced to retinal and no cofactor for rhodopsin can be produced in this way in vision. Furthermore, retinoic acid cannot transfer the blood-testis barrier (KURLANDSKY et al. 1995), and retinol is needed to transfer to the testis where it can be metabolized into retinoic acid.

II. Retinoid Receptor Ligands in the Embryo

All-*trans*-retinoic acid is present endogenously in the embryo and shows a specific distribution within areas such as the floor plate of the neural tube (WAGNER et al. 1990), in Hensens node (CHEN et al. 1992), and in the developing limb (THALLER and EICHELE 1987; DURSTON et al. 1989; SARTRE and

Kochhar 1989; Tamura et al. 1993; Scott et al. 1994). The limb bud was studied in particular detail and it was found that all-*trans*-retinoic acid formed a posterior-anterior concentration gradient which may be involved in the three-dimensional patterning of the limb (Thaller and Eichele 1987; Thaller et al. 1993; Helms et al. 1996; 1997). This view of all-*trans*-retinoic acid as a morphogen has been debated, however (Noji et al. 1991), and further studies in the mouse limb bud revealed a much smaller gradient as compared to the chick (Scott et al. 1994). Also the concentrations of the retinoids in mouse limb buds were much smaller – and showed a different pattern – compared to the chick (Scott et al. 1994). All-*trans*-didehydroretinoic acid was the predominant retinoic acid in chick limb bud, while this retinoid was essentially absent in the mouse. It is surprising that the pattern of retinoid metabolites present endogenously shows a great species variation. As discussed above, didehydroretinoids were found only in the chick embryo in significant amounts, and in the rabbit and human embryo to a minor degree. The predominant retinoid in mouse and rat embryos was all-*trans*-retinoic acid, while the 13-*cis* isomers of retinoic acid and 4-oxo-retinoic acid were prominently present in other species such as the human (Nau et al. 1994; Eckhoff and Nau 1990).

A further potentially important fact emerged from studies in the mouse embryos (Scott et al. 1994): the concentrations of all-*trans*-retinoic acid were much lower than those of its cellular binding protein CRABPI; in contrast, retinol exceeded its cellular binding protein CRBPI by a great factor. Thus it appears that all-*trans*-retinoic acid is predominately bound to CRABPI in mouse embryos, while most of retinol remains unbound to CRBPI. The exact significance of these findings remain unclear, but it is reasonable to speculate that CRABPI may protect the embryonic cells from excess retinoic acid and serve as reservoir for this ligand; on the other hand, much retinol is available for a direct – and presently unknown – action on embryonic structures as well as for the generation of all-*trans*-retinoic acid in specific areas. Such "hot spots" of retinoic acid synthesis have been identified in neural tissues, in particular in the spinal areas (McCaffery and Dräger 1994).

A number of enzymes were identified which can synthesize and further metabolize retinoic acid. The cytosolic alcohol dehydrogenase IV (ADH IV) was suggested to be inhibited by ethanol and retinoic acid deficiency of the embryo may be the basis of ethanol teratogenesis (Haselbeck et al. 1997; Duester 1998). Furthermore, a new microsomal cytochrome P450 was cloned (CYP 26 or CYP RAI) was recently cloned which shows an impressive specificity in regard to retinoic acid 4-hydroxylation (White et al. 1996). Both enzymes show a very specific expression pattern in the embryo.

Presently it is unknown how much all-*trans*-retinoic acid is received by the embryo via maternal sources, and how much is synthesized within the embryo itself. Studies in the adult rat have shown (Kurlandsky et al. 1995) that all-*trans*-retinoic acid present in well-perfused tissues such as the liver and brain is mostly derived from the central circulation. In other tissues all-*trans*-retinoic acid is derived both from the circulation and local synthesis. In the testis all-

trans-retinoic acid is predominately synthesized locally from retinol, as direct transfer form the circulation is apparently blocked by the blood-testis barrier (see above). Such detailed studies are not available for the embryo. It is clear, however, that both pathways are operating here. Knock-out of retinol binding protein (RBP), the plasma carrier protein for retinol, resulted in retinol and retinoic acid deficiency in the embryo as demonstrated in a culture system (BAVIK et al. 1996; WARD et al. 1997); it appears likely that the substrate retinol was lacking for local synthesis of all-*trans*-retinoic acid which resulted in severe defects in the embryo. Furthermore we recently showed that embryonic blood contains a protein which can bind all-*trans*-retinoic acid (unpublished results) (Fig. 2). This protein is probably RBP, which is synthesized in the yolk sac (SOPRANO et al. 1988), and thus may contain binding sites which are not yet occupied by retinol and are available for retinoic acid binding. In contrast, binding sites in adult plasma RBP are fully occupied by retinol, and little- if any – binding of all-*trans*-retinoic acid can take place. Thus, it seems likely that all-*trans*-retinoic acid from maternal circulation can be carried to the embryo via yolk sac RBP (Fig. 2). Both local sources and placental transfer may contribute to embryonic all-*trans*-retinoic acid. This view is supported by placental transfer studies of labeled all-*trans*-retinoic acid which had been administered to the dam: a very specific distribution within the embryonic structures was found indicating a direct placental transfer of all-*trans*-retinoic acid. It was also suggested that the cellular binding proteins CRBP I and II as well as CRABP I and II may be involved in placental transfer of retinoids (SAPIN et al. 1997).

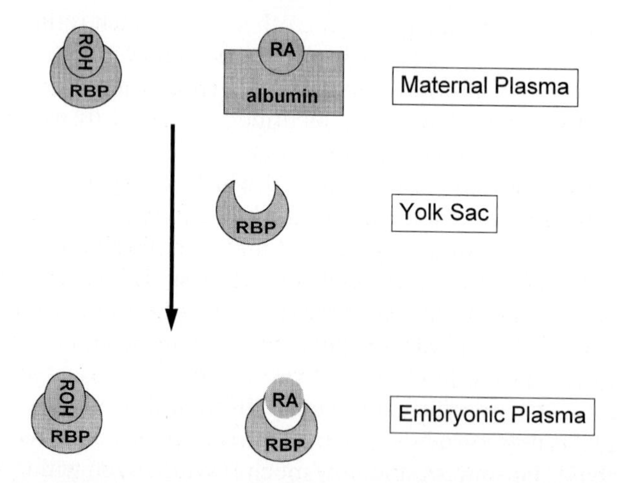

Fig. 2. Proposed scheme of retinol (*ROH*) placental transfer. Maternal plasma retinol binding protein RBP is fully occupied by retinol and all-*trans* retinoic acid is carried by albumin. RBP synthesized in the yolk sac has free binding sites which can subsequently be used to carry both retinol and all-*trans*-retinoic acid into the embryonic plasma

C. Retinoid Receptors and Binding Proteins in the Embryo

I. Retinoid Binding Proteins in the Embryo

The cytosolic binding proteins for retinoic acid and retinol (CRABPs and CRBPs, respectively) also show a very specific expression pattern within the embryo (MADEN et al. 1991; GUSTAFSON et al. 1993; DEKKER et al. 1994; SCOTT et al. 1994). These binding proteins may be involved in the metabolism of retinol and retinoic acid, as well as in the control of local concentrations of "free" all-*trans*-retinoic acid which is available for nuclear receptor binding. This view is supported by the fact that CRBP is relatively high in tissues sensitive to retinoid deficiency (e.g., the embryonic heart), while CRABP is high in tissues which are very sensitive to excess of retinoic acid (e.g., the CNS). On the other hand, knock-out experiments have suggested that these cytosolic binding proteins may be "dispensable" for embryonic development (GORRY et al. 1994): experiments with null mutant mice (both CRABP and CRBP were eliminated) revealed few major developmental defects. This discrepancy between the very specific distribution of CRABP and CRBP on the one hand, and their apparent "dispensability" on the other remains a mystery (see LI and NORRIS 1996); possibly the function of these proteins have not yet been uncovered, or other proteins such as FABPs (fatty acid binding proteins) can substitute here. It is clear, however, that CRABP can provide a "sink" for extensive accumulation of all-*trans*-retinoic acid in the embryo in the presence of low plasma levels (TZIMAS et al. 1996).

II. Retinoid Receptors in the Embryo

Numerous studies, mostly by immunohistochemistry, suggest that both the RARs and the RXRs show a very specific spatial and temporal distribution within the embryo (DOLLÉ et al. 1989; 1990; RUBERTE et al. 1990; 1991; 1993; YAMAGATA et al. 1994; MANGELSDORF et al. 1994). RARα is ubiquitously expressed throughout the embryo, while RARβ and RARγ show a temporally and spatially restricted expression. Often, RARβ and RARγ exhibit a mutually exclusive pattern of expression, which suggests that these receptors may have distinct developmental functions, and may therefore mediate distinct developmental effects following exposure to appropriate retinoid ligands. As one example, RARβ was found in the closed portion of the neural tube, while RARγ was present only in the open neural folds (mouse embryo, GD 8.5) (DOLLÉ et al. 1990; RUBERTE et al. 1991; 1993). These findings support an important role of RARβ and RARγ in the formation and closure of the neural tube; this view was strengthened by studies in mutant mice with spontaneous neural tube defects (CHEN et al. 1995). During later gestational periods (GD 13.5) RARβ is excluded from the cartilaginous condensations of the limb and was found in the interdigital mesenchyme, whereas RARγ is strongly expressed in the digits (DOLLE et al. 1990). This distribution pattern suggests that RARβ and RARγ may be involved in the formation of the digits and interference with retinoid signaling may cause interdigital webbing. RARβ was also present in endoderm-derived tissues of the lungs, liver and the genitourinary tract (DOLLÉ et al. 1989; 1990). RARγ was also prominently expressed in all mesenchyme giving rise to bones and facial cartilage, and in differentiating squamous keratinizing epithelia, as well as in the skin (RUBERTE et al. 1990).

From the RXR family, RXRβ showed the widest distribution pattern within the embryo, while RARα and RXRγ exhibited a more restricted pattern of expression (MANGELSDORF et al. 1994).

D. Experimental Models for the Study of the Significance of Retinoid Receptors in Teratogenesis

I. "Loss of Function" Approach: Studies with Null Mutant Mice

Null mutant mice were generated for various receptors, but functional redundancy between various receptors made it difficult to work out the precise function of each single receptor. Mice lacking all isoforms of RARβ develop normally and are susceptible to retinoic acid (LUO et al. 1995). While mice with a single mutation in the RARγ gene had normal offspring, null mutant mice for all RARγ isoforms exhibited growth deficiency, early lethality and male sterility as well as some cartilage malformations and homeotic transformations of the axial skeleton (LOHNES et al. 1993; 1995). Also, administration of all-*trans*-retinoic acid produced all expected malformations in RARγmutant mice except lumbosacral truncations (LOHNES et al. 1993). It therefore appears that RARγ is dispensable for induction of many developmental effects except for the generation of lumbosacral truncations. More recent studies suggest that RARγ null mutant mice are markedly resistant in regard to the induction of both anterior and posterior neural tube defects by administration of all-*trans*-retinoic acid (IULIANELLA and LOHNES 1997); the severity of limb defects induced by all-*trans*-retinoic acid was not altered in these mice. Related results were observed in regard to RXRα and limb defects: while mice with a mutation in the RXRα gene appear normal in limb development, these mice were resistant to limb malformations normally induced by all-*trans*-retinoic acid (SUCOV et al. 1995). Thus, considerable redundancy of putative retinoid receptor participation was found for some, but not other retinoic acid induced teratogenic effects (apparently other RARs can substitute for RARγ induced limb defects, but not for retinoic acid induced neural tube defects in the experimental protocol used). These results also indicate that a combination of "loss of function" and "gain of function" approaches (see below) may be a promising way to dissect various signaling pathways controlled by the retinoid receptors.

Functional redundancy was also found with other RARs and RXRs (KREZEL et al. 1996). RARα1 null mutants were normal, while disruption of the whole RARα gene resulted in early postnatal lethality and testis degeneration (LUFKIN et al. 1993).

A number of double mutants have been generated and multiple malformations have been found (LOHNES et al. 1994; 1995; KASTNER et al. 1994; MENDELSOHN et al. 1994b; GYSELINCK et al. 1997) which resembled those in the vitamin A deficiency syndrome (WILSON et al. 1953), exhibiting abnormalities of the eyes, respiratory tract, heart, urogenital system and diaphragm; cleft palate formation was also observed. Especially RXRα appears to play an important role in the generation of some of these defects (KASTNER et al. 1994). Marked effects, described as "synergy." were observed in RXRα-RAR double mutant mice which indicates that the RXRα is the main RXR implicated in transducing the developmental signal of the retinoid system in normal development (KASTNER et al. 1997); data described below indicate that this is also true for teratogenesis. No synergy was observed with RAR-RXRβ or RAR-RXRγ double mutants, suggesting that RXRβ and RXRγ may be of lower importance for development than RXRα. These results are supported by the findings, that RXRβ and RXRγ null mutant mice are viable, while RXRα null fetuses die in utero and display eye and heart malformations (SUCOV et al. 1994; KASTNER et al. 1994).

Male null mutant mice where either the RARα, the RARγ or the RXRβ was deleted are sterile, and receptors were shown to be involved in development of the testis and spermatogenesis (LOHNES et al. 1993; LUFKIN et al. 1993; KASTNER et al. 1996). The importance of RXRβ is highlighted also by the presence of 9-*cis*-retinoic acid in epididymal fluids (NEWCOMER and ONG 1990). Recently, also the 9-*cis*-retinol dehydrogenase was found in the testis. These results indicate a very special role of 9-*cis*-retinoic acid and RXRβ in sperm maturation (KASTNER et al. 1996).

Interesting results were also obtained in RARβ-RXRβ, RARβ-RXRγ or RXRβ-RXRγ double knock-out mice: the offspring of these mice showed impaired locomotion possibly originating from impairment of dopamine signaling (KREZEL et al. 1998). These results also indicate that more subtle effects may be uncovered in null mutant mice where previously an apparent dispensability and redundancy have been noted.

II. "Gain of Function" Approach: Administration of Selective Retinoid Receptor Ligands

All-*trans*-retinoic acid binds to all three RARs, and it is impossible to evaluate which of the three RARs are involved in teratogenic effects induced by this substance, and if the specific expression of the three RAR receptors in the mouse embryo could possibly be related to specific teratogenic effects produced (Fig. 3). We have therefore administered three synthetic retinoids, each preferentially binding to either the RARα (Am580) (HASHIMOTO and SHUDO 1991), the RARβ (CD2019) or the RARγ (CD437) (DARMON et al. 1989;

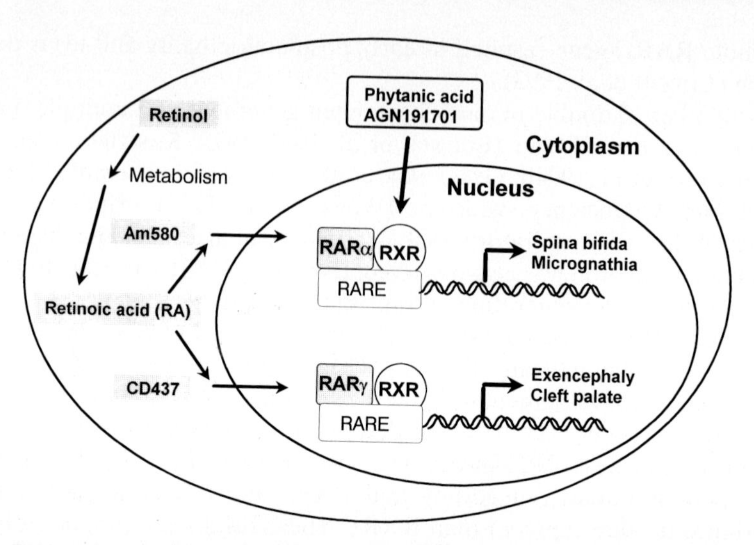

Fig. 3. Retinoid ligands can either bind to the RARs or RXRs. Administration of two ligands binding to both heterodimer partners results in a synergistic teratogenic response

BERNARD et al. 1992) (Fig. 1), and evaluated not only the potencies of the retinoids, but in particular the specific teratogenic effects observed.

The α ligand was the most potent, followed by the β ligand; the γ ligand was the least potent (ELMAZAR et al. 1996). These findings are compatible with the binding affinities and transactivating potencies of the three retinoids to the respective receptors. However, previous experience with other synthetic retinoids has shown that the correlation between teratogenic potency and receptor affinity is poor. As one example, some arotinoids are orders of magnitude more potent than all-*trans*-retinoic acid, but binding affinities are in the same range (CRETTAZ et al. 1990; LEHMANN et al. 1991). It may be that kinetic properties play role also, and all-*trans*-retinoic acid is much more rapidly eliminated as compared to synthetic retinoids. Also, more persistent binding on the receptor may result in higher efficacy, so that the residence time of receptor occupancy and not only receptor binding affinity must be considered (PIGANTELLO et al. 1997). These results are compatible with our finding that the area under the concentration-time curve (AUC), rather than the applied dose or maximal concentrations are the decisive correlates for teratogenic potency (NAU 1990; 1993; TZIMAS et al. 1997). We have therefore determined the pharmacokinetics following administration of the three selective receptor ligands (MARTIN et al. 1992; ARAFA et al. 1996, and unpublished results). While the most potent Am580 reached the highest maternal as well as embryonic AUC values, also administration of the other two ligands resulted in significant exposure; thus, the low teratogenic potency of the γ ligand is the result of the intrinsic activity, and not of low exposure of the embryo.

Table 1. Retinoid receptor pathways and teratogenic effects

Retinoid administered	Receptor pathway proposed	Teratogenic effect
RXR agonist[a]	RXR-RXR	No or low
RAR agonist[b]	RAR-RXR	High
RAR+RXR agonists[b]	RAR-RXR	Synergism
RAR agonist + RAR antagonist[c]	Antagonism	Decrease

[a] KOCHHAR et al. (1996); ELMAZAR et al. (1997).
[b] ELMAZAR et al. 1997
[c] APFEL et al. (1992); ECKHARDT and SCHMITT (1994); ELMAZAR et al. (1997).

The α ligand induced the most varied defects, compatible with the ubiquitous expression of the RARα receptor: ear, eye, mandible, palate, limb defects (phocomelia, ectrodactyly) and spina bifida were particularly pronounced (ELMAZAR et al. 1996) (Table 1). The β ligand induced defects of the urinary system with greater frequency than expected from its relative potency, which is compatible with the presence of the RARβ in these tissues. The γ ligand preferentially induced ossification deficiencies as well as defects of the sternebrae and the vertebral body, in addition to limb defects; these defects could be related to the presence of the RARγ in the mesenchyme, giving rise to bone and cartilage. RARγ was also shown to be expressed in the open neural tube (RUBERTE et al. 1991; 1993; CHEN et al. 1995), which may be related to exencephaly induced by CD437 (ELMAZAR et al. 1996) (see Fig. 3).

Similar results were obtained with another ligand (CD666) with relative selectivity for RARγ (ELMAZAR and NAU, unpublished observation). The more restricted teratogenic effects produced by the two RARγ ligands opens the possibility that retinoids can be developed with desirable properties in dermatology and oncology, but relatively low teratogenic side effects. The wide distribution of retinoid receptors in the organism will, however, make it very difficult to develop a therapeutically interesting retinoid without any teratogenic potential. Topical application of retinoids for treatment of skin disease may be one way to limit systemic (NAU 1993; BUCHAN et al. 1994), and thus embryonic exposure and teratogenic side effects.

Retinoids which are selective for the RXRs are much less teratogenic than those selective for the RARs (JIANG et al. 1995a; KOCHHAR et al. 1996; ELMAZAR et al. 1997) (Table 1). Excessive doses of RXR ligands must be administered during organogenesis to observe teratogenic effects. It may be that RXR ligands lead preferentially to the formation of RXR-RXR-homodimers (ZHANG et al. 1992) which may not play crucial roles in embryonic development. RXR ligands may therefore find important therapeutic applications such as in cancer chemotherapy – possibly because of effects on AP-1 pathways and apoptosis (SCHADENDORF et al. 1996) – and as insulin sensitizers – possibly because of interaction with PPARγ, the peroxisome proliferator activated receptor γ (MUKHERJEE et al. 1997). In *Xenopus* embryos, however, RXR-selective ligands do produce malformations (MINUCCI et al. 1996; 1997).

E. Synergistic Teratogenic Action Following Combined Administration of RAR and RXR Ligands

To further dissect the complex pattern of retinoid-induced developmental defects, a RXR-selective agonist (AGN191701) was coadministered with an RAR-selective agonist, either Am580 (RARα agonist) or CD437 (RARγ agonist) (ELMAZAR et al. 1997 and unpublished observation). The RXR agonist alone did not induce any developmental effects, which is in line with results on other RXR agonists (see above). However, coadministration of this RXR ligand with a RAR agonist resulted in a strong synergistic response in regard to a number of teratogenic effects (Fig. 4). Synergistic responses were observed for spina bifida, craniofacial and urogenital malformations as well as limb defects, among others (ELMAZAR et al. 1997); Interestingly, Am580-induced exencephaly and cleft palate were not increased – in some instances even reduced – by the coadministration of the RXR ligand. On the other hand, CD437-induced cleft palate and exencephaly were potentiated by coadministration of the RXR ligand, in addition to spina bifida and limb defects (Fig. 5). These results suggest that RAR-RXR-heterodimers – both liganded – were involved in the generation of defects such as spina bifida and a number of further malformations (Table 1). Both RARα and RARγ may serve as a heterodimer partner for RXR. On the other hand only RARγ-RXR heterodimers, but not RARα-RXR heterodimers appear to be involved in the induction of exencephaly and cleft palate (ELMAZAR et al. 1997 and unpub-

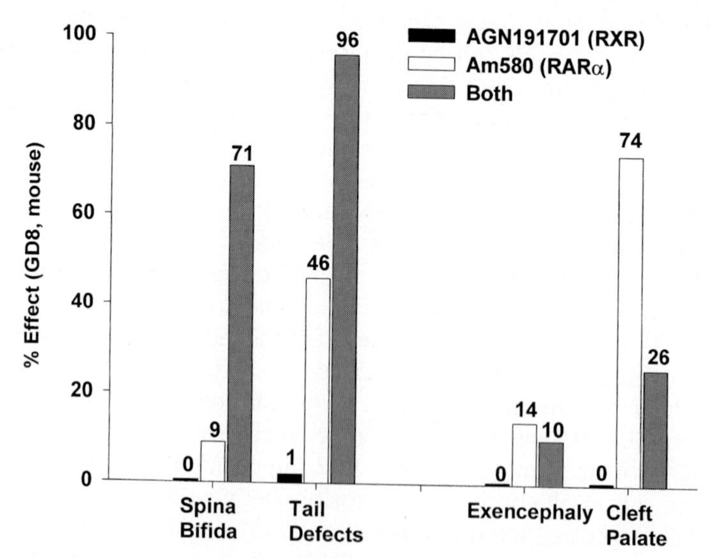

Fig. 4. Coadministration of a selective RARα and RXR ligand on day 8 of gestation in the mouse results in potentiation of the induction of spina bifida, but not of exencephaly and cleft palate

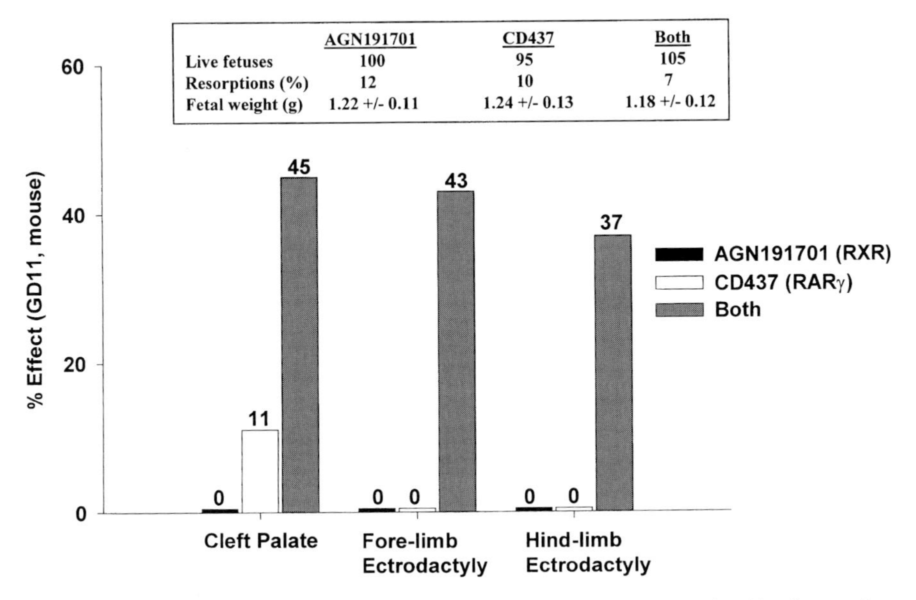

Fig. 5. Coadministration of a selective RARγ and RXR ligand on day 11 of gestation in the mouse results in synergism of limb defect induction; also cleft palate formation is potentiated

lished observation) (Table 1). No pharmacokinetic interaction was observed between the coadministered substances (Fig. 6).

These results suggest that RXR plays a central role in the induction of important teratogenic effects; The RAR-heterodimer partner determines which particular teratogenic effects are produced (Table 1). The fact that RARγ, but not RARα can serve as a heterodimer partner for RXR in the production of exencephaly may find its explanation by the fact that RARγ is specifically expressed in the open neural folds just before closure (see above).

Previously, a synergistic response between RAR- and RXR-selective ligands was seen in HeLa cells (DURAND et al. 1992; 1994), in F9 and P19 cells (ROY et al. 1995), in NB4 human leukemia cells (CHEN et al. 1996), and in cervical carcinoma cells (LOTAN et al. 1995). The results obtained in mice presented above clearly show that such heterodimers are functional also in vivo in regard to retinoid excess and teratogenesis (ELMAZAR et al. 1997).

Further proof for the involvement of retinoid receptors may be gained by the use of RAR-antagonists (APFEL et al. 1992; STANDEVEN et al. 1996). A particular antagonist (CD 2366) reduced some of the teratogenic effects produced by the RAR ligands; in particular those teratogenic effects were reduced which synergistically increased by coadministration of a RAR with a RXR ligand (ELAMAZAR et al. 1997).

These results show that the complex patterns of retinoid-induced developmental defects can be dissected by administration of selective retinoid

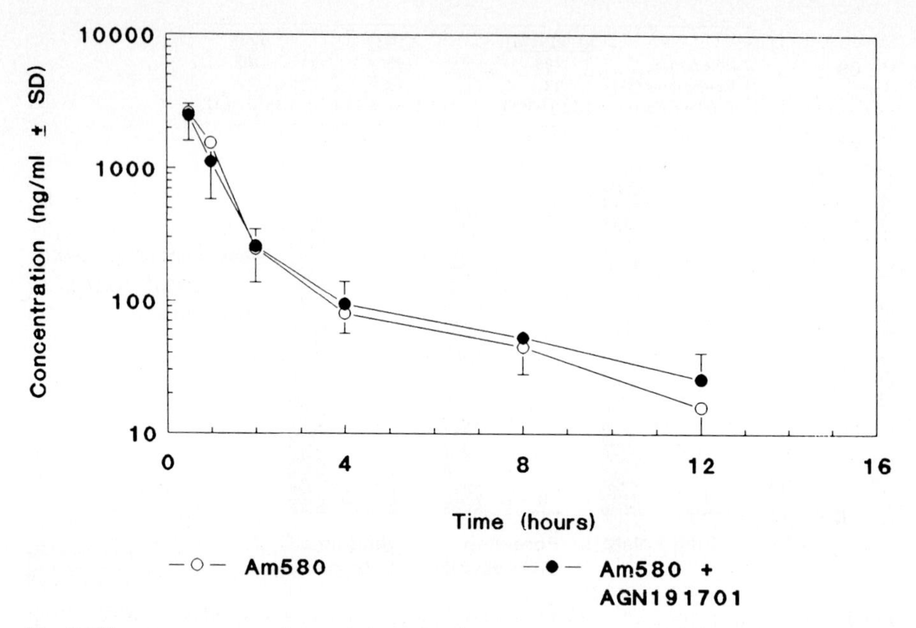

Fig. 6. Plasma concentrations of the selective RARα ligand Am580 is not altered by the coadministration of the RXR ligand, which suggests a pharmacodynamic and not phamacokinetic interaction between these ligands

receptor ligands, certain combinations of such agonists as well as coadministration of antagonists.

F. Molecular Pathways of Vitamin A and Retinoid Teratogenesis

During embryogenesis the molecular code present in the genome is used to generate mechanisms for control of cell proliferation, differentiation and morphogenesis. Numerous processes must be strictly defined in developmental time and location to allow the appropriate three-dimensional structure of the developing embryo during organogenesis, but also during all other developmental periods. Interference with such processes during organogenesis – when most drastic structural defects can be induced – may be expected to be most vulnerable to the action of substances such as retinoids via alteration of gene expression. The complex retinoid signaling pathways involving retinoid receptors, retinoid ligands and their strict spatiotemporal expression appear to play crucial roles by the control of the expression of retinoid-responsive target genes. Such target genes may be transcription factors on their own, resulting in complex patterns of retinoid-induced cascades of gene expression. Many of such pathways are induced by retinoids and their receptors, but some such as the AP-1 pathways – of particular importance in regard to apoptosis – may also be transrepressed. Altogether, hundreds of genes have been shown to be directly or indirectly influenced by the retinoid system. However, in the context of this review, it is far from clear which alteration of gene expression is causally related to teratogenesis induced by vitamin A deficiency or retinoid excess.

Some members of the homeobox gene family are directly controlled by the retinoid signaling system and have been shown to be involved in development of the vertebrae, the limbs, the hindbrain and the branchial arches (MADEN et al. 1998). These Hox genes are arranged spatially on chromosomes in clusters. The 3′ genes are expressed earlier and at the more anterior location, and are turned on at lower concentrations of all-*trans*-retinoic acid as compared to the 5′ genes. These Hox genes are therefore turned on sequentially, and the type of malformation produced depends on the concentration of the administered retinoid and the period of administration. This inappropriate influence on the normal pattern of gene expression is thus expected to result in major developmental defects.

In addition to Hox genes, numerous other genes have been found to be directly or indirectly influenced by retinoids/retinoid receptors: growth factors such as FGF, TGFα and β, EGF, IGF, PDGF, NGF, interleukines and interferons; growth factor receptors; intracellular matrix molecules which are of particular importance in regard to cell-cell communication and migration (collagen IV, laminin B1, metalloproteinases) and cell adhesion molecules such as NCAM; other transcription factors such as AP-1; synthesis of hormones such as GH, PTH, oxytocin and progesterone; enzymes of the signal transduction pathways such as protein kinase C (GUDAS et al. 1994). A causal relationship between such events and teratogenic effects are in most cases not known. Also retinoid binding proteins and receptors are autoregulated by retinoids (TAKEYAMA et al. 1996). In particular, induction of the RARβ gene expression has been related to teratogenesis (SOPRANO et al. 1994). If the RARβ-induction is causally related to retinoid-induced developmental effects, for example, via an imbalance of retinoid receptor pathways, also needs to be demonstrated, because other receptors such as RARα and RARγ isoforms are also upregulated. Furthermore, antisense oligonucleotides designed for inactivation of RARβ2, but not of RARγ1 inhibited chondrogenesis in limb bud cells in vitro (JIANG et al. 1995b; MONTOYAMA et al. 1994).

G. Conclusions

Many results obtained during the past few years speak for the hypothesis that an interference with the complex pattern of retinoid signaling pathways may be at the basis of teratogenesis and other developmental effects induced by excess as well as deficiency of vitamin A and retinoids. In addition, other compounds such as organohalogens (TCDD, PCBs) (MORSE and BROUWER 1995; FIORELLA et al. 1995), ketokonazol and derivatives (LEE et al. 1995) as well as some antiepileptic agents (NAU et al. 1995) which can interact with the retinoid system on various levels such as metabolic inhibition and induction, may also produce developmental effects via interference with the retinoid signaling pathways. The range of compounds with a potential to interfere with retinoids and their receptor pathways could possibly be further extended considering other receptors which can function as heterodimer partners of RXR. Cross-talk between these receptors as well as other nuclear receptors (Ah and steroid hormone receptors) have been shown on various molecular levels. The relevance of such interaction in regard to developmental effects will be the subject of much future research, as will be the underlying specific molecular events. Such research will not only improve our knowledge on the mechanism of normal as well as abnormal development, but may also lead the way to preventive measures.

Acknowledgements. The support by CIRD-Galderma (France) and the Deutsche Forschungsgemeinschaft (Bonn, Germany) is gratefully acknowledged. Retinoids were provided by Drs. U. Reichert and B. Shroot (CIRD-Galderma), R.A. Heyman (Ligand Pharm.) and R.A.S. Chandrarathna (Allergan). We also thank T. Arnhold for discussions and preparation of the figures, and Ms. U. Oberjatzas for preparation of the manuscript.

References

Achkar CC, Derguini F, Blumberg B, Langston A, Levin AA, Speck J, Evans RM, Bolado Jr J, Nakanishi K, Buck J, Gudas LJ (1996) 4-oxoretinol, a new natural ligand and transactivator of the retinoic acid receptors. Proc Natl Acad Sci USA 93:4879–4884

Allenby G, Bocquel MT, Saunders M, Kazmer S, Speck J, Rosenberger M, Lovey A, Kastner P, Grippo JF, Chambon P, Levin AA (1993) Retinoic acid receptors and retinoid X receptors: interactions with endogenous retinoic acids. Proc Natl Acad Sci USA 90:30–34

Apfel C, Bauer F, Crettaz M, Forni L, Kamber M, Kaufmann F, LeMotte P, Pirson W, Klaus M (1992) A retinoic acid receptor α antagonist selecitively counteracts retinoic acid effects. Proc Natl Acad Sci USA 89:7129–7133

Arafa HMM, Hamada FMA, Elmazar MMA, Nau H (1996) Fully automated determination of selective retinoic acid receptor ligands in mouse plasma and tissue by reversed-phase liquid chromatography coupled on-line with solid-phase extraction. J Chromatogr A 729:125–136

Arnhold T, Tzimas G, Wittfoht W, Plonait S, Nau H (1996) Identification of 9-*cis*-retinoic acid, 9,13-di-*cis*-retinoic acid, and 14-hydroxy-4,14-*retro*-retinol in human plasma after liver consumption. Life Sci 59: PL 169–177

Bavik C, Ward SJ, Chambon P (1996) Developmental abnormalities in cultured mouse embryos deprived of retinoic acid by inhibition of yolk-sac retinol binding protein synthesis. Proc Natl Acad Sci USA 93:3110–3114

Bernard BA, Bernardon JM, Delescluse C, Martin B, Lenoir MC, Maignan J, Charpentier B, Pilgrim WR, Reichert U, Shroot B (1992) Identification of synthetic retinoids with selectivity for human nuclear retinoic acid receptor β. Biochem Biophys Res Commun 186:977–983

Biesalski HK (1997) Bioavailability of vitamin A. Eur J Clin Nutr 51:S71–S75

Blumberg B, Bolado Jr J, Derguini F, Craig AG, Moreno TA, Chakravarti D, Heyman RA, Buck J, Evans RM (1996) Novel retinoic acid receptor ligands in *Xenopus* embryos. Proc Natl Acad Sci USA 93:4873–4878

Brinkmann E, Dehne L, Bijosono Oei H, Tiebach R, Baltes W (1995) Separation of geometrical retinol isomers in food samples by using narrow-bore high-performance liquid chromatography. J Chromagr A 693:271–279

Buchan P, Eckhoff C, Caron D, Nau H, Shroot B, Schaefer H (1994) Repeated topical administration of all-trans-retinoic acid and plasma levels of retinoic acids in man. J Am Acad Dermatol 30:428–434

Chambon P (1996) A decade of molecular biology of retinoic acid receptors. FASEB J 10:940–954

Chen JY, Clifford J, Zusi C, Starrett J, Tortolani D, Ostrowski J, Reczek PR, Chambon P, Gronemeyer H (1996) Two distinct actions of retinoid-receptor ligands. Nature 382:819–822

Chen WH, Morriss-Kay GM, Copp A (1995) Genesis and prevention of spinal neural tube in the *curly tail* mutant mouse: involvement of retinoic acid and its nuclear receptors RAR-β and RAR-γ. Development 121:681–691

Chen Y, Huang L, Russo AF, Solursh M (1992) Retinoic acid is enriched in Hensen's node and is developmentally regulated in the early chicken embryo. Proc Natl Acad Sci USA 89:10056–10059

Coberly S, Lammer E, Alashari M (1996) Retinoic acid embryopathy: case report and review of literature. Pediat Path Labor Med 16:823–836

Creech Kraft J, Schuh T, Juchau MR, Kimelman D (1994) The retinoid X receptor ligand, 9-*cis*- retinoic acid, is a potential regulator of early *Xenopus* development. Proc Natl Acad Sci USA 91:3067–3071

Crettaz M, Baron A, Siegenthaler G, Hunziker W (1990) Ligand specificities of recombinant retinoic acid receptors RAR-α and RAR-β. Biochem J 272:391–397

Darmon M, Rocher M, Cavey MT, Martin B, Rabilloud T, Delescluse C, Baily J, Eustache J, Jamoulle JC, Nedoncelle P, Shroot B (1989) Affinity of retinoids for nuclear receptors correlates with biological activity by absence of correlation with CRABP binding. In: U. Reichert, B. Shroot (eds) Pharmacology of retinoids in the skin, vol 3. Karger, Basel, pp 56–64

Dekker EJ, Vaessen M-J, van der Berg C, Timmermans A, Godsave S, Holling T, Nieuwkoop P, Geurts van Kessel A, Durston A (1994) Overexpression of a cellular retinoic acid binding protein (xCRABP) causes anteroposterior defects in developing *Xenopus* embryos. Development 120:973–985

De Luca LM, Darwiche N, Jones CS Scita G (1995) Retinoids in differentiation and neoplasia. Sci Amer Sci Med July/August:28–37

Dickman ED, Thaller C, Smith SM (1997) Temporally-regulated retinoic acid depletion produces specific crest, ocular and nervous system defects. Development 124:3111–3121

Dollé P, Ruberte E, Kastner P, Petkovich M, Stoner CM, Gudas LJ, Chambon P (1989) Differential expression of genes encoding α, β and γ retinoic acid receptors and CRABP in the developing limbs of the mouse. Nature 342:702–705

Dollé P, Ruberte E, Leory P, Morriss-Kay G, Chambon P (1990) Retinoic acid receptors and cellular retinoid binding proteins I. A systematic study of their differential pattern of transcription during mouse organogenesis. Development 110:1133–1151

Duester G (1998) Alcohol dehydrogenase as a critical mediator of retinoic acid synthesis from vitamin A in mouse embryo. J Nutr 128:459S-462 s

Durand B, Saunders M, Gaudon C, Roy B, Losson R, Chambon P (1994) Activation function 2 (AF-2) of retinoic acid receptor and 9-*cis* retinoic acid receptor: Presence of a conserved autonomous constitutive activating domain and influence of the nature of the response element of AF-2 activity. EMBO J 13:5370–5382

Durand B, Saunders M, Leroy P, Leid M, Chambon P (1992) *All*-trans and 9-*cis* retinoic acid induction of CRABPII transcription is mediated by RAR-RXR heterodimers bound to DR1 and DR2 repeated motifs. Cell 71:73–85

Durston AJ, Timmermans JPM, Hage WJ, Hendriks HFJ, de Vires NJ, Heideveld M, Nieuwkoop PD (1989) Retinoic acid causes an anteroposterior transformation in the developing central nervous system. Nature 340:140–144

Eckhardt K, Schmitt G (1994) A retinoic acid receptor α antagonist counteracts retinoid teratogenicity in vitro and reduced incidence and/or severity of malformations in vivo. Toxicol Lett 70:299–308

Eckhoff C, Collins MD, Nau H (1991) Human plasma all-trans-, 13-cis-, and 13-cis-4-oxoretinoic acid profiles during subchronic vitamin A supplementation: Comparison to retinol and retinyl ester plasma levels. J Nutr 121:1016–1025

Eckhoff C, Nau H (1990) Identification and quantitation of all-trans, 13-cis-, and 13-cis-4- oxoretinoic acid in human plasma. J Lipid Res 31:1445–1454

Elmazar MMA, Reichert U, Shroot B, Nau H (1996) Pattern of retinoid-induced teratogenic effects: Possible relationship with relative selectivity for nuclear retinoid receptors RARα, RARβ and RARγ. Teratology 53:158–167

Elmazar MMA, Rühl R, Reichert U, Shroot B, Nau H (1997) RARα-mediated teratogenicity in mice is potentiated by an RXR agonist and reduced by an RAR antagonist: Dissection of retinoid receptor-induced pathways. Toxicol Appl Pharmacol 146:21–28

Fiorella PD, Olson JR, Napoli JL (1995) 2,3,7,8-tetrachlorodibenz-p-dioxin induces diverse retinoic acid metabolites in multiple tissues of the Sprague-Dawley rat. Toxicol Appl Pharmacol 134:222–228

Ghyselinck NB, Dupé V, Dierich A, Messaddeq, Garnier JM, Rochette-Egly C, Chambon P, Mark M (1997) Role of retinoic acid receptor beta (RARβ) during mouse development. Int J Dev Biol 41:425–447

Giguère V, Ong ES, Segui P, Evans RM (1987) Identification of a receptor for the morphogen retinoid acid. Nature 330:624–629

Gorry P, Lufkin T, Dierich A, Rochette-Egly C, Decimo D, Dolle P, Mark M, Durand B, Chambon P (1994) The cellular retinoic acid binding protein I is dispensable. Proc Natl Acad Sci USA 91:9032–9036

Gudas LJ (1994) Retinoids and vertebrate development. J Biol Chem 269:15399–15402

Gustafson A-L, Dencker L, Eriksson U (1993) Non-overlapping expression of CRBP I and CRABP I during pattern formation of limbs and craniofacial structures in the early mouse embryo. Development 117:451–460

Gyselinck et al. (1997)

Haselbeck RJ, Ang HL, Deltour L, Duester G (1997) Retinoic acid and alcohol/retinol dehydrogenase in the mouse adrenal gland: a potential endocrine source of retinoic acid during development. Endocrinology 138:3035–3041

Hashimoto Y, Shudo K (1991) Retinoids and their nuclear receptors. Cell Biol Rev 25:209–230

Helms JA, Chang HK, Eichele G, Thaller C (1996) Retinoic acid signaling is required during early chick limb development. Development 122:1385–1394

Helms JA, Kim CH, Hu D, Minkoff R, Thaller C, Eichele G (1997) Sonic *hedgehog* participates in craniofacial morphogenesis and is down-regulated by teratogenic doses of retinoic acid. Dev Biol 187:25–35

Heyman RA, Mangelsdorf DJ, Dyck JA, Stein RB, Eichele G, Evans RM, Thaller C (1992) 9-*cis* retinoic acid is a high affinity ligand for the retinoid X receptor. Cell 68:397–406

Imai S, Okuno M, Moriwaki H, Muto Y, Murakami K, Shudo K, Suzuki Y, Kojima S (1997) 9,13-di-*cis* Retinoic acid induces the production of PA and activation of latent TGF-β via RARα in a human liver stellate cell line, LI90. FEBS Lett 411:102–106

Iulianella A, Lohnes D (1997) Contribution of retinoic acid receptor gamma to retinoid-induced craniofacial and axial defects. Dev Dyn 209:92–104

Jiang H, Penner JD, Beard RL, Chandraratna RAS, Kochhar DM (1995a) Diminished teratogenicity of retinoid X receptor-selective synthetic retinoids. Biochem Pharmacol 50 669–676

Jiang H, Soprano DR, Li SW, Soprano KJ, Penner JD, Gyda III M, Kochhar DM (1995b) Modulation of limb bud chondrogenesis by retinoic acid and retinoic acid receptors. Int J Dev Biol 39:617–627

Kastner P, Grondona JM, Mark M, Gansmuller A, LeMeur M, Decimo D, Vonesch JL, Dollé P, Chambon P (1994) Genetic analysis of RXRα developmental function: Convergence of RXR and RAR signaling pathways in heart and eye morphogenesis. Cell 78:987–1003

Kastner P, Mark M, Chambon P (1995) Nonsteroid nuclear receptors: What are genetic studies telling us about their role in real life? Cell 83:859–869

Kastner P, Mark M, Ghyselnick N, Krezel W, Dupé V, Grondona JM, Chambon P (1997) Genetic evidence that the retinoid signal is transduced by heterodimeric RXR/RAR functional units during mouse development. Development 124:313–326

Kastner P, Mark M, Leid M, Gansmuller A, Chin W, Grondona JM, Décimo D, Krezel W, Dierich A, Chambon P (1996) Abnormal spematogenesis in RXRβ mutant mice. Gen Dev 10:80–92

Kochhar DM, Jiang H, Penner JD, Beard RL, Chandraratna RAS (1996) Differential teratogenic response of mouse embryos to receptor selective analogs of retinoic acid. Chem Biol Interact 100:1–12

Krezel W, Dupé V, Mark M, Dierich A, Kastner P, Chambon P (1996) RXRγ null mice are apparently normal and compound RXR$\alpha^{+/-}$/RXR$\beta^{-/-}$/RXR$\gamma^{-/-}$ mutant mice are viable. Proc Natl Acad Sci USA 93:9010–9014

Krezel W, Ghyselinck N, Samad TA, Dupé V, Kastner P, Borrelli E, Chambon P (1998) Impaired locomotion and dopamine signaling in retinoid receptor mutant mice. Science 279:863–867

Kurlandsky SB, Gamble MV, Ramakrishnan R, Blaner WS (1995) Plasma delivery of retinoic acid to tissues in the rat. J Biol Chem 270:17850–17857

Lammer EJ, Chen DT, Hoar RM, Agnish ND, Benke PJ, Braun JT, Curry CJ, Fernhoff PM, Grix Jr AW, Lott IT, Richard JM, Sun SC (1985) Retinoic acid embryopathy. New Engl J Med 313:837–841

Lee JS, Newman RA, Lippman SA, Fossella FV, Calayag M, Raber MN, Krakoff IH, Hong WKPhase I evaluation of all-trans retinoic acid with and without ketoconazole in adults with solid tumours. J Clin Oncol 13:1501–1508

Lehmann JM, Dawson MI, Hobbs PD, Husmann M, Pfahl M (1991) Identification of retinoids with nuclear receptor subtype-selective activities. Cancer Res 51: 4804–4809

Lemotte PK, Keidel S, Apfel CM (1996) Phytanic acid is a retinoid X receptor ligand. Eur J Biochem 236:328–333

Levin AA, Sturzenbecker LJ, Kazmer S, Bosakowski T, Huselton C, Allenby G, Speck J, Kratzeisen C, Rosenberger M, Lovey A, Grippo JF (1992) 9-*cis* retinoic acid stereoisomer binds and activates the nuclear receptor RXRα. Nature 355:359–361

Li E, Norris AW (1996) Structure/function of cytoplasmic vitamin A-binding proteins. Annu Rev Nutr 16:205–234

Lohnes D, Kastner P, Dierich A, Mark M, LeMeur M, Chambon P (1993) Function of retinoic acid receptor γ in the mouse. Cell 73:643–658

Lohnes D, Mark M, Mendelsohn C, Dollé P, Decimo D, LeMeur M, Dierich A, Gorry P, Chambon P (1995) Developmental roles of retinoic acid receptors. J Steroid Biochem Mol Biol 53:475–486

Lohnes D, Mark M, Mendelsohn C, Dollé P, Dierich A, Gorry P, Gansmuller A, Chambon P (1994) Function of the retinoic acid receptors (RARs) during development (I) Craniofacial and skeletal abnormalities in RAR double mutants. Development 120:2733–3748

Lotan R, Dawson MI, Zou CC, Jong L, Lotan D, Zou CP (1995) Enhanced efficacy of combinations of retinoic acid- and retinoid X receptor-selective retinoids and α-interferon in inhibition of cervical carcinoma cell proliferation. Cancer Res 55:232–236

Lufkin T, Lohnes D, Mark M, Dierich A, Gorry P, Gaub MP, LeMeur M, Chambon P (1993) High postnatal lethality and testis degeneration in retinoic acid receptor α mutant mice. Proc Natl Acad Sci USA 90:7225–7229

Luo J, Pasceri P, Conlon RA, Rossant J, Giguère (1995) Mice lacking all isoforms of retinoic acid receptor β develop normally and are susceptible to the teratogenic effects of retinoic acid. Mech Dev 53:61–71

Maden M (1994) Vitamin A in embryonic development. Nutr Rev 52: S3-S12

Maden M, Gale E, Zile MH (1998) The role of vitamin A in the development of the central nervous system. J Nutr 128:471S-475 s

Maden M, Summerbell D, Maignan J, Darmon M, Shroot B (1991) The respecification of limb pattern by new synthetic retinoids and their interaction with cellular retinoic acid-binding protein. Differentiation 47:49–55

Mangelsdorf DJ, Umesono K, Evans RM (1994) The retinoid receptors. In: MB Sporn, AB Roberts, DS Goodman (eds) The retinoids: biology, chemistry, and medicine. Raven, New York, pp 319–349

Martin B, Bernadon J-M, Cavey M-T, Bernard B, Carlavan I, Charpentier B, Pilgrim WR, Shroot B, Reichert U (1992) Selective synthetic ligands for human nuclear retinoic acid receptors. Skin Pharmacol 5:57–65

McCaffery, Dräger U (1994) Hot spots of retinoic acid synthesis in the developing spinal cord. Proc Natl Acad Sci USA 91:7194–7197

Mendelsohn C, Lohnes D, Décimo D, Lufkin T, LeMeur M, Chambon P, Mark M (1994b) Function of the retinoic acid receptors (RARs) during development (II) Multiple abnormalities at various stages of organogenesis in RAR double mutants. Development 120:2749–2771

Minucci S, Leid M, Toyama R, Saint-Jeannet JP, Peterson VJ, Horn V, Ishmael JE, Bhattacharyya N, Dey A, Dawid IB, Ozato K (1997) Retinoid X Receptor (RXR) within the RXR-retinoic acid receptor heterodimer binds its ligand and enhances retinoid-dependent gene expression. Mol Cell Biol 17:644–655

Minucci S, Saint-Jeannet JP, Toyama R, Scita G, DeLuca LM, Taira M, Levin AA, Ozato K, Dawid IB (1996) Retinoid X receptor-selective ligands produce malformations in *Xenopus* embryos. Dev Biol 93:1803–1807

Morriss-Kay GM, Sokolova N (1996) Embryonic development and pattern formation. FASEB J 10:961–968

Morse DC, Brouwer A (1995) Fetal, neonatal, and long-term alterations in hepatic retinoid levels following maternal polychlorinated biphenyl exposure in rats. Toxicol Appl Pharmacol 131:175–182

Motoyama J, Eto K (1994) Antisense retinoic acid receptor β-1 oligonucleotide enhances chondrogenesis of mouse limb mesenchymal cells in vitro. FEBS Lett 338:319–322

Mukherjee R, Davies PJA, Crombie DL, Bischoff ED, Cesario RM, Jow L, Hamnn LG, Boehm MF, Mondon CE, Nadzan AM, Paterniti Jr JR, Heyman RA (1997) Sensitization of diabetic and obese mice to insulin by retinoid X receptor agonist. Nature 386:407–410

Napal S, Chandraratna RAS (1996) Retinoids as anti-cancer agents. Curr Pharm Design 2:295–316

Nau (1990) Correlation of transplacental and maternal pharmacokinetics of retinoids during organogenesis with teratogenicity. Methods Enzymol 190:437–448

Nau H (1993) Embryotoxicity and teratogenicity of topical retinoic acid. Skin Pharmacol 6 [Suppl 1]:35–44

Nau H (1994) Toxicokinetics and structure – activity relationships in retinoid teratogenesis. Ann Oncol 5 [Suppl 9]:S39–S43

Nau H, Chahoud I, Dencker L, Lammer E, Scott WJ (1994) Teratogenicity of vitamin A and retinoids. In: R Blomhoff (ed) Vitamin A in health and disease. Dekker, New York, pp 615–664

Nau H, Tzimas G, Mondry M, Plum C, Spohr H-L (1995) Antiepileptic drugs alter endogenous retinoid concentrations: a possible mechanism of teratogenesis of anticonvulsant therapy. Life Sci 57:53–60

Newcomer ME, Ong DE (1990) Purification and crystallization of retinoic acid-binding protein from rat epididymis. Identify with the major androgen-dependent epididymal proteins. J Biol Chem 265:12876–12879

Noji S, Nohno T, Koyama E, Muto K, Ohyama K, Aoki Y, Tamura K, Ohsugi K, Ide H, Taniguchi S, Saito T (1991) Retinoic acid induces polarizing activity but is unlikely to be a morphogen in the chick limb bud. Nature 350:83–86

Olson JA (1987) Recommended dietary intakes (RDI) of vitamin A in humans. Am J Clin Nutr 45:704–716

Petkovich M, Brand NJ, Krust A, Chambon P (1987) A human retinoic acid receptor which belongs to the family of nuclear receptors. Nature 330:444–450

Pignatello MA, Kaufmann FC, Levin AA (1997) Multiple factors contribute to the toxicity of the aromatic retinoid, TTNPB (Ro 13-7410): binding affinities and disposition. Toxicol Appl Pharmacol 142:319–327

Repa JJ, Hanson KK, Clagett-Dame M (1993) All-*trans*-retinol is a ligand for the retinoic acid receptors. Proc Natl Acad Sci USA 90:7293–7297

Rosa FW, Wilk AL, Kelsey (1986) Teratogen update: vitamin A congeners. Teratology 33:355–364

Roy B, Taneja R, Chambon P (1995) Synergistic activation of retinoic acid (RA)-responsive genes and induction of embryonal carcinoma cell differentiation by an RA receptor α (RARα)-, RAR-β-, or RAR-γ-selective ligand in combination with a retinoid X-receptor- specific ligand. Mol Cell Biol 15:6481–6487

Ruberte E, Dollé P, Chambon P, Morriss-Kay G (1991) Retinoic acid receptors and cellular retinoid binding proteins. II. Their differential pattern of transcription during early morphogenesis in mouse embryos. Development 111:45–60

Ruberte E, Dollé P, Krust A, Zelent A, Morriss-Kay G, Chambon P (1990) Specific spatial and temporal distribution of retinoic acid receptor gamma transcripts during mouse embryogenesis. Development 108:213–222

Ruberte E, Friederrich V, Chambon P, Morriss-Kay G (1993) Retinoic acid receptors and cellular retinoid binding proteins. III. Their differential transcript distribution during mouse nervous system development. Development 118:267–282

Sani BP, Venepally PR, Levin AA (1997) Didehydroretinoic acid: retinoid receptor-mediated transcriptional activation and binding properties. Biochem Pharmacol 53:1049–1053

Sapin V, Ward SJ, Bronner S, Chambon P (1997) Differential expression of transcripts encoding retinoid binding proteins and retinoic acid receptors during placentation of the mouse. Dev Dyn 208:199–210

Satre MA, Kochhar DM (1989) Elevations of the endogenous levels of the putative morphogen retinoic acid in embryonic mouse limb-buds associated with limb dysmorphogenesis. Dev Biol 133:529–536

Schadendorf D, Kern MA, Artuc M, Pahl H, Rosenbach T, Fichtner I, Nürnberg W, Süting S, v. Stebut E, Worm M, Makki A, Jurgovsky K, Kolde G, Henz BM (1996) Treatment of melanoma cells with the synthetic retinoid CD437 induces apoptosis via activation of AP-1 in vitro, and causes growth inhibition in xenografts in vivo. J Cell Biol 135:1889–1898

Scott WJ, Walter R, Tzimas G, Sass JO, Nau H, Collins MD (1994) Endogenous status of retinoids and their cytosolic binding proteins in limb buds of chick vs. mouse embryos. Dev Biol 165:397–409

Smith SM, Dickman ED, Power SC, Lancman J (1998) Retinoids and their receptors in vertebrate embryogenesis. J Nutr 128:467S–470S

Soprano DR, Gyda III M, Jiang H, Harnish DC, Ugen K, Satre M, Chen L, Soprano KJ, Kochhar DM (1994) A sustained elevation in retinoic acid receptor-β2 mRNA and protein occurs during retinoic acid-induced fetal dysmorphogenesis. Mech Dev 45:243–253

Soprano DR, Wyatt ML, Dixon JL, Soprano KJ, Goodman DWS (1988) Retinol-binding protein synthesis and secretion by the rat visceral yolk sac: Effect of retinol status. J Biol Chem 263:2934–2938

Standeven AM, Johnson AT, Escobar M, Chandraratna RAS (1996) Specific antagonist of retinoid toxicity in mice. Toxicol Appl Pharmacol 138:169–175

Sucov HM, Dyson E, Gumeringer CL, Price J, Chien KR, Evans RM (1994) RXR α mutant mice establish a genetic basis for vitamin A signaling in heart morphogenesis. Genes Dev 8:1007–1018

Sucov HM, Izpisua-Belmonte JC, Ganan Y, Evans RM (1995) Mouse embryos lacking RXRα are resistant to retinoic acid-induced limb defects. Development 121:3997–4003

Takeyama K, Kojima R, Ohashi T, Sato T, Mano H, Masushige S, Kato S (1996) Retinoic acid differentially up-regulates the gene expression of retinoic acid receptors α and β isoforms in embryo and adult rats. Biochem Biophys Res Commun 222:395–400

Tamura K, Hashimoto Y, Shudo K, Ide H (1993) Distribution of retinoids applied exogenously to chick limb buds: An autoradiographic analysis. Dev Growth Diff 35:593–599

Thaller C, Eichele G (1987) Identification and spatial distribution of retinoids in the developing chick limb bud. Nature 327:625–628

Thaller C, Hofmann C, Eichele G (1993) 9-*cis*-retinoic acid, a potent inducer of digit pattern duplications in the chick wing bud. Development 118:957–965

Tzimas G, Collins MD, Bürgin H, Hummler H, Nau H (1996) Embryotoxic doses of Vitamin A to rabbits result in low plasma but high embryonic concentrations of all-*trans*-retinoic acid: Risk of Vitamin A exposure in humans. J Nutr 126:2159–2171

Tzimas G, Thiel R, Chahoud I, Nau H (1997) The area under the concentration-time curve of all-*trans*-retinoic acid is the most suitable pharmacokinetic correlate to the embryotoxicity of this retinoid in the rat. Toxicol Appl Pharmacol 143:436–444

van der Saag PT (1996) Nuclear retinoid receptors: Mediators of retinoid effects. Eur J Clin Nutr 50:S24–S28

Wagner M, Thaller C, Jessell T, Eichele G (1990) Polarizing activity and retinoid synthesis in the floor plate of the neural tube. Nature 345:819–822

Ward SJ, Chambon P, Ong DE, Bavik C (1997) A retinol-binding protein receptor-mediated mechanism for uptake of vitamin A to postimplantation rat embryos. Biol Reprod 57:751–755

Watkins PA, Howard AE, Gould SJ, Avignan J, Mohalik SJ (1996) Phytanic acid activation in rat liver peroxisomes is catalyzed by long-chain acyl-CoA synthetase. J Lip Res 37:2288–2295

Wellik DM, Norback DH, DeLuca HF (1997) Retinol is specifically required during midgestation for neonatal survival. Am J Physiol 272: E25-E29

White JA, Guo YD, Baetz K, Beckett-Jones B, Bonasoro J, Hsu KE, Dilworth FJ, Jones G, Petkovich M (1996) Identification of the retinoic acid-inducible all-*trans*-retinoic acid 4- hydroxylase. J Biol Chem 271:29922–29927

Wilson JG, Roth CB, Warkany (1953) An analysis of the syndrome of malfunctions induced by maternal vitamin A deficiency. Effects of restoration of vitamin A at various times during gestation. Am J Anat 92:189–217

Yamagata T, Momoi MY, Yanagisawa M, Kumagai H, Yamakado M, Momoi T (1994) Changes of the expression and distribution of retinoic acid receptors during neurogenesis in mouse embryos. Dev Brain Res 77:163–176

Zhang XK, Lehmann J, Hoffmann B, Dawson MI, Cameron J, Graupner G, Hermann T, Tran P, Pfahl M (1992) Homodimer formation of retinoid X-receptor induced by 9-*cis* retinoic acid. Nature 358:587–591
Zhang XK, Pfahl M (1993) Regulation of retinoid and thyroid hormone action through homodimeric and heterodimeric receptors. Trends Endocrinol Metab 4:156–162
Zile MH (1998) Vitamin A and embryonic development: an overview. J Nutr 128:455S–458S

Section E
Skin

CHAPTER 17

Vitamin A Homeostasis in Human Epidermis: Native Retinoid Composition and Metabolism

R.K. RANDOLPH and G. SIEGENTHALER

A. Introduction

Vitamin A, or retinol, has long been known to play a critical role in epithelial homeostasis. Its importance in the maintenance of healthy epithelium was first recognized by MORI (1922) with the discovery that the cornea became keratinized in vitamin A deficient animals. Soon thereafter WOLBACH and HOWE (1925) documented the conversion of mucosal epithelial tissues to squamous, metaplastic, hyperkeratinized epithelia – epithelia resembling the epidermis – in vitamin A deficient rats. In the authors' words, ". . . replacement of many different epithelia by stratified keratinizing epithelium actually characterize[s] fat-soluble A avitaminosis. The specific pathology is the widespread keratinization." Indeed, even normal or orthokeratotic skin appeared to require vitamin A, as FRAZIER and HU (1931) and GOODWIN (1934) later described the cutaneous hyperkeratosis that accompanies vitamin A deficiency in human beings. In addition, the hyperkeratosis observed in vitamin A deficiency could be reversed by restoring vitamin A to the diet (FRAZIER and HU 1931; GOODWIN 1934). These early observations serve as the foundation for the idea that all epithelial tissues, including the epidermis, require adequate vitamin A nutritional status to sustain normal growth and differentiation.

Although these early studies clearly established the need for vitamin A in sustaining normal differentiation of the epidermis, the nature of the requirement for vitamin A became evident somewhat later during in vitro investigations. In 1953 FELL and MELLANBY observed that embryonic chick skin lost its keratinized phenotype and began to express the characteristics of a mucosal epithelium when cultured in medium containing vitamin A. Similar observations were made in studies of cultured human (BARNETT and SZABO 1973) and mouse skin (YUSPA and HARRIS 1974). More recently, the expression of molecular markers of differentiation in cultured human epidermal keratinocytes was shown to be suppressed by medium containing vitamin A (FUCHS and GREEN 1981). These in vitro results, when considered alone, might be interpreted as contrary to the supportive role of vitamin A in epidermal homeostasis in vivo. An alternative interpretation is that the supply of vitamin A in vivo, while necessary, must be limited or fall within a critical range. Evidence favoring this possibility came with the demonstration that optimum epidermal morphology in an in vitro model of skin is achieved only with nanomolar concentrations of extracellular vitamin A; concentrations above or below this result in abnormal keratinocyte differentiation (ASSELINEAU et al. 1989). Thus, normal epidermis not only requires vitamin A, it requires a critical quantity of vitamin A.

Qualitatively, retinoic acid is the form of vitamin A that is important for normal epidermal homeostasis (KANG et al. 1995). Retinoic acid exerts its activity in target cells by regulating the expression of retinoid-responsive genes (GUDAS et al. 1994; MANGELSDORF et al. 1994). Retinoic acid-mediated gene expression entails a sequence of events that is initiated by the reversible binding of an active retinoid such as retinoic acid to nuclear retinoic acid receptors and orphan retinoid "X" receptors (GIGUÈRE et al. 1987; PETKOVICH et al. 1987). These ligand-receptor complexes in turn interact with retinoid-response elements in the enhancer regions of retinoid-responsive genes or with other transcription factors, thus effecting transcriptional regulation (MANGELSDORF et al. 1994). The function of these receptors in the vitamin A signaling pathway in epidermis has been summarized recently (FISHER and VOORHEES 1996) and is not addressed here.

Since the nuclear retinoid receptors, are thought to be evenly distributed throughout the epidermis (TAVAKKOL et al. 1992), a retinoid message is likely modulated by the concentration of ligand retinoic acid available for binding to the receptors. Epidermal retinoic acid is potentially derived from two sources. First, retinoic acid can be generated in situ following the uptake and local metabolism of plasma retinol which is transported bound to retinol-binding protein (RBP; KANAI et al. 1968). Second, in addition to retinol, retinoic acid is also present in the circulation. Plasma retinoic acid is transported bound to albumin (LEHMAN et al. 1972; SMITH et al. 1973) and is taken up by a variety of cell types (KURLANDSKY et al. 1996). Retinol is constitutively present in human plasma at 1–2 μM (SOPRANO and BLANER 1994), whereas retinoic acid is present at 4–14 nM (DELEENHEER et al. 1982; ECKHOFF and NAU 1990). A clear understanding of how the plasma forms of vitamin A contribute to the maintenance of normal epidermal differentiation and function must thus consider the processes which control the accessibility of plasma retinol and retinoic acid to cells of the epidermis. These would include the uptake, transport, storage, and metabolism of extracellular retinol and retinoic acid.

The purpose of this review is to summarize the literature addressing native retinoid composition and metabolism at the cellular and tissue levels in human epidermis. Finally, to draw together all of the data, a model for vitamin A homeostasis in human epidermis is presented. We do not discuss the role of vitamin A in epidermal differentiation but refer readers to excellent reviews (VAHLQUIST and TÖRMÄ 1988; FUCHS 1990; DARMON 1991; VAHLQUIST 1994).

B. Total Vitamin A Content of the Epidermis

Despite the fact that the epidermis is an avascular tissue – the blood supply resides in underlying dermis – it contains significant quantities of vitamin A.

The earliest estimate of the vitamin A content of human epidermis was by WILLIAMS (1943) some 50 years ago as 390 ng vitamin A/g wet weight. In more recent years, with the advent of high performance liquid chromatography, several laboratories have identified and quantitated the vitamin A present in normal human epidermis. Table 1 summarizes these data. Since most of these measurements were obtained by analyses of saponified tissues, retinoid quantities are presented as total tissue retinoid alcohols. There are two vitamin A families represented; the vitamin A_1, or retinol-related retinoids and the vitamin A_2, or 3,4-didehydroretinol-related retinoids. There is good agreement between different laboratories on total epidermal vitamin A content. Total retinol content in normal epidermis averages 335 ng/g wet weight. This value compares closely to the concentration range of retinol in plasma (185 ng/ml–370 ng/ml; KANAI and GOODMAN 1968) and human dermis (~300 ng/g; VAHLQUIST et al. 1982). Vitamin A_2 exhibits a variable presence at 10%–50% of vitamin A_1 levels. Interestingly, vitamin A_2 is not known to be present in human plasma and has not been reported in any human tissues other than skin. The concentration of A_2 in human dermis, approximately 15 ng/g wet weight (see bottom of Table 1), is less than that in the epidermis.

The vitamin A content from various regions of the dermis is shown in the bottom of Table 1. The quantity of vitamin A_1 in papillary dermis, the portion of dermis containing the microvasculature of the skin and located immediately beneath the epidermis, is approximately half that of overlying epidermis. Vitamin A_1 in reticular dermis, the deeper layer of dermis containing larger blood vessels, is approximately the same as that in the plasma. The vitamin A_2 concentration in all dermal tissues of the skin is less than that in epidermis.

The literature contains a single report of a comparison of epidermal retinoid content from different body sites (VAHLQUIST et al. 1982). These data demonstrate that retinol content of epidermis is relatively constant at all body sites excepting the sole of the foot, which contains approximately one-half of the tissue retinol concentration relative to other body sites. There are no significant differences between body sites for the epidermal content of vitamin A_2. These same investigators have also reported an evaluation of epidermal vitamin A content of back skin between genders (VAHLQUIST et al. 1982). Their results show no differences in A_1 or A_2 content at this body site between men and women. There is a single report of the retinoid content of skin surface lipids (VAHLQUIST et al. 1982). Retinol was reported to be present at a concentration of 580 ng/g wet weight. No vitamin A_2 was detected.

C. Vitamin A Metabolites in the Epidermis

The total vitamin A content of epidermis clearly shows that plasma retinol has access to and is taken up by epidermal cells. The unique presence of 3,4-didehydroretinol in epidermis, and the other metabolites of vitamin A in this tissue (Table 2), points to active retinol metabolism. Despite the relatively good

Table 1. Vitamin A content of human epidermis and dermis

Source of tissue	Total retinol (A_1) (ng/g tissue wet weight)	Total 3,4-didehydroretinol (A_2) (ng/g tissue wet weight)	A_1/A_2 ratio	Reference
Breast skin	350[a]	70[a]	5	Rollman and Vahlquist (1981)
Breast skin	180	20	9	Vahlquist (1982)
Skin from various sites	294	70	4	Vahlquist et al. (1982)
Breast skin	240[a]	NR		Törmä and Vahlquist (1984)
Back skin	256[a]	119	2	Berne et al. (1984a)
Breast skin	714[a]	68	10	Berne et al. (1984b)
Neonatal foreskin	270	NR		Tang et al. (1994)
Buttock skin	346	NR		Kang et al. (1995)
Buttock skin	510	Trace		Duell et al. (1996b)
Back skin	193	51	4	Vahlquist et al. (1996)
Mean for epidermis ± SD	335 ± 163	66 ± 32	6 ± 3	
Papillary dermis from various body sites	186	20	9	Vahlquist et al. (1982)
Reticular dermis from various body sites	393	9	44	Vahlquist et al. (1982)
Blood plasma	580	0		Vahlquist et al. (1982)

NR, Not reported.
[a] Calculation based on 20% of wet tissue weight as protein.

agreement among laboratories with regard to total vitamin A content of the epidermis, approximately 7 pmole/mg protein, there is disparity on the relative abundance of retinoid metabolites. The reports by KANG et al. (1995) and DUELL et al. (1996) record that unesterified retinol is the most abundant epidermal retinoid, accounting for 88%–97% of total tissue vitamin A. This is in contrast to all other reports in human epidermis (Table 2), which find that the most abundant form of epidermal vitamin A is long chain fatty acid esters of retinol and 3,4-didehydroretinol. In these papers, vitamin A esters account for 43%–60% of total tissue retinoid. This discrepancy may be explained by the tissue composition of the biopsies that were analyzed. The "epidermal" samples analyzed by KANG et al. (1995) and DUELL et al. (1996) consisted of 200 μm keratomed biopsies of tape-stripped skin. This thickness of biopsy from skin lacking the stratum corneum (from tape-stripping) may have contained quantities of underlying dermal tissue and thus blood plasma (DIGIOVANNA et al. 1985). The presence of plasma retinol in samples would result in overestimating "epidermal" unesterified retinol. For this reason, the data from KANG et al. (1995) and DUELL et al. (1996) regarding epidermis vitamin A ester content may be underestimated.

The mole ratio of vitamin A_1 to A_2 esters (approximately 3) is less than that for total tissue vitamin A alcohols (approximately 6; Table 1). Thus, a greater proportion of tissue vitamin A_2 is esterified as compared to the proportion of tissue vitamin A_1 comprised as ester.

The balance of total epidermal retinoid content consists largely of the unesterified alcohols of A_1 and A_2 present at approximately 600 and 60 n*M*, respectively. This estimate of tissue concentrations is taken from the reports in Table 2 that show the presence of significant quantities of vitamin A ester (i.e., excluding studies from KANG et al. 1995; DUELL et al. 1996b). The higher retinol concentration derived from studies of KANG et al. (1995) and DUELL et al. (1996b) is approximately 1 μM and may represent an overestimation (see above). The conversion of tissue vitamin A content to tissue concentration in this and subsequent calculations assumes a protein density in epidermal keratinocytes as 0.16 pg protein/μm^3 cell volume (SUN and GREEN 1976). These tissue concentrations are important in the discussion of metabolism in the epidermis (see below) since the retinoid alcohols serve as substrates for the enzymes of metabolism.

The native vitamin A metabolite present in epidermis at the lowest concentration is retinoic acid. The retinoic acid in human epidermis exists as two stereoisomers; approximately 70% is 13-*cis*-retinoic acid and the balance is all-*trans*-retinoic acid (DUELL et al. 1996a). Under normal conditions, 9-*cis*-retinoic acid is not detected. Retinoic acid was first reported in epidermis by VAHLQUIST (1982), although its concentration was so low as to preclude quantitation (see Table 2). Another laboratory reports unquantifiable trace amounts of retinoic acid (<4 n*M*) in some samples of epidermal tissue (KANG et al. 1995, 1996) and 20 n*M* concentrations in others (DUELL et al. 1996a). These results are not contradictory considering that the values fall at the lower

Table 2. Relative retinoid composition of human epidermis

Source of epidermis	Total retinoid (pmol/mg protein)	Retinol (% of total)	Retinyl ester (% of total)	3,4-Didehydro-retinol (% of total)	3,4-Didehydro-retinyl ester (% of total)	Retinoic acid (n*M*)	Reference
Breast skin	3.5	30	60	3	7	<35	VAHLQUIST (1982)
	9.3	24	50	8.5	17	NR	TÖRMÄ and VAHLQUIST (1990a)
Neonatal foreskin	4.7	29	71	NR	NR	NR	TANG et al. (1994)
Tape-stripped buttock skin	6.1	88	12	NR	NR	<4	KANG et al. (1995)
	NR	NR	NR	NR	NR	20	DUELL et al. (1996a)
	9	97	2	1	NR	NR	DUELL et al. (1996b)
Back skin	7	30	43	10	17	NR	VAHLQUIST et al. (1996)
Buttock skin	NR	NR	NR	NR	NR	<4	KANG et al. (1996)
Mean ± SD	7.0 ± 2.4	53.8 ± 35.6	33.4 ± 25.1	5.6 ± 4.3	13.7 ± 5.8		

NR, Not reported.

limit of detection for most chromatography systems. Moreover there is the analytical challenge offered by intact epidermal tissue and its limited availability in biopsy material. From these few reports, one is led to the conclusion that retinoic acid is present, in at least some samples of epidermis, and its concentration is in the low nanomolar range, i.e., 20 n*M* or lower. This would place the retinoic acid concentration in epidermis, at its highest, near that in plasma which ranges between 4 n*M*-14 n*M* (DeLeenHeer et al. 1982; Eckhoff and Nau 1990).

D. Retinoid Composition of Epidermal Compartments

The native retinoid composition of epidermis considered thus far has considered the epidermis as a whole tissue. While this information gives insight into vitamin A homeostasis in the epithelium of skin, of more focused utility is the retinoid composition in the different epidermal strata. This is important since the epidermis contains keratinocytes in a wide variety of growth and states of differentiation, ranging from the undifferentiated dividing cells of the basal layer to the suprabasal cells in various stages of differentiation. As a result, it is likely that the cells in the different strata have distinct requirements for vitamin A.

A single laboratory has reported the retinoid alcohol and ester content in cell suspensions isolated from different regions of human epidermis (Törmä and Vahlquist 1990a). These data are summarized in Table 3. Using mechanical and enzymatic means, three cell populations were prepared as suspensions and were analyzed for retinoid content. The three cell populations included the basal layer, a "mixed" intermediate population consisting of basal, spinous, and granular layer cells, and a population consisting of cells from the stratum granulosum and stratum corneum. The authors have presented the data as "semiquantitative" since recovery of unesterified retinoid alcohols from non-saponified samples is less than ideal. Despite this analytical shortcoming, the data serve to give interesting insight to the relationship between retinoid

Table 3. Retinoid composition of epidermal strata

Epidermal cell suspension	Retinol	Retinyl ester	3,4-Didehydro-retinol (pmol/mg protein)	3,4-Didehydro-retinyl ester	Total
Basal	0.4	0.6	0.2	0.1	1.3
Basal + spinous + granular	0.7	1.1	0.3	0.6	2.7
Granular + corneal	1.2	2.9	0.4	0.9	5.4
Total	2.3	4.6	0.9	1.6	9.4

Data taken from Törmä and Vahlquist 1990a.

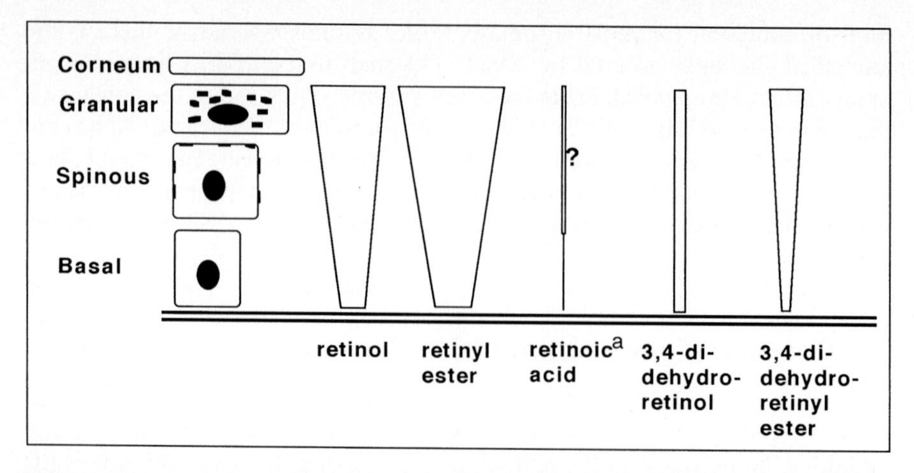

Fig. 1. Vitamin A composition of human epidermis. The trapezoidal shapes represent the relative composition of the indicated retinoid in keratinocytes as a function of location in the epidermis and were taken from the data in Tables 2 and 3. [a]The distribution of retinoic acid in the epidermis is not known. The representation presented here is speculative, and is based on in vitro metabolism studies (see text; From SIEGENTHALER et al. 1990a)

content and degree of differentiation. There is a trend toward increasing total retinoid content of cells with increasing suprabasal location. The majority of this increase is accounted for as an increase in retinyl ester. In addition, retinol and 3,4-didehydroretinyl ester are increased in more differentiated cells. The largest relative increase occurs for 3,4-didehydroretinyl ester, which increases about ninefold. The content of unesterified 3,4-didehydroretinol is relatively constant in all cell populations. The ratio of total A_1 to A_2 remains essentially the same, approximately 3, in all cell groups. The distribution of retinoic acid in the epidermis is not known. These semiquantitative data are also expressed diagrammatically in Fig. 1.

E. Retinoid Composition of Cultured Human Keratinocytes

Several laboratories have reported the native retinoid composition of human epidermal keratinocytes maintained in culture medium containing nanomolar concentrations of serum retinol. These data are summarized in Table 4. The total retinol to 3,4-didehydroretinol ratio is 2.4. This value compares to the range of A_1 to A_2 ratios in intact epidermis, between 3 and 6 (see above). Long chain fatty acid esters account for 78% of total cellular retinoid. The bulk of the remaining retinoid is distributed between the two unesterified retinoid alcohols. Retinoic acid is also present in cultured keratinocytes; its concentration is approximately 25 n*M*. Approximately 85% of the retinoic acid in cultured keratinocytes is all-*trans*-retinoic acid with balance accounted for as 13-*cis*-retinoic acid (RANDOLPH and SIMON 1995). The carboxylic acid metabolite of 3,4-didehydroretinol, 3,4-didehydroretinoic acid, is also found at low nanomolar concentrations.

Overall, the retinoid composition of keratinocytes cultured in the presence of serum (Table 4) qualitatively resembles that in intact epidermis. However, there are some quantitative differences. Total cell retinoid content in cultured keratinocytes is about 70% higher (~12 pmole/mg) than that in the epidermis (~7 pmol/mg; Table 1). The proportion of total cellular retinoid as vitamin A esters (including esters of A_1 and A_2) is considerably greater in cultured keratinocytes (~90%) than that in epidermal tissue (~55%). Corresponding to this, the concentration of the retinoid alcohols in cultured cells is lower than that in epidermis. And in addition, the concentration of retinoic acid, although in the low nanomolar range in the cultured cells (~25 n*M*), is higher than that in intact epidermis (<20 n*M*).

F. Extracellular Retinoid Transport

I. Retinol

The working model for this discussion is that both the quality and quantity of vitamin A are important in epidermal homeostasis. All of the vitamin A and it metabolites that are present in the epidermis are derived either directly from retinoids in the blood supply of the dermis, or indirectly via metabolism of the blood-borne retinoids. Plasma retinol is the most abundant native retinoid in plasma. It is transported primarily bound to its specific plasma transport protein, RBP (KANAI and GOODMAN 1968). Retinol and RBP are constitutively present in human plasma at a concentration of 1–2 μM (KANAI and GOODMAN 1968). The relatively low molecular weight of RBP (21 kDa), allows this transport protein access to the interstitial fluid compartment of most tissues including the skin. Consistent with the transport of retinol from the plasma to the extravascular compartment in bound form, the concentration of RBP and retinol in skin blister fluid is 0.7 μM (TÖRMÄ and VAHLQUIST 1983), about half their respective plasma concentrations. Since skin blister fluid is thought to approximate the plasma ultrafiltrate that is produced in the microvasculature of the dermis (ROSSING and WORM 1981), a 0.7 μM concentration of RBP-retinol likely represents the upper limit of extracellular retinol concentrations to which the epidermis would be exposed.

The concentration of RBP in intact epidermis is somewhat lower than this, approximately 50 n*M* (TÖRMÄ and VAHLQUIST 1983). Interestingly, most of the RBP present in intact epidermis is in the apo-form, i.e., lacking bound retinol (SIEGENTHALER and SAURAT 1987). This suggests that RBP-retinol permeating into the epidermis gives up its bound retinol to cells. Consistent with this, organ cultured skin and epidermis (TÖRMÄ and VAHLQUIST 1984), freshly isolated keratinocytes (TÖRMÄ and VAHLQUIST 1991), and cultured keratinocytes

(CREEK et al. 1993; BÄVIK et al. 1995), take up retinol from RBP. Retinol uptake by keratinocytes is thought to occur passively via diffusion across the membrane following its dissociation from RBP (CREEK et al. 1993). The extent of retinol uptake from RBP is determined by the balance between intracellular metabolism and extracellular supply (NOY and BLANER 1991; LEVIN 1993).

The actual supply of retinol to the epidermis is not known but can be estimated. The estimated supply of retinol is equal to the product of the retinol concentration in dermal plasma ultrafiltrate and the flow of the ultrafiltrate. The concentration of RBP-retinol in dermal interstitial fluid is assumed to be similar to that in skin blister fluid, 0.7 μM (TÖRMÄ and VAHLQUIST 1983). The flow of interstitial fluid through the skin can be estimated as follows. Since the concentration of albumin in skin blister fluid (ROSSING and WORM 1981), interstitial fluid and lymph (REED et al. 1993) are similar, the flux of plasma ultrafiltrate through dermal interstitium approximates the lymph flow from the skin (REED et al. 1993). The flow of lymph from the skin is assumed to be proportional to cardiac output to the skin, approximately 8.5% (SCHEUPLEIN 1979). Total lymph flow in an adult is approximately 5 l/day (BERNE and LEVY 1988). Adult body surface area is assumed to be 2 m^2. Utilizing these physiological constants and assumptions, the supply of retinol to epidermis in 1 cm^2 of skin is as follows: 5\l lymph/day × 0.085 × 0.7 μmol retinol/l ÷ 20000 cm^2 = 14.9 pmol retinol/cm^2/day. If this daily supply of retinol was contained within a 100 μm thickness of epidermis, and there was no loss, the total retinol content of epidermis would increase by 1.49 nmol/cm^3 epidermal volume in 1 day. This estimated daily supply of extracellular retinol compares to the total vitamin A content of epidermis of approximately 1 nmol/cm^3 (from Table 1). The comparable predicted supply, and the steady state content of retinol implies that there must be mechanisms that control the quantity of retinol delivered to and retained within epidermal cells. Although no such mechanisms have been described, they may include limited uptake and release or metabolism of retinol.

II. Retinoic Acid

As with retinol, retinoic acid is accessible (potentially) to the epidermis from the circulatory supply of underlying dermis. Retinoic acid is present constitutively in human plasma over a concentration range of 4–14 nM (DE LEEN HEER et al. 1982; ECKHOFF and NAU 1990), and is transported bound to albumin (SMITH et al. 1973). If albumin-bound retinoic acid is accessible to the epidermis, it would have to occur via extravascular transport, i.e., via the interstitial fluid. The concentration of retinoic acid in skin interstitial fluid is not known. However, albumin, the transport protein for retinoic acid in the plasma, is present in blister fluid at about one-half (300 μM) its concentration in plasma (600 μM; ROSSING and WORM 1981). If one assumes that the albumin permeating into the dermal interstitium retains its bound retinoic acid, and no metabolism of or production of retinoic acid by other cell types in the dermis, the

Table 4. Retinoid composition of cultured keratinocytes

Reference	Total retinoid (pmol/mg protein)	Retinol	Retinyl ester	3,4-Didehydro-retinol	3,4-Didehydro-retinyl ester (% of total retinoid)	Retinal	Retinoic acid	3,4-Didehydro-retinoic acid	A_1/A_2 ratio
RANDOLPH and SIMON (1993)	13	1	77	0.5	23	ND	0.5	0.3	3.3
ROLLMAN et al. (1993)	10	2	56	2.4	38	ND	ND	ND	1.4
KURLANDSKY et al. (1994)	NR	2	90	NR	NR	15	2.0	NR	
Mean ± SD	11.5 ± 2.4	2.7 ± 2.1	78.7 ± 16.5	1.0 ± 1.2	30.5 ± 10.6	15±	1.4 ± 0.8	0.4 ± 0.1	2.4 ± 1.3

ND, Not detected; NR, not reported.

Table 5. Content of cellular retinoid binding proteins in human epidermis

Reference	Source of epidermis	CRBP	Total CRABP[a]	CRABP I	CRABP II
		pmol/mg protein			
PUHVEL and SAKAMOTO (1984)	Facial skin		18.3		
SIEGENTHALER et al. (1984)	Abdominal skin		2.7		
DIGIOVANNA et al. (1985)	Lower limb skin	0.4	3.7		
GATES et al. (1987)	Neonatal foreskin	7.0	61.0		
SIEGENTHALER and SAURAT (1987)	Skin from various sites	1.0	2.4		
HIRSCHEL-SCHOLZ et al. (1989a)	Buttock skin		2.5		
HIRSCHEL-SCHOLZ et al. (1989b)	Buttock skin		2.5		
SIEGENTHALER and SAURAT (1991)	Skin from various sites	0.2	9.0		
SIEGENTHALER et al. (1992)	Skin from various sites			3.8[b]	5.2[b]
KANG et al. (1995)	Buttock skin		3.3	0.4	2.9
VAHLQUIST et al. (1996)	Back skin	0.2		NR	5.1
Mean (adult) ± SD		0.5 ± 0.4	5.5 ± 5.6		4.4 ± 1.3

[a] Method did not distinguish between CRABP types I and II and thus represents total CRABP.
[b] Values calculated from the relative proportions reported in the indicated reference, and the total CRABP value reported by SIEGENTHALER and SAURAT (1991).

extracellular concentration of retinoic acid in the dermis should be approximately 5 n*M*, about one-half its concentration in plasma. Since the concentration of albumin in skin blister fluid (Rossing and Worm 1981), interstitial fluid and lymph (Reed et al. 1993) are similar, the flux of plasma retinoic acid (albumin-bound) through dermal interstitium is assumed to approximate the lymph flow from the skin (Reed et al. 1993). Utilizing the same parameters described above for calculating retinol supply to the skin, the estimated supply of retinoic acid (originating from the plasma) to epidermis in 1 cm^2 of skin is as follows: 5 l lymph/day × 0.085 × 5 nmol retinoic acid/l ÷ 20 000 cm^2 = 0.11 pmol retinoic acid/cm^2/day. In order to predict the quantitative significance of this supply of retinoic acid, it is instructive to compare it with the concentration of retinoic acid in the epidermis. If the calculated daily supply of retinoic acid was contained in a 100 μm thickness of epidermis, its concentration would increase by about 11 pmol/cm^3 per day or about 11 n*M* per day! This increment compares to the steady state retinoic acid concentration of epidermis (<20 n*M*). The implication is, as for retinol, that there must be mechanisms in the skin that limit the access to or for metabolizing retinoic acid in the epidermis.

G. Retinoid Metabolism in Epidermis

I. Retinoid-Binding Proteins

1. Cellular Retinol-Binding Protein

The very low concentration of retinoic acid in the epidermis may derive not only by transport from the plasma, as discussed above, but via metabolism of retinol in situ. The substrate for retinoic acid synthesis is retinol that originates in the plasma (Blaner and Olson 1994; Napoli 1996). In many tissues, intracellular retinol is bound to cellular retinol-binding protein, type I (CRBP; Ong et al. 1994), a specific binding protein for retinol that resides in the cytoplasm. Cellular retinol-binding protein is known to function as a modulator of retinol metabolism in two ways. First, CRBP presents bound retinol to specific enzymes of metabolism via protein-protein interactions. For example, CRBP-retinol is restricted from esterification by the acyl CoA:retinol acyltransferase (ARAT) (Ong et al. 1988; Yost et al. 1988) reaction but is available for esterification by the lecithin:retinol acyltransferase (LRAT; Ong et al. 1988; Yost et al. 1988; Blaner and Olson 1994) reaction. Likewise, retinoic acid synthesis is thought to be initiated by the interaction of a specific retinol dehydrogenase with CRBP-retinol (Napoli 1994, 1996). Second, the extent to which CRBP in cells exists in holo-form, i.e., possessing bound retinol, versus in apo-form, i.e., lacking bound retinol, influences the activities of enzymes of esterification and oxidation (Napoli 1994, 1996). For example, a high proportion of total CRBP as apo-CRBP signifies limited availability of retinol and inhibits synthesis of retinyl ester by LRAT. This situation favors the presentation of

CRBP-retinol to enzymes for retinoic acid synthesis. On the other hand, a high proportion of total CRBP in holo-form is associated with an abundance of retinol and stimulates retinyl esterification. The significance of CRBP in a tissue is that it coordinates and directs the metabolism of retinol toward pathways of esterification or retinoic acid synthesis in accord with the respective expression of active enzymes in these pathways and with the abundance of retinol.

Cellular retinol-binding protein has been quantitated in epidermal tissue by several laboratories. The results are summarized in Table 5. The values reported for adult skin epidermis average 0.5 pmol CRBP/mg protein. This value corresponds to an epidermal CRBP concentration of approximately 80 n*M*. The concentration of CRBP in cultured keratinocytes, 50–60 n*M* (SIEGENTHALER et al. 1988), is less than that in the epidermis. The CRBP concentration in neonatal epidermis is very high, 1.4 μM (GATES et al. 1987), compared to adult epidermis. The reason for this higher level of CRBP in neonatal skin is not known.

The concentration of CRBP in a tissue is generally thought to be a good predictor of the concentration of retinol in that tissue (NOY and BLANER 1991). This is not the case, however, with adult epidermis (Tables 2, 5). Even considering the RBP concentration in epidermis, 50 n*M* (TÖRMÄ and VAHLQUIST 1983), with the CRBP concentration, 80 n*M* (Table 5), the concentration of total retinol-binding activity (CRBP plus RBP) is only about one-fourth of the unesterified retinol concentration in epidermis (600 n*M*). In this regard, the retinol content, and retinol-binding activity in the epidermis is very different than the situation in other tissues. For example, in liver the retinol concentration is approximately one-half the concentration of CRBP (NAPOLI 1994, 1996). At minimum, this excess of retinol over total binding protein suggests that epidermal CRBP is almost certainly saturated with retinol, a situation that favors retinol esterification by LRAT over retinoic acid synthesis (discussed below).

That a significant proportion of the unesterified retinol in the epidermis is presumably not bound to binding proteins (~470 n*M*) raises some important possibilities with regard to its metabolism; namely, that some metabolism could occur without the mediation of CRBP in the role of substrate presentation. These issues are discussed in the following sections addressing metabolism.

Cellular retinol-binding protein is expressed in the cytoplasm of keratinocytes in all strata of the epidermis. More CRBP is present in differentiating suprabasal keratinocytes than in the basal layer (SIEGENTHALER and SAURAT 1991; BUSCH et al. 1992; Fig. 2). Differentiating keratinocytes in culture likewise express higher levels of CRBP than their undifferentiated counterparts (SIEGENTHALER et al. 1988; SIEGENTHALER and SAURAT 1991). Following an expression pattern different than the protein, CRBP mRNA (KANG et al. 1995) appears to be uniformly distributed across the epidermis. TÖRMÄ et al. (1996) also report expression of CRBP mRNA in human epidermis although

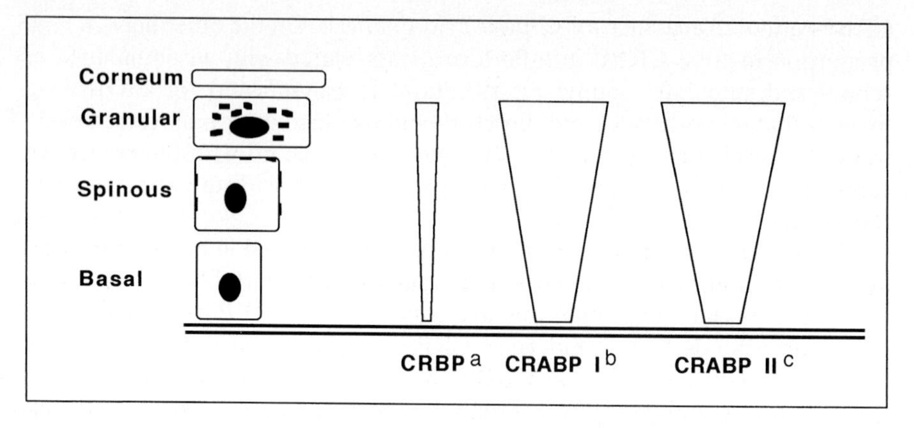

Fig. 2. Cellular retinoid binding proteins in human adult epidermis. The relative abundance of the indicated cellular retinoid binding proteins in keratinocytes is shown as a function of location in adult epidermis. Representations for each binding protein were taken from the indicated references. [a]From SIEGENTHALER and SAURAT 1991; BUSCH et al. 1992. [b]from BUSCH et al. 1992. [c]from SIEGENTHALER et al. 1988, 1992, and unpublished observations

no information is given regarding its specific distribution. The presence of CRBP in all of the living layers of the epidermis are consistent with both retinol esterification and retinoic acid synthesis, at least from the standpoint of substrate presentation, throughout the epidermis.

Consistent with a role for CRBP in modulating the production of retinoic acid from retinol, the expression of CRBP in the epidermis is regulated by retinoids. Increasing the content of retinol in the epidermis by topical treatment of skin with retinol, increases CRBP mRNA levels throughout the epidermis, presumably via the synthesis of retinoic acid (KANG et al. 1995). Concomitant with this increased CRBP expression, topical retinol also increases the epidermal content of retinyl ester (DUELL et al. 1996a). This response suggests that the epidermis responds to increased retinol levels by increasing the quantity of CRBP available to bind retinol. The resulting increased concentration of CRBP-retinol stimulates the esterification of retinol by LRAT (discussed further below), decreasing retinol levels and the synthesis of retinoic acid. Topical treatment of skin with retinoic acid increases the expression of CRBP mRNA in a manner similar to that observed for retinol (FISHER et al. 1995). Consequently, keratinocytes throughout the epidermis possess the potential to respond to and control their supply of retinol, and the synthesis of retinoic acid, by regulating CRBP expression.

2. Cellular Retinoic Acid Binding Proteins

Keratinocytes also contain cytosolic binding proteins for retinoic acid. Two cytosolic binding proteins for retinoic acid, cellular retinoic acid binding proteins (CRABP) type I and type II, are known (for review see ONG et al. 1994).

Several functions have been proposed for both of these binding proteins. The first function suggested for CRABP I was that it acts to convey retinoic acid from the cytosol to specific sites in the nucleus for gene regulation (TAKASE et al. 1986; BARKAI and SHERMAN 1987). In addition, CRABP I was shown subsequently to facilitate retinoic acid metabolism (FIORELLA and NAPOLI 1991; BOYLAN and GUDAS 1992). This is thought to occur in a manner similar to that for CRBP and retinol (discussed above); CRABP I-bound retinoic acid may be directed toward specific enzymes of metabolism by protein-protein interactions. The functions of CRABP II are less well understood and are inferred from its expression pattern under conditions known or thought to involve relative abundance or deficiency of retinoic acid. From these data it has been suggested that CRABP II functions as a cytosolic reservoir for retinoic acid or as a shuttle for its transport to the nucleus for signaling (ONG et al. 1994; BUCCO et al. 1995; SIEGENTHALER 1996). CRABP I and II may thus have both redundant and distinct functions.

The earliest quantitations of cytosolic retinoic acid-binding activity in epidermis did not distinguish between CRABP I and II. As a result, these early values were reported as total CRABP (Table 5). In neonatal foreskin, total CRABP is 61 pmol/mg protein (GATES et al. 1987). The total quantity of CRABP in adult epidermis is considerably lower, averaging 6.1 pmol/mg protein. In adult epidermis, this value for total retinoic acid-binding protein activity translates into a concentration of 1 μM. It is noteworthy that the concentration of total CRABP in the epidermis is greater than 50-fold in excess of the epidermal concentration of retinoic acid (Table 2). Total CRABP in cultured, differentiating keratinocytes (~3.5 μM; SIEGENTHALER et al. 1988; SIEGENTHALER and SAURAT 1991) is higher than that in epidermis. This excess in total CRABP versus retinoic acid concentration is conserved in cultured keratinocytes.

At the protein level, two disparate values have been published regarding the relative abundance of the two CRABPs in epidermis (Table 5). One report shows approximately equal quantities of CRABP I and II (SIEGENTHALER et al. 1992). This is in contrast to another study that shows a predominance of CRABP type II over type I by tenfold (KANG et al. 1995). The reason for this discrepancy is related to the different methodologies employed in each laboratory. In the study by KANG et al. (1995), ion exchange chromatography was utilized to quantitate the two CRABP isoforms. Since the isoelectric points of plasma albumin, pI = 4.9 (COHN et al. 1947), and CRABP II, pI = 4.86 (SIEGENTHALER et al. 1992) are similar, this technique does not resolve CRABP II and albumin. In addition, albumin is known to bind retinoic acid (SMITH et al. 1973), and is abundant in the epidermis (ROSSING and WORM 1981). Consequently, the coelution of CRABP II and albumin during ion exchange chromatography would overestimate the quantity of CRABP II present. The electrophoretic method utilized by SIEGENTHALER et al. (1992), resolves both of the respective CRABPs from albumin and RBP, both of which also bind retinoic acid, and thus is a more accurate method to measure the two proteins. The quantitation of CRABP II by VAHLQUIST et al. (1996) also employs the electrophoretic technique, and reports an epidermal CRABP II content that is very similar to that reported by SIEGENTHALER et al. (1992).

By immunohistochemical analysis, CRABP I is present throughout the epidermis and is more abundant in differentiating keratinocytes of suprabasal strata than in those resident in the basal layer (BUSCH et al. 1992; Fig. 2). In contrast to this expression pattern for CRABP I at the protein level, CRABP I mRNA is not detected or is not consistently detected in the epidermis, either by northern blot analysis (ELDER et al. 1992a,b; ELLER et al. 1992) or by in situ hybridization (DIDIERJEAN et al. 1991; ELLER et al. 1994). Cultured keratinocytes also express CRABP I mRNA at very low levels (ELLER et al. 1992; SANQUER et al. 1993). In contrast to epidermis however, CRABP I protein is not detected in cultured keratinocytes (SIEGENTHALER et al. 1992, 1996; VETTERMANN et al. 1997). These data suggest that the CRABP I gene is regulated differently in keratinocytes in the in vivo versus in vitro environments.

The localization of CRABP II in epidermis has not been published. The distribution of CRABP II mRNA is known, however. In neonatal foreskin epidermis, CRABP II mRNA is expressed at high levels in differentiating keratinocytes but is not detected in basal keratinocytes (ELLER et al. 1991). In adult skin, a different pattern of CRABP II expression is evident. CRABP II mRNA is present at low levels throughout the epidermis (TAVAKKOL et al. 1992; KANG et al. 1995). Perhaps predicting increased translation in the suprabasal strata of the epidermis, CRABP II protein levels increase substantially as cultured keratinocytes differentiate (SIEGENTHALER et al. 1988, 1992). Preliminary results by one of the authors are consistent with this expression pattern (Siegenthaler, unpublished observation).

The incremental abundance of CRABP I in keratinocytes in progressively later stages of differentiation suggests that it plays an important role in the differentiation process with regard to retinoic acid. It is not clear, however, what this role might be. Intriguingly, CRABP I, at the protein level, decreases in the skin treated with topical retinoid (SAURAT et al. 1994). Since no retinoic acid-response elements are known in the CRABP I gene, regulation by retinoid must occur by indirect means. This down-regulation is opposite to what is expected if CRABP I functions in the epidermis to facilitate the metabolism of excess retinoic acid as has been suggested to be the case in other tissues (FIORELLA and NAPOLI 1991; BOYLAN and GUDAS 1992). Interestingly, CRABP I is not expressed in cultured keratinocytes (SIEGENTHALER et al. 1992; SIEGENTHALER 1996; VETTERMANN et al. 1997) which exhibit a high capacity retinoic acid metabolism pathway (RANDOLPH and SIMON 1997, and discussed below). Other functions for CRABP I, such as trafficking retinoic acid to and from the nucleus, remain a possibility.

Exhibiting clear retinoid-responsive transcriptional regulation (DURAND et al. 1992), the expression of CRABP II increases dramatically at both the mRNA and protein levels in epidermis exposed to topical retinoic acid (HIRSCHEL-SHOLZ et al. 1989a,b; ÅSTRÖM et al. 1991; ELDER et al. 1992a,b 1993;

TAVAKKOL et al. 1992; SAURAT et al. 1994), retinol (SAURAT et al. 1994; KANG et al. 1995), or retinal (SAURAT et al. 1994), and in epidermis of individuals taking oral retinoids (HIRSCHEL-SHOLZ et al. 1989b). This regulation in response to exogenous retinoids suggests that CRABP II may function under these circumstances to buffer keratinocyte nuclear retinoic acid receptors against sudden or extreme changes in the availability of exogenous retinoic acid.

It is not clear, however, whether this is the function of CRABP II in the normal differentiation process. This follows from the fact that the cellular content of CRABP II increases as keratinocytes differentiate both in vivo and in vitro, conditions in which cellular retinoic acid concentration is very low, and yet appropriate for support of the differentiation program. Indeed, the capacity of keratinocyte cytosolic extracts to catalyze retinoic acid synthesis, the cellular content of CRABP II, and the stage of differentiation in vitro are positively correlated (SIEGENTHALER and SAURAT 1991; SIEGENTHALER 1996; VETTERMANN et al. 1997). All three of these parameters increase in concert when keratinocytes are stimulated to differentiate in culture by increasing medium calcium concentration. Since this differentiation-linked increase in retinoic acid synthesis appears to be a normal event in keratinocyte differentiation, it does not seem reasonable that cells would simultaneously increase CRABP II levels to sequester retinoic acid from the nucleus. Rather, CRABP II seems to play a facilitating role, enhancing the access of newly synthesized retinoic acid to retinoid receptors in the nucleus during differentiation. In fact, such a role for CRABP II in breast cancer cells has been recently reported (JING et al. 1997). Identifying and clarifying this and other functions of CRABP II in the epidermis will undoubtedly be the focus of continuing investigation.

II. Vitamin A Storage, Retinyl Ester Synthesis and Hydrolysis

Vitamin A is present in intact epidermis and in cultured keratinocytes, at least in part, as esters of long chain fatty acids (Tables 2, 3). Despite the lack of agreement regarding the retinyl ester content of epidermis (discussed above), high-affinity retinol esterification is observed in epidermal microsomes (TÖRMÄ and VAHLQUIST 1990a; KURLANDSKY et al. 1996), and in freshly isolated basal keratinocytes (KURLANDSKY et al. 1996). Topical application of pharmacological quantities of retinol to skin results in the formation of large amounts of retinyl ester in epidermis (KANG et al. 1995). Moreover, short-term culture of fresh, intact human epidermis with micromolar concentrations of RBP-retinol, the physiological substrate, results in the synthesis of substantial (50%–75% of total tissue retinoid) quantities of vitamin A ester (TÖRMÄ and VAHLQUIST 1990a,b). Thus, while there is some question regarding the precise quantity of retinyl ester (and retinol) in the epidermis (Table 2), it is clear that retinol esterification is a major pathway of retinol metabolism.

The retinyl ester present in epidermis is probably the result of synthesis by LRAT. Several observations are in support of this conclusion. First, in the initial description of retinol esterification in epidermis (TÖRMÄ and VAHLQUIST 1990a), the acyl Coenzyme A-independent esterifying activity present in epidermal microsomes is sufficient to account for all of the retinyl ester mass in the tissue. Second, the epidermal concentration of retinol (600 nM; TÖRMÄ and VAHLQUIST 1990a) and CRBP (80 n*M*; Table 5) suggest that about fourth of the retinol is bound to CRBP and as a result, is unavailable for esterification by ARAT (ONG et al. 1988; YOST et al. 1988) which is also present in epidermis (TÖRMÄ and VAHLQUIST 1990a). Third, the remaining unbound retinol, at nanomolar concentrations, would support an LRAT-catalyzed reaction ($K_m = 1\,\mu M$; ONG et al. 1994) significantly more than an ARAT-catalyzed reaction (K_m = 7–10 μM; TÖRMÄ and VAHLQUIST 1990a). Fourth, consistent with LRAT activity, nanomolar concentrations of retinol support robust esterification in cultured keratinocytes (RANDOLPH and SIMON 1993; ROLLMAN et al. 1993; KURLANDSKY et al. 1994). And finally, an LRAT-like activity has been described in cultured human keratinocytes (KURLANDSKY et al. 1996).

The retinol esterification reaction in keratinocytes has potential to regulate the synthesis of retinoic acid because both reactions utilize retinol as a substrate (RANDOLPH and SIMON 1995, 1996; KURLANDSKY et al. 1996). It is for this reason that the distribution of retinol esterifying activity across the strata of differentiating cells of the epidermis is important. The majority (~75%) of total epidermal LRAT activity resides in the basal cells of the epidermis (KURLANDSKY et al. 1996). The remaining LRAT activity is located in differentiating suprabasal cells. ARAT activity is distributed evenly throughout the epidermis (TÖRMÄ and VAHLQUIST 1990a). Consistent with a more significant role for LRAT activity in keratinocytes, the uptake of retinol by freshly isolated keratinocytes parallels the localization of LRAT activity; retinol uptake is greater in basal cells than in suprabasal cells (TÖRMÄ and VAHLQUIST 1991). This distribution of retinol esterifying activity is also consistent with the retinyl ester content of different epidermal strata (TÖRMÄ and VAHLQUIST (1990a; Table 4). As plasma retinol is transported from the blood supply in the dermis to the epidermis, it is taken up by and esterified in basal keratinocytes. Some retinol is not taken up in the basal layer and permeates to the suprabasal layers of the epidermis (retinol is present throughout the epidermis; Table 4), where it is taken up and esterified in differentiating keratinocytes. Thus, retinyl ester continues to accumulate in keratinocytes with increasing suprabasal location (TÖRMÄ and VAHLQUIST 1990a). This trend toward increasing retinyl ester content and fairly constant retinol content with increasingly suprabasal position suggests that the rate of retinol transport, uptake and esterification exceeds the rate of retinol utilization, i.e., retinoic acid synthesis, throughout the differentiation process.

Before the retinol stored as retinyl ester can be utilized for retinoic acid synthesis, the ester must undergo hydrolysis. The retinyl ester hydrolase that catalyzes this reaction in keratinocytes has not been identified. Hydrolysis of retinyl ester has been observed, however, in cultured keratinocytes. Studies by RANDOLPH and SIMON (1993, 1996) show that keratinocytes mobilize retinyl ester for retinoic acid synthesis with a half-life of 72h. Retinyl ester is the preferred source of substrate retinol for retinoic acid synthesis in cultured keratinocytes, even in the presence of extracellular retinol. Thus, in cultured keratinocytes the balance between the retinyl ester synthesis and hydrolysis reactions can influence the availability of at least some retinol for retinoic acid synthesis. Whether retinyl ester depots in the epidermis serve as a source of retinol for retinoic acid synthesis is not known.

In addition to serving as a source of retinol for retinoic acid synthesis, and a means for buffering cells from high retinol concentrations, epidermal retinyl ester may have other functions. This possibility must be considered since retinyl ester is present in the granular layer of the epidermis (Table 4), the population of cells that gives rise to the stratum corneum. Although the fate of retinyl ester contained in terminally differentiating keratinocytes is unknown, it may act as a protective barrier to ultraviolet light in the 300- to 400-nm wavelength range, or as an antioxidant, thus protecting the epithelium against potentially harmful oxidizing agents in the environment.

III. Retinoic Acid Synthesis

There are at least two sources for the low quantities of retinoic acid in the epidermis. As discussed above, plasma retinoic acid may be transported to the epidermis bound to albumin from the blood supply of the dermis. Is it possible that retinoic acid can also arise from synthesis in situ from retinol? If it does occur, what is its quantitative significance? Based on the affinity of retinol dehydrogenase enzymes (from other tissue sources) for retinol ($K_m < 1\,\mu M$; NAPOLI 1994, 1996), the concentration of retinol in epidermis (600 nM; TÖRMÄ and VAHLQUIST 1990a) is sufficient to support retinoic acid synthesis. Indeed, cytosolic extracts of human epidermis or cultured keratinocytes convert retinol into retinoic acid in the presence of NAD (Fig. 3) (SIEGENTHALER et al. 1990a,b).

Experimentally increasing the epidermal concentration of retinol by its topical application to skin results in the production of detectable levels of retinoic acid in human (KANG et al. 1995) or mouse (CONNOR and SMIT 1987) epidermis. However, under these conditions when epidermal retinol increases 100-fold or greater, retinoic acid levels increase only to the limit of detection in human epidermis (<4 nM; KANG et al. 1995), and less than ten fold in mouse epidermis (CONNOR and SMIT 1987). Similar to these in vivo results, less than 2% of total cell retinoid is converted to retinoic acid in cultured keratinocytes, even when the concentration of retinol is increased to the micromolar range (RANDOLPH and SIMON 1993; KURLANDSKY et al. 1994). Thus, while it is clear that the epidermis, and epidermal keratinocytes possess the potential to synthesize retinoic acid under experimental conditions, extrapolating the retinol concentration downward to that found in vivo, approx. 600 nM, leads to the conclusion that retinoic acid synthesis in the epidermis is at most, parsimonious.

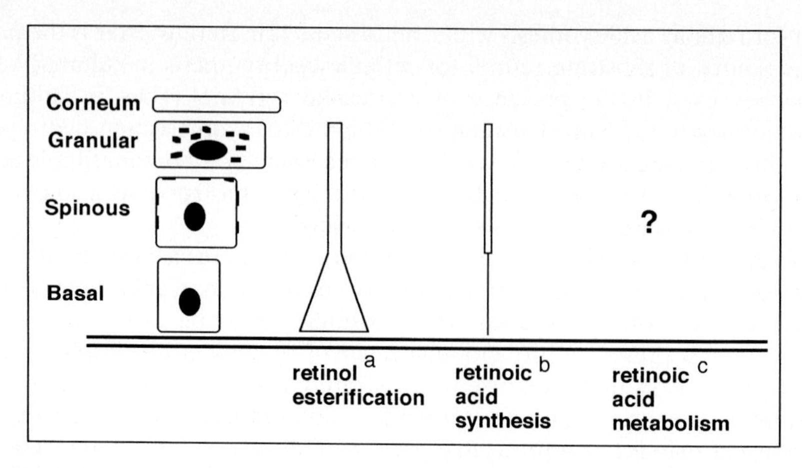

Fig. 3. Vitamin A metabolism in human epidermis. The relative activity of the indicated metabolic reactions in keratinocytes is shown as a function of location in the epidermis. Representations for each metabolic pathway are taken from the indicated references. [a]From KURLANDSKY et al. 1996. [b]The location of the retinoic acid synthesis pathway in the epidermis is not known. The representation presented here is speculative, and is based on in vitro metabolism studies (see text; SIEGENTHALER et al. 1990a). [c]The distribution of pathways for retinoic acid metabolism in the epidermis is not known

It must be emphasized that the fact that retinoic acid synthesis in the epidermis proceeds at a very low rate does not preclude a significant role in retinoic acid signaling in the epidermis. It is possible, for example, that physiologically relevant retinoic acid synthesis occurs in a small proportion of epidermal cells, and that its quantitative importance is diluted in experiments that deal with whole epidermal tissue.

The requirement for micromolar concentrations of substrate retinol to drive retinoic acid synthesis in the epidermis suggests that mechanisms in addition to or other than limiting concentrations of substrate retinol are operating to restrict retinoic acid synthesis under these experimental conditions. The prime candidates for this are the enzymes in the biosynthetic pathway. Since none of the enzymes of retinoic acid synthesis in epidermis have been purified or characterized, it remains to be determined as to whether and how this pathway is controlled.

Despite a dearth of information regarding the enzymatic pathway of retinoic acid synthesis in the epidermis, there is a hint that it is regulated at the level of enzyme expression in the context of differentiation. Cytosolic extracts from differentiating and undifferentiated cultured human and murine keratinocytes have been compared for their ability to catalyze the synthesis of retinoic acid (SIEGENTHALER et al. 1990a,b). Interestingly, only extracts from differentiating cells possess the capacity to synthesize retinoic acid from retinol. In contrast, extracts from both types of cells exhibit the capacity to synthesize retinoic acid from retinal, the intermediate in retinoic acid synthesis that is produced in the first oxidative reaction of retinol. These results suggest that the first enzyme in the retinoic acid synthesis pathway, retinol dehydrogenase, is not present in extracts from undifferentiated cells. The second enzyme in retinoic acid synthesis, retinal dehydrogenase, is present and similarly active in both cell populations. Thus, at least in cultured keratinocytes, a cytosolic pathway of retinoic acid synthesis possesses the potential for regulation at the level of enzyme activity. Moreover, the cytosolic potential for retinoic acid synthesis increases as cells undergo differentiation. It is not known whether similar regulation also occurs in intact cells or in vivo.

The retinoic acid synthesis pathway in keratinocytes is subject to negative feedback autoregulation, at least in vitro. Exposure of cultured keratinocytes to nanomolar concentrations of retinoic acid in the culture medium decreases retinoic acid production by approximately 50% (Randolph and Simon 1996). Under these conditions, declining retinoic acid synthesis occurs without comparable decreases in the cellular concentration of retinol. The down regulation is limited, however, since retinoic acid synthesis persists, even in the presence of pharmacological concentrations of retinoic acid.

In addition to the cytosolic retinol oxidizing activity described above in the epidermis (Siegenthaler et al. 1990a,b), cytosolic (Deuster 1996) and microsomal (Napoli 1996) retinol oxidizing activities have been described in other tissues. It is not known the extent to which either of these pathways for retinoic acid synthesis are present in the epidermis. Since the microsomal activity is thought to recruit CRBP-retinol as a preferred substrate in vivo, and since CRBP-retinol is present in epidermis, its activity needs to be determined. Likewise, the presence in the epidermis of a substantial quantity of retinol that is presumably not bound to CRBP raises a question as to its disposition as a substrate for cytosolic enzymes which do not require or recruit CRBP-bound retinol. This clearly is an area of retinol metabolism in the epidermis that requires much additional attention.

In summary, a low capacity retinoic acid synthesis pathway is present in human epidermis and keratinocytes. Retinoic acid synthesis in the epidermis is tightly regulated and proceeds at a slower rate than predicted by retinol concentration. Expression of retinol dehydrogenase enzyme activity, and retinoic acid synthetic potential, may be linked to differentiation. Retinoic acid production in keratinocytes is subject to autoregulated via negative feedback. The characterization of the biochemical pathway(s) of retinoic acid synthesis in the epidermis will most certainly be the subject of future investigation.

IV. Retinoic Acid Metabolism

The retinoic acid concentration of epidermis depends on the balance between input via the blood supply of the dermis, synthesis in situ, and output via metabolism. What is known about metabolism of retinoic acid in intact human epidermis comes from pharmacological studies. In the work published by DUELL et al. (1992), skin was treated with a cream containing 0.1% retinoic acid. After 4 days under occlusion, the skin was biopsied and the retinoid metabolites were extracted and analyzed. Nanogram quantities of 4-oxo-retinoic acid are recovered from epidermis of skin treated in this manner. In parallel experiments, epidermal microsomes prepared from the retinoic acid-treated skin exhibit a markedly enhanced rate of 4-oxo-retinoic acid synthesis compared to microsomes from untreated skin. The reaction present in microsomes from retinoic acid-treated skin is inhibited 68% by imidazole compounds, and 50% by CO, and is dependent upon NADPH, and molecular oxygen. In addition, it appears to be specific for all-*trans*-retinoic acid (DUELL et al. 1996b). This activity in epidermis resembles in many but not all respects, the cytochrome P450-linked activity observed in liver (BLANER and OLSON 1994; NAPOLI 1996).

The qualitative importance of cytochrome P450-linked retinoic acid metabolism to maintaining low epidermal retinoic acid concentrations is evident from in vivo studies in which an inhibitor of cytochrome P450 function is applied topically to skin. For example, KANG et al. (1996) demonstrate that application of a cream containing liarozole to skin increases the content of retinoic acid in the epidermis from undetectable levels to 60 n*M* 18 h after application. Epidermal retinoic acid concentrations decline thereafter but are still elevated (~20 n*M*) 48 h after application of the inhibitor. These experiments demonstrate that retinoic acid is indeed transported from the plasma or synthesized in the epidermis (discussed above), even when its concentration is below the limits of detection. Furthermore, a cytochrome P450-linked pathway is responsible for metabolism of endogenous as well as pharmacological retinoic acid in the epidermis.

Cytochrome P450-mediated metabolism of retinoic acid has also been demonstrated in cultured keratinocytes (RANDOLPH and SIMON 1997). Retinoic acid that is taken up by keratinocytes is rapidly metabolized to very polar compounds, and their production is inhibited by ketoconazole (HODAM and CREEK 1996; RANDOLPH and SIMON 1997). The polar metabolites of retinoic acid are rapidly released from cells to the medium. Similar to the retinoic acid metabolism catalyzed by epidermal microsomes from retinoic acid-treated skin, retinoic acid metabolism in keratinocytes is robust and exhibits sufficiently high capacity so as to limit retinoic acid-stimulated retinol esterification (RANDOLPH and SIMON 1997). In contrast to the sensitivity of retinoic acid metabolism in vivo to induction by retinoic acid, retinoic acid metabolism in cultured keratinocytes is not further increased by exposing the cells to pharmacological concentrations of retinoic acid. One reason for this difference is that the pathway for retinoic acid metabolism is maximally induced by the tissue culture environment and as such, cannot be induced further. Whatever the explanation, retinoic acid metabolism in epidermal keratinocytes, both in vivo and in vitro, has significant potential to limit the accumulation of intracellular retinoic acid.

While it is certain that retinoic acid metabolism is an important pathway in maintaining low concentrations of retinoic acid in the epidermis, the current literature does not reveal the quantitative contribution of this pathway under normal conditions. However, since the estimated supply of plasma retinoic acid to the epidermis compares to its concentration (discussed above), it is likely that substantial retinoic acid metabolism occurs in the epidermis or upper dermis.

V. 3,4-Didehydroretinol Synthesis and Metabolism

The synthesis of 3,4-didehydroretinol in epidermis was first demonstrated by TÖRMÄ and VAHLQUIST (1985), following its earlier identification in the epidermis (VAHLQUIST 1980). Since 3,4-didehydroretinol is not present in human plasma, its unique presence in the epidermis of skin is thought to be due to its synthesis from retinol. Epidermal keratinocytes in culture synthesize 3,4-didehydroretinol from retinol in the medium in sufficient quantity to explain all that is present in intact epidermis (RANDOLPH and SIMON 1993; ROLLMAN et al. 1993). Moreover, the synthesis of 3,4-didehydroretinol in keratinocytes is not reversible (RANDOLPH and SIMON 1995). This suggests that the vitamin A_2 content of epidermis arises by synthesis in keratinocytes. The potential for 3,4-didehydroretinol synthesis by other types of human cells has not been reported.

The 3,4-didehydroretinol that is present in epidermis appears in both esterified and unesterified forms, similar to retinol. Approximately 70% of total vitamin A_2 is present as ester with the remaining 30% in unesterified form (Table 2). This relative distribution of vitamin A_2 ester and alcohol also compares well to that seen in cultured keratinocytes (RANDOLPH and SIMON 1993; ROLLMAN et al. 1993). The synthesis of very small amounts of 3,4-didehydroretinoic acid is observed in cultured keratinocytes (RANDOLPH and SIMON 1993). This metabolite has not yet been observed in vivo, however.

Despite the unique presence of 3,4-didehydroretinol in the epidermis, and the similarity of its relative composition as ester and alcohol to that of retinol, little is known about its synthesis or metabolism. The production of 3,4-didehydroretinol from retinol in intact epidermis is partially inhibited by pharmacological concentrations of retinoic acid, by ketoconazole, and by citral (TÖRMÄ et al. 1991). In other in vitro experiments, despite the fact that 3,4-didehydroretinol binds to CRBP with affinity similar to that of retinol, the CRBP-bound form is not an efficient substrate for retinoic acid synthesis (BOERMAN and NAPOLI 1995). This observation raises the interesting possibility that 3,4-didehydroretinol might compete with retinol for binding to CRBP, and thus function as a nonproductive substrate for retinoic acid synthesis. This situation would position 3,4-didehydroretinol in the role of limiting endogenous retinoic acid production, a possibility that warrants investigation.

Even less is known about the function of vitamin A_2 in epidermis. Studies in vitro (DUELL et al. 1992; ALLENBY et al. 1993; TÖRMÄ et al. 1994; REPA et al.

1996) demonstrate that 3,4-didehydroretinoic acid, the vitamin A_2 counterpart to retinoic acid, appears to function similarly to retinoic acid in binding to nuclear receptors and effecting regulation of gene transcription. In other reports, increased abundance of 3,4-didehydroretinol in the epidermis is correlated with hyperproliferative disorders (ROLLMAN and VAHLQUIST 1981). It is not known, however, whether the relationship between epidermal 3,4-didehydroretinol and cellular proliferation is a consequence or a cause. The function of vitamin A_2, as its metabolism, needs further study.

H. Summary and Working Model for Retinoid Metabolism in Epidermis

This review began with historical evidence for and the assertion that both the quality and quantity of vitamin A and its metabolites are required to maintain epithelial homeostasis. The epidermis derives its vitamin A from the blood supply of underlying dermis. This source of vitamin A consists of micromolar concentrations of retinol and nanomolar concentrations of retinoic acid. When the flux of extracellular fluids are taken into account in estimating extracellular supply, one concludes that the epidermis is literally afloat on a sea of retinoid.

In the face of this constitutive and substantial supply of extracellular retinoid, cells of the epidermis control the cellular concentration of retinoic acid in the epidermis via multiple retinoic acid-responsive mechanisms acting in concert. The known means of regulation are summarized in Fig. 4, and

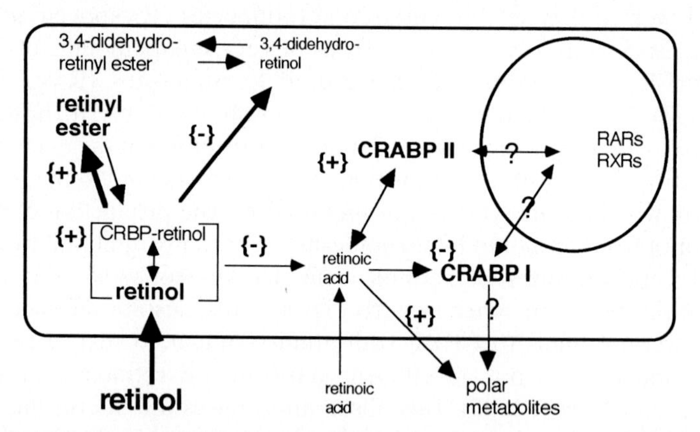

Fig. 4. Vitamin A homeostasis in epidermal keratinocytes. The major pathways of native retinoid metabolism in epidermal keratinocytes are depicted. *Weight of arrows and size of font* approximate the relative activity and the concentration in each pathway in general; {+}, {–}, direction of regulation by retinoic acid; ?, functions that are proposed but unconfirmed. The diagram is meant to convey the general pathways that are present in keratinocytes and is not intended to depict a particular stage of keratinocyte differentiation

include: (a) regulating cellular concentration of retinol via uptake, esterification and storage as ester; (b) limiting retinoic acid synthesis via restrained expression of enzyme activity; (c) maintaining a low cellular concentration of retinoic acid via metabolism, and (d) controlling access of retinoic acid to the nucleus via the buffering or transport function of CRABP I and CRABP II in keratinocyte cytosol. Among these mechanisms and the current data, it is not yet possible to identify one critical regulating event that is pivotal. While this insight may come with continued research, it seems likely that the epidermis orchestrates its essential, but sparing, vitamin A hormone message via the integrated function of these, and perhaps other yet to be defined mechanisms.

I. Future Prospects

The data summarized here regarding the composition and metabolism of native retinoids in the epidermis represents an interesting and engaging cross section of scientific investigation. It serves as a good foundation upon which a complete model of vitamin A homeostasis in the skin can be constructed. Indeed, many promising questions warrant continued inquiry. Issues that merit attention include: characterizing the pathway(s) of retinoic acid synthesis in the epidermis, including its expression in relation to differentiation; clarifying the relative roles of endogenous (synthesis in situ) versus exogenous retinoic acid to retinoic acid signaling in the epidermis; determining at the cellular level how the retinoic acid signal is controlled in the respective epidermal strata; characterizing the metabolism and function of the enigmatic vitamin A_2 family of retinoids, and finally, quantitating the impact of dermal cells such as fibroblasts, and other epidermal cells, such as Langerhans cells, and melanocytes, on the composition and quantity of plasma retinoids that are available to keratinocytes.

There is good potential for the native pathways of retinoid metabolism in the epidermis to be harnessed for therapeutic benefit. For example, the readily accessible and sizeable pool of retinyl ester in the epidermis could supply substrate retinol to drive local production of retinoic acid in lieu of exogenous retinoid therapy. Certainly this clinical potential and the goal of a complete understanding of vitamin A's role in epithelial homeostasis are sufficient to warrant continued exploration.

References

Allenby G, Bocquel MT, Saunders M, Kazmer S, Speck J, Rosenberger M, Lovey A, Kastner P, Grippo JF, Chambon P, Levin AA (1993) Retinoic acid receptors and retinoid X-receptors-interactions with endogenous retinoic acids. Proc Natl Acad Sci USA 90:30–34

Asselineau D, Bernard BA, Bailly C, Darmon M (1989) Retinoic acid improves epidermal morphogenesis. Dev Biol 133:322–335

Åström A, Tavakkol A, Pettersson U, Cromie M, Elder JT, Voorhees JJ (1991) Molecular cloning of two human cellular retinoic acid-binding proteins. J Biol Chem 266:17662–17666

Barkai U, Sherman MI (1987) Analysis of the interactions between retinoid-binding proteins and embryonal carcinoma cells. J Cell Biol 112:965–979

Barnett ML, Szabo G (1973) Effect of vitamin A on epithelial morphogenesis in vitro. Exp Cell Res 76:118–126

Bävik CO, Peterson PA, Eriksson U (1995) Retinol-binding protein mediates uptake of retinol to cultured human keratinocytes. Exp Cell Res 216:358–362

Berne B, Nilsson M, Vahlquist A (1984b) UV irradiation and cutaneous vitamin A: an experimental study in rabbit and human skin. J Invest Dermatol 83:401–404

Berne B, Vahlquist A, Fischer T, Danielson BG, Berne C (1984a) UV treatment of uraemic pruritis reduces the vitamin A content of the skin. Eur J Clin Invest 14:203–206

Berne RM, Levy MN (1988) The cardiovascular system. In: Berne RM, Levy MN (eds) Physiology, 2nd edn. Mosby, St Louis, p 506

Blaner WS, Olson JA (1994) Cellular-retinoid binding proteins. In: Sporn MB, Roberts AB, Goodman DS (eds) The retinoids: biology, chemistry, and medicine, 2nd edn. Raven, New York, p 283

Boerman MHEM, Napoli J (1995) Characterization of a microsomal retinol dehydrogenase: a short-chain alcohol dehydrogenase with integral and peripheral membrane forms that interacts with holo-CRBP (type I). Biochemistry 34:7027–7037

Boylan JF, Gudas LJ (1992) The level of CRABP-1 expression influences the amounts and types of all-*trans*-retinoic acid metabolites in F9 teratocarcinoma stem cells. J Biol Chem 267:21486–21491

Bucco RA, Melner MH, Gordon DS, Leers-Sucheta S, Ong DE (1995) Inducible expression of cellular retinoic acid-binding protein II in rat ovary: gonadotropin regulation during luteal development. Endocrinology 136:2730–2740

Busch C, Siegenthaler G, Vahlquist A, Nordlinder H, Sundelin J, Saksena P, Eriksson U (1992) Expression of cellular retinoid-binding proteins during normal and abnormal epidermal differentiation. J Invest Dermatol 99:795–802

Cohn EJ, Hughes WL Jr, Weare JHJ (1947) J Am Chem Soc 69:1753

Connor MJ, Smit MH (1987) The formation of all-*trans*-retinoic acid from all-*trans*-retinol in hairless mouse skin. Biochem Pharmacol 36:919–924

Creek KE, St Hilaire P, Hodam JR (1993) A comparison of the uptake, metabolism and biologic effects of retinol delivered to human keratinocytes either free or bound to serum retinol-binding protein. J Nutr 123:356–361

Darmon M (1991) Retinoic acid in skin and epithelia, Semin. Dev Biol 2:219–228

Deuster G (1996) Involvement of alcohol dehydrogenase, short-chain dehydrogenase/reductase, aldehyde dehydrogenase, and cytochrome P450 in the control of retinoid signaling by activation of retinoic acid synthesis. Biochemistry 35:12221–12227

De LeenHeer AP, Lambert WE, Claeys I (1982) All-*trans*-retinoic acid: Measurement of reference values in human serum by high performance liquid chromatography. J Lipid Res 23:1362–1367

Didierjean L, Durand B, Saurat JH (1991) Cellular retinoic acid-binding protein type 2 mRNA is overexpressed in human psoriatic skin as shown by in situ hybridization. Biochem Biophys Res Commun 180:204–208

DiGiovanna JJ, Fletcher R, Chader GJ (1985) Qualitative and quantitative analysis of cytosol retinoid binding proteins in human skin. J Invest Dermatol 85:460–464

Duell EA, Åström A, Griffiths CEM, Chambon P, Voorhees JJ (1992) Human skin levels of retinoic acid and cytochrome P-450-derived 4-hydroxyretinoic acid after topical application of retinoic acid in vivo compared to concentrations required to stimulate retinoic acid receptor-mediated transcription in vitro. J Clin Invest 90:1269–1274

Duell EA, Derguini F, Kang S, Elder JT, Voorhees JJ (1996a) Extraction of human epidermis treated with retinol yields retro-retinoids in addition to free retinol and retinyl esters. J Invest Dermatol 107:178–182
Duell EA, Kang S, Voorhees JJ (1996b) Retinoic acid isomers applied to human skin in vivo each induce a 4-hydroxylase that inactivates only *trans* retinoic acid. J Invest Dermatol 106:316–320
Durand B, Saunders M, Leroy P, Leid M, Chambon P (1992) All-*trans* and 9-*cis* retinoic acid induction of CRABP II transcription is mediated by RAR-RXR heterodimers bound to DR1 and DR2 repeated motifs. Cell 71:73–85
Eckhoff C, Nau H (1990) Identification and quantitation of all-*trans*- and 13-*cis*-retinoic acid and 13-*cis*-4-oxo-retinoic acid in human plasma. J Lipid Res 31:1445–1454
Elder JT, Åström A, Pettersson U, Tavakkol A, Griffiths CEM, Krust A, Kastner P, Chambon P, Voorhees JJ (1992a) Differential regulation of retinoic acid receptors and binding proteins in human skin. J Invest Dermatol 98:673–679
Elder JT, Åström A, Pettersson U, Tavakkol A, Griffiths CEM, Krust A, Kastner P, Chambon P, Voorhees JJ (1992b) Retinoic acid receptors and binding proteins in human skin. J Invest Dermatol 98:36S–41S
Elder JT, Cromie MA, Griffiths CEM, Chambon P, Joorhees JJ (1993) Stimulus-selective induction of CRABP-II mRNA: a marker for retinoic acid action in human skin. J Invest Dermatol 100:356–359
Eller MS, Oleksiak MF, McQuaid TJ, McAffee SG, Gilchrest BA (1992) The molecular cloning and expression of two CRABP cDNAs from human skin. Exp Cell Res 199:328–336
Eller MS, Harkness DD, Bhawan J, Gilcrest BA (1994) Epidermal differentiation enhances CRABP II expression in human skin. J Invest Dermatol 103:785–790
Fell HB, Mellanby E (1953) Metaplasia produced in cultures of chick ectoderm by vitamin A. Am J Physiol 119:470–488
Fiorella PD, Napoli JL (1991) Expression of cellular retinoic acid-binding protein (CRABP) in Eschericia coli. Characterization and evidence that holo-CRABP is a substrate in retinoic acid metabolism. J Biol Chem 266:16572–16579
Fisher GJ, Reddy AP, Datta SC, Kang S, Yi JY, Chambon P, Voorhees JJ (1995) All-*trans*-retinoic acid induces cellular retinol-binding protein in human skin in vivo. J Invest Dermatol 105:80–86
Fisher GJ, Voorhees JJ (1996) Molecular mechanisms of retinoid actions in skin. FASEB J 10:1002–1013
Frazier CN, Hu CK (1931) Cutaneous lesions associated with a deficiency in vitamin A in man. Arch Intern Med 48:507–514
Fuchs E (1990) Epidermal keratinization: the bare essentials. J Cell Biol 111:2807–2814
Fuchs E, Green H (1981) Regulation of terminal differentiation of cultured human keratinocytes by vitamin A. Cell 25:617–625
Gates RE, Mayfield C, Allred LE (1987) Human nonnatal keratinocytes have very high levels of cellular vitamin A-binding proteins. J Invest Dermatol 88:37–41
Goodwin GP (1934) A cutaneous manifestation of vitamin A deficiency. BMJ 2:113–114
Gudas LJ, Sporn MB, Roberts AB (1994) Cellular biology and biochemistry of the retinoids. In: Sporn MB, Roberts AB, Goodman DS (eds) The retinoids: biology, chemistry, and medicine, 2nd edn. Raven, New York, p 443
Guiguère V, Ong ES, Segui P, Evans RM (1987) Identification of a receptor for the morphogen retinoic acid. Nature 330:624–629
Hirschel-Scholz S, Siegenthaler G, Saurat JH (1989a) Ligand-specific and non-specific in vivo modulation of human epidermal cellular retinoic acid binding protein (CRABP). Eur J Clin Invest 19:220–227
Hirschel-Scholz S, Siegenthaler G, Saurat JH (1989b) Isotretionoin differs from other synthetic retinids in its modulation of human cellular retinoic acid binding protein (CRABP). Br J Dermatol 120:639–644

Hodam JR, Creek KE (1996) Uptake and metabolism of ^{3}H-retinoic acid delivered to human foreskin keratinocytes either bound to serum albumin or added directly to the culture medium. Biochim Biophys Acta 1311:102–110

Jing Y, Waxman S, Mira-y-Lopez R (1997) The cellular retinoic acid binding protein II is a positive regulator of retinoic acid signaling in breast cancer cells. Cancer Res 57:1668–1672

Kanai M, Raz A, Goodman DS (1968) Retinol-binding protein: the transport protein for vitamin A in human plasma. J Clin Invest 47:2025–2044

Kang S, Duell EA, Fisher GJ, Datta SC, Wang ZQ, Reddy AP, Tavakkol A, Jong YY, Griffiths EM, Elder JT, Voorhees JJ (1995) Application of retinol to human skin in vivo induces epidermal hyperplasia and cellular retinoid binding proteins characteristic of retinoic acid but without measurable retinoic acid levels or irritation. J Invest Dermatol 105:549–556

Kang S, Duell EA, Kim KJ, Voorhees JJ (1996) Liarozole inhibits human epidermal retinoic acid 4-hydroxylase activity and differentially augments human skin responses to retinoic acid and retinol in vivo. J Invest Dermatol 107:183–187

Kurlandsky SB, Duell EA, Kang S, Voorhees JJ, Fisher GJ (1996) Auto-regulation of retinoic acid biosynthesis through regulation of retinol esterification in human keratinocytes. J Biol Chem 271:15346–15352

Kurlandsky SB, Gamble MV, Rajesekhar R, Blaner WS (1995) Plasma delivery of retinoic acid to tissues in the rat. J Biol Chem 270:17850–17857

Kurlandsky SB, Xiao JH, Duell EA, Voorhees JJ, Fisher GJ (1994) Biological activity of all-*trans*-retinol requires metabolic conversion to all-*trans*-retinoic acid and is mediated through activation of nuclear retinoid receptors in human keratinocytes. J Biol Chem 269:32821–32827

Lehman ED, Spivey HO, Thayer RH, Nelson EC (1972) The binding of retinoic acid to serum albumin in plasma, Fed Proc 31:A672

Levin MS (1993) Cellular retinol-binding proteins are determinants of retinol uptake and metabolism in stably transfected Caco-2 cells. J Biol Chem 268:8267–8276

Mangelsdorf DJ, Umesono K, Evans RM (1994) The retinoid receptors. In: Sporn MB, Roberts AB, Goodman DS (eds) The retinoids: biology, chemistry, and medicine, 2nd edn. Raven, New York, p 319

Mori S (1922) The changes in the para-ocular glands which follow the administration of diets low in fat-soluble vitamin A; with notes of the effect of the same diets on salivary glands and mucus of the larynx and trachea, Bull Johns Hopkins Hosp 33:357–359

Napoli J (1994) Retinoic acid homeostasis: prospective roles of ß-carotene, retinol, CRBP, and CRABP. In: Blomhoff R (ed) Vitamin A in health and disease. Dekker, New York, pp 135–188

Napoli J (1996) Retinoic acid biosynthesis and metabolism. FASEB J 10:993–1001

Noy N, Blaner WS (1991) Interactions of retinol with binding proteins: Studies with rat cellular retinol-binding protein and with rat retinol-binding protein. Biochemistry 30:6380–6386

Ong DE, MacDonald PN, Gubitosi AM (1988) Esterification of retinol in rat liver: possible participation by cellular retinol-binding protein and cellular retinol-binding protein II. J Biol Chem 263:5789–5796

Ong DE, Newcomer ME, Chytil F (1994) Cellular retinoid binding proteins. In: Sporn MB, Roberts AB, Goodman DS (eds) The retinoids: biology, chemistry, and medicine, 2nd edn. Raven, New York, p 283

Petkovich M, Brand NJ, Krust A, Chambon P (1987) A human retinoic acid receptor which belongs to the family of nuclear receptors. Nature 330:444–450

Puhvel SM, Sakamoto M (1984) Cellular retinoic acid-binding proteins in human epidermis and sebaceous follicles. J Invest Dermatol 82:79–84

Randolph RK, Simon M (1993) Characterization of retinol metabolism in cultured human epidermal keratinocytes. J Biol Chem 268:9198–9205

Randolph RK, Simon M (1995), Metabolic regulation of active retinoid concentrations in cultured human epidermal keratinocytes by exogenous fatty acids. Arch Biochem Biophys 106:168–175
Randolph RK, Simon M (1996) All-*trans*-retinoic acid regulates retinol and 3,4-didehydroretinol metabolism in cultured human epidermal keratinocytes. J Invest Dermatol 106:168–175
Randolph RK, Simon M (1997) Metabolism of all-*trans*-retinoic acid by cultured human epidermal keratinocytes. J Lipid Res 38:1374–1383
Reed RK, Ishibashi M, Townsley MI, Parker JC, Taylor AE (1993) Blood-to-tissue clearance vs. lymph analysis in determining capillary transport characteristics for albumin in skin. Am J Physiol 264:H1394–H1401
Repa JJ, Plum LA, Tadikonda PK, Clagett-Dame M (1996) All-*trans* 3,4-didehydroretinoic acid equals all-*trans* retinoic acid in support of chick neuronal development. FASEB J 10:1078–1084
Rollman O, Vahlquist A (1981) Cutaneous vitamin A levels in seborrheic keratosis, actinic keratosis, and basal cell carcinoma. Arch Dermatol Res 270:193–196
Rollman O, Wood EJ, Olsson MJ, Cunliffe WJ (1993) Biosynthesis of 3,4-didehydroretinol from retinol by human skin keratinocytes in culture. Biochem J 293:675–682
Rossing N, Worm AM (1981) Interstitial fluid: exchange of macromolecules between plasma and skin interstitium, Clin Phys 1:275–284
Sanquer S, Eller MS, Gilchrest BA (1993) Retinoids and state of differentiation modulate CRABP II gene expression in a skin equivalent. J Invest Dermatol 100:148–153
Saurat JH, Didierjean L, Masgrau, Piletta PA, Jaconi S, Chatellard-Gruaz D, Gumowski D, Masouyé I, Salomon D, Siegenthaler G (1994) Topical retinal on human skin: biological effects and tolerance. J Invest Dermatol 103:770–774
Scheuplein RJ (1979) Fitzpatrick TB, Eisen AZ, Wolff K, Freedberg IM, Austen KF (eds) Dermatology in general medicine, vol 2. McGraw-Hill, New York, p 213
Siegenthaler G (1996) Extra- and intracellular transport of retinoids: a reappraisal. Horm Res 45:122–127
Siegenthaler G, Saurat JH, Ponec M (1988) Terminal differentiation in cultured human keratinocytes is associated with increased levels of cellular retinoic acid-binding protein. Exp Cell Res 178:114–126
Siegenthaler G, Gumowski-Sunek D, Saurat JH (1990a) Metabolism of natural retinoids in psoriatic epidermis. J Invest Dermatol 95:47S–48S
Siegenthaler G, Saurat JH, Ponec M (1990b) Retinol and retinal metabolism. Relation to the state of differentiation of cultured human keratinocytes. Biochem J 268:371–378
Siegenthaler G, Saurat JH (1991) Natural retinoids: Metabolism and transport in human epidermal cells. In: Saurat JH (ed) Retinoids: 10 years on. Karger, Basil, pp 56–68
Siegenthaler G, Saurat JH, Morin C, Hotz R (1984) Cellular retinol- and retinoic acid-binding proteins in the epidermis and dermis of normal human skin. Br J Dermatol 111:647–654
Siegenthaler G, Tomatis I, Chatellard-Gruaz D, Jaconi S, Eriksson U, Saurat JH (1992) Expression of CRABP-I and -II in human epidermal cells. Alteration of relative protein amounts is linked to state of differentiation. Biochem J 187:383–389
Smith JE, Milch PO, Muto Y, Goodman DS (1973) The plasma transport and metabolism of retinoic acid in the rat. Biochem J 132:821–827
Soprano D, Blaner WS (1994) Plasma retinol-binding protein. In: Sporn MB, Roberts AB, Goodman DS (eds) The retinoids: biology, chemistry, and medicine, 2nd edn. Raven, New York, p 257
Sun TT, Green H (1976) Differentiation of the epidermal keratinocyte in cell culture: formation of the cornified envelope. Cell 9:511–521

Takase S, Ong DE, Chytil F (1986) Transfer of retinoic acid from its complex with cellular retinoic acid-binding protein to the nucleus. Arch Biochem Biophys 47:328–334

Tang G, Webb AR, Russell Rm, Holick MF (1994) Epidermis and serum protect retinol but not retinyl esters from sunlight-induced photodegradation, Photodermatol Photoimmunol Photomed 10:1–7

Tavakkol A, Griffiths CEM, Keane KM, Palmer RD, Voorhees JJ (1992) Cellular localization of mRNA for cellular retinoic acid-binding protein II and nuclear retinoic acid receptor-g1 in retinoic acid-treated human skin. J Invest Dermatol 99:146–150

Törmä H, Vahlquist A (1983) Vitamin A transporting proteins in human epidermis and blister fluids. Arch Dermatol Res 275:324–328

Törmä H, Vahlquist A (1985) Biosynthesis of 3-dehydroretinol (Vitamin A_2) from all-*trans*-retinol (Vitamin A_1) in human epidermis. J Invest Dermatol 85:498–500

Törmä H, Vahlquist A (1990a) Vitamin A esterification in human epidermis: a relation to keratinocyte differentiation. J Invest Dermatol 94:132–138

Törmä H, Vahlquist A (1990b) Biosynthesis of 3,4-didehydroretinol and fatty acyl esters of retinol and 3,4-didehydroretinol by organ-cultured human skin. In: Packer L (ed) Methods in enzymology, vol 190. Academic, New York, p 210

Törmä H, Vahlquist A (1991) Retinol uptake and metabolism to 3,4-didehydroretinol in human keratinocytes at various stages of differentiation, Skin Parmacol 4:154–157

Törmä H, Stenström E, Andersson E, Vahlquist A (1991) Synthetic retinoids affect differently the epidermal synthesis of 3,4-didehydroretinol. Skin Pharmacol 4:246–253

Törmä H, Asselineau, Andersson E, Martin B, Reiniche P, Chambon P, Shroot B, Darmon M, Vahlquist A (1994) Biologic activities of retinoic acid and 3,4-didehydroretinoic acid in human keratinocytes are similar and correlate with receptor affinities and transactivation properties. J Invest Dermatol 102:49–54

Vahlquist A (1980) The identification of dehydroretinol (vitamin A_2) in human skin. Experientia 36:317–318

Vahlquist A (1982) Vitamin A in human skin: I. Detection and identification of retinoids in normal epidermis. J Invest Dermatol 79:89–93

Vahlquist A (1994) Role of retinoids in normal and diseased skin. In: Blomhoff R (ed) Vitamin A in health and disease. Dekker, New York, pp 365–424

Vahlquist A, Andersson E, Coble B-I, Rollman O, Törmä H (1996) Increased concentrations of 3,4-didehydroretinol and retinoic acid-binding protein (CRABP II) in human squamous cell carcinoma and keratoacanthoma but not in basal cell carcinoma of the skin. J Invest Dermatol 106:1070–1074

Vahlquist A, Lee JB, Michalsson G, Rollman O (1982) Vitamin A in human skin: II. Concentrations of carotene, retinol and dehydroretinol in various components of normal skin. J Invest Dermatol 79:94–97

Vahlquist A, Törmä H (1988) Retinoids and keratinization. Int J Dermatol 27:81–95

Vanden Bossche H, Willemsens G, Janssen PAJ (1988) Cytochrome-P-450-dependent metabolism of retinoic acid in rat skin microsomes: inhibition by ketoconazole. Skin Pharmacol 1:176–185

Vettermann O, Siegenthaler G, Winter H, Schweizer J (1997) Retinoic acid signaling cascade in differentiating murine epidermal keratinocytes: alterations in papilloma and carcinoma derived cell lines. Mol Carcinog 20:58–67

Williams RI (1943) The significance of the vitamin A content of tissues, Vit Horm 1:229–247

Wolbach SB, Howe PR (1925) Tissue changes following deprivation of fat soluble A vitamin. J Exp Med 42:753–777

Yost RW, Harrison EH, Ross AC (1988) Esterification by rat liver microsomes of retinol bound to cellular retinol-binding protein. J Biol Chem 263:18693–18701

Yuspa SH, Harris CC (1974) Altered differentiation of mouse epidermal cells treated with retinyl acetate in vitro. Exp Cell Res 86:95–105

CHAPTER 18

New Concepts for Delivery of Topical Retinoid Activity to Human Skin

J.-H. SAURAT, O. SORG, and L. DIDIERJEAN

A. Introduction

Retinoic acid is widely used for topical therapy of several skin diseases; it also improves the aspect of chronic solar damage (WEISS et al. 1988). Topical retinoic acid induces irritation of the skin, which precludes its use in some skin diseases that respond to systemic retinoids (GRIFFITHS et al. 1995; WEISS et al. 1988).

Irritation might be explained in part by an overload of the retinoic acid dependent pathways with non-physiological amounts of exogenous retinoic acid in the skin. The globally attainable concentrations of retinoic acid in the different layers from therapeutically efficient formulations have been determined in human skin *in vivo*. A steep concentration gradient with high concentrations in the epidermis, up to 8300 ng/g wet weight, and relatively low amounts in the dermis (490 ng/g wet weight) is achieved 100 min following application of retinoic acid 0.1% in isopropanol (330 nmol), corresponding to 15% of the applied dose of 100 mg (SCHAEFER and ZESCH 1975) (Table 1).

Retinoid content analysis in human skin treated with 0.1% retinoic acid (1670 nmol) for 4 days under occlusion showed the following values: retinoic acid 824 ng/g (2750 n*M*), 13-*cis* RA 745 ng/g (2480 n*M*) and 4-hydroxy-retinoic acid 93 ng/g wet weight (310 n*M*) (DUELL et al. 1992).

A recent study on percutaneous absorption of retinoic acid in human showed that about 2% of a single dose of 100 mg (330 μmol) in 0.05% formulation is absorbed; the same result is obtained after 28 days of daily application (LATRIANO et al. 1997). Due to the similarity of chemical structure, a similar pharmacokinetic behavior is expected for topical 13-*cis*-retinoic acid (SCHAEFER 1993). Such high tissue concentrations of retinoic acid are in excess of the concentrations needed to saturate nuclear receptors (ASTROM et al. 1990; DUELL et al. 1992).

Although major advances have been made in the analysis of the molecular events resulting from topical application of retinoic acid (FISHER et al. 1997), it is still not established if all the therapeutic activities of topical retinoic acid are mediated by nuclear receptors, and if irritation is necessary for obtaining some of these activities.

One possibility is that significant biological activity may still be achieved with much lower concentrations of topical retinoic acid. Alternatively, instead

Table 1. Retinoid contents (ng/g wet weight) of the skin after various treatments

Species	Part of skin	Topical treatment	RA	9-*cis*-RA	13-*cis*-RA	4-OH-RA	ROL	13-*cis*-ROL	3-DHR	14-HRR	RE[a]	RAL	Reference
Human	Epidermis	RA 0.1% (330 nmol) 100 min	8300	–		–		–		–		–	SCHAEFER and ZESH (1975)
		Vehicle	UQ	–		–		–		–		–	
	Dermis	RA 0.1% (330 nmol) 100 min	490	–		–		–		–		–	
		Vehicle	UQ	–		–		–		–		–	
Human[b]	Epidermis	RA 0.1% (1670 nmol) 4 days	824	–	745	93	–		–		–		DUELL et al. (1992)
		Vehicle	ND	–	ND	ND	–		–		–		
Human[c]	Epidermis	13-*cis*-RA (2.7 nmol) 48 h	–		70	–		–		–		–	LEHMAN et al. (1988)
		Vehicle	–		ND	–		–		–		–	
	Dermis	13-*cis*-RA (2.7 nmol) 48 h	–		4	–		–		–		–	
		Vehicle	–		ND	–		–		–		–	
Human[b]	Epidermis	Liarozole 3% 18 h	19	–		–		–		–		–	KANG et al. (1996)
		Vehicle	UQ	–		–		–		–		–	
Human[b]	Epidermis	ROL 0.4% (1400 nmol) 24 h	Trace	–		–	21 000	4000	–		10 500	–	KANG et al. (1995)
		Vehicle	UQ	–		–	300	15	–		40	–	
Human[b]	Epidermis	ROL 0.3% (1050 nmol) 4 days	–		–		2 500	–	48	18	–		DUELL et al. (1996)
		Vehicle	–		–		500	–	3	UQ	–		

		14-HRR 0.3% (1000 nmol) 4 days	–		–		450	–	5	80	–		
		Vehicle	–		–		500	–	3	UQ	–		
Human[b]	Epidermis	ROL-Pal 0.6% (1140 nmol) 3 days	–		–		975	140	–	70	385	–	Duell et al. (1997)
		Vehicle	–		–		175	12	–	UQ	UQ	–	
Hairless mouse	Epidermis	ROL 100 nmol 2 h[d]	873	–		–	33000	–		–		–	Connor and Smit (1987)
		Vehicle[d]	UQ	–		–	350	–		–		–	
	Dermis	ROL 100 nmol 2 h[d]	56	–		–	810	–		–		–	
		Vehicle[d]	UQ	–		–	49	–		–		–	
Mouse	Tail skin	RA 0.05% (833 nmol) 14 days[d]	4080	136	409	–	161	–		–		UQ	Didierjean et al. (1996)
		RA 0.05% (833 nmol) 14 days	310	24	62	–	107	–		–		UQ	
		Vehicle[d]	UQ	UQ	UQ	–	70	–		–		UQ	
Mouse	Tail skin	ROL 0.05% (875 nmol) 14 days[d]	UQ	UQ	UQ	–	672	–		56	2452	UQ	Sass et al. (1996)
		Vehicle[d]	UQ	UQ	UQ	–	70	–		UQ	219	UQ	
		RAL 0.05% (880 nmol) 14 days[d]	13	UQ	13	–	749	–		UQ	4515	350	
		Vehicle[d]	UQ	UQ	UQ	–	70	–		UQ	219	UQ	

3-DHR, 3-Dehydroretinol; 14-HRR, 14-hydroxy-4,14-retro-retinol; ND, not determined; 4-OH-RA, 4-hydroxy-all-*trans*-retinoic acid; RA, all-*trans*-retinoic acid; RAL, all-*trans*-retinal; RE, all-*trans*-retinyl esters; ROL, all-*trans*-retinol; UQ, under limit of detection.

[a] Most of the fatty acids esterified to retinol are palmitate and linoleate, but other fatty acids also form retinyl esters; however, authors do not always assay all the esters, neither do they indicate what esters they identified; thus this parameter is not the same for all studies.

[b] Application under occlusion.

[c] Studies performed *in vitro*.

[d] No washing or stratum corneum removing procedure were performed before retinoid analysis.

of treating the skin with the ligand of nuclear receptors itself, delivery could be distinctly targeted with "proligands."

B. The "Proligand-Nonligand" Concept

We have explored the possibility of delivering retinoid activity to human skin topically with a natural retinoid that does not bind to nuclear receptors. Such a precursor should be handled by enzymes of keratinocytes in the epidermis and transformed into either retinoic acid or storage forms such as retinol and retinyl esters. Epidermis is a differentiating, non-homogeneous tissue, made of keratinocytes that are not yet committed to terminal differentiation (basal cells layers) and a population of differentiating cells (suprabasal cell layers); the need for, and concentration of, retinoic acid is not identical in all layers and a gradient of retinoic acid has been considered to be a key event in the maturation of keratinocytes (RÉGNIER and DARMON 1989). This is supported by the fact that the conversion rate of retinol into retinoic acid by human keratinocytes depends on the state of keratinocyte differentiation, differentiating keratinocytes being able to convert at a higher rate than non-differentiated keratinocytes (SIEGENTHALER et al. 1990b).

That enzymes transforming the precursors into retinoic acid have distinct activities at different stages of differentiation indicates the possibility of targeting retinoic acid to epidermal cells in a differentiation related manner. This would be one approach to reduce side effects.

The use of retinoic acid precursors, such as retinol, retinyl esters, retinal and β-carotene, should therefore be considered in this context.

Validation of the concept would imply to demonstrate that: (a) epidermal cells distinctly metabolize the precursor into retinoic acid, (b) topical application of the precursor results in biological effects and (c) tolerability is better than that of retinoic acid.

Human keratinocytes transform retinol into retinal and then into retinoic acid by two enzymatic steps involving dehydrogenases; the first step is rate limiting and reversible; retinal can be converted enzymatically into either retinoic acid or retinol by human keratinocytes, both *in vivo* (SIEGENTHALER et al. 1990a; SIEGENTHALER and SAURAT 1991) and *in vitro* (SIEGENTHALER et al. 1990b). Epidermal cells have a weak capacity to transform retinol into retinal (SIEGENTHALER et al. 1990a; SIEGENTHALER and SAURAT 1991). β-Carotene is not converted into retinoids by epidermal cells (G. Siegenthaler, unpublished observations). We therefore hypothesized that retinal should be an interesting precursor for topical use because: (a) it bypasses the first limiting step of retinol oxidation into retinoic acid and (b) only the epidermal cells capable of retinal oxidation at pertinent stage of differentiation would generate active ligand(s).

In the following, we review our observations on topical retinal and works by other groups that are relevant to the "nonligand concept."

C. All-*trans*-Retinol

Retinol is the circulating form of vitamin A and is present in the plasma as a RBP/transthyretin complex, specific receptors for which may be present on epidermal cell membranes (Törmä and Vahlquist 1984). Retinol does not bind to nuclear receptors for retinoids (Blumberg et al. 1996), but is a precursor for molecules that can bind and activate these receptors. Mouse epidermis contains enzymes that can interconvert retinol and retinal and can oxidize retinal to retinoic acid (Kishore and Boutwell 1980), as do human keratinocytes in cultures (Siegenthaler et al. 1990b). Thus retinol appears as a precursor of gene modulating retinoids, and is widely used in cosmetics at low concentrations, although its metabolic pathways operative in skin physiology and pharmacology are not fully understood.

In normal physiological conditions, retinol is delivered to tissues bound to RBP (Soprano and Blaner 1994) and sequestered in the cell by CRBP (Ong et al. 1994); human keratinocytes express CRBP 1 (Bush et al. 1992; Siegenthaler et al. 1988; Törmä et al. 1994) and human epidermis contains truncated non-functional RBPs (Siegenthaler 1996). Concentrations of retinol and 3-dehydroretinol in the skin of various mammals in normal conditions have been reported in a review by Vahlquist et al. (1985). Application of pharmacological doses of retinol to human skin should circumvent this normal control.

Topical application of retinol on the skin of various mammals results in a dramatic increase in retinol concentrations, indicating penetration in the skin (Table 1). Application for 4 days of a cream containing 0.3% retinol (1050 nmol) on human skin under occlusion promoted an increase in viable epidermal (freed of stratum corneum) retinol concentration from 500 to 2500 ng/g wet weight (Duell et al. 1996) (Table 1). Using 0.4% retinol (1400 nmol) for 24 h, retinol reached much higher values, as did retinyl esters, whereas retinoic acid was still undetectable (Kang et al. 1995) (Table 1).

Törmä and Vahlquist had determined total retinol and 3-dehydroretinol content in human epidermal keratinocytes at different stages of differentiation (Törmä and Vahlquist 1990). Assays were performed after alkaline hydrolysis, thus giving the sum of free alcohol and esters for retinol and 3-dehydroretinol. In these conditions, retinol concentrations were shown to increase from basal keratinocytes (0.26 ng/mg protein) to granular/corneal keratinocytes (0.87 ng/mg protein), whereas 3-dehydroretinol concentrations were constant (about 0.21 ng/mg protein) (Törmä and Vahlquist 1990).

Connor and Smit (1987) used hairless mice to evaluate the penetration of retinol in the skin, as well as the concentration of retinoic acid that can be formed following topical retinol application. In this model, an application on dorsal skin of 100 nmol retinol (29 μg) dissolved in 200 μl acetone resulted in an increase in retinol concentration from 350 to 33,000 ng/g wet weight 2 h later in the epidermis, and from 49 to 810 ng/g wet weight in the dermis. In the same conditions, retinoic acid values raised from less than 3 (detection limit) to 873 and from less than 3 to 56 for the epidermis and dermis, respectively

(Table 1). The very high content of epidermal retinol after topical retinol treatment results probably from the absence of either stratum corneum removal or skin washing before retinoid extraction. In these conditions, probably most of the measured retinol was still on the surface, and only a small fraction did penetrate the skin. By applying 100 nmol of either retinol or retinoic acid for 4 h, they showed that the retinoic acid content of the epidermis was about 100 times higher after application of retinoic acid; a similar ratio was obtained for the dermis, with absolute values about 20 times lower than in the epidermis (CONNOR and SMIT 1987). It is difficult to compare these different studies, since they were not performed in the same conditions, and in the same species. CONNOR and SMIT used hairless mice, they applied a dose of 100 nmol of retinol, they assayed it 2 h later, and they did not remove the retinol remained on the surface. Conversely, KANG et al. (1995) used human epidermis freed of stratum corneum, they applied 1400 nmol of retinol, and assayed it 24 h later. Thus it appears that retinol penetration through the skin depends on the species, the dose, its mode and duration of application, as well as the treatment of the surface before skin analysis.

We studied by HPLC the metabolites detectable in mouse tail skin upon topical application of retinol 0.05% (875 nmol). As compared to vehicle-treated controls, retinol-treated mouse skin showed no detectable retinoic acid, 672 ng/g wet tissue of retinol, and 56 ng/g wet tissue of 14-hydroxy-retro-ROL (14-HRR) that was the predominant polar metabolite of retinol (SASS et al. 1996a) (Table 1). The detection of 14-HRR, a metabolite not previously reported in skin, that promotes growth of B lymphocytes and activation of T lymphocytes, outlines the distinct potentialities of topical retinol (DUELL et al. 1996).

We had developed an occlusive patch test for assessing topical retinoid action in human skin *in vivo* and found upregulation of CRABP2 by topical retinol, indicating a retinoid type of biological response (HIRSCHEL-SCHOLZ et al. 1989; SAURAT et al. 1994). KANG and coworkers (1995) have used this system to compare the activity of retinol with retinoic acid and to study retinoid metabolism after application of retinol. They reported that retinol treatment of human skin (a) results in an increase in retinol concentration with a maximal value at 24 h; (b) is much less erythemogenic than retinoic acid; (c) is similar to retinoic acid in its ability to induce epidermal hyperplasia, and to increase the expression of both cellular retinoic acid- and retinol-binding protein mRNAs; (d) is actively metabolized and stored as retinyl esters; and (e) does not results in significant retinoic acid accumulation (DUELL et al. 1996; KANG et al. 1995).

However, the relevance of the experimental conditions, i.e., applying the retinoid under an occlusive dressing that (a) enhances penetration, (b) exposes the compound to complex surface events including bacterial metabolism, and (c) is likely to modify epidermal physiology, is still a matter of debate. This question was partially addressed in a recent study published by DUELL and coworkers (1997). By comparing the efficiency of various topical retinoids under both occlusive and non-occlusive conditions on the activity of retinoic acid 4-hydroxylase, they showed no difference in activity between unoccluded and occluded retinol treated sites. Thus, retinol at 0.25% appears as a good precursor for retinoid action in the skin when administered topically without occlusion, since it does induce cellular and molecular events similar to those observed with 0.025% retinoic acid, while lacking the erythemogenic property of the latter (Duell et al. 1997). However, although these experiments contribute to our general understanding of the retinoid action in the skin, the activity of retinoic acid 4-hydroxylase is an indirect parameter, being an index of the metabolism of retinoic acid, and is thus not suitable for assessing the therapeutic value of a topical agent.

In this context, the study performed by Connor and Smit in hairless mice ten years ago is noteworthy in that it suggested a lack of correlation between the biological effects of retinol and its ability to generate retinoic acid. Indeed the epidermal and dermal levels of retinoic acid assayed after retinol application were 100 times lower than those found after retinoic acid application (Table 1) and could not account for the relative small difference in potencies between both retinoids, when assessed by epidermal hyperplasia in hairless mouse (Connor et al. 1986). This suggests that retinol might induce biological actions independently of its biotransformation into retinoic acid.

We have made identical observations in mouse tail skin treated with either retinol or retinoic acid, the former containing no detectable retinoic acid whereas the latter showing up to 310 ng/g wet weight (Didierjean et al. 1996). Thus, retinoic acid was statistically more active than retinol in time and dose-dependent reduction in the parakeratotic scales regions, in inducing increase in epidermal thickness (Fig. 1B), in reducing keratin 65-kDa and 70-kDa and inducing 50-kDa mRNAs (Didierjean et al. 1996) (Fig. 2). However, we observed early effects of low-dose 0.005% retinol that were superior to that induced by the same concentration of retinoic acid; thus the number of BrDU positive cells, as well as epidermal thickness, showed a peak after 2 days, and this effect was not found in retinoic acid treated animals (Fig. 1A).

Altogether these observations strongly suggest that topical retinol induces biological responses independently of its biotransformation into retinoic acid. This point is also addressed in more details in Sect. E.

D. All-*trans*-Retinyl Esters

In human epidermis about 70% of vitamin A occurs as fatty acyl esters (Vahlquist et al. 1985). Retinyl esters presumably function as cellular stores of vitamin A from which it can be mobilized in situations of increased demand or decreased supply from the blood (Törmä and Vahlquist 1987). Human epidermis is able to produce retinyl esters when incubated with retinol bound to CRBP, suggesting the involvement of the enzyme lecithin:retinol acyltransferase (LRAT) (Törmä and Vahlquist 1985), although the other esterifying

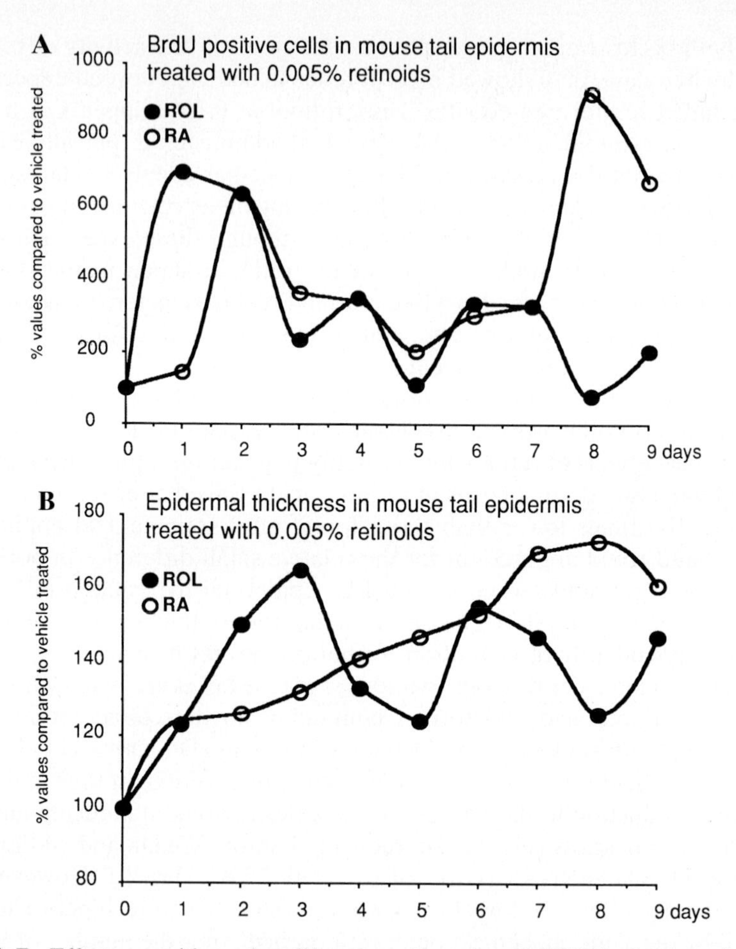

Fig. 1A,B. Effects on epidermal proliferation of low-dose (0.005%) of one to nine daily topical applications on mouse tail skin of either RA or ROL. **A** Keratinocyte DNA synthesis as shown by *in vivo* BrdU incorporation. Twenty hours after the last treatment mice were injected intraperitoneally with 250 μg BrdU per gram body weight dissolved in sterile physiological saline. The mice were killed, and one part of the tail skin was fixed in Dubosq-Brasil liquid, embedded in paraffin, and cut in 5-μm serial sections. BrdU was revealed by immunohistochemistry using anti-BrdU monoclonal antibody followed by staining by the Streptavidin-biotin peroxidase complex (ABC/horseradish peroxidase) method. The number of BrdU-positive cells in epidermis was counted in 100 interfollicular zones in samples of retinoids treated-skin and then expressed as percentage of BrdU-positive cells counted in 100 interfollicular zones in samples of vehicle-treated skin. **B** Epidermal thickness as assessed by morphometry. Sections were stained with hematoxylin-eosin and thickness was measured at the middle of 100 interfollicular zones from the basal membrane zone to the last nucleated cell layer. The thickness of retinoid-treated epidermis was expressed as percentage of thickness of vehicle-treated skin

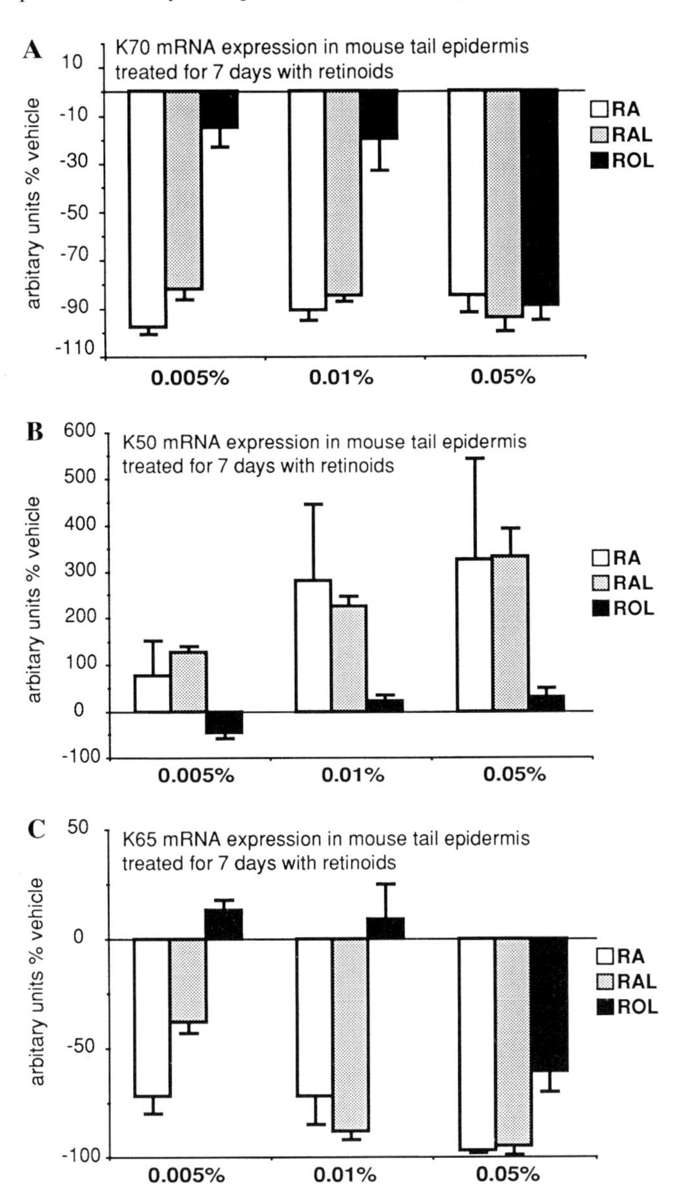

Fig. 2A–C. Effects on epidermal keratins mRNA expression of 0.005, 0.010, and 0.05% of either RA, RAL, or ROL after 7 days topical applications on mouse tail skin. Triplicate experiments of RNA slots blots of 1 μg total RNA hybridized by random-priming. Keratin 70-kDa (**A**), keratin 50-kDa (**B**), and keratin 65-kDa (**C**) mRNAs arbitrary units (pixels as measured by scanning) of RA, RAL, and ROL-treated samples expressed as percentage of vehicle-treated samples (mean ± SD). (From Didierjean et al. 1996)

enzyme, acyl CoA:retinol acyltransferase (ARAT) is expressed in mouse epidermis (TÖRMÄ and VAHLQUIST 1987). In human keratinocytes, LRAT has been shown to play a pivotal role in regulating the biosynthesis of retinoic acid according to retinoid availability in the culture medium (KURLANDSKY et al. 1996). Thus retinyl esters appear as precursors of biologically active retinoids, since they can be stored in relative high amounts in the tissues, and can release retinol upon demand. In order to represent a suitable way of vitamin A delivery to the skin, they should also penetrate the skin easily. Retinyl palmitate, a stable and abundant vitamin A ester, can be directly incorporated into the anhydrous phase of an oil-in-water emulsion, and is thus widely used in cosmetic formulations (IDSON 1990).

A single application of 0.6% retinyl palmitate (1140 nmol) resulted 3 days later in an increase from 175 to 975 ng/g wet weight of epidermal retinol in human. In these conditions, retinol was the most abundant retinoid extracted from the epidermis (DUELL et al. 1997) (Table 1).

The simultaneous study of percutaneous absorption and metabolism of a topical agent can be performed in an *in vitro* flow-through diffusion cell system (COLLIER et al. 1989). By using this system with human skin, BOEHNLEIN and coworkers reported that 18% of the applied dose of retinyl palmitate was recovered in skin and receptor fluid, and 44% of the absorbed dose was metabolized to retinol (BOEHNLEIN et al. 1994).

In a recent study comparing the efficacy of various topical retinoids under occlusion, retinyl palmitate was shown to be less effective than retinol in inducing retinoic acid 4-hydroxylase activity (DUELL et al. 1997). However, as discussed previously, the relevance of this parameter as an index of topical retinoid activity is questionable.

Taken together, the results presented in this section demonstrate the bioavailability of topical retinyl palmitate and indicate that it can exert some biological, retinoid-like, activity in the skin. More studies are needed to evaluate the amplitude of these effects, as well as the dose-effect ratio compared to other topical natural retinoids.

E. All-*trans*-Retinal

A β-carotene cleavage enzyme catalyses the conversion of all-*trans*-β-carotene to two molecules of retinal, while several other carotenoids and apocarotenoids are also cleaved to yield retinal (LAKSHMAN et al. 1989). Retinal is also formed from retinol (NAPOLI and RACE 1990). It is converted enzymatically into either retinyl esters, retinol or retinoic acid by human keratinocytes; the process is dependent on the differentiation stage of keratinocytes since it is higher in differentiating than in non-differentiated keratinocytes (BAUDRIMONT et al. 1994; CHATTELARD-GRUAZ et al. 1994; SIEGENTHALER et al. 1990b). This indicates that retinal would fulfill one criterion in the concept that is to target multipotential vitamin A activity into distinct compartments of the epidermis. Indeed, topically applied retinal would (a) be a precursor of either retinol, retinyl esters or retinoic acid (b) bypass the first, rate-limiting, step of retinol oxidation into retinoic acid and (c) be handled only by the epidermal cells having enzymatic activities at pertinent stage of differentiation, resulting in a controlled delivery of vitamin A metabolites into the cells. Retinal does not bind to retinoid nuclear receptors (CRETTAZ et al. 1990). Therefore its biological activity should result from its enzymatic transformation into retinoic acid by epidermal keratinocytes and should be qualitatively similar to that of retinoic acid. In order to exert quantitatively comparable activity to that resulting from direct application of retinoic acid, topical retinal should generate similar amounts of retinoic acid. To test this hypothesis we compared the effects of topical retinal to that of retinoic acid in the mouse tail model (DIDIERJEAN et al. 1983; WRENCH and DIDIERJEAN 1985) and analyzed by HPLC the retinoid metabolites resulting from their respective topical application (DIDIERJEAN et al. 1996). Retinoic acid and retinal induced a similar time and dose-dependent reduction in the parakeratotic scales regions and increase in epidermal thickness, a similar reduction in keratin 70-kDa mRNAs and a similar increase in filaggrin mRNAs. Loricrin, keratin 65- and 50-kDa mRNA, and BrdU-positive cells were increased by both retinoids but loricrin and keratin 65-kDa mRNA significantly more by retinoic acid whereas keratin 50- kDa mRNAs and BrdU positive cells significantly more by retinal. CRABP2 mRNA was increased 4-fold by all-*trans*-RA and 2.5-fold by RAL; CRBP1 mRNA was increased 1.3-fold by both retinoids. Only retinol was detectable by HPLC in vehicle treated-skin; retinal treated-skin contained 11-fold more retinol than vehicle-treated skin, together with low amounts of all-*trans*- and 13-*cis*-retinoic acid. In contrast, retinoic acid-treated skin showed 25-fold more all-*trans*- and 13-*cis*-retinoic acid, and 7-fold less retinol than retinal-treated skin (DIDIERJEAN et al. 1996 and Saurat et al., unpublished observations)

From these observations we concluded that (a) topical retinal exerts significant biological activity, similar to that of all-*trans*-RA in this model, (b) murine tail skin transforms retinal into retinoic acid *in vivo* and (c) the effects of retinal are not exclusively due to its transformation into retinoic acid because tissue levels of retinoic acid in retinal-treated skin are much lower than those in retinoic acid-treated skin, whereas the biological effects of retinoic acid and retinal were identical. Several explanations for that have been considered:

- A low cell loading of retinoic acid may be sufficient for full activity on the parameters studied in this model and only a small percentage of the retinoic acid recovered from retinoic acid treated skin for HPLC analysis may correspond to intracellular, biologically significant retinoic acid.
- The uptake of retinal by murine epidermal cells might be more effective than that of retinoic acid. *In vitro*, the uptake of [^{3}H]retinal by cultured human keratinocytes reached 30% at 8h (CHATTELARD-GRUAZ et al. 1994).

- Finally a possible explanation of our observations might be that metabolites other than retinoic acid are responsible for some biological effects of retinal, which might not be mediated by retinoic acid receptors. Since the major metabolites produced from retinal, both *in vitro* and *in vivo*, are retinol and several retinyl esters, the respective role of these metabolites, as well as that of retinal itself, in inducing selective topical activities, remain to be analyzed.

Since no published data were available concerning the topical use of retinal in humans, we have performed pilot studies to explore its biological effects and tolerability (SAURAT et al. 1994). Biological activity was shown by the induction of cellular retinoic acid binding protein type 2 (CRABP 2) mRNA and protein; the rank order for CRABP-2 increase was retinoic acid > retinal > 9 *cis* retinoic acid > retinol > β-carotene. This suggested that the CRABP 2 induction by topical retinal is a result of transformation by epidermal enzymes into active ligand(s). In volunteers treated 1–3 months by 0.5, 0.1, and 0.05% retinal, there was a dose-dependent and significant increase in: epidermal thickness, keratin 14 immunoreactivity, and Ki67-positive cells. The area of distribution of involucrin, transglutaminase, and filaggrin immunoreactivity was also increased in a dose dependent manner, and keratin 4 immunoreactivity was induced in the upper epidermis. In a pilot clinical tolerance studies 229 patients received topical retinal at different concentrations; the 1% preparation was tolerated by up to 70% of the treated subjects, tolerance of the 0.5% preparation was slightly better, whereas both 0.1 and 0.05% preparations applied on facial skin, were well tolerated and allowed prolonged use, up to 3 years, in patients with inflammatory dermatoses. These findings indicated that retinal has biological activity, is well tolerated and may be used as a topical agent on human skin.

In a double-blind study on 120 patients treated for 44 weeks comparing retinal and retinoic acid 0.05%, the tolerability was good to very good in 96% of patients treated with retinal or vehicle and only 65% in patients treated with retinoic acid (AGACHE et al. 1998).

To determine whether topical application of a large quantity of retinal on human skin is associated with detectable alteration in constitutive levels of plasma retinoids resulting from metabolism of retinal in the skin, plasma retinoids [retinol, retinoic acid, retinal, retinyl palmitate/oleate, 13-*cis*-retinoic acid and 4-oxo-13-*cis*-retinoic acid] were analyzed by high-pressure liquid chromatography (HPLC) in 10 healthy male volunteers kept under a poor vitamin A diet before, during and after daily topical application for 14 days of 7 mg retinal to 40% of the body surface (SASS et al. 1996b). The introduction of a restricted vitamin A diet before retinal application resulted in a decrease in the plasma levels of retinol, retinoic acid and retinyl palmitate/oleate. Retinal topical application did not induce an alteration in the retinoid metabolites plasma levels. No retinal was detectable in any of the plasma samples. These results indicated that skin metabolism of topically applied retinal does not result in detectable alteration in constitutive levels of plasma retinoids in humans. Since there was no increase in plasma retinyl esters during topical retinal treatment (despite a previous reduction in plasma retinyl esters due to the diet), it is likely that the retinyl esters produced from the 7 mg retinal administered daily remained stored in the epidermis. The proportion of retinal transformed into retinol is not likely to be delivered systemically since epidermal enzymes would metabolize it to retinyl esters. At any rate, the constitutive levels of plasma retinol (about 500 ng/ml) were not altered by the small amount that may originate from skin. In addition to providing important safety data pertinent to the potential use of retinal as a topical agent in humans, these results supported the concept of targeting vitamin A metabolites in the skin upon retinal topical treatment.

F. All-*trans*-Retinoyl-β-Glucuronide

Although retinoyl glucuronide does not fulfill the criteria for the concept of precursor, we review here some data on its proposed use as a topical natural retinoid.

Conjugation reactions, such as the formation of sulfates, glucuronides and glutathione derivatives of biologically active compounds, are classically considered as detoxification processes. In essence, lipophilic molecules are rendered less active by the formation of water-soluble metabolites, which can then be excreted in the urine. However, in the past decade, it has become clear that some conjugate metabolites are as active (or even more) than their parent compound. This is true for retinoids too. For instance, retinoyl glucuronide and its 9-*cis*- and 13-*cis*-isomers were identified as major plasma metabolites following administration of their corresponding retinoic acid isomers, in several mammalian species including human (Barua et al. 1991; Barua and Olson 1986; Creech et al. 1991; Sass et al. 1994). The percutaneous absorption of retinoyl glucuronide was studied in rat skin. After a single application of radiolabeled retinoyl glucuronide, about 1% of the applied dose appeared in the blood within 2–4 h, before decreasing to about 0.2% until 7 days (Barua and Olson 1996). Concerning its biological activity, retinoyl glucuronide was shown to be at least as active as retinoic acid in supporting rat growth, differentiation of vaginal epithelia, or inhibiting the DNA synthesis of mouse mammary gland (Olson et al. 1992). In human, topical retinoyl glucuronide was used in a pilot preliminary study for the treatment of acne, and did not induce skin irritation (Gunning et al. 1994).

These results show that retinoyl glucuronide might be another candidate for increasing the vitamin A content of the skin, although its therapeutic potential requires further investigation, especially in the light of a recent study demonstrating its high teratogenic potential in mouse limb bud mesenchymal cells *in vitro* (Sass et al. 1997).

G. Conclusions and Perspectives

Topical retinoic acid induces irritation of the skin which might be explained, in part, by an overload of the retinoic acid dependent pathways with non-physiological amounts of exogenous retinoic acid in the skin. The concept that biological activity may be achieved with natural retinoids that do not bind to nuclear receptors is now supported by many observations. In this context, retinal appears as a key compound due to its pivotal position in the metabolism of retinoids, as well as to the fact that it occurs (if any) in very low amounts in tissues. Results indicate that it delivers more activity than retinol and retinyl esters.

The questions pending are the following:

1. Are the biological activities resulting from the topical application of natural retinoids that do not bind to nuclear receptors all due to retinoic acid that is formed in minute amounts? It should be understood that the difference in the biological effects induced by 0.05% topical retinoic acid or retinal, is not in the order of magnitude of the difference in tissue loading in retinoic acid resulting from either application. Therefore, this indicates that either most of topically applied retinoic acid does not reach nuclear receptors, or that the biological effects are not all due to receptor binding. A great deal of work is required to dissect the pleomorphous effects of each topical natural retinoid and thus answer this question.

2. Another question is related to the potential of natural retinoids that do not bind nuclear receptors in the context of topical skin therapy. Some advantages over retinoic acid, in addition to the lack of, or low irritancy, may emerge. We have proposed that natural retinoids that do not bind to nuclear receptors may be considered for maintenance, prevention and adjuvant to skin therapy. We think that many insults to the skin, environmental or therapeutic, result in depletion of epidermal stores in retinol and retinyl esters, which result in a state of relative tissue vitamin A deficiency. This has been well shown for UV irradiation (BERNE et al. 1984a,b) and may be applicable for topical steroids (unpublished observations) as well. With aging, delivery of retinol to epidermis from endogenous stores may be altered; although no data are available on this issue, the concept is compatible with morphological and functional observations in sun exposed aging skin. Retinol binding protein is degraded in the epidermis and looses its functional properties (SIEGENTHALER and SAURAT 1987); the effect of aging and environmental insults in this carrier has not been analyzed. Decrease in retinol and retinyl esters has also been reported in precancerous lesions of the skin in humans (ROLLMAN and VAHLQUIST 1981). Relative tissue vitamin A deficiency cannot apparently be compensated by topical retinoic acid since it is not transformed into retinol and retinyl esters. In fact, an increase in both retinol (1.8-fold) and retinyl esters (3.3-fold) has been observed in mouse skin treated with topical retinoic acid; the mechanism of this increase is under study. It is much lower, however, than that resulting from topical application of retinal [retinol (7.5-fold) and

retinyl esters (9.8-fold)] or topical application of retinol [retinol (4.2-fold) and retinyl esters (4.5-fold)] (manuscript in preparation). The link between reconstitution of skin stores in vitamin A and any beneficial effect at the molecular and clinical levels remains to be analyzed.

References

Agache P, Vienne MP, Morel P, Lauze C, Bousquet V, Lagarde JM, Dupuy P (1998) Profilometric evaluation of photodamage after retinal and retinoic acid topical treatment (in press)

Astrom A, Pettersson U, Krust A, Chambon P, Voorhees JJ (1990) Retinoic acid and synthetic analogs differentially activate retinoic acid receptor dependent transcription. Biochem Biophys Res Commun 173:339–345

Barua AB, Gunning DB, Olson JA (1991) Metabolism in vivo of all-*trans* [11–^{3}H]-retinoic acid after an oral dose in the rat. Biochem J 277:527–531

Barua AB, Olson JA (1986) Retinoyl-β-glucuronide: an endogenous compound of human blood. Am J Clin Nutr 43:481–485

Barua AB, Olson JA (1996) Percutaneous absorption, excretion and metabolism of all-*trans*-retinoyl beta-glucuronide and of all-*trans*-retinoic acid in the rat. Skin Pharmacol 9:17–26

Baudrimont I, Betbeder AM, Gharbi A, Pfohl-Leszkowicz A, Dirheimer G, Creppy EE (1994) Effect of superoxide dismutase and catalase on the nephrotoxicity induced by subchronical administration of ochratoxin A in rats. Toxicology 89:101–111

Berne B, Nilsson M, Vahlquist A (1984a) UV irradiation and cutaneous vitamin A: an experimental study in rabbit and human skin. J Invest Dermatol 83:401–404

Berne B, Vahlquist A, Fischer T, Danielson BG, Berne C (1984b) UV-treatment of uraemic pruritus reduces the vitamin A content of the skin. Eur J Clin Invest 14:203–206

Blumberg B, Bolado J Jr, Derguini F, Craig AG, Moreno TA, Chakravarti D, Heyman RA, Buck J, Evans RM (1996) Novel retinoic acid receptor ligands in Xenopus embryos. Proc Natl Acad Sci USA 93:4873–4878

Boehnlein J, Sakr A, Lichtin JL, Bronaugh RL (1994) Characterization of esterase and alcohol dehydrogenase activity in skin. Metabolism of retinyl palmitate to retinol (vitamin A) during percutaneous absorption. Pharm Res 11:1155–1159

Bush C, Siegenthaler G, Vahlquist A, Nordlinder H, Sundelin J, Saksena P, Eriksson U (1992) Expression of cellular retinoid-binding proteins during normal and abnormal epidermal differentiation. J Invest Dermatol 99:795–802

Chattelard-Gruaz D, Jaconi S, Saurat JH, Siegenthaler G (1994) Metabolism of retinal in living cultured human keratinocytes. J Invest Dermatol 102:599

Collier SW, Sheikh NM, Lichtin JL, Stewart RF, Bronaugh RL (1989) Maintenance of skin viability during in vitro percutaneous absorption/metabolism studies. Toxicol Appl Pharmacol 99:522–533

Connor MJ, Ashton RE, Lowe NJ (1986) A comparative study of the induction of epidermal hyperplasia by natural and synthetic retinoids. J Pharmacol Exp Ther 237:31–35

Connor MJ, Smit MH (1987) The formation of all-*trans* retinoic acid from all-*trans* retinol in hairless mouse skin. Biochem Pharmacol 36:919–924

Creech KJ, Slikker W, Bailey JR, Roberts LG, Fischer B, Wittfoht W, Nau H (1991) Plasma pharmacokinetics and metabolism of 13-*cis*- and all-*trans*-retinoic acid in the cynomolgus monkey and the identification of 13-*cis*- and all-*trans*-retinoyl-β-glucuronides. Drug Metab Dispos 19:317–324

Crettaz M, Baron A, Siegenthaler G, Hunziker W (1990) Ligand specificities of recombinant retinoic acid receptors RARα and RARβ. Biochem J 272:391–397

Didierjean L, Carraux P, Grand D, Sass JO, Nau H, Saurat JH (1996) Topical retinal increases skin content of retinoic acid and exerts biological activity in mouse skin. J Invest Dermatol 107:714–719
Didierjean L, Wrench R, Saurat JH (1983) Expression of cytoplasmic antigens linked to orthokeratosis during the development of parakeratosis in newborn mouse tail epidermis. Differentiation 23:250–255
Duell EA, Åström A, Griffiths CEM, Chambon P, Voorhees JJ (1992) Human skin levels of retinoic acid and cytochrome P-450-derived 4-hydroxyretinoic acid after topical application of retinoic acid in vivo compared to concentrations required to stimulate retinoic acid receptor-mediated transcription in vitro. J Clin Invest 90:1269–1274
Duell EA, Derguini F, Kang S, Elder JT, Voorhees JJ (1996) Extraction of human epidermis treated with retinol yields retro-retinoids in addition to free retinol and retinyl esters. J Invest Dermatol 107:178–182
Duell EA, Kang S, Voorhees JJ (1997) Unoccluded retinol penetrates human skin in vivo more effectively than unoccluded retinyl palmitate or retinoic acid. J Invest Dermatol 109:301–305
Fisher GJ, Wang ZQ, Datta SC, Varani J, Kang S, Voorhees JJ (1997) Pathophysiology of premature skin aging induced by ultraviolet light. N Engl J Med 337:1419–1428
Griffiths CE, Kang S, Ellis CN, Kim KJ, Finkel LJ, Ortiz-Ferrer LC, White GM, Hamilton TA, Voorhees JJ (1995) Two concentrations of topical tretinoin (retinoic acid) cause similar improvement of photoaging but different degrees of irritation. A double- blind, vehicle-controlled comparison of 0.1% and 0.025% tretinoin creams. Arch Dermatol 131:1037–1044
Gunning DB, Barua AB, Lloyd RA, Olson JA (1994) Retinoyl-β-glucuronide: a non-toxic retinoid for the topical treatment of acne. J Dermatol Treat 5:181–185
Hirschel-Scholz S, Siegenthaler G, Saurat JH (1989) Isotretinoin differs from other synthetic retinoids in its modulation of human cellular retinoic acid binding protein (CRABP). Br J Dermatol 120:639–644
Idson B (1990) Vitamins in cosmetics, an update. I. Overview and vitamin A. Drug Cosmet. Ind. 22:363–410
Kang S, Duell EA, Fisher GJ, Datta SC, Wang ZQ, Reddy AP, Tavakkol A, Yi JY, Griffiths CEM, Elder JT, Voorhees JJ (1995) Application of retinol to human skin in vivo induces epidermal hyperplasia and cellular retinoid binding proteins characteristics of retinoic acid but without measurable retinoic acid levels or irritation. J Invest Dermatol 105:549–556
Kang S, Duell EA, Kim KJ, Voorhees JJ (1996) Liarozole inhibits human epidermal retinoic acid 4-hydroxylase activity and differentially augments human skin responses to retinoic acid and retinol in vivo. J Invest Dermatol 107:183–187
Kishore GS, Boutwell RK (1980) Enzymatic oxidation and reduction of retinal by mouse epidermis. Biochem Biophys Res Commun 94:1381–1386
Kurlandsky SB, Duell EA, Kang S, Voorhees JJ, Fisher GJ (1996) Auto-regulation of retinoic acid biosynthesis through regulation of retinol esterification in human keratinocytes. J Biol Chem 271:15346–15352
Lakshman MR, Mychkovsky I, Attlesey M (1989) Enzymatic conversion of all-*trans*-β-carotene to retinal by a cytosolic enzyme from rabbit and rat intestinal mucosa. Proc Natl Acad Sci USA 86:9124–9128
Latriano L, Tzimas G, Wong F, Wills RJ (1997) The percutaneous absorption of topically applied tretinoin and its effect on endogenous concentrations of tretinoin and its metabolites after single doses or long-term use. J Am Acad Dermatol 36: S37-S46
Lehman PA, Slattery JT, Franz TJ (1988) Percutaneous absorption of retinoids: influence of vehicle, light exposure, and dose. J Invest Dermatol 91:56–61
Napoli JL, Race KR (1990) Microsomes convert retinol and retinal into retinoic acid and interfere in the conversions catalyzed by cytosol. Biochim Biophys Acta 1034:228–232

Olson JA, Moon RC, Anders MW, Fenselau C, Shanes B (1992) Enhancement of biological activity by conjugation reactions. J Nutr 122:615–624

Ong DF, Newcomer ME, Chytil F (1994) Cellular retinoid binding proteins. In: Sporn MB, Roberts AB Goodman D (eds) The retinoids: biology, chemistry and medicine. Raven, New York, pp 283–317

Régnier M, Darmon M (1989) Human epidermis reconstructed in vitro: a model to study keratinocyte differentiation and its modulation by retinoic acid. In Vitro Cell. Dev Biol 25:1000–1008

Rollman O, Vahlquist A (1981) Cutaneous vitamin A levels in seborrheic keratosis, actinic keratosis, and basal cell carcinoma. Arch Dermatol Res 270:193–196

Sass JO, Didierjean L, Carraux P, Plum C, Nau H, Saurat JH (1996a) Metabolism of topical retinal and retinol by mouse skin in vivo: predominant formation of retinyl esters and identification of 14-hydroxy-4,14-retro-retinol. Exp Dermatol 5:267–271

Sass JO, Masgrau E, Piletta PA, Nau H, Saurat JH (1996b) Plasma retinoids after topical use of retinal on human skin. Skin Pharmacol 9:322–326

Sass JO, Tzimas G, Nau H (1994) 9-*cis*-Retinoyl-β-glucuronide is a major metabolite of 9-*cis*-retinoic acid. Life Sci 54:69–74

Sass JO, Zimmermann B, Rühl R, Nau H (1997) Effects of all-*trans*-retinoyl-β-glucuronide and all-*trans*-retinoic acid on chondrogenesis and retinoid metabolism in mouse limb bud mesenchymal cells in vitro. Arch Toxicol 71:142–150

Saurat JH, Didierjean L, Masgrau E, Piletta PA, Jaconi S, Chatellard-Gruaz D, Gumowski D, Masouyé I, Salomon I, Siegenthaler G (1994) Topical retinal on human skin: biological effects and tolerance. J Invest Dermatol 103:770–774

Schaefer H (1993) Penetration and percutaneous absorption of topical retinoids. Skin Pharmacol 6 [Suppl 1]:17–23

Schaefer H, Zesch A (1975) Penetration of vitamin A acid into human skin. Acta Derm. Venereol [Suppl] 74:50–55

Siegenthaler G (1996) Extra-and intracellular transport of retinoids: a reappraisal. Horm. Res 45:122–127

Siegenthaler G, Gumowski-Sunek D, Saurat JH (1990a) Metabolism of natural retinoids in psoriatic epidermis. J Invest Dermatol 95:47S–48S

Siegenthaler G, Saurat JH (1987) Loss of retinol-binding properties for plasma retinol-binding protein in normal human epidermis. J Invest Dermatol 88:403–408

Siegenthaler G, Saurat JH (1991) Natural retinoids: Metabolism and transport in human epidermal cells. In: Saurat JH (ed) Retinoids: 10 years on. Karger, Basel, pp 56–68

Siegenthaler G, Saurat JH, Ponec M (1988) Terminal differentiation in cultured human keratinocytes is associated with increased levels of cellular retinoic acid-binding protein. Exp Cell Res 178:114–126

Siegenthaler G, Saurat JH, Ponec M (1990b) Retinol and retinal metabolism. Relationship to the state of differentiation of cultured human keratinocytes. Biochem J 268:371–378

Soprano DR, Blaner WS (1994) Plasma retinol-binding protein. In: Sporn MB, Roberts AB, Goodman D (eds) The retinoids: biology, chemistry and medicine. Raven, New York, pp 257–281

Törmä H, Lontz W, Liu W, Rollman O, Vahlquist A (1994) Expression of cytosolic retinoid-binding protein genes in human skin biopsies and cultured keratinocytes and fibroblasts. Br J Dermatol 131:243–249

Törmä H, Vahlquist A (1984) Vitamin A uptake by human skin in vitro. Arch Dermatol Res 276:390–395

Törmä H, Vahlquist A (1985) Retinol esterification by cultured human skin. In: Saurat JH (ed) Retinoids. New trends in research and therapy. Karger, Basel, pp 194–197

Törmä H, Vahlquist A (1987) Retinol esterification by mouse epidermal microsomes: evidence for acyl-CoA: retinol acyltransferase activity. J Invest Dermatol 88:398–402

Törmä H, Vahlquist A (1990) Vitamin A esterification in human epidermis: a relation to keratinocyte differentiation. J Invest Dermatol 94:132–138
Vahlquist A, Törmä H, Rollman O, Berne B (1985) Distribution of natural and synthetic retinoids in the skin. In: Saurat JH (ed) Retinoids. New trends in research and therapy. Karger, Basel, pp 159–167
Weiss JS, Ellis CN, Headington JT, Tincoff T, Hamilton TA, Voorhees JJ (1988) Topical tretinoin improves photo aged skin. A double blind vehicle-controlled study. J Am Med Assoc 259:527–532
Wrench R, Didierjean L (1985) Antibodies to orthokeratotic keratinocytes in monitoring the drug-induced inhibition of parakeratotic differentiation in adult and infant mice. Arch Dermatol Res 277:201–208

CHAPTER 19

Retinoid Receptor-Selective Agonists and Their Action in Skin

B. Shroot, D.F.C. Gibson, and X.-P. Lu

A. The Molecular Aspects of Retinoid Signaling

I. The Role of Retinoid Receptors

Vitamin A (Retinol) and its biologically active metabolites, for example, all *trans* retinoic acid, (AtRA) are essential for the health of the individual. These natural compounds are involved in a wide range of biological processes, including vertebrate development and morphogenesis, cell proliferation and tissue differentiation, vision, and reproduction (Chambon 1996; Mangelsdorf et al. 1995; Marshall et al. 1996; Morriss-Kay and Sokolova 1996). These biological actions of vitamin A and related compounds (encompassed by the term "retinoids") are mediated by two families of nuclear receptors, termed retinoic acid receptors (RAR) and retinoid X receptors (RXR). The effects of RA are mediated through binding to RAR isoforms (α, β, γ) (Brand et al. 1988; Giguere et al. 1987; Krust et al. 1989), while the 9-*cis* isomer of RA binds to the RXR family (with α, β, γ isoforms also; Fleischhauer et al. 1992; Leid et al. 1992; Mangelsdorf et al. 1990), in addition to the RAR family (Allenby et al. 1993). The RARs and RXRs are members of the ever expanding superfamily of nuclear receptors, that includes the receptors for estrogen, thyroid hormone, and vitamin D_3 (Giguere 1994). In addition to RARs and RXRs, RA can also bind with high affinity to cytoplasmic retinoic acid binding proteins (CRABP 1 and 2). These binding proteins are thought to play a role in both retinoid transport and metabolism.

RARs and RXRs are nuclear transcription factors that are comprised of five functional domains. These are termed the N-terminal A/B domain, the DNA-binding domain (DBD), the hinge region (D domain), the ligand-binding domain (LBD), and the C-terminal dimerization domain (AF-2 or Tau-4). Mechanistically, RAR can form heterodimers with RXR (Fig. 1). These complexes then bind to their cognate DNA-response element (retinoic acid response element, RARE) within the promoters of responsive genes. In the response elements, the ligand-receptor complex recognizes direct repeat (DR) sequences of DNA spaced by 3, 4, or 5 base pairs (termed DR3, 4, or 5). Two models of RAR action have been established. First, upon ligand binding, RAR/RXR heterodimers function as transactivators of RARE-containing target genes. Alternatively, RAR or RXR may also interact in a ligand-dependent manner with other transcription factors, such as AP-1.

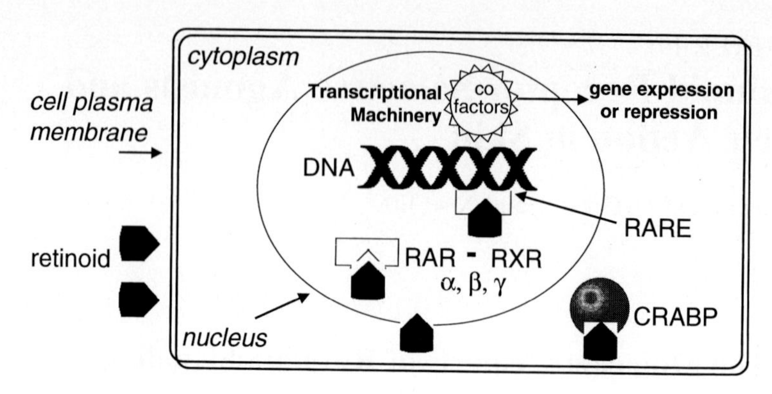

Fig. 1. Retinoid action at the molecular level

In this case, interaction leads to transrepression of AP-1-mediated gene expression, thus generating distinct regulatory signals to the cells (PFAHL 1993). In a broader sense, RXR acts as a central part of nuclear hormone receptor signaling. Most of the other receptors, including the vitamin D_3 receptor (VDR), the thyroid hormone receptor (TR), and the peroxisome proliferator-activated receptor (PPAR) bind and activate specific genes as heterodimers with RXR (GIGUERE 1994). The control of gene expression by these ligand-activated receptor complexes is sensitive to the specific DNA-receptor context (LA-VISTA PICARD et al. 1996; MANGELSDORF and EVANS 1995).

The complexity of retinoid signaling is further enhanced by the recent discovery that RAR and RXR signaling pathways interact with the cellular transcription machinery through nuclear receptor coactivators (NcoA) as well as nuclear receptor corepressors (NcoR) in both a ligand-dependent or independent manner (GLASS et al. 1997; HORWITZ et al. 1996). In this way, the nuclear hormone signaling pathway interacts with many other cellular signaling pathways, such as AP-1, NF-κB, and cAMP, ultimately leading to a fine control of gene expression in response to wide ranges of both physiological as well as stress signals (MONTMINY 1997). These concepts of RAR/RXR signal transduction pathways, generated by elegant molecular experiments, have been confirmed by genetic studies, in which RAR or RXR or RAR/RXR gene knock-out mice are observed to have virtually all of the phenotypic defects observed in Vitamin A-deficient animals (KASTNER et al. 1995).

This broad mechanism of action leads to drawbacks if the goal of the retinoid use is therapeutic. Indeed, many side effects that have been observed can be linked to the nonselective activation of retinoid signaling pathways in vivo. The design of receptor-selective ligands, in concert with a better understanding of the mechanism of retinoid action provides novel approaches for new retinoid or retinoid-associated therapies.

II. RAR Receptor-Selective Retinoids

Retinoic acid and its conformationally restricted synthetic analogs are ligands that bind to RARs. The three RAR subtypes identified to date (α, β, and γ) differ in the amino acid sequence at both the A/B domain and ligand-binding domain. It is clear now that such differences allow a given RAR subtype to interact with conformationally restricted retinoids in a more specific manner (CHEN et al. 1995a; FANJUL et al. 1994; GRAUPNER et al. 1991; KLEIN et al. 1996; LEHMANN et al. 1992). This specificity may be achieved by the induction of a distinct conformational change upon the interaction of the ligand-receptor complex with the DNA and general transcription machinery (KEIDEL et al. 1994; LU et al. 1993). For example, the naphthoic acid derivative CD271 (adapalene) activates RARγ on ApoAI-RARE, while all *trans*-RA activates RARγ equally well on all RAREs tested (Lu, unpublished data). This suggests that CD271 transactivates RAREs differentially depending on the context of receptor subtype and structure of RARE. Therefore the ability of a retinoid to activate specific gene transcription should be considered in the context of the receptor subtype involved, the receptor heterodimer formed, and the structure of the RARE within the promoter of the gene involved (CHIBA et al. 1997; LA-VISTA PICARD et al. 1996; NAGPAL et al. 1992; ROY et al. 1995). Factors that determine the retinoid regulation of gene transcription are:

- Receptor type (RAR, RXR)
- Receptor type isoform (α, β, γ)
- DNA interactions (response element)
- Transcription factors (coactivators, corepressors)

Finally, certain RAR antagonists can behave in a similar fashion to RAR subtype-selective retinoids, in that reporter gene expression is activated in a specific receptor-DNA context only (CHEN et al. 1996). This suggests a further mechanism by which retinoid signaling through selective retinoids can be restricted. Consequently, such selective action by a specifically designed retinoid promises a pattern of biological activity that may well lack much of the undesired side effects.

Recent reports have suggested another aspect of retinoid signaling that may not involve RARs per se. It appears that the certain receptors in the context of heterodimerization to RXR (in this case, the mouse homologue of FXR) can be activated by a panagonist of RAR, but not RXR-selective agonist, albeit at higher concentration (ZAVACKI 1997). The inability of an RXR-selective ligand to activate the heterodimer is in contrast to the activation of RXR/LXR and RXR/NGFI-B (Leblanc and Stunnenberg 1995). This suggests that the RXR-FXR heterodimer contains a novel ligand binding pocket that the panagonist can recognize. This concept further broadens the possible biological effects of retinoids in vivo, as well as adding to the complexity of retinoid action. For example, topical application of a retinoid should produce a sufficient concentration to activate the genes controlled by FXR/RXR pathway in the skin.

III. RXR Receptor Selective Retinoids

As has previously been discussed, RXR forms heterodimers with a range of nuclear hormone receptors. With the discovery of RXR-selective retinoids (Lehmann et al. 1992) it has been possible to investigate the control of transcription in RXR-containing heterodimers. It is generally believed that in response to an RXR ligand, RXR acts as a silent partner in RAR/RXR, TR/RXR, VDR/RXR heterodimers on DR-5, DR-4, and DR-3 hormone response elements (HREs), respectively. In contrast, RXR acts as a permissive partner in PPAR/RXR, LXR/RXR, and NGFI-B/RXR heterodimers on DR-1, DR-4, and DR-5 HRE's (Kurokawa et al. 1994; Mukherjee et al. 1997; Perlmann and Jansson 1995). Therefore, it appears that RXR not only functions as an auxiliary factor for high-affinity DNA binding of RAR and other receptors, but also as a true hormone receptor capable of mediating specific signals in vivo. This idea has been supported by recent observations (Vivat et al. 1997), which described an RXRα mutation with an inactivated ligand-dependent transactivation AF2 domain. This mutant receptor, still able to form heterodimers, was introduced into mice to replace the normal RXRα. These mice do exhibit some phenotypic aspects observed in RXR–/– animals, although to a much lesser extent (Perlmann and Evans 1997). This does indicate that liganded RXR does play a role in retinoid signaling, and such results suggest that an RXR-selective ligand could induce specific responses in vivo.

Many of the ligands that exert effects upon the epidermis (e.g., retinoids and vitamin D_3) bind to nuclear receptors as heterodimers with RXR. This has indicated that RXR occupies a central position in keratinocyte biology, and with the recent development of specific RXR ligands, the potential role of RXR in skin therapy can now be investigated. In keratinocytes, SR11237 (see Fig. 4), an RXR selective ligand, can synergize with vitamin D_3 to activate DR3 and induce a responsive gene, 24-hydroxylase (Li et al. 1997). Results of this nature indicate that RXR ligands could be used in concert with vitamin D_3 and related analogues to achieve the desired therapeutic effect with minimal calcemic risk. The advantage of using an RXR specific ligand is reflected by reports that RA actually inhibits vitamin D_3 induced gene expression (Gibson et al. 1998). Experiments of this nature indicate that the concept that RXR is a silent partner in RXR/VDR signaling is dependent upon the responsive gene. This further illustrates the importance of the promoter/gene/transcription factor environment.

There is evidence that in addition to heterodimers, RXR can also form homodimers. In the presence of 9-cisRA, RXR/RXR homodimers bind to DR-1 RXRE's in the promoter regions of the CRBP-II and $ApoA_I$ genes (ZHANG et al. 1992). Therefore, RXR receptors might play a dual role, as both a cofactor and as an independent signaling system. Interestingly, an antagonist of the RXR homodimer appears to function as an agonist in the context of reporter gene expression experiments using PPAR/RXR on DR-1 or RAR/RXR on DR-5 HRE's (LALA et al. 1996; MUKHERJEE et al. 1997). Although the antagonist does bind to RXR within the RAR/RXR heterodimer, it utilizes the AF-2 domain of RAR for transactivation. The term "phantom ligand effect" has been proposed for this mechanism (SCHULMAN et al. 1997). Again, such a gene transactivation mechanism could be expected to be selective.

B. Vitamin A and the Skin

I. Proliferation and Differentiation Within the Epidermis

Within the epidermis, the processes of proliferation and differentiation are tightly controlled, resulting in a tissue at steady-state in which the cells that are desquamated and lost are equal to those that are produced in the germinative basal layer. Cells within the basal layer are highly proliferative and display an undifferentiated phenotype characterized by the coexpression of the K5/K14 keratins. Mitosis is restricted to the basal layer, and as keratinocytes migrate into the suprabasal layers, they undergo a well-defined program of terminal differentiation, characterized by the coordinated induction of specific proteins (FUCHS 1990). The initial change is the loss of K5/K14 expression and the synthesis of the K1/K10 keratins (FUCHS 1988). With subsequent movement into more suprabasal layers, cornified envelope substrate proteins are synthesized, including loricrin (MEHREL et al. 1990), filaggrin (DALE et al. 1985), and involucrin (WARHOL et al. 1985), together with type 1 transglutaminase (THACHER and RICE 1985), the calcium dependent enzyme responsible for the cross-linking of the proteins. Additionally, these cells contain numerous lamellar bodies (ELIAS and MENON 1991). During the final stages of the differentiation process, the substrate protein cross-linking forms the resistant cornified envelopes (RICE et al. 1992). Lipids are secreted out of the lamellar bodies into the extracellular spaces between the cornified envelopes, resulting in the cutaneous permeability barrier. Many of the biochemical processes observed during differentiation can be replicated in vitro by exposing keratinocytes to elevated calcium (GIBSON et al. 1996).

II. Vitamin A and the Skin: The Historical Aspects

The dramatic effects of vitamin A upon the skin have been known for many years. Hyperkeratinization is the result of vitamin A deficiency in animals, with many epithelial sites becoming stratified and keratinized (WOLBACH and HOWE 1925). Conversely, an excess of vitamin A appears to inhibit keratinization, even inducing a mucous metaplasia in certain tissues (FELL and MELLANBY 1953). Human data, although obviously harder to obtain, appear to match these concepts. Vitamin A deficient soldiers were shown to have dry, scaly hyperkeratinized skin (FRAZIER and HU 1931), and the effects of hypervitaminosis A appear to include symptoms of inhibited keratinization (KAMM et al. 1984).

III. RARs and RXRs and the Development of the Skin

Both RARs and RXRs are spatially and temporally expressed during mammalian embryogenesis (HOFMANN and EICHELE 1994). RARs and RXRs are both expressed in premigratory and migratory neural crest cells that give rise to components of the epidermis. It is believed that RA functions as a morphogen through complex interactions with the Sonic Hedgehog, Hox, and TGF-β/BMP signaling pathways (CHANG et al. 1997; GAZIT et al. 1993; MARSHALL et al. 1996). The involvement of RA and RARs in the formation of epidermis as well as hair follicles, sebaceous glands, and other associated appendages have been previously reviewed (HARDY 1992; VIALLET et al. 1992), and it is clear that both vitamin A and RA have effects upon skin differentiation during development by switching between differential development pathways. For example, in explant organ cultures maintained in media containing high levels of vitamin A, the primitive epidermal layer on a fragment of 7-day developing chick skin becomes a mucous-secreting epithelium rather than a stratified squamous keratinizing tissue (FELL and MELLANBY 1953).

When the receptor types are specifically studied, it appears that both RARα and RARγ are expressed within the developing epidermis and dermis. In contrast, in mature epidermis, the major receptor forms are RARγ, together with RXRα (Fig. 2). Although RARβ is expressed in many other adult epithelial tissues, is not expressed in the epidermis at detectable levels within the developing or the adult epidermis. Interestingly, it can be detected within the mesenchymal cells of the dermis during development, and is inducible in adult dermis (ELDER et al. 1991; VIALLET and DHOUAILLY 1994; ZELENT et al. 1989). Recently, RXRα and RXRβ were found in dermal fibroblasts (TSOU et al. 1997). The importance of mesenchymal inductive signals controlling epidermal function is well established (CUNHA et al. 1992), and the differential expression of retinoid receptors within the epidermis and dermis does suggest retinoids might be part of such dermal-epidermal interactions. The involvement of RARs in these tissue interactions has been elegantly demonstrated in heterotopic and homotopic dermal-epidermal recombinant experiments in which RA-pretreated mesenchyme induces appendage morphogenesis within the untreated overlying epithelium (HARDY 1989). Since RARγ1, the main RARγ isoform in skin, does not activate the DR-5 type of RARE found in

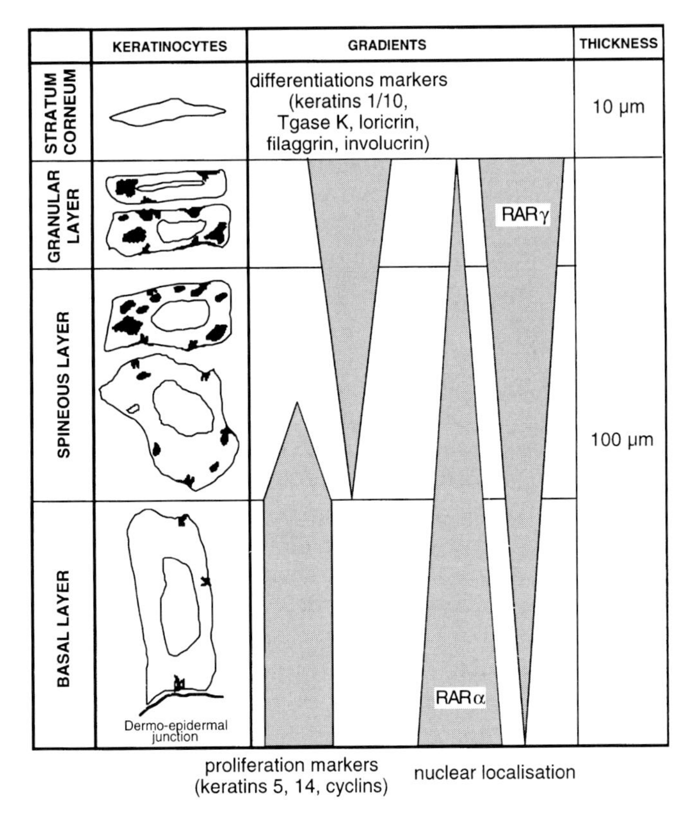

Fig. 2. A view of the epidermis, illustrating the changes in proliferation/differentiation markers and RARα and RARγ as keratinocytes move out of the basal and through the suprabasal layers

the RARβ2 promoter (Husmann et al. 1991), the inductive effect of RA in these experiment suggests an important role for RARβ or RARβ /RARα in mediating appendage differentiation during skin development. This may be important in the context of skin regeneration (including appendages) and diseases in which hair follicles are lost. In this context, a RARβ-selective ligand might be useful.

IV. Retinoid Effects upon Keratinocyte Proliferation and Differentiation

This historical data has been supported somewhat by more recent in vitro data. When keratinocytes are treated in vitro, RA inhibits terminal differentiation, suppressing the expression of K1/K10 (Kopan et al. 1987), profilaggrin (Fleckman et al. 1984), loricrin (Hohl et al. 1991), involucrin (Gibson et al. 1998), transglutaminase (Gibson et al. 1998; Rubin et al. 1989) and cornified envelope formation (GREEN and WATT 1982). Unfortunately, the situation in vitro is far from clear. Computer analysis of 2-D gels from RA treated and untreated cultures has indicated that only one third of differentiation-specific markers are down-regulated by RA (RABILLOUD et al. 1988), and in addition, some markers of the "undifferentiated state" (for example, K5/K14) can also be down-regulated by RA (GILFIX and ECKERT 1985).

The effects of RA upon the proliferation of cultured keratinocytes are highly dependent upon culture conditions. Undifferentiated, rapidly proliferating keratinocytes can be growth inhibited by RA (GIBSON et al. 1998; TONG et al. 1986; VARANI et al. 1989). Conversely, RA can stimulate the growth of quiescent keratinocytes (VARANI et al. 1989). There is evidence that the proliferative response of cultured keratinocytes to RA is related to cellular fatty acid composition (MARCELO and DUNHAM 1997).

In the light of these in vitro data, and the ability of RA to block epidermal differentiation, it is not immediately clear how the skin can exist in a differentiated form in the face of endogenous circulating retinoids? It has been suggested that retinoids are both rapidly metabolized by dermal fibro blasts (RANDOLPH and SIMON 1997) and "filtered" by the dermis (ASSELINEAU et al. 1989), leading to a gradient of RA within the epidermis. Topical application of RA promotes keratinocyte proliferation, resulting in epidermal hyperplasia (TUR et al. 1995). Specifically, the thickness of the granular layer is increased and the stratum corneum is compacted (LEVER et al. 1990), indicating effects upon differentiation in addition to proliferation. The increase in the number of granular layers results in an increased expression of "granular markers" such as transglutaminase (ROSENTHAL et al. 1992), loricrin and filaggrin (TUR et al. 1995). CRABPII is also induced (ELDER et al. 1992). RA treatment not only leads to a decrease in stratum corneum thickness, but also to cell loosening and fragility (ELIAS 1987), changes in lipid levels, and a subsequent decrease in the permeability barrier. The effects of RA in vitro (decreasing differentiation markers) and in vivo (increasing differentiation markers) are in direct contrast, and might reflect the lack of dermal interactions in submerged keratinocyte cultures. The validity of this concept has been proven with experiments using air-exposed 3-D cultures (ASSELINEAU et al. 1989). As outlined, in vivo RA treatment upregulates differentiation markers, including transglutaminase, CRABPII, and cornifin. However, all three of these proteins are upregulated in psoriatic lesions and this expression is down-regulated by RA treatment (FUJIMOTO et al. 1993; NAGPAL et al. 1997b). These data indicate that the regulation of differentiation in vivo is different to that in culture and abnormal states (psoriasis).

V. Modulation of Inflammation and the Immune Response in Skin by RA

In addition to keratinocytes and melanocytes, the epidermis also contains Langerhans cells, dendritic cells and the innervating peripheral neuron cells.

This means that the skin is an organ capable of responding to signals from both outside and from within the body.

A prominent feature during the pathogenesis of many skin disorders is chronic inflammation, characterized by the infiltration of activated T lymphocytes. Among the factors that are critical for controlling those processes, transcription factors (such as AP-1, NF-κB, GATAs, Oct, NF-IL6, NF-ATs, and STATs) appear to be essential (O'Shea 1997). Interestingly, the members of the nuclear hormone receptor superfamily are known to interact with many of these transcription factors, suggesting that ligands for these nuclear receptors may be an effective means of modulating inflammation, as well as the immune response in general (Alroy et al. 1995; Barnes and Karin 1997; Matikainen et al. 1997; Stocklin et al. 1996). Therefore, transcriptionally inactive antiAP-1 retinoids (Chen et al. 1995a; Fanjul et al. 1994) are likely to be anti-inflammatory candidates, as the regulation of transcription of many proinflammatory cytokines as well as chemokines are controlled through AP-1 alone or together with other factors. This is supported by the finding that a transcriptionally inactive synthetic analog of a glucocorticoid, with antiAP-1 activity, shows some anti-inflammatory effects in vivo (Vayssiere et al. 1997).

Many genes involved in immunomodulation are also regulated through AP-1, NF-κB, and other transcription factors. Retinoids have been reported to down regulate the activation of NF-κB in endothelial cells (Gille et al. 1997), cross talk with Oct factors (de Grazia et al. 1994), and modulate NF-IL6 (Kagechika et al. 1997; Nagpal et al. 1997a). In addition, it is known that both retinol and RA are essential factors for the survival of thymocytes and the development of T cells in vivo (Buck et al. 1991). Therefore, synthetic retinoids that are able to interact with transcription factors through the RAR represent novel effectors of the immune system. Given the role of the immune system in the skin, such compounds should prove extremely useful.

VI. Retinoid Receptors and Skin Function

As this outline has clearly shown, the effects of RA (as an example of a natural retinoid) upon the epidermis are widespread. In addition to affecting proliferation and differentiation, topical application of some retinoids leads to erythema, peeling and dryness (Orfanos et al. 1997), that often results in a cessation of treatment. More seriously, retinoids are known to be teratogenic. With these side effects in mind, synthetic retinoids have been made that are selective, in terms of RAR binding, in the hope of achieving an optimal balance between efficacy and tolerance.

With the development of more sophisticated molecular techniques, has come the opportunity for more information concerning the role of retinoids and RARs in vivo. Surprisingly, within double RARα/RARγ knock-out animals, no dramatic phenotypic defects were observed in the skin, although metaplasias were ready observed in other epithelial sites. While this result appears to contradict the apparent central role of RA within the skin, it may be explained, in part, by the existence RARβ in these knockout animals (Kastner et al. 1995). Experiments using transgenic mice have revealed further information. When a dominant-negative RARα is introduced into undifferentiated basal keratinocytes, epidermal differentiation is inhibited, with an absence in the formation of the spinous and horny layers and suprabasal expression of keratin 5 and 14 (Saitou et al. 1995). When a dominant-negative RARα was introduced into the differentiating suprabasal keratinocytes compartment, there was a disruption of lipid lamellar structure and barrier function (Attar et al. 1997; Imakado et al. 1995). While these data appear to confirm the central role of RA signaling in epidermal proliferation, differentiation and function, they also shed light on the complexity of the entire process. Specifically, if the mutant RARα receptor was disrupting RA signaling by sequestration of RXRs, other pathways controlled by receptors that also dimerize with RXR should also be affected. However, vitamin D_3 was still able to induce VDR-mediated responses (at least in the induction of 25-hydroxyvitamin D_3 hydroxylase) in RARα transgenic mice (Attar et al. 1997). This suggests that RXR sequestration is not the complete story. It is interesting to note that one of the RAR mutants used, RARα 403, shows a fourfold increased affinity for binding to the corepressor SMRT in solution (Chen and Evans 1995). This does imply that a highly expressed RARα 403 in transgenic epidermis may interfere with the interactions between corepressors or coactivators and variety of nuclear receptor signaling pathways (VDR, TR, LXR, or PPAR signaling in particular; Soderstrom et al. 1997).

Therefore, in the face of a wealth of data concerning the effects, both in vitro and in vivo, of retinoids upon the skin, the mechanisms of RAR/RXR signaling are becoming more apparent. This information has encouraged the rationale design of retinoids that are selective in terms of RAR binding, in the hope of achieving an optimal balance between efficacy and tolerance.

VII. The Development of Synthetic Retinoids and Their Action in Skin

Natural retinoid acids, including AtRA, 9-*cis*-RA, 13-*cis*-RA, and other retinoids such as 4-oxo-RA, 3, 4-didehydroretinol, 14-hydroxy-4, 14-retroretinol (Fig. 3) have been show be biologically active. In particular, AtRA, 13-*cis*-RA, and the first generation analogues (Etretinate and Acitretin) have been used successfully in controlling the symptoms of many skin diseases, including acne, rosacea, photodamage, skin-related malignances, and other miscellaneous skin disorders (Orfanos et al. 1997). However, the side effects observed with retinoid therapy (topical or oral routes), including mucocutaneous symptoms, hair loss, elevation of serum triglycerides, bone fragility, and malformation of the fetus does significantly limit retinoid use. Therefore, syn-

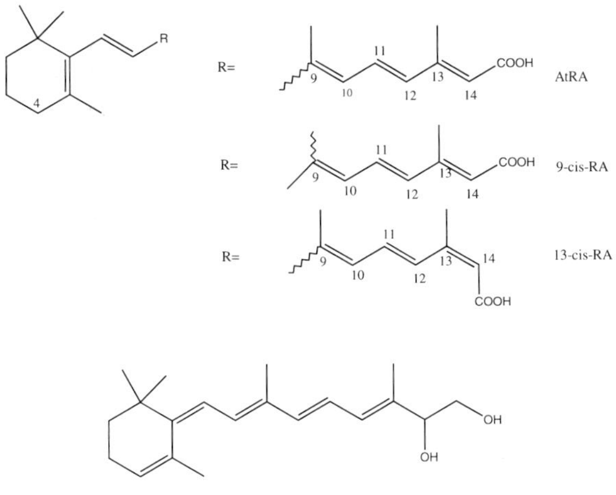

Fig. 3. The chemical structures of all-*trans* RA, 9-*cis* RA, 13-*cis* RA, and 14-hydroxy-4,14-*retro*-retinol

thetic retinoids were designed to separate the beneficial effects from the side effects.

In general, synthetic retinoids that are RAR panagonists appear to affect epidermal differentiation and proliferation in same manner as all-*trans* RA. RARγ-selective agonists, in particular, have been shown to be as active and as irritating as RA (Chen et al. 1995b). RXR-selective agonists, in line with RXR being a silent partner with RAR upon binding to the RARE, do not show any activity comparable with that of an RAR agonist (Geudimenico et al. 1994). However, RAR selective retinoids can exhibit therapeutic profiles that differ significantly from all-*trans* RA. Two such substances, Adapalene (CD271) and Tazarotene (AG-190168) (Fig. 4) have been developed and are presently used for acne and psoriasis, respectively.

1. Adapalene (CD271)

Adapalene is a synthetic retinoid that is used for the topical treatment of acne vulgaris. Although it is a RARγ/RARβ selective compound in terms of binding and transactivation (Shroot and Michel 1997), adapalene does appear to act in a similar fashion as RA. In vitro, adapalene inhibits keratinocyte differentiation as assayed by transglutaminase expression (tenfold more potent than RA), is as potent as RA in the inhibition of proliferation of HeLa cells (Shroot and Michel 1997), and in the modulation of the expression of K10 and filaggrin in cultures of reconstructed skin (Asselineau et al. 1992). In the epidermis of the Rhino mouse, adapalene shows a potent activity, reducing the number of comedones and increasing epidermal thickness (Shroot and Michel 1997). Clinically, in terms of lesion counts and global facial acne grade,

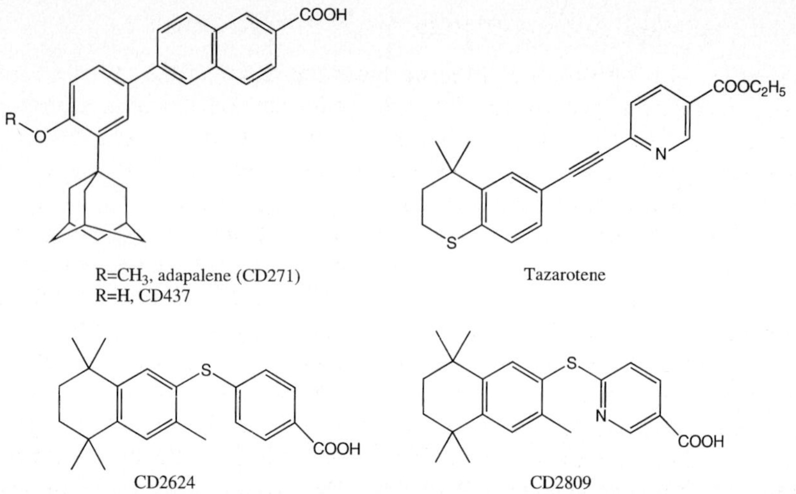

$R=CH_3$, adapalene (CD271)
R=H, CD437

Tazarotene

CD2624

CD2809

Fig. 4. The chemical structures of CD271 (adapalene), CD437, tazarotene, and the RXR ligands CD2624, CD2809, and SR11237

the effects of adapalene are comparable to those of tretinoin (CUNLIFFE et al. 1997). Most importantly, the signs of irritation (burning, dryness, scaling) were significantly lower with adapalene treatment. Interestingly, one explanation for this lowered irritation observed with adapalene may be the lack of binding to CRABP.

2. Tazarotene (AGN 190168)

Tazarotene is a synthetic RARβ/RARγ selective retinoid that is therapeutically effective in the treatment of psoriasis (ESGLEYES-RIBOT et al. 1994; WEINSTEIN 1996). Tazarotene acts in many ways as RA, including down-regulating transglutaminase in cultured keratinocytes (NAGPAL et al. 1996b) and up-regulating filaggrin (ESGLEYES-RIBOT et al. 1994) and inducing hyperplasia in the epidermis (THACHER et al. 1997). Skin-derived antileukoproteinase (SKALP) and migration inhibitory factor-related protein (MRP-8) are over-expressed in psoriasis as compared with normal skin, and as such, they represent markers of hyperproliferative keratinocyte differentiation. In both psoriasis and three-dimensional raft cultures, tazatotene down-regulated SKALP and MRP-8 (NAGPAL et al. 1996b). In addition, 3 novel genes (tazarotene-induced genes 1–3, TIG1–3) are up-regulated by tazarotene in skin raft cultures and psoriatic lesions (DUVIC et al. 1997; NAGPAL et al. 1996a 1997b). TIG2 has been shown to be 164 amino-acid protein that is membrane associated. These TIG's may have a role in the maintenance of normal skin homeostasis, and their induction by tazarotene may be central to this retinoid's therapeutic effects. Additionally, TIG1–3 represent the only known genes that are retinoid-inducible in psoriasis.

C. Apoptosis: An Emerging Role for Retinoids

As previously reviewed, the epidermis contains a population of proliferative keratinocytes that give rise to daughter cells that are destined to complete a program of terminal differentiation. As part of this differentiation, the cell nucleus is lost and the cornified cell is eventually sloughed off. Programmed cell death, or apoptosis, is a physiological method of cell death, that is a central part of development and homeostasis (Jacobson et al. 1997), and as such, keratinocyte terminal differentiation can be considered a form of apoptosis. Bcl-2 or Bcl-x, cell death inhibitor genes, appear not to be involved in keratinocyte terminal differentiation, but rather influence keratinocyte survival under stress situations (Pena et al. 1997). Interestingly, it has been reported that the expression of both Bcl-2 and Bcl-x are significantly increased in psoriatic lesional skin and basal cell carcinoma (Wrone-Smith et al. 1995). Other effectors of cell death such as Fas and Fas ligand are expressed within the skin (Freiberg et al. 1996), but their role in differentiation are far from clear.

A central aspect of terminal differentiation is lipid synthesis, resulting in an increased content of ceramides, cholesterol, and free fatty acids (Ponec 1994). Interestingly, it has been shown that both ceramide (Bose et al. 1995) and an excess of cholesterol (Brown and Goldstein 1997) are involved with apoptosis, and it may be that this lipid synthesis during differentiation controls both function (barrier formation) and ultimate death (apoptosis). Nuclear receptors also play a role in this lipid synthesis. Ligands for PPAR and LXR have been associated with both lipid and cholesterol metabolism (Janowski et al. 1996; Spiegelman and Flier 1996), and RXR ligands are able to activate PPAR/RXR and LXR/RXR heterodimers. This suggests that ligands for PPAR, LXR, and RXR may prove to be useful in promoting terminal differentiation and barrier formation.

The physiological balance between proliferation and differentiation is disturbed in a variety of skin diseases, resulting in hyperproliferation (e.g., psoriasis; Gottlieb et al. 1995). An increase in the number of dividing cells also is a hallmark of transformed phenotype. The known ability of retinoids to induce apoptosis in proliferating cells therefore indicates a therapeutic use within skin, and beyond. The synthetic retinoid CD437 (RARγ selective; Fig. 4) induces apoptosis in a variety of tumor cells in vitro and in vivo (Schadendorf et al. 1996; Shao et al. 1995). Mechanistically, CD437 was shown to down regulate Bcl-Xl expression in a breast cancer cell line (Hsu et al. 1997), and is known to activate the JNK kinase pathway in Jurkat cells (Lu et al. unpublished result). Recently, analogs of CD437 were shown to induce apoptosis in a range of cancer cells in a selective manner (Lu et al. 1997), and such apoptosis involved proteolysis of transcription factors (Piedrafita and Pfahl 1997). Although it has been debated whether RAR is involved in such events in vivo, analogs of CD437 that have been modified to form the methyl ester are unable to induce apoptosis, and are also unable to transactivate RAR. In a related report, two RAR antagonists (CD 2366 and CD2665) were unable to inhibit the antiproliferative effects of CD437 in cultures of non-small-cell lung carcinoma cells (Sun et al. 1997b), indicating that a nonretinoid receptor pathway may be involved. In a more detailed study, a range of retinoids were tested against non-small-cell lung carcinoma cells (Sun et al. 1997a). An RARα-specific compound (TD 550), an RARβ-specific compound (Ch 55), three RARγ-specific compounds (CD437, CD2325, SR11364), and an RARβ/RARγ-specific compound (CD271) were all active. Both RXR-specific compounds and an RAR antagonist (CD2665) were inactive. Whatever the mechanism, these reports do indicate that selective retinoids can induce apoptosis in a selective manner. A recent report, however, does suggest that RXR-specific ligands may have an advantage over RAR ligands, in terms of teratogenic potential. When evaluated using an assay of limb bud mesenchymal chrondrogenesis, RXR ligands (CD2809, CD2624; Fig. 4) were less teratogenic than RAR compounds (Beard et al. 1996).

D. Future Prospects

Many synthetic analogs were evaluated in the decade following the discovery that retinoic acid controlled gene expression through binding to receptors belonging to the nuclear steroid thyroid hormone family (Dawson and Hobbs 1994). The efforts to identify new drugs, which have an improved risk to benefit ratio as compared to all-*trans*-RA, have been successful at least as far as receptor type selective ligands for RAR and RXR are concerned. However, research into receptor subtype selective substances, while providing a plethora of pharmacological tools and even new drugs with perceived advantages, has yet to discover drugs with a radically changed profile. This review has clearly underlined the complexity of retinoid regulated transduction pathways. There is an undeniable possibility that new actions related to activators or repressors acting through orphan receptors may not be related to the ligand directly, but to posttranscriptional events occurring within or beyond the cell nucleus.

In parallel, the study of the genetic defects associated with skin diseases will reveal more specific targets that might be under the control of retinoid subtype receptors. The clinical use of natural isomers has indicated the potential that this class of molecules possess for treating skin diseases. Further research will reveal if these properties can be appropriately harnessed through finer control of the signaling pathway.

References

Allenby G, Bocquel M-T, Saunders M, Kazmer S, Speck J, Rosenberger M, Lovey A, Kastner P, Grippo JF, Chambon P, Levin AA (1993) Retinoic acid receptors and retinoid X receptors: interactions with endogenous retinoic acids. Proc Natl Acad Sci USA 90:30–34

Alroy I, Towers TL, Freedman LP (1995) Transcriptional repression of the interleukin-2 gene by vitamin D3: direct inhibition of NFATp/AP-1 complex formation by a nuclear hormone receptor. Mol Cell Biol 15:5789–5799

Asselineau D, Bernard BA, Bailly C, Darmon M (1989) Retinoic acid improves epidermal morphogenesis. Dev Biol 133:322–335

Asselineau D, Cavey MT, Shroot B, Darmon M (1992) Control of epidermal differentiation by a retinoid analogue unable to bind cytosolic retinoic acid binding proteins (CRABPs), J Invest Derm 98:128–134

Attar PS, Wertz PW, McArthur M, Imakado S, Bickenbach JR, Roop DR (1997) Inhibition of retinoic signaling in transgenic mice alters lipid processing and disrupts epidermal barrier function. Mol Endocrinol 11:792–800

Barnes PJ, Karin M (1997) Nuclear factor-Kappa B: a pivotal transcription factor in chronic inflammatory diseases. N Engl J Med 336:1066–1071

Beard RL, Colon DF, Song TK, Davies PJA, Kochhar DM, Chandraratna RAS (1996) Synthesis and structure-activity relationships of retinoid X receptor selective diaryl sulfide analogs of retinoic acid. J Med Chem 39:3556–3563

Bose R, Verheij M, Haimouitz-Friedman A, Scotto K, Kuks Z, Kolesnick R (1995) Ceramide synthase mediates daunorubicin-induced apoptosis: an alternative mechanism for generating death signals. Cell 82:405–414

Brand N, Petkovich A, Krust A, Chambon P, de The H, Marchio A, Tiollais P, Dejean A (1988) Identification of a second human retinoic acid receptor. Nature 332:850–853

Brown MS, Goldstein JL (1997) The SREBP pathway: regulation of cholesterol metabolism by proteolysis of a membrane-bound transcription factor. Cell 89:331–340

Buck J, Derguini F, Levi E, Nakanishi K, Hammerling U (1991) Intracellular signaling by 14-hydroxy-4,14-retro-retinol. Science 254:1654–1656

Chambon P (1996) A decade of molecular biology of retinoic acid receptors. FASEB J 10:940–954

Chang BE, Blader P, Fisher N, Ingham PW, Strahle U (1997) Axial (HNF3beta) and retinoic acid receptors are regulators of the zebrafish sonic hedgehog promoter. EMBO J 16:3955–3964

Chen JD, Evans RM (1995) A transcriptional co-repressor that interacts with nuclear hormone receptors. Nature 377:454–457

Chen JY, Penco S, Ostrowski J, Balaguer P, Pons M, Starrett JE, Reczek P, Chambon P, Gronemeyer H (1995a) RAR-specific agonist/antagonist which dissociate transactivation and AP-1 transrepression inhibit anchorage- independent cell proliferation. EMBO J 14:1187–1197

Chen JY, Clifford J, Zusi C, Starrett J, Tortolani D, Ostrowski J, Recek P, Chambon P (1996) Two distinct actions of retinoid-receptor ligands. Nature 382:819–822

Chen S, Ostrowski J, Whiting G, Roalsvig T, Hammer L, Currier SJ, Honeyman J, Kwasniewski B, Yu K-L, Sterzycki R, Kim CV, Starrett J, Mansuri M, Reczek RR (1995b) Retinoic acid receptor gamma mediates topical retinoid efficacy and irritation in animal models, J Invest Derm 104:779–783

Chiba H, Clifford J, Metzger D, Chambon P (1997) Distinct retinoid X receptor-retinoic acid receptor heterodimers are differentially involved in the control of expression of retinoid target genes in F9 embryonal carcinoma cells. Mol Cell Biol 17:3013–3020

Cunha GR, Young P, Hamamoto S, Guzman R, Nandi S (1992) Developmental response of adult mammary epithelial cells to various fetal and neonatal mesenchymes, Epith Cell Biol 1:105–118

Cunliffe WJ, Caputo R, Dreno B, Forstrom L, Heenen M, Orfanos CE, Privat Y, Robledo Aguilar A, Poncet M, Vershoore M (1997) Efficacy and safety comparison of adapalene (CD271) gel and tretinoin gel in the topical treatment of acne vulgaris. A European multicentre trial, J Dermatol Treat 8:173–178

Dale BA, Resing KA, Landsdale-Eccles JD (1985) A keratin filament associated protein. Ann NY Acad Sci 455:330–342

Dawson MI, Hobbs PD (1994) The synthetic chemistry of retinoids. In. Sporn MB, Roberts AB, Goodman DS (eds) The retinoids. Raven, New York, pp 5–178
DeGrazia U, Felli MP, Vacca A, Farina AR, Maroder M, Cappabianca L, Meco D, Farina M, Screpanti L, Frati L (1994) Positive and negative regulation of the composite octamer motif of the interleukin 2 enhancer by AP-1, Oct-2, and retinoic acid receptor. J Exp Med 180:1485–1497
Duvic M, Nagpal S, Asano AT, Chandraratna RAS (1997) Molecular mechanisms of tazarotene action in psoriasis. J Am Acad Dermatol 37:518–524
Elder JT, Astrom A, Pettersson U, Tavakkol A, Griffiths CE, Krust A, Kastner P, Chambon P, Voorhees JJ (1992) Differential regulation of retinoic acid receptors and binding proteins in human skin, J Invest Derm 98:673–679
Elder JT, Fisher GJ, Zhang QY, Eisen D, Krust A, Kastner P, Chambon P, Voorhees JJ (1991) Retinoic acid receptor gene expression in human skin, J Invest Derm 96:425–433
Elias PM (1987) Retinoid effects on the epidermis, Dermatologica 175 [Suppl]:28–36
Elias PM, Menon GK (1991) Structural and lipid biochemical correlates of the epidermal permeability barrier. Adv Lipid Res 24:1–26
Esgleyes-Ribot T, Chandraratna RAS, Lew-Kaya DA DA, Sefton J, Duvic M (1994) Response of psoriasis to a new topical retinoid, AGN 190168. J Am Acad Dermatol 30:581–590
Fanjul A, Dawson MI, Hobbs PD, Jong L, Cameron JF, Harlev E, Graupner G, Lu X-P, Pfahl M (1994) A new class of retinoids with selective inhibition of AP-1 inhibits proliferation. Nature 372:107–111
Fell HB, Mellanby E (1953) Metaplasia produced in cultures of chick ectoderm by vitamin A, J Physiol 119:470–488
Fleckman P, Haydock P, Blomquist C, Dale BA (1984) Profilaggrin and the 67-kD keratin are coordinately expressed in cultured human epidermal keratinocytes. J Cell Biol 99:315A
Fleischhauer K, Park JH, DiSanto JP, Marks M, Ozato K, Yang SY (1992) Isolation of a full-length cDNA clone encoding a N-terminal variant form of the human retinoid-X-receptor β, Nucleic Acid Res 20:1801
Frazier CN, Hu CK (1931) Cutaneous lesions associated with a deficiency in vitamin A in man. Arch Intern Med 48:507–514
Freiberg RA, Spencer DM, Choate KA, Peng PD, Shreiber SL, Crabtree GR, Khavari PA (1996) Specific triggering of the fas signal transduction pathway in normal human keratinocytes. J Biol Chem 271:31666–31669
Fuchs E (1988) Keratins as biochemical markers of epithelial differentiation. Trends Genet 4:277–281
Fuchs E (1990) Epidermal differentiation: the bare essentials. J Cell Biol 111:2807–2814
Fujimoto W, Marvin K, George M, Celli G, Darwiche N, DeLuca L, Jetten A (1993) Expression of cornifin in squamous differentiating epithelial tissues, including psoriasis and retinoic acid-treated skin, J Invest Derm 101:268–274
Gazit D, Ebner R, Kahn AJ, Derynck R (1993) Modulation of expression and cell surface binding of members of the transforming growth factor-beta superfamily during retinoic acid-induced osteoblastic differentiation of multipotential mesencymal cells. Mol Endocrinol 7:189–198
Geudimenico GJ, Stim TB, Corbo M, Janssen B, Mezick JA (1994) A pleiotropic response in induced in F9 embryonal carcinoma cells and Rhino mouse skin by all-*trans* retinoic acid, RAR agonist but not by SR 11237, a RXR selection agonist, J Invest Derm 102:676–680
Gibson DFC, Ratman AV, Bikle DD (1996) Evidence for separate control mechanisms at the message, protein, and enzyme activation level for transglutaminase during calcium-induced differentiation of normal and transformed human keratinocytes, J Invest Derm 106:154–161
Gibson DFC, Bikle DD, Harris J (1998) All-*trans* retinoic acid blocks the antiproliferative prodifferentiating actions of 1,25-dihydroxyvitamin D3 in normal human keratinocytes. J Cell Physiol 174:1–8

Giguere V, Ong S, Segui P, Evans RM (1987) Identification of a receptor for the morphogen retinoic acid. Nature 330:624–629

Giguere V (1994) retinoic acid receptors and cellular retinoid binding proteins: complex interplay in retinoid signaling, Endocrine Rev 15:61–79

Gilfix BM, Eckert RL (1985) Coordinate control by vitamin A of keratin gene expression in human keratinocytes. J Biol Chem 260:14026–14029

Gille J, Lani L, Paxton L, Lawley TJ, Wright Caugham S, Swerlick RA (1997) Retinoic acid inhibits the regulated expression of vascular cell adhesion molecule-1 by cultured dermal microvascular endothelial cells. J Clin Invest 99:492–500

Glass CK, Rose DW, Rosenfeld MG (1997) Nuclear receptor coactivators. Curr Opin Cell Biol 9:222–232

Gottlieb SL, Gilleaudeau P, Johnson R, Estes L, Woodworth TG, Gottlieb AB, Krueger JG (1995) Response of psoriasis to a lymphocyte-selective toxin (DAB389IL-2) suggests a primary immune, but not keratinocyte, pathogenic basis, Nature Med 1:442–447

Graupner G, Malle G, Maigan J, Lang G, Prunieras M, Pfahl M (1991) 6′-substituted naphthalene-2-carboxylic acid analogs, a new class of retinoic acid receptors subtype-specific ligands. Biochem Biophys Res Commun 179:1554–1561

Green H, Watt FM (1982) Regulation by vitamin A of envelope cross-linking in cultured keratinocytes derived from different human epithelia. Mol Cell Biol 2:1115–1117

Hardy MH (1989) The use of retinoids as probes for analyzing morphogenesis of glands from epithelial tissues, In Vitro Cell. Dev Biol 25:454–459

Hardy MH (1992) Diverse roles for retinoids in modifying skin, skin appendages, and oral mucosa in mammals and birds. In: Morriss-Kay G (ed) Retinoids in normal development and teratogenesis. Oxford University Press, New York, pp 181–198

Hofmann C, Eichele G (1994) Retinoids in development. In: Sporn M, Roberts AB, Goodman DS (eds) The retinoids. Raven, New York, pp 387–442

Hohl D, Lichti U, Breitkneutz, Steinart PM, Roop DR (1991) Transcription of the human loricrin gene in vitro is induced by calcium and cell density and suppressed by retinoic acid, J Invest Derm 96:414–418

Horwitz KB, Jackson TA, Bain DL, Richer JK, Takimoto GS, Tung L (1996) Nuclear receptor coactivators and corepressors. Mol Endocrinol 10:1167–1177

Hsu CK, Rishi AK, Li XS, Dawson MI, Reichert U, Shroot B, Fontana JA (1997) Bcl-X(L) expression and its downregulation by a novel retinoid in breast cancer cells. Exp Cell Res 232:17–24

Husmann M, Lehmann J, Hoffman B, Hermann T, Tzukerman M, Pfahl M (1991) Antagonism between retinoid receptors. Mol Cell Biol 11:4097–4103

Imakado S, Bickenbach JR, Bundman DS, Rothnagel PS, Attar PS, Eang XJ, Walczak VR, Wisiewski S, Pote J, Gordon JS, Heyman RA, Evans RM, Roop DR (1995) Targeting expression of a dominant-negative retinoic acid receptor mutant in the epidermis of transgenic mice results in a loss of barrier function, Gene Dev 9:317–329

Jacobson MD, Weil M, Raff MC (1997) Programmed cell death in animal development. Cell 88:347–354

Janowski BA, Willy PJ, Devi TR, Falck JR, Mangelsdorf DJ (1996) An oxysterol signalling pathway mediated by the nuclear receptor LXR alpha. Nature 383:728–731

Kagechika H, Kawachi E, Fukasawa H, Saito G, Iwanami N, Unemiya H, Hashimoto Y, Shudo K (1997) Inhibition of IL-1 induced IL-6 production by synthetic retinoids. Biochem Biophys Res Commun 231:243–248

Kamm JJ, Ashenfelter KO, Ehrmann CW (1984) Preclinical and clinical toxicity of selected retinoids. In: Sporn M, Roberts AB, Goodman DS (eds) The retinoids, vol 2. Academic, Orlando, p 288

Kastner P, Mark M, Chambon P (1995) Nonsteroid nuclear receptors: what are genetic studies telling us about their role in real life. Cell 83:859–869

Keidel S, LeMotte P, Apfel C (1994) Different agonist and antagonist-induced conformational changes in retinoic acid receptors analyzed by protease mapping. Mol Cell Biol 14:287–298

Klein ES, Pino ME, Johnson AT, Davies PJA, Nagpal S, Thatcher SM, Krasinski G, Chandraratna RAS (1996) Identification and functional separation of retinoic acid receptor neutral antagonists and inverse agonists. J Biol Chem 271:22692–22696
Kopan R, Traska G, Fuchs E (1987) Retinoids as important regulators of terminal differentiation: examining keratin expression in individual epidermal cells at various stages of keratinization. J Cell Biol 105:427–440
Krust A, Kastner P, Petkovich M, Zelent A, Chambon P (1989) A third human retinoic acid receptor, hRARγ. Proc Natl Acad Sci USA 86:5310–5314
Kurokawa R, DiRenzo J, Boehm M, Sugarman J, Gloss B, Rosenfeld MG, Heyman RA, Glass CK (1994) Regulation of retinoid signalling by receptor polarity and allosteric control of ligand binding. Nature 371:528–531
Lala DS, Mukherjee R, Schulman IG, Canan Koch SS, Dardashti LJ, Nadzan AM, Croston GE, Evans RM, Heyman RA (1996) Activation of specific heterodimers by an antagonist of RXR homodimers. Nature 383:450–453
La-Vista Picard N, Hobbs PD, Pfahl M, Dawson MI, Pfahl M (1996) The receptor-DNA complex determines the retinoid response: a mechanism from the diversification of the ligand signal. Mol Cell Biol 16:4137–4146
Leblanc BP, Stunnenberg HG (1995) 9-*cis*-retinoic acid signaling: changes in partners causes some excitement. Genes Dev 9:1811–1816
Lehmann JM, Jong L, Fanjul A, Cameron JF, Lu XP, Haefner P, Dawson MI, Pfahl M (1992) Retinoids selective for retinoid X receptor response pathways. Science 258:1944–1946
Leid M, Kastner P, Lyons R, Nakshari H, Saunders M, Zacharewski T, Chen J-Y, Staub A, Garnier J-M, Mader S, Chambon P (1992) Purification, cloning, and RXR identity of the HeLa cell factor with which RAR or TR heterodimerizes to bind target sequences efficiently. Cell 68:377–395
Lever L, Kimar P, Marks R (1990) Topical retinoic acid for treatment of solar damage. Br J Dermatol 122:91–98
Li X-Y, Xiao J-H, Feng X, Qin L, Voorhees JJ (1997) Retinoid X receptor-specific ligands synergistically upregulate 1,25-dihydroxyvitamin D3-dependent transcription in epidermal keratinocytes in vitro and in vivo, J Invest Derm 108:506–512
Lu XP, Eberhardt NL, Pfahl M (1993) DNA bending by retinoid X receptors containing retinoid and thyroid hormone receptor complexes. Mol Cell Biol 13:6509–6519
Lu XP, Fanjul A, Picard N, Pfahl M, Rungta D, Nared-Hood K, Carter B, Piedrafita J, Tang S, Fabbrizio E, Pfahl M (1997) Novel retinoid-related molecules as apoptosis inducers and effective inhibitors of human lung cancer cells in vivo. Nat Med 3:686–690
Mangelsdorf DJ, Ong ES, Dyck JA, Evans RM (1990) Nuclear receptor that identifies a novel retinoic acid response pathway. Nature 345:224–229
Mangelsdorf DJ, Evans RM (1995) The RXR heterodimers and orphan receptors. Cell 83:841–850
Mangelsdorf DJ, Thummel C, Beato M, Herrlich P, Schutz G, Umesono K, Blumberg B, Kastner P, Mark M, Chambon P, Evans RM (1995) The nuclear receptor superfamily, the second decade. Cell 83:835–839
Marcelo CL, Dunham WR (1997) Retinoic acid stimulates essential fatty acid-supplemented human keratinocytes in culture, J Invest Derm 108:758–762
Marshall H, Morrison A, Studer M, Popperl H, Krumlauf R (1996) Retinoids and hox genes. FASEB J 10:969–978
Matikainen S, Ronni T, Lehtanen A, Sareneva T, Melen K, Nordling S, Levy DE, Julkunen I (1997) Retinoic acid induces signal transducer and activator of transcription (STAT)1, STAT2, and p48 expression in myeloid leukemia cells and enhances their responsiveness to interferons. Cell Growth Differ 8:687–698
Mehrel T, Hohl D, Rothnagel JA, Langley MA, Bundman D, Cheng C, Lichti U, Bisher ME, Steven AC, Steinart PM, Yuspa SH, Roop DR (1990) Identification of a major keratincyte cell envelope protein, loricrin. Cell 61:1103–1112

Montminy M (1997) Something new to hang your HAT on. Nature 387:654–655
Morriss-Kay GM, Sokolova N (1996) Embryonic development and pattern formation. FASEB J 10:961–968
Mukherjee R, Davies PJA, Crombie DL, Bischoff ED, Cesario RM, Jow L, Hamann LG, Boehm MF, Mondon CE, Nadzan AM, Paterniti Jr JR, Heyman RA (1997) Sensitization of diabetic and obese mice to insulin by retinoid x receptor agonists. Nature 386:407–410
Nagpal S, Saunders M, Kastner P, Durand B, Nakshatri H, Chambon P (1992) Promoter context – and response element – dependent specificity of the transcriptional activation and modulating functions of retinoic acid receptors. Cell 70:1007–1019
Nagpal S, Patel S, Asano AT, Johnson AT, Duvic M, Chandraratna RAS (1996a) Tazarotene-induced gene 1(TIG1), a novel retinoic acid receptor-responsive gene in skin, J Invest Derm 106:269–274
Nagpal S, Thacher SM, Patel S, Friant S, Malhotra M, Shafer J, Krasinski G, Asano AT, Teng M, Duvic M, Chandraratna RAS (1996b) Negative regulation of two hyperproliferative keratinocyte differentiation markers by a retinoic acid receptor-specific retinoid: insight into the mechanism of retinoid action in psoriasis. Cell Growth Differ 7:1783–1791
Nagpal S, Cai J, Zheng T, Patel S, Masood R, Lin GY, Friant S, Johnson A, Smith DL, Chandraratna RAS, Gill PS (1997a) Retinoid antagonism of NF-IL6: Insight into the mechanism of antiproliferative effects of retinoids in Kaposi's sarcoma. Mol Cell Biol 17:4159–4168
Nagpal S, Patel S, Jacobe H, DiSepio D, Ghosn C, Malhotra M, Teng M, Duvic M, Chandraratna RAS (1997b) Tazarotene-induced gene 2 (TIG2), a novel retinoid-responsive gene in skin, J Invest Derm 109:91–95
Orfanos CE, Zouboulis CC, Almond-Roesler B, Geilen CC (1997) Current use and future potential role of retinoids in dermatology. Drugs 53:358–388
O'Shea JJ (1997) Jaks, STATs, cytokine signal transduction and immunoregulation: are we there yet? Immunity 7:1–11
Pena JC, Fuchs E, Thompson CB (1997) Bcl-x expression influences keratinocyte cell survival but not terminal differentiation. Cell Growth Differ 8:619–629
Perlmann T, Evans RM (1997) Nuclear receptors in Sicily. All in the famiglia. Cell 90:391–397
Perlmann T, Jansson (1995) A novel pathway for vitamin A signaling mediated by RXR heterodimerization with NGFI-B and NURR1 genes. Genes Dev 9:769–782
Pfahl M (1993) Nuclear receptor/AP-1 interactions, Endocrine Rev 14:651–656
Piedrafita FJ, Pfahl M (1997) Retinoid-induced apoptosis and Sp1 cleavage occur independently of transcription and require caspase activation. Mol Cell Biol 17:6348–6358
Ponec M (1994) Lipid biosynthesis. In: Leigh M, Lane B, Watt FM (ed) The keratinocyte handbook. Cambridge University Press, Cambridge, pp 351–363
Rabilloud T, Asselineau D, Bally C, Miquel C, Tarroux P, Darmon M (1988) Study of the human epidermal differentiation by two-dimensional electrophoreses. In: Schaefer-Nielson (ed) Electrophoreses '88. Verlag Chemie, Munich, pp 95–101
Randolph RK, Simon M (1997) Robust metabolism of retinoic acid by dermal fibroblasts has potential to reduce the availability of plasma retinoic acid to the epidermis, J Invest Derm 108:661 (abstract no 739)
Rice RH, Mehrpouyan M, O'Callahan W, Parenteau NL, Rubin AL (1992) Keratinocyte transglutaminase: differentiation marker and member of an extended family, Epith Cell Biol 1:128–137
Rosenthal D, Griffiths C, Yuspa S, Roop D, Voorhees J (1992) Acute or chronic topical retinoic acid treatment of human skin in vivo alters the expression of epidermal transglutaminase, loricrin, involucrin, filaggrin, keratins 6 and 13 but not 1:10, and 14, J Invest Derm 98:343–350
Roy B, Taneja R, Chambon P (1995) Synergistic activation of retinoic acid (RA)-responsive genes and induction of embryonal carcinoma cell differentiation by an

RA receptor α (RAR α)-, RAR β-, or RAR γ in combination with a retinoid X receptor-specific ligand. Mol Cell Biol 15:6481–6487
Rubin AL, Parenteau NL, Rice RH (1989) Coordination of keratinocyte programming in human SCC-13 carcinoma and normal epidermal cells. J Cell Physiol 138:208–214
Saitou M, Sugai S, Tanaka T, Shimouchi K, Fuchs E, Narumiya S, Kakizuka A (1995) Inhibition of skin development by targeted expression of a dominant-negative retinoic acid receptor. Nature 374:159–162
Schadendorf D, Kern MA, Artuc M, Pahl HL, Rosenbach T, Fichtner I, Nurnberg W, Stuting S, Stebut Ev, Worm M, Makkii A, Jurgovsky K, Kolde G, Henz BM (1996) Treatment of melanoma cells with the synthetic retinoid CD437 induces apoptosis via activation of AP-1 in vitro, and causes growth inhibition in xenografts in vivo. J Cell Biol 135:1889–1898
Schulman IG, Li C, Schwabe JWR, Evans RM (1997) The phantom ligand effect: allosteric control of transcription by the retinoid X receptor. Genes Dev 299–308
Shao ZM, Dawson MI, Li XS, Rishi AK, Sheikh MS, Han QX, Ordonez JV, Shroot B, Fontana JA (1995) p53 independent G0/G1 arrest and apoptosis induced by a novel retinoid in human breast cancer cells. Oncogene 11:493–504
Shroot B, Michel S (1997) Pharmacology and chemistry of adapalene. J Am Acad Dermatol 36:S96–S103
Soderstrom M, Vo A, Heinzel T, Lavinsky RM, Yang WM, Seto E, Peterson DA, Rosenfeld MG, Glass CK (1997) Differential effects of nuclear receptor co-repressor (N-CoR) expression leading to retinoic acid receptor-mediated repression supports the existence of dynamically regulated corepressor complexes. Mol Endocrinol 11:682–692
Spiegelman BM, Flier JS (1996) Adipogenesis and obesity: rounding out the big picture. Cell 87:373–389
Stocklin E, Wissler M, Gouilleux F, Grover B (1996) Functional interactions between stat5 and the glucocorticoid receptor. Nature 383:726–728
Sun SY, Yue P, Dawson MI, Shroot B, Michel S, Lamph WW, Heyman RA, Teng M, Chandraratna RAS, Shudo K, Hong WK, Lotan R (1997a) Differential effects of synthetic nuclear retinoid receptor-selective retinoids on the growth of human non-small cell lung carcinoma cells. Cancer Res 57:4931–4939
Sun SY, Yue P, Shroot B, Hong WK, Lotan R (1997b) Induction of apoptosis in human non-small cell lung carcinoma cells by the novel synthetic retinoid CD437. J Cell Physiol 173:279–284
Thacher SM, Rice RH (1985) Keratinocyte-specific transglutaminase of cultured human epidermal cells: relation to cross-linked envelope formation and terminal differentiation. Cell 40:685–695
Thacher SM, Standeven AM, Athanikar J, Kopper S, Castilleja O, Escobar M, Beard RL, Chandraratna RAS (1997) Receptor specificity of retinoid-induced epidermal hyperplasia: effect of RXR-selective agonists and correlation with topical irritation. J Pharmacol Exp Ther 282:528–534
Tong P, Mayes D, Wheeler L (1986) Extracellular calcium alters the effects of retinoic acid on DNA synthesis in cultured murine keratinocytes. Biochem Biophys Res Commun 138:483–488
Tsou HC, Xie XX, Yao YJ, Ping XL, Peacocke M (1997) Expression of retinoid X receptors in human dermal fibroblasts. Exp Cell Res 236:493–500
Tur E, Hohl D, Jetten A, Panizzon R, Frenk E (1995) Modification of late epidermal differentiation in photoaged skin treated with topical retinoic acid cream. Dermatology 191:124–128
Varani J, Nickoloff BJ, Dixit VM, Mitra RS, Voorhees JJ (1989) All-*trans* retinoic acid stimulates growth of adult human keratinocytes cultured in growth-factor deficient medium, inhibits production of thrombospondin and fibronectin, and reduces adhesion, J Invest Derm 93:449–454
Vayssiere BM, Dupont S, Choquart A, Petit F, Garcia T, Marchondeau C, Gronmeyer H, Resche-Rigon M (1997) Synthetic glucocorticoids that dissociate transactiva-

tion and AP-1 transrepression exhibit antiinflammatory activity in vivo. Mol Endocrinol 11:1245–1255
Viallet JP, Ruberte E, Krust A, Zelent A, Dhouailly D (1992) Expression of retinoic acid receptors and dermal-epidermal interactions during mouse skin morphogenesis. In: Morriss-Kay G (ed) Retinoids in normal development and teratogenesis. Oxford University Press, New York, pp 199–210
Viallet JP, Dhouailly (1994) Retinoic acid and mouse skin morphogenesis. I. Expression pattern of retinoic acid receptor genes during hair vibrissa follicle, plantar, and nasal gland development, J Invest Derm 103:116–121
Vivat V, Zechel C, Wurtz J-M, Bourguet W, Kagechika H, Umemiya H, Shudo K, Moras D, Gronmeyer H, Chambon P (1997) A mutation mimicking ligand-induced conformational changes yields a constitutive RXR that senses allosteric effects in heterodimers. EMBO J 16:5697–5709
Warhol MJ, Roth J, Lucocq JM, Pinkus GS, Rice RH (1985) Immunoultrastructural localization of involucrin in squamous epithelium and cultured keratinocytes. J Histochem Cytochem 33:141–149
Weinstein GD (1996) Safety, efficacy and duration of therapeutic effect of tazarotene used in the treatment of plaque psoriasis. Br J Dermatol 135(S49):32–36
Wolbach SB, Howe PR (1925) Tissue changes following deprivation of fat soluble A vitamin. J Exp Med 42:753–777
Wrone-Smith T, Johnson T, Nelson B, Boise LH, Thompson CB, Nunez G, Nickoloff BJ (1995) Discordant expression of Bcl-x and Bcl-2 by keratinocytes in vitro and psoriatic keratinocytes in vivo. Am J Pathol 146:1079–1088
Zavacki AM (1997) Activation of orphan receptor RIP14 by retinoids. Proc Natl Acad Sci USA 96:7909–7914
Zelent A, Krust A, Petkovich M, Kastner P, Chambon P (1989) Cloning of murine alpha and beta retinoic acid receptors and a novel receptor gamma predominantly expressed in skin. Nature 339:714–717
Zhang X-K, Lehmann J, Hoffmann B, Dawson MI, Cameron J, Gaupner G, Hermann T, Tran P, Pfahl M (1992) Homodimer formation of retinoid X receptor induced by 9-*cis* retinoic acid. Nature 358:587–591

Section F
Special Effects

CHAPTER 20
Retinoids in Mammalian Vision

J.C. SAARI

A. Introduction

I. Overview of the Visual Process

The vertebrate visual process is remarkable in its properties. Visual pigments characterized from various species vary in their maximum *wavelength sensitivity* from 380 nm to 620 nm, yet each has the same chromophore, 11-*cis*-retinal (or its 3-dehydro- or 3-hydroxy-derivative), which absorbs at 383 nm (393 nm, 3-dehydroretinal). The *dynamic range* of the visual system is enormous, allowing detection of contrast over an approx. 10^{10}-fold range of background intensities. This is accomplished, in part, by employing highly sensitive rod photoreceptors at low levels of illumination and less sensitive cone photoreceptors at higher levels of illumination. In addition, photoreceptors are capable of *reducing their sensitivity* in response to increasing background illumination (adaptation), contributing to the dynamic range of the visual system. The *extreme sensitivity* of the visual system to photostimulation requires a low background of dark noise, a condition met by enveloping the chromophore within the interior of a protein (opsin) to form a structure with an isomerization activation barrier of 45 kcal/mole (BIRGE 1990), ensuring that thermal isomerization of rhodopsin is a rare event (but see BARLOW et al. 1993). The resulting $t^1/_2$ of the complex for thermal excitation is 420 years (BAYLOR et al. 1984). Some of the *sensitivity* of rod photoreceptors results from their participation in summation pools, in which the output from hundreds of rods converges on one downstream neuron (a ganglion cell). Thus, the human retina is capable of detecting a single photon absorbed by each of five closely spaced rods within 20 ms (HECHT et al. 1942). In contrast, the output from each cone photoreceptor in the foveal region of the primate retina is directed to a single retinal ganglion cell, resulting in lower sensitivity but higher acuity. Finally, the visual system is capable of detecting both *differences in intensity and wavelength* of illumination, providing us with the monochrome world of twilight and night and the polychrome world of daylight. Because an individual visual pigment is "color blind," the visual system relies on comparing the output of two or more channels with overlapping spectral sensitivities. These channels correspond to the visual pigments of our cone photoreceptors, with their maximum sensitivities in the blue, green, and red regions of the spectrum. The reader is referred to Rodieck (1998) for a detailed discussion of the visual system.

It is notable that the foundations for this highly developed sensory system are the interactions of 11-*cis*-retinal with rod and cone opsins and the photoisomerization that occurs upon absorption of light. This chapter focuses on recent developments in our understanding of the molecular roles of retinoids in visual processes in the retina. Phototransduction has been reviewed recently by several authors (Baylor 1996; Lamb 1996; Polans et al. 1996) and is therefore not discussed in detail here. Similarly, the role of retinoic acid in the development of the retina has been summarized recently (Dräger and McCaffery 1997). The reader is referred to several recent reviews of the visual cycle, written from a different perspective (Crouch et al. 1996; Rando 1996; Saari 1994).

II. Anatomy

Discussion of molecular events in the visual cycle requires an introduction to the anatomy of the visual system, because metabolic events occur in two different cell types of the retina. Early anatomists noted the banded structure of the vertebrate retina, and the names they gave to the anatomic features remain in use today (Fig. 1). We now appreciate that this apparently simple arrangement of neuronal layers belies a bewildering complexity of interacting neurons in which three or four types of photoreceptors send their signals to at least ten types of bipolar cells. These cells, in turn, contact approximately 25

Fig. 1A–C. Anatomy. **A** Cross-section of an eye illustrating the location of the retina (*gray*). The boxed area is shown in higher magnification in **B**. **B** Cross-section of monkey retina; Richardson's stain. *CH*, Choroid; *RPE*, retinal pigment epithelium; *OS*, photoreceptor outer segments; *IS*, photoreceptor inner segments; *ELM*, external limiting membrane; *ONL*, outer nuclear layer (photoreceptor nuclei); *OPL*, outer plexiform layer (region of synaptic contacts of photoreceptors and inner retinal neurons); *INL*, inner nuclear layer (nuclei of inner retinal neurons); *IPL*, inner plexiform layer (area of synaptic contacts of inner retinal neurons); *GCL*, ganglion cell layer; *NFL*, nerve fiber layer (axonal processes of ganglion cells. Müller (glial) cells (not apparent in this photomicrograph) span nearly the entire thickness of the retina with their end feet in the NFL and their apical processes extending beyond the ELM. The visual pigments are localized to the photoreceptor outer segments. In this photo, cone photoreceptor cells are easily identified by their large inner segments. Nutrients for the photoreceptor cells are supplied by the choriocapillarus, in which red blood cells are seen within capillary lumina. Regeneration of bleached visual pigments involves movement of the retinoid chromophore of the visual pigments through an extracellular compartment bounded by the tight junctional complexes of the RPE, the selectively permeable junctional complexes of the ELM, and the plasma membranes of photoreceptor, RPE, and Müller cells. *Boxed area* shown in higher magnification in **C**. (Photomicrograph courtesy of Mr. Daniel E. Possin) **C** Electron photomicrograph of the junction of photoreceptor inner and outer segments. The disc membranes of the rod photoreceptor outer segment, which contain the visual pigment rhodopsin, are clearly separated from the plasma membrane of the cell. (Photo courtesy of Dr. Ann H. Milam)

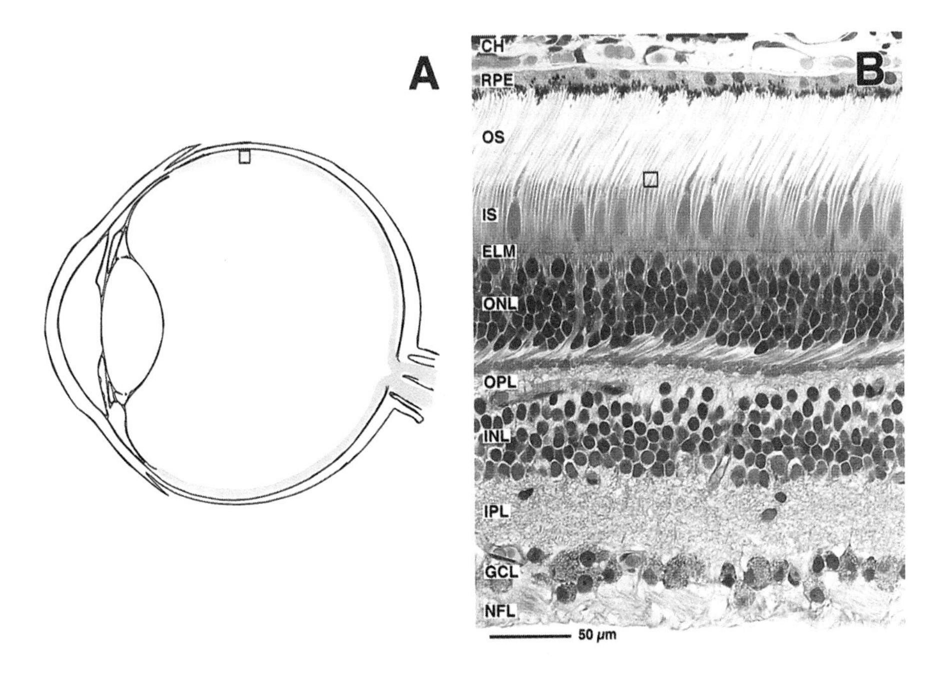
A
B
CH
RPE
OS
IS
ELM
ONL
OPL
INL
IPL
GCL
NFL
50 μm

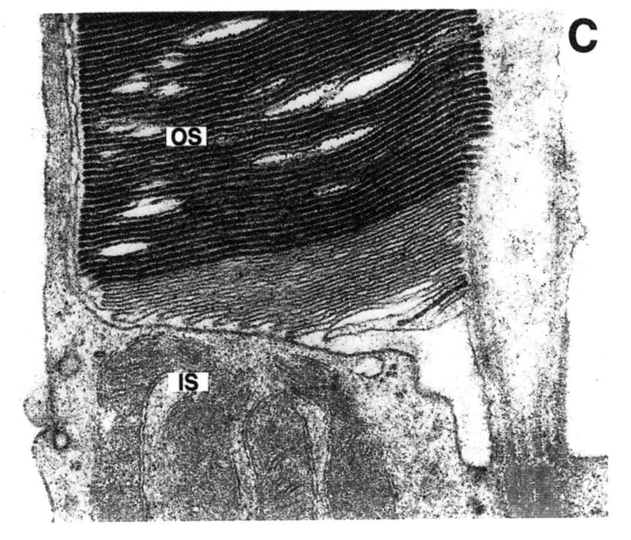
C
OS
IS

ganglion cells types whose axons leave the eye via the optic nerve and go to the brain. In addition, two types of horizontal cells and approximately 40 amacrine cells establish lateral contacts with retinal neurons (DACEY 1996). Müller (glial) cells, which are radially oriented and span nearly the entire thickness of the retina, are important in several retinal functions, and appear to be a site of retinoid metabolism. The microcircuits that result from the interactions of these cells are responsible for complex processing within the retina of the signals received from the photoreceptors.

The photoreceptor cell is a neuron whose role is to respond to light. One hallmark of this specialization is the modification of a cilium into an outer segment in which as many as 2000 flattened discs are stacked (Fig. 1C). The visual pigment (rhodopsin in rod cells) is an integral membrane protein associated with the disc membranes in high concentration (~3 mM). Other components of the phototransduction cascade are found associated with the disc or plasma membranes.

III. Nourishment of the Outer Retina

Neurons within the inner retina are nourished by blood vessels that course through the inner nuclear layer and the ganglion cell layer, whereas the photoreceptor cells of the outer retina must depend on the choroidal vasculature for their supply of nutrients, including glucose, oxygen, and vitamin A (Fig. 1). However, the choroidal vasculature and the photoreceptor cells are separated by a single layer of cells called the retinal pigment epithelium (RPE). These cells are connected by tight junctional complexes, which form a component of the blood retinal barrier. Thus, nutrients such as vitamin A, which are transported by high molecular weight carriers, do not have immediate access to the photoreceptors and must depend on other mechanisms to cross the RPE (discussed in more detail below).

Early investigators noted that regeneration of visual pigments after exposure to illumination required contact between the RPE and the neural retina in eye cup preparations (KÜHNE 1879). We now recognize that several reactions of the visual cycle occur within the RPE and that bleaching and regeneration of visual pigment require movement of the retinoid chromophore from photoreceptor cells to RPE and back again. The extracellular compartment through which the retinoids move is bounded by the tight junctional complexes of the RPE, the selectively permeable junctional complexes of the external limiting membrane and the plasma membranes of photoreceptor, Müller (glia) and RPE cells (Fig. 1B). This translocation process is discussed in more detail in subsequent sections.

RPE plays another role important for retinal function. The components of the photoreceptor outer segments, including proteins, lipids, and retinal, are continually renewed. Assembly of disc membranes occurs at the base of the outer segments. The length of the outer segment is kept approximately constant by a process that involves daily shedding of packets of discs, which are engulfed and phagocytosed by the RPE. Many of the components of the shed discs, including vitamin A and unsaturated lipids, are sent back to the photoreceptor cell for reutilization.

IV. Overview of the Visual Cycle

Both 11-*cis*-retinal and all-*trans*-retinal play critical roles in the visual process. 11-*cis*-retinal binds strongly to opsin via a protonated Schiff base and locks the photoreceptor in a stable, inactive form. In addition, 11-*cis*-retinal primes the receptor for activation because it is readily isomerized to all-*trans*-retinal by visible light. Conformational changes in the receptor, triggered by photoisomerization of 11-*cis*-retinal to all-*trans*-retinal, produce metarhodopsin II (meta II), which activates the phototransduction cascade, hyperpolarizes the cells, and ultimately results in a change in neurotransmitter release that is communicated to other retinal neurons and eventually to the brain. Complexes of all-*trans*-retinal with opsin are likely to set the sensitivity of the receptor, which is reduced by background or bleaching illumination. The activity of meta II must be quenched in order to allow continued processing of information from rapidly changing illumination conditions. Phosphorylation of meta II by rhodopsin kinase initiates the quenching process, and binding of arrestin to the phosphorylated receptor further reduces the activity of the receptor. The final quenching of meta II, however, requires the reduction of all-*trans*-retinal to all-*trans*-retinol by all-*trans*-retinol dehydrogenase. This eliminates the formation of active complexes with all-*trans*-retinal and commits the retinoid to the regeneration pathway.

Regeneration of the 11-*cis*-configuration requires several enzymatic reactions and takes place in the RPE (Fig. 1). Thus, all-*trans*-retinol must leave the photoreceptor cell and diffuse through an extracellular compartment to the RPE cells, where esterification, isomerization, and oxidation to 11-*cis*-retinal take place. To complete the process, 11-*cis*-retinal must leave the RPE and diffuse to the photoreceptor cell, where spontaneous association with opsin produces rhodopsin.

B. Structure and Function of Visual Pigments

I. Interaction of 11-*cis*-Retinal and Opsin

11-*cis*-retinal has long been known to bind to rod and cone opsins with very high affinity. The most dramatic consequence of this interaction is a bathochromic shift of the absorption maximum of the chromophore from 383 nm to 440, 500, 540 or 570 nm for the human S-cone pigment, rod rhodopsin, M-cone pigment, and L-cone pigments, respectively (S (short), M (middle), and L (long)) refer to cone pigments maximally sensitive to blue, green, and red light, respectively). This "fine tuning" of the absorption maximum of the complex produces a visual pigment that absorbs light in the visible range of the spectrum. In addition, the extinction coefficient of the chromophore is increased from 26355 liter/mole-cm in hexane to 40600 liter/mole-cm when bound to opsin, and the quantum efficiency for isomerization is increased from 0.23 to 0.63. The interactions with the protein responsible for these alterations in the properties of 11-*cis*-retinal have long intrigued investigators and recent structural studies have provided a more detailed understanding of these retinal-protein interactions.

A number of studies and experimental approaches have established that the opsin polypeptide crosses the membrane seven times (Fig. 2) in a compact cluster of α-helices (Fig. 3). Three of the helices (4, 6, and 7) are oriented approximately perpendicular to the plane of the membrane, whereas the remaining four (1, 2, 3, and 5) are tilted by 23° to 30°, forming a tighter cluster on the cytoplasmic side and splaying out on the extracellular (intradiscal) side (UNGER et al. 1997). Buried within this cluster of α-helices is the 11-*cis*-retinal, which is covalently attached to lysine 296 of helix 7 via a

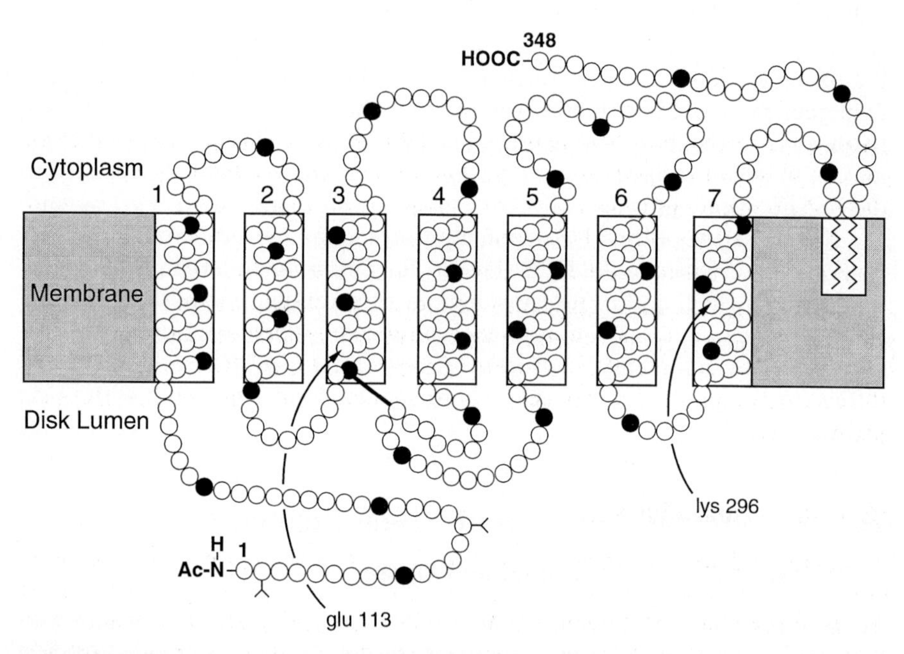

Fig. 2. Diagrammatic representation of the rhodopsin polypeptide chain. The seven transmembrane helices are shown traversing the disc membrane. Lys-296 is the site of attachment of 11-*cis*-retinal to the protein via a protonated Schiff base and Glu-113 provides the counterion. The polypeptide loops on the cytoplasmic side of the membrane are responsible for interaction with the G protein (transducin) and with rhodopsin kinase. Fatty acylation (*wavy lines*) of two adjacent cysteines is believed to provide an additional point of attachment of the receptor to the membrane. *Black bar*, a disulfide bond between two cysteines

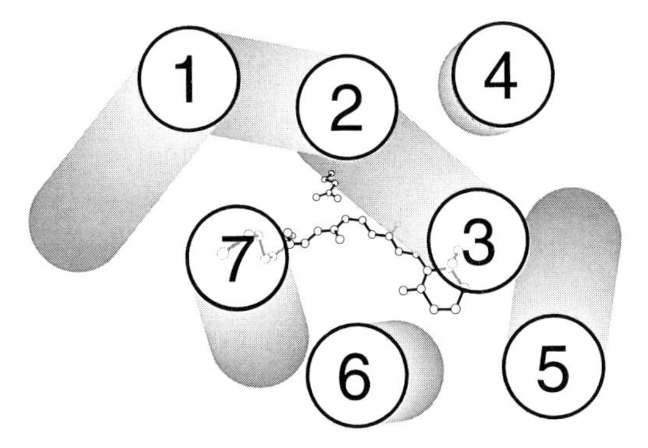

Fig. 3. View of the cluster of seven transmembrane helices from the cytoplasmic side of the disc membrane, generated from the electron density profiles of UNGER et al. (1997). The cylinders represent the courses of the helices over 20 Å. Lys-296, which forms a Schiff base with 11-*cis*-retinal, is about midway down helix 7. Asp-113 is toward the lumenal end of helix 3. The helical bundle is more tightly packed on the cytosolic side and splays out towards the intradiscal (lumenal) side. Exit and entry of retinal via the cytosolic side of the membrane during bleaching and regeneration would involve substantial rearrangement of the helical bundles. An approximate localization of 11-*cis*-retinal in Schiff base linkage to Lys-296 is shown

protonated Schiff base. The counterion for the positive charge of the Schiff base is provided by glutamate 113 of helix 3 (NATHANS 1990; SAKMAR et al. 1989; ZHUKOVSKY and OPRIAN 1989). Protonation of the Schiff base results in a bathochromic shift of the absorption maximum of the complex from ~383 nm to ~440 nm. The additional bathochromic shift of the absorption maximum of rod visual pigments (rhodopsins) to approximately 500 nm (493 nm in humans) is likely to be related to the orientation and distance of the counterion (glu113) from the protonated Schiff base (ERICKSON and BLATZ 1968; JÄGER et al. 1997) and perhaps to interactions with other amino acid side chains.

II. Cone Visual Pigments

As mentioned above, humans have three cone photoreceptor visual pigments in addition to rhodopsin. Human color vision results from the simultaneous stimulation of pairs of these visual pigments with overlapping wavelength sensitivities. Fine tuning of the sensitivity of the L/M visual pigment pair has been particularly well studied. The mechanisms responsible for the ~30 nm difference in spectral sensitivity between M- and L-cone pigments remain controversial; however, it is clear that substitutions of hydroxylated amino acids for nonpolar amino acids at a small number of positions in the M-cone sequence are responsible for generation of an L-cone spectral sensitivity. Analysis of the amino acid sequence of naturally occurring visual pigments in new world monkeys led to the hypothesis that substitutions in the M-cone pigment at positions 180 (Ala to Thr), 277 (Phe to Tyr), and 285 (Ala to Tyr) produce additive shifts in the absorption maxima of 5.3, 9.7, and 15.5 nm, respectively, accounting nearly completely for the bathochromic shift (NEITZ et al. 1991). However, others have challenged this theory and suggested that nonadditive interactions involving additional residues are also important (ASENJO et al. 1994; WILLIAMS et al. 1992).

III. Identification of the Activated Photoproducts

Photoisomerization of 11-*cis*-retinal to all-*trans*-retinal produces three activities in the previously quiescent visual pigment: (a) catalysis of GDP/GTP exchange on a heterotrimeric G protein (also known as transducin), (b) desensitization of the response of the photoreceptor cell, and (c) generation of a conformation of opsin that is a substrate for rhodopsin kinase. Identification of the bleaching intermediate(s) (Fig. 4) responsible for these activities has

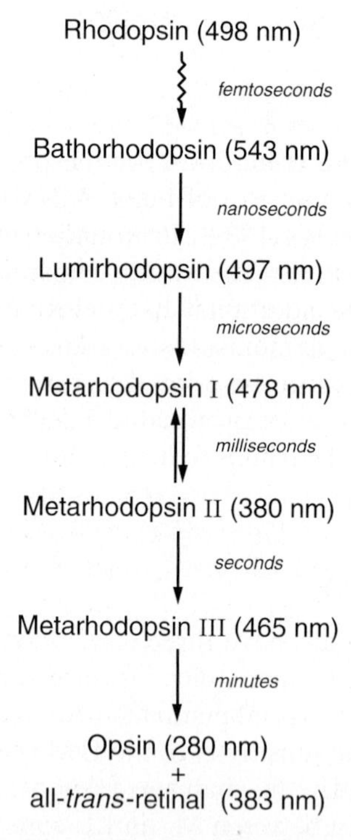

Fig. 4. Scheme depicting the bleaching intermediates of bovine rhodopsin. Conversion of 11-*cis*-retinal in rhodopsin to all-*trans*-retinal is largely complete in bathorhodopsin. The intermediate responsible for activation of the G protein is metarhodopsin II. Hydrolysis of the Schiff base between opsin and retinal is likely to occur at metarhodopsin III

challenged investigators for decades. Exchange of GDP for GTP on the G-protein is catalyzed by meta II, a nonprotonated Schiff base of all-*trans*-retinal and lysine 296 of rod opsin (ARNIS and HOFMANN 1993). This exchange reaction initiates events in the phototransduction cascade that eventually lower the concentration of cGMP and decrease the conductivity of the plasma membrane of the photoreceptor (BAYLOR 1996; LAMB 1996; POLANS et al. 1996). The structure of the specie responsible for the desensitization of the visual system is not known with certainty, but it cannot be meta II because its lifetime is too short (CONE and COBBS 1969). Recent studies of congenital stationary night blindness provide further information that the desensitizing specie cannot be opsin (PALCZEWSKI and SAARI 1997). Meta III is a likely candidate as are non-covalent complexes of all-*trans*-retinal and opsin (JÄGER et al. 1996; BUCZYLKO et al. 1996). Finally, the conformational changes that produce a substrate for rhodopsin kinase have been reviewed recently (PALCZEWSKI 1997).

C. Visual Pigment Regeneration

The term *visual cycle* was used by WALD (1968) to describe the cyclical process of conversion of 11-*cis*-retinal to all-*trans*-retinal by light and the regeneration of the 11-*cis*-configuration by a dark process (Fig. 5). In contrast to phototransduction, a system that has been well characterized over the past decade, the molecular characterization of visual cycle processes and enzymes is in its infancy. As of this writing, there is only one published molecular characterization of a visual cycle enzyme (11-*cis*-retinol dehydrogenase, SIMON et al. 1995), and reactions such as the transfer of retinoid between photoreceptors and RPE and the recombination of opsin with 11-*cis*-retinal remain only vaguely described. This discussion features a brief overview of the cycle and more detailed discussion of recent developments.

The description of the visual cycle which follows is derived from a consideration of the enzymology of individual reactions and metabolism of retinol in isolated RPE or ROS. The visual cycle begins with the absorption of a photon by 11-*cis*-retinal bound to opsin (Fig. 5). All-*trans*-retinal generated in this photoisomerization is reduced to all-*trans*-retinol in the outer segment by all-*trans*-retinol dehydrogenase (RDH). All-*trans*-retinol leaves the outer segment and diffuses to the RPE, where esterification by lecithin:retinol acyltransferase (LRAT) (BARRY et al. 1989; SAARI and BREDBERG 1989) produces all-*trans*-retinyl palmitate and smaller amounts of retinyl oleate and stearate. All-*trans*-retinyl ester is believed to be the substrate for the isomerase, which simultaneously cleaves the ester bond and isomerizes the polyene chain at the 11–12 double bond to generate 11-*cis*-retinol (DEIGNER et al. 1989). Esterification of 11-*cis*-retinol produces 11-*cis*-retinyl ester, which accumulates in the dark in some retinas. Alternatively, 11-*cis*-retinol can be oxidized to 11-*cis*-retinal by 11-*cis*-retinol dehydrogenase (11-RDH). Retinyl esters are hydrolyzed by retinyl ester hydrolase (REH). Finally, 11-*cis*-retinal leaves the

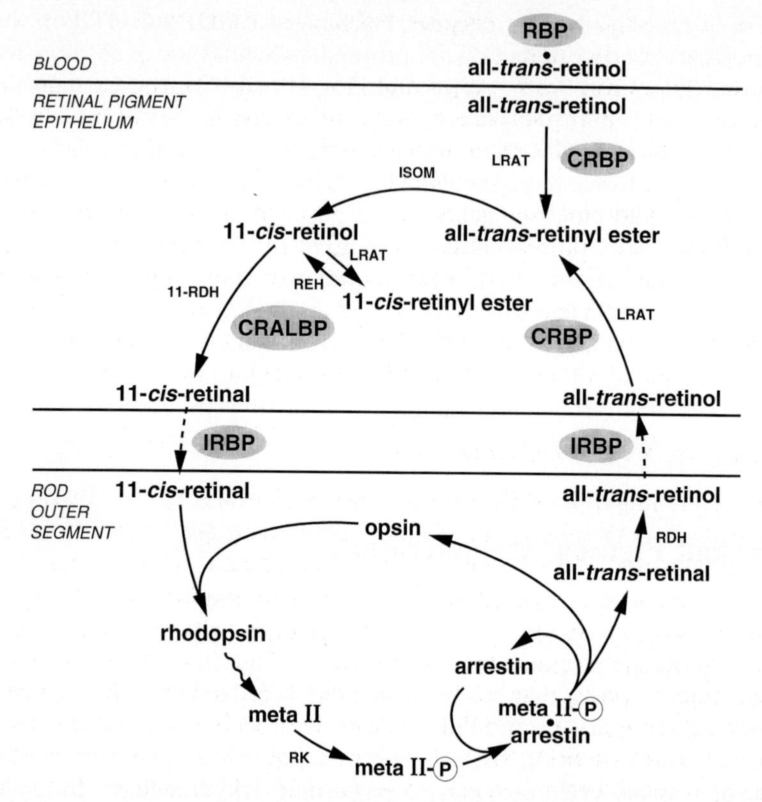

Fig. 5. The visual cycle of vertebrate rod photoreceptors. Reactions occur within two types of cells, the rod photoreceptor outer segments (*ROS*) and the retinal pigment epithelium (*RPE*). The double line represents the extracellular compartment separating these cells. All-*trans*-retinol is taken up from the blood where it is carried by the retinol-binding protein (*RBP*). Rhodopsin is a covalent complex of 11-*cis*-retinal and opsin, a protein. The visual cycle begins with photoisomerization of 11-*cis*-retinal (*wavy line*), which produces metarhodopsin II, the bleaching intermediate responsible for transducin activation. The activity of meta II is quenched by phosphorylation, catalyzed by rhodopsin kinase (*RK*), and binding of arrestin. Reduction of all-*trans*-retinal, catalyzed by all-*trans*-retinol dehydrogenase (*RDH*), results in dissociation of the complex (see text for further details) and recycling of arrestin and opsin. All-*trans*-retinol leaves the ROS and moves to the RPE where it is esterified by lecithin:retinol acyltransferase (*LRAT*). Movement across the extracellular compartment is thought to involve the interphotoreceptor matrix retinoid-binding protein (*IRBP*), and esterification may involve cellular retinol-binding protein (*CRBP*). 11-*cis*-Retinol is produced from all-*trans*-retinyl ester by concerted cleavage of the ester bond and isomerization of the 11-12 double bond by isomerohydrolase. 11-*cis*-Retinol can be esterified by LRAT for storage or oxidized to 11-*cis*-retinal by 11-*cis*-retinol dehydrogenase (*11-RDH*). Cellular retinal-binding protein (*CRALBP*) influences the metabolism of 11-*cis*-retinol by 11-RDH and LRAT in vitro. Retinyl esters are hydrolyzed by retinyl ester hydrolase (REH). Finally, 11-*cis*-retinal leaves the RPE, and moves to the outer segment where conjugation with opsin occurs spontaneously

RPE, diffuses to the photoreceptor cell and spontaneously combines with opsin to generate rhodopsin.

Bleaching intermediates responsible for activation of phototransduction are quenched by phosphorylation, which is catalyzed by rhodopsin kinase (RK), and by binding of arrestin to phosphorylated opsin (PALCZEWSKI 1997). Reduction of all-*trans*-retinal by RDH catalyzes the final step in the quenching of active photointermediates because all-*trans*-retinol cannot activate opsin (HOFMANN et al. 1992). The relationship of opsin phosphorylation and arrestin binding to retinal reduction requires further study. In vitro, these quenching reactions reduce the rate of reduction of all-*trans*-retinal by RDH, possibly by decreasing the rate of release of the substrate from the complex with opsin (PALCZEWSKI et al. 1994) (discussed in more detail below).

I. All-*trans*-Retinol Dehydrogenase

Molecular cloning of an RDH from mouse, bovine, and human retinas established that it is a member of the short chain dehydrogenase/reductase superfamily of enzymes (HESELEER et al. 1998). The enzyme is highly expressed in cone photoreceptors, where abundant amounts of all-*trans*-retinal are generated during daylight hours. The high expression of RDH in cone photoreceptor outer segments suggests that the faster rate of visual pigment regeneration in these photoreceptors may be due, in part, to their increased capacity for reduction of all-*trans*-retinal. Recent work has established that this enzyme catalyzes the rate-limiting step of the visual cycle in rod dominated mouse retina (SAARI et al. 1997). Is there a rod-specific RDH in addition to the RDH described here, or is the cone enzyme expressed in rods at very low levels? This question remains unanswered, but the existence of several related dehydrogenases in retina suggests that there may be photoreceptor cell-specific isozymes. Interestingly, mRNA encoding the enzyme was also found in several other tissues, including heart, liver, kidney, and pancreas. Perhaps the enzyme plays different roles in different tissues, similar to 11-*cis*-retinol dehydrogenase (see discussion below).

All-*trans*-retinal cannot be reduced by RDH until the Schiff base attaching it to opsin is hydrolyzed. The substrate for the hydrolysis reaction is not known with certainty, but it is likely to be a bleaching intermediate with a protonated Schiff base. Meta I has a protonated Schiff base but its lifetime is too short. All-*trans*-retinal is attached to meta II via a nonprotonated Schiff base whose rate of hydrolysis is likely to be slow relative to that of the protonated Schiff base (BRUYLANTS and FEYTMANTS-DE MEDICIS 1970). The Schiff base in meta III is protonated and its life time is long (min), making it tae likely substrate for the hydrolytic release of all-*trans*-retinal.

Phosphorylation of meta II in the presence of arrestin reduces the rate of reduction of endogenous all-*trans*-retinal in outer segment preparations, presumably by reducing the rate of hydrolysis of the Schiff base (PALCZEWSKI et al. 1994). However, it is not known whether phosphorylation alone or phosphorylation and capping with arrestin is responsible for this reduction in dehydrogenase rate. Finally, it is not known at what stage in the regeneration process that phosphates are removed by protein phosphatase 2B.

11-*cis*-retinol dehydrogenase (11-RDH). Molecular characterization of this enzyme followed its detection as a component interacting with RPE65, the putative RBP receptor (SIMON et al. 1995). It was the first retinoid-metabolizing dehydrogenase demonstrated to be a member of the short chain dehydrogenase/reductase superfamily. The original report suggested that this enzyme was expressed uniquely in RPE. However, subsequent studies have shown expression of the enzyme, in lower amounts, in several tissues including breast and liver. A recent study demonstrated 9-*cis*-retinol as a substrate for the enzyme, contrary to previous accounts (SUZUKI et al. 1993), giving rise to the suggestion that it is involved in the pathway for formation of 9-*cis*-retinoic acid (MERTZ et al. 1997). However, when both 11-*cis*- and 9-*cis*-retinols are present, the 11-*cis*-isomer is preferentially processed (SAARI et al. 1998). Thus, 11-RDH could play different roles in different tissues, depending on the presence of other hydrophobic substrates. In the RPE, 11-*cis*-retinol is present in substrate amounts and would be preferentially processed. In nonvisual tissues, where 11-*cis*-retinoids have not been reported, another substrate such as 9-*cis*-retinol, could be processed.

None of the other enzymes of the visual cycle have been characterized in molecular detail as of this writing.

Little information is available regarding the subcellular localization of the visual cycle enzymes in RPE cells. Most investigators of the regeneration cycle have used a microsomal membrane fraction from RPE, which consists of membranes from many parts of the cell. One report (OKAJIMA et al. 1994) of retinoid metabolism in RPE suggests that there are several nonequilibrating pools of retinoids.

II. Movement of Retinoids Between RPE and Photoreceptors

Early investigators of the visual cycle noted the movement of retinoids from neural retina to RPE during the course of bleaching and back again during dark recovery (DOWLING 1960). We now appreciate that several of the critical enzymatic reactions of the visual cycle occur within the RPE and that the RPE must be in apposition to retina before regeneration of bleached visual pigments can occur. During prolonged and full bleaching the retinoid must leave the disc membrane, diffuse through the cytosol in the outer segment, pass through the plasma membrane, diffuse across an extracellular compartment in the retina, and enter the RPE cell where enzymatic processing occurs. During dark adaptation the reverse of this process must occur. The role of retinoid-binding proteins in this process is discussed in a subsequent section.

III. Regeneration in Rods Compared to Cones

Visual pigment regeneration in rods and cones may differ in minor details, and perhaps even fundamentally. (a) Human cone visual pigments regenerate about three times more rapidly than rod visual pigments (ALPERN 1971; ALPERN et al. 1971). This observation could be explained, in part, by differences in the structures of the outer segments of the two photoreceptors. In rods the bulk of the rhodopsin in the disc membranes is separated from the extracellular milieu by the plasma membrane, whereas in cones the lamellar system of membranes is formed by invaginations of the plasma membrane. Thus, cones have a high effective surface area for release and uptake of retinoids, perhaps resulting in a more rapid recycling of visual pigment. It is also possible that an increased amount of RDH in cones, or a more efficient isozyme, could result in more rapid regeneration, because this enzyme catalyzes the rate-limiting step of the cycle (at least in rodents) (SAARI et al. 1997a). Mutagenesis studies have demonstrated that replacement of Glu-122 in rhodopsin by one of the residues present in the corresponding sequence position in L- or M-pigments converted the rates of meta II decay and regeneration into those of the cone pigments. Conversely, substitution of a glutamate into the M-cone pigment sequence changed the kinetics of decay and regeneration into those of rhodopsin (IMAI et al. 1997). Thus, differences in the rates of regeneration and cycling may also be related to intrinsic structural differences in rod and cone visual pigments.
(b) Isolated amphibian rods resensitize only with exogenous 11-*cis*-retinal, whereas isolated amphibian cones resensitize with either 11-*cis*-retinol or 11-*cis*-retinal (JONES et al. 1989). This finding suggests that the complement of enzymes in amphibian cones and rods is different. It is not known whether this finding extends to mammalian rods and cones. (c) Isolated toad rods recover their sensitivity only if 11-*cis*-retinal is applied to their outer segments. Toad cones recover their sensitivity if 11-*cis*-retinal is applied to either the inner or outer segment (JIN et al. 1994). The authors speculate that Müller cells are involved in cone visual pigment regeneration, perhaps as a source of 11-*cis*-retinoid, because the cone inner segment is in close proximity to the apical end of the Müller cells. Müller cells have previously been shown to be a site of retinoid metabolism (see discussion of Müller cells below).

IV. Müller Cells

The Müller glial cells were long considered to play only a structural role in the retina, perhaps serving as guides for the migrating retinal cells during differentiation. However, it is clear from recent investigations that Müller cells play important functional roles in retinal physiology, including acting as sinks for extracellular K^+ released by photoreceptors during stimulation and processing neurotransmitters released by retinal neurons (NEWMAN and

REICHENBACH 1996). In addition, several lines of evidence indicate that Müller cells are involved in retinoid metabolism. For instance, several retinoid-binding proteins have been localized to Müller cells, including cellular retinol-binding protein (CRBP) (BOK et al. 1984), cellular retinal-binding protein (CRALBP) (BUNT-MILAM and SAARI 1983), and cellular retinoic acid-binding protein (CRABP) in some species (MILAM et al. 1990). The presence of CRALBP is particularly interesting because this binding protein in retina is known to carry 11-*cis*-retinol and 11-*cis*-retinal as endogenous ligands (SAARI et al. 1982). Cultured chicken Müller cells have been demonstrated to produce 11-*cis*-retinol from exogenous all-*trans*-retinol (DAS et al. 1992), and cultured rabbit Müller cells are able to synthesize all-*trans*-retinoic acid from exogenous retinol (EDWARDS et al. 1992).

V. Control of the Visual Cycle

A frequent theme in biochemistry is the control of substrate flux through a biological pathway by allosteric or covalent control of a rate-limiting step. Pathways controlled by these mechanisms generally involve substrate concentrations in excess of the K_m for the rate-limiting step. In adult tissues and cells, free retinol concentrations are estimated to be greater than 10 nM (NAPOLI et al. 1991), well below the K_m values reported for retinoid metabolizing enzymes, which are typically 0.1–5 μM. Thus, allosteric regulation of such a retinoid metabolizing enzyme would not be particularly effective. However, the situation in the photoreceptor outer segment is different, because full bleaching of the visual pigment produces ~3 mM all-*trans*-retinal, assuming that the OS is a homogeneous compartment. Concentrations of all-*trans*-retinal within the disc membrane are likely to be even higher. Based on this reasoning, one might expect the flux through the visual cycle to be regulated to ensure that adequate amounts of 11-*cis*-retinal are available during periods of intense visual pigment bleaching. However, it is also possible that flux of retinoids through the cycle is governed simply by availability of substrate.

Two experimental approaches have been used to explore control of the visual cycle: retinal densitometry and retinoid analysis. Retinal densitometry in humans has been used to demonstrate that the rate of cone visual pigment regeneration can be described by an equation that includes only the concentration of opsin; in other words, the reaction is pseudo first order in opsin concentration. An interpretation of this finding is that 11-*cis*-retinal is present in a constant amount and thus does not appear in the rate equation. This suggests that reactions that produce 11-*cis*-retinal in the RPE are fast and that the rate-limiting step is likely to be associated with either conjugation of 11-*cis*-retinal with opsin or perhaps reduction of all-*trans*-retinal (see discussion below). Other studies have demonstrated that the time constant of human cone visual pigment regeneration is sensitive to the type of bleaching regimen, suggesting that a change in the rate-limiting step has occurred (RUSHTON and

Henry 1968). In contrast to these studies of human cones, the time constant for regeneration of the rodent retina is proportional to the extent of visual pigment bleached (Perlmann 1978). The explanation for this species difference is not known but may be related to a comparison of cones with rods or to the two different types of bleaching regimens used.

A second approach has involved measurement of the levels of retinoid intermediates during steady state bleaching of visual pigments. The assumption is that the retinoid that accumulates is the substrate for the slowest reaction. Mice were subjected to constant illumination that produced a steady-state level of ~35% visual pigment bleached, and visual cycle retinoids were quantified by HPLC (Saari et al. 1998). Surprisingly, all-*trans*-retinal was the only retinoid to accumulate under these conditions, and its amount was approximately equal to the amount of the visual pigment bleached. The rate of decay of all-*trans*-retinal in the dark was approximately equal to the rate of regeneration of visual pigment. These results indicate that during steady state cycling all steps subsequent to the reduction of all-*trans*-retinal are rapid and that reduction of all-*trans*-retinal appears to limit the rate of visual pigment recycling in the mouse retina.

Several groups have suggested that cellular retinoid-binding proteins play an important role in regulating retinoid metabolism based on studies in vitro (Ong et al. 1988; Napoli et al. 1991; Saari and Bredberg 1982; Saari et al. 1994). The significance of this hypothesis, which is based on in vitro experiments, will be critically examined when the phenotypes of mice lacking the genes encoding these proteins are thoroughly examined. For instance, CRABP-negative mice develop normally, suggesting that the role of the binding protein is not essential (Gorry et al. 1994; Lampron et al. 1995).

VI. Communication Between Neural Retina and RPE

Efficient coupling of visual pigment bleaching and visual cycle activity would appear to require communication between neural retina and RPE. The tacit assumption behind this statement is that visual cycle activity is not controlled simply by mass action, an assumption that has not been proved. The RPE could sense the rate of bleaching of visual pigment by monitoring the steady state extracellular concentration of K+ released by photoreceptors in response to stimulation. Alternatively, a light-sensitive retinoid-binding protein in the RPE could sense the photon flux impinging on the retina or the release of all *trans*-retinol from photoreceptors or another retinoid in the RPE. Candidate sensors for retinoids in the RPE include peropsin, an opsinlike protein (Sun et al. 1997); retinal G protein coupled receptor (RGR), another opsinlike protein (Shen et al. 1994); and cellular retinal-binding protein (CRALBP), a water-soluble retinoid binding protein (Saari 1994) (see discussion below). Peropsin localizes to the RPE apical processes, in close proximity to the photoreceptor outer segments. Its similarity to opsin suggests that it could bind a retinoid and act as a light sensing receptor; however, attempts to demonstrate retinal binding have not been successful (Sun et al. 1997). RGR binds all-*trans*-retinal to produce a photosensitive complex with absorption maxima of 370 and 467 nm (Hao and Fong 1996). The protein localizes to an intracytoplasmic compartment of RPE and Müller cells and thus could sense retinoids taken up by these cells. Alternately, RGR could function as a photopigment, sensing the photon flux. Finally, CRALBP (see discussion below) could serve as a sensing molecule, signaling the flow of substrate through the regeneration pathway in RPE. A regulatory role has been proposed for yeast SEC14p, a protein related to CRALBP (Skinner et al. 1995).

D. Cellular and Extracellular Retinoid-Binding Proteins

A systematic search in the 1970s for retinoid-binding proteins in retinal extracts resulted in detection of CRBP, CRABP, CRALBP, and IRBP (reviewed in Chader 1989; Saari 1994; Pepperberg et al. 1993). These four water-soluble retinoid-binding proteins are expressed at high levels in retina. Cellular retinol-binding protein (CRBP) and cellular retinoic acid-binding protein (CRABP) occur in a number of other tissues as well. Cellular retinal-binding protein (CRALBP) was originally thought to be expressed uniquely in retina. However, recent evidence shows that it is expressed in several other tissues. Interphotoreceptor retinoid-binding protein (IRBP) appears to be expressed only in retina. The function of these proteins remains controversial; however, their presence in a tissue with a substantial flux of retinoid metabolism and their complement of endogenous retinoids suggest that they are associated with retinoid metabolism or function. Recent developments in our understanding of the function of these proteins is summarized in the sections below.

I. Cellular Retinal-Binding Protein

CRALBP is the founding member of a small family of proteins whose common feature is the binding of hydrophobic substances. Members of this family include α-tocopherol transfer protein (Sato et al. 1993), yeast sec14 protein or phosphatidyl inositol transfer protein (Salama et al. 1990), and a cytosolic protein tyrosine-phosphatase (Gu et al. 1992). CRALBP was originally detected because of its ability to bind 11-*cis*- but not all-*trans*-retinal (Futterman et al. 1977). Subsequent studies demonstrated that CRALBP isolated from RPE was saturated with 11-*cis*-retinal, whereas the protein isolated from bovine neural retina was saturated with a mixture of 11-*cis*-retinol and 11-*cis*-retinal (Saari et al. 1982). Later studies established that the 11-*cis*-retinol bound to CRALBP was preferentially oxidized to 11-*cis*-retinal by enzymes of RPE, whereas the retinoid complexed with bovine serum albumin was preferentially esterified (Saari et al. 1994) by the same enzymes. These in vitro findings suggested that CRALBP could influence the direction of metabolism of visual cycle retinoids in vivo. The ultimate test of this hypothesis will come from an examination of the phenotype of mice with targeted disruption of the CRALBP gene or perhaps from detailed analysis of the visual function of retinitis pigmentosa patients carrying mutations of the CRALBP gene (see discussion below).

Immunocytochemistry was used to demonstrate the localization of CRALBP to Müller and RPE cells of retina (Bunt-Milam and Saari 1983). Subsequent studies revealed CRALBP in the developing pigmented ciliary and iris epithelium of rat (Eisenfeld et al. 1985), in adult bovine ciliary body (Martin-Alonso et al. 1993), and in corneal epithelium (Sarthy 1996). A recent study demonstrated CRALBP in a subpopulation of oligodendrocytes of optic nerve and brain (Saari et al. 1997b). CRALBP from bovine brain was capable of binding exogenous [^{3}H]11-*cis*-retinal; however, no retinoids were found associated with CRALBP purified from this tissue. How can one reconcile the presence of CRALBP in a number of extra-retinal tissues and its specificity for 11-*cis*-retinoids, which do not occur outside the retina and pineal? Does CRALBP from brain bind another hydrophobic ligand that was not detected by the analytical procedures employed? What is the function of CRALBP in brain or other tissues? Only further experiments will answer these questions. In particular, it will be interesting to discover whether myelination and oligodendrocyte morphology are normal in mice lacking a functional CRALBP gene.

II. Interphotoreceptor Retinoid-Binding Protein

IRBP is a large (144 kDa) glycoprotein synthesized in photoreceptor cells and secreted into the extracellular space between photoreceptor and RPE cells, where it is retained by the restrictive pore size of junctional complexes at the external limiting membrane of the retina and the retinal pigment epithelium (Bunt-Milam et al. 1977). There is evidence to suggest that IRBP is involved in the diffusion of retinoids between photoreceptors and RPE cells. The protein is located in precisely the extracellular compartment through which the retinoids must move (Bunt-Milam and Saari 1983), and the major endogenous retinoids associated with IRBP are 11-*cis*-retinal in the dark and all *trans*-retinol in the light (Adler and Spencer 1991). Studies have demonstrated that IRBP is required for efficient regeneration of bleached visual pigment in skate retina in vitro (Duffy et al. 1993).

The mechanism of delivery of retinoids to RPE remains unsettled. Does IRBP shuttle bound retinoids across the extracellular compartment, or does retinoid transfer take place through simple aqueous diffusion of the retinoid, with IRBP acting as a buffer (Ho et al. 1989)? The large size and cigar shape of IRBP (Saari et al. 1985) would appear ill-suited to a shuttle protein, and transfer of the retinol through the aqueous phase is rapid and actually retarded by IRBP in vitro (Ho et al. 1989). If IRBP is necessary for transport of retinoid from photoreceptors to RPE, one would anticipate delayed visual pigment regeneration and a buildup of all-*trans*-retinol in photoreceptors and 11-*cis*-retinal in RPE during steady state illumination in mice lacking a functional IRBP. Analysis of visual cycle activity in such mice will contribute to the resolution of these questions. Similarly, the mechanism of release of 11-*cis*-retinal from RPE has not been determined. IRBP appears necessary for release of 11-*cis*-retinal from RPE in cultured RPE cells (CARLSON and BOK 1992) or delivery of retinol in an eye cup model system (OKAJIMA et al. 1989), suggestive of a receptor-mediated process. However, specific binding of IRBP to RPE membranes has not been reported, and the molecular components of this putative receptor have not been detected.

III. Cellular Retinol-Binding Protein

CRBP is a small (16.5 kDa), water-soluble protein found in nearly all tissues. In retina, it is found in RPE and Müller cells (BOK et al. 1984; SAARI et al. 1984) and purifies fully saturated with all-*trans*-retinol (SAARI et al. 1982). All-*trans*-retinol bound to CRBP is a substrate for lecithin:retinol acyltransferase (LRAT) (ONG et al. 1988) and an NADP-dependent retinol dehydrogenase (KAKKAD and ONG 1988; POSCH et al. 1992), suggesting that it may function as a substrate carrier protein and influence the metabolism of retinol (NAPOLI et al. 1991). An additional functional role may involve the uptake of retinol from blood (discussed in more detail below).

IV. Retinol-Binding Protein

Vitamin A circulates in the blood bound to retinol-binding protein (RBP). As noted previously, the basal plasma membrane of the RPE cells is bathed by the choroidal circulation and is the site of uptake of plasma vitamin A.

The mechanism of vitamin A uptake by cells, including RPE, remains controversial despite decades of research. Evidence has been provided for specific binding of RBP to RPE basal membranes (BOK and HELLER 1976; HELLER 1975), suggesting that vitamin A uptake is receptor-mediated. Indeed, a candidate RBP receptor was isolated following its chemical cross-linking to RBP (BAVIK et al. 1991). This component was called p63 and later RPE65 or p61 by other groups (HAMEL et al. 1993; NICOLETTI et al. 1995). However, detailed analysis of the putative receptor showed it to be localized generally to internal membranes of the RPE cell, including those near the apical plasma membrane (ERIKSSON 1997), and others have assigned different functions to the protein (HAMEL et al. 1993). The association of RPE65 with RBP and with 11-*cis*-retinol dehydrogenase suggests it is involved in retinoid metabolism (SIMON et al. 1995); however, its ability to bind retinoids has not been reported, and its biological function remains controversial (discussed in more detail below).

An alternative mechanism for uptake of retinol involves diffusion of free all-*trans*-retinol through the aqueous media, with the direction of flow determined by the relative affinities and amounts of retinoid-binding proteins in extracellular and intracellular compartments (Noy and Blaner 1991), similar to the model advocated for uptake of plasma steroid hormones. Free all-*trans*-retinol is postulated to be in equilibrium with that bound to RBP and the plasma membrane in the extracellular compartment and with that bound to cellular retinol-binding protein (CRBP) and membranous compartments within the cell. The affinities of CRBP and RBP for all-*trans*-retinol are similar (20 and 13 n*M*, respectively, but see Napoli et al. 1991) whereas the off rate for CRBP is faster than that for RBP (Noy and Blaner 1991). Thus, the partition of retinol between blood and the interior of the cell is postulated to be determined by the levels of CRBP within the cell. An unexplained aspect of this model is that the large amount of membrane present should act as a non-saturable sink for the vitamin.

E. Diseases Associated with Retinoid Metabolism or Function in the Retina

Recently, several inherited retinal diseases have been shown to be associated with mutations in genes encoding enzymes or proteins related to the visual cycle. These diseases are discussed because they have clarified our understanding of visual cycle and quenching reactions and have emphasized the importance of the visual cycle in human visual physiology.

I. Congenital Stationary Night Blindness

Characteristic features of this group of diseases include delayed dark adaptation and complaints of night blindness. Oguchi disease and fundus albipunctatus represent the opposite ends of the spectrum of CNSBs. Oguchi patients exhibit severely delayed dark adaptation and consequent night blindness, with normal kinetics of visual pigment regeneration (Ripps 1982). In two studies, mutations in the gene encoding arrestin (Fuchs et al. 1995) or rhodopsin kinase (Yamamoto et al. 1997) were linked to this condition. A loss of function of either of these two proteins would severely reduce the patient's ability to quench the activated forms of photopigment that result in desensitization of the visual system. Thus, complete resensitization of the visual system would require regeneration of all the bleached visual pigment and would be consistent with the observed phenotypes of delayed dark adaptation and night blindness. In contrast, patients with fundus albipunctatus display delayed dark adaptation and visual pigment regeneration (Ripps 1982). The genetic basis for this condition has not been reported as of this writing; however, defects in a visual cycle enzyme or protein would produce the described phenotype.

II. Autosomal Recessive Retinitis Pigmentosa

Mutations in the CRALBP gene have been found in three family members with early onset of night blindness (3- and 4-years-old), progressive loss of vision, and legal blindness in their late twenties (MAW et al. 1997). Clinical examination of these patients (in their thirties and forties) revealed macular degeneration and an extinguished electroretinogram (ERG), suggesting that the photoreceptors had degenerated. CRALBP has been shown to affect the rates of metabolism of 11-*cis*-retinol by enzymes of the visual cycle *in vitro* (SAARI et al. 1994), decreasing the rate of esterification and modestly enhancing the rate of oxidation of 11-*cis*-retinol. Is the proposed role of CRALBP compatible with the phenotype observed in these patients? Because the disease is recessive, and because the patients begin life with vision, one is led to conclude that CRALBP is not essential for the regeneration of visual pigments. Perhaps a role in protection of 11-*cis*-retinoids is also consistent with the observed phenotype of gradual decline in visual function. A more detailed analysis of visual pigment and dark adaptation kinetics in young patients may resolve this question.

Mutations in the gene encoding RPE65 have been linked to an early onset type of ARRP in which both rod and cone photoreceptors are affected simultaneously (GU et al. 1997; MARLHENS et al. 1997). Patients in one study displayed a childhood-onset form of retinal dystrophy (severe visual impairment at ages 3–7 years) (GU et al. 1997) associated with mutations that were likely to result in nonfunctional RPE65 protein (folding, inactive splice site, or frame shifts). The role of RPE65 in vision is not known; however, the observation that these patients have vision early in life again suggests that RPE65 is not essential to visual function. In contrast to these results, a second study described mutations in the gene encoding RPE65 that were associated with a severe form of disease called Leber's congenital amaurosis (MARLHENS et al. 1997). Patients in this study were examined clinically at ages 20 and 13 years, but apparently had severe visual defects at birth. This phenotype would be consistent with an essential role of RPE65 in visual function. Targeted disruption of the RPE65 gene in mice and analysis of the phenotype should resolve the ambiguities apparent in these two studies.

III. Retinoids and Lipofuscin

Fluorescent deposits, known as lipofuscins, accumulate in a number of tissues with advancing age and in some pathological states. In retina these deposits are particularly noticeable in RPE and are believed to result from incomplete digestion or recycling of photoreceptor discs phagocytosed by RPE. Recently, the structure of a fluorescent component of RPE lipofuscin called A_2-E has been determined to be a condensation product of two moles of retinal and one mole of ethanolamine (SEKAI et al. 1996). Although the authors suggest that the detergentlike properties of A_2-E and its ability to generate reactive oxygen species are likely to contribute to human macular degeneration, it is important to remember that A_2-E is only one of many components in RPE lipofuscin. Addition of retinal to rod outer segment lipids resulted in the production of several fluorescent products with chromatographic migrations similar to those of fluorescent lipofuscin components (WASSELL and BOULTON 1997). The physiological relevance of this observation is not known.

F. Summary

Although the role of retinoids in the visual process would appear to be highly specialized and therefore of little general significance, a more careful analysis indicates that a number of fundamentally important general features of retinoid function have emerged from studies of the visual system. First, the fundamental importance of retinoid-protein interactions was first revealed from studies of the profound effects that opsins have on the chemical properties of 11-*cis*- and all-*trans*-retinals. Second, the substantial flux of retinoid metabolism associated with visual cycle activity has provided a fertile ground for the study of retinoid metabolizing enzymes. Third, while the significance of retinoid binding proteins remains unclear, they are particularly abundant in the retina. and continued study should provide a clearer picture of their importance. Finally, the ability to detect inherited defects in visual pigment regeneration and visual system resensitization with sensitive, noninvasive techniques (ERG, retinal densitometry) offers the potential for a more precise dissection of the significance of visual cycle enzymes and proteins in the visual process.

Acknowledgements. The author thanks Gregory Garwin, Daniel Possin, Jing Huang, and Julie Huber for help with preparation of this article. Studies from the author's laboratory were supported in part by United States Public Health Service grants EY02317, EY01730; by an award from The Royalty Research Fund of the University of Washington; and by a Senior Scientific Investigator Award from Research to Prevent Blindness, Inc. (New York, N.Y.).

References

Adler AJ, Spencer SA (1991) Effect of light on endogenous ligands carried by interphotoreceptor retinoid-binding protein. Exp Eye Res 53:337–346
Alpern M (1971) Rhodopsin kinetics in the human eye. J Physiol (Lond) 217:447–471
Alpern M, Maaseidvaag F, Ohba N (1971) The kinetics of cone visual pigments in man. Vision Res 11:539–549
Arnis S, Hofmann KP (1993) Two different forms of metarhodopsin II: Schiff base deprotonation precedes proton uptake and signaling state. Proc Natl Acad Sci USA 90:7849–7853
Asenjo AB, Rim J, Oprian DD (1994) Molecular determinants of human red/green color discrimination. Neuron 12:1131–1138
Barlow RB, Birge RR, Kaplan E, Tallent JR (1993) On the molecular origin of photoreceptor noise. Nature 366:64–66

Barry RJ, Canada FJ, Rando RR (1989) Solubilization and partial purification of retinyl ester synthetase and retinoid isomerase from bovine ocular pigment epithelium. J Biol Chem 264:9231–9238
Baylor D (1996) How photons start vision. Proc Natl Acad Sci USA 93:560–565
Baylor D, Nunn GJ, Schnapf JL (1984) The photocurrent, noise and spectral sensitivity of rods of the monkey Macaca fascicularis. J Physiol (Lond) 357:575–607
Bavik CO, Eriksson U, Allen RA, Peterson PA (1991) Identification and partial characterization of a retinal pigment epithelial membrane receptor for plasma retinol-binding protein. J Biol Chem 266:14978–14985
Birge RR (1990) Nature of the primary photochemical events in rhodopsin and bacteriorhodopsin. Biochim Biophys Acta 1016:293–327
Bok D, Heller J (1976) Transport of retinol from the blood to the retina: an autoradiographic study of the pigment epithelial cell surface receptor for plasma retinol-binding protein. Exp Eye Res 22:395–402
Bok D, Ong DE, Chytil F (1984) Immunocytochemical localization of cellular retinol binding protein in the rat retina. Invest Ophthalmol Vis Sci 25:877–883
Bruylants A, Feytmants-De Medicis E (1970) Cleavage of the carbon-nitrogen double bond. In: Patai S (ed) The chemistry of the carbon-nitrogen double bond. Interscience, London, p 465
Buczylko J, Saari JC, Crouch RK, Palczewski K (1996) Mechanisms of opsin activation. J Biol Chem 271:20621–20630
Bunt-Milam AH, Saari JC (1983) Immunocytochemical localization of two retinoid-binding proteins in vertebrate retina. J Cell Biol 97:703–712
Bunt-Milam AH, Saari JC, Klock IB, Garwin GG (1977) Zonulae adherentes pore size in the external limiting membrane of the rabbit retina. Invest Ophthalmol Vis Sci 26:1377–1380
Carlson A, Bok D (1992) Promotion of the release of 11-*cis*-retinal from cultured retinal pigment epithelium by interphotoreceptor retinoid-binding protein. Biochemistry 31:9056–9062
Chader GJ (1989) Interphotoreceptor retinoid-binding protein (IRBP): a model protein for molecular biological and clinically relevant studies. Invest Ophthalmol Vis Sci 30:7–22
Cone RA, Cobbs III WH (1969) Rhodopsin cycle in the living eye of the rat. Nature 221:820–622
Crouch RK, Chader GJ, Wiggert B, Pepperberg DR (1996) Retinoids and the visual process. Photochem Photobiol 64:613–621
Dacey DM (1996) Circuitry for color coding in the primate retina. Proc Natl Acad Sci USA 93:582–588
Das SR, Bhardwaj N, Kjeldbye H, Gouras P (1992) Müller cells of chicken retina synthesize 11-*cis*-retinol. Biochem J 285:907–913
Deigner PS, Law WC, Canada FJ, Rando RR (1989) Membranes as the energy source in the endergonic transformation of vitamin A to 11-*cis*-retinol. Science 244:968–971
Dowling JE (1960) Chemistry of visual adaptation in the rat. Nature 188:114–118
Dräger U, McCaffery P (1997) Retinoic acid and development of the retina. Prog Retinal Eye Res 16:323–351
Duffy M, Sun Y, Wiggert B, Duncan T, Chader GJ, Ripps H (1993) Interphotoreceptor retinoid binding protein (IRBP) enhances rhodopsin regeneration in the experimentally detached retina. Exp Eye Res 57:771–782
Edwards RB, Adler AJ, Dev S, Claycomb RC (1992) Synthesis of retinoic acid from retinol by cultured rabbit Müller cells. Exp Eye Res 54:481–490
Eisenfeld AJ, Bunt-Milam AH, Saari JC (1985) Localization of retinoid-binding proteins in developing rat retina. Exp Eye Res 41:299–304
Erickson JO, Blatz PE (1968) N-Retinylidene-1-amino-2-propanol: a Schiff base analog for rhodopsin. Vision Res: 1367–1375
Eriksson U (1997) A membrane receptor for plasma retinol-binding protein (RBP) is expressed in the retinal pigment epithelium. Prog Retinal Eye Res 16:375–399

Fuchs S, Nakazawa Y, Maw M, Tamai M, Oguchi Y, Gal A (1995) A homozygous 1-base pair deletion in the arrestin gene is a frequent cause of Oguchi disease in Japanese. Nat Genet 10:360–362
Futterman S, Saari JC, Blair S (1977) Occurrence of a binding protein for 11-*cis*-retinal in retina. J Biol Chem 252:3267–3271
Gorry P, Lufkin T, Dierich A, Rochette-Egly C, D'ecimo D, Doll'e P, Mark M, Durand B, Chambon P (1994) The cellular retinoic acid-binding protein I is dispensable. Proc Natl Acad Sci USA 91:9032–9036
Gu M, Warshawsky I, Majerus PW (1992) Cloning and expression of a cytosolic megakaryocyte protein-tryosine-phosphatase with sequence homology to retinal-binding protein and yeast SEC14p. Proc Natl Acad Sci USA 89:2980–2984
Gu S-M, Thompson DA, Srikumari CRS, Lorenz B, Finckh U, Nicoletti A, Murthy KR, Rathmann M, Kumaramanickavel G, Denton MJ, Gal A (1997) Mutations in RPE65 cause autosomal recessive childhood-onset severe retinal dystrophy. Nat Genet 17:194–197
Haeseleer F, Huang J, Buczylko J, Possin DE, Lebioda L, Sokal I, Palczewski K, Saari JC (1998) Characterization of a putative photoreceptor all-*trans*-retinol dehydrogenase (RDH). Invest Ophthalmol Vis Sci (to be published)
Hamel CP, Tsilou E, Pfeffer BA, Hooks JJ, Detrick B, Redmond TM (1993) Molecular cloning and expression of RPE65, a novel retinal pigment epithelium-specific microsomal protein that is post-transcriptionally regulated in vitro. J Biol Chem 268:15751–15757
Hao W, Fong HKW (1996) Blue and ultraviolet light-absorbing opsin from the retinal pigment epithelium. Biochemistry 35:6251–6256
Hecht S, Shlaer S, Pirenne M (1942) Energy, quanta and vision. J Gen Physiol 25:819–840
Heller J (1975) Interactions of plasma retinol-binding protein with its receptor. J Biol Chem 250:3613–3619
Ho M-TP, Massey JB, Pownall HJ, Anderson RE, Hollyfield JG (1989) Mechanism of vitamin A movement between rod outer segments, interphotoreceptor retinoid-binding protein, and liposomes. J Biol Chem 264:928–935
Hofmann KP, PulverMüller A, Buczylko J, Van Hooser P, Palczewski K (1992) The role of arrestin and retinoids in the regeneration pathway of rhodopsin. J Biol Chem 267:15701–15706
Imai H, Kojima D, Oura T, Tachibanaki S, Terakita A, Shichida Y (1997) Single amino acid residue as a functional determinant of rod and cone visual pigments. Proc Natl Acad Sci USA 94:2322–2326
Jäger S, Lewis JW, Zvyaga TA, Szundi I, Dakmar TP, Kliger DS (1997) Chromophore structural changes in rhodopsin from nanoseconds to microseconds following pigment photolysis. Proc Natl Acad Sci USA 94:8557–8562
Jäger S, Palczewski K, Hofmann KP (1996) Opsin/all-*trans*-retinal complex activates transducin by different mechanisms than photolyzed rhodopsin. Biochemistry 35:2901–2908
Jin J, Jones GJ, Carter Cornwall M (1994) Movement of retinal along cone and rod photoreceptors. Visual Neurosci 11:389–399
Jones GJ, Crouch RK, Wiggert B, Carter Cornwall M, Chader GJ (1989) Retinoid requirements for recovery of sensitivity after visual-pigment bleaching in isolated photoreceptors. Proc Natl Acad Sci USA 86:9606–9610
Kakkad BP, Ong DE (1988) Reduction of retinal bound to cellular retinol-binding protein (Type II) by microsomes from rat small intestine. J Biol Chem 263:12916–12919
Kühne W (1879) Chemische Vorgänge in der Netzhaut In: Hermann L, (ed) Handbuch der Physiologie, Erster Theil FCW Vogel, Leipzig: (In English as: Chemical processes in the retina. Vision Res 17:1269–1316 [1977])
Lamb TD (1996) Gain and kinetics of activation in the G-protein cascade of phototransduction. Proc Natl Acad Sci USA 93:566–570

Lampron C, Rochette-Egly C, Gorry P, Dolle P, Mark M, Lufkin T, LeMeur M, Chambon P (1995) Mice deficient in cellular retinoic acid binding protein II (CRABPII) or in both CRABPI and CRABPII are essentially normal. Development 121:539–548

Marlhens F, Bariel C, Griffoin J-M, Zrenner E, Amalric P, Eliaou C, Liu S-Y, Harris E, Redmond TM, Arnaud B, Mireille C, Hamel CP (1997) Mutations in RPE65 cause Leber's congenital amaurosis. Nat Genet 17:139–141

Martin-Alonso M-P, Ghosh S, Hernando N, Crabb JW, Coca-Prados M (1993) Differential expression of the cellular retinal-binding protein in bovine ciliary epithelium. Exp Eye Res 56:659–669

Maw MA, Kennedy B, Knight A, Bridges R, Roth KE, Mani EJ, Mukkadan JK, Nancarrow D, Crabb JW, Denton MJ (1997) Mutation of the gene encoding cellular retinal-binding protein in autosomal recessive retinitis pigmentosa. Nat Genet 17:198–200

Mertz JR, Shang E, Piantedosi R, Wei S, Wolgemuth DJ, Blaner WS (1997) Identification and characterization of a stereospecific human enzyme that catalyzes 9-*cis*-retinol oxidation. J Biol Chem 272:11744–11749

Milam AH, De Leeuw AM, Gaur VP, Saari JC (1990) Immunolocalization of cellular retinoic acid-binding protein to Müller cells and/or a subpopulation of GABA-positive amacrine cells in retinas of different species. J Comp Neurol 296:123–129

Napoli JL, Posch KP, Fiorella PD, Boerman MHEM (1991) Physiological occurrence, biosynthesis and metabolism of retinoic acid: evidence for roles of cellular retinol-binding protein (CRBP) and cellular retinoic acid-biding protein (CRABP) in the pathway of retinoic acid homeostasis. Biomed Pharmacother 45:131–143

Nathans J (1990) Determinants of visual pigment absorbance: identification of the retinylidene Schiff's base counterion in bovine rhodopsin. Biochemistry 29:9746–9752

Neitz M, Neitz J, Jacobs GH (1991) Spectral tuning of pigments underlying red-green color vision. Science 252:971–974

Newman E, Reichenbach A (1996) The Müller cell: a functional element of the retina. Trends Neurol Sci 19:307–312

Nicoletti A, Wong DJ, Kawase K, Gibson LH, Yang-Feng TL, Richards JE, Thompson DA (1995) Molecular characterization of the human gene encoding an abundant 61 kDa protein specific to the retinal pigment epithelium. Hum Mol Genet 4:461–525

Noy N, Blaner WS (1991) Interactions of retinol with binding proteins: studies with rat cellular retinol-binding protein and with rat retinol-binding protein. Biochemistry 30:6380–6386

Okajima-T-I, Wiggert B, Chader GJ, Pepperberg DR (1994) Retinoid processing in retinal pigment epithelium of toad (Bufo marinus) J Biol Chem 269:21983–21989

Okajima T-I, Pepperberg DR, Ripps H, Wiggert B, Chader GJ (1989) Interphotoreceptor retinoid-binding protein: role in delivery of retinol to the pigment epithelium. Exp Eye Res 49:629–644

Ong DE, MacDonald PN, Gubitosi AM (1988) Esterification of retinol in rat liver. J Biol Chem 263:5789–5796

Palczewski K (1997) GTP-binding-protein-coupled receptor kinases. Eur J Biochemistry 248:261–269

Palczewski K, Jäger S, Buczylko J, Crouch RK, Bredberg DL, Hofmann KP, Asson-Batres MA, Saari JC (1994) Rod outer segment retinol dehydrogenase: substrate specificity and role in phototransduction. Biochemistry 33:13741–13750

Palczewski K, Saari JC (1997) Activation and inactivation steps in the visual transduction pathway. Curr Opin Neurobiol 7:500–504

Pepperberg DR, Okajima T-I, Wiggert M, Ripps H, Crouch RK, Chader GJ (1993) Interphotoreceptor retinoid-binding protein (IRBP) Molecular biology and physiological role in the visual cycle of rhodopsin. Mol Neurobiol 7:61–84

Perlman I (1978) Kinetics of bleaching and regeneration of rhodopsin in abnormal (RCS) and normal albino rats in vivo. J Physiol (Lond) 278:141–159

Polans A, Baehr W, Palczewski K (1996) Turned on by Ca^{2+}! The physiology and pathology of Ca^{2+}-binding proteins in the retina. Trends Neurosci 19:547–554

Posch KC, Burns RD, Napoli JL (1992) Biosynthesis of all-*trans*-retinoic acid from retinal. J Biol Chem 267:19676–19682

Rando RR (1996) Polyenes and vision. Chem Biol 3:255–262

Ripps H (1982) Night blindness revisited: from man to molecules. Invest Ophthalmol Vis Sci 23:588–609

Rodieck RW (1998) The first steps in seeing. Sinauer, Sunderland

Rushton WAH, Henry GH (1968) Bleaching and regeneration of cone pigments in man. Vis Res 8:617–631

Saari JC (1994) Retinoids in photosensitive systems. In: Sporn MB, Roberts AB, Goodman DS (eds) The retinoids: biology, chemistry, and medicine, 2nd edn. Raven, New York, p 351

Saari JC, Bredberg L (1982) Enzymatic reduction of 11-*cis*-retinal bound to cellular retinal-binding protein. Biochim Biophys Acta 716:266–272

Saari JC, Bredberg DL (1989) Lecithin: retinol acyltransferase in retinal pigment epithelial microsomes. J Biol Chem 264:8636–8640

Saari JC, Bredberg L, Garwin GG (1982) Identification of the endogenous retinoids associated with three cellular retinoid-binding proteins from bovine retina and retinal pigment epithelium. J Biol Chem 257:13329–13333

Saari JC, Bredberg DL, Noy N (1994) Control of substrate flow at a branch in the visual cycle. Biochemistry 33:3106–3112

Saari JC, Bunt-Milam AH, Bredberg DL, Garwin GG (1984) Properties and immunocytochemical localization of three retinoid-binding proteins from bovine retina. Vision Res 24:1595–1603

Saari JC, Garwin GG, Haeseleer F, Jang G-F, Palczewski K (1999) Phase partition and HPLC assays of retinoid dehydrogenases. Meth Enzymol 316, in press

Saari JC, Garwin GG, Van Hooser JP, Palczewski K (1997a) (to be published) Reduction of all-*trans*-retinal limits regeneration of visual pigment in mice. Vision Res 38:1325–1333

Saari JC, Huang J, Possin DE, Fariss RN, Leonard J, Garwin GG, Crabb JW, Milam AH (1997b) Cellular retinal-binding protein is expressed in oligodendrocytes in optic nerve and brain. Glia 21:259–268

Saari JC, Teller DC, Crabb JW, Bredberg L (1985) Properties of an interphotoreceptor retinoid-binding protein from bovine retina. J Biol Chem 260:195–201

Sakmar TP, Franke RR, Khorana HG (1989) Glutamic acid-113 serves as the retinylidene Schiff base counterion in bovine rhodopsin. Proc Natl Acad Sci USA 86:8309–8313

Salama SR, Cleves AE, Malehorn DE, Whitters EA, Bankaitis VA (1990) Cloning and characterization of Kluyveromyces lactis SEC14, a gene whose product stimulates secretory function in Saccharomyces cerevisiae. J Bacteriol 172:4510–4521

Sarthy V (1996) Cellular retinal-binding protein localization in cornea. Exp Eye Res 63:759–762

Sato Y, Arai H, Miyata A, Tokita S, Yamamoto K, Tanabe T, Inoue K (1993) Primary structure of α-tocopherol transfer protein from rat liver. J Biol Chem 268:17705–17710

Sekai H, Decatur J, Nakanishi K, Eldred GE (1996) Ocular age pigment "A2-E": an unprecedented pyridinium bisretinoid. J Am Chem Soc 118:1559–1560

Shen D, Jiang M, Wenshan H, Tal L, Salazar M, Fong HKW (1994) A human opsin-related gene that encodes a retinal-binding protein. Biochemistry 33:13117–13125

Simon A, Hellman U, Wernstedt C, Eriksson U (1995) The retinal pigment epithelial-specific 11-*cis*-retinol dehydrogenase belongs to the family of short chain alcohol dehydrogenases. J Biol Chem 270:1107–1112
Skinner HB, McGee TP, McMaster CR, Fry MR, Bell RM, Bankaitis VA (1995) The Saccharomyces cerevisiae phosphatidylisonitol-transfer protein effects a ligand-dependent inhibition of choline-phosphate cytidylyltransferase activity. Proc Natl Acad Sci USA 92:112–116
Suzuki Y, Ishiguro S-I, Tamai M (1993) Identification and immunohistochemistry of retinol dehydrogenase from bovine retinal pigment epithelium. Biochim Biophys Acta 1163:201–208
Sun H, Gilbert DJ, Copeland NG, Jenkins NA, Nathans J (1997) Peropsin, a novel visual pigment-like protein located in the apical microvilli of the retinal pigment epithelium. Proc Natl Acad Sci USA 94:9893–9898
Unger VM, Hargrave PA, Baldwin JM, Schertler GFX (1997) Arrangement of rhodopsin transmembrane α-helices. Nature 389:203–206
Wald G (1968) Molecular basis of visual excitation. Science 162:230–239
Wassell J, Boulton M (1997) A role for vitamin A in the formation of ocular lipofuscin. Br J Ophthalmol 81:911–918
Williams AJ, Hunt DM, Bowmaker JK, Mollon JD (1992) The polymorphic photopigments of the marmoset: spectral tuning and genetic basis. EMBO J 11:2039–2045
Yamamoto S, Sippel KC, Berson EL, Dryja TP (1997) Defects in the rhodopsin kinase gene in the Oguchi form of stationary night blindness. Nat Genet 15:175–178
Zhukovsky EA, Oprian DD (1989) Effect of carboxylic acid side chains on the absorption maximum of visual pigments. Science 246:928–930

CHAPTER 21

Retinoids and Immunity

C.E. HAYES, F.E. NASHOLD, F. ENRIQUE GOMEZ, and K.A. HOAG

A. Introduction

Nearly 80 years ago at the University of Wisconsin, McCOLLUM and DAVIS (1913) discovered a lipophilic, growth-supporting substance, and STEENBOCK et al. (1921) reported that rats ingesting a diet devoid of this substance, vitamin A, often died of respiratory infections. The increased morbidity and mortality due to infection was quickly confirmed in vitamin A deficient (VAD) rabbits, pigeons, and humans (WERKMAN 1923; GREEN and MELLANBY 1928). Mortality from infection occurred before vitamin A deficiency signs (xerophthalmia, keratinization) appeared (BOYNTON and BRADFORD 1931; McCOY 1934). These early reports showing a causal relationship between vitamin A deficiency and predisposition to infection began to establish the cycle of malnutrition and infection (Fig. 1). GREENE (1933) documented poor antigen-specific immunoglobulin (Ig) responses in VAD animals, and BRESSE et al. (1942) discovered a causal relationship between infection and vitamin A depletion, completing the cycle. The challenge was then and is now to elucidate the molecular basis for the cycle of vitamin A deficiency, immune dysfunction, and infection, and to interrupt the cycle with nutritional intervention.

An eloquent review of the malnutrition – infection cycle stimulated research into the immune system's need for vitamin A (SCRIMSHAW et al. 1968). DRESSER (1968) discovered vitamin A's enhancing effect on Ig responses, foreshadowing further research into adaptive immunity. Adaptive responses require prior antigen exposure, involve B and T lymphocytes, and show immunological memory. COHEN and ELIN (1974) discovered vitamin A's stimulating effect on nonspecific resistance to infection, foreshadowing additional research into innate immunity. The innate responses require no prior antigen exposure and do not show immunological memory. These seminal studies established the principle that vitamin A is a pleiotropic immune system regulator.

This review was written for scientists who are interested in vitamin A and immunity, but who are not necessarily researchers in this field. The review emphasizes recent investigations into the immunobiological and molecular mechanisms that explain the immune system's need for vitamin A. Many excellent reviews summarize older research (DENNERT 1985; NAUSS 1986; HAYES et al. 1994; ROSS and HÄMMERLING 1994; SEMBA 1994; KOLB 1995; ROSS

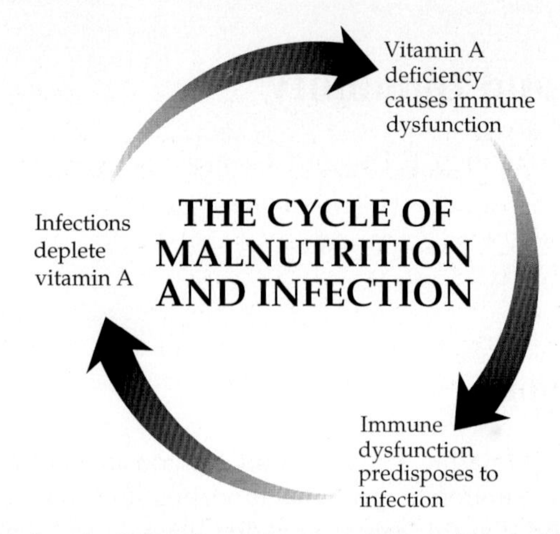

Fig. 1. The cycle of vitamin A deficiency, immune dysfunction, and infection

and Stephenson 1996). The chapter is organized into sections on the immunobiology, biochemistry, and molecular biology of vitamin A as an immune system regulator. For further information on the immune system, we suggest Immunobiology by Janeway and Travers (1996).

B. Innate Immunity

Vitamin A is required by the natural killer (NK) cells and phagocytic cells that mediate innate immunity. The epithelial barrier is an important innate defense against infection; refer to Chap. 8 for a discussion of vitamin A and the maintenance of this barrier.

I. Natural Killer Cells

NK cells lyse some target cells and secrete certain cytokines without antigen sensitization. Basal NK cell activity was reduced in VAD animals, and dietary retinoid repletion restored this activity (Goldfarb and Herberman 1981; Santom et al. 1986; Goettsch et al. 1992). The VAD rats had fewer NK cells and repletion restored the NK cell number to control levels (Ross 1996). However, vitamin A deficiency did not impair the basal lytic efficiency per NK cell or the ability of poly(I:C) to stimulate NK cell proliferation, interferon-gamma (IFN-γ) synthesis, and lytic efficiency in vivo (Ross 1996). In VAD mice, activated NK cell IFN-γ synthesis was actually higher than in controls (Cantorna et al. 1995). Thus, NK cell function is not retinoid-dependent. Further experiments are needed to elucidate how vitamin A supports basal NK cell numbers.

II. Phagocytic Cells

Mononuclear phagocytes (macrophages) and polymorphonuclear neutrophils ingest invading organisms in infection sites and necrotic cells in wounds and inflamed sites. They produce reactive oxygen metabolites and degradative enzymes which kill ingested organisms, extracellular organisms, and tumor cells. Finally, phagocytes secrete cytokines, and macrophages present peptides to initiate adaptive T cell-mediated responses.

Macrophage and neutrophil function are retinoid-dependent. Phagocytic cell numbers were unchanged, but phagocytosis and intracellular killing mechanisms were lower in VAD rats than in controls (ONGSAKUL et al. 1985; WIEDERMANN et al. 1996a; TWINING et al. 1997). The neutrophils from VAD rats had abnormal, hypersegmented nuclei (TWINING et al. 1996), poor in vitro chemotactic ability, and reduced production of oxidative molecules (Twining et al. 1997). However, psoriasis patients ingesting a retinoid had fewer neutrophils in their psoriatic lesions, possibly owing to less recruitment (ORFANOS and BAUER 1983). Retinoid effects on neutrophil chemotaxis and function remain to be clarified.

Retinoid effects on macrophage function also need clarification. Retinoid supplementation in vivo increased macrophage phagocytic and tumoricidal activity (TACHIBANA et al. 1984; MORIGUCHI et al. 1985; HATCHIGIAN et al. 1989), and cytostatic activity against *Mycobacterium tuberculosis* in vitro (CROWLE and ROSS 1989). In contrast, RHODES and OLIVER (1980) observed retinoid inhibition of Fc receptor expression and opsonized cell phagocytosis in human peripheral blood macrophage cultures. Finding a biochemical explanation for these results, and for the decreased chemotaxis, phagocytosis, and oxidative burst in retinoid-depleted macrophages and neutrophils requires additional investigation.

C. Adaptive Immunity

The lymphocyte-mediated adaptive immune mechanisms provide protection beyond the innate defenses. Foreign antigen binding to a lymphocyte's receptors activates proliferation and differentiation of that lymphocyte in a clonal selection process. The B cell receptors bind native antigens, but the T cell receptors only bind antigenic peptides bound to a multiprotein complex displayed on antigen-presenting cells (APC).

I. Antibody-Mediated Immunity

The B lymphocytes produce Ig and provide antibody-mediated immunity (AMI). Multivalent, T-independent antigens such as bacterial cell walls can stimulate B cells to produce IgM. A monovalent, T-dependent antigen can only stimulate B cells to produce Ig when it binds the B cell antigen receptor and at the same time, the B cell receives stimulatory signals from antigen-activated, T helper (Th) lymphocytes. Via membrane-bound proteins and soluble cytokines such as interleukin (IL)-4 and others, the T helper 2 (Th2) cells deliver the most potent signals driving B cell clonal expansion, Ig class switching, and Ig variable region diversification (ABBAS et al. 1996).

GREENE (1933) first documented defective Ig responses to a T-dependent antigen in VAD animals, and DRESSER (1968) found evidence that vitamin A (retinol) stimulated Ig responses in vivo. Many similar reports have been reviewed (DENNERT 1985; ROSS and HÄMMERLING 1994; SEMBA 1994). Here we review studies on antigen-specific Ig responses of particular isotypes and evidence relevant to the target cells for retinoid action.

Vitamin A deficiency diminishes nearly all Ig isotype responses to T-dependent antigens. Depressed antigen-specific serum IgM and IgG responses correlated with deficiency in mice immunized with hemocyanin (SMITH et al. 1987; SMITH and HAYES 1987), in rats immunized with tetanus toxin (ROSS 1996) or β-lactoglobulin (WIEDERMANN et al. 1993b), and in children vaccinated against tetanus (SEMBA et al. 1992). Defective mucosal IgA responses were associated with vitamin A deficiency in children (PALMER 1978; KARALLIEDDE et al. 1979), in rats immunized with dinitrophenyl-bovine gamma globulin (SIRISINHA et al. 1980) or with cholera vaccine (WIEDERMANN et al. 1993a), and in mice innoculated with influenza A virus (STEPHENSEN et al. 1993; GANGOPDAHYAY et al. 1996; STEPHENSEN et al. 1996). The serum IgE response was diminished in VAD rats immunized with β-lactoglobulin (WIEDERMANN et al. 1993b). Where repletion was done, the Ig responses were restored. The foregoing Ig responses that are decreased in VAD animals depend on Th2 stimulation to B cells. In contrast, VAD rats responding to influenza A produced more flu-specific IgG2a than vitamin A-sufficient (VAS) rats (STEPHENSEN et al. 1996); this IgG2a response depends on T helper 1 (Th1) cell stimulation to B cells.

Although Ig responses are generally poor in VAD animals, these animals seem to have no fundamental B cell defect. Most studies reported normal or nearly normal B cell numbers in VAD animals, indicating adequate B lymphopoiesis (ROSS and HÄMMERLING 1994). Memory B cell formation was undisturbed in VAD rats (ROSS 1996). Normal IgM responses were obtained in VAD rats immunized with the T-independent antigens (WIEDERMANN et al. 1993b; ROSS 1996). Normal IgG responses to a T-dependent protein antigen were obtained in VAD rats, if a B cell mitogen was administered with the antigen (ARORA and ROSS 1994). Finally, B cells from VAD mice produced excellent IgG1 responses when VAS Th2 cells provided costimulatory signals (CARMAN et al. 1989). Together, this evidence strongly suggests that B lymphopoiesis, activation, Ig synthesis and secretion, Ig isotype diversification, and memory formation are not explicitly retinoid-dependent.

A Th defect apparently accounts for depressed AMI in VAD animals. The T-dependent but not T-independent Ig responses were depressed. Moreover, the deficiency reduced the Ig-secreting B cell frequency (SMITH and HAYES

1987; WIEDERMANN et al. 1993a; GANGOPADHYAY et al. 1996), but had no effect on response kinetics or the Ig secretion rate per Ig-secreting cell (SMITH and HAYES 1987). The failure to clonally expand the Ig-secreting B cells suggests a Th cell defect, because in the absence of a B cell mitogen, Th2-derived cytokines are required to support B cell clonal expansion (Fig. 2).

CARMAN et al. (1989) provided definitive evidence for a Th2 cell dysfunction in VAD mice. The VAD Th2 cells provided inadequate signals to support VAS B cell IgG1 synthesis. Moreover, Th2 cells were infrequent in VAD mice, and retinyl acetate addition restored the Th2 frequency to control levels (Fig. 2). Thus, the VAD mice had an insufficiency of Th2 cells to drive B cell proliferation and differentiation. Excessive IFN-γ synthesis (see below) may

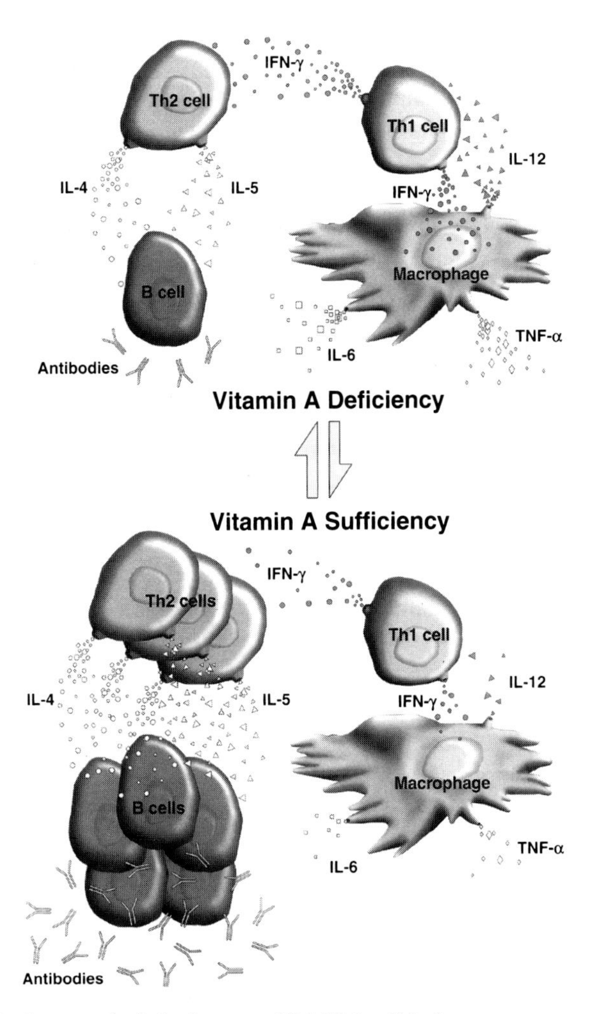

Fig. 2. Vitamin A control of the immune Th1-Th2 cell balance

partially account for this Th2 cell insufficiency, because IFN-γ inhibits Th2 cell development (ABBAS et al. 1996). However, the Th2 insufficiency in VAD rodents is not entirely due to excessive IFN-γ synthesis, because neutralizing the IFN-γ did not restore Th2 cells (CANTORNA et al. 1994). Adding all-*trans*-retinoic acid (atRA) was necessary to fully restore Th2 cells. Thus for unknown reasons, Th2 cell development is retinoid-dependent.

II. Cell-Mediated Immunity

The T lymphocytes direct cell-mediated immunity (CMI). T cells are classified according to their function; surface CD4 molecules demarcate Th subsets, while CD8 molecules denote cytotoxic T lymphocytes (CTL). Th1 cells secrete IFN-γ, tumor necrosis factor-beta (TNF-β), and IL-2, which serves as an autocrine growth factor, while the Th2 cells secrete IL-3, IL-4, IL-5, IL-6, IL-10, and IL-13, with IL-4 serving as an autocrine growth factor (ABBAS et al. 1996). Th1 cells support B cell synthesis of complement-fixing and opsonizing Ig such as rodent IgG2a, macrophage activation, and CTL responses. The Th2 cells support B cell synthesis of most Ig isotypes and eosinophil differentiation and activation, and they inhibit inflammatory responses. There is a regulatory relationship between Th1 and Th2 cells (Fig. 2). While Th1 cell IFN-γ secretion persists, Th2 cell development and function are inhibited; when IFN-γ secretion subsides, Th2 cells develop and their IL-4, IL-10, and IL-13 production limits further Th1 cell activation and function.

JAYALAKSHMI and GOPALAN (1958) first reported poor DTH in VAD children. This was also seen in VAD rodents (SMITH et al. 1987; AHMED et al. 1991; WIEDERMANN et al. 1993b), but not consistently in humans (ROSS and HÄMMERLING 1994). The DTH reaction is very complex, but IL-2-stimulated proliferation of IL-2 receptor (IL-2R) positive Th1 cells is thought to be the driving force. The poor DTH may reflect retinoid-dependent IL-2R expression (Sect. H). Mitogen-driven T cell proliferation in VAD animals has been reviewed (NAUSS 1986; ROSS and HÄMMERLING 1994). Antigen-driven T cell proliferation is a more specific measure of Th function. Studies have reported increases (WIEDERMANN et al. 1993b), no change (CARMAN et al. 1989), and decreases (FRIEDMAN and SKLAN 1989) for this parameter in VAD animals.

The APC have a critical role in T cell activation. The macrophage capacity to stimulate antigen-specific T cell proliferation was unaltered in VAD rodents (CARMAN et al. 1989; WIEDERMANN et al. 1993b). However, atRA treatment in vivo increased CD11c and major histocompatibility complex (MHC) molecule expression on human Langerhans cells and enhanced their capacity to stimulate MHC-mismatched T cells (MEUNIER et al. 1994). In contrast, atRA treatment in vitro decreased the murine dendritic cell capacity to stimulate MHC-mismatched T cells (BEDFORD and KNIGHT 1989) and depressed the murine macrophage capacity to stimulate Th1 cell IFN-γ synthesis (CANTORNA et al. 1994). Further detailed analyses of explicit APC cell types and functions are required to resolve these apparent discrepancies.

The VAD rodents had excessive inflammation (WIEDERMAN et al. 1996b) and Th1 cell IFN-γ synthesis, and atRA inhibited the Th1 cell IFN-γ synthesis (CARMAN et al. 1991, 1992; WIEDERMANN et al. 1993b; CANTORNA et al. 1994, 1995, 1996; CANTORNA and HAYES 1996) (Fig. 2). In VAD mice, IFN-γ and the inducible IL-12 p40 subunit were constitutively synthesized, establishing a microenvironment very conducive for Th1 cell development but inhibitory for Th2 cells (CANTORNA et al. 1995). This effect contributed strongly to the Th1-Th2 imbalance seen in VAD mice (CARMAN et al. 1989; CANTORNA et al. 1994) (Fig. 2).

$CD8^+$ CTL function may also be retinoid-dependent. A 40% decrease in CTL activity was reported in VAD chickens infected with Newcastle disease virus (SIJTSMA et al. 1990). Conversely, treating mice with retinoids enhanced the rejection of a transplantable, MHC-mismatched tumor (DENNERT 1985), which may indicate retinoid-dependent CTL function, since CTL mediate MHC-mismatched tumor rejection. Blocking retinoic acid receptor (RAR)β_2 synthesis in hematopoietic cells compromised the rejection of MHC-mismatched cardiac grafts, due to a decrease in CTL activity (CHEN et al. 1997). The mechanism accounting for the apparent retinoid-dependence of CTL remains to be explored.

D. Immunity to Infections

Excellent reviews have summarized the relationship between vitamin A deficiency, poor immunity, and increased morbidity and mortality due to infections (SOMMER 1993; SEMBA 1994; ROSS and HÄMMERLING 1994). As SOMMER (1993) poignantly described, preexisting vitamin A deficiency increased a young child's relative risk of death from infection three to tenfold, and, conversely, vitamin A supplementation decreased a deficient child's mortality risk about 35%. Globally, about 190 million pre-school-age children are at risk for hypovitaminosis A, and their death rate attributable to this deficiency was estimated to be one child per minute (HUMPHREY et al. 1992). Research reviewed here aims to define human populations and contemporary disease epidemics where vitamin A treatment may be an effective and inexpensive (U.S.$0.02 per capsule) strategy to limit morbidity, mortality, and disease transmission.

I. Measles Infection

ELLISON (1932) reported that high-dose vitamin A therapy halved the measles mortality rate in VAD children. This result has been confirmed (SOMMER 1993; SEMBA 1994). Also, vitamin A therapy reduced morbidity, duration, complications, and hospitalization consequential to measles infection (BARCLAY et al. 1987; HUSSEY and KLEIN 1990). Since measles fatality rates for infants in the developing countries were nearly 50%, the World Health Organization (WHO) recommended administering vitamin A to these infants at the time of measles vaccination (Semba et al. 1995c). When this was carried out for 6 month old infants, the high-dose vitamin A reduced the effectiveness of the measles vaccine (Semba et al. 1995c). The oral vitamin A supplement did not interfere with vaccination in 9 month old infants (Benn et al. 1997; Semba et al. 1997). It is assumed that some type of immune system enhancement underlies the value of high-dose vitamin A therapy in VAD children with measles, but this has not been investigated. It would be valuable to learn what effect this therapy has on measles virus-specific CTL responses, virus-specific neutralizing Ig titer, and viral titer in vivo.

II. Human Immunodeficiency Virus Type 1 Infection

Studies are in progress to determine whether vitamin A treatment may be beneficial in the human immunodeficiency virus type 1 (HIV-1) global pandemic. In populations where severe vitamin A deficiency was prevalent, HIV-1^+ individuals who were vitamin A deficient faced a five- to sixfold increased risk of death (Semba et al. 1993, 1995a; Rwangawoba et al. 1996) and a four- to fivefold increased risk of maternal-fetal transmission (Semba et al. 1994, 1995b; Coutsoudis et al. 1995; Greenberg et al. 1997) over those who were not deficient. In a population where vitamin A deficiency was not prevalent, there was no correlation between serum retinol and maternal-fetal HIV-1 transmission (Burger et al. 1997).

The prevalence of hypovitaminosis A among HIV-1^+ individuals is uncertain, because disparate serum retinol values have been taken as deficient (Semba 1997). According to WHO guidelines, serum retinol levels lower than 0.35 μmol/l (10 μg/dl) are severely deficient, 0.35–0.7 μmol/l is mildly deficient, and levels higher than 0.7 μmol/l are sufficient (Burger et al. 1997). In mice (whose serum retinol levels are similar to humans), immune impairments occurred at about 0.7 μmol/l (Smith et al. 1987). Using serum retinol below 1.05 μmol/l as the cutoff, the prevalence of hypovitaminosis A among HIV-1^+ individuals in the United States was below 5% in homosexual men (many of whom took vitamin supplements), 29% in injection drug users, and 22% in pregnant women (Semba 1997). The prevalence was greater in economically disadvantaged countries, 37% in Ugandan adults, 63% in pregnant Malawi women, and 75% in lactating Haitian women (Semba 1997). Others reported similar hypovitaminosis A rates (Karter et al. 1995; Tang et al. 1997) and β-carotene deficiency (Sappey et al. 1994; Phuapradit et al. 1996). Although the cutoff point for deficiency may be debated, it is clear that economically disadvantaged, HIV-1^+ persons, especially pregnant and lactating women are at some significant risk for hypovitaminosis A.

Controlled vitamin A supplementation trials are in progress to learn how supplementation may affect HIV-1 morbidity, mortality, and maternal-fetal transmission (Semba 1997). Unlike expensive therapeutics, this intervention strategy could be applied globally, should it prove to be effective. However, there is a possible risk of treating pregnant women with vitamin A, in that excessive vitamin A levels can cause developmental defects in the fetus. The outcomes of broad-based supplementation trials will certainly be important. In addition, it would be helpful to have more mechanistic information regarding possible impairments in protective antiviral immune responses as a function of vitamin A status, and how these measures may improve with supplementation. The HIV-1-specific CTL responses and mucosal HIV-1-neutralizing IgA responses would be most important to study. This knowledge may reveal new strategies to combat the HIV-1 pandemic.

III. Parasitic Infections

There is substantial evidence that low serum retinol levels (<1.05 μmol/l) are common in children infected with parasites, but no causal relationship between the parasitic infection and the deficiency has been demonstrated. Vitamin A deficiency was seen during infection with *Plasmodium* species (malaria) (WOLDE-GEBRIEL et al. 1993; DAVIS et al. 1994; BINKA et al. 1995; DAS et al. 1996; FRIIS et al. 1997), *Schistosoma mansoni* (schistosomiasis) (FRIIS et al. 1996; FRIIS et al. 1997), *Ascaris lumbricoides* (CURTALE et al. 1994) and *Onchocerca vovulus* (river blindness) (STOREY 1993). AHMED et al. (1993) reported that less than 1% of a vitamin A supplement given to ascaris-infected children was recovered in their stools, suggesting good absorption. Although vitamin A supplementation improved the serum retinol levels, it did not reduce mortality due to malaria (BINKA et al. 1995) or ascariasis (TANUMIHARDJO et al. 1996). In a mouse model of *Trichinella spiralis* infection, the VAD and non-VAD mice did not differ in the control of this helminthic parasite (CARMAN et al. 1992). Taken together, these results seem to indicate no beneficial effects of vitamin A on immunity to parasites. Direct effects of vitamin A supplementation on parasite growth have not been studied.

E. Autoimmunity

I. Rheumatoid Arthritis

Susceptibility to rheumatoid arthritis, a debilitating disease characterized by chronic joint destruction, depends on genetic and environmental factors (FELDMANN et al. 1996). A major arthritis susceptibility locus maps in the human MHC. The MHC class II molecules are thought to present arithogenic peptides to autoimmune T lymphocytes, thereby initiating autoimmune-mediated pathology.

Vitamin A deficiency exacerbated arthritis in rodents (WIEDERMANN et al. 1995; WIEDERMANN et al. 1996a; CANTORNA and HAYES 1996). The VAD mice innoculated with *Borrelia burgdorferi*, the causative organism in Lyme arthritis, had elevated amounts of IL-12, IFN-γ, and TNF, that peaked immediately before acute arthritis symptoms appeared (CANTORNA and HAYES 1996). The VAD rats innoculated with *Staphylococcus aureus* had a higher arthritis incidence, decreased phagocytosis and complement lysis activity, and increased IFN-γ synthesis accompanying arthritis development (Wiedermann et al. 1995; Wiedermann et al. 1996a). Thus, failing innate immune defense mechanisms and heightened pro-inflammatory cytokine synthesis were associated with vitamin A deficiency in these infectious arthritis models.

Vitamin A has shown some value in treating arthritis. Feeding 13-*cis*-retinoic acid (13-cRA) to rats reduced both developing and established adjuvant arthritis and streptococcal cell wall-induced arthritis (Brinkerhoff et al. 1985). Retinoid inhibition of collagenase gene expression may partially explain suppression of joint destruction (Vincenti et al. 1996). Retinoids may also decrease arthritic inflammation through combined effects on innate immune system cells and on epithelial, endothelial, and connective tissue cells in the inflammation site.

Two clinical trials have tested retinoids in arthritis patients. Etretinate treatment reduced joint swelling and pain in patients with psoriatic arthritis (Ciompi et al. 1988). However, *N*-(4-hydroxyphenyl)-retinamide (4-HPR) provided no benefit to patients with severe, long-standing rheumatoid arthritis (Gravallese et al. 1996), perhaps reflecting the biological activities of 4-HPR, or the advanced stage of disease or the disease etiology. Additional experimentation directed at retinoid mechanisms of action in arthritis is needed.

II. Multiple Sclerosis

Multiple sclerosis (MS) is an autoimmune, demyelinating disease of the central nervous system (CNS) (Steinman 1996). The cause of the disease is uncertain, but heritable susceptibility genes, including one in the human MHC, are contributing factors. Poorly defined environmental factors also play a disease-determining role, and nutrition may be one of them (Hayes et al. 1997).

Experimental autoimmune encephalomyelitis (EAE) is a useful and widely studied model for MS. This disease is induced by immunizing animals with CNS proteins such as myelin basic protein, or by passively transferring myelin-specific Th1 cells into naive animals. When 13-cRA was fed to Lewis rats after immunization but before EAE onset, it suppressed the inflammatory cell infiltration into the CNS and the disease took a delayed and less severe course (Massacesi et al. 1987). Incubating the myelin-specific T cells with 13-cRA or atRA in vitro inhibited IL-2 synthesis and proliferation, and the capacity of these cells to transfer EAE in rodents (Massacesi et al. 1991; Racke et al. 1995). The atRA incubation in vitro also decreased IFN-γ and TNF-α mRNA, and increased IL-4 mRNA (Racke et al. 1995) (Fig. 2). As a treatment for established EAE, 4-HPR provided little benefit (Racke et al. 1995). A myelin-specific Th1 cell clone capable of transferring EAE showed atRA inhibition of IFN-γ synthesis when the CD28-costimulation pathway was activated (Cantorna et al. 1996). Thus, atRA may protect against EAE by diminishing Th1 cell inflammatory cytokine biosynthesis and/or stimulating nonencephalitogenic Th2 cells directly or indirectly (Fig. 2).

F. Lymphocyte Turnover

Retinoid effects on hematopoietic cell differentiation are described in Chap. 9. Here we briefly review new information on thymocytes and T cells.

Lymphocyte turnover is tightly controlled, and vitamin A may be one of the regulators. Bone marrow-derived lymphopoietic cells migrate into the thymus where they generate mature T lymphocytes. Thymic atrophy occurred during vitamin A deficiency in some species, and retinoid repletion restored the intrathymic small lymphocyte numbers (Ross and Hämmerling 1994). However, retinoic acid (RA) addition reduced the accumulation of $CD4^+CD8^+$ T cells in vitro (Meco et al. 1994; Foerster et al. 1996).

The decrease in $CD4^+CD8^+$ T cell development may reflect retinoid downregulation of *Bcl-2* gene expression (Agarwal et al. 1997; Bruel et al. 1997). The *Bcl-2* gene is required for thymocyte maturation. During positive selection through the T cell antigen receptor, $CD4^+CD8^+$ thymocytes increase their *Bcl-2* expression, blocking apoptosis. Knocking out *Bcl-2* impeded lymphopoiesis (Matsuzaki et al. 1997); increasing *Bcl-2* expression promoted T cell differentiation in the absence of T cell receptor rearrangement, but did not fully substitute for positive selection (Linette et al. 1994). Thus, retinoid downregulation of *Bcl-2* may have promoted $CD4^+CD8^+$ thymocyte apoptosis.

In contrast, retinoid addition inhibited activation-induced apoptosis of mature T cells in vitro (Iwata et al. 1992). Activation-induced T cell apoptosis ensures that T cell-mediated immune reactions are self-limiting (Abbas et al. 1996). Engagement of Fas by Fas ligand can initiate apoptosis; Fas protein is expressed on pluripotent stem cells, multipotent progenitors, pro-T and pre-T cells, most thymocytes and a subset of mature T lymphocytes (Li et al. 1996). Retinoid addition decreased Fas ligand gene transcription through an RAR-RXR-dependent mechanism, but did not change Fas signaling (Bissonnette et al. 1995; Yang et al. 1995a; Yang et al. 1995b). The existing data suggest retinoids may have profound effects on developing T cells, but much work is needed to elucidate the pathways involved.

G. Vitamin A Metabolism

I. Active Metabolites for Immune Function

Excellent discussions of vitamin A supplementation and repletion studies have been published (Ross and Hämmerling 1994; Ross 1996). Retinyl esters have supported all immune functions examined in vivo. However, the RA isomers, which cannot be reduced to retinal or retinol (Chap. 2), may be more active metabolites for immune function. RA isomers given in vivo supported innate immunity (phagocytic and cytostatic function; NK cell number), AMI (Ig responses), and several measures of CMI (DTH; T cell proliferation; CTL activity against MHC-incompatible tumors) (Ross and Hämmerling 1994;

Ross 1996). Quantitative comparisons of retinoids in support of IgG responses (CHUN et al. 1992) and as inhibitors of Th1 cell IFN-γ synthesis (CANTORNA et al. 1996) yielded a ranking of RA > retinal >> retinol. In addition, immune system cells express RAR and retinoid X receptor (RXR) isoforms (Sect. H). Together, these data suggest that RA isomers are immune regulators.

Retinol metabolism to an active metabolite may be defective in the VAD animal. The VAD mice fed a small daily amount of retinyl acetate did not make an IgG response to an antigen given 24 hr after repletion began, although at the time of immunization, the serum retinol had reached a normal level (CHUN et al. 1992). Failing Ig responses were also seen in VAD rats given a single retinol dose to normalize serum retinol (PASATIEMPO et al. 1992). That lymphoid cells did not utilize the retinol available in the blood may indicate that they depend on active metabolite synthesis in some distant organ, and this metabolism may be inefficient in the VAD animal. In VAD rats, the liver lacked lecithin:retinol acyltransferase activity (RANDOLPH and ROSS 1991), and therefore a large retinyl acetate dose did not contribute to hepatic retinyl palmitate stores (BHAT and LACROIX 1983). Hepatic retinyl palmitate, which may be the substrate for active metabolite synthesis, was low and increased only slightly in the low retinyl acetate repletion protocol (CHUN et al. 1992). A striking new finding was the sequestration of retinol in the bone marrow of VAD rats (TWINING et al. 1996); possibly retinol is directed first to bone marrow to support functions such as hematopoiesis, and second to liver for storage.

The research on metabolites that support immune function may have important implications for human vitamin A repletion and vaccination protocols. Single dose repletion protocols coupled to vaccination may be ineffective. Adequate repletion may require an initial dose to upregulate the necessary enzymes, followed by a second dose to support active metabolite synthesis and delivery to lymphoid cells before vaccination.

Based on in vitro studies, 14-hydroxy-4,14-*retro*-retinol (14-HRR), has been suggested as an active vitamin A metabolite for immune function (BUCK et al. 1991; Chap. 3). It is theorized to bind to a cytoplasmic receptor that translocates to the nucleus to control gene expression. The suggested 14-HRR requirement to sustain lymphopoiesis appears inconsistent with the fact that animals maintained in the vitamin A-deficient state do not become B or T cell deficient (ROSS and HÄMMERLING 1994). To verify the 14-HRR hypothesis, it will be important to identify the enzymes involved in 14-HRR synthesis, to isolate the putative 14-HRR receptor, and to test 14-HRR in VAD animals for its ability to support retinoid-dependent lymphocyte functions.

II. Metabolism During Infection or Inflammation

Significantly decreased serum retinol is often correlated with infection (SEMBA 1994) and was previously attributed to vitamin A malabsorption (BREESE et al. 1942) and vitamin A excretion (LAWRIE et al. 1941 and references therein).

Adults with pneumonia or sepsis, and children with acute diarrhea, excreted significantly higher amounts of retinol and retinol binding protein than uninfected controls (STEPHENSEN et al. 1994; ALVAREZ et al. 1995). These studies confirmed accelerated retinol losses as a result of infection. New evidence indicates that decreased hepatic retinol export may also contribute to depleted serum retinol during infection. Serum retinol dropped significantly but transiently when VAS animals were infected with *Borrelia burgdorferi* (CANTORNA and HAYES 1996) or influenza A virus (STEPHENSEN et al. 1996). Serum retinol and retinol binding protein also decreased transiently in surgical patients (LOUW et al. 1992) and lipopolysaccharide-injected rats (ROSALES et al. 1996) undergoing the acute-phase response. The lipopolysaccharide-injected rats had diminished liver and kidney retinol binding protein and transthyretin, and reduced liver retinol binding protein mRNA (ROSALES et al. 1996). This evidence points to decreased hepatic retinol export during an acute-phase response induced by surgery or infection. A review of the recommended vitamin A intake during infection is clearly needed, as is further study of vitamin A malabsorption and metabolism during infection, and the consequences for immune system function.

H. Mechanisms of Vitamin A Action in the Immune System

The molecular mechanism of atRA action is thought to be transcriptional regulation of gene expression (Chap. 5). The RA isomers bind to nuclear RAR and RXR, members of the steroid receptor superfamily. These receptors regulate gene expression through RA responsive elements (RARE). The expression of RAR and RXR in immune system cells suggests that RAR and RXR function as transcriptional regulators in these cells. Despite the appeal of this hypothesis, to date no RARE has been defined in a gene expressed by an immune system cell.

There are three RAR and three RXR isotypes (α, β, and γ) encoded separately in the genome. Within an isotype, there are sometimes isoforms (α_1, α_2) generated by alternative transcript splicing. Splenocytes reportedly expressed RARα, RARβ, RXRα, and RXRβ, but not RARγ or RXRγ (DE THE et al. 1989; WAN et al. 1992; MANGELSDORF et al. 1992; ZHUANG et al. 1995). The immature $CD3^+CD4^-CD8^-$ thymocytes expressed RARα and RARγ1, but not RARβ, whereas the mature splenic T lymphocytes expressed only RARα (FRIEDMAN et al. 1993; MECO et al. 1994). The RARβ was detected in peritoneal macrophages and myeloid lineage cells, but not in splenic B or T lymphoctyes (CHEN et al. 1997; DE THE et al. 1989).

There is now considerable evidence for retinoid control of the immune response via regulated cytokine expression (Fig. 2). As reviewed in Sects. B and C, cells of the innate and adaptive immune systems mutually regulate one another through cell-to-cell interactions and a network of secreted cytokines.

The cytokine network allows the innate system to orchestrate an appropriate adaptive immune response, and permits the adaptive immune system to retain immunological memory without perpetuating unnecessary immune reactions.

The proinflammatory cytokines are released immediately following wounding or infectious insult. Monocyte/macrophages release IL-1 and TNF-α following exposure to platelet-derived transforming growth factor-β (wounding) or direct bacterial stimulation (infection). The IL-1 and TNF-α elicit tissue fibroblast IL-6 release, triggering local inflammation and immunocyte recruitment, and the hepatic acute-phase response (BAUMANN and GAULDIE 1994). These cytokines aid Th sensitization, but their prolonged expression, especially IL-1 and TNF-α, can damage tissue. Research in vitro has shown atRA-dependent inhibition of TNF-α in B cells, macrophages (BLOMHOFF et al. 1992; MEHTA et al. 1994), and immune Th1 cells (RACKE et al. 1995), and inhibition of IL-6 (BLOMHOFF et al. 1992; GROSS et al. 1993) (Fig. 2). Regarding macrophage IL-1 synthesis, some researchers observed atRA-dependent increases (MORIGUCHI et al. 1985; TRECHSEL et al. 1985; GOETTSCH et al. 1992), in one case at the transcript level (MATIKAINEN et al. 1991), others found no effect (MATIKAINEN et al. 1991; GROSS et al. 1993), and others reported suppression (GROSS et al. 1993). It will be interesting to learn whether any of these retinoid activities reflect proinflammatory cytokine gene expression control.

Retinoids also control some Th1 cell cytokine responses. The atRA effects on IL-2 synthesis are unclear. CARMAN et al. (1992) reported no change in antigen-specific T cell IL-2 production in VAD mice, while RACKE et al. (1995) demonstrated an atRA-dependent decrease in antigen-specific T cell IL-2 mRNA. However, reports have consistently found atRA-dependent IL-2R increases on T cells (SIDELL and RAMSDELL 1988; FEGAN et al. 1995) and at the mRNA level (SIDELL et al. 1993, 1997).

There is a negative effect of atRA on IFN-γ synthesis in all cells that produce it (CANTORNA et al. 1995) (Fig. 2). The IFN-γ is upregulated in VAD animals and downregulated by atRA addition (see Sect. C); the atRA acted on an unidentified component of the CD28 costimulation pathway (CANTORNA et al. 1996). CIPPITELLI et al. (1996) mapped a region of the human IFN-γ promoter that was required for atRA inhibition of phorbol ester-stimulated IFN-γ synthesis in Jurkat lymphoma cells; this region did not harbor an RARE, and did not bind RAR-RXR complexes.

Retinoid-dependence of Th2 cytokine synthesis has been reported. Decreased IL-4, IL-5, and IL-10 responses to antigen stimulation were seen in VAD animals (CARMAN et al. 1992; CANTORNA et al. 1994). Dose-dependent increases in mRNA were observed for IL-4 (RACKE et al. 1995) and IL-5 (CANTORNA et al. 1994), but not IL-10 (CANTORNA et al. 1994), when atRA was added to immune lymphocytes in vitro. Regarding IL-5, atRA addition increased the IL-5 secreting cell frequency, without altering the secretion rate (CANTORNA et al. 1994). Thus, it is not entirely clear whether atRA alters Th2 cell frequency or cytokine gene expression or both (Fig. 2).

I. Summary and Future Research Directions

Vitamin A deficiency is associated with high morbidity and mortality due to infectious disease, and may tend to exacerbate inflammatory diseases. Research suggests that failing innate defense mechanisms, inadequate Th2 support for B lymphocyte Ig responses, and excessive inflammatory Th1 cell cytokine synthesis characterize the VAD animal. In contrast, vitamin A supplementation restores innate immunity and the Th1-Th2 balance. It facilitates Th2 development thereby enhancing B cell Ig responses, and depresses many of the proinflammatory macrophage functions and the excessive inflammatory Th1 cell functions (Fig. 2). Thus, vitamin A supplementation is correlated with strong natural defenses against infections, increases in vaccination efficiency, and diminution of inflammatory disease.

Despite the progress in understanding the immunobiological role vitamin A plays in supporting the immune system, detailed molecular mechanisms for the identified effects are largely unknown. Further biochemical details are needed regarding the alterations in vitamin A metabolism brought about by vitamin A deficiency and/or infection. A mechanistic explanation is needed for retinoid support of basal NK cell number, and for the decreased chemotaxis, phagocytosis, and oxidative burst in retinoid-depleted macrophages and neutrophils. Detailed studies on APC are needed to understand how retinoids inhibit their proinflammatory TNF-α and IL-6 synthesis and excessive Th1 stimulating capacity. For unknown reasons, Th2 cell and perhaps CTL development and/or cytokine synthesis are retinoid-dependent, suggesting additional areas to be explored. In Th1 cells, the biochemical mechanism for retinoid inhibition of IFN-γ transcription requires further study. The second messenger hypothesis for 14-HRR action requires in vivo verification, as well as identification of key enzymes and 14-HRR receptors. The gene expression hypothesis awaits identification of an RARE in a gene expressed by an immune system cell, and proof of RAR-RXR involvement in the mechanism. It will be extremely interesting to see the biochemical and molecular details come into focus with further research into vitamin A and immunity.

Acknowledgements. We thank the United States National Institutes of Health (grant DK46820) and the United States National Multiple Sclerosis Society (grant PP0509) for their generous research support. We also wish to acknowledge many colleagues at the University of Wisconsin–Madison for valuable discussions. Finally, we thank Ms. P Salzwedel and Ms. C Kunen for their effort in support of our research.

References

Abbas AK, Murphy KM, Sher A (1996) Functional diversity of helper T lymphocytes. Nature 383:787–793

Agarwal N, Mehta K (1997) Possible involvement of Bcl-2 pathway in retinoid X receptor alpha-induced apoptosis of HL60 cells. Biochem Biophys Res Commun 230:251–253

Ahmed F, Jones DB, Jackson AA (1991) Effect of vitamin A deficiency on the immune response to epizootic diarrhoea of infant mice (EDIM) rotavirus infection in mice. Br J Nutr 65:475–485

Ahmed F, Mohiduzzaman M, Jackson AA (1993) Vitamin A absorption in children with ascariasis. Br J Nutr 69:817–825

Alvarez JO, Salazar-Lindo E, Kohatsu J, Miranda P, Stephensen CB (1995) Urinary excretion of retinol in children with acute diarrhea. Am J Clin Nutr 61:1273–1276

Arora D, Ross AC (1994) Antibody response against tetanus toxoid is enhanced by lipopolysaccharide or tumor necrosis factor-alpha in vitamin A-sufficient and deficient rats. Am J Clin Nutr 59:922–928

Barclay AJG, Foster A, Sommer A (1987) Vitamin A supplements and mortality related to measles: a randomised clinical trial. BMJ 294:294–296

Baumann H, Gauldie J (1994) The acute phase response. Immunol Today 15:74–80

Bedford PA, Knight SC (1989) The effect of retinoids on dendritic cell function. Clin Exp Immunol 75:481–486

Benn CS, Aaby P, Bale C, Olsen J, Michaelsen KF, George N, Whittle H (1997) Randomised trial of effect of vitamin A supplementation on antibody response to measles vaccine in Guinea-Bissau, West Africa. Lancet 350:101–105

Bhat PV, Lacroix A (1983) Metabolism of [11–^{3}H]retinyl acetate in liver tissues of vitamin A-sufficient, -deficient and retinoic acid-supplemented rats. Biochim Biophys Acta 752:451–459

Binka FN, Ross DA, Morris SS, Kirkwood BR, Arthur P, Dollimore N, Gyapong JO, Smith PG (1995) Vitamin A supplementation and childhood malaria in northern Ghana. Am J Clin Nutr 61:853–859

Bissonnette RP, Brunner T, Lazarchik SB, Yoo NJ, Boehm MF, Green DR, Heyman RA (1995) 9-*cis* Retinoic acid inhibition of activation-induced apoptosis is mediated via regulation of Fas ligand and requires retinoic acid receptor and retinoid x receptor activation. Mol Cell Biol 15:5576–5585

Blomhoff HK, Smeland EB, Erikstein B, Rasmussen AM, Skrede B, Skjønsberg C, Blomhoff R (1992) Vitamin A is a key regulator for cell growth, cytokine production, and differentiation in normal B cells. J Biol Chem 267:23988–23992

Boynton LC, Bradford WL (1931) Effect of vitamins A and D on resistance to infection. J Nutr 4:323–329

Bresse BB, Watkins E, McCoord AB (1942) The absorption of vitamin A in tuberculosis. JAMA 119:3–4

Brinckerhoff CE, Sheldon LA, Benoit MC, Burgess DR, Wilder RL (1985) Effect of retinoids on rheumatoid arthritis, a proliferative and invasive non-malignant disease. In: Retinoids, differentiation and disease, Ciba Foundation Symposium 113. Pitman, London, p 191

Bruel A, Karsenty E, Schmid M, McDonnell T, Lanotte M (1997) Altered sensitivity to retinoid-induced apoptosis associated with changes in the subcellular distribution of Bcl-2. Exp Cell Res 233:281–287

Buck J, Derguini F, Levi E, Nakanishi K, Hammerling U (1991) Intracellular signaling by 14-hydroxy-4,14-retro-retinol. Science 254:1654–1656

Burger H, Kovacs A, Weiser B, Grimson R, Nachman S, Tropper P, van Bennekum AM, Elie MC, Blaner WS (1997) Maternal serum vitamin A levels are not associated with mother-to-child transmission of HIV-1 in the United States, J Acquir Immune Defic Syndr Hum Retrovirol 14:321–326

Cantorna MT, Nashold FE, Hayes CE (1994) In vitamin A deficiency multiple mechanisms establish a regulatory T helper cell imbalance with excess Th1 and insufficient Th2 function. J Immunol 152:1515–1522

Cantorna MT, Nashold FE, Hayes CE (1995) Vitamin A deficiency results in a priming environment conducive for Th1 cell development. Eur J Immunol 25:1673–1679

Cantorna MT, Hayes CE (1996) Vitamin A deficiency exacerbates murine Lyme arthritis. J Infect Dis 174:747–751

Cantorna MT, Nashold FE, Chun TY, Hayes CE (1996) Vitamin A down-regulation of IFN-γ synthesis in cloned mouse Th1 lymphocytes depends on the CD28 costimulatory pathway. J Immunol 156:2674–2679
Carman JA, Smith SM, Hayes CE (1989) Characterization of a helper T lymphocyte defect in vitamin A-deficient mice. J Immunol 142:388–393
Carman JA, Hayes CE (1991) Abnormal regulation of IFN-γ secretion in vitamin A-deficiency. J Immunol 147:1247–1252
Carman JA, Pond L, Nashold F, Wassom DL, Hayes CE (1992) Immunity to Trichinella spiralis infection in Vitamin A-deficient mice. J Exp Med 175:111–120
Chen H, Bérard J, Luo H, Landers M, Vinet B, Bradley WEC, Wu J (1997) Compromised allograft rejection response in transgenic mice expressing antisense sequences to retinoic acid receptor $\beta 2$. J Immunol 159:623–634
Chun TY, Carman JA, Hayes CE (1992) Retinoid repletion of vitamin A-deficient mice restores IgG responses. J Nutr 122:1062–1069
Ciompi ML, Bazzichi L, Marotta G, Pasero G (1988) Tigason (etretinate) treatment in psoriatic arthritis, Int J Tissue Reactions 10:25–27
Cippitelli M, Ye J, Viggiano V, Sica A, Ghosh P, Gulino A, Santoni A, Young HA (1996) Retinoic acid-induced transcriptional modulation of the human interferon-γ promoter. J Biol Chem 271:26783–26793
Cohen BE, Elin RJ (1974) Vitamin A-induced nonspecific resistance to infection. J Infect Dis 129:597–600
Coutsoudis A, Bobat RA, Coovadia HM, Tsai WY, Stein ZA (1995) The effects of vitamin A supplementation on the morbidity of children born to HIV-infected women, Am J Publ Health 85:1076–1081
Crowle AJ, Ross EJ (1989) Inhibition by retinoic acid of multiplication of virulent tubercle bacilli in cultured human macrophages. Infect Immun 57:840–844
Curtale F, Vaidya Y, Muhilal, Tilden RL (1994) Ascariasis, hookworm infection and serum retinol amongst children in Nepal, Panminerva Medica 36:19–21
Das BS, Thurnham DI, Das DB (1996) Plasma α-tocopherol, retinol, and carotenoids in children with falciparum malaria. Am J Clin Nutr 64:94–100
Davis TM, Binh TQ, Danh PT, Dyer JR, St John A, Garcia-Webb P, Anh TK (1994) Serum vitamin A and E concentrations in acute falciparum malaria: modulators or marker of severity? Clin Sci 87:505–511
de The H, Marchio A, Tiollais P, Dejean A (1989) Differential expression and ligand regulation of the retinoic acid receptor α and β genes. EMBO J 8:429–433
Dennert G (1985) Immunostimulation by retinoic acid. Retinoids, differentiation and disease. Pitman, London, p 117
Dresser DW (1968) Adjuvanticity of vitamin A. Nature 217:527–529
Ellison JB (1932) Intensive vitamin A therapy in measles. BMJ 2:708–711
Fegan C, Bailey-Wood R, Coleman S, Phillips SA, Neale L, Hoy T, Whittaker JA (1995) All *trans* retinoic acid enhances human LAK activity. Eur J Haematol 54:95–100
Feldmann M, Brennan FM, Maini RN (1996) Rheumatoid arthritis. Cell 85:307–310
Foerster M, Sass JO, Rühl R, Nau H (1996) Comparative studies on effects of all-*trans*-retinoic acid and all-*trans* retinoyl-β-D-glucuronide on the development of foetal mouse thymus in an organ culture system, Toxicology in Vitro 10:7–15
Friedman A, Sklan D (1989) Impaired T lymphocyte immune response in vitamin A depleted rats and chicks. Br J Nutr 62:439–449
Friedman A, Halevy O, Schrift M, Arazi Y, Sklan D (1993) Retinoic acid promotes proliferation and induces expression of retinoic acid receptor-α gene in murine T lymphocytes, Cellular Immunol 152:240–248
Friis H, Ndhlovu P, Kaondera K, Sandstrom B, Michaelsen KF, Vennervald BJ, Christensen NO (1996) Serum concentration of micronutrients in relation to schistosomiasis and indicators of infection: a cross-sectional study among rural Zimbabwean schoolchildren. Eur J Clin Nutr 50:386–391
Friis H, Mwaniki D, Omondi B, Muniu E, Magnussen P, Geissler W, Thiong'o F, Michaelsen KF (1997) Serum retinol concentrations and Schistosoma mansoni,

intestinal helminths, and malarial parasitemia: a cross-sectional study in Kenyan preschool and primary school children. Am J Clin Nutr 66:665–671
Gangopadhyay NN, Moldoveanu Z, Stephensen CB (1996) Vitamin A deficiency has different effects on immunoglobulin A production and transport during influenza A infection in BALB/c mice. J Nutr 126:2960–2967
Goettsch W, Hatori Y, Sharma RP (1992) Adjuvant activity of all-*trans*-retinoic acid in C57Bl/6 mice, Int J Immunopharmac 14:143–150
Goldfarb RH, Herberman RB (1981) Natural killer cell reactivity: regulatory interactions among phorbol ester, interferon, cholera toxin, and retinoic acid. J Immunol 126:2129–2135
Gravallese EM, Handel ML, Coblyn J, Anderson RJ, Sperling RI, Karlson EW, Maier A, Ruderman EM, Formelli F, Weinblatt ME (1996) N-[4-hydroxyphenyl] retinamide in rheumatoid arthritis: a pilot study, Arthritis & Rheumatism 39:1021–1026
Green HN, Mellanby E (1928) Vitamin A as an anti-infective agent. BMJ 2:691–696
Greenberg BL, Semba RD, Vink PE, Farley JJ, Sivapalasingam M, Steketee RW, Thea DM, Schoenbaum EE (1997) Vitamin A deficiency and maternal-infant transmissions of HIV in two metropolitan areas in the United States. AIDS 11:325–332
Greene MR (1933) The effects of vitamin A and D on antibody production and resistance to infection, Am J Hyg 17:60–101
Gross V, Villiger PM, Zhang B, Lotz M (1993) Retinoic acid inhibits interleukin-1-induced cytokine synthesis in human monocytes, J Leukocyte Biol 54:125–132
Hatchigian EA, Santos JI, Broitman SA, Vitale JJ (1989) Vitamin a supplementation improves macrophage function and bacterial clearance during experimental Salmonella infection. Proc Soc Exp Biol Med 191:47–54
Hayes CE, Nashold F, Chun TY, Cantorna M (1994) Vitamin A: a regulator of immune function. In: Livrea MA, Vidali G (eds) Retinoids: from basic science to clinical applications. Birkhäuser, p 215
Hayes CE, Cantorna MT, DeLuca HF (1997) Vitamin D and multiple sclerosis. Proc Soc Exp Biol Med 216:21–27
Humphrey JH, West KP Jr, Sommer A (1992) Vitamin A deficiency and attributable mortality among under-5-year-olds, Bull World Health Organ 70:225–232
Hussey GD, Klein M (1990) A randomized, controlled trial of vitamin A in children with severe measles, New Eng. J Med 323:160–164
Iwata M, Mukai M, Nakai Y, Iseki R (1992) Retinoic acids inhibit activation-induced apoptosis in T cell hybridomas and thymocytes. J Immunol 149:3302–3308
Janeway CA, Travers P (1996) Immunobiology: the immune system in health and disease, 2nd edn. Current Biology. Garland, London
Jayalakshmi VT, Gopalan C (1958) Nutrition and tuberculosis. I. An epidemiological study. Ind J Med Res 46:87–92
Karalliedde S, Dissanayake S, Wikramanayake TW (1979) Salivary immunoglobulin A in vitamin A deficiency. Ceylon Med J 24:69–70
Karter DL, Karter AJ, Yarrish R, Patterson C, Kass PH, Nord J, Kislak JW (1995) Vitamin A deficiency in non-vitamin-supplemented patients with AIDS: a cross-sectional study. J Acquir Immune Defic Syndr Hum Retrovirol 8:199–203
Kolb E (1995) The significance of vitamin A for the immune system. Berliner Munch Tierarztl Wochensch 108:385–390
Lawrie NR, Moore T, Rajagopal KR (1941) The excretion of vitamin A in urine. Biochem J 35:825–836
Li T, Ramirez K, Palacios R (1996) Distinct patterns of fas cell surface expression during development of T- or B-lymphocyte lineages in normal, scid, and mutant mice lacking or overexpressing p53, bcl-2, or rag-2 genes, Cell Growth and. Differentiation 7:107–114
Linette GP, Grusby MJ, Hedrick SM, Hansen TH, Glimcher LH, Korsmeyer SJ (1994) Bcl-2 is upregulated at the $CD4^{+}CD8^{+}$ stage during positive selection and promotes thymocyte differentiation at several control points. Immunity 1:197–205

Louw JA, Werbeck A, Louw MEJ, Kotze TJW, Cooper R, Labadarios D (1992) Blood vitamin concentrations during the acute-phase response. Crit Care Med 20:934–941

Mangelsdorf DJ, Borgmeyer U, Heyman RA, Zhou JY, Ong ES, Oro AE, Kakizuka A, Evans RM (1992) Characterization of three RXR genes that mediate the action of 9-*cis* retinoic acid, Genes. Development 6:329–344

Massacesi L, Abbamondi AL, Giorgi C, Sarlo F, Lolli F, Amaducci L (1987) Suppression of experimental allergic encephalomyelitis by retinoic acid. J Neurol Sci 80:55–64

Massacesi L, Castigli E, Vergelli M, Olivotto J, Abbamondi AL, Sarlo F, Amaducci L (1991) Immunosuppressive activity of 13-*cis*-retinoic acid and prevention of experimental autoimmune encephalomyelitis in rats. J Clin Invest 88:1331–1337

Matikainen S, Serkkola E, Hurme M (1991) Retinoic acid enhances IL-1β expression in myeloid leukemia cells and in human monocytes. J Immunol 147:162–167

Matsuzaki Y, Nakayama K, Nakayama K, Tomita T, Isoda M, Loh DY, Nakauchi H (1997) Role of bcl-2 in the development of lymphoid cells from the hematopoietic stem cell. Blood 89:853–862

McCollum EV, Davis M (1913) The necessity of certain lipins in the diet during growth. J Biol Chem 15:167–175

McCoy OR (1934) The effect of vitamin A deficiency on the resistance of rats to infection with Trichinella spiralis. Am J Hyg 20:169–180

Meco D, Scarpa S, Napolitano M, Maroder M, Bellavia D, De Maria R, Ragano-Caracciolo M, Frati L, Modesti A, Gulino A, Screpanti I (1994) Modulation of fibronectin and thymic stromal cell-dependent thymocyte maturation by retinoic acid. J Immunol 153:73–83

Mehta K, McQueen T, Tucker S, Pandita R, Aggarwal BB (1994) Inhibition by all-*trans*-retinoic acid of tumor necrosis factor and nitric oxide production by peritoneal macrophages. J Leukoc Biol 55:336–342

Meunier L, Bohjanen K, Voorhees JJ, Cooper KD (1994) Retinoic acid upregulates human Langerhans cell antigen presentation and surface expression of HLA-DR and CD11c, a beta 2 integrin critically involved in T-cell activation. J Invest Dermatol 103:775–779

Moriguchi S, Werner L, Watson RR (1985) High dietary vitamin A (retinyl palmitate) and cellular immune functions in mice. Immunology 56:169–177

Nauss KM (1986) Influence of vitamin A status on the immune system. In: Bauernfiend JC (ed) Vitamin A deficiency and its control Academic, Orlando, p 207

Ongsakul M, Sirisinha S, Lamb AJ (1985) Impaired blood clearance of bacteria and phagocytic activity in vitamin A-deficient rats. Proc Soc Exp Biol Med 178:204–208

Orfanos CE, Bauer R (1983) Evidence for anti-inflammatory activities of oral synthetic retinoids: experimental findings and clinical experience. Br J Dermatol 109(S25):55–60

Palmer S (1978) Influence of vitamin A nutriture on the immune response: findings in children with Down's syndrome. Int J Vitamin Nutr Res 48:188–216

Pasatiempo AMG, Abaza M, Taylor CE, Ross AC (1992) Effects of timing and dose of vitamin A on tissue retinol concentrations and antibody production in the previously vitamin A-depleted rats. Am J Clin Nutr 55:443–451

Phuapradit W, Chaturachinda K, Taneepanichskul S, Sirivarasry J, Khupulsup K, Lerdvuthisopon N (1996) Serum vitamin A and β-carotene levels in pregnant women infected with human immunodeficiency virus-1. Obstet Gynecol 87:564–567

Racke MK, Burnett D, Pak SH, Albert PS, Cannella B, Raine CS, McFarlin DE, Scott DE (1995) Retinoid treatment of experimental allergic encephalomyelitis: IL-4 production correlates with improved disease course. J Immunol 154:450–458

Randolph RK, Ross AC (1991) Vitamin A status regulates hepatic lecithin: retinol acyltransferase activity in rats. J Biol Chem 266:16453–16457

Rhodes J, Oliver S (1980) Retinoids as regulators of macrophage function. Immunology 40:467–472

Rosales FJ, Ritter SJ, Zolfaghari R, Smith JE, Ross AC (1996) Effects of acute inflammation on plasma retinol, retinol-binding protein, and its mRNA in the liver and kidneys of vitamin A-sufficient rats. J Lipid Res 37:962–971
Ross AC, Hämmerling UG (1994) Retinoids and the immune system. In: Sporn MB, Roberts AB, Goodman DS (ed) The retinoids: biology, chemistry, and medicine. Raven, New York, p 521
Ross AC (1996) Vitamin A deficiency and retinoid repletion regulate the antibody response to bacterial antigens and the maintenance of natural killer cells. Clin Immunol Immunopathol 80:S63–S72
Ross AC, Stephensen CB (1996) Vitamin A and retinoids in antiviral responses. FASEB J 10:979–985
Rwangabwoba JM, Humphrey J, Coberly J, Moulton L, Desormeaux J, Halsey NA (1996) Vitamin A status and development of tuberculosis and/or mortality. In: Abstracts of the XI International Conference on AIDS. XI International Conference on AIDS Society, Vancouver, Canada vol 2 p 103
Santom A, Sola SC, Giovarelli M, Martinetto P, Vietti D, Forni G (1986) Modulation of natural killer activity in mice by prolonged administration of various doses of dietary retinoids, Nat Immun Cell. Growth Regul 5:259–266
Sappey C, Leclercq P, Coudray C, Faure P, Micoud M, Favier A (1994) Vitamin, trace element and peroxide status in HIV seropositive patients: asymptomatic patients present a severe β-carotene deficiency. Clin Chim Acta 230:34–42
Scrimshaw NS, Taylor CE, Gordon JE (1968) Interactions of nutrition and infection. WHO Monograph Series No 57, World Health Organization, Geneva, p 3
Semba RD, Muhilal, Scott AL, Natadisastra G, Wirasasmita S, Mele L, Ridwan E, West KP Jr, Sommer A (1992) Depressed immune response to tetanus in children with vitamin A deficiency. J Nutr 122:101–107
Semba RD, Graham NMH, Caiaffa WT, Margolick JB, Clement L, Vlahov D (1993) Increased mortality associated with vitamin A deficiency during human immunodeficiency virus type 1 infection. Arch Int Med 153:2149–2154
Semba RD (1994) Vitamin A, immunity, and infection. Clin Infect Dis 19:489–499
Semba RD, Miotti PG, Chiphangwi JD, Saah AJ, Canner JK, Dallabetta GA, Hoover DR (1994) Maternal vitamin A deficiency and mother-to-child transmission of HIV-1. Lancet 343:1593–1597
Semba RD, Caiaffa WT, Graham NMH, Cohn S, Vlahov D (1995a) Vitamin A deficiency and wasting as predictors of mortality in human immunodeficiency virus-infected injection drug users. J Infect Dis 171:1196–1202
Semba RD, Miotti PG, Chiphangwi JD, Liomba G, Yang L-P, Saah AJ, Dallabetta GA, Hoover DR (1995b) Infant mortality and maternal vitamin A deficiency during human immunodeficiency virus infection. Clin Infect Dis 21:966–972
Semba RD, Munasir Z, Beeler J, Akib A, Muhilal, Audet S, Sommer A (1995c) Reduced seroconversion to measles in infants given vitamin A with measles vaccination. Lancet 345:1330–1332
Semba RD (1997) Vitamin A and human immunodeficiency virus infection. Proc Nutr Soc 56:459–469
Semba RD, Akib A, Beeler J, Munasir Z, Permaesih D, Muherdiyantiningsih, Komala, Martuti S, Muhailal (1997) Effect of vitamin A supplementation on measles vaccination in nine-month-old infants. Public Health 111:245–247
Sidell N, Ramsdell F (1988) Retinoic acid upregulates interleukin-2 receptors on activated human thymocytes. Cell Immunol 115:299–309
Sidell N, Chang B, Bhatti L (1993) Upregulation by retinoic acid of interleukin-2-receptor mRNA in human T lymphocytes. Cell Immunol 146:28–37
Sidell N, Kummer U, Aframian D, Thierfelder S (1997) Retinoid regulation of interleukin-2 receptors on human T-cells. Cell Immunol 179:116–125
Sijtsma SR, Rombout JH, West CE, Van der Zijpp AJ (1990) Vitamin A deficiency impairs cytotoxic T lymphocyte activity in Newcastle disease virus-infected chickens, Vet Immunol Immunopathol 26:191–201

Sirisinha S, Darip MD, Moongkarndi P, Ongsakul M, Lamb AJ (1980) Impaired local immune response in vitamin A-deficient rats. Clin Exp Immunol 40:127–135

Smith SM, Hayes CE (1987) Contrasting impairments in IgM and IgG responses in vitamin A-deficient mice. Proc Natl Acad Sci USA 84:5878–5882

Smith SM, Levy NS, Hayes CE (1987) Impaired immunity in vitamin A-deficient mice. J Nutr 117:857–865

Sommer A (1993) Vitamin A, infectious disease, and childhood mortality: a 2 cent solution?. J Infect Dis 167:1003–1007

Steenbock H, Sell MT, Buell MV (1921) Fat-soluble vitamins, VII, The fat-soluble vitamins and yellow pigmentation in animal fats with some observations on its stability to saponification. J Biol Chem 47:89–109

Steinman L (1996) Multiple sclerosis: a coordinated immunological attack against myelin in the central nervous system. Cell 85:299–302

Stephensen CB, Blount SR, Schoeb TR, Park JY (1993) Vitamin A deficiency impairs some aspects of the host response to influenza A virus infection in BALB/c mice. J Nutr 123:823–833

Stephensen CB, Alvarez JO, Kohatsu J, Hardmeier R, Kennedy JI Jr, Gammon RB Jr (1994) Vitamin A is excreted in the urine during acute infection. Am J Clin Nutr 60:388–392

Stephensen CB, Moldoveanu Z, Gangopadhyay NN (1996) Vitamin A deficiency diminishes the salivary immunoglobulin A response and enhances the serum immunoglobulin G response to influenza A virus infection in BALB/c mice. J Nutr 126:94–102

Storey DM (1993) Filariasis: nutritional interactions in human and animal hosts. Parasitol 107:S147–S158

Tachibana K, Sone S, Tsubura E, Kishino Y (1984) Stimulatory effect of vitamin A on tumoricidal activity of rat alveolar macrophages. Br J Cancer 49:343–348

Tang AM, Graham NM, Semba RD, Saah AJ (1997) Association between serum vitamin A and E levels and HIV-1 disease progression. AIDS 11:613–620

Tanumihardjo SA, Permaesih D, Muherdiyantiningsih, Rustan E, Rusmil K, Fatah AC, Wilbur S, Muhilal, Karyadi D, Olson JA (1996) Vitamin A status of Indonesian children infected with Ascaris lumbricoides after dosing with vitamin A supplements and albendazole. J Nutr 126:451–457

Trechsel U, Evêquoz V, Fleisch H (1985) Stimulation of interleukin 1 and 3 production by retinoic acid in vitro. Biochem J 230:339–344

Twining SS, Schulte DP, Wilson PM, Fish BL, Moulder JE (1996) Retinol is sequestered in the bone marrow of vitamin A-deficient rats. J Nutr 126:1618–1626

Twining SS, Schulte DP, Wilson PM, Fish BL, Moulder JE (1997) Vitamin A deficiency alters rat neutrophil function. J Nutr 127:558–565

Vincenti MP, White LA, Schroen DJ, Benbow U, Brinkerhoff CE (1996) Regulating expression of the gene for matrix metalloproteinase-1 (collagenase): mechanisms that control enzyme activity, transcription, and mRNA stability, Crit Rev Eukaryotic Gene Expression 6:391–411

Wan Y-JY, Wang L, Wu T-CJ (1992) Detection of retinoic acid receptor mRNA in tissues by reverse transcriptase-polymerase chain reaction. J Mol Endocrinol 9:291–294

Werkman CH (1923) Immunological significance of vitamins, II, Influence of lack of vitamins on resistance of rat, rabbit and pigeon to bacterial infection. J Infect Dis 32:255–262

Wiedermann U, Hanson LA, Holmgren J, Kahu H, Dahlgren UI (1993a) Impaired mucosal antibody response to cholera toxin in vitamin A deficient rats immunized with oral cholera vaccine. Infect Immun 61:3952–3957

Wiedermann U, Hanson LA, Kahu H, Dahlgren UI (1993b) Aberrant T cell function in vitro and impaired T cell dependent antibody responses in vivo in vitamin A-deficient rats. Immunology 80:581–586

Wiedermann U, Hanson LA, Bremell T, Kahu H, Dahlgren UI (1995) Increased translocation of Escherichia coli and development of arthritis in vitamin A-deficient rats. Infect Immun 63:3062–3068

Wiedermann U, Tarkowski A, Bremell T, Hanson LA, Kahu H, Dahlgren UI (1996a) Vitamin A deficiency predisposes to Staphylococcus aureus infection. Infect Immun 64:209–214

Wiedermann U, Chen X-J, Enerbäck L, Hanson LA, Kahu H, Dahlgren UI (1996b) Vitamin A deficiency increases inflammatory responses. Scand J Immunol 44:578–584

Wolde-Gebriel Z, West CE, Gebru H, Tadesse AS, Fisseha T, Gabre P, Aboye C, Ayana G, Hautvast JG (1993) Interrelationship between vitamin A, iodine and iron status in school children in Shoa Region, central Ethiopia. Br J Nutr 70:593–607

Yang Y, Mercep M, Ware CF, Ashwell JD (1995a) Fas and activation-induced Fas ligand mediate apoptosis of T cell hybridomas: inhibition of Fas ligand expression by retinoic acid and glucocorticoids. J Exp Med 181:1673–1682

Yang Y, Minucci S, Ozato K, Heyman RA, Ashwell JD (1995b) Efficient inhibition of activation-induced Fas ligand up-regulation and T cell apoptosis by retinoids requires occupancy of both retinoid X receptors and retinoic acid receptors. J Biol Chem 270:18672–18677

Zhuang Y-H, Sainio E-L, Sainio P, Vedeckis WV, Ylikomi T, Tuohimaa P (1995) Distribution of all-*trans*-retinoic acid in normal and vitamin A deficient mice: correlation to retinoic acid receptors in different tissues of normal mice. Gen Comp Endocrinol 100:170–178

Subject Index

A
absorption of retinoids 31 pp.
actinic keratosis 301
ACTR 168
acute
– promyelocytic
–– leukemia (*see* APL) 204
–– leukemic cells 281
– retinoid deficiency 413
adalpene (*see* CD 271)
adaptive immunity 591 pp.
ADH (alcohol dehydrogenase) 123, 127, 130, 372, 454
– isoenzymes 123
ADH1 123–125
ADH2 130, 132, 133
ADH4 123–125, 420
administration of selective retinoid receptor ligands 473
AER (apical ectodermal ridge) 380, 384
aerodigestive tract cancer 301
AHD (aldehyde dehydrogenase) 131, 419
AHD-2 419, 420
AHPN 242, 252, 255, 256, 305
– apoptosis 257 p.
alcohol dehydrogenase (*see* ADH) 123
aldehyde dehydrogenase (*see* AHD) 131
ALDH 372
all-trans-retinal 530 p.
all-trans-retinoic acid (*see* ATRA) 186
all-trans-retinol 525 p.
all-trans-retinol-β-glucuronide 82
all-trans-retinol dehydrogenase 573 p.
all-trans-retinoyl-β-glucuronide 533 p.
all-trans-retinyl esters 527 p.
ALRT1550 221
anhydroretinol 98
– biosynthesis of 110 pp.
– induced cell death 106
anterior amputation phenotype 406
antibody-mediated immunity 591 p.
AP-1 166, 167
– mediated growth signals 249
AP-2 400
apical ectodermal ridge (*see* AER) 380
APL (acute promyelocytic leukemia) 204, 284 pp., 287 pp., 302, 305, 310 pp.
– ATRA 310
–– differentiation therapy 287
– low-dose ATRA 287
– molecular characteristics 284 p.
– RARα 330 p.
– retinoid resistance 288
APL93 trial 288, 289
β-apo-carotenals 34
apolipoprotein E (apoE) 38
apoptosis 239 pp., 551 pp.
– AHPN 257 pp.
– 4-HPR 256 p.
– induction 169 pp.
–– by RAR-selective retinoids 173
– in neural crest 414
–– derived structures 413
– receptor-mediated induction of 252 pp.
– regulation of 250 pp.
apoptotic protease activating factor-1 (Apaf1) 169
aqueous solubility 3, 4
AR as reversible inhibitor of B-lymphocyte proliferation 103
ARAT 508, 530
Ascaris lumbricoides 597
associated CNS structures 411
ATBF1 400
ATRA (all-trans-retinoic acid) 186, 189, 195, 199, 200, 204, 207, 278, 281, 282, 305, 539, 548
– APL 286 pp., 310 pp.
–– differentiation therapy 287
– breast cancer 327
– cervical cancer 328 p.
– embryonal carcinoma 329
– HEC 327

– HNSCC 325, 326
– myeloid leukemia 329, 330
– NT2/D1 human embryonal carcinoma cell line 308 p.
– side effects 289
– signaling pathway 303
– skin cancer 328
– therapy 281, 282
ATRA syndrome 290
ATRA-induced
– HL-60 cell differentiation 197, 198
– transcriptional activity for RARα selective antagonists 201
autoimmunity 597 pp.
autosomal recessive retinitis pigmentosa 582

B
basal cell carcinoma 335
BEAS-2B cells 246
bile salts 33, 48
biosynthesis of retinoids 31 pp.
B-lymphocytes 99, 526
– growth of 101
bone
– marrow 44
– morphogenetic proteins (BMP) 360
Borrelia burgdorferi 597
breast cancer 327, 334
– ATRA 327

C
cAct-RIIa 452
cancer
– chemoprevention 306 pp.
– therapy 301 pp.
cardia bifida 452
cardiogenesis 450 p.
α-carotene 34
– mechanism for the oxidative cleavage of 35
β-carotene 34
carotenoid (*see* provitamin A) 32
CBP (CREB-binding protein) 164, 169, 217, 220, 225
CBP/p300 166 p., 168
CCTC-binding factor (CTCF) 248
CD 271 (adalpene) 173, 549, 550, 552
CD2366 477
CD437 171, 172, 242, 255, 256, 473, 475, 551, 552
Ced3 169
Ced4 169
Ced9 169
cell cycle
– regulation 244 pp.
– regulatory targets of retinoid action 246 pp.
cell-mediated immunity 594 p.
cellular
– retinal-binding proteins (*see* CRALBP) 11
– retinoic acid-binding proteins (CRABP) 11
– retinol-binding protein CRPB 11
–– type I (*see* CRBP I) 38
–– type II (*see* CRBP II) 36
ceramide 243
cerebrospinal fluid (*see* CSF) 79
cervical cancer 328 p., 336
– ATRA 328, 329
cervical dysplasia 301
CETP (cholesteryl ester transfer protein) 39, 81, 82
c-fos 245
channeling 14
chemical structure 4
chemotherapy of carcinogen-induced breast cancer 230 pp.
cholesteryl ester
– hydrolase 33
– transfer protein (*see* CETP) 39, 81, 82
chylomicron remnants 39–41
– uptake by hepatocytes 45
chylomicrons 37, 38
11-cis-retinal 567 p.
11-cis retinal (*see* RAL) 97
9-cis-retinoic acid 548
– biosynthetic pathways 134
– formation of 133 pp.
13-cis-retinoic acid 80 p., 548
9-cis-retinol dehydrogenases (9cRDH) 136
11-cis-retinol dehydrogenases (11cRDH) 136, 574
cis-retinol/3α-hydroxy sterol short-chain deydrogenase (*see* CRAD) 136
cis-retinol/3α-hydroxysteroid dehydrogenases (CRAD) 127
c-jun 245
CMC (critical micellar concentration) 5
c-myc 245
– regulation of 248
– transcription 248
cNR-1 452
CNS 453 p.
– CRABP 422 pp.
– CRBP 421 pp.
– endogenous RA 415
– RAR 426 p.

– RXR 426 p.
cone visual pigments 569
congenital stationary night blindness 581
COUP-TF 158, 241, 327
CRABP (cellular retinoic acid-binding proteins) 11, 172, 578
– epidermis (*see also there*) 504 pp.
– I 82, 101, 118, 119, 277, 283, 350, 421, 427, 469, 470
–– CNS 422 pp.
– II 119, 277, 283, 351, 352, 354, 421, 425 p., 427, 470, 526, 531, 532, 546
CRAD (cis-retinol/3α-hydroxy sterol short-chain deydrogenase) 127, 128
– CRAD1 136
– CRAD2 136
CRALBP (cellular retinal-binding proteins) 11, 576–579, 582
CRBP I (cellular retinol-binding protein type I) 38, 45, 49 p., 53, 75, 100, 101, 118–121, 132, 133, 277, 348–350, 421, 469, 531
– CNS 421 p.
CRBP II (cellular retinol-binding protein type II) 36–38, 119, 277, 348
CREB 166, 167
CREB-binding protein (*see* CBP) 164
critical micellar concentration (CMC) 5
CRPB (*see* cellular retinol-binding proteins) 11
β-cryptoxanthin 34
CSF (cerebrospinal fluid) 79
CTCF (CCTC-binding factor) 248
CYP26 139, 140, 142
cytochrome
– P450 138, 141, 142, 469
– P4502D1 (CYP2D1) 126

D
DHR (13, 14-dihydroxy-retinol) 109, 110, 242
differential expression of nuclear RAR and RBP 283
differentiation-proliferation switches 165 p.
differin 173
13, 14-dihydroxy-retinol (*see* DHR) 109
direct synthesis of retinoic acid from carotenoids 137
diseases (*see* syndromes)
dithiothreotol (DTT) 60–62
DR-1 162
DR-1-PPARE 162
DR-5 HRE 168
DR-5-RARE 162

E
early development 443 pp.
– retinoid metabolism 455 p.
EGF 239
embryo
– retinoid
–– development of vertebrate 371
–– acid
––– degrading enzymes 371 pp.
––– synthesizing enzymes 371 pp.
embryonal carcinoma 329
– ATRA 329
– cells 399
embryonic malformations 445
epidermal
– compartments 497
– proliferation 528
– strada 497
epidermis 491 pp.
– ARAT 508
– CRABP 504 pp.
–– I and II 505–507
– CRBP 502 pp., 508
– 3, 4-didehydroretinol 513
– inflammation 546
– LRAT 503
– RBP 502 p.
– retinoic acid
–– metabolism 511 p.
–– synthesis 509 p.
– retinoid metabolism 502 pp.
– vitamin A
–– content 492 pp.
–– differentiation 543
–– metabolites 493 pp.
–– proliferation 543
–– retinyl ester 507 p.
–– storage 507 p.
epididymal retinoic acid-binding protein (*see* E-RABP) 121
equilibrium dissociation constants 11
– retinoid-binding proteins 12, 13
– RAR 14
– RXR 14
ER LBD 163
E-RABP (epididymal retinoic acid-binding protein) 120, 121
ERAP160 162
ER-RAR receptors 203
estrogen 239
– receptor 241

ethanol-induced retinoid depletion 454 p.
extrahepatic metabolism 54

F
fenretinide (*see* 4-HPR) 170, 255
fetal alcohol syndrome (FAS) 454
fibroblast 104
– growth factor (FGF) 386
flip-flop 10, 21, 68
α-fodrin 251

G
GADD45 258
GAP-43 400
GATA-4/5/6 451
GCN5 168
glucuronides 82
glutathione-S-transferase gene (GSTP1-1) 249
GR-interacting proteins (GRIP) 162
growth control by retinoids 239 pp.
GSTP1-1 (glutathione-S-transferase gene) 249

H
hamster cheek-pouch carcinogenesis 332
HAT (histone acetyltransferase) 168, 169
HDL (high density lipoprotein) 39, 81
head and neck squamous cell carcinoma (*see* HNSCC) 325
heart 450
HEC (human esophageal cancer) 326
– ATRA 327
hematopoiesis 280 pp.
hematopoietic cells 277 pp.
heparan sulfate proteoglycans (HSPG) 39, 42
hepatic
– LRAT 50, 51
– vitamin A 48
hepatocellular cancer 301
hepatocytes 45 pp.
– retinoid metabolism 46
high
– density lipoprotein (*see* HDL) 39
– performance liquid chromatography (HPLC) 415, 416, 526, 531, 532, 577
– affinity RXR ligands 218
histone
– acetylation in receptor-mediated transactivation 168 p.
– acetyltransferase (*see* HAT) 168
HIV-1 596
HL-60 cells 106, 107, 246, 252, 253, 257, 283, 336
HNSCC (head and neck squamous cell carcinoma) 325, 336
– ATRA 325, 326
– nuclear retinoid receptors 325, 331 p.
homebox 400
hormone response elements (HREs, *see* RARE) 156
Hox 375, 377, 385, 410, 411
Hox/HOM-C genes 375
4-HPR 242, 252, 255, 304, 598, 600
– apoptosis 256 p.
14-HRR (14-hydroxy-retro-retinol) 83, 98, 101, 109, 170–172, 242, 526
human esophageal cancer (*see* HEC) 326
14-hydroxy-4, 14- retroretinol (*see* 14-HPR)

I
IGF-1 253
IL1α 239
IL6 239
immediate early genes (IEG) 245
immune
– response in skin 546 p.
– system 98
immunity 589 pp.
induced myeloid leukemik cell apoptosis 282
inflammation 600
inhibition of RA synthesis 414
insulin 239
interaction with membranes 6, 8
interferon γ 239
intermediate markers 307
interphotoreceptor matrix IPM 8, 9
intracellular signaling 106 p.
inverse agonism 201
ionization behavior of retinoic acid in membranes 10
IRBP 23, 120, 578, 579
– targeting of retinoids 22, 24

J
juvenile chronic myelogenous leukemia 302

K
K5/K14 keratins 543
keratinocytes 498 pp.
– retinoic acid 500, 502
– retinol 499, 500
kidney cancer 302

L
LDL (low density lipoprotein) 31, 35, 39, 81
– receptor 39–42
–– related protein (LRP) 39, 42
lecithin:retinol acyltransferase (LRAT) 37, 38, 527, 571
left-right asymmetry 451
leukemic myeloid leukemik cell differentiation 280
LG030282 221
LG100268 221, 223, 224, 227–230
LG100754 223–225
LGD1069 222–224, 231
ligand-binding
– affinities of RARs and RXRs 17
– domain (LBD) 154, 216
limb
– differentiation 386 p.
– regeneration 387 p.
lipofuscin 582
lipolysis-stimulated receptor (*see* LSR) 39
lipoprotein
– bound retinyl ester 81 p.
– lipase 38
liver 44 pp.
– retinoid metabolism 45
loss of anterior hindbrain 407
low density lipoprotein (*see* LDL) 31
LSR (lipolysis-stimulated receptor) 39, 41
lung cancer
– nuclear retinoid receptors 325
– prevention model 313 p.
lymphocyte turnover 599

M
malaria 597
mannose-6-phosphate/insulin-like growth factor II receptor 78, 242
MASH1 400
MASH2 400
Mauthner neuron 410, 411
MCF-7 breast cancer cells 204, 245, 247, 254
measles infection 595
membranes
– radius of curvature 8
– solubility of retinoids 7, 8
metabolism of retinoids 31 pp.
– intramucosal metabolism 36
α-methyl-effect 187, 188
3-Me-TTNPB 188
micronutrient vitamin A 443
microsomal retinol dehydrogenases (*see* RoDH) 125
migration inhibitory factor-related protein (MRP-8) 550
Müller cells 575 p.
multiple sclerosis 598
mutations
– binding proteins 355
– retinoid receptors 355 pp.
– spermatogenesis 360 p.
MX335 171, 174
mycosis fungoides 302
MYD118 258
myeloid leukemia 329 p.
– ATRA 329, 330
myeloproliferative
– disorders 290
– growth 283

N
N-(4-hydroxyphenyl)retinamide (4-HPR) 242
NADPH 512
natural killer cells 590
N-CoA 169
N-CoR (nuclear receptor corepressor) 161, 217, 220
N-CoR/SMRT 168, 169
nerve growth factor induced receptor (NGFIB) 218
neural
– crest cells 403 p.
– development 399 pp.
– retina 577
neuroblastoma cells 399
neuronal
– cells 402 p.
– differentiation, induction of 399 pp.
noninsulin dependent diabetes mellitus (NIDDM) 226, 228, 230, 231, 375
nonprovitamin A carotenoids 34
non-small cell lung cancer 333
NT2/D1 human embryonal carcinoma cell line 308 p.
NTERA-2cl.D1 teratocarcinoma 248
nuclear
– receptor corepressor (*see* N-CoR) 161
– retinoid receptors (*see* retinoid receptors) 153 pp.
null mutant mice 472
Nur77 241, 337

O
Onchocerca vovulus 597
opsin 567 p., 570

oral leukoplakia 301, 305, 307
4-oxo-retinal 97
4-oxo-retinol 97, 242, 548

P
P/CAF 169
P142 69, 73
p21$^{WAF/Cip1}$ 245, 258
p27^{Kip1} 245
p53 258
p57^{Kip2} 245
P63 68, 69, 73, 74
P450RA 139, 140
P450RAI 139, 140, 372
pancreatic lipase 33
parasitic infection 597
PARP (poly-ADP-ribose polymerase) 251, 257
peroxisomal proliferator-activated receptor (*see* PPAR) 217, 540
phagocytic cells 591
phantom ligand effect 222 pp.
plasma all-trans-retinoic acid 77 pp.
platelet derived growth factor (PDGF) 104
PML oncogenic domains (POD) 310
poly-ADP-ribose polymerase (PARP) 251
positive cooperativity in ligand-induced dissociation of RXR-tetramers 16
posteriorisation 405 p.
PPAR (peroxisomal proliferator-activated receptor) 217, 218, 540
– α 37
proligand-nonligand concept 524
promyelocytic myeloid leukemia (PML) 204, 330
provitamin A carotenoid 32, 34
– intestinal abbsorption 32
– metabolism 33 pp.
– uptake 33 pp.

R
radius of curvature of membranes 8
RAL (11-cis Retinal) 97, 98
RalDH (retinal dehydrogenases) 132, 372, 373, 420
– I 132, 372
– II 132, 133, 372, 421
RAR 11, 185, 240, 277 pp., 304, 539
– antagonists 197 pp., 206 p.
–– pharmacology 203
– cancer, aberrant expression and function of 323 pp.
– CNS 426 p.
– inversive agonists 201 pp., 206 p.
– mutations 356 p.
– neutral antagonists 201 pp.
– selective
–– agonists 190 pp.
––– pharmacology 203
–– ligands 185 pp
–– retinoids 541 p.
– signal transduction 373 p.
– teratogenesis (*see also there*) 465 pp.
– testis 352
RARα selective agonists 192, 193, 204, 205
RARβ selective agonists 194, 195, 205
RARγ
– selective
–– agonists 195, 196, 205
–– antagonist 200
– VP16 202
RARE (retinoid response elements) 156–158, 206, 207, 240, 324, 351, 384, 411, 416, 418, 451, 544, 549
– dependent transactivation 241
– mediated transcriptional activation 242
RARE/RXRE-dependent transactivation 240
rate of
– dissociation from retinoid
–– nuclear receptors 18, 19
–– binding proteins 17, 18
– uptake of retinol by cells 21, 22
Rb 246 pp.
RBP (retinol-binding protein) 10, 31, 45, 56 pp., 101, 120
– cellular synthesis 58 pp.
– embryo 471
– fractional rates of synthesis 66
– gene 63
– properties 63 p.
– receptor 67 pp.
RBP-TTR complex 63 pp.
receptor
– independent regulation of cell proliferation 242
– mediated
–– induction of apoptosis 252 p.
–– inhibition of apoptosis 254 p.
– selective agonists 429 p.
reporter cells 417 p.
retinal
– dehydrogenases (*see* RalDH) 132
– epithelium cells (RPE) 8, 9
– G protein coupled receptor 577
– oxidation of 130 pp.
– pigment epithelium 566, 572
retinoic acid 117 pp.

– degrading enzymes in embryos 371 pp.
– formation from retinal derived from carotenoids 36
– induced heparin binding protein (*see* RI-HB) 401
– as a novel myeloid differentiation factor 282
– oxidative metabolism 137 pp.
– receptors (*see* RAR) 11
– syndrome 289
– synthesis 121 pp.
– synthesizing enzymes in
–– CNS 419 p.
–– embryos 371 pp.
retinoid
– delivery to target tissues 55
– development of vertebrate embryos 371
– ligand knockout model 447 p.
– lipid interaction 7
– metabolism 278
– receptors 153 pp.
–– APL 330, 331
–– basal
––– cell carcinoma 335
––– transcription factors 158, 160
–– breast
––– cancer 327
––– tissue 333 p.
–– cervical cancer 328, 329, 336
–– coactivators 158, 159
–– corepressors 158, 159
–– embryonal carcinoma 329
–– embryo 471
–– esophageal cancer 326, 327
–– hamster cheek-pouch carcinogenesis 332
–– HNSCC 325, 326, 331
–– lung cancer 325
–– myeloid leukemia 329, 330
–– non-small cell lung cancer 333
–– pathway and teretogenic effects 475
–– promyelocytic leukemia 330
–– selective agonists 539 pp.
–– skin cancer 328, 335
–– synergistic teratogenic action 476
– responsive genes 362
– signal transduction in the vertebrate embryo 370 pp.
– target genes 375 pp.
– X receptors (*see* RXR) 11, 539
–– CNS 426 p.
–– mutations 357 p.
–– selective
––– ligands 186
––– retinoids 542 p.
–– signal transduction 373 p.
–– teratogenesis (*see also there*) 465 pp.
–– testis 352
––– dependent mechanisms 240
––– mediated control of cell proliferation 241 p.
retinol
– binding protein (*see* RBP) 10
– dehydratase 111, 118
– placental transfer 470
– RPB turnover from the circulation 66
retinoyl β-glucuronide 83, 142, 143
retinyl ester 33
– hydrolase (REH) 45
– hydrolase/carboxylesterase 49
retinyl β-glucuronide 83
retro-metabolites of retinol 83
retro-retinoids 97 pp.
– cellular effects 101 pp.
– general properties 99 p.
– metabolism 108 p.
rexinoids 218
rheumatoid arthritis 597
rhodopsin 570
RI-HB (retinoic acid-induced heparin binding protein) 401, 403
RIP140 162, 163, 220
river blindness 597
RO40-8757 247
rod
– photoreceptor outer segments (ROS) 572
– visual pigments 575
RoDH (microsomal retinol dehydrogenases) 125, 126, 128, 129, 372, 373
RPE (retinal pigment epithelium cells) 8, 9
– membrane receptor 69
RPE65 574, 580, 582
RXR 11, 240, 277 pp., 304
– dependent heterodimers 220
– RXR-response element (RXRE) 218
– pharmacology 226 pp.
– RXR-specific agonists 215 pp.
RXR-RAR heterodimers 221

S

schistosomiasis 597
self
– aggregation 4
– association 4, 5

β-series retinods 100
Sertoli cell blood-testis barrier 349
serum response element (SRE) 249
silencing mediator for retinoid and thyroid hormone receptors (*see* SMRT) 161
situs inversus 451, 452
SK-BR-3 breast cancer cells 204
Skin (*see also* epidermis) 539 pp.
– cancer 328, 335
–– ATRA 328
skin-derived antileukoproteinase (SKALP) 550
SMRT (silencing mediator for retinoid and thyroid hormone receptors) 161, 217, 220
solubility in membranes 7
sonic hedgehog (Shh) 377, 382, 383, 452 452
spermatogenesis 347 pp.
– mutations 360 p.
sqaumous cell cancer of the skin 302
SRC 400
SRC-1 167, 168, 220, 225
SRC-2 220
SRC-3 220
SSEA-1 401
stellate cells 51 pp.
– retinoid metabolism 52
surface active membranolytic properties 4
syndromes / diseases (names only)
– ATRA syndrome 290
– fetal alcohol syndrome (FAS) 454
– retinoic acid syndrome 289
– Sweets syndrome 290
synthetic retinoids 304
– skin 548

T

3T3-L1 differentiation 228
t(11;17) translocation 285
T47 breast cancer cells 139–141, 247, 248
T-lymphocytes 104
TAF (TBP-associated factors) 160, 161
Taf110 225
targeting of retinoids by IRBP 22
TATA box binding protein (*see* TBP) 160
Tazarotene (AGN 190168) 550
TBP (TATA box binding protein) 160, 220, 225
– associated factors (*see* TAF) 160
TCR-mediated apoptosis 255
TE-671-1A rhabdomyosarcoma 248
teratogenesis 465 pp.
– retinoid receptors
–– ligands
––– embryo 468 pp.
––– endogenous 467
–– RAR 465
––– antagonists 466
––– ligands 466
–– RXR 465
––– ligands 466
teratogenic effects of vitamin A deficiency and excess 444 p.
– gestation 445
teratogenicity of retinoids on the CNS 404
testis 349
– RAR 352
– RXR 352
– vitamin A deficiency 353
TGase II (transglutaminase II) 251, 254
TGFα 239
TGFβ 242
Th-lymphocytes 592, 593
thyroid hormone receptors 540
topical retinoid 521 pp.
– all-trans-retinal 530 p.
– all-trans-retinol 525 p.
– all-trans-retinoyl-β-glucuronide 533 p.
– all-trans-retinyl esters 527 p.
topoisomerase I 251
β-Trace 120
transcription intermediary factor (TIF) 359, 360
– TIF 1 163
transformation of anterior hindbrain 410 p.
transgenic embryos 416 p.
transglutaminase II (TGase II) 251
transport
– processes 19
– of retinoids 31 pp.
transthretin (*see* TTR) 349
Trichinella spiralis 597
triglyceride lipase 33
TTNPB 186, 187, 192,
TTR (transthretin) 348, 349
– fractional rates of synthesis 66
– properties 63 pp.
tumor necrosis factor α (TNFα) 281
TZD 227, 228

U

UPD-glucouronosyltransferase 143
uptake of retinol 33
– by target cells 20

V
vasculogenesis 452 p.
vertebrate limb
– development 378 pp.
– embryology 379 p.
– morphogenesis 369 pp.
very low density lipoprotein (*see* VLDL) 31
vision 563 pp.
visual
– cycle 567
–– control of 576
– pigments 567 pp.
–– regeneration 571 p.
– process 563 pp.
vitamin A
– deficiency 58, 59, 99, 324, 353, 354, 412, 443, 452, 468, 590
–– CNS 412
–– embryo 449
–– teratogenic effects (*see* teratogenic effects of vitamin A deficiency and excess) 444 p.
–– testis 353
– dependent transcription in Sertoli cells 362
– immune system 98, 99
vitamin D 239
vitamin D_3 receptor 540
VLDL (very low density lipoprotein) 31, 81

W
Wnt 401

X
xeroderma pigmentosum 301

Z
ZPA (zone of polarizing activity) 380, 381 pp.
– generation of 383 pp.

Printing: Saladruck, Berlin
Binding: H. Stürtz AG, Würzburg